Computation to Fight SARS-CoV-2 (CoVid-19)

Computation to Fight SARS-CoV-2 (CoVid-19)

Editors

Simone Brogi
Vincenzo Calderone

MDPI

Basel • Beijing • Wuhan • Barcelona • Belgrade • Novi Sad • Cluj • Manchester

Editors
Simone Brogi
Department of Pharmacy
University of Pisa
Pisa
Italy

Vincenzo Calderone
Department of Pharmacy
University of Pisa
Pisa
Italy

Editorial Office
MDPI
St. Alban-Anlage 66
4052 Basel, Switzerland

This is a reprint of articles from the Special Issue published online in the open access journal *Computation* (ISSN 2079-3197) (available at: https://www.mdpi.com/journal/computation/special_issues/computation_CoVid_19).

For citation purposes, cite each article independently as indicated on the article page online and as indicated below:

Lastname, A.A.; Lastname, B.B. Article Title. *Journal Name* **Year**, *Volume Number*, Page Range.

ISBN 978-3-7258-0048-3 (Hbk)
ISBN 978-3-7258-0047-6 (PDF)
doi.org/10.3390/books978-3-7258-0047-6

Contents

computation

Editorial

Computation to Fight SARS-CoV-2 (COVID-19)

Simone Brogi [1,2,*] and Vincenzo Calderone [1,*]

1 Department of Pharmacy, University of Pisa, Via Bonanno 6, 56126 Pisa, Italy
2 Bioinformatics Research Center, School of Pharmacy and Pharmaceutical Sciences, Isfahan University of Medical Sciences, Isfahan 81746-7346, Iran
* Correspondence: simone.brogi@unipi.it (S.B.); vincenzo.calderone@unipi.it (V.C.); Tel.: +39-050-2219613 (S.B.); +39-050-2219589 (V.C.)

Citation: Brogi, S.; Calderone, V. Computation to Fight SARS-CoV-2 (COVID-19). *Computation* **2023**, *11*, 185. https://doi.org/10.3390/computation11090185

Received: 8 September 2023
Accepted: 14 September 2023
Published: 18 September 2023

1. Introduction

In April 2020, during the last pandemic health emergency, we launched a Special Issue hosted by Computation—section Computational Biology, entitled "Computation to Fight SARS-CoV-2 (COVID-19)". The COVID-19 infective condition is caused by the etiological agent SARS-CoV-2 (2019-nCoV), a novel, highly virulent betacoronavirus. Therefore, SARS-CoV-2 is an important worldwide health hazard with high mortality and high contagiousness. Despite the introduction of vaccines and therapeutic options, the occurrence of resistant phenomena in SARS-CoV-2 keeps the public attentive to health concerns with a relevant interest in developing effective antiviral agents, vaccines, and social measures to mitigate the transmission of the virus and its variants. Accordingly, to advance our knowledge of the pathogenic mechanism of viruses and to identify novel drug candidates, computer science may play a significant role in the search for effective therapeutics to cure this infection. In addition, because of this global health alert, such computational methodologies could hasten the development of creative and focused solutions to the coronavirus emergency. This Special Issue showcases advancements in epidemiology, virus biology, and medication discovery to provide researchers with cutting-edge computational strategies for combating SARS-CoV-2. Based on this consideration, the Special Issue has attracted the attention of scientists in the field, and 27 research articles have been published on this topic. We divided this editorial article into two sections, one dedicated to the progress in the development of computer-based tools, mathematical models, and algorithms related to the socioeconomic impact, epidemiology, diffusion, and dynamics of SARS-CoV-2. The second section is devoted to computational approaches for understanding virus behavior, selecting possible vaccine candidates, and identifying promising antiviral agents.

1.1. Socioeconomic Impact, Epidemiology, Diffusion, and Dynamics of SARS-CoV-2

In this section, we analyze the articles published in the Special Issue that focused on the development of computational models related to the impact on society, with a particular focus on social, epidemiological, and management issues raised from the inception of SARS-CoV-2 diffusion. The first article, authored by Chowdhury and coworkers, explored the social media impact on awareness of COVID-19 related to the global health emergency, considering scholars at Bangladeshi University. To assess actual shifts in student isolation, psychological numbness, and trust and belief in social media coverage data, the authors used a cross-sectional design with a quantitative method. The authors presented to the students an online survey to determine the connection between knowledge of the pandemic health emergency and the activity on social media. Data were extracted from 189 completed surveys. Using exploratory factor analysis (EFA) and path analysis, the authors indicated that social media are changing the perspective of health problems, affording information and advice on undesired effects of SARS-CoV-2 infections, favoring psychological wellness, and having a beneficial effect on quarantine or lockdown. This interesting work indicated that social media are pivotal in improving knowledge about current and future pandemic conditions [1]. Examining the media's contribution to the COVID-19 pandemic in more detail, Yakunin and colleagues investigated the mass media influence in relation to the COVID-19 global health emergency, considering that this situation is an excellent example of a time when the media are crucial for educating the public about essential information. The purpose of this study was to undertake a comparative examination of the depiction of pandemic-related issues in Kazakhstani and Russian Federation internet media. The major objective of this study is to suggest a technique that would enable the analysis of the relationship between COVID-19 data from the World Health Organization and dynamic indicators from mainstream media. Three methods for numerically representing mass media dynamics were devised and used to complete this challenge, each of which was used in accordance with a manually chosen set of search terms. The result analysis indicated both parallel and divergent representations of the epidemiological situation in Russian and Kazakhstan periodicals. In particular, there was a correlation between publication activity in the two categories and absolute measures, e.g., the daily death rate and the daily rate of novel infections. However, the media typically fails to include viral reproduction and positive rates of confirmed cases. When the rigorousness of quarantine measures is considered, mainstream media in Russia and Kazakhstan exhibit very different correlations. According to an analysis of search terms, the issue of fake news and misinformation in Kazakhstan is worse during times when the epidemiological condition is deteriorating and crime and poverty are rising. The originality of this study lies in the formulation and application of a methodological approach that enables a comparison between media indices and objective COVID-19 statistical parameters [2]. Another important study was conducted by González-Parra and Arenas. To model the COVID-19 pandemic, the authors used a highly nonlinear mathematical model to investigate the Omicron wave, considering the effects of the vaccine. The developed model comprises asymptomatic and immunized individuals, which affects SARS-CoV-2 dynamics. In addition, the developed computational tool considered the decrease in vaccination immunity and efficacy against the Omicron strain. The results of the simulation suggested that even if the Omicron strain is less lethal, it may still result in more deaths, infections, and hospitalizations. Interestingly, the authors presented some scenarios that aid in understanding the Omicron wave and its repercussions. Overall, the described mathematical modeling approach, along with the simulation of the selected biological systems, explained the enormous Omicron wave under varied vaccination and transmissibility circumstances. These findings raise awareness that even though SARS-CoV-2 genotypes have a reduced mortality rate, they can nevertheless result in more deaths. Accordingly, the developed model can be useful for understanding the Omicron wave and the impact of novel highly transmissible strains [3]. In a different study authored by Afiahayati and colleagues, a computer-based technique was described for precisely estimating the number of cases of COVID-19 in the coming days, which could be extremely valuable in decision-making for providing proper advice to mitigate pandemic health emergencies. The researchers forecasted the total number of verified cases of COVID-19 in Indonesia using the flower pollination algorithm (FPA), a metaheuristic optimization algorithm. FPA is a robust and adaptive computational technique for optimizing the curve fitting of COVID-19 cases. A machine learning (ML) technique known as the recurrent neural network (RNN), which is popular for prediction, was used for analyzing and comparing the FPA performance. The best hyperparameters for the RNN and FPA were found after a thorough experiment (For the FPA and RNN, there are 24 and 72 different hyperparameter combinations, respectively) to be used for developing the COVID-19 predictive model. According to the outcomes, the FPA method outperformed the RNN in both long- and short-term predictions of the COVID-19 cases. Notably, in the last iteration for long-term forecasting, the FPA model (0.38%) had a substantially lower mean absolute percentage error (MAPE) than the RNN model (5.31%). In the last iteration for short-term forecasting of the cumulative COVID-19 instances in Indonesia, the MAPE for the FPA model (0.74%) was also lower than that for the RNN model (4.8%). The cutting-edge findings from this study could aid efforts to combat the global COVID-19 health emergency [4]. Intriguingly, a multivariate analysis using COVID-19 vaccine infodemic data in Brazil was described by da Penha Harb and colleagues. This study deals with the exposure of people to an enormous volume of data considering diverse media channels. Notably, this information and these data are not always official and true, and often, when false, they can affect data readability and disease control; incorrect information might worsen the pandemic's harmful impacts. Using the multivariate analytic technique, the research uncovered similar patterns of behavior in the selected population throughout 2021 in two analyses by including information on immunizations from all age groups and with people aged 64 years or older (13% of the population). The authors employed dendrograms as a cluster visualization method. To validate the formed clusters, two techniques were used: the cophenetic correlation coefficient, which produced good findings over 0.7, and the elbow technique, which confirmed the quantity of identified clusters. As a consequence of examining Brazilian states across all age categories, more homogeneous divisions were detected, according to the findings. In contrast, the second analysis produced more heterogeneous clusters, indicating that at the time of vaccinations, there may have been fear, skepticism, and a strong belief in the infodemic [5]. In another paper related to the social impact of vaccination campaigns, Shahzad and collaborators used an ML approach to categorize COVID-19 vaccine-related user responses. In fact, in the pandemic global health emergency, the availability of COVID-19 immunization offered hope for humanity. Unfortunately, people still believe that vaccinations have risks, and they express their beliefs and experiences on social media platforms despite determined vaccination efforts and recommendations from medical professionals and governments. Such opinions may be analyzed to identify societal trends and develop strategies for boosting vaccination acceptance. Accordingly, the authors described a method for sentiment analysis of worldwide opinions and impressions of COVID-19 immunization. The study was conducted using data from Twitter and considered five vaccines, including Moderna, AstraZeneca, Sinovac, Pfizer, and Sinopharm. For sentiment classification, the tweet datasets were divided into three categories (e.g., positive, negative, and natural) using different ML classifiers such as Random Forest (RF), Support Vector Machine (SVM), Logistic Regression (LR), Decision Tree (DT), and Naive Bayes (NB). Interestingly, compared with the other ML methods, the DT classifier has the best classification performance among all the selected datasets. The highest accuracy for COVID-19 vaccine tweets with sentiment annotation was 93.0%. This accuracy was also achieved for the vaccine dataset of AstraZeneca (90.94%), Pfizer (91.07%), Moderna (88.01%), Sinovac (92.8%), and Sinopharm (93.87%). The quantitative comparisons indicated that the proposed study was extremely accurate. Therefore, starting from this type of work, it will be possible to apply deep learning (DL) methods to achieve better accuracy [6]. De La Hoz-M and colleagues conducted a different type of study. They examined the published research on COVID-19 from its beginning and monitored the development of research over two years (February 2020–January 2022). Accordingly, using text mining, latent Dirichlet allocation modeling, trend analysis, and other techniques, the authors searched the PubMed database to extract topics and examine the time-based variations in research for each subject. Furthermore, the authors examined how these themes are covered in different nations (e.g., United States of America, United Kingdom, Italy, India, and China), and journals (7040 sources such as Sci. Rep., PLoS ONE, and Int. J. Environ. Res. Public Health, which represented the leading publishing journals on the COVID-19 outbreak), and 16 research topics and 126,334 peer-reviewed publications were found. The authors found eight topics (vaccine immunity, telemedicine, prevention, morbidity and mortality, mental health, ML, risk factors, academic parameters, and information synthesis methods) showing a rising trend, five (COVID-19 pathology complications, etiopathogenesis, epidemiology, diagnostic test, and political and health factors) showing a falling trend, and the remaining three (pharmacological factors, therapeutics, and others) varied over time with no clear patterns. In conclusion, the findings can offer new study guidelines and be helpful to academics and politicians in understanding research trends in the context of global events. The outcomes demonstrated that topic modeling is a rapid and effective technique for assessing the development of a sizable and rapidly evolving research topic, such as COVID-19 [7]. Another type of work was focused on the development of a computational method for classifying time-series data related to the number of cases of COVID-19 infection per day. Pangestu and collaborators assessed the effectiveness of community activity limitations in reducing the number of novel COVID-19 cases in West Java (WJ). For this investigation, the researchers used time-series clustering on daily positive case data for COVID-19 across 27 cities and regencies in WJ. Clustering was accomplished using the k-medoids clustering method with shape-based lock step metrics, notably the cross-correlation-based distance. During the worst outbreak, the researchers utilized daily new COVID-19 infection cases in the mentioned cities and regions, considering different periods. The findings indicated that four was the optimal number of clusters that could be generated from the data used considering the first period, whereas two was the optimal number of clusters considering the second period, with silhouette values of 0.2633 and 0.6363, respectively. Results showed that, except for Bogor and Depok, the activity restriction period was successful in reducing COVID-19 infection during the initial time frame and during the second period throughout the whole city and region of WJ. Interestingly, the authors discovered that geography, in addition to the activity restriction period effectiveness, had an impact on the cluster that was formed. A city's likelihood of experiencing an increase in COVID-19 instances depends on the distance from a COVID-19 hotspot area [8]. Curiously, Alinizzi and colleagues investigated traffic congestion during the COVID-19 pandemic period as a rising issue in addition to several socioeconomic problems. Accordingly, the period of traffic congestion hazard (HTC) in metropolitan areas depends on commuters' decisions regarding home-to-shopping center departure times. The decision to leave early or stay late to go shopping depended on both internal (related to commuters) and external (related to shopping centers) considerations. A useful method to evaluate the HTC time following curfew timings was developed in the selected study. The commuters' perception of time spent shopping was evaluated, along with the effects of eight internal (family size, nature of the job, involvement in extracurricular activities, education level, number of cars, number of children, age, and availability of a private driver) and three external (dimensions of the city, distance to shopping center, and approachability of favored shopping center in proximity) factors on their choice. Chi-square and Cramer's V tests, with an acceptable 20% response rate, identified family size and participation in other activities as the most important internal factors and accessibility to the preferred shopping area as the key external factor. The commuters' choices of leaving early or later depended in part on their age, the number of children they had, and the dimensions of the city. Except for educational level and the presence of household drivers, most of the characteristics showed significant relationships. The commuters' responses were divided into four categories using fuzzy synthetic evaluation (FSE): no delay, short delay, moderate delay, and long delay. The peak period of traffic congestion was successfully identified by hierarchical bottom-up aggregation. According to the survey findings, most commuters (approximately 65%) go shopping within 15 min of the curfew lifting; hence, HTC within the first hour of the no curfew time deserved consideration. Traffic regulation agencies can use the described method to determine the HTC period and conduct practical traffic management methods in accordance with the basic sociodemographic information of residents of an urban neighborhood. By implementing the proposed strategy in various locations and conducting traffic monitoring studies when the curfew is lifted for the duration of the pandemic, future research can confirm the results of the current study [9]. Sukandar and coworkers conducted an important study. They used the cumulative case function to generate dynamic operators that recover all state dynamics of a susceptible exposed infectious recovered (EIRs) model for the transmission of COVID-19. In particular, to accommodate immeasurable control and intervention mechanisms, this analysis considered known and unrecorded EIRs and a time-dependent infection rate. Cumulative cases were used to build and implement the dynamic operators. By implementing the generating operators, all infection processes hidden in this cumulative function can be completely recovered.

All recorded state dynamics were obtained by directly implementing the operators on the cumulative function. Furthermore, the unrecorded infection rate per day was calculated using the ratio of the infection mortality ratio (IFR) to the case mortality ratio (CFR). The generating operators were used to obtain the residual dynamics of the unrecorded states. The simulations were run using infection data from ten different countries provided by Worldometers. The increased amount of daily PCR testing was demonstrated to have a direct influence on lowering the effective reproduction ratio. Simulations of all state dynamics, infection rates, and effective reproduction ratios were performed for a number of countries during the first and second waves of transmission. With this method, daily transmission indicators are directly measured and can be utilized to successfully control the epidemic on a daily basis [10]. Aslam published a further article of interest. In this paper, artificial intelligence was used to predict death and ventilator support early in COVID-19 patients to reduce mortality and increase the chance for more effective and prompt therapies. COVID-19 hospitalized patients from King Abdulaziz Medical City in Riyadh were included in this study. To determine the influence of specific qualities on the prediction of death and ventilator support in patients affected by COVID-19, this work coupled a DL model with explainable artificial intelligence (EAI). Despite producing important results, the DL model is difficult to interpret. Data were collected from patient demographics, laboratory tests, and chest X-ray (CXR) results. Due to the unbalanced nature of the dataset, specificity, sensitivity, balanced accuracy, AUC metrics, and the Youden index were applied to assess the efficiency of the developed computational tool. In addition, the SMOTE (over- and under-sampled) datasets and the original datasets were used in the studies. With a balanced accuracy of 0.98 and an AUC of 0.998 to predict mortality employing the entire feature set, the developed model performs better than the baseline study. A maximum adjusted accuracy of 0.979 and an AUC of 0.981 were obtained to predict ventilator support. The described predictive computer-based tool could help physicians identify patients with COVID-19 who may need ventilator assistance or die early on, which would improve the use of hospital resources [11]. Demongeot and coworkers investigated other attractive aspects related to the pandemic scenario. In fact, it is infrequently investigated how to estimate daily reproduction counts during the contagiousness phase, and only their aggregate R0 is usually estimated to define an infectious disease contagiousness level. Using a deconvolution method on a set of novel COVID-19 infections, the researchers derived an equation for the discrete dynamics of epidemic propagation and determined the number of daily reproductions. Considering various nations and waves of the COVID-19 outbreak, results and estimations were obtained to determine how noise can affect the stability of the epidemic dynamics. Accordingly, it will be possible to enhance estimates of the distribution of daily reproduction numbers during the infectious period by accounting for heterogeneity owing to different host age classes [12]. Moreover, Bertrand and Pirch developed a susceptibility-exposed-infected-quarantined-recovered-deceased (SEIQRD) model for the propagation of COVID-19 using a flux-based finite element method. The model was largely based on susceptible-exposed-infected-recovered-deceased (SEIRD) models recently established, with the addition of a quarantined compartment of the living population. A least-squares mesoscale method is then used to solve the resulting first-order system of coupled PDEs. To establish an indicator that affects the predictions generated by the approach, a variety of data on governmental actions taken to control the spread throughout 2020 was used. When compared with actual disease-spreading data, the results of numerical tests showed that predictions of the virus's space-time behavior were remarkably accurate [13]. In an interesting paper, Zeng and coworkers investigated the possibility of understanding the relationship among components crucial to patient disease progression to decrease the effects of the epidemic. For this purpose, the authors developed an improved COVID-19 structured dataset from many sources, incorporating local weather information and study sentiment relevant to a particular nation using natural language processing. The researcher used both ML and DL methods on the 301,363 samples and 43 attributes in the expanded structured dataset to predict the likelihood that a patient will survive. To enhance the model performance, the authors imported alignment sequence data. When used on the expanded structured dataset, Extreme Gradient Boosting (XGBoost) predicted patient survival with 97% accuracy, with climatic conditions and age exhibiting the greatest significance. Similarly, a multilayer perceptron (MLP) application achieved 98% accuracy. Accordingly, it would be beneficial to add more potentially significant variables to the patient data that are already available, such as the current weather, to improve the prediction of the likelihood that the patient will survive [14]. Sarv Ahrabi and colleagues investigated the patient monitoring issue. They highlighted the importance of careful patient monitoring in keeping the condition entirely under control, in addition to extensive medical research. It is well established that the study of X-rays is useful because of its accessibility; however, viral testing is most commonly used to discover COVID-19. Many studies have used DL paradigms with the goal of enhancing COVID-19 radiography-based identification of lung infection. In this context, the authors compared notable methods for binary classification of infected photos using DL techniques, deriving a convolutional neural network (CNN) variation with optimized parameters that showed satisfactory performance on a recent dataset of images obtained from COVID-19 patients. Contrary to the other models provided, the effectiveness of the generated model is of great importance. In this method, a random selection of a few images from the dataset was used as a holdout set. The developed tool successfully identified the majority of COVID-19 X-rays, showing an outstanding general accuracy of 99.8%. Additionally, the relevance of the findings from evaluating other datasets with various features (which, notably, are not utilized in the training process) showed that the suggested approach is beneficial in terms of accuracy up to 93% [15]. Moreover, Gencoglu and collaborators reported an analysis of effective crisis management during unfavorable health occurrences. During this period, it is fundamental to comprehend the traits of public attention and sentiment. This is extremely relevant during a pandemic like COVID-19 because the primary role of risk management is disseminated across society rather than being centered on a single entity. While many studies have used Twitter data to describe or anticipate the COVID-19 outbreak, causal modeling of public attention has not been studied. In the mentioned work, a causal inference approach was utilized to pinpoint and measure causal links among Twitter activity, public opinion, and pandemic features such as infection and mortality rates. The outcomes showed that the suggested strategy may effectively capture epidemiological domain knowledge and recognize factors that influence public opinion. Notably, by separating events that correlate with public attention from those that cause public attention, this work could advance the discipline of infodemiology [16]. Finally, an interesting aspect that correlated SARS-CoV-2 diffusion and transmission with air pollution and related pollutants was investigated in two papers published in the Special Issue. In particular, Delnevo and collaborators explored the correlation between air pollution and the fatal effects of COVID-19. To this end, the authors considered a series of daily values of $PM_{2.5}$, PM_{10}, and NO_2 over time, and the Granger causality statistical hypothesis test was used to determine the presumption of causation. Surely, numerous additional investigations at a level commensurate with the size of this phenomenon (e.g., physical, chemical, and biological) would be required to fully comprehend the relationship between the spread of this lethal virus and air pollutants. However, as strictly viewed from a Granger causality standpoint, the outcomes acquired both during and after the government lockdown decisions demonstrate a definite association [17]. Subsequently, the same research team focused further attention on the correlation between the propagation of the virus and the presence of airborne particle pollutants ($PM_{2.5}$, PM_{10}, and NO_2). In this study, the authors have described a new metric for forecasting COVID-19 diffusion. An ML model was developed and trained using the following data: (i) all COVID-19 illnesses that occurred between February and July 2020 in Emilia Romagna (Italy), a region in Europe that is among the most polluted; (ii) the region-specific daily values of all particulates. The traditional ten-fold cross-validation approach was then utilized on the ML model, and the results showed an accuracy rate of 90%. Finally, the model was applied to forecast the potential reappearance of the virus in Emilia-Romagna between September and December 2020. The authors were unable to verify the accuracy of the forecasts at the time of writing this article. However, this COVID-19 prediction model is the only one of its kind in the world because the authors speculated on a scenario based on a novel premise [18].

1.2. Structural Modeling, Vaccines, and Antiviral Drug Discovery

The second section is dedicated to the development and applications of computer-aided procedures for antiviral drug discovery, vaccine candidate selection, and understanding the behavior of the virus, simulating the dynamics of drug targets. In this context, Liang and collaborators investigated SARS-CoV-2 spike glycoprotein (S-protein) dynamics by employing a coarse-grained approach using physic-informed ML. Coarse-grained methods are useful for modeling systems that are not possible to model utilizing classical all-atom molecular dynamics (MD). Accordingly, by employing learned interaction parameters, coarse-grained MD simulations attained the microsecond time scale with stability (simulation speed 40,000 times faster than the conventional MD). When compared with the usual iterative approach, the proposed framework more accurately matches all-atom reference structures. The increased efficiency improves the timeliness in developing long-time simulations of SARS-CoV-2 drug targets and creates opportunities for revealing protein processes and anticipating environmental changes [19]. Regarding the selection of vaccine candidate, Oluwagbemi and coworkers developed a computational protocol using bioinformatics and immunoinformatic techniques to generate a multi-epitope mRNA vaccine that protects against the SARS-CoV-2 S-protein variants that were present in African nations at the time of the study. In particular, predictions of T- and B-lymphocyte epitopes were performed using various immunoinformatic methods. To select epitopes that could trigger a long-lasting immune response, they were subjected to additional tests. Seven epitopes, a highly immunogenic adjuvant, an MHC I-targeting domain (MITD), a signal peptide, and linkers comprise the proposed vaccine. The proposed vaccine was also antigenic, nonallergenic, nontoxic, thermostable, and hydrophilic, according to the results. In 100 randomly chosen SARS-CoV-2 S-proteins, none of the seven epitopes showed alterations. The vaccine construction secondary structure was stabilized by 36.44% α-helices, 20.45% drawn filaments, and 33.38% random helices. The simulated vaccine exhibited a strong affinity for TLR-4, as revealed by molecular docking, indicating its capacity to activate both innate and adaptive immunity. Further in vitro and in vivo studies should be conducted after the results and performance of this computational research [20]. In an interesting paper, Singh and colleagues described a computational approach to find chemicals potentially able to recover the activity of interferon-stimulated genes (ISGs). ISGylation is a critical step in the process by which ISGs induce an antiviral response in host cells. In fact, numerous viruses, such as SARS-CoV-2, decrease host immune responses by negatively influencing the ISGylation process (de-ISGylation). SARS-CoV-2 papain-like protease (PLpro) interacts with host ISG15, resulting in de-ISGylation. Thus, blocking de-ISGylation to recover ISG activity may be an appealing method for enhancing the host immunological response to SARS-CoV-2. For this purpose, the authors evaluated in silico several phytochemicals derived from well-known immunomodulatory herbs, including *Andrographis paniculata*, *Tinospora cordifolia*, and *Ocimum sanctum*, for their influence on de-ISGylation induced by SARS-CoV2 PLpro. The authors used a crystallographic complex that reflects the SARS-CoV-2 PLpro and ISG15 protein interacting model (PDB ID: 6XA9). The ability of these phytochemicals to bind to the interface region between PLpro and ISG15 was evaluated. Molecular docking calculations revealed that 14-deoxy-15-isopropylidene-11,12-didehydroandrographolide (AG1), isocolumbin (GU1), and orientin (TU1) have satisfactory binding energies. According to MD parameters and MM/PBSA calculations, TU1, GU1, and AG1 may bind to the interface, targeting pivotal residues in the PLpro-ISG15 complex [21]. In the field of antiviral drug discovery, Brogi and collaborators published two papers within the Special Issue considering two different SARS-CoV-2 drug targets, the RNA-dependent RNA polymerase (RdRp) and the main protease (Mpro or 3CLpro). The first paper described a computational approach aimed at optimizing the alkaloid Quinadoline B (Q3) as a possible SARS-CoV-2 RdRp inhibitor. For this purpose, starting from a previously identified anti-viral fungal metabolite Q3 as a SARS-CoV-2 RdRp inhibitor, a computational combinatorial methodology was used to generate a chemical library based on the Q3 compound. The resulting chemical library (>900,000 different Q3 derivatives) was screened against RdRp to identify RdRp binders with higher affinity than the Q3-derived starting molecule. Using this method, along with the evaluation of the physchem profile, 26 derivatives were identified as potential RdRp inhibitors. In addition, the most promising derivatives were subjected to MD simulation to thoroughly examine the binding mechanism. Five compounds showed improved binding affinity for the RdRp enzyme and are therefore worth further study as potential antivirals. The described in silico strategy offers a practical computational method for hit-to-lead optimization, with implications for the search for anti-SARS-CoV-2 drugs and the overall drug optimization process [22]. Later, the same research group analyzed in silico the potential of peptide-based derivatives containing bifunctional warheads that could interact with prime and non-prime residues to covalently bind the Mpro of SARS-CoV-2 to develop novel antiviral agents. As a result, the authors proposed a computer-based protocol for discovering potential SARS-CoV-2 Mpro covalent inhibitors. They examined the possibility of a peptide-based scaffold with diverse warheads as a substantial alternative to aldehyde and nitrile electrophilic groups using multiple in silico methodologies. As warheads, we rationally generated four possible inhibitors, including difluorostatone and a Michael acceptor. Based on molecular and covalent docking, MD simulation, and free energy perturbation (FEP), the in silico investigation showed that the generated compounds might function as covalent inhibitors of Mpro and that the examined warheads could be employed to develop inhibitors that can covalently bind cysteine or serine proteases, including the Mpro of SARS-CoV-2. Notably, the abovementioned research provided a rigorous computational protocol for identifying and developing powerful antiviral agents [23]. Further considering Mpro and other possible drug targets, Muhammad and colleagues examined the interactions of eight natural eucalyptus compounds with SARS-CoV-2 Mpro to determine whether they could be used as herbal treatments for the emerging SARS-CoV-2 virus. Atomistic interactions were inspected using various in silico techniques, such as molecular docking, MD simulations, and MM/PBSA calculations. On the basis of the outcomes of molecular docking, all drugs examined showed significant binding energy for Mpro. Three computational hits, α-gurjunene, aromadendrene, and alloaromadendrene, with satisfactorily predicted affinity, were simulated using GROMACS to analyze the interactions between Mpro and inhibitor molecules at the molecular level. According to the results of our MD simulation, aromadendrene and α-gurjunene were found to be the most promising compounds, with binding energies of −18.99 kcal/mol and −17.91 kcal/mol, respectively. The outcomes indicated that eucalyptus could be a hypothetical therapeutic opportunity to inhibit the Mpro enzyme. Remarkably, this work is one of the first in which has been investigated the role of structural flexibility in Mpro interactions with herbal drugs [24]. Culletta and coworkers used different computational techniques to identify drugs against different established drug targets (3CLpro, PLpro, and different non-structural viral proteins). Homology modeling (for targets with no available experimental structures) and a structure-based pharmacophore modeling study were conducted for each drug target. Next, using the developed pharmacophore models, a virtual screening was conducted employing the chemical library provided by DrugBank. Each target's potential inhibitors were identified using XP docking, induced fit docking, and MM/GBSA calculations. After the docking study, 34 hits for the explored targets were selected (26 experimental drugs, 5 investigational drugs, and 3 approved drugs). The best binding energy for each molecule, as determined by MM/GBSA calculations, was used to make the final selection of candidate inhibitors. These chemicals were found able to interact with crucial residues of each target according to the molecular recognition analysis. The effectiveness of these drug candidates in successfully inhibiting COVID-19 can be further assessed. The findings of this study provide crucial information for anti-COVID-19 drug discovery efforts, identify the primary binding sites for the most significant SARS-CoV-2 proteins, and present a crucial path for the development of novel antivirals [25]. Finally, Qiao and coworkers developed two computational protocols to identify SARS-CoV-2 S-protein and 3CLpro inhibitors for possible COVID-19 treatment. Among the screened compounds showing a significant inhibitory profile in preventing the recognition of the S-protein of SARS-CoV-2 and ACE-2 in host cells, vancomycin, amphotericin B, and ergotamine were identified as the most promising compounds. On the other hand, the researchers also identified possible inhibitors of SARS-CoV-2 3CLpro. Among them, the most interesting drugs identified were dasatinib, rivaroxaban, montelukast, sildenafil, saquinavir, tadalafil, and vardenafil, which showed docking scores lower than −8.5 kcal/mol [26]. Aminpour and colleagues authored the last paper analyzed here. They used in silico methods to explore the possible mechanism of action of ivermectin and its derivatives as possible multitarget antivirals. The authors conducted computational work to estimate the binding affinity of possible antivirals for the S-protein of SARS-CoV-2, the CD147 receptor (secondary attachment point for the virus), and the α-7 nicotinic acetylcholine receptor (α-7nAChr) (important for viral penetration of neuronal tissue as well as an activation site for the cholinergic anti-inflammatory pathway controlled by the vagus nerve). For each compound's various docking locations and binding mechanisms, binding affinities were computed. Our findings show that ivermectin has a strong affinity for all three of these molecular targets, with some other drugs having even greater affinities. Interestingly, these findings point to potential molecular processes through which ivermectin could reduce the infectiousness and morbidity of the SARS-CoV-2 virus and activate an anti-inflammatory pathway controlled by α-7nAChR, which might reduce the production of cytokines by immune cells [27].

Finally, as Guest Editors, we express our gratitude to the Computation Editorial team for their generous support, all the authors and co-authors for their pertinent contributions to this Special Issue, and all the reviewers for their work in assessing the submissions. All of these combined efforts contributed to the research topic's outstanding success. In addition to being a significant source of knowledge and inspiration for researchers and students, we anticipate that this topic will boost the understanding of SARS-CoV-2 behavior, helping to find effective treatments and measures for reducing viral transmission. You can freely obtain this Special Issue by clicking on the following link: https://www.mdpi.com/journal/computation/special_issues/computation_COVID_19, (accessed on 13 September 2023).

Author Contributions: Conceptualization, S.B. and V.C.; data curation, S.B. and V.C.; writing—original draft preparation, S.B.; writing—review and editing, S.B. and V.C.; supervision, S.B. and V.C. All authors have read and agreed to the published version of the manuscript.

Conflicts of Interest: The authors declare no conflict of interest.

References

1. Chowdhury, S.; Rahman, M.; Doddanavar, I.A.; Zayed, N.M.; Nitsenko, V.; Melnykovych, O.; Holik, O. Impact of Social Media on Knowledge of the COVID-19 Pandemic on Bangladeshi University Students. *Computation* **2023**, *11*, 38. [CrossRef]
2. Yakunin, K.; Mukhamediev, R.I.; Zaitseva, E.; Levashenko, V.; Yelis, M.; Symagulov, A.; Kuchin, Y.; Muhamedijeva, E.; Aubakirov, M.; Gopejenko, V. Mass Media as a Mirror of the COVID-19 Pandemic. *Computation* **2021**, *9*, 140. [CrossRef]
3. González-Parra, G.; Arenas, A.J. Mathematical Modeling of SARS-CoV-2 Omicron Wave under Vaccination Effects. *Computation* **2023**, *11*, 36. [CrossRef]
4. Afiahayati; Wah, Y.; Hartati, S.; Sari, Y.; Trisna, I.; Putri, D.; Musdholifah, A.; Wardoyo, R. Forecasting the Cumulative COVID-19 Cases in Indonesia Using Flower Pollination Algorithm. *Computation* **2022**, *10*, 214. [CrossRef]
5. Harb, M.d.P.; Silva, L.; Ayass, T.; Vijaykumar, N.; Silva, M.; Francês, C.R. Dendrograms for Clustering in Multivariate Analysis: Applications for COVID-19 Vaccination Infodemic Data in Brazil. *Computation* **2022**, *10*, 166. [CrossRef]
6. Shahzad, A.; Zafar, B.; Ali, N.; Jamil, U.; Alghadhban, A.J.; Assam, M.; Ghamry, N.A.; Eldin, E.T. COVID-19 Vaccines Related User's Response Categorization Using Machine Learning Techniques. *Computation* **2022**, *10*, 141. [CrossRef]
7. De La Hoz, M.J.; Mendes, S.; Fernández-Gómez, M.J.; González Silva, Y. Capturing the Complexity of COVID-19 Research: Trend Analysis in the First Two Years of the Pandemic Using a Bayesian Probabilistic Model and Machine Learning Tools. *Computation* **2022**, *10*, 156. [CrossRef]

8. Pangestu, D.S.; Sukono, S.; Anggriani, N. Evaluation of the Effectiveness of Community Activities Restriction in Containing the Spread of COVID-19 in West Java, Indonesia Using Time-Series Clustering. *Computation* **2022**, *10*, 153. [CrossRef]
9. Alinizzi, M.; Haider, H.; Alresheedi, M. Assessing Traffic Congestion Hazard Period due to Commuters' Home-to-Shopping Center Departures after COVID-19 Curfew Timings. *Computation* **2022**, *10*, 132. [CrossRef]
10. Sukandar, K.K.; Louismono, A.L.; Volisa, M.; Kusdiantara, R.; Fakhruddin, M.; Nuraini, N.; Soewono, E. A Prospective Method for Generating COVID-19 Dynamics. *Computation* **2022**, *10*, 107. [CrossRef]
11. Aslam, N. Explainable Artificial Intelligence Approach for the Early Prediction of Ventilator Support and Mortality in COVID-19 Patients. *Computation* **2022**, *10*, 36. [CrossRef]
12. Demongeot, J.; Oshinubi, K.; Rachdi, M.; Seligmann, H.; Thuderoz, F.; Waku, J. Estimation of Daily Reproduction Numbers during the COVID-19 Outbreak. *Computation* **2021**, *9*, 109. [CrossRef]
13. Bertrand, F.; Pirch, E. Least-Squares Finite Element Method for a Meso-Scale Model of the Spread of COVID-19. *Computation* **2021**, *9*, 18. [CrossRef]
14. Zeng, W.; Gautam, A.; Huson, D.H. On the Application of Advanced Machine Learning Methods to Analyze Enhanced, Multimodal Data from Persons Infected with COVID-19. *Computation* **2021**, *9*, 4. [CrossRef]
15. Sarv Ahrabi, S.; Scarpiniti, M.; Baccarelli, E.; Momenzadeh, A. An Accuracy vs. Complexity Comparison of Deep Learning Architectures for the Detection of COVID-19 Disease. *Computation* **2021**, *9*, 3. [CrossRef]
16. Gencoglu, O.; Gruber, M. Causal Modeling of Twitter Activity during COVID-19. *Computation* **2020**, *8*, 85. [CrossRef]
17. Delnevo, G.; Mirri, S.; Roccetti, M. Particulate Matter and COVID-19 Disease Diffusion in Emilia-Romagna (Italy). Already a Cold Case? *Computation* **2020**, *8*, 59. [CrossRef]
18. Mirri, S.; Delnevo, G.; Roccetti, M. Is a COVID-19 Second Wave Possible in Emilia-Romagna (Italy)? Forecasting a Future Outbreak with Particulate Pollution and Machine Learning. *Computation* **2020**, *8*, 74. [CrossRef]
19. Liang, D.; Zhang, Z.; Rafailovich, M.; Simon, M.; Deng, Y.; Zhang, P. Coarse-Grained Modeling of the SARS-CoV-2 Spike Glycoprotein by Physics-Informed Machine Learning. *Computation* **2023**, *11*, 24. [CrossRef]
20. Oluwagbemi, O.; Oladipo, E.; Kolawole, O.; Oloke, J.; Adelusi, T.; Irewolede, B.; Dairo, E.; Ayeni, A.; Kolapo, K.; Akindiya, O.; et al. Bioinformatics, Computational Informatics, and Modeling Approaches to the Design of mRNA COVID-19 Vaccine Candidates. *Computation* **2022**, *10*, 117. [CrossRef]
21. Singh, P.; Bhat, S.S.; Punnapuzha, A.; Bhagavatula, A.; Venkanna, B.U.; Mohamed, R.; Rao, R.P. Effect of Key Phytochemicals from Andrographis paniculata, Tinospora cordifolia, and Ocimum sanctum on PLpro-ISG15 De-Conjugation Machinery—A Computational Approach. *Computation* **2022**, *10*, 109. [CrossRef]
22. Brogi, S.; Quimque, M.T.; Notarte, K.I.; Africa, J.G.; Hernandez, J.B.; Tan, S.M.; Calderone, V.; Macabeo, A.P. Virtual Combinatorial Library Screening of Quinadoline B Derivatives against SARS-CoV-2 RNA-Dependent RNA Polymerase. *Computation* **2022**, *10*, 7. [CrossRef]
23. Brogi, S.; Rossi, S.; Ibba, R.; Butini, S.; Calderone, V.; Campiani, G.; Gemma, S. In Silico Analysis of Peptide-Based Derivatives Containing Bifunctional Warheads Engaging Prime and Non-Prime Subsites to Covalent Binding SARS-CoV-2 Main Protease (Mpro). *Computation* **2022**, *10*, 69. [CrossRef]
24. Muhammad, I.A.; Muangchoo, K.; Muhammad, A.; Ajingi, Y.u.S.; Muhammad, I.Y.; Umar, I.D.; Muhammad, A.B. A Computational Study to Identify Potential Inhibitors of SARS-CoV-2 Main Protease (Mpro) from Eucalyptus Active Compounds. *Computation* **2020**, *8*, 79. [CrossRef]
25. Culletta, G.; Gulotta, M.R.; Perricone, U.; Zappalà, M.; Almerico, A.M.; Tutone, M. Exploring the SARS-CoV-2 Proteome in the Search of Potential Inhibitors via Structure-Based Pharmacophore Modeling/Docking Approach. *Computation* **2020**, *8*, 77. [CrossRef]
26. Qiao, Z.; Zhang, H.; Ji, H.-F.; Chen, Q. Computational View toward the Inhibition of SARS-CoV-2 Spike Glycoprotein and the 3CL Protease. *Computation* **2020**, *8*, 53. [CrossRef] [PubMed]
27. Aminpour, M.; Cannariato, M.; Preto, J.; Safaeeardebili, M.E.; Moracchiato, A.; Doria, D.; Donato, F.; Zizzi, E.A.; Deriu, M.A.; Scheim, D.E.; et al. In Silico Analysis of the Multi-Targeted Mode of Action of Ivermectin and Related Compounds. *Computation* **2022**, *10*, 51. [CrossRef]

computation

MDPI

Article

Impact of Social Media on Knowledge of the COVID-19 Pandemic on Bangladeshi University Students

Shanjida Chowdhury [1], Mahfujur Rahman [2], Indrajit Ajit Doddanavar [3], Nurul Mohammad Zayed [4], Vitalii Nitsenko [5,6,*], Olena Melnykovych [7] and Oksana Holik [7]

1 Southeast Business School, Southeast University, Dhaka 1208, Bangladesh
2 Department of Statistics, Cumilla University, Cumilla 3506, Bangladesh
3 Jain College of MCA & MBA, Belagavi, Afflated to Rani Channamma University, Belgaum 590014, Karnataka, India
4 Department of Business Administration, Daffodil International University, Dhaka 1216, Bangladesh
5 Department of Entrepreneurship and Marketing, Institute of Economics and Management, Ivano-Frankivsk National Technical Oil and Gas University, 76019 Ivano-Frankivsk, Ukraine
6 SCIRE Foundation, 00867 Warsaw, Poland
7 Department of Advertising and Journalism, State University of Trade and Economics, 02156 Kyiv, Ukraine
* Correspondence: vitaliinitsenko@onu.edu.ua; Tel.: +380-939983073

Citation: Chowdhury, S.; Rahman, M.; Doddanavar, I.A.; Zayed, N.M.; Nitsenko, V.; Melnykovych, O.; Holik, O. Impact of Social Media on Knowledge of the COVID-19 Pandemic on Bangladeshi University Students. *Computation* **2023**, *11*, 38. https://doi.org/10.3390/computation11020038

Academic Editor: Simone Brogi

Received: 27 January 2023
Revised: 12 February 2023
Accepted: 14 February 2023
Published: 16 February 2023

Abstract: This study aimed to examine the role and impact of social media on the knowledge of the COVID-19 pandemic in Bangladesh through disseminating actual changes in health safety, trust and belief of social media's coverage statistics, isolation, and psychological numbness among students. This study used a cross-sectional design in which a quantitative approach was adopted. Data from an online survey were collected in a short period of time during the early stages of COVID-19 to determine the relationship between social media activity and knowledge of the COVID-19 pandemic with accuracy. A total of 189 respondents were interviewed using structured questionnaires during the onset of the COVID-19 outbreak in Bangladeshi university students. Exploratory factor analysis (EFA) and path analysis were performed. Out of 189 respondents, about 80% were aged between 16 and 25 years, of which nearly 60.33% were students. This study explored four factors—knowledge and health safety, trust in social media news, social distancing or quarantine, and psychological effect—using factor analysis. These four factors are also found to be positively associated in path analysis. Validation of the model was assessed, revealing that the path diagram with four latent exogenous variables fit well. Each factor coefficient was treated as a factor loading (β = 0.564 to 0.973). The results suggested that the measurement models using four elements were appropriate. The coefficient of determination was 0.98, indicating that the model provided an adequate explanation. Social media is transforming the dynamics of health issues, providing information and warnings about the adverse effects of COVID-19, having a positive impact on lockdown or quarantine, and promoting psychological wellness. This comprehensive study suggested that social media plays a positive role in enhancing knowledge about COVID-19 and other pandemic circumstances.

Keywords: social media; COVID-19; psychological impact; social distancing; knowledge

1. Introduction

The coronavirus (COVID-19) pandemic is running rampant globally, creating a worldwide health crisis. It has already significantly impacted the everyday lives of people. This deadly virus has killed many millions, among which older adults have been the main victims. A novel strain of coronavirus, SARS-CoV-2, was first identified in December 2019 in Wuhan, a city in China's Hubei province, after a flare-up of pneumonia without apparent reason. Globally, there were 54,558,120 confirmed cases of COVID-19, including 1,320,148 deaths according to the latest update by the World Health Organization (WHO) on 17 November 2020, which declared a pandemic on 11 March 2020 [1,2]. Most countries have tried to save people's lives from the pandemic threat by putting in place various restrictive measures including lockdowns and social distancing. The economy in many countries has come tumbling down due to the COVID-19 [3]. The whole world has come to a standstill, focusing its efforts on resisting this disease. If adequate precautions are not implemented, developing countries with frail healthcare systems may suffer the most disastrous consequences from this epidemic [4,5]. This pandemic has become a serious socioeconomic, behavioral, psychosocial, governance, and technology challenge, particularly for the frontline healthcare sector. This health crisis has already been transformed into a global economic crisis [5].

Bangladesh declared a war against COVID-19, like many other countries, becoming one of the worst-affected nations by this havoc [6,7]. In many parts of the nation, formal measures such as closing schools, closing offices for a 1 month trial period, prohibiting people from leaving their homes after 6 p.m., taking legal action against those who do, banning gatherings in mosques, and restricting public gatherings were swiftly implemented. This nation has faced a number of challenges as a result of the pandemic, including maintaining social distancing, inadequate COVID-19 testing facilities, a lack of COVID-19 mitigation strategies, and limited financial support [8–10]. To battle the outbreak, Bangladesh was compelled to proclaim a state of emergency commencing 26 March 2020 [8,11,12]. The elected government then imposed social distancing, isolation, and home quarantine measures to reduce infection rates [13]. However, the lives of all individuals have been affected by COVID-19 from the social, professional, and personal aspects. A national lockdown strategy was implemented as a remedy, as in other countries [14]. This strategy significantly affected agricultural production, food supply, and demand. People in Bangladesh came to know about the infectious virus through television, radio, newspaper, social media, or personal experience, stimulating anxiety and agitation among citizens, as well as their friends and family. Social media has recently emerged as indispensable to reach people easily. Defining social media is a difficult task because it is a constantly changing field. According to Joosten [15], the term "social media" refers to any number of technological systems that are linked to cooperation and community. Again, Saydan and Dulek [16] defined social media as "social platforms where users share their information, manners, and interests via the internet or mobile systems" and big data applications as "social platforms where users share their information, manners, and interests via the internet or mobile systems" [17,18]. Today, social media plays an important role in shaping society and is perhaps one of the easiest ways to broadcast news or share an idea. During the COVID-19 pandemic, social media also educated people on how to prevent infectious disease and save lives. There were more than three billion active social media users before the outbreak of COVID-19. Since started the disease, a substantial rise in time spent on social media was observed, which facilitated the sharing of COVID-19 information. Online social media platforms such as Facebook, Instagram, and Twitter allow individuals to associate with one another across the globe, including the sharing of COVID-19 articles, papers, and reports. The young generation predominantly embraced web-based life during the pandemic. In addition to gathering COVID-19-related knowledge, previous studies [9,19] concluded that students obtained academic knowledge through the use of a variety of social media apps such as YouTube for self-learning, WhatsApp for exchanging papers, information, and presentations, and Zoom, Skype, and Google Meet for video conferencing to speed up learning. In addition to text messaging, video conferencing solutions have been widely used to promote communication between instructors and students. Students are becoming more assured in their capacity to use technology to learn, access, share, and generate relevant information, as well as gain knowledge about a subject. Social media has, therefore, been essential for spreading information throughout the pandemic. Despite the benefits of social media, a challenge during the pandemic has been the rapid spread of misinformation or fake news related to the virus outbreak [20,21]. People spent an average of more than 2 h a day on social media for news mainly related to COVID-19 [22,23], resulting in increased panic triggered by misinformation [22].

During the pandemic, direct dissemination of critical COVID-19 guidance through government offices became impossible [24]. Social media platforms, particularly Facebook and YouTube, as well as television and various websites, all played important roles in disseminating health messages and keeping people up to date on the pandemic [20,25]. Social media platforms provide direct access to an unprecedented amount of content and may amplify rumors and questionable information. Twitter is playing an increasingly important role in the dissemination of health information. There is mounting evidence that a highly mentioned paper on this social media platform may reflect the quality of the paper, which may then be subject to debate in journal clubs, as well as a post-publication social peer review process that may aid in retraction [20,26]. Information shared on social media, such as general health precautionary measures, mask use, maintaining social distancing, hand washing, and lockdowns, has had a positive impact during the pandemic [1,27]. Governments and public health authorities use social networking sites to inform citizens about COVID-19 testing locations and more affected areas, strictly taking responsibility for posting legitimate information related to COVID-19. By identifying and tracking user behavioral patterns, social media can transfer useful information about infectious diseases. Pandemic-related social media health campaigns can be effective in slowing disease spread by conveying positive attitudes.

This study contributes to investigating the effect of social media use on the knowledge of the COVID-19 outbreak in Bangladesh among university students. Furthermore, this study considers educational level as an indicator with a distinct influence on all the variables (predictors and dependent) investigated. This approach was infrequent in previous research [28–30]. Structural equation modeling (SEM) has also been rarely used in studies on the COVID-19 pandemic in Bangladesh. To the best of our knowledge, this is the first study to empirically establish the assumed effect of social media on the knowledge of COVID-19 in the context of Bangladesh using exploratory factor analysis (EFA).

The physical threat of virus spread also requires social distancing by refraining from regular activity. China's strict actions, including institutionalized quarantine, isolation, dedicated hospitals, and social distancing, were highly effective. Social media has a powerful role in influencing behavior. According to Radwan and Radwan [31], social media can have a positive impact on the public if used correctly. Therefore, this paper aims to identify the impact of social media on the individual, social, and societal levels during the pandemic. The remainder of the paper is organized as follows: Section 2 describes the methods, along with the basic statistics and final findings of the article; Section 3 discusses the results of path coefficients and exploratory factor analysis (EFA); Section 4 provides a discussion; lastly, Section 5 presents the concluding remarks.

2. Materials and Methods

2.1. The Data

This study obtained primary data by conducting a survey. A structured questionnaire was developed to collect data on the impact and role of social media during the pandemic. The questionnaire was propagated as a self-administered Google Form [32] to the target respondents. A total of 189 responses were recorded, with a response rate of 18%, which is acceptable for an online survey. This survey was conducted from 21 March to 15 April 2020. Criteria for collecting data were being a regular or frequent internet (or Facebook, WhatsApp, Twitter, YouTube, Instagram etc.) user, with an age >15.

2.2. Methods

Categorical data analysis mainly involves statistical tools such as logistic, ordinal, or multinomial regression; logit or probit models including the structural equation model are considered superior. This complex and widespread model is used massively in marketing [33,34], psychology [35,36], and education [37,38]. In the case of the perception of social media use, SEM can amalgamate many tools such as factor analysis, path diagrams, latent growth models, and MIMIC. In this study, our goal was to explore latent information from respondents related to social media's role in the COVID-19 pandemic. Each variable was recorded on a Likert scale (1 = strongly disagree, 2 = disagree, 3 = neutral, 4 = agree, and 5 = strongly agree). For the path diagram, we examined the measurement model through exploratory factor analysis, allowing us to classify various variables into a limited number of selected factors. Structural equation modeling (SEM) has several benefits compared to other statistical analyses [39]. Firstly, SEM implies a hypothesis-testing approach for data, which enables scholars to build hypothesis-based methods. Secondly, SEM solves measurement errors by obtaining unbiased estimates of the relationship between variables. This error is removed upon correcting correlation or regression estimates. Thirdly, SEM assembles both latent and observed variables. Lastly, it provides the direct effect, indirect effect, and total effect of multivariate relations.

In this study, SEM was combined with statistical techniques to exhibit a latent relationship between dependent and independent or observed and unobserved variables [40]. Exploratory factor analysis (EFA) was used to display latent relationships among factors or latent variables. Path analysis models were used to perceive observed variables. This enabled us to estimate and exhibit the relationship between observed variables. The models also represent an essential part of the historical development of SEM, and they employ the same underlying process of model testing and fitting as other SEM models. The relationship of indicators was determined with latent variables through EFA in the measurement model. This study scrutinized four factors with 11 variables. Therefore, exploratory factor analysis (EFA), was examined using four factors: knowledge and safeguarding health through media, self-detainment at home, social media's accuracy, and psychological monotony.

3. Results

For this study, data were collected from 189 respondents, involving university students from different private and public universities. Data were collected through a self-administered questionnaire within a limited time. The background characteristics of the respondents are shown in Table 1. Among the respondents, about 64% were male; most of the respondents (80%) were from the age group 16–25 years old, whereas about 6.7% were more than 36 years old. Approximately 54.4% of respondents were from the megacity, whereas more than one-fourth were from a rural area.

Table 1. Background characteristics of the respondents.

Variables	Category and Measurement	N	%
Sex	Male = 1	115	63.9
	Female = 2	65	36.1
Age group	16–25 = 1	144	80.0
	26–35 = 2	24	13.3
	>36 = 3	12	6.7
Residence	Mega city = 1	98	54.4
	Urban = 2	35	19.5
	Rural = 3	47	26.1

In univariate statistics, each variable under an item also exhibits a correlation. Each variable's mean response was greater than or equal to four (Table 2), indicating that the respondents were optimistic about being queried. The standard deviation was less than one, indicating that discrepancy was not observed in response to any question. Additionally, this study found an intercorrelation of the items ranging from 0.6 to 0.8, indicating the acceptability of the items.

Table 2. Mean score and correlation of social media use during COVID-19 pandemic.

Variable	Mean	SD	Item-Test Correlation
Teaches about symptoms of coronavirus (COVID-19)	4.44	0.70	0.714
Teaches about the spread of coronavirus (COVID-19)	4.35	0.77	0.702
Teaches precautionary steps to reduce the chances of getting infected	4.33	0.73	0.769
Teaches categories of risk related to coronavirus (COVID-19)	4.21	0.82	0.736
Teaches about being properly sanitized	4.24	0.73	0.720
Teaches about the minimum safe distance between two persons being 1 m (3 ft)	4.22	0.85	0.701
Teaches the difference between isolation and home quarantine	4.01	0.97	0.669
Fake news/information related to coronavirus is spreading	4.21	1.00	0.369
There has been a negative effect on mental health during the outbreak of COVID-19	3.97	1.11	0.503
Helps to create public awareness of health issues	4.30	0.73	0.642
Helps to maintain social distance from others	4.14	0.87	0.627
Highlights actual figures related to death or infection during the pandemic	3.56	1.25	0.593
Highlights COVID-19 without any biases as it is a global issue	3.95	1.13	0.613
Effectively broadcasts government initiatives to fight against COVID-19	4.03	0.91	0.712
Keeps one entertained during the home quarantine/lockdown period	4.17	0.97	0.506
Effectively presents the benefits of the "stay home and stay safe" slogan	4.21	0.91	0.663

Before conducting factor analysis, some precautionary steps were taken to perceive general knowledge about the dataset. This study accumulated 189 respondents to explore the impact of social media on the COVID-19 pandemic. For observing outliers and reducing unobserved variables, Cronbach's alpha was found to be 0.85. For sampling adequacy, the KMO measure and Bartlett tests (Table 3) both suggested that the sample size and correlation of items were acceptable, enabling further analysis.

Table 3. KMO and Bartlett's test.

Kaiser–Meyer–Olkin Measure of Sampling Adequacy	Bartlett's Test of Sphericity		
	Approx. Chi-Square	Degree of Freedom	*p*-Value
0.805	1016.23	55	0.001

For a more in-depth factor analysis, principal component analysis was employed in this study. According to the rules of thumb and Horst's parallel analysis, this study accepted four factors that explained 78.23% of the variance (Table 4).

Table 4. Factor loading explained through Factor analysis.

Component	Total Eigenvalue	% of Variance	Cumulative %
1	4.751	43.188	43.188
2	1.475	13.408	56.595
3	1.248	11.348	67.943
4	1.132	10.289	78.232

On rotated factor loadings, Kaiser normalization was applied due to its simplicity and preferable performance. Under factor loadings, the first factor comprised five variables associated with knowledge-related issues (knowledge of symptoms, spread of knowledge, precautionary steps for protection, knowledge of risk criteria, and sanitization) (Table 5). The second factor comprised two variables (the accuracy of social media facts and figures, and biases in news broadcasting). The third factor incorporated two variables (home quarantine and lockdown issues). The fourth and fifth factors comprised two variables (fake information issues and psychological effect of lockdown). For subsequent estimation, the researchers further analyzed the normality of residuals, anti-image correlation and covariance matrices, and scree plot (Figure 1). In the scree plot, factors with eigenvalues >1 were counted. Furthermore, this study used parallel analysis to more effectively validate the factor analysis.

Table 5. Rotated component matrix under factor analysis.

Variables	Component			
	1	2	3	4
Teaches about symptoms of coronavirus (COVID-19)	0.82			
Teaches about the spread of coronavirus (COVID-19)	0.847			
Teaches precautionary steps to reduce the chances of getting infected	0.868			
Teaches categories of risk related to coronavirus (COVID-19)	0.833			
Teaches about being properly sanitized.	0.774			
Highlights actual figures related to death or infection during the pandemic		0.889		
Highlights COVID-19 without any biases as it is a global issue		0.864		
Keeps one entertained during the home quarantine/lockdown period			0.903	
Effectively presents the benefits of the "stay home and stay safe" slogan			0.814	
Fake news/information related to coronavirus is spreading				0.875
There has been a negative effect on mental health during the outbreak of COVID-19				0.849

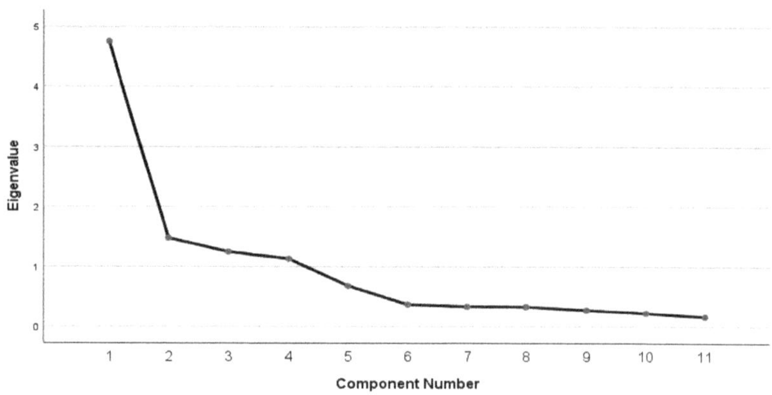

Figure 1. Scree plot after factor analysis.

Exploratory Factor Analysis (EFA)

Factor analysis identified four factors—(i) knowledge and health safety, (ii) trust in social media news, (iii) distance maintenance or quarantine, and (iv) psychological effect. The study conducted a path analysis using the latent relationships among the four factors, and the model was assessed according to its goodness of fit. As no previous studies analyzed the effect of social media on education, this study innovatively employed exploratory analysis by scrutinizing three factors and then establishing a structural equation model (SEM). Moreover, various cutoff points were examined using particular software.

For determining parameters, maximum-likelihood estimation was used as it is better than other traditional techniques and its estimates are unbiased, consistent, and efficient. A positive path coefficient between factors (standardized and unstandardized estimates) indicates their positive influence on each other through social media (Table 6). Therefore, social media plays a key role in enhancing knowledge, which ultimately helps to improve health and safety issues, as well as increase trust in social media news, quarantine maintenance, and psychological impact. In other words, social media can be used as a positive trigger for maintaining lockdown and raising the awareness of safety issues. Furthermore, trusting figures and abiding by lockdowns were treated as factor loadings (β = 0.564 to 0.973). This suggested that the measurement models of four factors fitted well. A squared factor loading shows the proportion of variance in the observed variable that is explained by the factor. A value of rotated loadings closer to 1 better explains the path coefficient from one variable to another. In other words, the measurement error is reduced. In Table 6, under the first factor (f1) related to knowledge of health and safety in acknowledging symptoms, the squared factor loading of knowledge about COVID-19 spread explained 86% of the variance for that variable under the first factor, with the remaining 14% representing the measurement error. Similarly, under f4 (psychological monotony), the squared loading of the negative effect on mental health during the outbreak of COVID-19 explained 97.3% of the variance. Each coefficient was positively and significantly associated at a 1% level of significance.

Table 6. Path coefficients of structural equation model.

Parameter	Items	B	*p*-Value
F1 (knowledge and safeguarding health through media)	Teaches about symptoms of coronavirus (COVID-19)	0.818 ***	0.001
	Teaches about spread of coronavirus (COVID-19)	0.854 ***	0.001
	Teaches precautionary steps to reduce the chances of getting infected	0.885 ***	0.001
	Teaches categories of risk related to coronavirus (COVID-19)	0.792 ***	0.001
	Teaches about being properly sanitized.	0.735 ***	0.001
F2 (social media's accuracy)	Highlights the actual figures related to death or infection during the pandemic	0.809 ***	0.001
	Highlights COVID-19 without any biases as it is a global issue	0.818 ***	0.001
F3 (self-detainment at home)	Keeps one entertained during the home quarantine/lockdown period	0.656 ***	0.001
	Effectively presents the benefits of the "stay home and stay safe" slogan	0.972 ***	0.001
F4 (psychological monotony)	Fake news/information related to coronavirus is spreading	0.564 ***	0.001
	There has been a negative effect on mental health during the outbreak of COVID-19	0.973 ***	0.002

*** The 1% level of significance for β-coefficient.

As another path diagram in tabulated form, we demonstrate the direct relationships between variables, as well as the total effect of each diagram. As each coefficient was positive, the overall path diagram indicated a positive relationship with statistical significance. In Table 7, the coefficients of the last two factors were >1. Hence, psychological monotony and lockdowns had a greater effect according to the responses to "keeps one entertained during the home quarantine/lockdown period" and "fake news/information related to coronavirus is spreading". Hence, our path diagram linking factors was statistically significant in terms of the total effect, model measurement, model fitness, and overall estimation.

Table 7. Total effect on factors of explanatory variables.

Paths	Coefficients	SE	*p*-Value
Teaches about symptoms of coronavirus (COVID-19) > f1	1	(constrained)	
Teaches about the spread of coronavirus (COVID-19) > f1	1.148	0.082	<0.001
Teaches precautionary steps to reduce the chances of getting infected > f1	1.126	0.082	<0.001
Teaches categories of risk related to coronavirus (COVID-19) > f1	1.134	0.094	<0.001
Teaches about being proper sanitized > f1	0.933	0.088	<0.001
Fake news/information related to coronavirus is spreading > f2	1	(constrained)	
There has been a negative effect on mental health during the outbreak of COVID-19 > f2	0.917	0.144	<0.001
Highlights the actual figures related to death or infection during pandemic > f3	1	(constrained)	
Highlights COVID-19 without any biases as it is a global issue > f3	1.390	0.144	<0.001
Keeps one entertained during the home quarantine/lockdown period > f4	1	(constrained)	
Effectively presents the benefits of the "stay home and stay safe" slogan > f4	1.920	0.238	<0.001

The chi-square (df = 38) ratio was 121.3, with a *p*-value of 0.19. This suggests that our hypothesized model fit the sample data well, and the null hypothesis of the model vs. saturated model was accepted [41,42]. In this model, the root-mean-square error of approximation (RMSEA) was 0.042 (<0.080); thus, the model adequacy was acceptable [37,41].

Additionally, the computed CLOSE (0.602) was significantly higher than 0.050, indicating no evidence to reject the fact that the RMSEA was greater than 0.500. Furthermore, the comparative fit index (CFI) and Tucker–Lewis index (TLI) [39,41,43] were 0.95 and 0.96, respectively. The coefficient of determination was also close to 1 (0.98). These measures all suggested that the model had a good fit. Subsequently, the skewness and kurtosis for normality, residuals, and basic statistics of variables were estimated [43]. There were no outliers, and each variable obeyed a normal distribution with an asymmetric shape according to skewness and kurtosis.

4. Discussion

This study aimed to evaluate the effects of social media on knowledge and safety issues, the accuracy of figures, isolation, and psychological monotony through exploratory factor analysis (EFA), in contrast to other theories such as knowledge, attitudes, and practices (KAP) [23,37,44–46] or other equation-based models [47–50]. Spreading information has a significant impact on people's behavior and can change how well government countermeasures work. Despite lockdown being the only active preventive measure against

COVID-19, it was impossible to reduce labor in several areas of Bangladesh [10,12,49]. This study elucidated the positive and negative effects of social media on knowledge of COVID-19 among university students. Dependency on social media and knowledge on health safety represented the first factor explaining awareness of COVID-19 symptoms, spread, and transmission, as well as sanitization. These items revealed that social media played a significant role in preparing mankind for COVID-19, in line with similar research [24,50–53].

Accordingly, models that predict viral propagation are beginning to take into consideration the population's behavioral reaction to public health measures and the communication dynamics underlying content consumption [5,6,20,49,53]. Experts and beginners alike use social media to share their sensible and irrational viewpoints with little moderation. Self-detainment or isolation was the solution to fight this pandemic, and social media presented us with several miniature movies or animations. To reduce depression, anxiety, or stress, social media can play a positive role through entertainment [30,54]. Moreover, a recent study discovered that internet-based smartphone use can improve the perceived quality of life through facilitating positive social media connections, online shopping, online conferencing, and constant interaction with friends and family living in different countries [13].

Social media has been a blessing in this tough time for millions. It has been an excellent platform to enhance interaction and study collaboratively. Teachers can now create interactive lectures, graphical contents, and motion pictures, as well as use diversified digital tools to enable students to grasp lessons swiftly. The media has also been active in transmitting the latest news, highlighting the unfavorable attitude of many Bangladeshis. Long-term lockdowns, as well as unfavorable news or information, may impact mental health. An earlier study looked at the impact of lockdowns on mental health during the severe acute respiratory syndrome (SARS) outbreak [55–60].

Facebook, Twitter, YouTube, WhatsApp, and similar social sites assist students in getting updated information on national and international issues. Education is no longer confined to textbooks. Even before the pandemic, schools and colleges assigned tasks or set question papers according to the curriculum, thus limiting a student's learning. With access to social media platforms through online learning, students have the opportunity to search several sources for well-researched solutions. Teachers are no longer confined to the traditional teaching system. They provide students with educational video links, access to important resources, and assignment-based tasks to evaluate their understanding. Students are forced to sit exams, while question patterns are designed to require interaction and cooperation among friends. Social media platforms such as Messenger and WhatsApp enable learners get instant information, reviews, or solutions to their problems or get in touch with professors.

Facebook has played a vital role during this crisis. Several educational groups have been created to understand students' problems and provide them with proper guidance. Ed-tech platforms in Bangladesh use these social media platforms to offer courses at an affordable cost. In addition, these ed-tech organizations are coming up with more interactive video lessons to make learning enjoyable, while providing educational content. The COVID-19 pandemic has led to the birth of many ed-tech companies, changing the traditional educational system in Bangladesh at an unprecedented pace [8,19,49].

Students from remote areas craving quality education can browse YouTube to connect with the best teachers. Watching YouTube videos can motivate students to learn more, share information, or increase their attention span. It has become much easier to read the desired books, access online notes, or arrange video calls with teachers.

Not many students are connected to Twitter, but those who have signed up can gain global knowledge. This platform is paving the way in building connections and providing an opportunity to stand up for oneself. Disseminating views is a great way to build confidence at a younger age. With students facing less academic pressure, they are taking their creativity to a whole new level. Moreover, by sharing their work on various social media platforms, students are getting a break from their daily study routine.

This study tried to examine alternative ways of indicating social media's effect on knowledge about the COVID-19 pandemic. Firstly, according to statistics, the majority of the respondents agreed strongly that social media is a driving force in safeguarding health, remaining home safe and sound, being aware of figures and reports spread through media, and adjusting with monotony. Knowledge through media helps to increase the washing of hands, as well as the use of a handkerchief or sanitizer, and to reduce touching of the eyes, nose, and mouth. Secondly, lockdowns or quarantines are novel concepts to most; however, they are vital in reducing transmission. Due to COVID-19′s infectivity, today's primary remedy is to stay home and follow social distancing measures. Thirdly, social media is not a government agency or solicit organization where people blindly rely on the news. Instead, people's confidence in statistics, data, and outcomes mainly derive from posting on social media, as well as interacting with Facebook, Twitter, or WhatsApp updates. Lastly, exploratory factor analysis (EFA) revealed a positive relationship between social media and health outcomes, suggesting that health safeguarding, lockdowns, reliance on media figures, and psychological resentment are fortified through social media.

5. Conclusions

Social media is the most convenient way to access information, share an opinion, and evaluate its justification. It has been vital in following COVID-19 and its traumatic death toll. This study investigated the impact of social media usage on the COVID-19 pandemic through an online-based questionnaire. The COVID-19 outbreak has become a global catastrophe. Our study tried to evaluate latent information accessed through social media on the perception of health issues, quarantine maintenance, data validation and accuracy, and psychological monotony related to COVID-19.

Empirical evidence through factor analysis portrayed a well-fitted model explaining 78% of the variance. Path analysis revealed that health-related issues and safety were significantly associated at the 1% level with quarantines, media figure accuracy, and the psychological effect. The hypothetical factors were positively related to each other through social media. Univariate analysis highlighted issues such as the health consciousness of social media users related to knowledge about COVID-19 and its symptoms, risk assessment, and social media accuracy. Factor analysis established four pillars: knowledge and safeguarding health through media, self-detainment at home, social media's accuracy, and psychological monotony. The results indicated a positive impact on health consciousness in terms of washing hands, using handkerchiefs or tissue, and staying safe at home as much as possible. Thus, social media has had a positive impact on improving humankind and conquering COVID-19. Individuals worldwide have faced enormous stress related to e health concerns stemming from the COVID-19 pandemic, which has escalated social media use. People used social media to seek accurate health information and stay in touch with coworkers, peers, friends, and family members. Precautionary health practices are regarded as the most effective preventive measures for COVID-19 transmission. Even though a vaccination program has begun, both vaccinated and unvaccinated people are advised to take preventive measures [2,24]. People use social media and other educational technology platforms to gain health-related information and to seek major emotional, informational, educational, and peer support. More individuals are utilizing social media, increasing access to health-related information. The introduction of facemasks, handwashing, and social seclusion foreshadowed peer, informational, and emotional support. Some limitations of this study were that it did not rely on theories such as knowledge, attitudes, and practices (KAP) [44–46] or psychometric scales. Secondly, this study investigated COVID-19 without considering mental condition, sleep quality, or crucial variables. Overall, this study mostly exhibited a positive effect of social media on the knowledge of students.

Author Contributions: Conceptualization, S.C., N.M.Z. and V.N.; data curation, O.M., I.A.D. and O.H.; formal analysis, S.C., N.M.Z., I.A.D. and O.H.; investigation, S.C., N.M.Z., O.M. and V.N.; methodology, S.C., N.M.Z. and I.A.D.; project administration, M.R. and O.M.; software, I.A.D. and M.R.; supervision, O.M. and O.H.; validation, M.R., N.M.Z. and V.N.; writing—original draft, S.C., N.M.Z. and V.N.; writing—review and editing, S.C., N.M.Z., M.R. and V.N. All authors have read and agreed to the published version of the manuscript.

Funding: This research did not receive any specific grant from funding agencies in the public, commercial, or not-for-profit sectors.

Institutional Review Board Statement: The research study entitled "Migrant organizations and the co-production of social protection" underlying this article includes human research participants. The study was prospectively approved by the legal office of the Technical University of Dortmund. Ethical approval was not mandatory for this study.

Informed Consent Statement: Informed consent was obtained from all subjects involved in the study.

Data Availability Statement: To protect the privacy of our research participants, research data will not be shared.

Conflicts of Interest: The authors declare no conflict of interest.

References

1. Hua, J.; Shaw, R. Corona virus (COVID-19) "infodemic" and emerging issues through a data lens: The case of China. *Int. J. Environ. Res. Public Health* **2020**, *17*, 2309. [CrossRef] [PubMed]
2. Zhang, H.; Shaw, R. Identifying Research Trends and Gaps in the Context of COVID-19. *Int. J. Environ. Res. Public Health* **2020**, *17*, 3370. [CrossRef] [PubMed]
3. Fernandes, N. Economic Effects of Coronavirus Outbreak (COVID-19) on the World Economy. *SSRN Electron. J.* **2020**. [CrossRef]
4. Arshad, A.; Shajeea, M.B.; Naseem, A.; Asadullah, A.A.; Ayman, I. The Outbreak of Coronavirus Disease 2019 (COVID-19)—An Emerging Global Health Threat. *J. Infect. Public Health* **2020**, *13*, 644–646. [CrossRef]
5. Lai, C.-C.; Wang, C.-Y.; Wang, Y.-H.; Hsueh, S.-C.; Ko, W.-C.; Hsueh, P.-R. Global epidemiology of coronavirus disease 2019 (COVID-19): Disease incidence, daily cumulative index, mortality, and their association with country healthcare resources and economic status. *Int. J. Antimicrob. Agents* **2020**, *55*, 105946. [CrossRef]
6. Shaman, J.; Karspeck, A.; Yang, W.; Tamerius, J.; Lipsitch, M. Real-time influenza forecasts during the 2012–2013 season. *Nat. Commun.* **2013**, *4*, 2837. [CrossRef]
7. Zayed, N.M.; Edeh, F.O.; Islam, K.M.A.; Nitsenko, V.; Polova, O.; Khaietska, O. Utilization of Knowledge Management as Business Resilience Strategy for Microentrepreneurs in Post-COVID-19 Economy. *Sustainability* **2022**, *14*, 15789. [CrossRef]
8. Anwar, S.; Nasrullah, M.; Hosen, M.J. COVID-19 and Bangladesh: Challenges and How to Address Them. *Front. Public Health* **2020**, *8*, 154. [CrossRef]
9. Edeh, F.O.; Zayed, N.M.; Nitsenko, V.; Brezhnieva-Yermolenko, O.; Negovska, J.; Shtan, M. Predicting Innovation Capability through Knowledge Management in the Banking Sector. *J. Risk Financ. Manag.* **2022**, *15*, 312. [CrossRef]
10. Zayed, N.M.; Edeh, F.O.; Islam, K.M.A.; Nitsenko, V.; Dubovyk, T.; Doroshuk, H. An Investigation into the Effect of Knowledge Management on Employee Retention in the Telecom Sector. *Adm. Sci.* **2022**, *12*, 138. [CrossRef]
11. Edeh, F.O.; Zayed, N.M.; Perevozova, I.; Kryshtal, H.; Nitsenko, V. Talent Management in the Hospitality Sector: Predicting Discretionary Work Behaviour. *Adm. Sci.* **2022**, *12*, 122. [CrossRef]
12. Zayed, N.M.; Edeh, F.O.; Darwish, S.; Islam, K.M.A.; Kryshtal, H.; Nitsenko, V.; Stanislavyk, O. Human Resource Skill Adjustment in Service Sector: Predicting Dynamic Capability in Post COVID-19 Work Environment. *J. Risk Financ. Manag.* **2022**, *15*, 402. [CrossRef]
13. IIslam, M.S.; Sujan, M.S.H.; Tasnim, R.; Ferdous, M.Z.; Masud, J.H.B.; Kundu, S.; Mosaddek, A.S.M.; Choudhuri, M.S.K.; Kircaburun, K.; Griffiths, M.D. Problematic internet use among young and adult population in Bangladesh: Correlates with lifestyle and online activities during the COVID-19 pandemic. *Addict. Behav. Rep.* **2020**, *12*, 100311. [CrossRef] [PubMed]
14. Bonaccorsi, G.; Pierri, F.; Cinelli, M.; Flori, A.; Galeazzi, A.; Porcelli, F.; Schmidt, A.L.; Valensise, C.M.; Scala, A.; Quattrociocchi, W.; et al. Economic and social consequences of human mobility restrictions under COVID-19. *Proc. Natl. Acad. Sci. USA* **2020**, *117*, 15530–15535. [CrossRef] [PubMed]
15. Joosten, T. *Social Media for Educators: Strategies and Best Practices*; John Wiley & Sons: Hoboken, NJ, USA, 2012.
16. Saydan, R.; Dülek, B. The impact of social media advertisement awareness on brand awareness, brand image, brand attitude and brand loyalty: A research on university students. *Int. J. Contemp. Econ. Adm. Sci.* **2019**, *9*, 470–494.
17. Souravlas, S.; Anastasiadou, S. Pipelined Dynamic Scheduling of Big Data Streams. *Appl. Sci.* **2020**, *10*, 4796. [CrossRef]
18. Souravlas, S.; Anastasiadou, S.; Katsavounis, S. More on Pipelined Dynamic Scheduling of Big Data Streams. *Appl. Sci.* **2021**, *11*, 61. [CrossRef]

19. Dutta, A. Impact of Digital Social Media on Indian Higher Education: Alternative Approaches of Online Learning during COVID-19 Pandemic Crisis. *Int. J. Sci. Res. Publ.* **2020**, *10*, 604–611. [CrossRef]
20. Cinelli, M.; Quattrociocchi, W.; Galeazzi, A.; Valensise, C.M.; Brugnoli, E.; Schmidt, A.L.; Zola, P.; Zollo, F.; Scala, A. The COVID-19 social media infodemic. *Sci. Rep.* **2020**, *10*, 16598. [CrossRef]
21. Pennycook, G.; McPhetres, J.; Zhang, Y.; Lu, J.G.; Rand, D.G. Fighting COVID-19 Misinformation on Social Media: Experimental Evidence for a Scalable Accuracy-Nudge Intervention. *Psychol. Sci.* **2020**, *31*, 770–780. [CrossRef]
22. Depoux, A.; Martin, S.; Karafillakis, E.; Preet, R.; Wilder-Smith, A.; Larson, H. The pandemic of social media panic travels faster than the COVID-19 outbreak. *J. Travel Med.* **2020**, *27*, taaa031. [CrossRef] [PubMed]
23. Ni, M.Y.; Yang, L.; Leung, C.M.C.; Li, N.; Yao, X.I.; Wang, Y.; Leung, G.M.; Cowling, B.J.; Liao, Q. Mental Health, Risk Factors, and Social Media Use During the COVID-19 Epidemic and Cordon Sanitaire Among the Community and Health Professionals in Wuhan, China: Cross-Sectional Survey. *JMIR Ment. Health* **2020**, *7*, e19009. [CrossRef] [PubMed]
24. Sharif, N.; Opu, R.R.; Alzahrani, K.J.; Ahmed, S.N.; Islam, S.; Mim, S.S.; Khan, F.B.; Zaman, F.; Dey, S.K. The positive impact of social media on health behavior towards the COVID-19 pandemic in Bangladesh: A web-based cross-sectional study. *Diabetes Metab. Syndr. Clin. Res. Rev.* **2021**, *15*, 102206. [CrossRef] [PubMed]
25. Ngai, C.S.B.; Singh, R.G.; Lu, W.; Koon, A.C. Grappling With the COVID-19 Health Crisis: Content Analysis of Communication Strategies and Their Effects on Public Engagement on Social Media. *J. Med. Internet Res.* **2020**, *22*, e21360. [CrossRef]
26. Eysenbach, G. Can Tweets Predict Citations? Metrics of Social Impact Based on Twitter and Correlation with Traditional Metrics of Scientific Impact. *J. Med. Internet Res.* **2011**, *13*, e123. [CrossRef]
27. Henrich, N.; Holmes, B. Communicating during a pandemic: Information the public wants about the disease and new vaccines and drugs. *Health Promot. Pract.* **2011**, *12*, 610–619. [CrossRef] [PubMed]
28. Ahmad, A.R.; Murad, H.R. The Impact of Social Media on Panic During the COVID-19 Pandemic in Iraqi Kurdistan: Online Questionnaire Study. *J. Med Internet Res.* **2020**, *22*, e19556. [CrossRef]
29. Doza, B.; Shammi, M.; Bahlman, L.; Islam, A.R.M.T.; Rahman, M. Psychosocial and Socio-Economic Crisis in Bangladesh Due to COVID-19 Pandemic: A Perception-Based Assessment. *Front. Public Health* **2020**, *8*, 341. [CrossRef]
30. Islam, M.M.; Islam, M.M.; Ahmed, F.; Rumana, A.S. Creative social media use for COVID-19 prevention in Bangladesh: A structural equation modeling approach. *Soc. Netw. Anal. Min.* **2021**, *11*, 1–14. [CrossRef]
31. Radwan, E.; Radwan, A. The Spread of the Pandemic of Social Media Panic during the COVID-19 Outbreak. *Eur. J. Environ. Public Health* **2020**, *4*, em0044. [CrossRef]
32. Paul, L. *Encyclopedia of Survey Research Methods*; Sage Publications, Inc.: Thousand Oaks, CA, USA, 2008. [CrossRef]
33. Jarvis, C.B.; MacKenzie, S.B.; Podsakoff, P.M. A Critical Review of Construct Indicators and Measurement Model Misspecification in Marketing and Consumer Research. *J. Consum. Res.* **2003**, *30*, 199–218. [CrossRef]
34. Williams, L.J.; Edwards, J.R.; Vandenberg, R.J. Recent Advances in Causal Modeling Methods for Organizational and Management Research. *J. Manag.* **2003**, *29*, 903–936. [CrossRef]
35. Martens, M.P. The Use of Structural Equation Modeling in Counseling Psychology Research. *Couns. Psychol.* **2005**, *33*, 269–298. [CrossRef]
36. Roger, M. *The SAGE Handbook of Quantitative Methods in Psychology*; Sage Publications, Inc.: Thousand Oaks, CA, USA, 2009. [CrossRef]
37. Hancock, G.R.; Stapleton, L.M.; Mueller, R.O. *The Reviewer's Guide to Quantitative Methods in the Social Sciences*, 2nd ed.; Routledge: New York, NY, USA, 2018. [CrossRef]
38. Kieffer, M.J. Converging Trajectories: Reading Growth in Language Minority Learners and Their Classmates, Kindergarten to Grade 8. *Am. Educ. Res. J.* **2011**, *48*, 1187–1225. [CrossRef]
39. Bollen, K.A. *Structural Equations with Latent Variables*; John Wiley & Sons, Inc.: Hoboken, NJ, USA, 2014. [CrossRef]
40. Howard, A.L. Handbook of Structural Equation Modeling. *Struct. Equ. Model. A Multidiscip. J.* **2013**, *20*, 354–360. [CrossRef]
41. Hu, L.T.; Bentler, P.M. Cutoff criteria for fit indexes in covariance structure analysis: Conventional criteria versus new alternatives. *Struct. Equ. Model. Multidiscip. J.* **1999**, *6*, 1–55. [CrossRef]
42. Tabri, N.; Elliott, C.M. Principles and Practice of Structural Equation Modeling. *Can. Grad. J. Sociol. Criminol.* **2012**, *1*, 59–60. [CrossRef]
43. Byrne, B.M. *Structural Equation Modeling with EQS: Basic Concepts, Applications, and Programming*, 2nd ed.; Routledge: New York, NY, USA, 2013; pp. 1–440. [CrossRef]
44. Hesaraki, M.; Akbarizadeh, M.; Ahmadidarrehsima, S.; Moghadam, M.P.; Izadpanah, F. Knowledge, attitude, practice and clinical recommendations of health care workers towards COVID-19: A systematic review. *Rev. Environ. Health* **2020**, *36*, 345–357. [CrossRef]
45. Kumar, B.; Pinky, S.D.; Nurudden, A. Knowledge, attitudes and practices towards COVID-19 guidelines among students in Bangladesh. *Soc. Sci. Humanit. Open* **2021**, *4*, 100194. [CrossRef]
46. Rahman, M.; Khan, S.J.; Sakib, M.S.; Halim, A.; Rahman, M.; Asikunnaby; Jhinuk, J.M. COVID-19 responses among university students of Bangladesh: Assessment of status and individual view toward COVID-19. *J. Hum. Behav. Soc. Environ.* **2021**, *31*, 512–531. [CrossRef]
47. Fedushko, S.; Ustyianovych, T. E-Commerce Customers Behavior Research Using Cohort Analysis: A Case Study of COVID-19. *J. Open Innov. Technol. Mark. Complex.* **2022**, *8*, 12. [CrossRef]

48. Hossain, M.A.; Hossain, K.M.A.; Saunders, K.; Uddin, Z.; Walton, L.M.; Raigangar, V.; Sakel, M.; Shafin, R.; Kabir, F.; Faruqui, R.; et al. Prevalence of Long COVID symptoms in Bangladesh: A prospective Inception Cohort Study of COVID-19 survivors. *BMJ Glob. Health* **2021**, *6*, e006838. [CrossRef] [PubMed]
49. Shammi, M.; Doza, B.; Islam, A.R.M.T.; Rahman, M. COVID-19 pandemic, socioeconomic crisis and human stress in resource-limited settings: A case from Bangladesh. *Heliyon* **2020**, *6*, e04063. [CrossRef] [PubMed]
50. Hu, Z.; Khokhlachova, Y.; Sydorenko, V.; Opirskyy, I. Method for Optimization of Information Security Systems Behavior under Conditions of Influences. *Int. J. Intell. Syst. Appl.* **2017**, *9*, 46. [CrossRef]
51. Limaye, R.J.; Sauer, M.; Ali, J.; Bernstein, J.; Wahl, B.; Barnhill, A.; Labrique, A. Building trust while influencing online COVID-19 content in the social media world. *Lancet Digit. Health* **2020**, *2*, e277–e278. [CrossRef]
52. O'Sullivan, E.; Cutts, E.; Kavikondala, S.; Salcedo, A.; D'Souza, K.; Hernandez-Torre, M.; Anderson, C.; Tiwari, A.; Ho, K.; Last, J. Social Media in Health Science Education: An International Survey. *JMIR Med. Educ.* **2017**, *3*, e6304. [CrossRef]
53. Kim, L.; Fast, S.M.; Markuzon, N. Incorporating media data into a model of infectious disease transmission. *PLoS ONE* **2019**, *14*, e0197646. [CrossRef]
54. Di Blasi, M.; Giardina, A.; Giordano, C.; Coco, G.L.; Tosto, C.; Billieux, J.; Schimmenti, A. Problematic video game use as an emotional coping strategy: Evidence from a sample of MMORPG gamers. *J. Behav. Addict.* **2019**, *8*, 25–34. [CrossRef]
55. Hawryluck, L.; Gold, W.L.; Robinson, S.; Pogorski, S.; Galea, S.; Styra, R. SARS Control and Psychological Effects of Quarantine, Toronto, Canada. *Emerg. Infect. Dis.* **2004**, *10*, 1206–1212. [CrossRef]
56. Nitsenko, V. What is the Government Really Pursuing by Introducing Quarantine Measures in the Conditions of COVID-19? The Case of Ukraine. *Ukr. Policymaker* **2021**, *9*, 69–77. [CrossRef]
57. Dorfman, N. "Social-Distancing" as a Chance to Revise the Paradoxes of Humanistic Philosophy: Personality Vs. Identity in Online Artistic Practices of the Pandemic. *Philos. Cosmol.* **2021**, *27*, 116–125. [CrossRef] [PubMed]
58. Pooran, S. Global Health Security in a New World Order: Winning the Battle but Losing the War. *Future Hum. Image* **2022**, *18*, 42–55. [CrossRef]
59. Malysheva, N.; Hurova, A. New Frontiers of Sustainable Human's Activities: Challenges for Legal Order of Space Mining Economy. *Adv. Space Law* **2021**, *8*, 76–85. [CrossRef]
60. Veljanovska Blazhevska, K. Populism Versus a Transparently Informed Public: The State of the Media Space in South East Europe. *Ukr. Policymaker* **2022**, *11*, 92–102. [CrossRef]

computation

MDPI

Article

Mass Media as a Mirror of the COVID-19 Pandemic

Kirill Yakunin [1,2,3], Ravil I. Mukhamediev [1,2,*], Elena Zaitseva [4,*], Vitaly Levashenko [4], Marina Yelis [1,2,*], Adilkhan Symagulov [1,2], Yan Kuchin [1,2,*], Elena Muhamedijeva [1], Margulan Aubakirov [5] and Viktors Gopejenko [6,7]

1 Institute of Information and Computational Technologies MES RK, Pushkin Street, 125, Almaty 050010, Kazakhstan; Yakunin.k@mail.ru (K.Y.); a.symagulov@satbayev.university (A.S.); muhamedijeva@gmail.com (E.M.)
2 Institute of Automation and Information Technologies, Satbayev University, Satpayev Street, 22A, Almaty 050013, Kazakhstan
3 School of Engineering Management, Almaty Management University, Rozybakiev Street, 227, Almaty 050060, Kazakhstan
4 Faculty of Management Science and Informatics, University of Zilina, Univerzitná 8215/1, 010 26 Žilina, Slovakia; vitaly@kifri.fri.uniza.sk
5 Department of Information Technology, Maharishi International University, 1000 N 4th Street, Fairfield, IA 52557, USA; margulan.aubakir@gmail.com
6 International Radio Astronomy Centre, Ventspils University of Applied Sciences, Inzhenieru Street, 101, LV-3601 Ventspils, Latvia; viktors.gopejenko@isma.lv
7 Department of Natural Science and Computer Technologies, ISMA University, Lomonosov Street, 1, LV-1011 Riga, Latvia
* Correspondence: ravil.muhamedyev@gmail.com (R.I.M.); elena.zaitseva@fri.uniza.sk (E.Z.); k.marina92@gmail.com (M.Y.); ykuchin@mail.ru (Y.K.)

Citation: Yakunin, K.; Mukhamediev, R.I.; Zaitseva, E.; Levashenko, V.; Yelis, M.; Symagulov, A.; Kuchin, Y.; Muhamedijeva, E.; Aubakirov, M.; Gopejenko, V. Mass Media as a Mirror of the COVID-19 Pandemic. *Computation* **2021**, *9*, 140. https://doi.org/10.3390/computation9120140

Academic Editors: Simone Brogi and Vincenzo Calderone

Received: 23 October 2021
Accepted: 2 December 2021
Published: 13 December 2021

Abstract: The media plays an important role in disseminating facts and knowledge to the public at critical times, and the COVID-19 pandemic is a good example of such a period. This research is devoted to performing a comparative analysis of the representation of topics connected with the pandemic in the internet media of Kazakhstan and the Russian Federation. The main goal of the research is to propose a method that would make it possible to analyze the correlation between mass media dynamic indicators and the World Health Organization COVID-19 data. In order to solve the task, three approaches related to the representation of mass media dynamics in numerical form—automatically obtained topics, average sentiment, and dynamic indicators—were proposed and applied according to a manually selected list of search queries. The results of the analysis indicate similarities and differences in the ways in which the epidemiological situation is reflected in publications in Russia and in Kazakhstan. In particular, the publication activity in both countries correlates with the absolute indicators, such as the daily number of new infections, and the daily number of deaths. However, mass media tend to ignore the positive rate of confirmed cases and the virus reproduction rate. If we consider strictness of quarantine measures, mass media in Russia show a rather high correlation, while in Kazakhstan, the correlation is much lower. Analysis of search queries revealed that in Kazakhstan the problem of fake news and disinformation is more acute during periods of deterioration of the epidemiological situation, when the level of crime and poverty increase. The novelty of this work is the proposal and implementation of a method that allows the performing of a comparative analysis of objective COVID-19 statistics and several mass media indicators. In addition, it is the first time that such a comparative analysis, between different countries, has been performed on a corpus in a language other than English.

Keywords: COVID-19; topic modeling; BigARTM; latent Dirichlet analysis; mass media analysis

1. Introduction

COVID-19 has highlighted the relative inefficiency and low productivity in the health sector, which in turn have contributed to increased social tension and a steady decline in the economic growth in most countries during the pandemic [1]. The healthcare system can be considered as one of the main factors determining the sustainable growth of welfare in many countries including Kazakhstan. However, healthcare systems in Kazakhstan and throughout the world face multiple problems, which cause an increased demand for health services, high public expectations, and higher expenses [2]. Not only economic but also social and medical efficiency is important in the healthcare system; "medical measures of therapeutic and preventive nature may be economically unprofitable, but medical and social effects require them" [3]. According to the authors of [4], a fundamental transformation of healthcare systems, based on Artificial Intelligence (AI) technology, is necessary. The economic impact of AI on healthcare in Europe is estimated at 200 billion euros [5]. The effect is associated with savings in time and an increase in the number of lives saved.

One of the technologies related to AI is Natural Language Processing (NLP) [6], which effectively uses machine learning techniques to process natural language texts and speech; it is used in healthcare to extract information from clinical records [7], to process speech messages, and to create question answering systems [8,9]. NLP methods can be used not only to address the direct healthcare objectives but also to assess how the mass media (media) reflect the public health situation during the pandemic. Mass media and social networks have a substantial influence on the informational environment of society. Nowadays, the media not only act as a source of information on current events, but often shape the information agenda and form the discourse of socially important topics [10,11]. The inadequate presentation of health authorities in the media may contribute to the spread of rumors and misinformation [12], and affect the mental health of the population [13]. Topic modeling in combination with sentiment analysis is often used to evaluate media texts [14–16].

The severity of the COVID-19 pandemic in Kazakhstan and Russia [17] is significantly higher than for an average nation. While Russia is the ninth largest country by population in the world, the total number of COVID-19 cases made Russia the third–fifth largest pandemic nation. Kazakhstan is in 63rd place in terms of population, and holds 36th–39th place according to the number of new cases. This makes these two nations an interesting target for this kind of research, especially since mass media in both countries is primarily in the same language (Russian).

In this paper, we aimed to achieve two research objectives: to identify the differences in the publication activities of the two countries regarding the COVID-19 pandemic and to assess the correlation of publication activity with the COVID-19 pandemic indicators.

The main contribution of this study was the development of a new method to compare and analyze real statistical data on COVID-19 (published by the WHO) and the responses of mass media specified in the study. These responses were evaluated by new indicators that are elaborated and introduced in the study. The indicators were developed based on three approaches used in the evaluation of mass media dynamics—automatically obtained topics, average sentiment, and dynamic indicators—according to manually selected search queries.

This study also represents the first time such a comparative analysis of COVID-19's representation in mass media was performed for languages other than English.

The obtained results showed the substantial differences in the representation of the pandemic in the media of Russia and Kazakhstan, as well as providing several insights on how the internet media tended to react to changes in epidemiological situations.

In this work, we used topic modeling for a comparative analysis of the corpus of media publications on the COVID-19 pandemic in two countries, and we also assessed the correlation of publication activities with the statistics of the pandemic. The work consisted of the following parts.

The first part of this study examined the existing research on media publications during the COVID-19 pandemic. The analysis showed the lack of comparative studies of the pandemic publication corpora in languages other than English.

In the second part, we considered the publication corpus and the method of processing this corpus; this made it possible to obtain three types of mass media dynamic indicators describing the different aspects of the representation of the COVID-19 situation by mass media.

In the third part, we described and discussed the obtained results. The main result of the experiments was a quantitative comparison of the coefficients of correlation between the mass media indicators and the indicators of the COVID-19 epidemiological situation, as well as the analysis of these coefficients. We also briefly described the system architecture of the proposed method of data collection and processing.

In the end, we briefly described the advantages and limitations of the proposed approach, and formulated the future research objectives.

2. Related Work

Evaluation of the media content is a focus of many research studies due to its practical importance for advertising companies, news agencies and governmental bodies. Based on the analysis of media content, it is possible to predict the possible popularity of news [18] and to plan PR strategies for the promotion of products and services [19,20]. Government sectors can use tools for the promotion of their opinions, as well as to improve the planning of publication activities (i.e., which topics and events should be emphasized) and to identify negative content.

According to the Edelman Trust Barometer [21], the trust in information presented by government and media channels remains low. The gap between the informed and general public is growing [21]. When people do not have reliable information or experience to comprehend what is going on, they become dependent on the information accessible via the mass media sources [22]. According to previous studies [23,24], mass media and social network sources employ manipulative techniques to form public opinion, or to focus the audience on specific topics. An additional factor affecting public perception is the increased availability of various news items on the Internet, which can create confusion due to the usage of personal, often unchecked sources of information, such as personal blogs, video streaming, and unverified news [25]. Hence, many researchers focus on the possibilities of assessing the negative effects of media and facilitating its positive effects [26].

During the COVID-19 pandemic, media messages have significantly affected people's emotions and their psychological stability [27]. During this period, more than 51% of news headlines in English-language media have had a negative sentiment and only about 30% of them were found to be positive [28]. Such information can cause anxiety, fear, anger, longing, sadness, etc. in a great number of people [29].

Public reactions to the implemented measures, assessed via the analysis of large volumes of documents, permits the adjustment of the restrictions imposed by government agencies. In particular, the positive attitudes of the population to the measures of the governments of South Korea [30] and Singapore [31] were revealed. There is some evidence of increasing confidence in traditional media in the United States [32] and in India [33]. During the pandemic, the amount of misinformation and rumors circulating on social media increased significantly; some of them could be detected automatically [34].

The analysis of mass media, social media, and publicly available datasets is important to encourage analytical efforts and to provide data for pandemic mitigation planning [35]. Such an analysis can also be used as one of the possible proxy indicators and even predictors of the economic situation in the country [36,37]. The list of analyzed indicators can include the level of inflation, unemployment, poverty, economic development, etc.

One of the main tools used to analyze large corpora of texts is topic modeling. The topic model determines the quantitative relationships between documents and topics, as well as between topics and words or phrases. Clusters of terms and phrases formed in the process of topic modeling, in particular, allow the solving of problems of synonymy and polysemy of terms [38]. To build a topic model of a document corpus, the following methods are generally used: Probabilistic Latent Semantic Analysis (PLSA), ARTM (Additive Regularization of Topic Models) [39] and, very commonly, Latent Dirichlet Allocation (LDA) [40].

Many studies that use LDA as a primary tool focus on identifying the list of topics prevalent in publications and further analyzing the sentiment of the messages [41]. For example, the authors of [42] identified several main topics on Twitter, including "news on new confirmed cases", "COVID-19-related deaths", "cases outside China (worldwide)", "COVID-19 outbreak in South Korea", "early signs of outbreak in New York", "Diamond Princess cruise", "economic impact", "preventive measures", "authorities", and "supply chain". However, topics related to treatment and symptoms are not as important on Twitter. In [43], the topics of the publications are summarized as follows: "work and life under pandemic conditions", "social problems", "understanding the nature of the virus", and "methods of prevention". The authors of the paper [44] determined that users in South Africa focus their attention on the following topics: "sale and consumption of alcohol", "staying at home", "daily tracking of statistics", "police brutality", "5G", "spread of disease", "testing", "doctors", and "conspiracy theories" about vaccines. An analysis of publications in different countries revealed common themes widely covered in the UK, India, Japan, and South Korea: "education", "economy", "USA", and "sports" [14].

LDA is a prevalent method of topic modeling (see Table 1). The most frequent language of the text corpora is English. Most of the publications are based on the social networks Twitter, Sina Weibo, and Facebook. The most frequently considered corpus type is publications on the situation in a particular country.

Table 1. Some examples of objectives and methods of research of publications about COVID-19.

Purpose of Research	Method	Data Source	The Language of the Corpus of Publications
The impact of news about COVID-19 on people's emotional state [27]	Statistical analysis	Online survey	English
Analysis of social media information during a pandemic [45]	LDA, Random Forest	Twitter, Sina Weibo	English, Chinese
Testing the hypothesis that COVID-19 is more likely to spread between regions with closer ties in social networks [46].	Statistical analysis	Facebook	data
Understanding the discourse and psychological reactions of Twitter users to COVID-19 [42].	LDA, sentiment analysis	Twitter	data
Identifying predominant themes and accompanying emotions [43]	LDA	Twitter	English
Identifying what topics are discussed by the public and how they affect the implementation of measures taken by the government [44]	LDA	Papers	English
Analysis of PubMed® publication topics and their evolution over time during the COVID-19 pandemic [47]	LDA	PubMed	English

Table 1. *Cont.*

Purpose of Research	Method	Data Source	The Language of the Corpus of Publications
Identification of the most representative themes and sentiment analysis [14]	Top2vec [48], RoBERTa [49]	The media	English
Assessing social media sentiment toward coronavirus [41]	LDA, sentiment analysis	Twitter	English
Analysis of Indian online users' tweets during the COVID-19 lockdown to identify texts containing fear, sadness, anger, and joy [50]	Sentiment analysis based on BERT	Twitter	English
Sentiment predictions on Covid-19 data [51]	Sentiment analysis based on LSTM	Twitter	English

Therefore, there is a certain lack of research on corpora of publications about COVID-19 in languages other than English. We did not identify the studies devoted to a comparative analysis of corpora of texts from the traditional internet media either. It is not clear how the sentiment of statements in social networks and mass media correlates with the objective indicators of the pandemic (the number of infected and sick people, the mortality rate, etc.).

One of the main aims of the research is to reduce the above-mentioned gap in studies; this paper performs a comparative analysis of the Russian-language media in Russia and Kazakhstan based on the corpus of texts we collected earlier.

We evaluated the correlations between the sentiment expressed in the media, and the number of publications on certain topics with objective indicators of the COVID-19 epidemic in Russia and Kazakhstan.

In this work, we define media as "traditional" mass media (newspapers, magazines, and TV-channels) presented in electronic form as well as purely electronic news websites and social networks represented by widespread services such as Twitter, Sina Weibo, Telegram, VK., etc. Attention is mainly paid to the mass media in the traditional sense, which continues to play an important role in shaping the opinions of the population. The media readership behavior in Kazakhstan and in Russia is very similar, although the lists of popular news sources are obviously different and generally have almost no intersections.

3. Methods and Data

The employed methods included the following steps (Figure 1): text corpus collection (a), text corpus processing (b), and correlation analysis using the objective data on the epidemiological situation (c).

(a) Data collection. Mass media and social network news publications were gathered using automatic scrapping tools.
(b) Text corpus processing was necessary to extract meaningful dynamic indicators of mass media publication activity. Three types of indicators were proposed; they were described in the section below: topics were obtained by a cascade of topic models, sentiment analysis, and analysis of full-text search queries.
(c) Correlation analysis. We performed the assessment of pairwise correlation between two groups of dynamic indicators—mass media indicators obtained in step (b) and COVID-10 epidemiological indicators. We used COVID-19 indicators, which were processed and prepared by the Center for Systems Science and Engineering at Johns Hopkins University (JHU CSSE) [52].

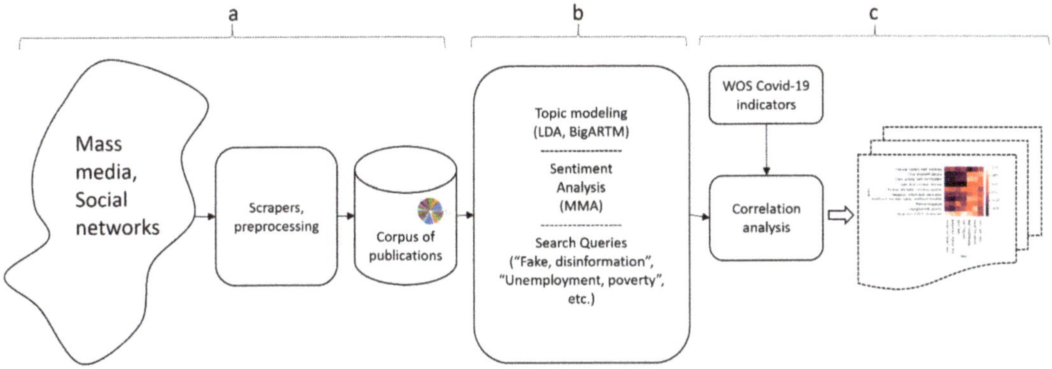

Figure 1. Main steps of the method for correlation analysis of mass media indicators and COVID-19 data.

Finally, the proposed method assumed the performance of manual analysis of the obtained correlation coefficients in order to draw conclusions about the similarities and differences of the pandemic reflection in different countries.

A corpus of news publications from media in Russia and in Kazakhstan was used for the research [53]. It included social networks (VK.com, YouTube, Instagram, and Telegram) and more than 20 news websites. The total numbers of news items were as follows: 4,233,990 documents, received from various sources in Kazakhstan, and 2,027,963 documents from various sources in Russia; the date span of the publications was from 2000 to 2021 (see Figure 2). The data mainly contained news publications from traditional news web sites or from official groups/channels of those web sites and resources on social networks. There was a small number of news publications from independent bloggers or slightly moderated social network groups.

The data were collected using the Python library Scrapy, for which a custom configurable Spider (crawler) was developed. The scraper was run regularly according to a source-by-source schedule, which ranged from hourly to daily execution based on source priority. The scheduling was implemented using Apache Airflow DAGs. The scraping process was started in late 2019 and subsequently worked according to the schedule; hence, the date and time of collection for each news publication were very close to its publication date.

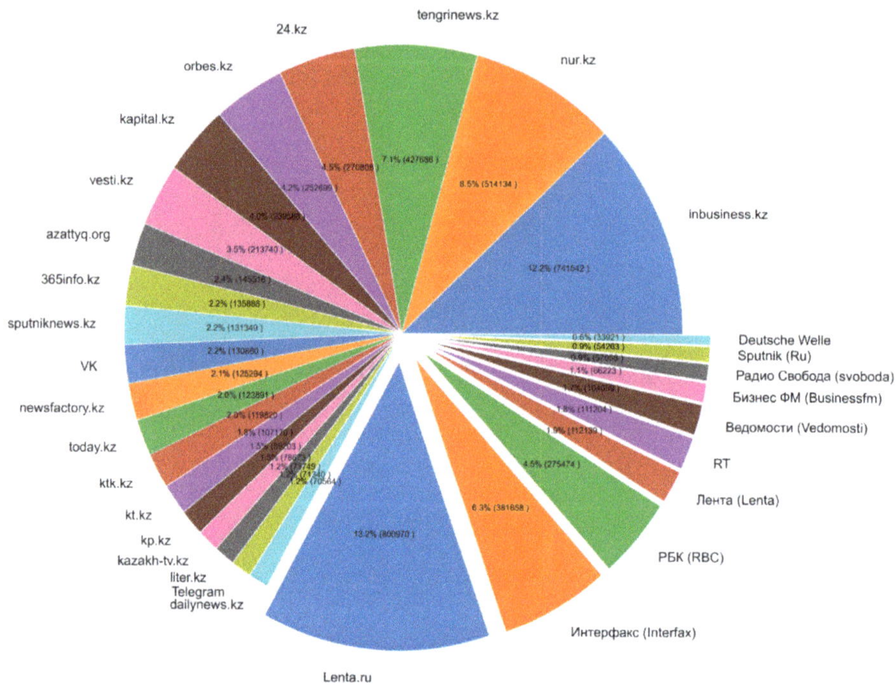

Figure 2. Major sources of the corpus.

3.1. COVID-19 Data

Data from JHU CSSE [52] were used is order to analyze the objectivity of the representation of the COVID-19 epidemiological situation in media in Kazakhstan and in Russia. The daily analyzed indicators are presented in Table 2.

Table 2. COVID-19 epidemiological indicators that were used in the analysis.

Indicator	Description
Number of new tests	Daily number of new tests for COVID-19
Positive rate	The share of COVID-19 tests that are positive, given as a rolling 7-day average (this is the inverse of tests per case indicator)
Number of new cases smoothed	New confirmed cases of COVID-19 (7-day smoothed)
Number of new deaths smoothed	New deaths attributed to COVID-19 (7-day smoothed)
Tests per case	Tests conducted per new confirmed case of COVID-19, given as a rolling 7-day average (this is the inverse of positive rate indicator)
Virus reproduction rate	Real-time estimate of the effective reproduction rate (R) of COVID-19 [54]
Stringency index	Government Response Stringency Index: composite measure based on 9 response indicators including school closures, workplace closures, and travel bans, rescaled to a value from 0 to 100 (100 = strictest response)

3.2. Methods

We proposed the use of three main approaches to analyzing media data:

- Topic-modeling approach;
- Sentiment analysis;

- Analysis by search queries.

The first approach applies Topic Modeling (TM). TM is a method that allows the automatic finding of the hidden latent structures of corpora based on the statistical characteristics of document collections. TM is often used in humanitarian research, since it allows the efficient representation of large volumes of textual data in the form of a distribution of a set of terms over documents (D) and a distribution of documents over topics (T) [55].

LDA [40,56] is often used as a method for building topic models. There is also a popular generalization of LDA, which employs a set of ARTM. We use an ARTM in the form of the BigARTM library [39] in this research.

The probabilistic thematic model is based on the assumption that each document is a set of words generated randomly and independently from the conditional probability distribution of words (w) in documents (d) [55]:

$$p(w|d) = \sum_{t \in T} p(w \mid t,\ d)\ p(t \mid d) = \sum_{t \in T} \varphi_{wt}\theta_{td} \quad (1)$$

which represents the sum of mixed conditional distributions on all T-set topics, where $p(w \mid t)$ is the conditional distribution of words (w) in topics (t), and $p(t \mid d)$ is the conditional distribution of topics in the documents (d), w defines the distribution of words and d represents the documents, φ_{wt} is a matrix representing distribution of words over topics and θ_{td} is a matrix representing the distribution of topics over documents. This ratio is true, based on the assumption that there is no need to maintain the order of documents in the corpus and the order of words in the documents. The LDA method assumes that the components φ_{wt} and θ_{td}. are generated by Dirichle's continuous multidimensional probability distribution. The aim of the algorithm is to search for parameters φ_{wt} and θ_{td} by maximizing the likelihood function with appropriate regularization:

$$\sum_{d \in M} \sum_{w \in d} n_{dw} ln \sum_{t \in T} \varphi_{wt}\theta_{td} + R(\varphi, \theta) \rightarrow max \quad (2)$$

where n_{dw} is the number of occurrences of the word w in the document d, and $R(\varphi, \theta)$ is a logarithmic regularizer. To determine the optimal number of topics (clusters) T, the method of maximizing the coherence value with the use of UMass metrics is often applied [57].

BigARTM is an open-source library for the simultaneous calculation of topic models on large text corpora, the implementation of which is based on the additive regularization approach (ARTM), in which the maximization of the logarithm of plausibility, restoring the original distribution of W words on documents D, is added to a weighted sum of regularizers, by many criteria:

$$R(\varphi, \theta) = \sum_{i=1} \tau_i R_i(\varphi, \theta) \quad (3)$$

This summand is a weighted linear combination of regularizers, with non-negative τ_i weights.

BigARTM offers a set of regularizers:

1. The smoothing regularizer, based on the assumption that the matrix columns φ and θ are generated by Dirichlet distributions with hyperparameters $\beta_0\beta_t$ and $\alpha_0\alpha_t$ (identical to the implementation of the LDA model, in which hyperparameters can only be positive);

$$R(\varphi, \theta) = \beta_0 \sum_{t \in T} \sum_{w \in W} \beta_{wt} ln \varphi_{wt} + \alpha_0 \sum_{d \in D} \sum_{w \in W} \alpha_{td} ln \theta_{td} \rightarrow max \quad (4)$$

In this way we can highlight the background topics, defining the vocabulary of the language, or calculate the general vocabulary in the section of each document.

2. By decreasing the regularizer coefficients, the reverse smoothing regularizer can be obtained:

$$(\varphi,\theta) = -\beta_0 \sum_{t\in T} \sum_{w\in W} \beta_{wt} ln\varphi_{wt} - \alpha_0 \sum_{d\in D} \sum_{w\in W} \alpha_{td} ln\theta_{td} \rightarrow max \quad (5)$$

This aims to identify the significant subject words, so-called lexical kernels, in addition to subject topics in each document, zeroing out small probabilities.

3. The decorrelator Phi regularizer makes topics more "different". The selection of topics allows the model to discard small, uninformative, duplicate, and dependent topics:

$$(\varphi,\theta) = -0.5 * \tau \sum_{t\in T} \sum_{s\in \frac{T}{t}} cov(\varphi_t \varphi_s) \rightarrow max,\ cov(\varphi_t \varphi_s) = \sum_{w\in W} \varphi_{wt} \quad (6)$$

This regularizer is independent of matrix θ. The estimation of differences in the discrete distributions is implemented by $\varphi_{wt} = p(w|t)$, in which the measure is the covariance of the current distribution of words in the topics φ_t versus the calculated distributions φ_s, where $s \in T/t$.

The BigARTM topic model was applied to a corpus of over a million texts (news) published from 1 January 2020 to 25 February 2021 from over 30 major internet media sources in Russia and in Kazakhstan. Concatenation of news titles and article bodies was used to extract the topics.

The analyzed media sources publish news articles in three different languages: Russian, English, and Kazakh. For the purpose of comparing Kazakhstani and Russian media, only the news in Russian was considered. Since all three languages use distinctly different alphabets, the filtering was based on simple character-frequency statistics. Next, the news in the Russian language was lemmatized using the PyMyStem3 library [58]. A list of stop words provided by the stop-words Python library [59] was applied.

A cascade of topic models was used in this research, since the preliminary experiments showed that a single topic model is not capable of providing the required details. First, a topic model with 200 topics was built; we refer to it as level-0 (initial) model. Then, experts manually chose and labelled the topics related to medicine, the pandemic, and healthcare. This labelling was used to filter a sub-corpus of documents that had relative weights, in relation to the selected topics (from θ-matrix), above a constant threshold, which was set to 0.05, empirically determined based on experiments. Then, a level-1 topic model (150 topics) was calculated based on the text document from the sub-corpus. However, the analysis of the level-1 topic model showed that the accuracy of results of medicine-related filtration was not high enough, and parts of the topics were irrelevant. Hence, the described process was re-iterated in order to obtain two more models (Level 2 and Level 3). Each time, the number of topics was chosen empirically based on quality metrics (perplexity, coherence, and contrast [39]), as presented in Table 3. Let us discuss the metrics used for the assessment of the models:

Table 3. Main information on the obtained topic models.

Topic Model	# Topics	# Documents	Membership Threshold	Perplexity	Contrast	Purity
Level-0	200	1679803	-	3165	0.48	0.203
Level-1	150	285564	0.05	1853	0.505	0.207
Level-2	100	241536	0.04	1895	0.509	0.244
Level-3	50	194392	0.1	1859	0.503	0.284

Perplexity is an indicator from information theory, which defines how well a probability model predicts a sample. Lower values indicate that the model predicts the sample better. The perplexity of a probability model (topic model) can be defined as:

$$PP(p) = 2^{H(p)} = 2^{-\sum_x p(x)log_2 p(x)} = \prod_x p(x)^{-p(x)}$$

where $H(p)$ is the information entropy of the distribution, and x represents an iterator over the samples (documents). Perplexity value does not have a minimum value; hence, it is usually used to compare different models over the same set of data or to detect the "elbow effect" to determine the optimal number of topics [60].

Contrast of the topics is defined by the formula $\frac{1}{|W_t|}\sum_{w \in w_t} p(t|w)$, where w_t is a topic kernel, i.e., the words from the topic with relation weight greater than or equal to a given threshold.

Purity is defined by the formula $\sum_{w \in w_t} p(w|t)$, where w_t is also a topic kernel [60].

Finally, the following list of topic models was used for analysis:

- The level-0 topic model, which mainly consisted of general topics, such as economy, medicine (in general), education, etc.
- The level-2 topic model comprised the topics related to medicine including the ones somehow related to medicine and the epidemiological situation, such as quarantine limitations in education, sport events and public life, the economic situation in the context of the pandemic, etc. (Figure 3)
- The level-3 topic model provided very high accuracy in classifying medicine- and healthcare-related topics and documents

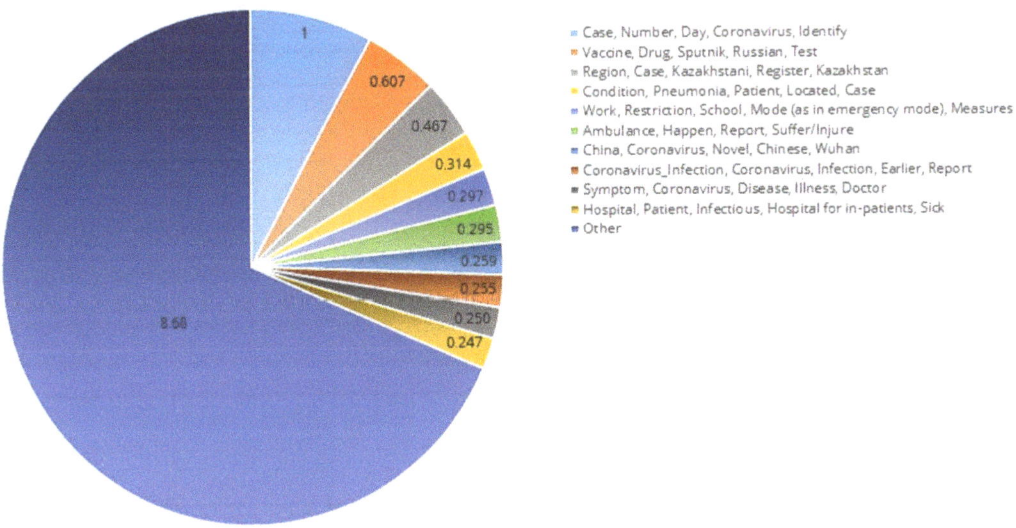

Figure 3. Distribution of topics by relative volume in the second level of the thematic model.

Table 3 illustrates number of topics, membership thresholds, and quality metrics for the obtained models.

The level-1 topic model did not provide the required level of accuracy; therefore, it was excluded from the analysis. It only served as an intermediate model, which allowed the obtaining of the more accurate models.

The obtained topic models were used in order to calculate the relative weight of each of the topics. The relative weight was calculated daily in order to be able to analyze the topics' dynamics within the publication activity. The relative weight of a topic is a ratio of a column of the θ-matrix, representing the given topic in relation to the sum of the whole θ-matrix. The relative weight ranges from 0 to 1 and shows the ratio of information related to the given topic in the information field described by the corpus under analysis. Figures 4–6 show the dynamics of weekly relative weight of 3 of 100 topics within the level-2 topic model.

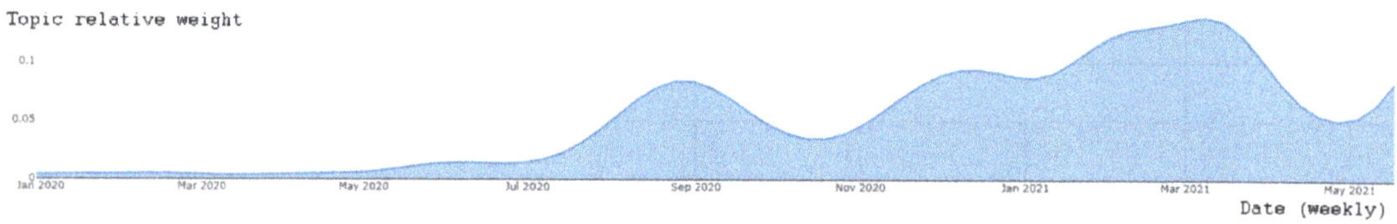

Figure 4. Weekly smoothed dynamics of "Vaccine, Drug, Sputnik-V, Russian, Test" topic relative weight.

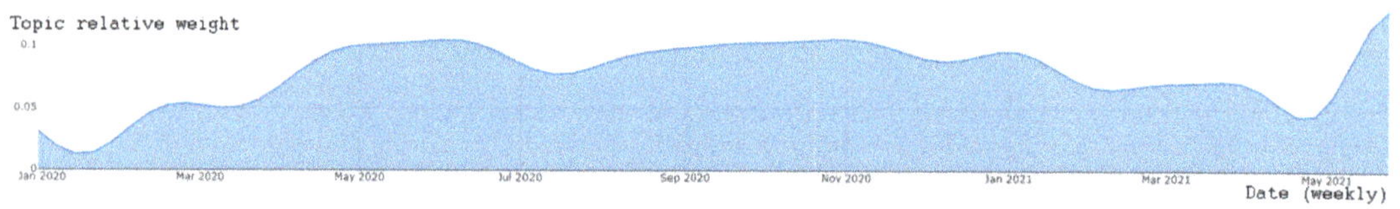

Figure 5. Weekly smoothed dynamics of "Case, Number, Day, Coronavirus, Reveal" topic relative weight.

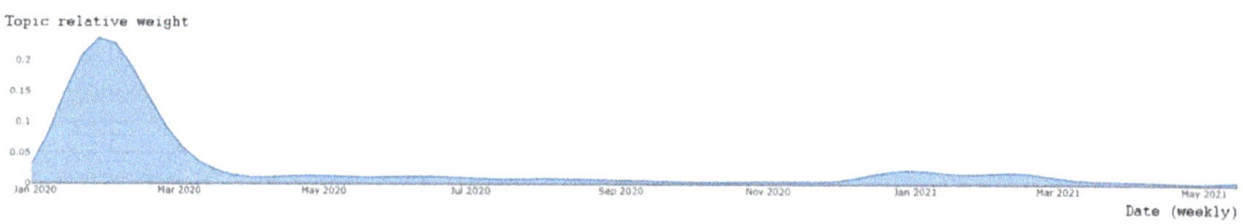

Figure 6. Weekly smoothed dynamics of "China, Coronavirus, New, Chinese, Wuhan" topic relative weight.

The topic model made it possible to exclude the personal bias from the process of analysis; it enhanced the model's utility in the task of assessing the reflection of epidemiological situation by mass media. However, this approach was found to have two main limitations:

1. It considers the dynamic weight of only single topic, while it might be possible that some combination of topics according to other criteria may be more representative;
2. Topic modelling cannot consider the expert opinion, and certain topics, which are considered to be important by experts, may not be distinguished automatically by the topic model, depending on its meta-parameters.

In order to resolve this limitations, two other approaches were proposed. Sentiment analysis was based on MMA (Mass Media Assessment) method [15], which required expert labelling of topics by sentiment. This approach allowed the analysis of some combinations of topics grouped by their sentiment. It also allowed the creation of effective classification models with low volume high-level manual labeling—in this case, labeling topics by sentiment in the range from −1 to +1. Then, the result for each document was obtained by a summation of expert labeling results weighted by document related to each topic. Another aggregation method could also be used, as described in [15]. On test data, this approach made it possible to achieve an ROC AUC of 0.93, which is comparable to modern deep learning classifiers.

It also should be noted that, in this case, the definition of sentiment differed from the conventional definition: we did not define sentiment as an author's opinion on some issue, but rather the general positivity or negativity of the described event for the society. Journalism ethics requires news publications to be neutral and objective; hence, the conventional definition of sentiment does not seem to apply to the problem.

The third approach aimed to perform an analysis of specific search queries constructed by experts. It allowed the testing of the specific hypothesis by manually defined search queries without relying on the topic model to distinguish the corresponding topics. The list was composed based on the assumption that the COVID-19 pandemic had significantly affected almost all areas of human activities [61], including healthcare, the economy (unemployment, crisis, poverty) [62], remote work and education [63], crime rate [64], and the abundance of fake news [65]. In the list, we attempted to encompass the most important areas that might have been affected by the COVID-19 pandemic according to common sense and literature review.

The translated list of queries used for the analysis is presented below:

- Fake, disinformation, anti-vax;
- Unemployment, poverty;
- Crisis, economic decline;
- Famine, starvation, homeless, poverty;
- Remote education;
- Freelance, remote work, brain drain;
- Criminal, robbery, theft, homicide;
- Crisis, lending, debt, microcredits;
- Healthcare, hospitals, issues, healthcare scandals;
- Vaccination, COVID-19 vaccines.

This list was composed manually in order to address the main hypothesis: what areas of human activity were most affected by the COVID-19 pandemic. The list was based on the opinion of populations of Kazakhstan and Russia as perceived by experts. The population was mainly concerned with the economic impact of the pandemic, including unemployment and poverty, the potential growth of criminal activity due to the economic decline, impacts on education and healthcare, and also vaccination and how fake news and disinformation can affect public opinion on the COVID-19 vaccination.

These queries (in Russian) were searched via ElasticSearch with the employment of a multi-match full-text search method, which returned a list of matching documents with relative weights. Then, a daily average of these relative weights was calculated for analysis.

ElasticSearch is a NoSQL in memory storage database, which uses the Apache Lucene engine for full text search and provides REST API to index (create), modify, and access (search) different types of data, including texts of arbitrary lengths [66]. ElasticSearch makes it possible to effectively perform full-text search queries on large volumes of textual data and is able to assess document relevance based on search queries using built-in algorithms implemented in the Apache Lucene engine [67].

Distribution of media by the top negative and top positive criteria is presented in Appendix A.

The next step was calculating the Pearson correlation coefficient and Spearman correlation coefficient between the three groups of indicators described above and the COVID-19 data (Table 2). The use of these two correlation coefficients was justified by the necessity to verify the results using two fundamentally different statistics (parametric and non-parametric). The Pearson correlation assesses the linear dependency between two variables, while the Spearman coefficient assesses how well the relationship of two variables can be described using a monotonic function (which may or may not be linear). These coefficients can be applied to assess the correlation between two time series. Usually, such research is performed under the hypothesis that one of the time series is an independent variable and another is a dependent variable. Such a method of analysis allows the performance of an automatic dependency search between hundreds of variables (time series);

however, the results must be manually analyzed and verified by experts, since correlation analysis may produce inadequate results, especially when there are too many different indicators under consideration. The following experimental procedure was proposed:

- Experiments should be performed for Russia and Kazakhstan separately;
- Experiments should be performed for each of the topic models' average daily sentiments using the level-2 topic model, and for each of the search queries separately;
- Experiments should be performed for each of seven COVID-19 indicators selected for analysis.

Below is a description of the system architecture, which implemented the proposed methods for data collection, processing and analysis. During the development of the data processing architecture, the following key needs were identified [68]:

- The possibility of simultaneous calculations with the employment of several machines;
- Ability to flexibly plan the various data processing tasks;
- Ability to monitor tasks in real time, including prompt notification of exceptions;
- Flexibility in using the tools and technologies.

The Apache Airflow open-source software platform was chosen to meet all these needs that were identified in the analysis.

The system components (see Figure 7) were organized as Docker containers [68].

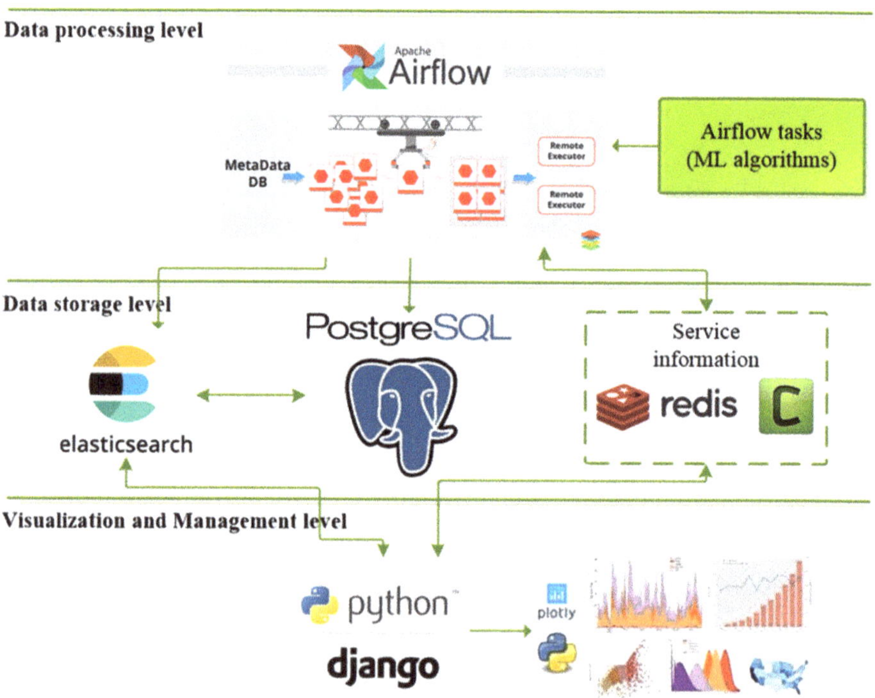

Figure 7. Multilayer system architecture.

The containers had access to the same virtual network, which provided the ability to exchange data using standard network protocols (TCP). This system implementation ensured the operation of subsystems as independent components, and each of them could be replaced if necessary. The interaction of the system layers—the visualization layer and the data processing layer—was carried out with the help of the storage system.

There were three types of storage in the system:

- PostgreSQL—performed the role of a persistent storage medium for structured data;
- ElasticSearch—in-memory NoSQL storage, dedicated to storing unstructured or poorly structured data, as well as fast search (including full-text), filtering, and streaming access;
- Redis—fast key-value storage, used for caching individual pages and elements, and for caching authorization sessions. Redis stored service data as well as page and element caches, which were often accessed.

The general scheme of component interaction was organized according to the ETL (extract, transform, load) principle: the user makes a request for data in ElasticSearch (if data are rarely used) or in Redis (if data are often used). Text processing algorithms were implemented as Airflow-tasks. The processing subsystem used Airflow-scheduler, which writes information about the distribution of tasks by workers to Redis; they, in turn, report to Redis about the status of their tasks. The subsystem interface was an HTML + CSS + JS website accessible via the HTTP protocol. The web application was implemented on the Python Django framework, the webserver was Gunicorn, and the reverse proxy was Nginx. The web application had access to both the persistent storage PostgreSQL and ElasticSearch.

4. Results and Discussion

The proposed method was applied to the obtained topic models over the course of 42 experiments in which analysis was performed across two countries, three topic models and seven indicators. Then, the results of the experiments were analyzed by experts. Table 4 illustrates an example of data obtained during the experiments, and shows the topics with top correlation.

Table 4. Correlation between the number of new deaths from COVID-19 and topics from the initial topic model.

Correlation (Pearson/Spearman)	Topic Name (Top-Words)	Topic Volume (Documents)
0.91/0.87	Vaccine, Vaccination, Drug, Coronavirus, Test, Sputnik-V, Russian	15,434
0.86/0.85	Petersburg, Saint Petersburg, Petersburg, Leningrad region, Moscow, report deaths, COVID	1495
0.77/0.69	Health, Product, Doctor, Alcohol, Organism, Nutrition, Healthy	8318
0.74/0.63	Tell, Photo, Arrive, Depart, Tourism, Return, Go	2693
0.67/0.49	Temperature, Degree, Night, Snow, Weather, Air, Strong	8196

The implemented research allowed us to propose some results and recommendations.

The numbers of daily deaths and daily new cases had higher maximum correlations in all of the experiments (typically 0.6–0.8). However, more informative relative indicators, such as the positive test rate, reproduction rate, and number of tests per positive result (an indicator, reversed to the "positive test rate" indicator) had lower maximum correlations (typically 0.4–0.6). Hence, the media in Kazakhstan and in Russia focused too much on absolute numbers, which can be argued to be biased and less informative. For example, the absolute number of new identified COVID-19 cases does not reflect the situation accurately, since the number of performed tests may vary drastically. Such types of analysis can lead to situations when, although the overall epidemiological situation is steady, media start to inflate the public opinion and cause panic due to an increase in the number of tests, which leads to an increase in the absolute number of new cases, and the reverse situation is also possible. However, media agencies in Kazakhstan and in Russia seemed to ignore the relative indicators.

The index of the stringency of quarantine restrictions seemed to have high correlation with topics in media in Russia (0.75, Figure 8), while media in Kazakhstan did not seem to focus on stringency, and the highest correlation was only 0.43 (Figure 9). This may indicate that in Kazakhstan, there was a divergence between restrictions due to the pandemic and their reflection in the media, which can be considered as a problem, since mass media are one of the main tools for government to broadcast information about current the situation and related restrictions.

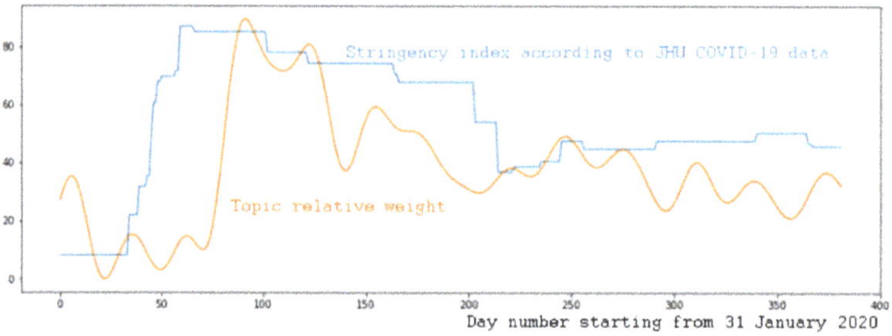

Figure 8. Blue line shows quarantine limitations index in Russia over time, orange line shows dynamic weight of topic with the highest correlation (0.75) over time. The topic is related to the reports on newly identified COVID-19 cases across the different regions of the country.

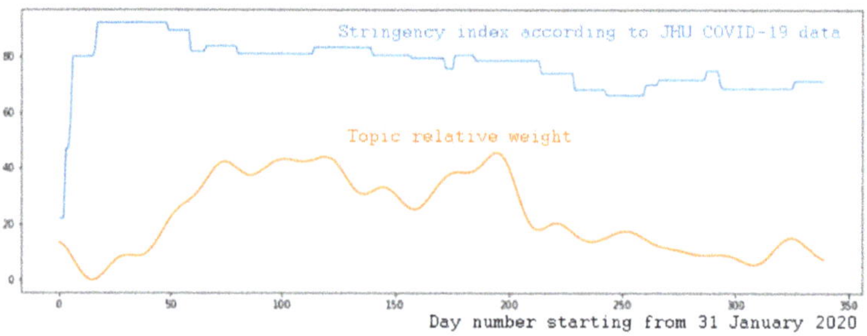

Figure 9. Blue line shows quarantine limitations index in Kazakhstan over time, orange line shows dynamic weight of topic with the highest correlation (0.43) over time. The topic is related to the reports on newly identified COVID-19 cases across the different regions of the country.

A topic from the Kazakhstani news corpus with main associated words being "oxygen, investigation, embezzlement" had the highest correlation with the number of daily deaths from COVID-19 (0.8). This correlation was one of the highest among all topics and COVID-19 indicators for Kazakhstan. It could be argued that this proves that criminal embezzlements of liquid oxygen, required for critical cases of lung damage induced by COVID-19, might be a reason for the increase in the number of deaths. Topics related to the coronavirus vaccination had the highest correlation with the number of deaths in Russia (0.82–0.84). This could be interpreted as an attempt to mitigate the risks of panic among the population by informing them about new methods of stopping the epidemic.

One counterintuitive outcome of the series of experiments was that there was a strong negative correlation between the number of deaths and newly identified cases and topics related to the economy and banking (−0.3, −0.4). Although the negative impact of the pandemic on the world economy is obvious, the deterioration of the epidemiological situation did not cause the sudden increases in information on politics- and economics-related topics; at the same time, neutral topics, such as culture, art, celebrities, and lifestyle, do not correlate considerably with epidemiological indicators.

Tables 5–7 show the correlation coefficients between COVID-19 indicators and sentiment in the media in Russia and in Kazakhstan. It is important to note that most topics that were labelled by experts to be positive according to the level-2 topic model were related to vaccination, COVID-related scientific research, and medical development. There were also several topics on positive COVID-19 dynamics occurring as a result of compliance with quarantine conditions, as well as a topic about governmental support of small businesses. Hence, this explains why there was a high correlation between positive news and overall average sentiment and such indicators as numbers of new deaths and numbers of new cases.

Table 5. Correlation between average sentiment and COVID-19 indicators.

Russia		Kazakhstan	
Indicator	**Correlation Coefficient (Pearson/Spearman)**	**Indicator**	**Correlation Coefficient (Pearson/Spearman)**
New deaths smoothed	0.81/0.77	New tests	0.55/0.54
New cases smoothed	0.66/0.67	New cases smoothed	0.51/0.76
Positive test rate	0.57/0.54	Positive test rate	0.23/0.62
New tests	0.36/0.49	New deaths smoothed	0.22/0.58
Reproduction rate	−0.007/−0.11	Reproduction rate	−0.18/−0.43
Stringency index	−0.12/−0.16	Stringency index	−0.53/−0.56
Tests per case	−0.13/−0.05	Tests per case	−0.54/−0.58

Table 6. Correlation between numbers of news stories with negative sentiment and COVID-19 indicators.

Russia		Kazakhstan	
Indicator	**Correlation Coefficient (Pearson/Spearman)**	**Indicator**	**Correlation Coefficient (Pearson/Spearman)**
Reproduction rate	0.71/0.72	Tests per case	0.42/0.36
Stringency index	0.65/0.57	Stringency index	0.41/0.75
Tests per case	0.08/0.26	Reproduction rate	0.16/0.42
Positive test rate	0.02/0.17	Positive test rate	0.14/−0.40
New tests	−0.11/0.10	New deaths smoothed	−0.29/−0.47
New cases smoothed	−0.18/0.08	New cases smoothed	−0.35/−0.47
New deaths smoothed	−0.32/−0.10	New tests	−0.56/−0.50

Table 7. Correlation between numbers of news stories with positive sentiment and COVID-19 indicators.

Russia		Kazakhstan	
Indicator	**Correlation Coefficient (Pearson/Spearman)**	**Indicator**	**Correlation Coefficient (Pearson/Spearman)**
New deaths smoothed	0.84/0.71	New cases smoothed	0.50/0.63
New cases smoothed	0.70/0.68	Positive test rate	0.36/0.67
Positive test rate	0.68/0.62	New deaths smoothed	0.29/0.54
New tests	0.47/0.48	New tests	0.20/0.13
Reproduction rate	0.25/0.14	Reproduction rate	−0.06/−0.43
Stringency index	0.05/−0.06	Stringency index	−0.21/−0.11
Tests per case	−0.14/−0.11	Tests per case	−0.5/−0.63

These data make it possible to draw some conclusions, which are presented below.

This results obtained using this approach support the hypothesis that the media in Russia reflected COVID-19 situation more accurately.

Moreover, the number of negative news stories in media in Russian strongly correlated with two very representative parameters—the virus reproduction rate and the quarantine stringency index—which also indicates that the mass media in Russia presented the situation in an objective and accurate manner.

Rankings according to the Pearson and Spearman correlation coefficients were identical to the data obtained for the Russian Federation and very similar to the data for Kazakhstan. This might indicate that, in the Russian Federation, mass media publication activity was responsive the changes in the epidemiological situation in a more linear way, as compared to the media in Kazakhstan.

Generally, there was a moderate correlation between the number of deaths, new cases, and new tests, and the number of positive news stories (which was also considerably higher in the Russian media). This might indicate that the media generally tend to smooth out the negative psychological effects caused by the pandemic situation, rather than inflating fear. Specifically, when the epidemiological situation deteriorated, the media tended to publish more information about the latest research on—and benefits of—vaccines.

Lastly, we consider the results of the manually constructed full-text search query analysis. There were several options for obtaining the time series from the query results. In this case, the average daily relative weights of the documents were used (the results are presented in Table 8).

Table 8. Correlation between average relative weights of queries and indicators with the highest correlations.

Russia			Kazakhstan		
Query	**Top Indicator (Pearson/Spearman)**	**Correlation Coefficient (Pearson/Spearman)**	**Query**	**Top Indicator (Pearson/Spearman)**	**Correlation Coefficient (Pearson/Spearman)**
Vaccination, COVID-19 vaccines	Positive rate/Positive rate	0.76/0.78	Healthcare, hospitals, issues, healthcare scandals	Stringency index/Stringency index	0.42/0.32
Healthcare, hospitals, issues, healthcare scandals	Positive rate/Reproduction rate	0.67/0.53	Crisis, lending, debt, microcredits	Tests per case/Stringency index	0.41/0.46
Crisis, lending, debt, microcredits	Reproduction rate/Stringency index	0.56/0.54	Vaccination, COVID-19 vaccines	Reproduction rate/Stringency index	0.38/0.52
Unemployment, poverty	Stringency index/Stringency ondex	0.56/0.48	Fake, disinformation, anti-vax	Tests per case/Reproduction rate	0.33/0.21
Crisis, economic decline	Tests per case/Stringency index	0.49/0.44	Crisis, economic decline	Tests per case/Stringency index	0.29/0.47
Freelance, remote work, brain drain	Stringency index/Stringency index	0.39/0.35	Remote education	New tests/Positive rate	0.29/0.38
Famine, starvation, homeless, poverty	Stringency index/Stringency index	0.35/0.47	Unemployment, poverty	Tests per case/Reproduction rate	0.27/0.29
Fake, disinformation, anti-vax	Tests per case/Tests per case	0.33/0.39	Freelance, remote work, brain drain	Stringency index/Stringency index	0.26/0.46
Remote education	Reproduction rate/Tests per case	0.11/0.39	Criminal, robbery, theft, homicide	New tests/New tests	0.20/0.15
Criminal, robbery, theft, homicide	New deaths/Positive rate	0.07/0.10	Famine, starvation, homeless, poverty	New tests/Positive rate	0.09/0.16

These experiments also demonstrated that the Russian media reflected the COVID-19 situation more objectively.

The rankings that were constructed according to the Pearson and Spearman correlation coefficients were also identical for the Russian Federation, while for Kazakhstan, there was considerable inconsistency.

In the cases of both sentiment and query correlation, the most inconsistent COVID-19 indicator, which accounted for the most of the differences in ranking, was the stringency index of Kazakhstan. According to the analysis, this might indicate that the stringency index in Kazakhstan changed non-linearly and it was less responsive to changes in the epidemiological situation as compared to the stringency index in Russia. Figures 8 and 9 illustrate this difference, since it is visible that the spread of the stringency index was much lower in Kazakhstan (while the epidemiological situation's spread seemed to be comparable).

In both countries, the Healthcare, Crisis, and Vaccination queries showed the highest correlation, while Crime and Famine/Starvation were ranked much lower, even in the media in Russia, which might indicate that the fears that the pandemic critically damaged the economy and led to severe problems such as crime and extreme poverty were not justified.

In Kazakhstan, the query about fake news and disinformation was ranked much higher than in Russia, which might indicate that these were significantly more acute problems for Kazakhstan.

The queries about remote education, freelancing, and unemployment showed moderate correlations in both countries.

The hypothesis that there might be a lag between COVID-19 indicators and mass media reaction was already addressed in a number of computational experiments. Different lags between −10 and +10 days were tested. The experiments showed that mass media and COVID-19 indicators steadily demonstrated maximum correlation at close to zero lag, while increases in the lag (either positive or negative) led to monotonic decreases in the maximum and average correlation coefficients. This regularity was observed in both countries. Although it might intuitively be assumed that mass media should react to COVID-19 indicators with some delay, in practice, this idea is not supported. Two explanations can be considered in this regard:

- Mass media received actual information rather promptly, and react to it operatively;
- There was some inherent lag in the analyzed COVID-19 indicators. For example, daily statistics may have actually contained some sort of aggregated information over several days due to imperfections in statistical data collection processes in Kazakhstan and Russia.

The main contribution of this work is the proposal of a model to perform a comparative analysis of the representation of the COVID-19 pandemic by mass media in two different countries, where English was not used as the language of communication, taking into account multiple points of view—automatically obtained topics, average sentiment, and dynamic indicators—according to manually selected search queries.

5. Conclusions and Future Research

The COVID-19 pandemic has had a great impact on the life of society in almost all countries of the world. The analysis of media texts allows us to evaluate the public reaction to the non-standard situation and the measures taken by national governments.

We proposed a method that, in this study, made it possible to analyze how statistical indicators related to COVID-19 were reflected in mass media. The method assumes the application of BigARTM or another topic model in order to obtain the topical structure of the corpus, which can then be used to calculate the topics' dynamics. Those dynamical indicators of publication activity can be compared with COVID-19 indicators, such as numbers of new cases, positive test rates, stringency indexes, and others, in order to perform the correlation analysis. In this study, sentiment analysis based on topic embeddings [14] was also conducted, as well as an analysis of correlation in which the daily average relevance weights were obtained from 10 full-text search queries constructed manually by experts.

The main advantage of the proposed method is that it combines the analysis the dynamics of unbiased and automatically obtained topics, sentiment analysis based on expert labelling, and manual queries. It can be argued that an such approach may produce more objective results and conclusions through the comparative analysis of the results of three groups of computational experiments.

The proposed method can potentially be used to obtain insights on how COVID-19 is presented in the media, and on which statistical indicators describe the media activities. For example, it was found that the media in Russia and in Kazakhstan focused on absolute values, while more informative relative indicators such as positive test rates and virus reproduction rates were generally ignored, since such indicators showed much lower correlations with publishing activities on several topics as well as with sentiment.

The method might also be applied to estimate how the stringency of quarantine limitations is reflected in the media. Such an analysis may help indicate reasons for deteriorations in the epidemiological situation, since quarantine restrictions must not only be introduced, but also enforced and broadcasted.

The limitations of the current study include the fact that only daily aggregations were used, while valuable insights could theoretically also have been obtained at different degrees of granularity. Time-lagged correlation was not considered in the study; however, this is not a limitation of the method.

One obvious methodological limitation, which is inherent to all correlation-based approaches, is the possibility of sporadic correlations. However, the proposed method attempts to avoid this problem by using small (daily) granularity, which makes the chances of sporadic random correlation much lower, and also by including three different groups of experiments into the analysis to make it possible to cross-check the findings.

It should also be noted that, in the study, the cross-national effects were not taken into account; thus, the lack of generalizability to a global perspective is a limitation of this study.

Directions of further research include:

- Attempt to build a model for the recognition of inaccuracies in official statistical indicators regarding the COVID-19 pandemic using mass media data as a reference;
- Conduct an analysis of the topical profile of the COVID-19 pandemic in different countries and explore how it evolved over time. Such an analysis can be used to assess its impact on the economy, education, politics, tourism, etc.

Author Contributions: Conceptualization, R.I.M. and K.Y.; methodology, R.I.M., K.Y. and E.Z.; software, K.Y.; validation, V.L., Y.K. and M.Y.; formal analysis R.I.M.; investigation, A.S., M.A., E.Z. and V.G.; resources, R.I.M.; data curation, K.Y., E.M. and M.Y.; writing—original draft preparation, R.I.M. and K.Y.; writing—review and editing, M.Y., A.S., V.G. and R.I.M.; visualization, R.I.M., E.M. and K.Y.; supervision, R.I.M.; project administration, Y.K. and V.G.; funding acquisition, R.I.M., V.L. and M.A. All authors have read and agreed to the published version of the manuscript.

Funding: This research was funded by the Science Committee of the Ministry of Education and Science of the Republic of Kazakhstan, Grant No. AP09259587, "Developing of methods and algorithms of intelligent GIS for multi-criteria analysis of healthcare data", and by the Slovak Research and Development Agency, no. APVV PP-COVID-20–0013, "Development of methods of healthcare system risk and reliability evaluation under coronavirus outbreak".

Institutional Review Board Statement: Not applicable.

Informed Consent Statement: Not applicable.

Data Availability Statement: Data are available at https://data.mendeley.com/datasets/2vz7vtbhn2/1 (accessed on 4 December 2021); https://data.mendeley.com/datasets/hwj24p9gkh/1 (accessed on 4 December 2021) under dataset License: CC-BY-SA.

Acknowledgments: The authors would like to express their sincere gratitude to the anonymous referees for their useful comments and to Yelena Popova for her professional support.

Conflicts of Interest: The authors declare no conflict of interest.

Appendix A

Visualization of the topic modeling and sentiment analysis results for the corpus of media publications.

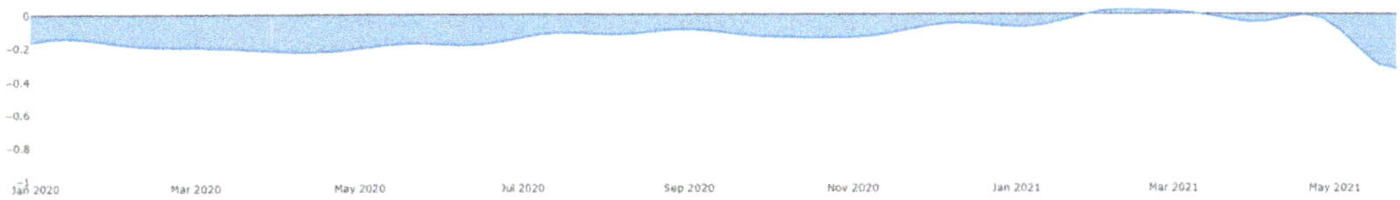

Figure A1. The dynamics of sentiment.

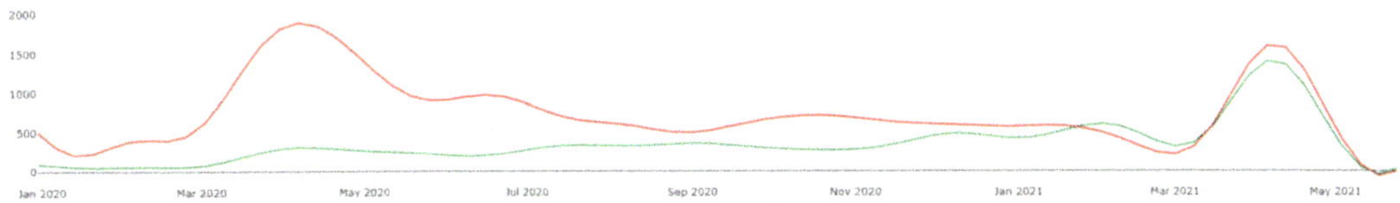

Figure A2. Number of positive and negative articles.

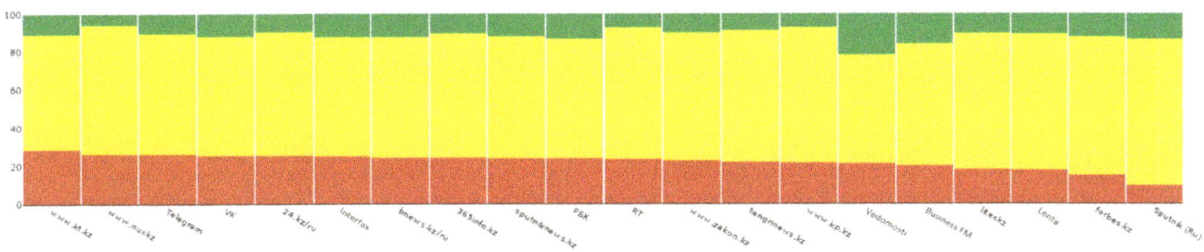

Figure A3. Negative media.

Sentiment Score	Publication date and time	Title	Source
-1,000	2020-03-22T12:44:00+00:00	Almaty woman infected with coronavirus had symptoms similar to ARVI	https://24.kz/ru/
-0,981	2021-04-08T14:28:00+00:00	Modular hospital in Atyrau is 100% full	https://www.zakon.kz/
-0,973	2020-03-23T05:58:00+00:00	Two of those infected with coronavirus in Almaty became infected in Kazakhstan	https://24.kz/ru/
-0,970	2020-03-23T05:39:00+00:00	Last two Almaty residents infected with coronavirus infected in Kazakhstan	https://bnews.kz/ru (baigenews.kz)
-0,957	2021-04-02T18:26:00+00:00	How the number of deaths from COVID has changed in Russia. Infographics	RBK News
-0,944	2021-01-21T04:25:00+00:00	298 Kazakhstanis with coronavirus are in serious condition	https://24.kz/ru/
-0,938	2021-04-24T04:39:00+00:00	93 Kazakhstanis with CVI are connected to mechanical ventilation	https://365info.kz/
-0,937	2021-04-10T05:15:00+00:00	More than 11 thousand people receive treatment from CVI in hospitals	https://365info.kz/
-0,934	2021-04-03T05:05:00+00:00	377 patients with COVID in serious condition in Kazakhstan	https://24.kz/ru/
-0,934	2021-04-18T05:05:00+00:00	171 patients with CVI in Kazakhstan are in critical condition	https://24.kz/ru/

Figure A4. Top negative news.

Sentiment Score	Publication date and time	Title	Source
1,000	2021-04-28 01:28:00+00:00	Found a way to kill coronavirus in a second	Lenta.ru
0,962	2021-02-06 08:50:00+00:00	A substance has been created that increases the effectiveness of anticoid vaccines by 10 times	https://bnews.kz/ru (baigenews.kz)
0,961	2021-04-22 18:47:10+00:00	Mass vaccination using the first batch of this drug begins before the end of the final phase of clinical trials.	VK
0,956	2021-01-27 09:40:00+00:00	Turkey has found a way to destroy coronavirus in a minute	https://www.zakon.kz/
0,955	2020-04-17 02:39:00+00:00	Scientists have established the death temperature of the coronavirus	Business FM
0,948	2021-04-14 10:28:03+00:00	The trials of the Kazakhstani QazVac vaccine are almost completed 50% of the third phase of clinical trials of the inactivated QazVac vaccine will be completed in ...	Telegram
0,945	2021-03-31 07:38:00+00:00	The sun has proven beneficial against coronavirus	Lenta.ru
0,944	2021-04-26 03:01:06+00:00	Aleksey Tsoi (Minister of Healthcare) was vaccinated with the Kazakh drug QazVac. The immunization process is the same as with Sputnik V - that is, the injection is injected into the shoulder. The difference ...	Telegram
0,943	2021-01-21 08:15:00+00:00	Hungary registers Sputnik V vaccine	Interfax
0,926	2021-02-06 09:17:00+00:00	Ten times more effective: scientists have invented help for vaccines against COVID-19	Sputnik (Ru)

Figure A5. Top positive news.

References

1. Baldwin, R.; di Mauro, B.W. Economics in the time of COVID-19: A new eBook. *VOX CEPR Policy Portal* **2020**, 2–3.
2. Atun, R. Transitioning health systems for multimorbidity. *Lancet* **2015**, *386*, 721–722. [CrossRef]
3. Orlov, E.M. The category of effectiveness in the health care system. *Basic Res.* **2010**, *10*, 70–75.
4. Panch, T.; Szolovits, P.; Atun, R. Artificial intelligence, machine learning and health systems. *J. Glob. Health* **2018**, *8*, 020303. [CrossRef] [PubMed]
5. *The Socio-Economic Impact of AI in Healthcare*. October 2020. Available online: https://www.medtecheurope.org/wp-content/uploads/2020/10/mte-ai_impact-in-healthcare_oct2020_report.pdf (accessed on 10 September 2021).
6. Mukhamediev, R.I.; Symagulov, A.; Kuchin, Y.; Yakunin, K.; Yelis, M. From Classical Machine Learning to Deep Neural Networks: A Simplified Scientometric Review. *Appl. Sci.* **2021**, *11*, 5541. [CrossRef]
7. Lee, J.; Yoon, W.; Kim, S.; Kim, D.; Kim, S.; So, C.H.; Kang, J. BioBERT: A pre-trained biomedical language representation model for biomedical text mining. *Bioinformatics* **2020**, *36*, 1234–1240. [CrossRef]
8. Daniel, J.; Willie, R.; Copley, C. Towards automating healthcare question answering in a noisy multilingual low-resource setting. In Proceedings of the 57th Annual Meeting of the Association for Computational Linguistics, Florence, Italy, 28 July–2 August 2019; pp. 948–953. [CrossRef]
9. Feng, L.; Xiaoli, W.; Qingfeng, W.; Jiaying, L.; Xueliang, Q.; Zhifeng, B. HQADeepHelper: A deep learning system for healthcare question answering. In *Companion Proceedings of the Web Conference 2020 (WWW '20)*; Association for Computing Machinery: New York, NY, USA, 2020; pp. 194–197. [CrossRef]
10. Draganescu, O. Forms of Influencing Young People through Media Discourse. *EIRP Proc.* **2019**, *13*. Available online: http://www.proceedings.univ-danubius.ro/index.php/eirp/article/view/1965/2250 (accessed on 2 December 2021).
11. Choudhary, V. Role of mass media in shaping public opinion. *Aut Aut Res. J.* **2020**, *XI*, 398–404.
12. Tasnim, S.; Hossain, M.; Mazumder, H. Impact of Rumors and Misinformation on COVID-19 in Social Media. *J. Prev. Med. Public Health* **2020**, *53*, 171–174. [CrossRef] [PubMed]
13. Gao, J.; Zheng, P.; Jia, Y.; Chen, H.; Mao, Y.; Chen, S.; Wang, Y.; Fu, H.; Dai, J. Mental health problems and social media exposure during COVID-19 outbreak. *PLoS ONE* **2020**, *15*, e0231924.
14. Ghasiya, P.; Okamura, K. Investigating COVID-19 News across Four Nations: A Topic Modeling and Sentiment Analysis Approach. *IEEE Access* **2021**, *9*, 36645–36656. [CrossRef]
15. Mukhamediev, R.I.; Yakunin, K.; Mussabayev, R.; Buldybayev, T.; Kuchin, Y.; Murzakhmetov, S.; Yelis, M. Classification of Negative Information on Socially Significant Topics in Mass Media. *Symmetry* **2020**, *12*, 1945. [CrossRef]
16. Kirill, Y.; Mihail, I.G.; Sanzhar, M.; Rustam, M.; Olga, F.; Ravil, M. Propaganda Identification Using Topic Modelling. *Proc. Comput. Sci.* **2020**, *178*, 205–212. [CrossRef]
17. Battineni, G.; Chintalapudi, N.; Amenta, F. Forecasting of COVID-19 epidemic size in four high hitting nations (USA, Brazil, India and Russia) by Fb-Prophet machine learning model. *Appl. Comput. Inform.* **2020**. [CrossRef]
18. Yakunin, K.; Murzakhmetov, S.; Mussabayev, R.; Muhamedyev, R. News popularity prediction using topic modelling. In Proceedings of the 2021 IEEE International Conference on Smart Information Systems and Technologies (SIST), Nur-Sultan, Kazakhstan, 28–30 April 2021. [CrossRef]
19. Tatar, A.; Antoniadis, P.; de Amorim, M.D.; Fdida, S. Ranking news articles based on popularity prediction. In Proceedings of the 2012 IEEE/ACM International Conference on Advances in Social Networks Analysis and Mining, Istanbul, Turkey, 26–29 August 2012; pp. 106–110.
20. Bandari, R.; Asur, S.; Huberman, B. The pulse of news in social media: Forecasting popularity. In Proceedings of the International AAAI Conference on Web and Social Media, Dublin, Ireland, 4–7 June 2012.
21. Edelman Trust Barometer. Available online: https://www.edelman.com/trust-barometer (accessed on 5 August 2021).
22. Miller, D. Promotional strategies and media power. In *Introduction to Media*; Briggs, A., Cobley, P., Eds.; Longman: London, UK, 1998; pp. 65–80. ISBN 0582 27798 1.
23. Bushman, B.; Whitaker, J. Media influence on behavior. In *Encyclopedia of Human Behavior*, 2nd ed.; Elsevier Inc.: Amsterdam, The Netherlands, 2012; pp. 571–575.
24. Stacks, D.; Li, Z.C.; Spaulding, C. Media effects. In *International Encyclopedia of the Social & Behavioral Sciences*, 2nd ed.; Elsevier Inc.: Amsterdam, The Netherlands, 2015; pp. 29–34.
25. Ko, H.; Hong, J.Y.; Kim, S.; Mesicek, L.; Na, I.S. Human-machine interaction: A case study on fake news detection using a backtracking based on a cognitive system. *Cogn. Syst. Res.* **2019**, *55*, 77–81. [CrossRef]
26. Bushman, B.J.; Whitaker, J.L. *Media Influence on Behavior. Reference Module in Neuroscience and Biobehavioral Psychology*; Elsevier Inc.: Amsterdam, The Netherlands, 2017.
27. Giri, S.P.; Maurya, A.K. A neglected reality of mass media during COVID-19: Effect of pandemic news on individual's positive and negative emotion and psychological resilience. *Personal. Individ. Differ.* **2021**, *180*, 110962. [CrossRef]
28. Aslam, F.; Awan, T.M.; Syed, J.H.; Kashif, A.; Parveen, M. Sentiments and emotions evoked by news headlines of coronavirus disease (COVID-19) outbreak. *Human. Soc. Sci. Commun.* **2020**, *7*, 1–9. [CrossRef]
29. Hamidein, Z.; Hatami, J.; Rezapour, T. How people emotionally respond to the news on COVID-19: An online survey. *Basic Clin. Neurosci.* **2020**, *11*, 171. [CrossRef]

30. Jo, W.; Chang, D. Political Consequences of COVID-19 and Media Framing in South Korea. *Front. Public Health* **2020**, *8*, 425. [CrossRef] [PubMed]
31. Ridhwan, K.M.; Hargreaves, C.A. Leveraging Twitter Data to Understand Public Sentiment for the COVID-19 Outbreak in Singapore. *Int. J. Inf. Manag. Data Insights* **2021**, *1*, 100021. [CrossRef]
32. Casero-Ripollés, A. Impact of Covid-19 on the media system. Communicative and democratic consequences of news consumption during the outbreak. *Prof. Inf.* **2020**, *29*, e290223. [CrossRef]
33. Tandoc, E.C., Jr. Tell me who your sources are: Perceptions of news credibility on social media. *J. Pract.* **2019**, *13*, 178–190. [CrossRef]
34. Song, X.; Petrak, J.; Jiang, Y.; Singh, I.; Maynard, D.; Bontcheva, K. Classification aware neural topic model for COVID-19 disinformation categorisation. *PLoS ONE* **2021**, *16*, e0247086. [CrossRef]
35. Sun, K.; Chen, J.; Viboud, C. Early epidemiological analysis of the coronavirus disease 2019 outbreak based on crowdsourced data: A population-level observational study. *Lancet Digit. Health* **2020**, *2*, e201–e208. [CrossRef]
36. Gabrielyan, D.; Masso, J.; Uuskula, L. Mining news data for the measurement and prediction of inflation expectations. In Proceedings of the CARMA 2020—3rd International Conference on Advanced Research Methods and Analytics, Valencia, Spain, 8–9 July 2020. [CrossRef]
37. Leombroni, M.; Vedolin, A.; Venter, G.; Whelan, P. Central bank communication and the yield curve. *J. Financ. Econ.* **2021**. [CrossRef]
38. Parkhomenko, P.A.; Grigor'yev, A.A.; Astrakhantsev, N.A. Review and experimental comparison of text clustering methods. *Proc. Inst. Syst. Program. RAS* **2017**, *29*, 161–200. [CrossRef]
39. Vorontsov, K.; Frei, O.; Apishev, M.; Romov, P.; Dudarenko, M. Bigartm: Open source library for regularized multimodal topic modeling of large collections. In Proceedings of the International Conference on Analysis of Images, Social Networks and Texts, Yekaterinburg, Russia, 9–11 April 2015; pp. 370–381.
40. Jelodar, H. Latent Dirichlet allocation (LDA) and topic modeling: Models, applications, a survey. *Multimed. Tools Appl.* **2019**, *78*, 15169–15211. [CrossRef]
41. Alsolamy, M.; Alotaibi, A.; Alabbas, A.; Abdullah, M. Topic based Sentiment Analysis for COVID-19 Tweets. *Int. J. Adv. Comput. Sci. Appl.* **2021**, *12*. [CrossRef]
42. Xue, J.; Chen, J.; Chen, C.; Zheng, C.; Li, S.; Zhu, T. Public discourse and sentiment during the COVID 19 pandemic: Using Latent Dirichlet Allocation for topic modeling on Twitter. *PLoS ONE* **2020**, *15*, e0239441. [CrossRef]
43. Tao, G.; Miao, Y.; Ng, S. COVID-19 topic modeling and visualization. In Proceedings of the 2020 24th International Conference Information Visualisation (IV), Melbourne, Australia, 7–11 September 2020; pp. 734–739.
44. Mutanga, M.B.; Abayomi, A. Tweeting on COVID-19 pandemic in South Africa: LDA-based topic modelling approach. *Afr. J. Sci. Technol. Innov. Dev.* **2020**, 1–10. [CrossRef]
45. Tsao, S.-F.; Chen, H.; Tisseverasinghe, T.; Yang, Y.; Li, L.; Butt, Z.A. What social media told us in the time of COVID-19: A scoping review. *Lancet Digit. Health* **2021**, *3*, e175–e194. [CrossRef]
46. Kuchler, T.; Russel, D.; Stroebel, J. JUE Insight: The geographic spread of COVID-19 correlates with the structure of social networks as measured by Facebook. *J. Urban Econ.* **2021**, 103314. [CrossRef]
47. Gupta, A.; Aeron, S.; Agrawal, A.; Gupta, H. Trends in COVID-19 Publications: Streamlining Research Using NLP and LDA. *Front Digit Health* **2021**, *3*, 686720. [CrossRef] [PubMed]
48. Angelov, D. Top2vec: Distributed representations of topics. *arXiv* **2020**, arXiv:2008.09470 2020.
49. Liu, Y.; Ott, M.; Goyal, N.; Du, J.; Joshi, M.; Chen, D.; Levy, O.; Lewis, M.; Zettlemoyer, L.; Stoyanov, V. Roberta: A robustly optimized bert pretraining approach. *arXiv* **2019**, arXiv:1907.11692 2019.
50. Chintalapudi, N.; Battineni, G.; Amenta, F. Sentimental Analysis of COVID-19 Tweets Using Deep Learning Models. *Infect. Dis. Rep.* **2021**, *13*, 329–339. [CrossRef]
51. Chakraborty, A.K.; Das, S.; Kolya, A.K. Sentiment analysis of covid-19 tweets using evolutionary classification-based LSTM model. In *Proceedings of Research and Applications in Artificial Intelligence*; Springer: Berlin/Heidelberg, Germany, 2021; pp. 75–86.
52. Dong, E.; Du, H.; Gardner, L. An interactive web-based dashboard to track COVID-19 in real time. *Lancet Infect. Dis.* **2020**, *20*, 533–534. [CrossRef]
53. Yakunin, K.; Kalimoldayev, M.; Mukhamediev, R.; Mussabayev, R.; Barakhnin, V.; Kuchin, Y.; Murzakhmetov, S.; Buldybayev, T.; Ospanova, U.; Yelis, M.; et al. KazNewsDataset: Single Country Overall Digital Mass Media Publication Corpus. *Data* **2021**, *6*, 31. [CrossRef]
54. Arroyo-Marioli, F.; Bullano, F.; Kucinskas, S.; Rondón-Moreno, C. Tracking R of COVID-19: A new real-time estimation using the Kalman filter. *PLoS ONE* **2021**, *16*, e0244474. [CrossRef]
55. Vorontsov, K.V.; Potapenko, A.A. Regularization, robustness and sparsity of probabilistic thematic models. *Comput. Res. Model.* **2012**, *4*, 693–706. [CrossRef]
56. Blei, D.M.; Ng, A.Y.; Jordan, M.I. Latent dirichlet allocation. *J. Mach. Learn. Res.* **2003**, *3*, 993–1022.
57. Mimno, D.; Wallach, H.; Talley, E.; Leenders, M.; McCallum, A. Optimizing semantic coherence in topic models. In Proceedings of the Conference on Empirical Methods in Natural Language Processing, Edinburgh, UK, 27–31 July 2011; pp. 262–272.
58. Segalovich, I. A fast morphological algorithm with unknown word guessing induced by a dictionary for a web search engine. *MLMTA* **2003**, *2003*, 273.

59. Stop-Words 2018.7.23. Available online: https://pypi.org/project/stop-words/ (accessed on 13 November 2021).
60. Krasnov, F.; Anastasiia, S. The number of topics optimization: Clustering approach. *Mach. Learn. Knowl. Extr.* **2019**, *1*, 416–426. [CrossRef]
61. Haleem, A.; Mohd, J.; Raju, V. Effects of COVID-19 pandemic in daily life. *Curr. Med. Res. Pract.* **2020**, *10*, 78. [CrossRef] [PubMed]
62. Fernandes, N. Economic Effects of Coronavirus Outbreak (COVID-19) on the World Economy. 2020. SSRN 3557504. Available online: https://papers.ssrn.com/sol3/papers.cfm?abstract_id=3557504 (accessed on 4 December 2021).
63. Galanti, T.; Guidetti, G.; Mazzei, E.; Zappalà, S.; Toscano, F. Work from Home During the COVID-19 Outbreak: The Impact on Employees' Remote Work Productivity, Engagement, and Stress. *J. Occup. Environ. Med.* **2021**, *63*, e426–e432. [CrossRef]
64. Campedelli, G.; Alberto, A.; Serena, F. Exploring the immediate effects of COVID-19 containment policies on crime: An empirical analysis of the short-term aftermath in Los Angeles. *Am. J. Crim. Justice* **2021**, *46*, 704–727. [CrossRef] [PubMed]
65. Apuke, D.; Bahiyah, O. Fake news and COVID-19: Modelling the predictors of fake news sharing among social media users. *Telemat. Inform.* **2021**, *56*, 101475. [CrossRef]
66. Divya, M.; Shiv Kumar, G. ElasticSearch: An advanced and quick search technique to handle voluminous data. *Compusoft* **2013**, *2*, 171–175.
67. Białecki, A.; Muir, R.; Ingersoll, G. Apache lucene. In Proceedings of the SIGIR 2012 Workshop on Open Source Information Retrieval, Portland, OR, USA, 16 August 2012; Volume 4, pp. 17–24.
68. Barakhnin, V.; Kozhemyakina, O.; Mukhamedyev, R.; Borzilova, Y.; Yakunin, K. The design of the structure of the software system for processing text document corpus. *Bus. Inform.* **2019**, *13*, 60–72. [CrossRef]

computation

MDPI

Article

Mathematical Modeling of SARS-CoV-2 Omicron Wave under Vaccination Effects

Gilberto González-Parra [1,*] and Abraham J. Arenas [2]

1 Department of Mathematics, New Mexico Tech, New Mexico Institute of Mining and Technology, Socorro, NM 87801, USA
2 Departamento de Matematicas y Estadistica, Universidad de Cordoba, Monteria 230002, Colombia; aarenas@correo.unicordoba.edu.co
* Correspondence: gilberto.gonzalezparra@nmt.edu

Abstract: Over the course of the COVID-19 pandemic millions of deaths and hospitalizations have been reported. Different SARS-CoV-2 variants of concern have been recognized during this pandemic and some of these variants of concern have caused uncertainty and changes in the dynamics. The Omicron variant has caused a large amount of infected cases in the US and worldwide. The average number of deaths during the Omicron wave toll increased in comparison with previous SARS-CoV-2 waves. We studied the Omicron wave by using a highly nonlinear mathematical model for the COVID-19 pandemic. The novel model includes individuals who are vaccinated and asymptomatic, which influences the dynamics of SARS-CoV-2. Moreover, the model considers the waning of the immunity and efficacy of the vaccine against the Omicron strain. This study uses the facts that the Omicron strain has a higher transmissibility than the previous circulating SARS-CoV-2 strain but is less deadly. Preliminary studies have found that Omicron has a lower case fatality rate compared to previous circulating SARS-CoV-2 strains. The simulation results show that even if the Omicron strain is less deadly it might cause more deaths, hospitalizations and infections. We provide a variety of scenarios that help to obtain insight about the Omicron wave and its consequences. The proposed mathematical model, in conjunction with the simulations, provides an explanation for a large Omicron wave under various conditions related to vaccines and transmissibility. These results provide an awareness that new SARS-CoV-2 variants can cause more deaths even if their fatality rate is lower.

Keywords: SARS-CoV-2 variant; Omicron wave; mathematical modeling; vaccination; scenarios; simulations

Citation: González-Parra, G.; Arenas, A.J. Mathematical Modeling of SARS-CoV-2 Omicron Wave under Vaccination Effects. *Computation* **2023**, *11*, 36. https://doi.org/10.3390/computation11020036

Academic Editors: Simone Brogi and Vincenzo Calderone

Received: 1 February 2023
Revised: 9 February 2023
Accepted: 10 February 2023
Published: 15 February 2023

1. Introduction

Over the course of the COVID-19 pandemic, at least 671 million confirmed cases and 6.83 million deaths have been reported (December 2022) [1]. These reported numbers are in the lower bounds, since there are asymptomatic and under-reported cases [2–7]. During 2019, 2020, 2021 and 2022, different strains of the SARS-CoV-2 virus have been found [8–13]. These strains have different characteristics related to contagiousness and severity. Thus, some SARS-CoV-2 variants affect the count of infected cases, hospitalizations and deaths [14,15]. Vaccination programs against SARS-CoV-2 started at the very end of 2019 and the beginning of 2020 in some countries [16–21]. For the year 2022, many countries have already implemented vaccination programs and some countries have also implemented booster programs [4,22–24]. The evolution of SARS-CoV-2 is affected by various factors that are difficult to quantify [25–28]. For instance, social behavior and vaccination status are major factors that influence the COVID-19 pandemic [27,29–37]. New SARS-CoV-2 strains also play a major role in the evolution of the COVID-19 pandemic and have generated different spatial-temporal waves in different countries [12,38–44]. These waves are mainly the product of different contagiousness of new SARS-CoV-2 strains and public health interventions.

The Omicron variant caused a new wave during 2022. The count of cases has been very large and has exceeded previews waves. Omicron was first detected in South Africa and Botswana in early November 2021, but using retrospective testing, it was found that Omicron was also present in England, Nigeria and the United States during November of 2021 [45–47]. Omicron has more than fifty mutations in comparison with the original circulating SARS-CoV-2 [46]. The Omicron strain carries an unusually high number of mutations, suggesting potential immune evasion [22]. A near-complete lack of neutralizing activity has been reported against Omicron in polyclonal sera from individuals vaccinated with two doses of the BNT162b2 COVID-19 vaccine and from convalescent individuals, as well as resistance to different monoclonal antibodies in clinical use [22]. In [48], results suggest that two doses of COVID-19 vaccines only offer modest protection against symptomatic Omicron infection. In [24], the authors showed that Omicron exhibits significant immune evasion compared to other strains. In addition, they found that the Omicron spike exhibits reduced receptor binding and cell–cell fusion, but increased cell-to-cell transmission [24].

Despite the fact that the Omicron strain has lower severity, it has caused a large number of hospitalizations and the average daily number of deaths has been substantial [49]. Some studies have reported a lower rate of hospitalization for the Omicron strain compared with infections caused by the Delta strain [50]. It has been found that booster vaccination and vaccination of individuals with a history of SARS-CoV-2 infection generated lower antibody titers than those against the Delta strain [51,52].

One important aspect for studies predicting health outcomes related to this pandemic is how deadly each of the SARS-CoV-2 strains are. There are two main ways to compute how deadly a disease is. The first is the infection–fatality ratio (IFR), which is given by the ratio of deaths to all infected individuals. The second is computing the case fatality ratio (CFR), which is given by the ratio of deaths to confirmed cases. Estimating the IFR is complex, since it requires knowing the total number of infected cases. Some studies have estimated the CFR as being from less than 0.1% to over 25% [53]. For COVID-19, the true level of transmission is frequently underestimated because a substantial proportion of people with the infection are undetected, either because they are asymptomatic or are not reported [53–55]. In places where testing is extensive, the estimation of CFR is more robust [56]. Another aspect that affects health outcomes is the immunity level of the population which is related to the herd immunity. The increase in population immunity makes it more difficult to compare Omicron's severity with previous circulating SARS-CoV-2 strains, since previous exposure to SARS-CoV-2 strains is expected to prevent to some extent severe outcomes from subsequent infection [57].

The main objective here is to obtain insight into the impact of the Omicron strain. In particular, our aim is to propose a mathematical approach that helps to provide an explanation of the large Omicron wave and the great number of deaths during this wave despite its lower fatality rate. We propose a mathematical modeling framework to study the Omicron wave and attain some additional insight into its evolution. Mathematical models are fruitful and have been used to investigate a variety of scenarios related to the behavior of the COVID-19 disease [6,58–74]. These models are used to study the impact of a variety of health interventions on epidemics. With in silico simulations of the mathematical models we can produce a variety of outcomes that are difficult to foresee due to the nonlinearity and complexity of the epidemics [75–77]. In addition, in some cases the mathematical analysis permits us to determine under what conditions the disease can disappear. Previous studies have investigated the dynamics of the COVID-19 pandemic under two SARS-CoV-2 variants, but some of them did not include vaccination and waning since they were designed for the early pandemic [78–85]. Recently, some researchers have studied the Omicron wave dynamics [86–89]. In [88], the authors analyzed a second wave of COVID-19 and in particular on the Omicron variant pandemic data in India. In [86], a stochastic and second-order model is proposed to deal with the Omicron wave. A mathematical model considering age structure, vaccine, antiviral treatment and influx of the Omicron variant in Korea was developed in [87]. The authors in [58] proposed a fractal–fractional age-structure model for the omicron SARS-CoV-2 variant and considered two age groups. They found that there is a high infection and recovery rate of the Omicron SARS-CoV-2 variant infection among the population under 50. In [90], a generalized SIR model was used to simulate and predict the dynamics of Omicron waves in Ukraine and in the whole world. Mathematical models also have bee used for within-host dynamics for SARS-CoV-2 and in particular the Omicron variant [91,92]

Over this pandemic, many SARS-CoV-2 strains have appeared and these have different characteristics [11,93–97]. Previous models have been used to investigate the influence of new SARS-CoV-2 strains that have a higher probability of transmission [6,38,79,80,83,98–101]. In particular, some interesting studies have considered the mathematical modeling of new SARS-CoV-2 strains and at the same time imperfect vaccination or waning [83,99,101]. Furthermore, some mathematical models have been proposed for studying SARS-CoV-2 waves [40,102,103]. The models have different underlying assumptions and, as any mathematical model of an epidemic, they have advantages and limitations. A variety of work has been carried out considering continuous and discrete models that have included vaccinated subpopulations where people have less probability to get infected, proliferate the virus, or die [78,81,101,104–108].

In this study, we build a mathematical model for the Omicron wave situation. Individuals who are asymptomatic and vaccinated are included in the model since they influence the evolution of the Omicron wave [109–115]. In this study, we use the fact that the Omicron strain has a higher transmissibility than the previously circulating SARS-CoV-2 strains and that the vaccine efficacy is lower for the Omicron strain [22,48]. In addition, we take into account that preliminary studies have found that Omicron has a lower case fatality rate compared to previous circulating SARS-CoV-2 strains. We perform in silico simulations with a variety of scenarios to attain insight into the Omicron wave, its potential consequences and to explain the Omicron wave situation. In this study, we perform a brief stability analysis of the developed model and we also identify the basic reproduction number $\mathcal{R}_0$ despite the fact that the in silico simulations are aimed more toward shorter dynamics [116,117]. The reproduction number $\mathcal{R}_0$ is strongly connected to the effective reproduction number $\mathcal{R}_t$, and therefore is useful in obtaining insight into the behavior of epidemics and pandemics. The motivation of this work is to provide additional knowledge-based support to health authorities and the population in general. Scientific studies that bring awareness of health issues are important to public health despite sometimes the scientific tools used not being very complex [118]. In summary, we propose a mathematical approach to provide an explanation of a large Omicron wave arising under various conditions as a function of vaccination status and transmissibility. These results provide awareness that new SARS-CoV-2 variants can cause more deaths even if their fatality rates are lower.

There are some certain previous studies and mathematical models related to the Omicron wave [119–121]. In [121], the authors implemented a stochastic, discrete-time-, individual-based transmission model of SARS-CoV-2 infection and COVID-19 disease. The model considers an age-structured, small-world network. Using sensitivity analysis due to many uncertainties they show that a new SARS-CoV-2 variant dominance is primarily driven by its infectivity, which does not necessarily lead to an increased public health burden. In [119], the authors used a model fitted to more than 2 years of epidemiological data from England to project potential dynamics of SARS-CoV-2 infections and deaths in England to December 2022. They considered several key uncertainties including behavioral changes and waning immunity. They concluded that for the particular case of England and under the assumption that no new variants emerge, SARS-CoV-2 transmission is expected to decline. The authors concluded that the projections depend largely on assumptions around waning immunity, social behavior and seasonality. Other interesting work related to Omicron waves is presented in [120]. In this work, a generalized SEIR model assuming gamma-distributed incubation and infectious periods is presented. The model includes susceptibility to Omicron. Their results suggest that even in those regions where the Delta variant is controlled before the beginning of the Omicron wave a significant Omicron wave can be expected. It is important to remark that for the particular case of England the Omicron wave was smaller than the Delta wave. In our paper, we provide additional insight regarding the Omicron wave.

This paper is organized as follows: In Section 2, we build the mathematical model for the Omicron wave dynamics. Section 3 is devoted to the stability analysis of the model. In Section 4, the numerical simulation results regarding the Omicron wave are presented, and the final section is devoted to discussion and conclusions.

2. Mathematical Model for the Omicron Wave Dynamics

We constructed a mathematical model that relies on nonlinear differential equations. The model includes the Omicron strain and one previous circulating strain of SARS-CoV-2. The mathematical model uses the fact that Omicron is more contagious than the previously circulating SARS-CoV-2 strain. The constructed model also encompass people who are vaccinated and asymptomatic. Moreover, the model assumes the waning of immunity for vaccinated and recovered individuals. All these are major components of the constructed model and a novelty in comparison with other models. The developed model assumes that the pre-existent circulating SARS-CoV-2 strain(s) has (have) lower contagiousness than the Omicron strain. The constructed model can be extended to other circulating SARS-CoV-2 strains if similar conditions hold.

The model encompass individuals in the susceptible ($S_i(t)$), symptomatic ($I_i(t)$), asymptomatic ($A_i(t)$) and recovered ($R_i(t)$) groups for each SARS-CoV-2 strain. In addition, the model comprise three type of subclasses for vaccinated individuals. The first is when susceptible individuals are vaccinated $V(t)$, the second when individuals who have recovered from strain 1 get vaccinated V_{1R}, and the last arises when individuals who have recovered from strain 2 get vaccinated V_{2R}. The individuals in the last two subpopulations have stronger immunity and protection against the SARS-CoV-2, as immunology studies have suggested [57]. The model is depicted in Figure 1.

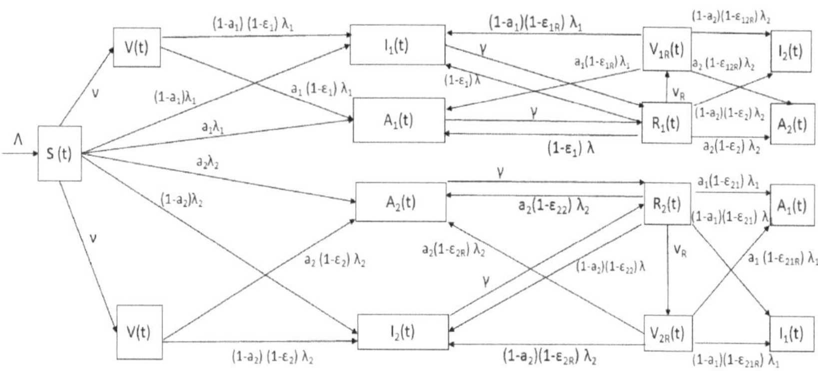

Figure 1. Diagram of the mathematical model (2) with classes and relevant parameters.

The flow of individuals from one subpopulation to another depends on the individual COVID-19 disease status. The model bears in mind partial cross-immunity against the other SARS-CoV-2 strain due to the adaptive immune response [75,122–124]. A susceptible individual can get infected with either strain and progress to the symptomatic classes (with either the previously circulating strain or Omicron) or to the asymptomatic classes ($A_1(t)$ or $A_2(t)$). The symptomatic and asymptomatic individuals stay in the infectious stage for a certain time with mean $1/\gamma$. The symptomatic and asymptomatic individuals then move to the recovered classes ($R_1(t)$ or $R_2(t)$ respectively). Then, individuals in the recovered class $R_1(t)$ can progress to the vaccinated class $V_{1R}(t)$ if they get vaccinated. However, they can also progress (with different probabilities) to infected subpopulations $I_1(t), I_2(t), A_1(t)$ or $A_2(t)$, depending on which strain they get and the symptoms. Analogously, individuals in the recovered class $R_2(t)$ can progress to the vaccinated class $V_{2R}(t)$ if they get vaccinated and to the infected subpopulations $I_1(t), I_2(t), A_1(t)$ or $A_2(t)$. Recovered individuals cannot go back to the susceptible population due to partial cross immunity and adaptive immune system that has memory [57,124–128]. Finally, symptomatic individuals can die due to COVID-19, but the model assumes that people who are asymptomatic cannot. The model, as any other epidemiological model, is obviously a simplification of reality. For instance, the hospitalization subpopulation is not considered explicitly, nor are presymptomatic individuals. It is important to remark that a large number of studies assume these simplifications in order to focus on some particular stages and/or parameters.

The model allows us to analyze the dynamics of the Omicron wave taking into account two SARS-CoV-2 strains. Studies have shown that exposure to small airborne particles is equally, or even more, likely to lead to infection with SARS-CoV-2 as the more widely recognized transmission via larger respiratory droplets and/or direct contact with infected people or contaminated surfaces [129]. Thus, we can model the transmission of SARS-CoV-2 by mass action, i.e., a term $\beta S\, I$, where β is the SARS-CoV-2 transmission rate [117]. The total population size is given by

$$\begin{aligned} N(t) &= S(t) + I_1(t) + A_1(t) + I_2(t) + A_2(t) + R_1(t) + R_2(t) \\ &\quad + V(t) + V_{1R}(t) + V_{2R}(t). \end{aligned} \tag{1}$$

The total population $N(t)$ does not include the cumulative deaths but we can compute them in the in silico simulations. The model is represented by the next differential equations

$$\begin{aligned}
\dot{S}(t) &= \Lambda - (\nu + d)S(t) - \lambda_1(t)S(t) - \lambda_2(t)S(t), \\
\dot{I}_1(t) &= (1-a_1)\lambda_1(t)\Big(S(t) + (1-\epsilon_1)V(t) + (1-\epsilon_{1R})V_{1R}(t) + (1-\epsilon_{21R})V_{2R}(t) \\
&\quad + (1-\epsilon_1)R_1(t) + (1-\epsilon_{21})R_2(t)\Big) - (d + d_1 + \gamma)I_1(t), \\
\dot{A}_1(t) &= a_1\lambda_1(t)\Big(S(t) + (1-\epsilon_1)V(t) + (1-\epsilon_{1R})V_{1R}(t) + (1-\epsilon_{21R})V_{2R}(t) \\
&\quad + (1-\epsilon_1)R_1(t) + (1-\epsilon_{21})R_2(t)\Big) - (d + \gamma)A_1(t), \\
\dot{I}_2(t) &= (1-a_2)\lambda_2(t)\Big(S(t) + (1-\epsilon_2)V(t) + (1-\epsilon_{2R})V_{2R}(t) + (1-\epsilon_{12R})V_{1R}(t) \\
&\quad + (1-\epsilon_2)R_1(t) + (1-\epsilon_{22})R_2(t)\Big) - (d + d_2 + \gamma)I_2(t), \\
\dot{A}_2(t) &= a_2\lambda_2(t)\Big(S(t) + (1-\epsilon_2)V(t) + (1-\epsilon_{2R})V_{2R}(t) + (1-\epsilon_{12R})V_{1R}(t) \\
&\quad + (1-\epsilon_2)R_1(t) + (1-\epsilon_{22})R_2(t)\Big) - (d + \gamma)A_2(t), \qquad (2) \\
\dot{R}_1(t) &= \gamma(I_1(t) + A_1(t)) - (d + \nu_r)R_1(t) - \lambda_1(t)(1-\epsilon_1)R_1(t) - \lambda_2(t)(1-\epsilon_2)R_1(t), \\
\dot{R}_2(t) &= \gamma(I_2(t) + A_2(t)) - (d + \nu_r)R_2(t) - \lambda_2(t)(1-\epsilon_{22})R_2(t) - \lambda_1(t)(1-\epsilon_{21})R_2(t), \\
\dot{V}(t) &= \nu S(t) - dV(t) - (1-\epsilon_1)\lambda_1(t)V(t) - (1-\epsilon_2)\lambda_2(t)V(t), \\
\dot{V}_{1R}(t) &= \nu_r R_1(t) - dV_{1R}(t) - (1-\epsilon_{1R})\lambda_1(t)V_{1R}(t) - (1-\epsilon_{12R})\lambda_2(t)V_{1R}(t), \\
\dot{V}_{2R}(t) &= \nu_r R_2(t) - dV_{2R}(t) - (1-\epsilon_{2R})\lambda_2(t)V_{2R}(t) - (1-\epsilon_{21R})\lambda_1(t)V_{2R}(t),
\end{aligned}$$

where $\lambda_1(t) = \beta_{I_1} I_1(t) + \beta_{A_1} A_1(t)$ and $\lambda_2(t) = \beta_{I_2} I_2(t) + \beta_{A_2} A_2(t)$ are the sources that produce infections in the different at risk subpopulations. The model comprises ten dependent variables, representing the different subpopulations. The parameters with their respective meaning and numerical values are shown in Table 1.

Table 1. Parameters for the Omicron wave mathematical model (2) with their respective meaning and numerical values.

Parameter	Symbol	Value
Inflow rate	λ	7.864180×10^3 people/day [130]
Natural death rate	d	0.00002378 1/day [130]
Infectious period	γ^{-1}	7 days [131]
Transmission rate	β_i	varied $\beta_1 \leq \beta_2$
Death rate (infected with previous circulating strains)	d_1	0.01 days^{-1} [106,132]
Death rate (infected with Omicron)	d_2	varied < 0.01 days^{-1} [49]
Vaccination rates	ν, ν_R	varied $\nu \geq \nu_R$ 1/day [130]
Proportion of asymptomatic	a_i	0.5 [133,134]

We will analyze some basic features of the model (2) in order to obtain a mathematical framework for the stability analysis. Some conditions of the model (2) are The initial conditions satisfy

$$S(0) > 0,\ I_1(0) \geq 0,\ A_1(0) \geq 0,\ I_2(0) \geq 0,\ A_2(0) \geq 0,\ R_1(0) \geq 0,\ R_2(0) \geq 0,\\ V(0) > 0,\ V_{1R}(0) \geq 0,\ V_{2R}(0) \geq 0. \tag{3}$$

The parameters satisfy

$$\Lambda,\ \beta_{I_2}, \beta_{I_1}, \beta_{A_2}, \beta_{A_1}, \alpha, \gamma, d, d_i, \nu_r \in \mathbb{R}^+, \text{ and } a_i, \epsilon_i, \epsilon_{iR}, \epsilon_{ij}, \epsilon_{ijR} \in [0,1]. \tag{4}$$

Positivity

By the classical theory of ordinary differential equations [135,136], it deduces that the system (2) is well-posed, and has a unique solution

$$\mathcal{Z}(t) := \Big(S(t), I_1(t), A_1(t), I_2(t), A_2(t), R_1(t), R_2(t), V(t), V_{1R}(t), V_{2R}(t)\Big)$$

satisfying the initial conditions given by (3). The dependent variables of the system (2) are subpopulations; therefore, we must show that if (3) holds, then the solutions of the mathematical model (2) are positive $\forall t > 0$.

Theorem 1. *Assume that (2) and (3) hold. Then the solution $\mathcal{Z}(t)$ of (2) is positive and uniformly bounded $\forall t > 0$.*

Proof. We define the following number

$$\mathcal{W} = \sup\{\rho > 0/\forall t \in [0,\rho],\ S(t) > 0, I_i(t) \geq 0, A_i(t) \geq 0, R_i(t) \geq 0, V(t) > 0, V_{1R}(t) \geq 0, V_{2R}(t) \geq 0\},$$

for $i = 1,2$. Suppose that $\mathcal{W} < \infty$. Since the solutions of the model (2) are continuous, it follows that

$$S(\mathcal{W}) = 0, \text{ or } I_2(\mathcal{W}) = 0, \text{ or } I_1(\mathcal{W}) = 0, \text{ or } A_2(\mathcal{W}) = 0, \text{ or } A_1(\mathcal{W}) = 0, \text{ or } R_1(\mathcal{W}) = 0, \text{ or}\\ R_2(\mathcal{W}) = 0, \text{ or } V(\mathcal{W}) = 0, \text{ or } V_{1R}(\mathcal{W}) = 0, \text{ or } V_{2R}(\mathcal{W}) = 0.$$

Thus, if $S(\mathcal{W}) = 0$, is obtained before the other variables, one obtains

$$\frac{dS(\mathcal{W})}{dt} = \lim_{t \to \mathcal{W}^-} \frac{S(\mathcal{W}) - S(t)}{\mathcal{W} - t} \leq 0.$$

Accordingly, from first the Equation of the model (2), one obtains that

$$\begin{aligned}\dot{S}(\mathcal{W}) &= \Lambda - (\nu + d)S(\mathcal{W}) - \lambda_1(\mathcal{W})S(\mathcal{W}) + \lambda_2(\mathcal{W})S(\mathcal{W})\\ &= \Lambda > 0,\end{aligned}$$

which is a contradiction. Therefore, $S(t) > 0$, for all $t \geq 0$. Now, similarly, if we assume that $V(\mathcal{W}) = 0$, occurs before any of the other variables are zero, one obtains

$$\frac{dV(\mathcal{W})}{dt} = \lim_{t \to \mathcal{W}^-} \frac{V(\mathcal{W}) - V(t)}{\mathcal{W} - t} \leq 0,$$

and using the eighth Equation (2), another contradiction follows

$$\dot{V}(\mathcal{W}) = \nu S(\mathcal{W}) - dV(\mathcal{W}) - (1-\epsilon_1)\lambda_1(t)V(\mathcal{W}) - (1-\epsilon_2)\lambda_2(t)V(\mathcal{W}) > 0.$$

We can use a similar process for the other dependent variables to obtain to similar contradictions. Therefore, $\mathcal{W} = +\infty$, and therefore

$$S(t) \geq 0,\ I_1(t) \geq 0,\ A_1(t) \geq 0,\ I_2(t) \geq 0,\ A_2(t) \geq 0,\ R_1(t) \geq 0, R_2(t) \geq 0,\ V(t) \geq 0,$$
$$V_{1R}(t) \geq 0, V_{2R}(t) \geq 0,$$

for $t > 0$. Next, using (2) one obtains

$$\dot{N}(t) = \Lambda - dN(t) - d_1 I_1(t) - d_2 I_2(t) \leq \Lambda - dN(t), \tag{5}$$

and using Gronwall inequalities one obtains that

$$N(t) \leq \frac{\Lambda}{d} + \left(N(0) - \frac{\Lambda}{d}\right)e^{-dt}, \tag{6}$$

for $t \geq 0$. Now, taking $N(0) \leq \frac{\Lambda}{d}$, then $N(t) \leq \frac{\Lambda}{d}$. On the other hand, from the first and eighth Eqs. of system (2) it follows that

$$\dot{S}(t) = \Lambda - (\nu + d)S(t) - \lambda_1(t)S(t) - \lambda_2(t)S(t) \leq \Lambda - (\nu + d)S(t),$$

and

$$\dot{V}(t) = \nu S(t) - dV(t) - (1-\epsilon_1)\lambda_1(t)V(t) - (1-\epsilon_2)\lambda_2(t)V(t) \leq \nu S(t) - dV(t).$$

Taking the limit, we have that $S(t) \leq \frac{\Lambda}{\nu + d}$ and $V(t) \leq \frac{\nu\Lambda}{d(\nu + d)}$ as $t \to \infty$. As a result, $\theta \in [0,1)$ implies that

$$0 < S(t) + \theta V(t) \leq \frac{\Lambda[d + \theta\nu]}{d(d+\nu)}, \text{ as } t \to \infty.$$

Therefore, we can consider the region

$$\mathcal{O} = \left\{ (S, I_1, A_1, I_2, A_2, R_1, R_2, V, V_{1R}, V_{2R}) \in \mathbf{R}_+^{10} \;\middle|\; \begin{array}{l} N(t) \leq \frac{\Lambda}{d},\ S(t) \leq \frac{\Lambda}{d(\nu + d)}, \\ 0 < S(t) + \theta V(t) \leq \frac{\Lambda[d + \theta\nu]}{d(d+\nu)}, \theta \in [0,1) \end{array} \right\} \tag{7}$$

which is positively invariant. Thus, the solutions of system (2) are bounded. Furthermore, if $N(0) > \frac{\Lambda}{d}$, then either the solution enters $\mathcal{O}$ for infinite time or $N(t) \to \frac{\Lambda}{d}$ asymptotically. $\square$

3. Stability Analysis

In the qualitative analysis of the model solutions, it is common to determine the stationary points that identify the disease-free and endemic equilibrium points. In this case, in the model (2) there is a disease-free point (F_1^*), which can be found by setting

$I_1 = I_2 = A_1 = A_2 = 0$, and indicates that SARS-CoV-2 becomes extinct. Now, it is of great importance to determine in epidemiological models the different parameters that delimit the different states of a disease. One in particular is the basic reproduction number $\mathcal{R}_0$, which measures the influence of introducing one infected individual into a total susceptible population [76,137].

3.1. Disease-Free Equilibrium Point and $\mathcal{R}_0$

The disease-free equilibrium (F_1^*) point of the model (2) is given by

$$F_1^* = \left(S^0, I_1^0, A_1^0, I_2^0, A_2^0, R_1^0, R_2^0, V^0, V_{1R}^0, V_{2R}^0\right) = \left(\frac{\Lambda}{d+\nu}, 0, 0, 0, 0, 0, 0, 0, \frac{\nu\Lambda}{d\,(d+\nu)}, 0, 0\right). \tag{8}$$

In order to obtain an expression for $\mathcal{R}_0$ in the model (2), we use the next generation matrix (NGM) method [116,137]. For this purpose, we determine the matrix $\mathcal{F}$ representing the new infection cases and the matrix $\mathcal{V}$ represents the progression between classes. The eigenvalue of the matrix $\mathcal{F}\mathcal{V}^{-1}$ with largest absolute value is the basic reproduction number $\mathcal{R}_0$. For further technicalities see [116,137]. Thus,

$$\mathcal{F} = \begin{bmatrix} (1-a_1)\mathbf{B}_{I_1} & (1-a_1)\mathbf{B}_{A_1} & 0 & 0 \\ a_1\mathbf{B}_{I_1} & a_1\mathbf{B}_{A_1} & 0 & 0 \\ 0 & 0 & (1-a_2)\mathbf{B}_{I_2} & (1-a_2)\mathbf{B}_{A_2} \\ 0 & 0 & a_2\mathbf{B}_{I_2} & a_2\mathbf{B}_{A_2} \end{bmatrix}, \tag{9}$$

and

$$\mathcal{V} = \begin{bmatrix} d+d_1+\gamma & 0 & 0 & 0 \\ 0 & d+\gamma & 0 & 0 \\ 0 & 0 & d+d_2+\gamma & 0 \\ 0 & 0 & 0 & d+\gamma \end{bmatrix}. \tag{10}$$

Then, one obtains

$$\mathcal{F}\mathcal{V}^{-1} = \begin{bmatrix} \dfrac{(1-a_1)\mathbf{B}_{I_1}}{d+d_1+\gamma} & \dfrac{(1-a_1)\mathbf{B}_{A_1}}{d+\gamma} & 0 & 0 \\ \dfrac{a_1\mathbf{B}_{I_1}}{d+d_1+\gamma} & \dfrac{a_1\mathbf{B}_{A_1}}{d+\gamma} & 0 & 0 \\ 0 & 0 & \dfrac{(1-a_2)\mathbf{B}_{I_2}}{d+d_2+\gamma} & \dfrac{(1-a_2)\mathbf{B}_{A_2}}{d+\gamma} \\ 0 & 0 & \dfrac{a_2\mathbf{B}_{I_2}}{d+d_2+\gamma} & \dfrac{a_2\mathbf{B}_{A_2}}{d+\gamma} \end{bmatrix},$$

which is the NGM, and the positive eigenvalues are given by

$$\mathcal{R}_{0_1} = \frac{\mathbf{B}_{A_1} a_1 (d + d_1 + \gamma) + \mathbf{B}_{I_1}(1 - a_1)(d + \gamma)}{(d + \gamma)(d + d_1 + \gamma)},$$
$$\mathcal{R}_{0_2} = \frac{\mathbf{B}_{A_2} a_2 (d + d_2 + \gamma) + \mathbf{B}_{I_2}(1 - a_2)(d + \gamma)}{(d + \gamma)(d + d_2 + \gamma)}, \tag{11}$$

or

$$\mathcal{R}_{0_1} = \frac{\mathbf{B}_{A_1} a_1}{d + \gamma} + \frac{\mathbf{B}_{I_1}(1 - a_1)}{d + d_1 + \gamma},$$
$$\mathcal{R}_{0_2} = \frac{\mathbf{B}_{A_2} a_2}{d + \gamma} + \frac{\mathbf{B}_{I_2}(1 - a_2)}{d + d_2 + \gamma},$$

where

$$\mathbf{B}_{I_1} = \frac{\beta_{I_1}\Lambda}{\nu + d}\left(1 + \frac{(1 - \epsilon_1)\nu}{d}\right), \ \mathbf{B}_{A_1} = \frac{\beta_{A_1}\Lambda}{\nu + d}\left(1 + \frac{(1 - \epsilon_1)\nu}{d}\right),$$
$$\mathbf{B}_{I_2} = \frac{\beta_{I_2}\Lambda}{\nu + d}\left(1 + \frac{(1 - \epsilon_2)\nu}{d}\right), \ \mathbf{B}_{A_2} = \frac{\beta_{A_2}\Lambda}{\nu + d}\left(1 + \frac{(1 - \epsilon_2)\nu}{d}\right).$$

The parameters $\mathcal{R}_{0_1}$ and $\mathcal{R}_{0_2}$ are related to the two different SARS-CoV-2 strains, respectively. Thus, one obtains the spectral radius of $\mathcal{F}V^{-1}$

$$\mathcal{R}_0 = \max\{\mathcal{R}_{0_1}, \mathcal{R}_{0_2}\}. \tag{12}$$

The parameter $\mathcal{R}_0$ allows us to determine if an outbreak would occur. When $\mathcal{R}_0 < 1$, and if the initial conditions of the model (2) are close enough to the equilibrium (F_1^*), then no outbreak would occur. However, when $\mathcal{R}_0 > 1$, an epidemic would occur. Thus, one obtains the next theorem.

Theorem 2. *When the basic reproduction number $\mathcal{R}_0 < 1$ ($\mathcal{R}_0 > 1$), the disease-free equilibrium point F_1^* of the model (2) and given in (8) is locally asymptotically stable (unstable).*

Proof. The proof follows from applying Theorem 2 in [137]. □

Global Stability of Disease-Free Equilibrium Point

Analyzing the behavior of the solutions of an epidemiological model represented by a system of differential equations such as (2) around the disease-free equilibrium point is important because it determines what public health measures are necessary in order to avoid endemic situations. Thus, we want to analyze whether the disease-free point F_1^* is a global attractor, i.e., it must be proven that if $\mathcal{R}_0 < 1$, the disease becomes extinct regardless of the initial conditions of the model (2). In other words, the point F_1^* is globally asymptotically stable (GAS). In order to prove the global stability of F_1^*, we apply the methodology used in [138]. The system (2) can be written as

$$\dot{Y}(t) = \mathbf{F}(Y, Z), \quad \dot{Z}(t) = \mathbf{I}(Y, Z), \quad \mathbf{I}(Y, \mathbf{0}) = \mathbf{0} \in \mathbb{R}^4, \tag{13}$$

with $Y = (S, V, R_1, R_2, V_{R1}, V_{R2})$ which denotes the vector of uninfected compartments, and $Z = (I_1, A_1, I_2, A_2)$ is the vector of infected compartments. Moreover, $\mathbf{F}(Y, \mathbf{0})$ is the right-hand side of $\dot{S}(t), \dot{V}(t), \dot{R}_1(t), \dot{R}_2(t), \dot{V}_{R1}(t), \dot{V}_{R2}(t)$, setting $I_1 = A_1 = I_2 = A_2 = 0$. Thus, F_1^* is rewritten as $Y^0 = (S^0, V^0, \mathbf{0})$, $\mathbf{0} \in \mathbb{R}^4$. The following result guarantees the GAS of F_1^*.

Theorem 3. *The point F_1^* given by (8) of system (2) is GAS in $\mathcal{O}$ if $\mathcal{R}_0 \leq 1$, and if the next conditions hold:*

- **Condition 1** : *Given* $\dot{Y}(t) = \mathbf{F}(Y, \mathbf{0})$, $\mathbf{0} \in \mathbf{R}^4$, *then* Y^0 *is GAS.*
- **Condition 2** : $\mathbf{I}(Y, Z) = \mathcal{J}Z - \mathring{\mathbf{I}}(Y, Z)$, *then* $\mathring{\mathbf{I}}(Y, Z) \geq 0$ *in* $\mathcal{O}$ *as* $t \to \infty$, *and* $\mathcal{J} = D_Z(\mathring{\mathbf{I}}, \mathbf{0})$ *is an* $M-$*matrix, i.e., the off-diagonal elements are non-negative.*

Proof. For the **Condition 1,** we write $\dot{Y}(t) = \mathbf{F}(Y, \mathbf{0})$, $\mathbf{0} \in \mathbf{R}^4$ as

$$\begin{aligned} \dot{S}(t) &= \Lambda - (\nu + d)S(t), \\ \dot{V}(t) &= \nu S(t) - dV(t), \\ \dot{R_1}(t) &= -(d + \nu_r)R_1(t), \\ \dot{R_2}(t) &= -(d + \nu_r)R_2(t), \\ \dot{V_{1R}}(t) &= \nu_r R_1(t) - dV_{1R}(t), \\ \dot{V_{2R}}(t) &= \nu_r R_2(t) - dV_{2R}(t). \end{aligned} \tag{14}$$

After some calculations using (14) one obtains

$$(S(t), V(t), R_1(t), R_2(t), dV_{1R}(t), V_{2R}(t)) \to \left(S^0, V^0, \mathbf{0}\right) \text{ as } t \to \infty. \tag{15}$$

On the other hand, for the **Condition 2,** from (9) and (10) we can obtain the matrix $\mathcal{J} = \mathcal{F} - \mathcal{V}$.

$$\mathcal{J} = \begin{bmatrix} (1-a_1)\mathbf{B}_{I_1} - (d + d_1 + \gamma) & (1-a_1)\mathbf{B}_{A_1} & 0 & 0 \\ a_1\mathbf{B}_{I_1} & a_1\mathbf{B}_{A_1} - (d+\gamma) & 0 & 0 \\ 0 & 0 & (1-a_2)\mathbf{B}_{I_2} - (d + d_2 + \gamma) & (1-a_2)\mathbf{B}_{A_2} \\ 0 & 0 & a_2\mathbf{B}_{I_2} & a_2\mathbf{B}_{A_2} - (d+\gamma) \end{bmatrix},$$

and $\mathcal{J}$ is an $M-$matrix. Next, from (15) and in view of (7) yields

$$\mathring{\mathbf{I}}(Y, Z) = \mathcal{J}Z - \mathbf{I}(Y, Z) = \begin{bmatrix} \left\{\frac{\Lambda(d + (1-\epsilon_1)\nu)}{d(d+\nu)} - (S + (1-\epsilon_1)V)\right\}(1-a_1)\lambda_1 - (1-a_1)\lambda_1\mathbf{W}_1 \\ \left\{\frac{\Lambda(d + (1-\epsilon_1)\nu)}{d(d+\nu)} - (S + (1-\epsilon_1)V)\right\}a_1\lambda_1 - a_1\lambda_1\mathbf{W}_1 \\ \left\{\frac{\Lambda(d + (1-\epsilon_2)\nu)}{d(d+\nu)} - (S + (1-\epsilon_2)V)\right\}(1-a_{12})\lambda_2 - (1-a_2)\lambda_2\mathbf{W}_2 \\ \left\{\frac{\Lambda(d + (1-\epsilon_2)\nu)}{d(d+\nu)} - (S + (1-\epsilon_2)V)\right\}a_2\lambda_2 - a_2\lambda_2\mathbf{W}_2 \end{bmatrix} \geq 0,$$

in Ω as $t \to \infty$, where

$$\mathbf{W}_1(t) = [(1-\epsilon_{1R})V_{1R}(t) + (1-\epsilon_{21R})V_{2R}(t) + (1-\epsilon_1)R_1(t) + (1-\epsilon_{21})R_2(t)],$$

and

$$\mathbf{W}_2(t) = (1-\epsilon_{2R})V_{2R}(t) + (1-\epsilon_{12R})V_{1R}(t) + (1-\epsilon_2)R_1(t) + (1-\epsilon_{22})R_2(t),$$

with $\mathbf{W}_1(t), \mathbf{W}_2(t) \to 0$, as $t \to \infty$. Thus, it is very clear that $\mathring{\mathbf{I}}(Y, Z) \geq \mathbf{0}$, with $\mathbf{0} \in \mathbf{R}^4$. □

The consequence of Theorem 3 from the epidemiological viewpoint is that COVID will not become endemic as long as $\mathcal{R}_0 < 1$, regardless of the initial conditions.

3.2. Endemic Equilibrium Point

The behavior of the solutions of the model (2) when $\mathcal{R}_0 > 1$ depends on the endemic points. We can find these endemic points by simply setting the right-hand side of the system (2) to zero and obtaining the algebraic solutions representing the endemic points as a function of the parameters of the mathematical model (2).

For the model (2), we want to determine the endemic points, which will be denoted by

$$E^* = (S^*, A_1^*, I_1^*, A_2^*, I_2^*, R_1^*, R_2^*, V^*, V_{1R}^*, V_{2R}^*), \tag{16}$$

and this vector is a solution of the following algebraic system:

$$\begin{aligned}
0 &= \Lambda - (\nu + d)S^* - \lambda_1^* S^* - \lambda_2^* S^*, \\
0 &= (1 - a_1)\lambda_1^* \Big(S^* + (1 - \epsilon_1)V^* + (1 - \epsilon_{1R})V_{1R}^* + (1 - \epsilon_{21R})V_{2R}^* + (1 - \epsilon_1)R_1^* \\
&+ (1 - \epsilon_{21})R_2^*\Big) - (d + d_1 + \gamma)I_1^*, \\
0 &= a_1\lambda_1^* \Big(S^* + (1 - \epsilon_1)V^* + (1 - \epsilon_{1R})V_{1R}^* + (1 - \epsilon_{21R})V_{2R}^* + (1 - \epsilon_1)R_1^* \\
&+ (1 - \epsilon_{21})R_2^*\Big) - (d + \gamma)A_1^*, \\
0 &= (1 - a_2)\lambda_2^* \Big(S^* + (1 - \epsilon_2)V^* + (1 - \epsilon_{2R})V_{2R}^* + (1 - \epsilon_{12R})V_{1R}^* + (1 - \epsilon_2)R_1^* \\
&+ (1 - \epsilon_{22})R_2^*\Big) - (d + d_2 + \gamma)I_2^*, \\
0 &= a_2\lambda_2^* \Big(S^* + (1 - \epsilon_2)V^* + (1 - \epsilon_{2R})V_{2R}^* + (1 - \epsilon_{12R})V_{1R}^* + (1 - \epsilon_2)R_1^* \\
&+ (1 - \epsilon_{22})R_2^*\Big) - (d + \gamma)A_2^*, \\
0 &= \gamma(I_1^* + A_1^*) - (d + \nu_r)R_1^* - \lambda_1^*(1 - \epsilon_1)R_1^* - \lambda_2^*(1 - \epsilon_2)R_1^*, \\
0 &= \gamma(I_2^* + A_2^*) - (d + \nu_r)R_2^* - \lambda_2^*(1 - \epsilon_{22})R_2^* - \lambda_1^*(1 - \epsilon_{21})R_2^*, \\
0 &= \nu S^* - dV^* - (1 - \epsilon_1)\lambda_1^* V^* - (1 - \epsilon_2)\lambda_2^* V^*, \\
0 &= \nu_r R_1^* - dV_{1R}^* - (1 - \epsilon_{1R})\lambda_1^* V_{1R}^* - (1 - \epsilon_{12R})\lambda_2^* V_{1R}^*, \\
0 &= \nu_r R_2^* - dV_{2R}^* - (1 - \epsilon_{2R})\lambda_2^* V_{2R}^* - (1 - \epsilon_{21R})\lambda_1^* V_{2R}^*,
\end{aligned} \tag{17}$$

where $\lambda_1^* = \beta_{I_1} I_1^* + \beta_{A_1} A_1^*$ and $\lambda_2^* = \beta_{I_2} I_2^* + \beta_{A_2} A_2^*$. We can see from the first Eq. of the system (17) that $S^* > 0$. Moreover, $\Lambda - (d + \nu)S^* > 0$, that is, $S^* \in \mathcal{O}$. Using the second, third, fourth and fifth Equation (17) we arrive to the next result,

$$I_1^* = \frac{(1 - a_1)(d + \gamma)A_1^*}{a_1(d + d_1 + \gamma)}, \; I_2^* = \frac{(1 - a_2)(d + \gamma)A_2^*}{a_2(d + d_2 + \gamma)}. \tag{18}$$

Thus

$$\lambda_1^* = \frac{d\mathcal{R}_{0_1}(d + \gamma)(\nu + d)A_1^*}{a_1\Lambda(d + (1 - \epsilon_1)\nu)}, \; \lambda_2^* = \frac{d\mathcal{R}_{0_2}(d + \gamma)(\nu + d)A_2^*}{a_2\Lambda(d + (1 - \epsilon_2)\nu)}. \tag{19}$$

Now, from the first Equation (17) it follows that

$$S^* = \frac{\Lambda}{(\nu + d)(1 + \lambda_1^* + \lambda_2^*)}. \tag{20}$$

Next, from the sixth and seventh Equation (17), and putting (18), it follows that

$$R_1^* = \frac{\gamma\left(1 + \dfrac{(1 - a_1)(d + \gamma)}{a_1(d + d_1 + \gamma)}\right)A_1^*}{(d + \nu_r + \lambda_1^*(1 - \epsilon_1) + \lambda_2^*(1 - \epsilon_2))}, \; R_2^* = \frac{\gamma\left(1 + \dfrac{(1 - a_2)(d + \gamma)}{a_2(d + d_2 + \gamma)}\right)A_2^*}{(d + \nu_r + \lambda_1^*(1 - \epsilon_{22}) + \lambda_2^*(1 - \epsilon_{21}))} \tag{21}$$

In the same way, from the ninth and tenth Equation (17), one obtains

$$V_{1R}^* = \frac{\nu_r R_1^*}{(d + \lambda_1^*(1 - \epsilon_{1R}) + \lambda_2^*(1 - \epsilon_{12R}))}, \; V_{2R}^* = \frac{\nu_r R_2^*}{(d + \lambda_1^*(1 - \epsilon_{2R}) + \lambda_2^*(1 - \epsilon_{21R}))}, \tag{22}$$

and finally

$$V^* = \frac{S^* \nu}{d + (1-\epsilon_1)\lambda_1^* + (1-\epsilon_2)\lambda_2^*}. \tag{23}$$

Thus, there are three endemic equilibrium points that can be obtained from Equation (18). Indeed, if $A_1^* = 0$ and $A_2^* > 0$, then one obtains from Equations (18)–(23) that $\lambda_1^* = 0$, $\lambda_2^* > 0, S^* > 0$, $R_1^* = 0, R_2^* > 0, V_{1R}^* = 0, V_{2R}^* > 0$, and $V^* > 0$. Thus, the first endemic point given by

$$E_1^* = (S^*, 0, 0, A_2^*, I_2^*, 0, R_2^*, V^*, 0, V_{2R}^*),. \tag{24}$$

Next, if $A_2^* = 0$ and $A_1^* > 0$, then one obtains from Equations (18)–(23) that $\lambda_2^* = 0$, $\lambda_1^* > 0, S^* > 0$, $R_2^* = 0, R_1^* > 0, V_{2R}^* = 0, V_{1R}^* > 0$, and $V^* > 0$. Therefore, the second endemic point is

$$E_2^* = (S^*, A_1^*, I_1^*, 0, 0, R_1^*, 0, V^*, V_{1R}^*, 0). \tag{25}$$

Finally, if $A_1^* > 0$ and $A_2^* > 0$ then we can obtain the third endemic point given by Equations (18)–(23).

Thus, the steady states are one of the endemic equilibrium points depending on the numerical values of $\mathcal{R}_{0_2}$ and $\mathcal{R}_{0_1}$. For instance, if $\mathcal{R}_{0_2} > \mathcal{R}_{0_1} > 1$ then both SARS-CoV-2 strains survive in the population. This is due to the fact that the mathematical model (2) does not consider full immunity either from vaccination or natural immunity [62,80]. Recent studies suggest that this is true for the COVID-19 pandemic situation [22,139–143]. We did not perform further stability analysis related to periodic solutions, backward bifurcations and global stability since the aim of this study is the short dynamics of the Omicron wave and obtaining further insight into it.

4. Simulations for the Omicron Wave

We performed in silico simulations of the Omicron wave model (2) for a variety of scenarios (in fact, infinitely many) in order to obtain insight into the Omicron wave situation and additional potential consequences of the Omircon strain on the dynamics of this pandemic. We varied the vaccine's efficacy against the Omicron strain in order to consider, as some articles have mentioned, that the efficacy of the vaccine is lower against the Omicron strain [22,48]. We also varied the transmissibility and severity of the Omicron strain since it has been revealed that both factors significantly differ in comparison to the previous circulating SARS-CoV-2 strains [22,24,48]. The in silico simulations allow us to explain, at least partially, the Omicron wave period. We focus here on the qualitative results of the in silico simulations since there are uncertainties that make it very difficult to have accurate forecasts as time has proven over the COVID-19 waves.

The dependence of the transmission rate on the natural daily variability in human behavior makes estimation of the transmission rate very difficult. Sensitivity analysis is one means researchers often use to approach the uncertainties in the COVID-19 pandemic. The numerical simulations presented in the present study show different potential situations in order to remark on the distinct possibilities regarding the transmission rates. For instance, when the Omicron variant arose, the scientific community did not know if it was more transmissible or deadly than the previous strain. The simulations also have the aim of corroborating the theoretical results in addition to potential explanations of what happened in the real world. The simulations allow us to present different scenarios regarding the real values of transmission rate and case fatality rate. This provides additional insight regarding the COVID-19 pandemic dynamics and future scenarios for new variants.

All numerical simulations were carried out in Python 3.8. Ordinary differential equations were solved using the scipy.odeint routine. The simulations were performed with a PC (Intel(R) Core(TM) i7-7820HQ CPU, 2.90 GHz) with 64 Mb RAM. Table 1 shows the numerical values of the parameters that were used for the in silico simulations. For some parameters, we used a wide range of values in order to consider a larger number of scenarios and potentially extreme cases that might arise due to uncertainty in the parameters. For the initial subpopulations, we took the values from the particular situation of the USA just before the start of the Omicron wave period [1]. Based on previous works, the Omicron wave started around mid-November [45]. The values of the initial conditions can be extracted from different data sources. Like the CDC, we considered the possibility that for every symptomatic infected case there would be one asymptomatic case, even though there is some uncertainty for this [1,4,133]. We chose the situation of the USA since the reported data are more reliable than in other countries and the population is large enough to observe the main effects on the Omicron wave dynamics. In the numerical simulated scenarios, there is an effective reproduction number $\mathcal{R}_t$ that decreases as the susceptible subpopulation decreases [80,144]. During the in silico simulations, we assumed that the parameters are time-invariant, despite that in reality some parameters might vary over time. Introducing time-varying parameters is a difficult task although some modelers have attempted it [101]. For the percentage of asymptomatic cases we considered 50%, which is a situation proposed by the CDC [133]. Making reasonable changes to this percentage does not affect the qualitative conclusions of this study.

4.1. *Efficacy of the Vaccine against the Omicron Strain*

The Omicron strain has been detected in many countries [145]. Preliminary data related to the efficacies of current vaccines against the Omicron strain are available. It has been revealed that these efficacies are different in comparison with other SARS-CoV-2 strains. In [145], the authors analyzed 133,437 PCR test results and found that during the proxy Omicron period the vaccine efficacy against hospitalization was 70%, which is much lower than the 93% efficacy for the comparator period. In [52], the authors carried out a narrative review from 32 scientific articles supporting the idea that Omicron evades antibodies induced by primary vaccination or by SARS-CoV-2 infection. We use this information in order to set the efficacies of the vaccines for the numerical simulations. Based on several scientific articles, we assume that the current vaccines have less efficacy against the Omicron strain [24,51,128,146–148]. On the other hand, it has been revealed that the Omicron pseudovirus infects cells more efficiently than other SARS-CoV-2 strains [128]. Furthermore, those who received two doses of vaccine have lower neutralizing activity against Omicron [22].

Table 2 shows the different efficacies of the vaccines for a variety of status related to COVID-19. Some of these efficacies are high if the individuals already had the disease in good agreement with previous studies [149,150]. Due to a short time study of less than one year, the model does not consider a particular subpopulation for the cases where individuals contracted the disease twice, which is very unlikely. However, the model can also be used as an approximation for longer times, since it considers that once individuals have been infected with SARS-CoV-2, the likelihood to get infected again is lower due to memory cells and adaptive immunity [151–153]. The model considers implicitly the waning of the effectiveness of the vaccine as well as natural immunity since vaccinated and recovered people can get infected but with lower probability [153–155].

Table 2. Values of the assumed efficacies for the SARS-CoV-2 vaccines used in the in silico simulations.

Parameter	From	To	Value
ϵ_1	V	I_1, A_1	[0.8,0.95]
ϵ_1	R_1	I_1, A_1	[0.8,0.95]
ϵ_2	V	I_2, A_2	[0.37,0.6]
ϵ_{1R}	V_{1R}	I_1, A_1	[0.98,0.99]
ϵ_{12R}	V_{1R}	I_2, A_2	[0.95,0.98]
ϵ_{22}	R_2	I_2, A_2	[0.9,0.95]
ϵ_{21}	R_2	I_1, A_1	[0.37,0.6]
ϵ_{21R}	V_{2R}	I_1, A_1	[0.98,0.99]

4.2. Numerical Simulations towards Steady States

We present three in silico simulations of the model (2) in order to provide additional support to the theoretical results and observe the long-term behavior. For these scenarios we used initial conditions where the number of infected cases is very small since we just want to compare with the theoretical results and since $\mathcal{R}_0$ is defined for almost fully susceptible populations [116,137]. We varied the transmissibility of circulating SARS-CoV-2 strains and we considered that the Omicron strain has a higher likelihood to be transmitted than the previously circulating strain. This allows us to foresee the long-term qualitative effects of the Omicron wave.

Figure 2 displays the evolution of the symptomatic subpopulations $I_1(t)$ and $I_2(t)$. We chose the transmission rate such that $\mathcal{R}_{0_1} < 1$ and $\mathcal{R}_{0_2} < 1$. Both symptomatic (the asymptomatic cases were also treated but are not shown) subpopulations approach the disease-free steady state F_1^*. In order to obtain manageable and useful graphs for the steady states we use a large natural death rate for faster dynamics only in this subsection. Figure 3 displays the long-term behavior when $\mathcal{R}_{0_2} > 1 > \mathcal{R}_{0_1}$ and the initial infected subpopulations are small. Note that the Omicron strain becomes the prevalent one and the previous circulating one vanishes. In Figure 4 we consider the case where the initial number of infected people with the previously circulating strain is large in order to resemble reality when Omicron was introduced. It can be seen that despite having a large vaccination rate, the system (2) still approaches the endemic steady state E_1^* due to the higher transmissibility of the Omicron strain. Figure 5 depicts the case where $\mathcal{R}_{0_2} > \mathcal{R}_{0_1} > 1$ and it can be seen that the previously circulating strains and the Omicron strain become endemic. The explanation for this is due to the fact that people who got either of the SARS-CoV-2 strains can get the other strain. After this long-term dynamics results, the next subsection is devoted to the transient dynamics of the Omicron wave.

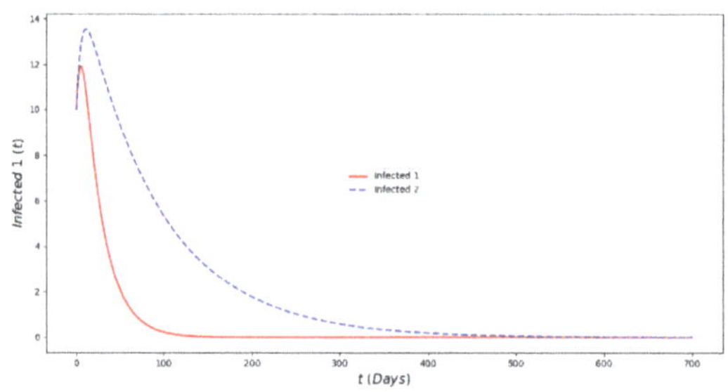

Figure 2. In silico simulation of the Omicron wave model (2) when $\mathcal{R}_{0_2} \approx 0.95 > \mathcal{R}_{0_1} \approx 0.82$. The previously circulating and Omicron strains disappear, while the system approaches the point F_1^*. We use a large natural death rate for faster dynamics.

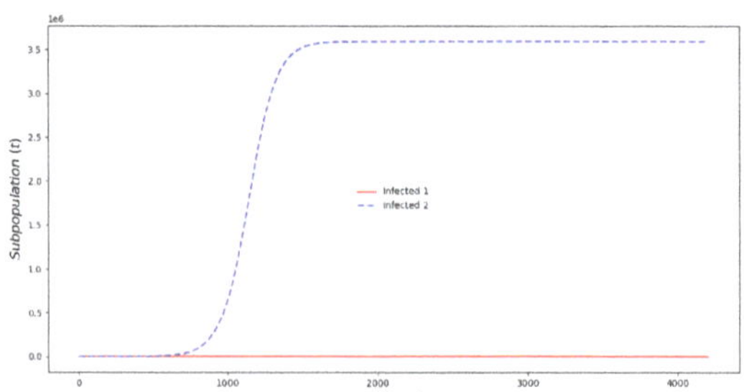

Figure 3. In silicosimulations of the Omicron wave model (2) when $\mathcal{R}_{0_2} \approx 1.04 > \mathcal{R}_{0_1} \approx 0.9$. The Omicron strain becomes prevalent and the system approaches the point E_1^*.

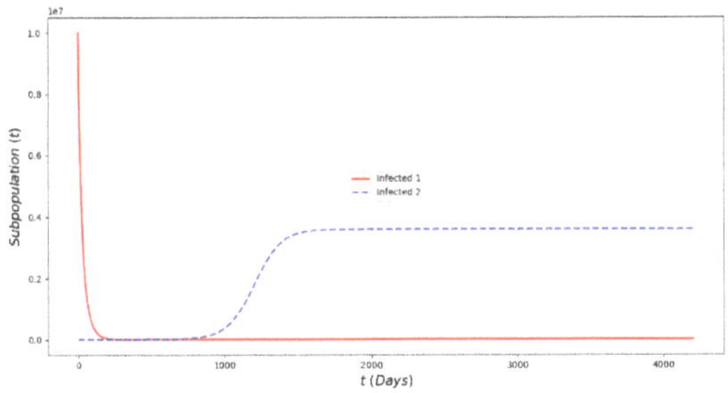

Figure 4. In silico of the Omicron wave model (2) when $\mathcal{R}_{0_2} \approx 1.04 > \mathcal{R}_{0_1} \approx 0.9$. The Omicron strain becomes prevalent and the system approaches the endemic steady state E_1^* despite the fact that the initial prevalence of the non-Omicron strain has a very large prevalence.

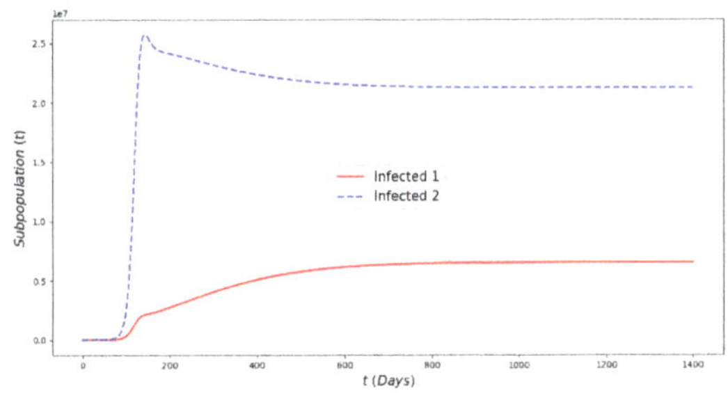

Figure 5. In silico simulations of the Omicron wave model (2) when $\mathcal{R}_{0_2} \approx 1.5 > \mathcal{R}_{0_1} \approx 1.4$. The previously circulating and Omicron strains become prevalent and the system approaches the endemic steady state E^*.

4.3. Numerical Simulations to Assess Critical Outcomes

For the in silico simulations we considered various efficacies of the vaccine against the Omicron strain, transmissibility and severity of the Omicron strain. In the analysis, we focus on the qualitative results and the effects of the aforementioned factors.

Figure 6 displays the paths of each of the subpopulations and some cumulative numbers. This is a particular case where we can see the evolution of the Omicron wave for one scenario. This is not a suitable way to understand the effects of the Omicron strain since there is no comparison with other scenarios. Thus, the next simulations consider variations of the vaccine's efficacy against Omicron and also Omicron infectivity.

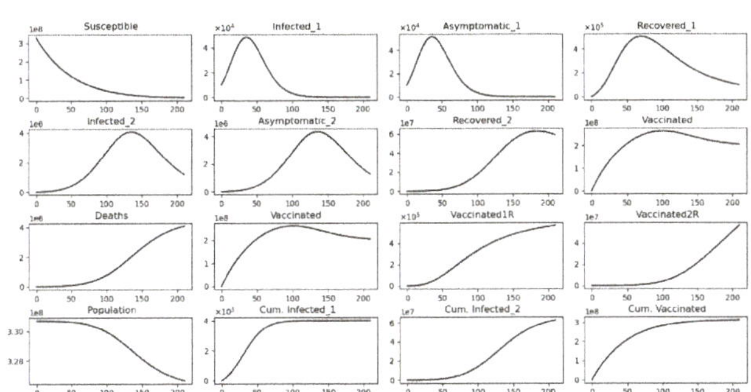

Figure 6. In silico simulation of the Omicron wave model (2) when $\mathcal{R}_{0_2} \approx 0.95 > \mathcal{R}_{0_1} \approx 0.87$. The two strains vanish and the system approaches the disease-free equilibrium point F_1^*.

Figure 7 displays different outcomes regarding the final cumulative infected population with each strain. It can be seen that when the Omicron infectivity rate increases, the final cumulative number of people infected with Omicron increases, but the final cumulative number of people infected with the previously circulating strain decreases. This is due to a competition for the susceptible people among the strains. The model does not consider co-infection. Furthermore, the final cumulative number of infected people with the previously circulating SARS-CoV-2 strain increases if the vaccine's efficacy against Omicron increases. The opposite situation occurs for the final cumulative number of infected people with Omicron. However, the changes to final cumulative numbers for people infected with Omicron are much larger, which partially explains the large number of infected cases that have been recorded for the Omicron wave.

Figure 8 displays the final cumulative number of deaths when we vary the vaccine's efficacy against the Omicron strain and the infectivity of the Omicron strain. As can be observed, the final cumulative number of deaths increases as Omicron's infectivity increases despite assuming the same case fatality rate. This is a major result to bring awareness to, given that even if the Omicron strain is less deadly the final cumulative deaths can increase as has indeed occurred [1,49]. We also performed in silico simulations assuming standard incidence in the model (2) and the results are qualitatively similar.

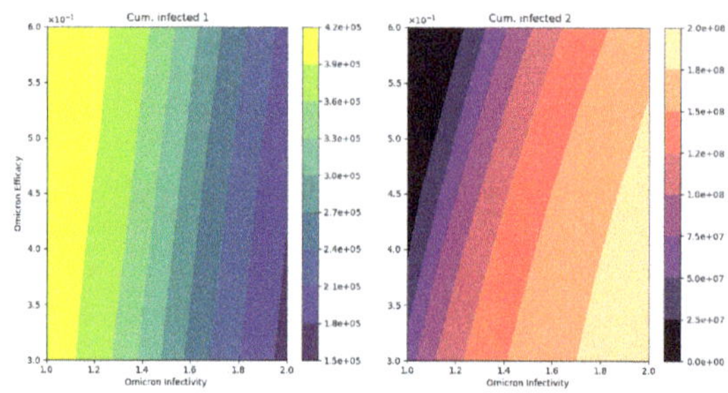

Figure 7. In silico simulation of the Omicron wave model (2). The outcomes regarding the final cumulative infected people for each strain. As the Omicron infectivity rate increases, the final cumulative number of people infected with Omicron increases but the final cumulative number of people infected with the previously circulating strain decreases.

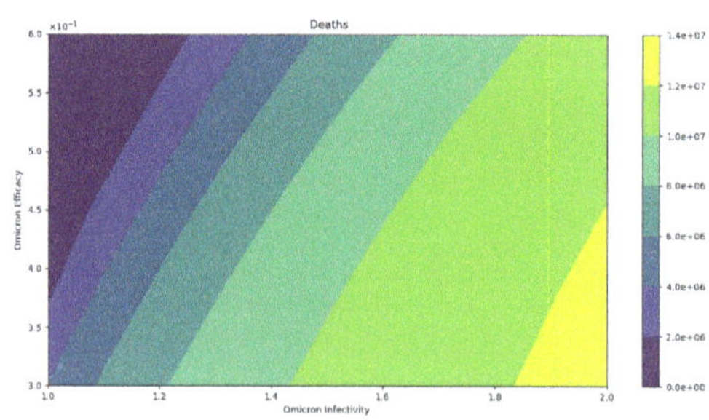

Figure 8. In silico simulation of the Omicron wave model (2). The outcomes regarding total deaths. As the Omicron infectivity rate increases, the final total number of deaths increases. As can be observed, the number of deaths increases despite assuming the same case fatality rate for the two strains.

4.4. Comparison of the Omicron Wave with the Non-Omicron Situation

Finally, we present additional in silico simulations to compare the non-Omicron with the Omicron situation. In the analysis we focus on the qualitative results related to infected people and total number of deaths since these are the crucial health outcomes of the pandemic.

Figure 9 displays the infected subpopulations over a period of six months. The total number of infected people is larger under the Omicron wave in comparison with the situation where no Omicron is introduced, as reflected in reality. Notice that initially the number of people infected with Omicron is much smaller, also as reflected in the real world.

Figure 10 displays the number of deaths over a period of six months assuming a smaller death rate for people infected with Omicron (25% of previous circulating strain). The total number of deaths is larger under the Omicron wave in comparison with the situation with no Omicron despite a large number of the population being vaccinated and a relative acceptable vaccine efficacy. These results are in good agreement with the results that have occurred during the Omicron wave [1,49].

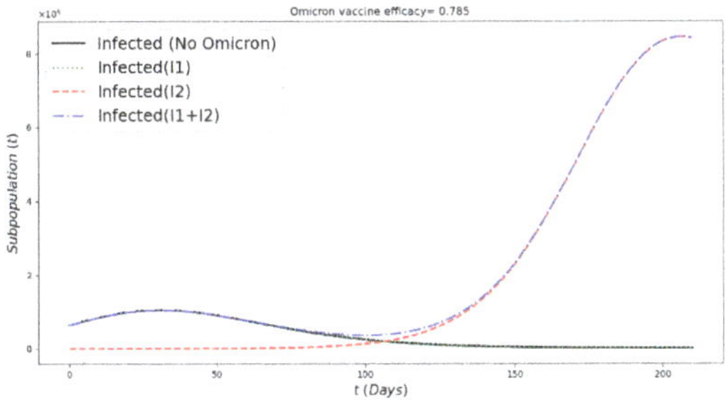

Figure 9. In silico simulation of the Omicron wave model (2) when $\mathcal{R}_{0_1} \approx 0.81$, $\mathcal{R}_{0_2} \approx 1.74$ and vaccine efficacy against Omicron is approximately 79%. More infected cases during the Omicron wave, despite a large number of the population being vaccinated and a relative acceptable vaccine efficacy.

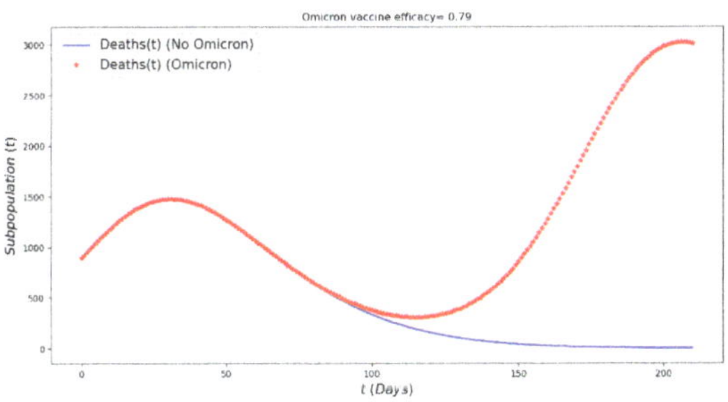

Figure 10. In silico simulation of the Omicron wave mathematical model (2) when $\mathcal{R}_{0_1} \approx 0.81$, $\mathcal{R}_{0_2} \approx 1.74$ and vaccine efficacy against Omicron is approximately 79%. More deaths during the Omicron wave, despite a lower case fatality rate for Omicron, large number of population vaccinated and a relative acceptable vaccine efficacy.

4.5. Discussion of Numerical Simulation Results

The numerical simulation results presented here agree with those obtained in previous work related to the Omicron wave. For instance, in [121] the authors found that a new SARS-CoV-2 variant's (for example, Omicron) dominance is primarily driven by its infectivity, which does not always lead to an increased public health burden. This has been shown in our work through the theoretical results and the numerical simulations. In [119], the authors considered several key uncertainties and concluded that in the particular case of England and under the assumption that no new variants emerge, SARS-CoV-2 transmission is expected to decline. This also agrees with our results, since the basic effective reproductive number depends on the transmission rates. The authors mentioned that the projections depend largely on assumptions around waning immunity, social behavior and seasonality. In our work, we presented sensitivity analysis to assess the effects of uncertainty of some factors related to the Omicron variant and the results agree with the aforementioned work. It is important to remark that during the Omicron wave, people from each region have different levels of immunity protection. This was investigated in in [120]. Their results suggested that even in those regions where the Delta variant is controlled before the beginning of the Omicron wave a significant Omicron wave can be expected. This has been shown in our study and under some mathematical conditions that we have found. Thus, all these results provide additional insight into the understanding of new SARS-CoV-2 variants.

Previous studies have modeled the Omicron wave [119–121]. In [121], the authors implemented a stochastic, discrete-time- and individual-based transmission model of SARS-CoV-2 infection and COVID-19 disease. The model considers an age-structured, small-world network. Using sensitivity analysis, they show that a new SARS-CoV-2 variant dominance is primarily driven by its infectivity, which does not necessarily lead to an increased public health burden. In [119] the authors used a model fitted to more than 2 years of epidemiological data from England to project potential dynamics of SARS-CoV-2 infections and deaths to December 2022. They considered several key uncertainties including behavioral change and waning immunity. They concluded that for the particular case of England and under the assumption that no new variants emerge, SARS-CoV-2 transmission is expected to decline. The authors concluded that the projections depend largely on assumptions of waning immunity, social behavior and seasonality. Other interesting work related to Omicron waves is presented in [120]. In this work, a generalized SEIR model assuming gamma-distributed incubation and infectious periods is presented. The model includes susceptibility to Omicron. Their results suggest that even in those regions where the Delta variant is controlled before the beginning of the Omicron wave a significant Omicron wave can be expected. For the particular case of England, the Omicron wave was smaller than the Delta wave. In our paper, we provide additional insight regarding the Omicron wave.

5. Conclusions

Mathematical models are fruitful for the study of various epidemics and infectious diseases. The models allow us to learn about the evolution of epidemics and also to grasp the potential effects of public health control strategies on the epidemics. Forecasting epidemics is frequently a complex task. Mathematical models are able to provide results that sometimes are difficult to anticipate without mathematical tools.

We constructed a mathematical model to investigate the evolution of the Omicron wave. The Omicron strain has caused a new wave with a large amount of infected cases and deaths worldwide. In some countries, the average number of deaths during this Omicron wave has only slightly increased in comparison with previous circulating SARS-CoV-2 waves. We used a mathematical model to study and approximate the Omicron wave situation in the USA, but it can be extended to other countries. This study uses the facts that the Omicron strain exhibits a higher intrinsic transmissibility than the previously circulating SARS-CoV-2 strain but is less deadly. The numerical simulation results show that despite the fact that the Omicron strain is less deadly it can nevertheless cause more deaths and hospitalizations. This result is of paramount importance for public health, since many people might think that since the Omicron strain is less deadly then the number of deaths will be fewer during the Omicron wave. The spread of the Omicron strain depends on several factors, which vary according to the region; therefore, the Omicron wave situation can be different in other countries or regions. In summary, we used a mathematical model in conjunction with numerical simulations to provide an explanation of a large Omicron wave under various conditions related to the variant's transmissibility. These results provide awareness that new SARS-CoV-2 variants can cause more deaths even if their fatality rate is lower. In fact, we can mention that in the USA the peak of number of deaths during the Omicron wave was comparable to that during the Delta wave despite the fact that during the former wave people already had immunity protection due to vaccination programs [1]. In addition, in Brazil and Colombia, the numbers of infected cases were larger than those during the Delta wave. These facts point out the different potential outcomes of new SARS-CoV-2 variants with different transmissibility and fatality rates.

From a mathematical analysis viewpoint, we studied first the local stability using the well-known NGM method. We computed the basic reproduction number $\mathcal{R}_0$ and found that it is the largest of the two parameters $\mathcal{R}_{0_1}$ and $\mathcal{R}_{0_2}$. This theoretical result reveals that the COVID-19 pandemic can become extinct if $\mathcal{R}_0 < 1$. This is achievable if the vaccination rate is increased (this implies that people are willing to get vaccinated) and/or the transmission rate is decreased such that $\mathcal{R}_0 < 1$. We also performed global stability analysis for the disease-free steady state. The numerical simulations provided additional support to the theoretical analysis and showed qualitative effects of the Omicron strain on the US population. This study is more designed for a relatively short time horizon. However, we provide long-term mathematical analysis to obtain a better picture of the dynamics. Interesting and deeper mathematical analysis can be be carried out regarding the endemic states, global stability, periodic solutions and bifurcations.

We provided a variety of scenarios that help to obtain insight into the Omicron wave and its consequences. The numerical simulations showed the Omicron wave outcomes under different conditions related to the vaccines and transmissibility. The results show that the final cumulative number of infected people can be greater with respect to previous waves despite a large number of people being vaccinated. These results are in good agreement with what has occurred during the Omicron wave. For instance, this happened in Brazil and Colombia [1,49].

The results presented here help to support public health policies and, most importantly, to bring awareness to people about the Omicron strain or future highly contagious SARS-CoV-2 strains. At this time, China is suffering one of the largest waves in spite of the fact that in the past they were able to control the spread of SARS-CoV-2. As in any mathematical model, we need to be aware of the limitations in order to understand potential misunderstandings or mistaken conclusions. For instance, the constructed mathematical model assumes homogeneous mixing and constant proportional vaccination rates which obviously is not the case in the real world. One way to better approximate reality would be to describe the vaccination using real data which would give a more complex model since it would then become non-autonomous (see [105]). In addition, more detailed models can include age structure and seasonality. However, despite the usual limitations of mathematical models, this study provides useful means of explaining and obtaining deeper insight about the Omicron wave. As shown by the simulations, the appearance of the Omicron strain or highly contagious SARS-CoV-2 strains changes the dynamics of the pandemic and can increase the number of deaths despite a lower mortality rate.

As in any mathematical model of the real world there are limitations in the results and conclusions. The proposed model is just an approximation of the reality during the Omicron wave. During this wave several SARS-CoV-2 variants were circulating. The model assumes the existence of just two main variants. The model assumes a constant transmission rate for each of the Delta and Omicron variants, but the reality is that these rates change continuously depending on many complex factors. The proposed model does not consider explicitly people hesitant to be vaccinated. The model does not consider the spatial effects of the diffusion of SARS-CoV-2. This has been a common weakness of many models. The model considers only one vaccinated population without any distinction between the number of doses received by individuals. The model does not include human behavioral changes, but considers a variety of transmission rates in the sensitivity analysis.

Finally, we would like to point out that the results presented here are helpful to obtain further insight into the Omicron wave and the effect of new highly transmissible strains and new vaccines. Various graphical illustrations show the impact of vaccines and transmissibility on the Omicron wave. From the results, it can be seen that the COVID-19 pandemic can be eliminated under some circumstances and following the recommendations of the World Health Organization (WHO).

Author Contributions: Conceptualization, G.G.-P.; Formal analysis, G.G.-P. and A.J.A.; Investigation, G.G.-P. and A.J.A.; Methodology, G.G.-P. and A.J.A.; Software, G.G.-P.; Supervision, G.G.-P. and A.J.A.; Validation, G.G.-P. and A.J.A.; Visualization, G.G.-P. and A.J.A.; Writing—original draft, G.G.-P. and A.J.A.; Writing—review and editing, G.G.-P. and A.J.A. All authors have read and agreed to the published version of the manuscript.

Funding: This research was funded by the National Institute of General Medical Sciences (P20GM103 451) via NM-INBRE, and is gratefully acknowledged by the first author. Support from the University of Córdoba, Colombia, is acknowledged by the second author.

Data Availability Statement: Data are contained within the article. Codes are available upon request.

Acknowledgments: The authors are grateful to the anonymous reviewers for their valuable comments and suggestions which improved the quality and the clarity of the paper. The first author has benefited from discussions and English grammar corrections with Roy J. Little.

Conflicts of Interest: The authors declare no conflict of interest.

References

1. Johns Hopkins University and Medicine. Available online: https://coronavirus.jhu.edu (accessed on 1 November 2022).
2. Arvisais-Anhalt, S.; Lehmann, C.U.; Park, J.Y.; Araj, E.; Holcomb, M.; Jamieson, A.R.; McDonald, S.; Medford, R.J.; Perl, T.M.; Toomay, S.M.; et al. What the Coronavirus Disease 2019 (COVID-19) Pandemic Has Reinforced: The Need for Accurate Data. *Clin. Infect. Dis.* **2020**, *72*, 920–923. [CrossRef] [PubMed]
3. Burki, T. COVID-19 in Latin America. *Lancet Infect. Dis.* **2020**, *20*, 547–548. [CrossRef] [PubMed]
4. A Weekly Summary of U.S. COVID-19 Hospitalization Data. Available online: https://gis.cdc.gov/grasp/covidnet/COVID19_5.html (accessed on 1 November 2022).
5. Do Prado, M.F.; de Paula Antunes, B.B.; Bastos, L.D.S.L.; Peres, I.T.; Da Silva, A.D.A.B.; Dantas, L.F.; Baião, F.A.; Maçaira, P.; Hamacher, S.; Bozza, F.A. Analysis of COVID-19 under-reporting in Brazil. *Rev. Bras. Ter. Intensiv.* **2020**, *32*, 224. [CrossRef] [PubMed]
6. Ivorra, B.; Ferrández, M.R.; Vela-Pérez, M.; Ramos, A. Mathematical modeling of the spread of the coronavirus disease 2019 (COVID-19) taking into account the undetected infections. The case of China. *Commun. Nonlinear Sci. Numer. Simul.* **2020**, *88*, 105303. [CrossRef]
7. Sarnaglia, A.J.; Zamprogno, B.; Molinares, F.A.F.; de Godoi, L.G.; Monroy, N.A.J. Correcting notification delay and forecasting of COVID-19 data. *J. Math. Anal. Appl.* **2021**, *514*, 125202. [CrossRef]
8. Faria, N.R.; Mellan, T.A.; Whittaker, C.; Claro, I.M.; Candido, D.d.S.; Mishra, S.; Crispim, M.A.; Sales, F.C.; Hawryluk, I.; McCrone, J.T.; et al. Genomics and epidemiology of the P. 1 SARS-CoV-2 lineage in Manaus, Brazil. *Science* **2021**, *372*, 815–821. [CrossRef]
9. Fraser, B. COVID-19 strains remote regions of Peru. *Lancet* **2020**, *395*, 1684. [CrossRef]
10. Lemieux, J.E.; Li, J.Z. Uncovering Ways that Emerging SARS-CoV-2 Lineages May Increase Transmissibility. *J. Infect. Dis.* **2021**, *223*, 1663–1665. [CrossRef]
11. Plante, J.A.; Liu, Y.; Liu, J.; Xia, H.; Johnson, B.A.; Lokugamage, K.G.; Zhang, X.; Muruato, A.E.; Zou, J.; Fontes-Garfias, C.R.; et al. Spike mutation D614G alters SARS-CoV-2 fitness. *Nature* **2020**, *592*, 116–121. [CrossRef]
12. Leung, K.; Shum, M.H.; Leung, G.M.; Lam, T.T.; Wu, J.T. Early transmissibility assessment of the N501Y mutant strains of SARS-CoV-2 in the United Kingdom, October to November 2020. *Eurosurveillance* **2021**, *26*, 2002106. [CrossRef]
13. Torjesen, I. COVID-19: Delta variant is now UK's most dominant strain and spreading through schools. *BMJ* **2021**, *373*, n1445. [CrossRef] [PubMed]
14. Le Page, M. Threats from new variants. *New Sci.* **2021**, *249*, 8–9. [CrossRef] [PubMed]
15. van Oosterhout, C.; Hall, N.; Ly, H.; Tyler, K.M. COVID-19 evolution during the pandemic–Implications of new SARS-CoV-2 variants on disease control and public health policies. *Virulence* **2021**, *12*, 507. [CrossRef] [PubMed]
16. Benest, J.; Rhodes, S.; Quaife, M.; Evans, T.G.; White, R.G. Optimising Vaccine Dose in Inoculation against SARS-CoV-2, a Multi-Factor Optimisation Modelling Study to Maximise Vaccine Safety and Efficacy. *Vaccines* **2021**, *9*, 78. [CrossRef] [PubMed]
17. Dinleyici, E.C.; Borrow, R.; Safadi, M.A.P.; van Damme, P.; Munoz, F.M. Vaccines and routine immunization strategies during the COVID-19 pandemic. *Hum. Vaccines Immunother.* **2020**, *17*, 400–407. [CrossRef] [PubMed]
18. Haque, A.; Pant, A.B. Efforts at COVID-19 Vaccine Development: Challenges and Successes. *Vaccines* **2020**, *8*, 739. [CrossRef]
19. Koirala, A.; Joo, Y.J.; Khatami, A.; Chiu, C.; Britton, P.N. Vaccines for COVID-19: The current state of play. *Paediatr. Respir. Rev.* **2020**, *35*, 43–49. [CrossRef]
20. Lurie, N.; Saville, M.; Hatchett, R.; Halton, J. Developing Covid-19 vaccines at pandemic speed. *N. Engl. J. Med.* **2020**, *382*, 1969–1973. [CrossRef]
21. Yamey, G.; Schaferhoff, M.; Hatchett, R.; Pate, M.; Zhao, F.; McDade, K.K. Ensuring global access to COVID-19 vaccines. *Lancet* **2020**, *395*, 1405–1406. [CrossRef]

22. Gruell, H.; Vanshylla, K.; Tober-Lau, P.; Hillus, D.; Schommers, P.; Lehmann, C.; Kurth, F.; Sander, L.E.; Klein, F. mRNA booster immunization elicits potent neutralizing serum activity against the SARS-CoV-2 Omicron variant. *Nat. Med.* **2022**, *28*, 477–480. [CrossRef]
23. Rana, P.; Jha, D.; Chauhan, S. Dynamical Analysis on Two Dose Vaccines in the Presence of Media. *J. Comput. Anal. Appl.* **2022**, *30*, 260–280.
24. Zeng, C.; Evans, J.P.; Qu, P.; Faraone, J.; Zheng, Y.M.; Carlin, C.; Bednash, J.S.; Zhou, T.; Lozanski, G.; Mallampalli, R.; et al. Neutralization and stability of SARS-CoV-2 Omicron variant. *BioRxiv* **2021**. [CrossRef]
25. Mandal, S.; Bhatnagar, T.; Arinaminpathy, N.; Agarwal, A.; Chowdhury, A.; Murhekar, M.; Gangakhedkar, R.R.; Sarkar, S. Prudent public health intervention strategies to control the coronavirus disease 2019 transmission in India: A mathematical model-based approach. *Indian J. Med. Res.* **2020**, *151*, 190. [PubMed]
26. Reis, R.F.; de Melo Quintela, B.; de Oliveira Campos, J.; Gomes, J.M.; Rocha, B.M.; Lobosco, M.; dos Santos, R.W. Characterization of the COVID-19 pandemic and the impact of uncertainties, mitigation strategies, and underreporting of cases in South Korea, Italy, and Brazil. *Chaos Solitons Fractals* **2020**, *136*, 109888. [CrossRef] [PubMed]
27. Wang, X.; Pasco, R.F.; Du, Z.; Petty, M.; Fox, S.J.; Galvani, A.P.; Pignone, M.; Johnston, S.C.; Meyers, L.A. Impact of social distancing measures on coronavirus disease healthcare demand, central Texas, USA. *Emerg. Infect. Dis.* **2020**, *26*, 2361. [CrossRef]
28. Pinky, L.; Dobrovolny, H.M. SARS-CoV-2 coinfections: Could influenza and the common cold be beneficial? *J. Med. Virol.* **2020**, *92*, 2623–2630. [CrossRef]
29. Bedson, J.; Skrip, L.A.; Pedi, D.; Abramowitz, S.; Carter, S.; Jalloh, M.F.; Funk, S.; Gobat, N.; Giles-Vernick, T.; Chowell, G.; et al. A review and agenda for integrated disease models including social and behavioural factors. *Nat. Hum. Behav.* **2021**, *5*, 834–846. [CrossRef]
30. Block, P.; Hoffman, M.; Raabe, I.J.; Dowd, J.B.; Rahal, C.; Kashyap, R.; Mills, M.C. Social network-based distancing strategies to flatten the COVID-19 curve in a post-lockdown world. *Nat. Hum. Behav.* **2020**, *4*, 588–596. [CrossRef]
31. Jentsch, P.C.; Anand, M.; Bauch, C.T. Prioritising COVID-19 vaccination in changing social and epidemiological landscapes: A mathematical modelling study. *Lancet Infect. Dis.* **2021**, *28*, 1097–1106. [CrossRef]
32. Kucharski, A.J.; Russell, T.W.; Diamond, C.; Liu, Y.; Edmunds, J.; Funk, S.; Eggo, R.M.; Sun, F.; Jit, M.; Munday, J.D.; et al. Early dynamics of transmission and control of COVID-19: A mathematical modelling study. *Lancet Infect. Dis.* **2020**, *20*, 553–558. [CrossRef]
33. Qazi, A.; Qazi, J.; Naseer, K.; Zeeshan, M.; Hardaker, G.; Maitama, J.Z.; Haruna, K. Analyzing situational awareness through public opinion to predict adoption of social distancing amid pandemic COVID-19. *J. Med. Virol.* **2020**, *92*, 849–855. [CrossRef] [PubMed]
34. Morato, M.M.; Pataro, I.M.; da Costa, M.V.A.; Normey-Rico, J.E. A parametrized nonlinear predictive control strategy for relaxing COVID-19 social distancing measures in Brazil. *ISA Trans.* **2020**, *124*, 197–214. [CrossRef]
35. Ran, L.; Chen, X.; Wang, Y.; Wu, W.; Zhang, L.; Tan, X. Risk factors of healthcare workers with corona virus disease 2019: A retrospective cohort study in a designated hospital of Wuhan in China. *Clin. Infect. Dis.* **2020**, *71*, 2218–2221. [CrossRef] [PubMed]
36. Yang, H.; Duan, G. Analysis on the epidemic factors for the corona virus disease. *Zhonghua Fang Xue Zhi [Chin. J. Prev. Med.]* **2020**, *54*, E021.
37. Zhang, X.; Tan, Y.; Ling, Y.; Lu, G.; Liu, F.; Yi, Z.; Jia, X.; Wu, M.; Shi, B.; Xu, S.; et al. Viral and host factors related to the clinical outcome of COVID-19. *Nature* **2020**, *583*, 437–440. [CrossRef] [PubMed]
38. Dyson, L.; Hill, E.M.; Moore, S.; Curran-Sebastian, J.; Tildesley, M.J.; Lythgoe, K.A.; House, T.; Pellis, L.; Keeling, M.J. Possible future waves of SARS-CoV-2 infection generated by variants of concern with a range of characteristics. *Nat. Commun.* **2021**, *12*, 5730. [CrossRef]
39. Fiorentini, S.; Messali, S.; Zani, A.; Caccuri, F.; Giovanetti, M.; Ciccozzi, M.; Caruso, A. First detection of SARS-CoV-2 spike protein N501 mutation in Italy in August, 2020. *Lancet Infect. Dis.* **2021**, *21*, e147. [CrossRef] [PubMed]
40. Mohammadi, H.; Rezapour, S.; Jajarmi, A. On the fractional SIRD mathematical model and control for the transmission of COVID-19: The first and the second waves of the disease in Iran and Japan. *ISA Trans.* **2021**, *124*, 103–114. [CrossRef] [PubMed]
41. Nakhaeizadeh, M.; Chegeni, M.; Adhami, M.; Sharifi, H.; Gohari, M.A.; Iranpour, A.; Azizian, M.; Mashinchi, M.; Baneshi, M.R.; Karamouzian, M.; et al. Estimating the Number of COVID-19 Cases and Impact of New COVID-19 Variants and Vaccination on the Population in Kerman, Iran: A Mathematical Modeling Study. *Comput. Math. Methods Med.* **2022**, *2022*, 6624471. [CrossRef]
42. Rahimi, F.; Abadi, A.T.B. Implications of the Emergence of a New Variant of SARS-CoV-2, VUI-202012/01. *Arch. Med. Res.* **2021**, *52*, 569–571. [CrossRef]
43. Shereen, M.A.; Khan, S.; Kazmi, A.; Bashir, N.; Siddique, R. COVID-19 infection: Emergence, transmission, and characteristics of human coronaviruses. *J. Adv. Res.* **2020**, *24*, 91–98. [CrossRef] [PubMed]
44. Shil, P.; Atre, N.M.; Tandale, B.V. Epidemiological findings for the first and second waves of COVID-19 pandemic in Maharashtra, India. *Spat.Spatio Temporal Epidemiol.* **2022**, *41*, 100507. [CrossRef] [PubMed]
45. Mallapaty, S. Where did Omicron come from? Three key theories. *Nature* **2022**, *602*, 26–28. [CrossRef]
46. Martin, D.; Lytras, S.; Lucaci, A.; Maier, W.; Gruning, B.; Shank, S. Selection analysis identifies significant mutational changes in Omicron that are likely to influence both antibody neutralization and Spike function (Part 1 of 2). *Virological. Org.* **2021**, *5*, 1–2 .
47. Viana, R.; Moyo, S.; Amoako, D.G.; Tegally, H.; Scheepers, C.; Althaus, C.L.; Anyaneji, U.J.; Bester, P.A.; Boni, M.F.; Chand, M.; et al. Rapid epidemic expansion of the SARS-CoV-2 Omicron variant in southern Africa. *Nature* **2022**, *603*, 679–686. [CrossRef]

48. Buchan, S.A.; Chung, H.; Brown, K.A.; Austin, P.C.; Fell, D.B.; Gubbay, J.; Nasreen, S.; Schwartz, K.L.; Sundaram, M.E.; Tadrous, M.; et al. Effectiveness of COVID-19 vaccines against Omicron or Delta infection. *medRxiv* **2022**. [CrossRef]
49. Iuliano, A.D.; Brunkard, J.M.; Boehmer, T.K.; Peterson, E.; Adjei, S.; Binder, A.M.; Cobb, S.; Graff, P.; Hidalgo, P.; Panaggio, M.J.; et al. Trends in Disease Severity and Health Care Utilization during the Early Omicron Variant Period Compared with Previous SARS-CoV-2 High Transmission Periods—United States, December 2020–January 2022. Available online: https://stacks.cdc.gov/view/cdc/113758 (accessed on 1 July 2022).
50. Ledford, H. How severe are Omicron infections? *Nature* **2021**, *600*, 577–578. [CrossRef]
51. Planas, D.; Saunders, N.; Maes, P.; Benhassine, F.G.; Planchais, C.; Porrot, F.; Staropoli, I.; Lemoine, F.; Pere, H.; Veyer, D.; et al. Considerable escape of SARS-CoV-2 variant Omicron to antibody neutralization (preprint). *Nature* **2022**, *602*, 671–675. [CrossRef] [PubMed]
52. Minka, S.; Minka, F. A tabulated summary of the evidence on humoral and cellular responses to the SARS-CoV-2 Omicron VOC, as well as vaccine efficacy against this variant. *Immunol. Lett.* **2022**, *243*, 38–43. [CrossRef] [PubMed]
53. Statista. Available online: https://www.who.int/news-room/commentaries/detail/estimating-mortality-from-covid-19 (accessed on 1 July 2022).
54. Kim, G.U.; Kim, M.J.; Ra, S.H.; Lee, J.; Bae, S.; Jung, J.; Kim, S.H. Clinical characteristics of asymptomatic and symptomatic patients with mild COVID-19. *Clin. Microbiol. Infect.* **2020**, *26*, 948-e1. [CrossRef]
55. Nishiura, H.; Kobayashi, T.; Miyama, T.; Suzuki, A.; Jung, S.m.; Hayashi, K.; Kinoshita, R.; Yang, Y.; Yuan, B.; Akhmetzhanov, A.R.; et al. Estimation of the asymptomatic ratio of novel coronavirus infections (COVID-19). *Int. J. Infect. Dis.* **2020**, *94*, 154–155. [CrossRef]
56. Luo, G.; Zhang, X.; Zheng, H.; He, D. Infection fatality ratio and case fatality ratio of COVID-19. *Int. J. Infect. Dis.* **2021**, *113*, 43–46. [CrossRef]
57. Bhattacharyya, R.P.; Hanage, W.P. Challenges in Inferring Intrinsic Severity of the SARS-CoV-2 Omicron Variant. *N. Engl. J. Med.* **2022**, *386*, e14. [CrossRef]
58. Addai, E.; Zhang, L.; Asamoah, J.K.K.; Preko, A.K.; Arthur, Y.D. Fractal–fractional age-structure study of omicron SARS-CoV-2 variant transmission dynamics. *Partial. Differ. Equations Appl. Math.* **2022**, *6*, 100455. [CrossRef]
59. Ahmed, H.M.; Elbarkouky, R.A.; Omar, O.A.; Ragusa, M.A. Models for COVID-19 Daily Confirmed Cases in Different Countries. *Mathematics* **2021**, *9*, 659. [CrossRef]
60. Benlloch, J.M.; Cortés, J.C.; Martínez-Rodríguez, D.; Julián, R.S.; Villanueva, R.J. Effect of the early use of antivirals on the COVID-19 pandemic. A computational network modeling approach. *Chaos Solitons Fractals* **2020**, *140*, 110168. [CrossRef]
61. Garrido, J.M.; Martínez-Rodríguez, D.; Rodríguez-Serrano, F.; Sferle, S.M.; Villanueva, R.J. Modeling COVID-19 with Uncertainty in Granada, Spain. Intra-Hospitalary Circuit and Expectations over the Next Months. *Mathematics* **2021**, *9*, 1132. [CrossRef]
62. Gonzalez-Parra, G.; Arenas, A.J. Nonlinear Dynamics of the Introduction of a New SARS-CoV-2 Variant with Different Infectiousness. *Mathematics* **2021**, *9*, 1564. [CrossRef]
63. González-Parra, G.; Díaz-Rodríguez, M.; Arenas, A.J. Mathematical modeling to study the impact of immigration on the dynamics of the COVID-19 pandemic: A case study for Venezuela. *Spat. Spatio-Temporal Epidemiol.* **2022**, *43*, 100532. [CrossRef]
64. Kong, J.D.; Tchuendom, R.F.; Adeleye, S.A.; David, J.F.; Admasu, F.S.; Bakare, E.A.; Siewe, N. SARS-CoV-2 and self-medication in Cameroon: A mathematical model. *J. Biol. Dyn.* **2021**, *15*, 137–150. [CrossRef]
65. Law, K.B.; Peariasamy, K.M.; Gill, B.S.; Singh, S.; Sundram, B.M.; Rajendran, K.; Dass, S.C.; Lee, Y.L.; Goh, P.P.; Ibrahim, H.; et al. Tracking the early depleting transmission dynamics of COVID-19 with a time-varying SIR model. *Sci. Rep.* **2020**, *10*, 21721. [CrossRef] [PubMed]
66. Mbogo, R.W.; Orwa, T.O. SARS-COV-2 outbreak and control in Kenya-Mathematical model analysis. *Infect. Dis. Model.* **2021**, *6*, 370–380. [CrossRef] [PubMed]
67. Mugisha, J.Y.; Ssebuliba, J.; Nakakawa, J.N.; Kikawa, C.R.; Ssematimba, A. Mathematical modeling of COVID-19 transmission dynamics in Uganda: Implications of complacency and early easing of lockdown. *PLoS ONE* **2021**, *16*, e0247456. [CrossRef] [PubMed]
68. Mumbu, A.R.J.; Hugo, A.K. Mathematical modelling on COVID-19 transmission impacts with preventive measures: A case study of Tanzania. *J. Biol. Dyn.* **2020**, *14*, 748–766. [CrossRef]
69. Oliveira, J.F.; Jorge, D.C.; Veiga, R.V.; Rodrigues, M.S.; Torquato, M.F.; da Silva, N.B.; Fiaccone, R.L.; Cardim, L.L.; Pereira, F.A.; de Castro, C.P.; et al. Mathematical modeling of COVID-19 in 14.8 million individuals in Bahia, Brazil. *Nat. Commun.* **2021**, *12*, 333. [CrossRef] [PubMed]
70. Sperrin, M.; Grant, S.W.; Peek, N. Prediction models for diagnosis and prognosis in COVID-19. *BMJ* **2020**, *369*, m1464. [CrossRef]
71. Sweilam, N.; Al-Mekhlafi, S.; Baleanu, D. A hybrid stochastic fractional order Coronavirus (2019-nCov) mathematical model. *Chaos Solitons Fractals* **2021**, *145*, 110762. [CrossRef]
72. Tilahun, G.T.; Alemneh, H.T. Mathematical modeling and optimal control analysis of COVID-19 in Ethiopia. *J. Interdiscip. Math.* **2021**, *24*, 2101–2120. [CrossRef]
73. Wang, B.G.; Wang, Z.C.; Wu, Y.; Xiong, Y.; Zhang, J.; Ma, Z. A mathematical model reveals the influence of NPIs and vaccination on SARS-CoV-2 Omicron Variant. *Nonlinear Dyn.* **2022**, *111*, 3937–3952. [CrossRef]
74. Wintachai, P.; Prathom, K. Stability analysis of SEIR model related to efficiency of vaccines for COVID-19 situation. *Heliyon* **2021**, *7*, e06812. [CrossRef]

75. Brauer, F. Mathematical epidemiology: Past, present, and future. *Infect. Dis. Model.* **2017**, *2*, 113–127. [CrossRef] [PubMed]
76. Hethcote, H.W. Mathematics of infectious diseases. *SIAM Rev.* **2005**, *42*, 599–653. [CrossRef]
77. Mehta, S.R.; Smith, D.M.; Boukadida, C.; Chaillon, A. Comparative Dynamics of Delta and Omicron SARS-CoV-2 Variants across and between California and Mexico. *Viruses* **2022**, *14*, 1494. [CrossRef]
78. Forde, J.E.; Ciupe, S.M. Modeling the influence of vaccine administration on COVID-19 testing strategies. *Viruses* **2021**, *13*, 2546. [CrossRef]
79. Gonzalez-Parra, G.; Martínez-Rodríguez, D.; Villanueva-Micó, R.J. Impact of a new SARS-CoV-2 variant on the population: A mathematical modeling approach. *Math. Comput. Appl.* **2021**, *26*, 25. [CrossRef]
80. González-Parra, G.; Arenas, A.J. Qualitative analysis of a mathematical model with presymptomatic individuals and two SARS-CoV-2 variants. *Comput. Appl. Math.* **2021**, *40*, 199. [CrossRef]
81. Gumel, A.B.; Iboi, E.A.; Ngonghala, C.N.; Elbasha, E.H. A primer on using mathematics to understand COVID-19 dynamics: Modeling, analysis and simulations. *Infect. Dis. Model.* **2021**, *6*, 148–168. [CrossRef]
82. Massard, M.; Eftimie, R.; Perasso, A.; Saussereau, B. A multi-strain epidemic model for COVID-19 with infected and asymptomatic cases: Application to French data. *J. Theor. Biol.* **2022**, *545*, 111117. [CrossRef]
83. Ramos, A.; Vela-Pérez, M.; Ferrández, M.; Kubik, A.; Ivorra, B. Modeling the impact of SARS-CoV-2 variants and vaccines on the spread of COVID-19. *Commun. Nonlinear Sci. Numer. Simul.* **2021**, *102*, 105937. [CrossRef]
84. Shim, E. Projecting the impact of SARS-CoV-2 variants and the vaccination program on the fourth wave of the COVID-19 pandemic in South Korea. *Int. J. Environ. Res. Public Health* **2021**, *18*, 7578. [CrossRef]
85. Yang, H.M.; Junior, L.P.L.; Castro, F.F.M.; Yang, A.C. Evaluating the impacts of relaxation and mutation in the SARS-CoV-2 on the COVID-19 epidemic based on a mathematical model: A case study of São Paulo State (Brazil). *Comput. Appl. Math.* **2021**, *40*, 1–27. [CrossRef]
86. Khan, M.A.; Atangana, A. Mathematical modeling and analysis of COVID-19: A study of new variant Omicron. *Phys. A Stat. Mech. Its Appl.* **2022**, *599*, 127452. [CrossRef]
87. Ko, Y.; Mendoza, V.M.; Mendoza, R.; Seo, Y.; Lee, J.; Lee, J.; Kwon, D.; Jung, E. Multi-faceted analysis of COVID-19 epidemic in Korea considering omicron variant: Mathematical modeling-based study. *J. Korean Med. Sci.* **2022**, *37*, e209. [CrossRef]
88. Muniyappan, A.; Sundarappan, B.; Manoharan, P.; Hamdi, M.; Raahemifar, K.; Bourouis, S.; Varadarajan, V. Stability and Numerical Solutions of Second Wave Mathematical Modeling on COVID-19 and Omicron Outbreak Strategy of Pandemic: Analytical and Error Analysis of Approximate Series Solutions by Using HPM. *Mathematics* **2022**, *10*, 343. [CrossRef]
89. Hussein, T.; Hammad, M.H.; Surakhi, O.; AlKhanafseh, M.; Fung, P.L.; Zaidan, M.A.; Wraith, D.; Ershaidat, N. Short-Term and Long-Term COVID-19 Pandemic Forecasting Revisited with the Emergence of OMICRON Variant in Jordan. *Vaccines* **2022**, *10*, 569. [CrossRef]
90. Nesteruk, I. Epidemic waves caused by SARS-CoV-2 omicron (B.1.1.529) and pessimistic forecasts of the COVID-19 pandemic duration. *MedComm* **2022**, *3*, e122. [CrossRef]
91. Afonyushkin, V.N.; Akberdin, I.R.; Kozlova, Y.N.; Schukin, I.A.; Mironova, T.E.; Bobikova, A.S.; Cherepushkina, V.S.; Donchenko, N.A.; Poletaeva, Y.E.; Kolpakov, F.A. Multicompartmental Mathematical Model of SARS-CoV-2 Distribution in Human Organs and Their Treatment. *Mathematics* **2022**, *10*, 1925. [CrossRef]
92. Haq, I.U.; Yavuz, M.; Ali, N.; Akgül, A. A SARS-CoV-2 fractional-order mathematical model via the modified euler method. *Math. Comput. Appl.* **2022**, *27*, 82. [CrossRef]
93. Pachetti, M.; Marini, B.; Giudici, F.; Benedetti, F.; Angeletti, S.; Ciccozzi, M.; Masciovecchio, C.; Ippodrino, R.; Zella, D. Impact of lockdown on Covid-19 case fatality rate and viral mutations spread in 7 countries in Europe and North America. *J. Transl. Med.* **2020**, *18*, 338. [CrossRef]
94. van Dorp, L.; Acman, M.; Richard, D.; Shaw, L.P.; Ford, C.E.; Ormond, L.; Owen, C.J.; Pang, J.; Tan, C.C.; Boshier, F.A.; et al. Emergence of genomic diversity and recurrent mutations in SARS-CoV-2. *Infect. Genet. Evol.* **2020**, *83*, 104351. [CrossRef]
95. Li, Q.; Wu, J.; Nie, J.; Zhang, L.; Hao, H.; Liu, S.; Zhao, C.; Zhang, Q.; Liu, H.; Nie, L.; et al. The impact of mutations in SARS-CoV-2 spike on viral infectivity and antigenicity. *Cell* **2020**, *182*, 1284–1294. [CrossRef] [PubMed]
96. Grubaugh, N.D.; Hanage, W.P.; Rasmussen, A.L. Making sense of mutation: What D614G means for the COVID-19 pandemic remains unclear. *Cell* **2020**, *182*, 794–795. [CrossRef] [PubMed]
97. Zhu, W.; Yang, J.; Lu, S.; Lan, R.; Jin, D.; Luo, X.l.; Pu, J.; Wu, S.; Xu, J. Beta-and Novel Delta-Coronaviruses Are Identified from Wild Animals in the Qinghai-Tibetan Plateau, China. *Virol. Sin.* **2020**, *36*, 402–411. [CrossRef]
98. Gupta, R.K. Will SARS-CoV-2 variants of concern affect the promise of vaccines? *Nat. Rev. Immunol.* **2021**, *21*, 340–341. [CrossRef]
99. León, U.A.P.d.; Avila-Vales, E.; Huang, K. Modeling the transmission of the SARS-CoV-2 delta variant in a partially vaccinated population. *Viruses* **2022**, *14*, 158. [CrossRef]
100. Khyar, O.; Allali, K. Global dynamics of a multi-strain SEIR epidemic model with general incidence rates: Application to COVID-19 pandemic. *Nonlinear Dyn.* **2020**, *102*, 489–509. [CrossRef]
101. Mancuso, M.; Eikenberry, S.E.; Gumel, A.B. Will vaccine-derived protective immunity curtail COVID-19 variants in the US? *Infect. Dis. Model.* **2021**, *6*, 1110–1134. [CrossRef]
102. Muñoz-Fernández, G.A.; Seoane, J.M.; Seoane-Sepúlveda, J.B. A SIR-type model describing the successive waves of COVID-19. *Chaos Solitons Fractals* **2021**, *144*, 110682. [CrossRef]

103. Santra, P.; Ghosh, D.; Mahapatra, G.; Bonyah, E. Mathematical Analysis of Two Waves of COVID-19 Disease with Impact of Vaccination as Optimal Control. *Comput. Math. Methods Med.* **2022**, *2022*, 2684055. [CrossRef]
104. Amaku, M.; Covas, D.T.; Coutinho, F.A.B.; Azevedo, R.S.; Massad, E. Modelling the impact of delaying vaccination against SARS-CoV-2 assuming unlimited vaccine supply. *Theor. Biol. Med. Model.* **2021**, *18*, 14. [CrossRef]
105. Islam, M.R.; Oraby, T.; McCombs, A.; Chowdhury, M.M.; Al-Mamun, M.; Tyshenko, M.G.; Kadelka, C. Evaluation of the United States COVID-19 vaccine allocation strategy. *PLoS ONE* **2021**, *16*, e0259700. [CrossRef] [PubMed]
106. Paltiel, A.D.; Schwartz, J.L.; Zheng, A.; Walensky, R.P. Clinical Outcomes Of A COVID-19 Vaccine: Implementation Over Efficacy: Study examines how definitions and thresholds of vaccine efficacy, coupled with different levels of implementation effectiveness and background epidemic severity, translate into outcomes. *Health Aff.* **2021**, *40*, 42–52.
107. Gonzalez-Parra, G. Analysis of Delayed Vaccination Regimens: A Mathematical Modeling Approach. *Epidemiologia* **2021**, *2*, 271–293. [CrossRef] [PubMed]
108. Eikenberry, S.E.; Mancuso, M.; Iboi, E.; Phan, T.; Eikenberry, K.; Kuang, Y.; Kostelich, E.; Gumel, A.B. To mask or not to mask: Modeling the potential for face mask use by the general public to curtail the COVID-19 pandemic. *Infect. Dis. Model.* **2020**, *5*, 293–308. [CrossRef]
109. Al-Qahtani, M.; AlAli, S.; AbdulRahman, A.; Alsayyad, A.S.; Otoom, S.; Atkin, S.L. The prevalence of asymptomatic and symptomatic COVID-19 in a cohort of quarantined subjects. *Int. J. Infect. Dis.* **2020**, *102*, 285–288. [CrossRef]
110. Bai, Y.; Yao, L.; Wei, T.; Tian, F.; Jin, D.Y.; Chen, L.; Wang, M. Presumed asymptomatic carrier transmission of COVID-19. *JAMA* **2020**, *323*, 1406–1407. [CrossRef]
111. Buitrago-Garcia, D.; Egli-Gany, D.; Counotte, M.J.; Hossmann, S.; Imeri, H.; Ipekci, A.M.; Salanti, G.; Low, N. Occurrence and transmission potential of asymptomatic and presymptomatic SARS-CoV-2 infections: A living systematic review and meta-analysis. *PLoS Med.* **2020**, *17*, e1003346. [CrossRef]
112. Clarke, C.; Prendecki, M.; Dhutia, A.; Ali, M.A.; Sajjad, H.; Shivakumar, O.; Lightstone, L.; Kelleher, P.; Pickering, M.C.; Thomas, D.; et al. High prevalence of asymptomatic COVID-19 infection in hemodialysis patients detected using serologic screening. *J. Am. Soc. Nephrol.* **2020**, *31*, 1969–1975. [CrossRef]
113. Gandhi, M.; Yokoe, D.S.; Havlir, D.V. Asymptomatic Transmission, the Achilles' Heel of Current Strategies to Control COVID-19. *N. Engl. J. Med.* **2020**, *382*, 2158–2160.
114. Johansson, M.A.; Quandelacy, T.M.; Kada, S.; Prasad, P.V.; Steele, M.; Brooks, J.T.; Slayton, R.B.; Biggerstaff, M.; Butler, J.C. SARS-CoV-2 Transmission From People Without COVID-19 Symptoms. *JAMA Netw. Open* **2021**, *4*, e2035057. [CrossRef]
115. Teixeira, S.C. Mild and asymptomatic cases of COVID-19 are potential threat for faecal–oral transmission. *Braz. J. Infect. Dis.* **2020**, *24*, 368. [CrossRef] [PubMed]
116. Van den Driessche, P.; Watmough, J. *Further Notes on the Basic Reproduction Number*; Springer: Berlin/Heidelberg, Germany, 2008; pp. 159–178.
117. Van den Driessche, P. Reproduction numbers of infectious disease models. *Infect. Dis. Model.* **2017**, *2*, 288–303. [CrossRef] [PubMed]
118. Abdel-Hamid, T.; Ankel, F.; Battle-Fisher, M.; Gibson, B.; Gonzalez-Parra, G.; Jalali, M.; Kaipainen, K.; Kalupahana, N.; Karanfil, O.; Marathe, A.; et al. Public and health professionals' misconceptions about the dynamics of body weight gain/loss. *Syst. Dyn. Rev.* **2014**, *30*, 58–74. [CrossRef]
119. Barnard, R.C.; Davies, N.G.; Jit, M.; Edmunds, W.J. Modelling the medium-term dynamics of SARS-CoV-2 transmission in England in the Omicron era. *Nat. Commun.* **2022**, *13*, 4879. [CrossRef] [PubMed]
120. Bartha, F.A.; Boldog, P.; Tekeli, T.; Vizi, Z.; Dénes, A.; Röst, G. Potential severity, mitigation, and control of Omicron waves depending on pre-existing immunity and immune evasion. In Proceedings of the Trends in Biomathematics: Stability and Oscillations in Environmental, Social, and Biological Models: Selected Works from the BIOMAT Consortium Lectures, Rio de Janeiro, Brazil, 1–5 November 2021; pp. 407–419.
121. Le Rutte, E.A.; Shattock, A.J.; Chitnis, N.; Kelly, S.L.; Penny, M.A. Modelling the impact of Omicron and emerging variants on SARS-CoV-2 transmission and public health burden. *Commun. Med.* **2022**, *2*, 93. [CrossRef]
122. Amador Pacheco, J.; Armesto, D.; Gómez-Corral, A. Extreme values in SIR epidemic models with two strains and cross-immunity. *Math. Biosci. Eng.* **2019**, *16*, 1992–2022. [CrossRef]
123. Meskaf, A.; Khyar, O.; Danane, J.; Allali, K. Global stability analysis of a two-strain epidemic model with non-monotone incidence rates. *Chaos Solitons Fractals* **2020**, *133*, 109647. [CrossRef]
124. Shayak, B.; Sharma, M.M.; Gaur, M.; Mishra, A.K. Impact of reproduction number on multiwave spreading dynamics of COVID-19 with temporary immunity: A mathematical model. *Int. J. Infect. Dis.* **2021**, *104*, 649–654. [CrossRef]
125. Altmann, D.M.; Boyton, R.J.; Beale, R. Immunity to SARS-CoV-2 variants of concern. *Science* **2021**, *371*, 1103–1104. [CrossRef]
126. Clemente-Suárez, V.J.; Hormeño-Holgado, A.; Jiménez, M.; Benitez-Agudelo, J.C.; Navarro-Jiménez, E.; Perez-Palencia, N.; Maestre-Serrano, R.; Laborde-Cárdenas, C.C.; Tornero-Aguilera, J.F. Dynamics of population immunity due to the herd effect in the COVID-19 pandemic. *Vaccines* **2020**, *8*, 236. [CrossRef]
127. Ehrhardt, M.; Gasper, J.; Kilianová, S. SIR-based mathematical modeling of infectious diseases with vaccination and waning immunity. *J. Comput. Sci.* **2019**, *37*, 101027. [CrossRef]

128. Garcia-Beltran, W.F.; Denis, K.J.S.; Hoelzemer, A.; Lam, E.C.; Nitido, A.D.; Sheehan, M.L.; Berrios, C.; Ofoman, O.; Chang, C.C.; Hauser, B.M.; et al. mRNA-based COVID-19 vaccine boosters induce neutralizing immunity against SARS-CoV-2 Omicron variant. *Cell* **2022**, *185*, 457–466. [CrossRef]
129. Tang, J.W.; Bahnfleth, W.P.; Bluyssen, P.M.; Buonanno, G.; Jimenez, J.L.; Kurnitski, J.; Li, Y.; Miller, S.; Sekhar, C.; Morawska, L.; et al. Dismantling myths on the airborne transmission of severe acute respiratory syndrome coronavirus-2 (SARS-CoV-2). *J. Hosp. Infect.* **2021**, *110*, 89–96. [CrossRef]
130. Wordometer. Available online: https://www.worldometers.info/coronavirus/country/us/ (accessed on 1 July 2022).
131. Li, Q.; Guan, X.; Wu, P.; Wang, X.; Zhou, L.; Tong, Y.; Ren, R.; Leung, K.S.; Lau, E.H.; Wong, J.Y.; et al. Early transmission dynamics in Wuhan, China, of novel coronavirus–infected pneumonia. *N. Engl. J. Med.* **2020**, *382*, 1199–1207. [CrossRef]
132. Quah, P.; Li, A.; Phua, J. Mortality rates of patients with COVID-19 in the intensive care unit: A systematic review of the emerging literature. *Crit. Care* **2020**, *24*, 285. [CrossRef]
133. Centers for Disease Control and Prevention. Available online: https://www.cdc.gov/coronavirus/2019-nCoV/index.html (accessed on 1 November 2022).
134. Oran, D.P.; Topol, E.J. Prevalence of Asymptomatic SARS-CoV-2 Infection: A Narrative Review. *Ann. Intern. Med.* **2020**, *173*, 362–367. [CrossRef]
135. Lambert, J.D. *Computational Methods in Ordinary Differential Equations*; Wiley: New York, NY, USA, 1973.
136. Fred Brauer, J.A.N. *The Qualitative Theory of Ordinary Differential Equations: An Introduction*; Dover Publications: Mineola, NY, USA , 1989.
137. van den Driessche, P.; Watmough, J. Reproduction numbers and sub-threshold endemic equilibria for compartmental models of disease transmission. *Math. Biosci.* **2002**, *180*, 29–48. [CrossRef]
138. Castillo-Chavez, C.; Feng, Z.; Huang, W. On the computation of ro and its role on. *Math. Approaches Emerg. Reemerging Infect. Dis. Introd.* **2002**, *1*, 229.
139. Bedston, S.; Akbari, A.; Jarvis, C.I.; Lowthian, E.; Torabi, F.; North, L.; Lyons, J.; Perry, M.; Griffiths, L.J.; Owen, R.K.; et al. COVID-19 vaccine uptake, effectiveness, and waning in 82,959 health care workers: A national prospective cohort study in Wales. *Vaccine* **2022**, *40*, 1180–1189. [CrossRef]
140. Dolgin, E. COVID vaccine immunity is waning-how much does that matter. *Nature* **2021**, *597*, 606–607. [CrossRef]
141. Dzinamarira, T.; Tungwarara, N.; Chitungo, I.; Chimene, M.; Iradukunda, P.G.; Mashora, M.; Murewanhema, G.; Rwibasira, G.N.; Musuka, G. Unpacking the Implications of SARS-CoV-2 Breakthrough Infections on COVID-19 Vaccination Programs. *Vaccines* **2022**, *10*, 252. [CrossRef] [PubMed]
142. Leung, K.; Wu, J.T. Managing waning vaccine protection against SARS-CoV-2 variants. *Lancet* **2022**, *399*, 2–3. [CrossRef] [PubMed]
143. Rabiu, M.; Iyaniwura, S.A. Assessing the potential impact of immunity waning on the dynamics of COVID-19 in South Africa: An endemic model of COVID-19. *Nonlinear Dyn.* **2022**, *109*, 203–223. [CrossRef] [PubMed]
144. Yan, P.; Chowell, G. Beyond the Initial Phase: Compartment Models for Disease Transmission. In *Quantitative Methods for Investigating Infectious Disease Outbreaks*; Springer: Berlin/Heidelberg, Germany, 2019; pp. 135–182.
145. Collie, S.; Champion, J.; Moultrie, H.; Bekker, L.G.; Gray, G. Effectiveness of BNT162b2 vaccine against omicron variant in South Africa. *N. Engl. J. Med.* **2021**, *386*, 494–496. [CrossRef] [PubMed]
146. Nemet, I.; Kliker, L.; Lustig, Y.; Zuckerman, N.; Erster, O.; Cohen, C.; Kreiss, Y.; Alroy-Preis, S.; Regev-Yochay, G.; Mendelson, E.; et al. Third BNT162b2 vaccination neutralization of SARS-CoV-2 Omicron infection. *N. Engl. J. Med.* **2022**, *386*, 492–494. [CrossRef]
147. Basile, K.; Rockett, R.J.; McPhie, K.; Fennell, M.; Johnson-Mackinnon, J.; Agius, J.; Fong, W.; Rahman, H.; Ko, D.; Donavan, L.; et al. Improved neutralization of the SARS-CoV-2 Omicron variant after Pfizer-BioNTech BNT162b2 COVID-19 vaccine boosting. *bioRxiv* **2021**. [CrossRef]
148. Pilishvili, T.; Gierke, R.; Fleming-Dutra, K.E.; Farrar, J.L.; Mohr, N.M.; Talan, D.A.; Krishnadasan, A.; Harland, K.K.; Smithline, H.A.; Hou, P.C.; et al. Effectiveness of mRNA Covid-19 vaccine among US health care personnel. *N. Engl. J. Med.* **2021**, *385*, e90. [CrossRef]
149. Hall, V.; Foulkes, S.; Insalata, F.; Kirwan, P.; Saei, A.; Atti, A.; Wellington, E.; Khawam, J.; Munro, K.; Cole, M.; et al. Protection against SARS-CoV-2 after Covid-19 Vaccination and Previous Infection. *N. Engl. J. Med.* **2022**, *386*, 1207–1220. [CrossRef]
150. Kojima, N.; Klausner, J.D. Protective immunity after recovery from SARS-CoV-2 infection. *Lancet Infect. Dis.* **2022**, *22*, 12–14. [CrossRef]
151. Adamo, S.; Michler, J.; Zurbuchen, Y.; Cervia, C.; Taeschler, P.; Raeber, M.E.; Baghai Sain, S.; Nilsson, J.; Moor, A.E.; Boyman, O. Signature of long-lived memory CD8+ T cells in acute SARS-CoV-2 infection. *Nature* **2022**, *602*, 148–155. [CrossRef]
152. De Sanctis, J.B.; Garmendia, J.V.; Hajdúch, M. Overview of Memory NK Cells in Viral Infections: Possible Role in SARS-CoV-2 Infection. *Immuno* **2022**, *2*, 52–67. [CrossRef]
153. Gurevich, M.; Zilkha-Falb, R.; Sonis, P.; Magalashvili, D.; Menascu, S.; Flechter, S.; Dolev, M.; Mandel, M.; Achiron, A. SARS-CoV-2 memory B and T cell profiles in mild COVID-19 convalescent patients. *Int. J. Infect. Dis.* **2022**, *115*, 208–214. [CrossRef] [PubMed]

154. Walensky, R.P.; Walke, H.T.; Fauci, A.S. SARS-CoV-2 variants of concern in the United States challenges and opportunities. *JAMA* **2021**, *325*, 1037–1038. [CrossRef] [PubMed]
155. Martínez-Rodríguez, D.; Gonzalez-Parra, G.; Villanueva, R.J. Analysis of key factors of a SARS-CoV-2 vaccination program: A mathematical modeling approach. *Epidemiologia* **2021**, *2*, 140–161. [CrossRef]

computation

MDPI

Article

Forecasting the Cumulative COVID-19 Cases in Indonesia Using Flower Pollination Algorithm

Afiahayati [1,*], Yap Bee Wah [2,3,4], Sri Hartati [1], Yunita Sari [1], I Nyoman Prayana Trisna [5], Diyah Utami Kusumaning Putri [1], Aina Musdholifah [1] and Retantyo Wardoyo [1]

1 Department of Computer Science and Electronics, Faculty of Mathematics and Natural Sciences, Universitas Gadjah Mada, Yogyakarta 55281, Indonesia
2 Institute for Big Data Analytics and Artificial Intelligence (IBDAAI), Universiti Teknologi MARA (UiTM), Shah Alam 40450, Malaysia
3 Centre of Statistical and Decision Sciences Studies, Faculty of Computer and Mathematical Sciences, Universiti Teknologi MARA (UiTM), Shah Alam 40450, Malaysia
4 UNITAR Graduate School, UNITAR International University, Jalan SS6/3, SS6, Petaling Jaya 47301, Malaysia
5 Information Technology Study Program, Faculty of Engineering, Universitas Udayana, Badung 80362, Indonesia
* Correspondence: afia@ugm.ac.id

Abstract: Coronavirus disease 2019 (COVID-19) was declared as a global pandemic by the World Health Organization (WHO) on 12 March 2020. Indonesia is reported to have the highest number of cases in Southeast Asia. Accurate prediction of the number of COVID-19 cases in the upcoming few days is required as one of the considerations in making decisions to provide appropriate recommendations in the process of mitigating global pandemic infectious diseases. In this research, a metaheuristics optimization algorithm, the flower pollination algorithm, is used to forecast the cumulative confirmed COVID-19 cases in Indonesia. The flower pollination algorithm is a robust and adaptive method to perform optimization for curve fitting of COVID-19 cases. The performance of the flower pollination algorithm was evaluated and compared with a machine learning method which is popular for forecasting, the recurrent neural network. A comprehensive experiment was carried out to determine the optimal hyperparameters for the flower pollination algorithm and recurrent neural network. There were 24 and 72 combinations of hyperparameters for the flower pollination algorithm and recurrent neural network, respectively. The best hyperparameters were used to develop the COVID-19 forecasting model. Experimental results showed that the flower pollination algorithm performed better than the recurrent neural network in long-term (two weeks) and short-term (one week) forecasting of COVID-19 cases. The mean absolute percentage error (MAPE) for the flower pollination algorithm model (0.38%) was much lower than that of the recurrent neural network model (5.31%) in the last iteration for long-term forecasting. Meanwhile, the MAPE for the flower pollination algorithm model (0.74%) is also lower than the recurrent neural network model (4.8%) in the last iteration for short-term forecasting of the cumulative COVID-19 cases in Indonesia. This research provides state-of-the-art results to help the process of mitigating the global pandemic of COVID-19 in Indonesia.

Keywords: COVID-19; forecasting; flower pollination algorithm; recurrent neural network

Citation: Afiahayati; Wah, Y.B.; Hartati, S.; Sari, Y.; Trisna, I.N.P.; Putri, D.U.K.; Musdholifah, A.; Wardoyo, R. Forecasting the Cumulative COVID-19 Cases in Indonesia Using Flower Pollination Algorithm. *Computation* **2022**, *10*, 214. https://doi.org/10.3390/computation10120214

Academic Editors: Simone Brogi and Vincenzo Calderone

Received: 27 October 2022
Accepted: 30 November 2022
Published: 7 December 2022

1. Introduction

COVID-19 was declared as a global pandemic by the World Health Organization (WHO) on 12 March 2020. It is an ongoing pandemic and as of 19 January 2021, more than 95.5 million cases have been confirmed, with more than 2.03 million deaths attributed to COVID-19 across 190 countries around the world [1,2]. The coronavirus was first identified in December 2019 in Wuhan, China. COVID-19 has spread globally, with America, Europe, and countries in Asia reporting high numbers of cases. The government of China quickly implemented policies such as lockdown, physical distancing, mandatory masks, and quarantine to mitigate the spread of the virus. China has successfully controlled the pandemic rapidly and effectively, but many countries around the world are still struggling to control the spread of the virus. The virus spread to Southeast Asia on 13 January 2020, when a 61-year-old woman from Wuhan tested positive in Thailand [3]. Indonesia, a country with a population of 273 million, is the worst-hit nation in the region, with a rapid increase in cases since the first case reported in March 2020.

In the beginning, the COVID-19 pandemic has not only disrupted the normal way of life of the community, business, and government operations, but also the economy. COVID-19 has affected all levels of society and all areas of life. Hospitals and doctors are struggling to provide care for the COVID-19 patients, and businesses are affected due to lockdowns. The COVID-19 pandemic has also forced many activities to be carried out online, and new standard operating procedures (SOPs) were enforced by the government to ensure safety protocols for the public and for business operations. The COVID-19 pandemic is also causing an economic recession. The governments of many countries are allowing some economic movement while still enforcing strict health safety protocols for the public and business owners to follow. In any health disease crises, prediction of the number of cases is of utmost importance because it helps the relevant authorities to take strategic actions to mitigate the effect of the rise in numbers or control the spread of the disease.

Accurate forecasts are needed to provide useful information in the process of mitigating the global pandemic infectious disease. Thus, forecasting the number of COVID-19 cases in the upcoming few days will be most useful for considerations in making decisions, including the provision of personal equipment (PPE), preparation of economic policies, preparation of health facilities, lockdown policies, and opening of schools or businesses.

Currently, there are two approaches to forecasting COVID-19 cases. The first approach is forecasting COVID-19 using mathematical and statistical models. The mathematical and statistical model approach requires knowledge of epidemiology and statistical assumptions regarding the distribution of the data. Mathematical and statistical model approaches include the autoregressive integrated moving average (ARIMA) [4–6], seasonal ARIMA (SARIMA) [4], the susceptible-infected-recovered (SIR) model [5,7], the logistic growth model [7], and the Richards model, which is an extension of a simple logistic growth model [8].

The second approach is forecasting COVID-19 using artificial intelligence. One of the artificial intelligence approaches is machine learning. Machine learning is a computational method with sophisticated algorithms which can learn the pattern of data to solve forecasting problems. Some machine learning forecasting algorithms for forecasting COVID-19 include multi-layer perceptron, random forest, support vector regression, the Elman neural network [9–11], and the recurrent neural network (RNN) [9,10,12,13]. Sahid et al. [9] concluded that RNN outperformed support vector regression and ARIMA. Hao et al.'s [10] experimental results showed that RNN is more suitable for the prediction of the cumulative confirmed cases compared to death and cured cases.

RNN utilized network architecture which is suitable for processing sequential data. Qiu, Wang, and Zhou [14] applied RNN with long short-term memory (LSTM) architecture and attention mechanism for stock price forecasting. Uras et al. [15] applied RNN with LSTM architecture for Bitcoin closing price forecasting. Yao and Guan [16] applied RNN with an improved LSTM for natural language processing. RNN is also widely applied for speech recognition [17] and to solve fuzzy non-linear programming [18]. Hewamalage, Bergmeir, and Bandara's [19] experimental studies concluded that RNN is a good algorithm for obtaining reliable forecasts.

Another artificial intelligence approach for forecasting is a metaheuristics optimization algorithm. The flower pollination algorithm (FPA) is a robust and adaptive metaheuristics optimization algorithm which is inspired by how flower pollination occurs. The FPA solves the balance of global and local search and uses Lévy flight distribution for better global search performance. The FPA is a method that aims for optimization. The

FPA outperformed other nature-inspired methods such as the genetic algorithm and particle swarm optimization [20]. The FPA has been deployed to estimate transportation energy demand [21], to forecast Organization of the Petroleum Exporting Countries (OPEC) petroleum consumption [22], to forecast electricity energy consumption [23], and to solve combined economic and emission dispatch problems [24]. FPA was created by Yang [20] in 2014 and has been reported to perform better than other metaheuristic algorithms.

In this paper, the FPA was used to determine the optimal coefficients of the variables in the forecasting function of cumulative confirmed COVID-19 cases in Indonesia. In other words, the FPA was used to perform optimization for curve fitting of cumulative confirmed COVID-19 cases. We compare the performance of the FPA with a machine learning method which is popular for forecasting, the recurrent neural network (RNN). Experimental results showed that the FPA performed better than the RNN in long-term (two weeks) and short-term (one week) forecasting. This research provides state-of-the-art results to help the process of mitigating the global pandemic of COVID-19 in Indonesia. This paper is structured as follows: after this introduction, the second section covers related works on forecasting COVID-19 cases. This is followed by the explanation of the data and the methodology in the third and fourth section. The results and discussion are presented in the fifth section, and the conclusion is provided in the last section.

2. Related Works

In this section, some related works related to forecasting of COVID-19 cases are presented. As explained in the first section, there are two approaches on forecasting COVID-19 cases. The first one, the mathematical and statistical model approach, is presented here [4–8,25,26].

Mishra et al. [4] applied the ARIMA, SARIMA, and Prophet model to forecast the cumulative deaths, cumulative cases, and new cases of COVID-19 in India. The model was used to forecast the COVID-19 cases for next 15–20 days starting on 1 September 2020.

Abuhasel, Khadr, and Alquraish [5] applied SIR and ARIMA models to analyze and forecast the daily COVID-19 cases in the Kingdom of Saudi Arabia. The deterministic SIR model was applied to analyze the COVID-19 spread in Saudi Arabia, while the ARIMA model was used to forecast the daily COVID-19 cases. The two models were applied to the daily data from March 3 until 30 June 2020.

Ali et al. [6] applied the ARIMA model to forecast the cumulative confirmed cases, recovered cases, and deaths in Pakistan from COVID-19. The training data to develop the ARIMA model were from 27 February until 24 June 2020, and then the ARIMA model was used to forecast the next 10 days (25 June 2020 to 4 July 2020).

Malavika et al. [7] developed mathematical model approaches to forecast COVID-19 in India. The SIR models were applied to forecast the maximum number of active cases and peak time, the logistics growth curve model was applied for short-term prediction and the time interrupted regression model was used to analyze the effect of lockdown and other policies. The models were used to forecast the COVID-19 epidemic in India by the end of May 2020.

Zuhairoh and Rosadi [8] applied the Richards model, which is an extension of a simple logistic growth model, to forecast daily cases of COVID-19 in South Sulawesi Province, Indonesia. In addition to forecasting, the objective of this research was to predict when this pandemic would reach the peak of its spread, and when it would end. The data used in this paper were compiled as of 24 June 2020.

Anastassopoulou et al. [25] developed a mathematical model approach to estimate the fatality ratio (death rate) and recovery case ratio based on time series of positive case data, death rate, and recovered cases from COVID-19 in Hubei, China. The model was based on data distribution from Middle East respiratory syndrome (MERS) and severe acute respiratory syndrome (SARS) cases that occurred previously. The model was applied to forecast the COVID-19 cases by the end of February 2020.

Petropoulos and Makridakis [26] applied a simple time series model from the exponential smoothing family to forecast the global number of positive cases, the number of deaths, and the number of patients who have been cured of COVID-19 infection. The model was used to forecast the COVID-19 cases from February until March 2020.

The second approach in forecasting the COVID-19 cases is using artificial intelligence, especially machine learning methods [9–13]. Shahid, Zameer, and Muneeb [9] applied four different machine learning methods and the well-known ARIMA method to forecast the confirmed cases, recovered cases, and death cases in 10 major countries affected by COVID-19. The machine learning methods were RNN with bidirectional LSTM (Bi-LSTM) architecture, RNN with LSTM architecture, RNN with gated recurrent unit (GRU) architecture, and support vector regression (SVR). The data used in this research were from 22 January until 10 May 2020 for training, and from 11 May until 27 June 2020 for testing. The RNN model outperformed the SVR and ARIMA for forecasting COVID-19. The models' ranking, from the best to the worst performance, was: RNN Bi-LSTM, RNN LSTM, RNN GRU, SVR, and ARIMA.

Hao et al. [10] applied three machine learning methods to forecast the cumulative confirmed cases, cumulative deaths, and cumulative cured cases in Wuhan, Hubei Province, China. The machine learning methods were the Elman neural network, RNN-LSTM, and support vector machine (SVM). The data used in this research were from 23 January 2020 to 6 April 2020. Based on the experimental results, the RNN-LSTM model is more suitable for the prediction of the cumulative confirmed cases compared to death and cured cases.

Balli [11] applied four different machine learning time series methods to forecast the weekly cumulative confirmed COVID-19 cases for the United States of America (USA), Germany, and the world. The machine learning methods were linear regression, multi-layer perceptron, random forest, and support vector machine. The data used in this research were from between 20 January and 18 September 2020. The data consist of weekly cumulative confirmed cases for 35 weeks. SVM outperformed other methods for forecasting the COVID-19 cases.

Hawas [12] developed an RNN to forecast the data of COVID-19′s daily infections in Brazil. The training data to develop the RNN model were from 7 April until 6 May 2020, and then the RNN model was used to forecast the next 54 days (7 May 2020 until 29 June 2020). In this research, there were two alternative timesteps used for the RNN, 30 and 40.

Shastri et al. [13] developed an RNN to forecast the confirmed cases and death cases of COVID-19 in India and USA. In this research, variants of LSTM architecture of RNN are developed, including stacked LSTM, bi-directional LSTM, and convolutional LSTM. The data of confirmed cases used in this research, for both India and USA, were from 7 February until 7 July 2020, while the data of death cases for India were from 12 March until July 2020, and for USA were from 26 February until 7 July 2020. The training data constituted 80% of the total, while the validation data were 20%.

In the COVID-19 research area, machine learning was used for another task beside forecasting. Machine learning has been applied to COVID-19 patient data. Zoabi et al. [27] used gradient-boosting machine model built with decision-tree base-learner for prediction of COVID-19 positive case based on symptoms while Kim et al. [28] evaluated several machine learning models to predict the need for intensive care. Recently, Ahmad et al. [29] proposed Shallow Single-Layer Perceptron Neural Network (SSLPNN) and Gaussian Process Regression (GPR) model for classification and prediction of confirmed COVID-19 cases. Elzeki et al. [30] proposed a computer-aided model using deep learning to classify positive COVID-19 based on Chest X-ray image data.

The results of closely related works are summarized in Table 1. In this research, a metaheuristics optimization algorithm, the FPA, is used to forecast the cumulative confirmed COVID-19 cases in Indonesia. The FPA is a robust and adaptive method to perform optimization for curve fitting of COVID-19 cases. The performance of the FPA was evaluated and compared with a machine learning method which is popular for forecasting, the RNN.

Table 1. Summarization of closely related works.

Authors	Methods	Forecasting of COVID-19 Cases	Results
Mishra et al. [4]	ARIMA, SARIMA, and Prophet model.	The cumulative deaths, cumulative cases, and daily confirmed cases in India.	The best root-mean-square error (RMSE) of forecasting for the cumulative cases from 23 August 2020 to 1 September 2020: 82090.21.
Abuhasel, Khadr, and Alquraish [5]	SIR and ARIMA models.	The daily confirmed cases in the Kingdom of Saudi Arabia.	The best RMSE of forecasting for the next 10 days: 341.
Ali et al. [6]	ARIMA model.	The cumulative confirmed cases, recovered cases, and deaths in Pakistan.	The best RMSE of forecasting for the cumulative confirmed cases from 25 June 2020 till 4 July 2020: 413.9.
Petropoulos and Makridakis [26]	A simple time series model from the exponential smoothing family.	The global number of cumulative positive cases, the number of deaths, and the number of recovered cases.	The absolute percentage error of forecasting for the cumulative confirmed cases: a. 01/02/2020 till 10/02/2020: 388%; b. 11/02/2020 till 20/02/2020: 7.7%; c. 21/02/2020 till 01/03/2020: 6.2%; d. 02/03/2020 till 11/03/2020: 12.1%.
Zuhairoh and Rosadi [8]	The Richards model.	The daily confirmed cases in South Sulawesi Province, Indonesia.	They provided the prediction that the peak of the COVID-19 pandemic in South Sulawesi Province, Indonesia, would be the middle of June 2020 until the end of July 2020, with 10,000–12,000 cases per day.
Shahid, Zameer, and Muneeb [9]	RNN with bidirectional LSTM (Bi-LSTM) architecture, RNN with LSTM architecture, RNN with GRU architecture, support vector regression, and ARIMA method.	The confirmed cases, recovered cases, and death cases in 10 major countries.	The best RMSE of forecasting for the daily confirmed cases from 11 May 2022 to 27 June 2022 (48 days): a. China: 180.63; b. Italy: 3612.81; c. USA: 273,851.39.
Hao et al. [10]	Elman neural network, RNN-LSTM, and SVM.	The cumulative confirmed cases, cumulative deaths, and cumulative cured cases in Wuhan, Hubei Province, China.	The best MSE of forecasting for the cumulative confirmed cases from 24 March 2022 to 6 April 2022: 0.0320.
Balli [11]	Linear regression, multi-layer perceptron, random forest, and support vector machine.	The weekly cumulative confirmed cases in USA, Germany, and the world.	The best RMSE of forecasting for the weekly cumulative cases from 24 May 2022 to 18 September 2022 (17 weeks): a. Germany: 329,196; b. USA: 9.531,6776; c. Global: 25,825.8366.
Hawas [12]	RNN.	The daily confirmed cases in Brazil.	R2 of forecasting for the daily confirmed cases from 7 May 2020 to 29 June 2020: 0.665.
Shastri et al. [13]	RNN (stacked LSTM, bi-directional LSTM, and convolutional LSTM).	The daily confirmed cases and death cases in India and USA.	The best MAPE of forecasting for the daily cases from 8 June 2020 to 7 July 2020: a. India: 2.17; b. USA: 2.00.

3. Data

This research used cumulative daily cases data from Indonesia, which are available publicly from the Ministry of Health, Indonesia at https://kawalcovid19.id/ (accessed on 11 February 2021). Firstly, this research used data compiled since the first case reported in March 2020. This research used data from 2 March 2020, the date of the first reported case, until 24 August 2020. The data from that period are used for training and validation of models to determine the appropriate hyperparameters. After validation, the next step is testing. A detailed explanation related to the partition of training, validation, and testing data is explained in Section 4.4. The pattern of cumulative COVID-19 cases in Indonesia is presented in Figure 1.

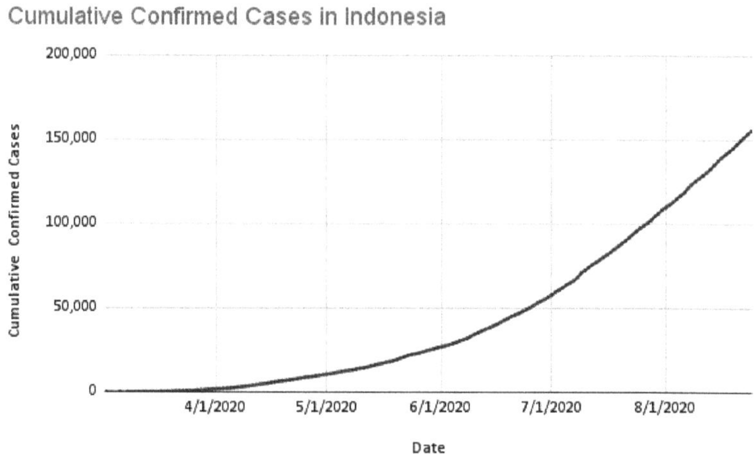

Figure 1. Cumulative confirmed cases of COVID-19 in Indonesia.

4. Methods

4.1. Forecasting Using Flower Pollination Algorithm

The flower pollination algorithm (FPA) is a nature-inspired metaheuristic algorithm proposed by Yang [20]. The FPA is based on the flower pollination process of flowering plants. Flower pollination can occur by self-pollination or cross-pollination. Self-pollination refers to pollination that occurs from a different flower, or from the same flower, of a single plant. When there is no reliable pollinator available, it is usually aided by wind. Self-pollination is also referred to as abiotic pollination. Cross-pollination, on the other hand, refers to pollination from a flower of a different plant. Cross-pollination is aided by a pollinator, such as bees, bats, birds, and flies, who can fly a long distance. The pollinators may demonstrate as Lévy flight behavior. They jump or fly with distance steps that obey Lévy distribution. Cross-pollination is also referred to as biotic pollination. Cross-pollination is considered to be global pollination, while self-pollination is considered to be local pollination.

There are four rules for the FPA, based on the above flower pollination process of flowering plants:

1. Rule 1—biotic, cross-pollination, or pollination between flowers is global pollination following Lévy Distribution. This first rule is represented mathematically in Equation (1), where x_i^t is the pollen i or solution vector x_i at iteration t, $g*$ is the current best solution found among all solutions at the current iteration, and $L(\pi)$ is the strength of the pollination (step size). Lévy flight is used to mimic it; therefore, $L(\pi)$ is derived from a Lévy distribution with a value greater than 0. Lévy distribution is represented in Equation (2). Lévy distribution uses the standard gamma function $\Gamma(\pi)$, which is valid for large steps $s > 0$.

$$x_i^{t+1} = x_i^t + \gamma L(\lambda)\left(x_i^t - g*\right), \tag{1}$$

$$L \sim \frac{\lambda\Gamma(\lambda)sin(\pi\lambda/2)}{\pi}\frac{1}{s^{1+\lambda}}, (s \gg s_0 \gg 0), \tag{2}$$

2. Rule 2—abiotic, self-pollination, or pollination of flowers from the same plants. Local pollination is represented mathematically in Equation (3). x_j^t and x_k^t are two pollens of the same plant but from different flowers. ε is a random value from a uniform distribution in range [0,1].

$$x_i^{t+1} = x_i^t + \epsilon\left(x_j^t - x_k^t\right), \tag{3}$$

3. Rule 3–flower constancy or equivalent to a reproduction probability proportional to the likeness of the two flowers involved is often developed by the pollinators.
4. Rule 4—a probability $P \in [0,1]$ is used to switch between local pollination and global pollination.

In this study, the FPA was used to forecast cumulative cases of COVID-19. The FPA was used to obtain the best solution $g*$ from the set of solutions x. Each x consists of a multilinear regression coefficient θ_l, where $l = 1, 2, \ldots, N$ and bias θ_0 to predict the cumulative daily cases of COVID-19 for day D'_T based on the previous N days, so that $x = \{\theta_0, \theta_1, \theta_2, \ldots, \theta_N\}$. The θ_l will be used as sum-product for D_{T-l} and then the results are summed by θ_0. Formally, the multilinear regression in this research is represented in Equation (4):

$$D'_T(x) = \theta_0 + \sum_{l=1}^{n} \theta_l \cdot D_{T-l}, \tag{4}$$

The objective function for each solution x is to minimize the difference between predicted cumulative case D'_T and actual cumulative case D_T. In this research, root-mean-square error (RMSE) is used to measure the difference. RMSE is presented in Equation (5), where m is equal to the length of the time series record:

$$RMSE(x) = \sqrt{\sum_{i=1}^{m} \frac{\left(D'_i(x) - D_i\right)^2}{m}}, \tag{5}$$

Based on the objective function that has been determined, the fitness function for each solution to be evaluated is represented mathematically in Equation (6). The best solution for each generation is $g*$, and will be used as the final solution:

$$fitness(x) = \frac{1}{RMSE(x) + 1}, \tag{6}$$

For each generation t, n solutions as a population are generated. From initial generation t_0, the best solution in the population will be stated as $g*$. In generation t, where $t = 1, 2, \ldots, MaxGeneration$, if there is one solution that is better than $g*$, that solution will replace the existing $g*$. The alteration of $g*$ is performed iteratively in each generation; therefore, a dynamic approach is required. The solutions in generation t are formed from the pollination of the solutions in generation $t-1$ (either global pollination or local pollination, as stated in Equations (1) and (3), respectively). The switch between global or local pollination in generation t is controlled by switch probability P, as stated in Rule 4.

4.2. *Forecasting Using Recurrent Neural Network*

The second method applied is the recurrent neural network (RNN). RNN is a kind of neural network architecture which is suitable for processing sequential data. The advantage of the RNN architecture is that it is more flexible and can be attuned according to the number of sequences in input or output. The RNN uses iterative function cycles to store information [31]. The RNN architecture is constructed in a form such that the network will remember the previous information and apply it to calculate the current output. In the RNN, the nodes between the hidden layers are connected periodically, and the hidden layer's input includes not only the output of the input layer, but also the output of the hidden layer at the last time, thus RNN can preserve, learn, and record historical information in sequence data [32].

The RNN has a similar forward pass process to that of a multilayer perceptron with a single hidden layer. The difference lies in the fact that RNNs accept activations from both the current external input and also the hidden layer activations from previous timesteps [31] As shown in Figure 2, the structure of the RNN includes the input layer, hidden layer, output layer, the weights of input layer to hidden layers, the weights of hidden layers to output layers, and learnable weights for the previously hidden state. These recurrent connections serve to pass values over timestep or sequence.

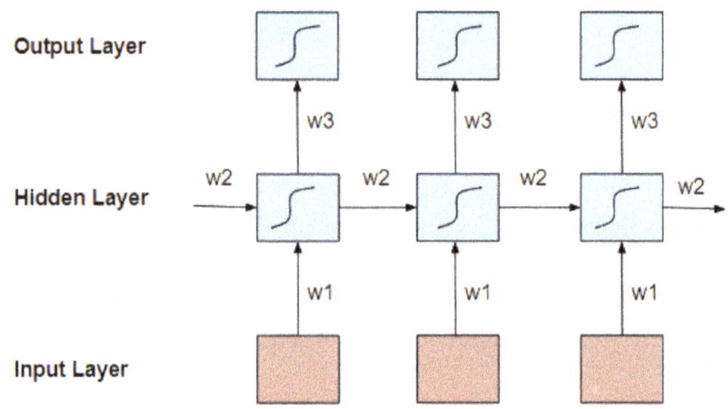

Figure 2. Unfolded recurrent neural network.

With this architecture, the current output in the RNN depends on the previous state. In a simple RNN, hidden units will receive the input in the current state and the output from the previous hidden state. The current hidden unit and the output can be defined mathematically in Equations (7) and (8), respectively:

$$h_t = \sigma(W_1 x_t + W_2 h_{t-1} + b), \tag{7}$$

$$o_t = W_3 h_t + b_2, \tag{8}$$

For Equation (7), h_t is the hidden state and x_t is the input at the current timestep. W_1 is the learnable weight from the input layer to the hidden layer, while W_2 are learnable weights for the previously hidden state's input. σ is an activation function and b_1 is the bias for the hidden layer. The activation function σ can be switched depending on the situation. The purpose of using the activation function is to ensure that the model is a non-linear machine. Common activation function choices are sigmoid, tanh, and ReLU functions. For Equation (8), o_t is the output state, h_t is the hidden state, W_3 is the learnable weight from hidden layer to the output layer, and b_2 is the bias for the output layer.

The complete sequence of hidden activations can be calculated by starting at the first timestep and then recursively applying Equation (7), incrementing time at each step. For the initially hidden unit at the start of the timestep, the value of the previously hidden state unit can either be manually adjusted to a certain value or set to zero. It is known that RNN stability and performance can be improved by using non-zero initial values. As for the weights, the norm is to randomize the weight without known information about the data. However, they can be set to particular values to help avoid overfitting [31].

In neural networks, the error of the prediction with respect to the target is calculated after the output is obtained. This error is normally in the form of a partial derivative of a differentiable loss function, where the derivative with respect to the weights can be used to improve the weights. There are two well-known algorithms that can be used to calculate the loss derivatives for RNNs: real-time recurrent learning (RTRL) and backpropagation through time (BPTT). BPTT is known to be simpler and more efficient in computation time, particularly since its process is similar to normal backpropagation in the neural network [31].

In this research, the data of cumulative COVID-19 daily cases are represented sequentially. Each sequence consists of data from the previous N. This sequence will be fed to RNN architectures to predict the cumulative COVID-19 cases of day D'_T. The experiments are conducted using several combinations of hyperparameters, such as the number of hidden layers, the dimension of neurons, learning rates, and dropout ratio, to attempt to determine the best model with minimum RMSE. ReLU is used as the activation function and Adam is used as the optimizer.

4.3. Model Performance Measurement

In order to measure the performance of the forecasting model, two performance measurements are used in this research, which are root-mean-square error (RMSE) and mean absolute percentage error (MAPE). RMSE is represented mathematically in Equation (5). The smaller the RMSE values are, the more accurate the forecasting model is; conversely, the larger the RMSE values are, the more inaccurate the model is [33]. RMSE value is the error number, which doesn't provide any information about the percentage of error compared to the actual value. Meanwhile, MAPE is a widely used evaluation metric for forecasting methods presenting the percentage of error. MAPE is represented mathematically in Equation (9), where A_t is actual value, F_t is forecast value, and n is the length of time series recorded.

$$MAPE = \frac{1}{n}\sum_{t=1}^{n}\left|\frac{A_t - F_t}{A_t}\right|, \quad (9)$$

The code of both forecasting models, the FPA and RNN, are available to be accessed publicly at http://ugm.id/covidforecasting (accessed on 29 November 2022). The code is written in Python programming language.

4.4. Training, Validation, and Testing Data

This study involved two phases, which are Phase 1: Development of FPA and RNN Model, and Phase 2: Evaluation of the Forecast Performance of the FPA and RNN Model Developed in Phase 1.

In Phase 1: Model Development, the data period is from 2 March 2020, to 10 July 2020. The dataset from 2 March 2020, to 10 July 2020, is divided into a ratio of 80:20; 80% for training data and 20% for validation data. Therefore, the training data are from 2 March 2020, to 4 June 2020, while the validation data are from 15 June 2020, to 10 July 2020, represented in Table 2. The validation process is carried out to determine the appropriate hyperparameters for the model.

Table 2. Period for developing the FPA and RNN.

Sample	Period
Training (n = 105)	2 March–14 June 2020
Validation (n = 26)	15 June–10 July 2020

In Phase 2: Model Evaluation, after the appropriate hyperparameters for the FPA and RNN model are obtained, the testing process is conducted. The FPA and RNN model is tested for short- and long-term forecast of the cumulative COVID-19 cases. We refer to some references [4–6,10,26] conducting forecasting for the next 7–14 days. Therefore, we used one-week forecast for the short-term and two-week forecast for the long-term forecasting.

1. Long-term forecast, which forecasts the cumulative cases of COVID-19 over the next 14 days (2-week forecast);
2. Short-term forecast, which forecasts the cumulative COVID-19 cases for the next 7 days (1-week forecast).

In order to obtain more comprehensive results of the performance of the models, the testing (forecast) process is conducted in several rounds or iterations. Long-term testing is conducted in 5 iterations, while short-term testing is conducted in 10 iterations. The model is updated with the relevant training data in each iteration using the hyperparameters defined in the validation sample in Phase 1. Table 3 presents the period of training data and testing data for long-term testing, while Table 4 presents the period of training data and testing data for short-term testing.

Table 3. Period for long-term testing (forecast).

Iteration	Types of Data	Period
Iteration 1	Training Data	2 March–15 June 2020
	Testing Data	16 June–29 June 2020
Iteration 2	Training Data	2 March–29 June 2020
	Testing Data	30 June–13 July 2020
Iteration 3	Training Data	2 March–13 July 2020
	Testing Data	14 July–27 July 2020
Iteration 4	Training Data	2 March–27 July 2020
	Testing Data	28 July–10 August 2020
Iteration 5	Training Data	2 March–10 August 2020
	Testing Data	11 August–24 August 2020

Table 4. Period for short-term testing (forecast).

Iteration	Types of Data	Period
Iteration 1	Training Data	2 March–15 June 2020
	Testing Data	15 June–22 June 2020
Iteration 2	Training Data	2 March–22 June 2020
	Testing Data	23 June–29 June 2020
Iteration 3	Training Data	2 March–29 June 2020
	Testing Data	30 June–6 July 2020
Iteration 4	Training Data	2 March–6 July 2020
	Testing Data	7 June–13 July 2020
Iteration 5	Training Data	2 March–13 July 2020
	Testing Data	14 July–20 July 2020
Iteration 6	Training Data	2 March–20 July 2020
	Testing Data	21 July– 27 July 2020
Iteration 7	Training Data	2 March–27 July 2020
	Testing Data	28 July–3 August 2020
Iteration 8	Training Data	2 March–3 August 2020
	Testing Data	4 August–10 August 2020
Iteration 9	Training Data	2 March–10 August 2020
	Testing Data	10 August–17 August 2020
Iteration 10	Training Data	2 March–17 August 2020
	Testing Data	18 August–24 August 2020

5. Results and Discussion

5.1. Hyperparameter

The validation process is conducted to obtain appropriate hyperparameters for the FPA and RNN model. The experiments engage several combinations of hyperparameters and choose the best one, providing the model with minimum RMSE. As explained in Section 4.4., the training data is from 2 March 2020, until 14 June 2020, while the validation data is from 15 June 2020, until 10 July 2020.

The combinations of hyperparameters for the FPA and RNN model are:

1. FPA Model:
   a. Length of the input timestep: 5 or 7;
   b. Switch probability between global pollination or local pollination: 0.3, 0.5, or 0.8;
   c. Population size (number of generated solutions): 50, 100, 150, or 200.
2. RNN Model:
   a. Length of the input timestep: 5 or 7;
   b. Dimension of neurons in LSTM cell: 10, 30, 50;

c. Learning rates: 0.001 or 0.01;
d. The number of hidden layers: 1 or 2;
e. Dropout ratio for each hidden layer: 20%, 50%, or no dropout.

In total, there are 24 combinations of hyperparameters for the FPA and 72 combinations of hyperparameters for the RNN. The best hyperparameters will be used in the testing process. For the RNN model, we use one and two hidden layers. Deeper RNN architecture required more data for training. In our research, the training data are limited enough (105 days). Therefore, if we use three or more hidden layers for the RNN, the model will have high possibility to be trapped in overfitting and it may not provide better results.

5.2. *Results and Performance Analysis*

The experiments of the validation process used 24 hyperparameter combinations for the FPA and 72 hyperparameter combinations for RNN in order to determine the best hyperparameters for this forecasting model. Based on the observation of the RMSE value for each generation, the number of generations to run the FPA is 100. The RMSE value for 100 generations reached convergence. While the number of epochs to run RNN is 1000, the RMSE value at 1000 epochs also reached convergence.

The complete 96 experiment results of the validation process are presented in Supplementary Tables S1 and S2, while the best hyperparameters, with the lowest RMSE values, are:

1. FPA Model:
   a. Length of the input timestep: 5;
   b. Switch probability between global pollination or local pollination: 0.3;
   c. Population size (number of generated solutions): 100;
   d. RMSE value: 292.66.
2. RNN Model:
   a. Length of the input timestep: 7;
   b. Dimension of neurons in LSTM cell: 10;
   c. Learning rate: 0.01;
   d. The number of hidden layers: 1;
   e. Dropout ratio for each hidden layer: no dropout;
   f. RMSE value: 502.95.

These parameters were then used to generate the FPA and RNN model for the testing process. In this validation process, the RMSE value from the FPA model (292.66) is significantly lower than that of the RNN model (502.95).

5.2.1. Long-Term Forecasting

There are two types of testing processes: (1) long-term forecasting for 5 iterations (different time periods); and (2) short-term forecasting for 10 iterations (different time periods). In this testing process, two performance measurements are calculated, RMSE and MAPE.

The long-term forecasting results are explained in this section, while the short-term forecasting is explained in the next section. The results for long-term forecasting are presented in Table 5. The FPA and RNN models are not overfitted, because the MAPE value for testing data is lower than the training data for all iterations. The FPA model has the lowest MAPE in the last iteration.

Table 5. Long-term forecasting results.

Iteration	Data	FPA		RNN	
		RMSE	MAPE (%)	RMSE	MAPE (%)
Iteration 1	Training Data	185.80	4.12	177.35	2.57
	Testing Data	289.05	0.45	567.92	1.11
Iteration 2	Training Data	394.88	7.74	760.54	10.91
	Testing Data	997.16	1.07	1802.61	2.62
Iteration 3	Training Data	431.61	9.50	632.57	4.62
	Testing Data	1927.70	2.10	4639.33	4.93
Iteration 4	Training Data	550.55	10.90	1082.42	5.03
	Testing Data	752.77	0.53	2459.13	2.13
Iteration 5	Training Data	562.86	7.09	1620.60	6.98
	Testing Data	621.37	0.38	7715.96	5.31

Figure 3 represents a bar chart of RMSE for long-term forecasting in testing data. Figure 4 represents a bar chart of MAPE for long-term forecasting in testing data. Based on Table 5 and the clustered bar chart in Figure 3, the RMSE value of training and testing provided by the FPA model is lower than that of the RNN model for all iterations. It can be observed in Table 5 that the MAPE is higher for the FPA for the training sample at iteration 1, 3, and 4. However, the MAPE for the testing sample is lower for the FPA model compared to the RNN model, which has high MAPE values, as shown in Table 5 and Figure 3. This shows that the FPA provided more reliable long-term forecasts.

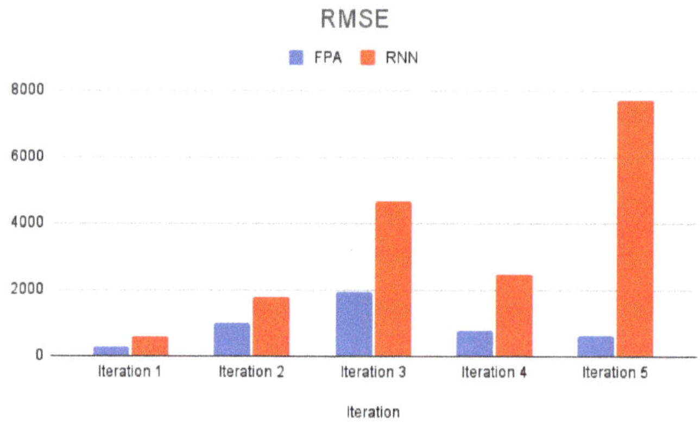

Figure 3. Bar chart of RMSE for long-term forecasting in testing data.

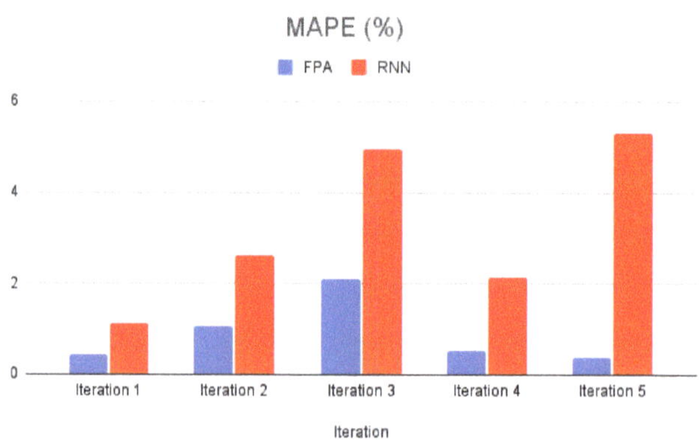

Figure 4. Bar chart of MAPE for long-term forecasting in testing data.

In total, until the last iteration (iteration five), we have training data consisting of 169 records (2 March 2020 until 17 August 2020). After we observed the forecasting results, it was determined that the RNN model provides more accurate forecasting in the beginning of training data (day 1–50), but less accurate in the following days. In contrast with the RNN model, the FPA model provides more accurate forecasting results than the RNN starting at day 51. The MAPE value represents the proportion between the error and the actual numbers. The RNN model provides more accurate forecasting results in the beginning, when training data contain less than 10,000 cumulative cases, but provides less accurate forecasting results in the following days, when training data contains more than 10,000 cumulative cases, reaching a total of 140,000 cases on the last day. For this reason, the RNN model has a higher RMSE value but lower MAPE value than the FPA model for iteration 3, 4, and 5 of the training data. The FPA model provides more accurate forecasting after learning, for some iterations; therefore, the FPA model provides a lower RMSE value but a higher MAPE value than the RNN model in training data. The RNN model requires more data for training. Unlike the training data, the testing data for each iteration only consists of 14 days. The next analyses will focus on forecasting results of testing data.

Figure 5 represents the trend for the actual data and long-term forecasting results using the FPA for each iteration. The x-axis represents the date and the y-axis represents the cumulative COVID-19 cases. The actual number of cumulative COVID-19 cases are represented with a blue line (real), the forecasting result for iteration 1 is represented with a red line (testing 1), iteration 2 is represented with a yellow line (testing 2), iteration 3 is represented with a green line (testing 3), iteration 4 is represented with an orange line (testing 4), and iteration 5 is represented with a black line (testing 5).

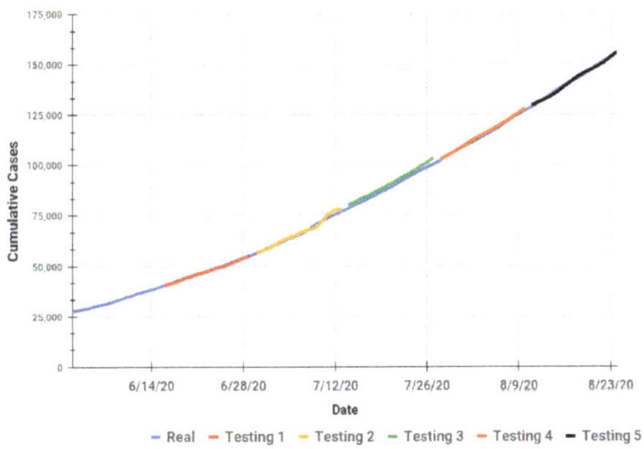

Figure 5. Actual and long-term forecasting for cumulative COVID-19 cases using the FPA model.

The RMSE and MAPE value for testing data in iteration 3 are the highest compared to other iterations. As we can see from Figure 5, the forecasting in iteration 3 (testing 3) learns the pattern from iteration 1 and iteration 2. The trend of data in iteration 3 has a steeper slope than the previous iterations. This may be the reason why the error value in iteration 3 is the highest one.

Figure 6 presents the trend of actual data and forecasting results for the RNN model. The forecasting results of the FPA model are better than those of the RNN model. This is also confirmed with the RMSE and MAPE results in Table 4, which shows that the overall RMSE and MAPE values of the FPA model are lower than those of the RNN model. The RNN model is a deep neural network which requires more data for training. The FPA model is better than RNN for long-term forecasting.

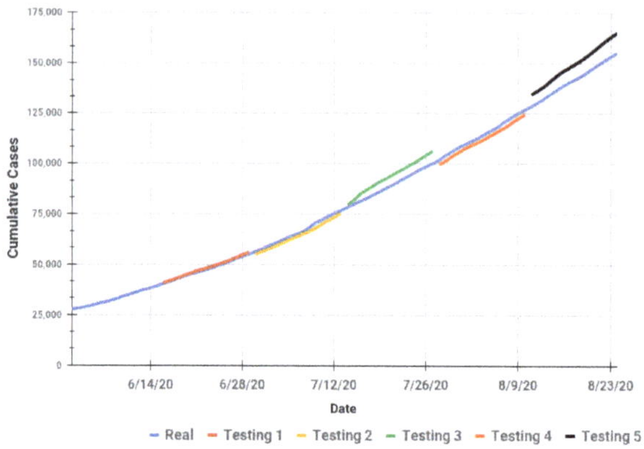

Figure 6. Actual and long-term forecasting for cumulative COVID-19 case using the RNN model.

5.2.2. Short-Term Forecasting

The results for short-term forecasting are presented in Table 6. The FPA model has lower RMSE for both training and testing data for all iterations except iteration 7 (RMSE for the FPA is higher than for the RNN). On the other hand, the RNN model, overall, has lower MAPE for training data, except iteration 7. The FPA has lower MAPE for testing data, except at iteration 7. These results show the RNN model may be overfitted in the iterations where MAPE is higher in testing than the training sample. Overfitting occurs when the performance of the model is good for the training but not the testing data.

Table 6. Short-term forecasting results.

Iteration	Data	FPA		RNN	
		RMSE	MAPE (%)	RMSE	MAPE (%)
Iteration 1	Training Data	372.30	4.57	1612.35	3.79
	Testing Data	1179.31	0.74	7240.29	4.80
Iteration 2	Training Data	167.45	5.49	582.25	7.92
	Testing Data	195.66	0.30	1878.50	3.64
Iteration 3	Training Data	172.18	3.31	148.22	3.37
	Testing Data	346.66	0.48	306.01	0.39
Iteration 4	Training Data	189.55	5.72	508.51	4.50
	Testing Data	735.18	0.82	1331.62	1.75
Iteration 5	Training Data	243.84	5.76	627.81	2.59
	Testing Data	739.52	0.88	2467.42	2.89
Iteration 6	Training Data	651.86	14.74	1031.12	5.51
	Testing Data	2184.86	2.22	3277.67	3.41
Iteration 7	Training Data	477.98	10.00	542.96	5.69
	Testing Data	1589.29	1.34	221.98	0.13
Iteration 8	Training Data	395.56	6.41	1166.66	2.91
	Testing Data	1228.22	0.99	5895.64	4.85
Iteration 9	Training Data	260.75	3.03	285.52	2.24
	Testing Data	373.41	0.21	622.63	0.42
Iteration 10	Training Data	372.30	4.57	1612.35	3.79
	Testing Data	1179.31	0.74	7240.29	4.80

Based on Table 6, the RMSE value of training data for iteration 4, 5, 6, 7, 8, 9, and 10 provided by the FPA model is lower than the RNN model, but the MAPE value provided by the FPA model is higher than the RNN model. This is the same as what occurred for the long-term forecasting. The RNN model provides more accurate forecasting results in the beginning, when training data contain less than 10,000 cumulative cases, but provides less accurate forecasting results in the following days, when training data contains more than

10,000 cumulative cases, reaching a total of 140,000 cases on the last day. For this reason, the RNN model has a higher RMSE value but lower MAPE value than the FPA model for iteration 4,5,6,7,8,9, and 10 in training data. The FPA model provides more accurate forecasting after learning for some iterations; therefore, the FPA model provides lower RMSE value but higher MAPE value than the RNN model in training data.

Figure 7 represents a bar chart of RMSE for short-term forecasting in testing data. Figure 8 represents a bar chart of MAPE for short-term forecasting in testing data. Figure 9 represents the trend for the actual data and short-term forecasting results of the FPA model for each iteration. The x-axis represents the date, and the y-axis represents the cumulative COVID-19 cases. The actual number of cumulative COVID-19 cases are represented with a blue line (real), the forecasting result for iteration 1 is represented with a red line (testing 1), iteration 2 is represented with a yellow line (testing 2), iteration 3 is represented with a green line (testing 3), iteration 4 is represented with an orange line (testing 4), iteration 5 is represented with a brown line (testing 5), iteration 6 is represented with a purple line (testing 6), iteration 7 is represented with a gray line (testing 7), iteration 8 is represented with a dark blue line (testing 8), iteration 9 is represented with a pink line (testing 9) and iteration 10 is represented with a black line (testing 10).

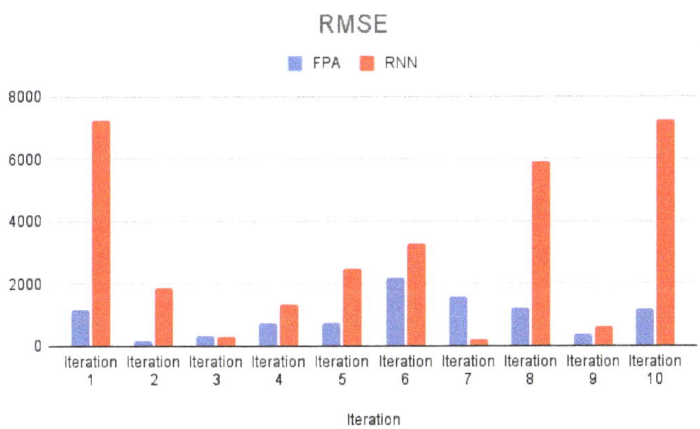

Figure 7. Bar chart of RMSE for short-term forecasting in testing data.

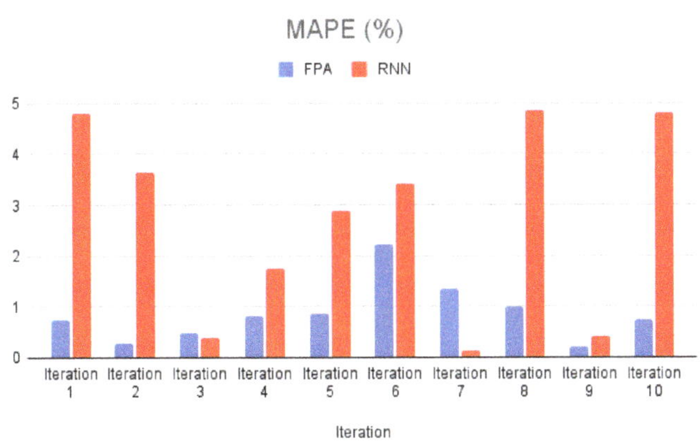

Figure 8. Bar chart of MAPE for short-term forecasting in testing data.

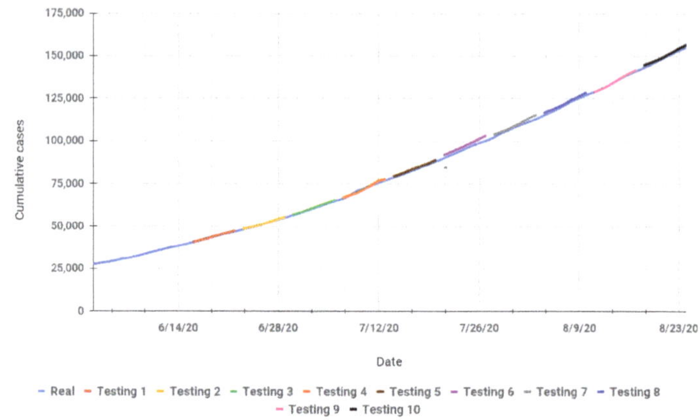

Figure 9. Actual and short-term forecasting for cumulative COVID-19 cases using the FPA model.

Based on Figures 7–9, as with long-term forecasting, the FPA model for short-term forecasting has the highest RMSE and MAPE value for testing data in iteration 6, which is iteration 3 for long-term forecasting. The trend of data in iteration 6 has a steeper slope than the previous iterations. The MAPE value for testing data in iteration 7 provided by the FPA model (1.34 %) is higher than that of the RNN model (0.13%). The FPA model learned a new pattern of data in iteration 6, with a steeper slope; therefore, the FPA model has the highest MAPE in iteration 6 (2.22%). The MAPE decreases in iteration 7 (1.34%) and in the next iterations. This does not occur in long-term forecasting. The FPA model can learn a new pattern of data better in long-term forecasting, which is the training data updated for 2 weeks.

Figure 10 represents the trend of actual data and forecasting results for the RNN model. Based on Figure 8, the highest MAPE value is in iteration 8, which is confirmed with the forecasting result in Figure 10 (testing 8). The forecasting results of the RNN model for the long-term model are better than for the short-term model. The training model in the RNN may not be adequately up to date with the addition of 1 week of data for each iteration. The RNN could not calculate the pattern of the data with the addition of only a few data (1 week of data).

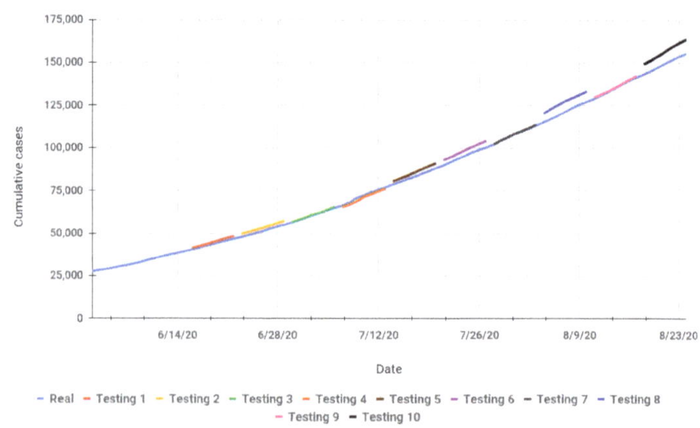

Figure 10. Actual and short-term forecasting for cumulative COVID-19 cases using the RNN model.

Overall, the forecasting results of the FPA model are better than the RNN model, both for long-term forecasting and short-term forecasting. The FPA model is better than the

RNN model in the presence of limited training data. The RNN model requires more data for training and to learn the pattern of data. The FPA model is better than the RNN model for forecasting the cumulative COVID-19 cases in Indonesia.

6. Conclusions

In this research, we presented forecasts of the cumulative COVID-19 cases in Indonesia using the FPA, a natured-inspired algorithm, to determine the optimal coefficients of the variables in the forecasting function of COVID-19 cases. We compared the performance of the FPA with a machine learning method which is popular for forecasting, the RNN. Several comprehensive experiments were conducted to determine the optimal hyperparameters for the FPA and RNN. The best hyperparameters were used to develop a model for forecasting. Long-term and short-term forecasting were conducted using different iterations with data added as more cases were reported. The FPA model has lower MAPE value than the RNN model for both long-term and short-term forecasting. These results show that the FPA model is better than the RNN model for forecasting cumulative COVID-19 cases. The FPA model was able to provide more reliable forecasts. This research provides state-of-the-art results to aid the process of mitigating the global pandemic of COVID-19 in Indonesia. In future, this forecasting model will be extended for COVID-19 active cases and deaths. Then, the forecasting results will be provided online and updated each day by developing an online dashboard for users; therefore, it will be more useful.

Supplementary Materials: The following supporting information can be downloaded at: https://www.mdpi.com/article/10.3390/computation10120214/s1, Table S1: The experiment results of validation process using 72 hyperparameter combinations for RNN.; Table S2: The experiment results of validation process using 24 hyperparameter combinations for the FPA.

Author Contributions: Conceptualization, A.; Methodology, A., Y.S., I.N.P.T. and R.W.; software, A. and I.N.P.T.; validation, A. and I.N.P.T.; formal analysis, A., Y.B.W., S.H., Y.S., A.M. and R.W.; investigation, Y.B.W. and A.M.; data curation, D.U.K.P.; writing—original draft, A., S.H., Y.S. and D.U.K.P.; writing—review and editing, A., Y.B.W. and S.H.; visualization, A.; supervision, S.H.; project administration, A.; funding acquisition, A. All authors have read and agreed to the published version of the manuscript.

Funding: This work is supported in part by the Department of Computer Science and Electronics, Faculty of Mathematics and Natural Sciences, Universitas Gadjah Mada, Schema C Research Grant No. 224/J01.1.28/PL.06.02/2020.

Data Availability Statement: The dataset and the code of the forecasting model are available to be accessed publicly at http://ugm.id/covidforecasting (accessed on 29 November 2022).

Acknowledgments: This work was carried out under the scheme of research collaboration between Intelligent System Laboratory, Department of Computer Science and Electronics, Universitas Gadjah Mada (UGM), Indonesia and Advanced Analytics Engineering Centre (AAEC), Faculty of Computer and Mathematical Sciences, Universiti Teknologi MARA (UiTM), Malaysia. We would like to thank Nurbaizura Borhan and Puan Aida Wati Zainan Abidin from UiTM, Malaysia, and Ilona Usuman, Anifuddin Azis, and Sri Mulyana from UGM for providing their suggestions regarding this research.

Conflicts of Interest: The authors declare no conflict of interest.

Abbreviations

The following abbreviations are used in this manuscript:

ARIMA	Autoregressive integrated moving average
Bi-LSTM	Bidirectional LSTM
BPTT	Backpropagation through time
COVID-19	Coronavirus disease 2019
FPA	Flower pollination algorithm
GRU	Gated recurrent unit
LSTM	Long short-term memory

MAPE	Mean absolute percentage error
MERS	Middle East respiratory syndrome
OPEC	Organization of the Petroleum Exporting Countries
PPE	Personal equipment
ReLU	Rectified linear activation function
RMSE	Root-mean-square error
RNN	Recurrent neural network
RTRL	Real-time recurrent learning
SARIMA	Seasonal ARIMA
SARS	Severe acute respiratory syndrome
SIR	Susceptible-infected-recovered
SOP	Standard operating procedures
SVM	Support vector machine
SVR	Support vector regression
USA	United States of America
WHO	World Health Organization

References

1. WHO. Coronavirus Disease (COVID-19). Available online: https://www.who.int/emergencies/diseases/novel-coronavirus-2019/situation-reports/ (accessed on 11 February 2021).
2. Worldometer Coronavirus. Available online: https://www.worldometers.info/coronavirus/ (accessed on 11 February 2021).
3. Wuhan Pneumonia: Thailand Confirms First Case of Virus Outside China. Available online: https://www.scmp.com/news/hong-kong/health-environment/article/3045902/wuhan-pneumonia-thailand-confirms-first-case (accessed on 13 January 2020).
4. Mishra, P.; Al Khatib, A.M.; Sardar, I.; Mohammed, J.; Ray, M.; Manish, K.; Rawat, D.; Pandey, S.A.; Dubey, A.; Feys, J.; et al. Modelling and Forecasting of COVID-19 in India. *J. Infect. Dis. Epidemiol.* **2020**, *6*, 162.
5. Abuhasel, K.A.; Khadr, M.; Alquraish, M.M. Analyzing and forecasting COVID-19 pandemic in the Kingdom of Saudi Arabia using ARIMA and SIR models. *Comput. Intell.* **2020**, *38*, 770–783. [CrossRef] [PubMed]
6. Ali, M.; Khan, D.M.; Aamir, M.; Khalil, U.; Khan, Z. Forecasting COVID-19 in Pakistan. *PLoS ONE* **2020**, *15*, e0242762. [CrossRef]
7. Malavika, B.; Marimuthu, S.; Joy, M.; Nadaraj, A.; Asirvatham, E.S.; Jeyaseelan, L. Forecasting COVID-19 epidemic in India and high incidence states using SIR and logistic growth models. *Clin. Epidemiol. Glob. Health* **2021**, *9*, 26–33. [CrossRef]
8. Zuhairoh, F.; Rosadi, D. Real-time Forecasting of the COVID-19 Epidemic using the Richards Model in South Sulawesi, Indonesia. *Indones. J. Sci. Technol.* **2020**, *5*, 456–462. [CrossRef]
9. Shahid, F.; Zameer, A.; Muneeb, M. Predictions for COVID-19 with deep learning models of LSTM, GRU and Bi-LSTM. *Chaos Solitons Fractals* **2020**, *140*, 110212. [CrossRef]
10. Hao, Y.; Xu, T.; Hu, H.; Wang, P.; Bai, Y. Prediction and analysis of corona virus disease 2019. *PLoS ONE* **2020**, *15*, e0239960. [CrossRef] [PubMed]
11. Balli, S. Data analysis of COVID-19 pandemic and short-term cumulative case forecasting using machine learning time series methods. *Chaos Solitons Fractals* **2021**, *142*, 110512. [CrossRef]
12. Hawas, M. Generated time-series prediction data of COVID-19's daily infections in Brazil by using recurrent neural networks. *Data Brief* **2020**, *32*, 106175. [CrossRef]
13. Shastri, S.; Singh, K.; Kumar, S.; Kour, P.; Mansotra, V. Time series forecasting of COVID-19 using deep learning models: India-USA comparative case study. *Chaos Solitons Fractals* **2020**, *140*, 110227. [CrossRef]
14. Qiu, J.; Wang, B.; Zhou, C. Forecasting stock prices with long-short term memory neural network based on attention mechanism. *PLoS ONE* **2020**, *15*, e0227222. [CrossRef] [PubMed]
15. Uras, N.; Marchesi, L.; Marchesi, M.; Tonelli, R. Forecasting Bitcoin closing price series using linear regression and neural networks models. *PeerJ Comput. Sci.* **2020**, *6*, e279. [CrossRef] [PubMed]
16. Yao, L.; Guan, Y. An improved LSTM structure for natural language processing. In Proceedings of the 2018 IEEE International Conference of Safety Produce Informatization (IICSPI), Chongqing, China, 10–12 December 2018; pp. 565–569.
17. Graves, A.; Mohamed, A.R.; Hinton, G. Speech recognition with deep recurrent neural networks. In Proceedings of the 2013 IEEE International Conference on Acoustics, Speech and Signal Processing, Vancouver, BC, Canada, 26–31 May 2013; pp. 6645–6649.
18. Mansoori, A.; Effati, S.; Eshaghnezhad, M. An efficient recurrent neural network model for solving fuzzy non-linear programming problems. *Appl. Intell.* **2017**, *46*, 308–327. [CrossRef]
19. Hewamalage, H.; Bergmeir, C.; Bandara, K. Recurrent neural networks for time series forecasting: Current status and future directions. *Int. J. Forecast.* **2019**, *37*, 388–427. [CrossRef]
20. Yang, X.S. *Nature-Inspired Optimization Algorithms*; Elsevier: Amsterdam, The Netherlands, 2014; pp. 155–173.
21. Korkmaz, E.; Akgüngör, A.P. Flower pollination algorithm approach for the transportation energy demand estimation in Turkey: Model development and application. *Energy Sources Part B Econ. Plan. Policy* **2018**, *13*, 429–447. [CrossRef]

22. Chiroma, H.; Khan, A.; Abubakar, A.I.; Saadi, Y.; Hamza, M.F.; Shuib, L.; Gital, A.Y.; Herawan, T. A new approach for forecasting OPEC petroleum consumption based on neural network train by using flower pollination algorithm. *Appl. Soft Comput.* **2016**, *48*, 50–58. [CrossRef]
23. Volkan, A.; BARIŞÇI, N. Short-term load forecasting model using flower pollination algorithm. *Int. Sci. Vocat. Stud. J.* **2017**, *1*, 22–29.
24. Abdelaziz, A.Y.; Ali, E.S.; Abd Elazim, S.M. Flower pollination algorithm to solve combined economic and emission dispatch problems. *Eng. Sci. Technol. Int. J.* **2016**, *19*, 980–990. [CrossRef]
25. Anastassopoulou, C.; Russo, L.; Tsakris, A.; Siettos, C. Data-based analysis, modelling and forecasting of the COVID-19 outbreak. *PLoS ONE* **2020**, *15*, e0230405. [CrossRef]
26. Petropoulos, F.; Makridakis, S. Forecasting the novel coronavirus COVID-19. *PLoS ONE* **2020**, *15*, e0231236. [CrossRef]
27. Zoabi, Y.; Deri-Rozov, S.; Shomron, N. Machine learning-based prediction of COVID-19 diagnosis based on symptoms. *NPJ Digit. Med.* **2021**, *4*, 3. [CrossRef] [PubMed]
28. Kim, H.J.; Han, D.; Kim, J.H.; Kim, D.; Ha, B.; Seog, W.; Lee, Y.K.; Lim, D.; Hong, S.O.; Park, M.J.; et al. An Easy-to-Use Machine Learning Model to Predict the Prognosis of Patients with COVID-19: Retrospective Cohort Study. *J. Med. Internet Res.* **2020**, *22*, e24225. [CrossRef] [PubMed]
29. Ahmad, F.; Almuayqil, S.N.; Humayun, M.; Naseem, S.; Khan, W.A.; Junaid, K. Prediction of covid-19 cases using machine learning for effective public health management. *Comput. Mater. Contin.* **2021**, *66*, 2265–2282. [CrossRef]
30. Elzeki, O.M.; Shams, M.; Sarhan, S.; Abd Elfattah, M.; Hassanien, A.E. COVID-19: A new deep learning computer-aided model for classification. *PeerJ Comput. Sci.* **2021**, *7*, e358. [CrossRef] [PubMed]
31. Graves, A. *Supervised Sequence Labelling with Recurrent Neural Networks;* Springer: Berlin, Germany, 2012; pp. 5–13.
32. Li, W.; Liao, J. A comparative study on trend forecasting approach for stock price time series. In Proceedings of the 2017 11th IEEE International Conference on Anti-Counterfeiting, Security, and Identification (ASID), Xiamen, China, 27–29 October 2017; pp. 74–78.
33. Klimberg, R.K.; Sillup, G.P.; Boyle, K.J.; Tavva, V. Forecasting performance measures—What are their practical meaning? *Adv. Bus. Manag. Forecast.* **2010**, *7*, 137–147.

computation

MDPI

Article

Dendrograms for Clustering in Multivariate Analysis: Applications for COVID-19 Vaccination Infodemic Data in Brazil

Maria da Penha Harb [1,*], Lena Silva [2], Thalita Ayass [1], Nandamudi Vijaykumar [3], Marcelino Silva [4] and Carlos Renato Francês [1]

1 Institute of Technology, Federal University of Para, Belem 66075-110, Brazil
2 Center for Exact Sciences and Technology, University of Amazon, Belem 66060-902, Brazil
3 National Institute for Space Research, São José dos Campos 12227-010, Brazil
4 Institute of Engineering and Geosciences, Federal University of West Para, Santarem 68040-255, Brazil
* Correspondence: mpenha@ufpa.br

Abstract: Since December 2019, with the discovery of a new coronavirus, humanity has been exposed to a large amount of information from different media. Information is not always true and official. Known as an infodemic, false information can increase the negative effects of the pandemic by impairing data readability and disease control. The paper aims to find similar patterns of behavior of the Brazilian population during 2021 in two analyses: with vaccination data of all age groups and using the age group of 64 years or more, representing 13% of the population, using the multivariate analysis technique. Infodemic vaccination information and pandemic numbers were also used. Dendrograms were used as a cluster visualization technique. The result of the generated clusters was verified by two algorithms: the cophenetic correlation coefficient, which obtained satisfactory results above 0.7, and the elbow method, which corroborated the number of clusters found. In the result of the analysis with all age groups, more homogeneous divisions were perceived among Brazilian states, while the second analysis resulted in more heterogeneous clusters, recalling that at the start of vaccinations they could have had fear, doubts, and significant belief in the infodemic.

Keywords: dendrogram; infodemic; COVID-19; Google Trends; multivariate analysis

Citation: Harb, M.d.P.; Silva, L.; Ayass, T.; Vijaykumar, N.; Silva, M.; Francês, C.R. Dendrograms for Clustering in Multivariate Analysis: Applications for COVID-19 Vaccination Infodemic Data in Brazil. *Computation* **2022**, *10*, 166. https://doi.org/10.3390/computation10090166

Academic Editors: Simone Brogi and Vincenzo Calderone

Received: 8 August 2022
Accepted: 5 September 2022
Published: 19 September 2022

1. Introduction

Humanity has been seriously affected by the COVID-19 pandemic that originated in late 2019 in Wuhan, Hubei province in China, caused by the SARS-CoV-2 virus [1,2]. It was declared a pandemic on 11 March 2020, by the World Health Organization (WHO) [3]. According to the WHO data [4], by May 2022, more than 533 million cases of the disease had already been confirmed, causing more than 6 million deaths worldwide.

Both the impact of an entirely new and unknown disease, which was already immense, and the lack of information associated with it allowed false and dubious information to quickly appear and spread across various social media platforms, newspapers, and magazines [5]. Thus, the COVID-19 pandemic was accompanied by a massive and widespread wave of disinformation about the disease that can be described as an infodemic [6].

The WHO explains that infodemics are an excessive amount of information about a problem, making it extremely hard to identify a solution. They can spread both misinformation and disinformation (accidentally and deliberately false information, respectively) as well as rumors during a health emergency. Infodemics can severely affect or damage an effective public health response and create confusion and mistrust among people [7].

The battle against the COVID-19 pandemic and infodemic continues. A long-awaited step was the development of effective vaccines, which was highly anticipated, and several vaccines are now available. The development and availability of vaccines are not the only obstacles to overcome from a public health perspective. The increasing acceptance of the vaccine by the population is also paramount to designing public health measures and reaching a considerable proportion of vaccinated population to achieve herd immunity [8].

However, there is a process of rejection or delay in accepting vaccines, which can be affected by the variables of trust, compliance, and convenience, and directly influences the historical context of vaccination [9]. Such resistance is known as vaccine hesitancy, and before the pandemic, in 2019, the WHO identified it as one of the major threats to global health [10]. In addition to fake news, other factors impact the drop in vaccine coverage, such as social, cultural, religious, and economic issues in which the population is involved, and this can influence whether the population goes to vaccination centers [11].

Following the evolution of the pandemic, until May 2022, Brazil was the third-most affected country in the total number of cases and second in the total number of deaths [12], ranked 34th in vaccination (counting both first and second vaccine doses) [13]. Brazil applied its first vaccine against COVID-19 on 17 January 2021. In three months, 5.32% of the population had received one of the two necessary doses and only 2% were fully immunized [13]. The pace was very slow compared to countries like Israel, the United Kingdom, and Chile [14]. This was due to problems and delays in vaccine purchases (and when a campaign started, doses ran out), doubts about the effectiveness of the results, and false and dubious information circulated on the internet and social networks and in speeches by the President of Brazil, who was considered a denialist and anti-vaccine [14,15].

Because of the above, the purpose of this work is to carry out a spatial analysis of COVID-19 vaccination by the Federation Units in Brazil from the contribution of the cluster analysis technique, applying multivariate techniques using indicators that cover the infodemic data of vaccination from COVID-19 in Google Trends (GT) searches, internet proliferation in states, and data on deaths and the number of COVID-19 cases.

Cluster analysis makes it possible to group cases or variables into a homogeneous group based on their similarity. Each object is like the others in the group, maximizing homogeneity within the group and maximizing heterogeneity between groups. Dendrograms are used to cluster the states and, thus, generate new divisions of the regions, different from the geographic regions that already exist in Brazil: first carried out for an analysis of all age groups available and second with an analysis of the vaccination of the elderly 64 years or older.

Brazil is the fifth largest country in the world geographically and the sixth in population; it has 5570 municipalities divided into 27 federative units (26 states and one Federal District (DF)), which are grouped into five geographic macro-regions (Central-West, Northeast, North, Southeast, and South). After the results obtained from the techniques used, the new division of the groupings of the states will be visualized through the behavior of the population.

2. Related Work

The internet is revolutionizing the way epidemic intelligence is collected and offers solutions to some of these challenges. Freely available sources of information can allow us to detect disease outbreaks earlier with reduced cost and greater transparency in reporting [16]. Search engines have become pervasive in recent years, retrieving information easily on a variety of topics, ranging from customer service to general information. In addition to these research interests, there is a growing interest in obtaining health advice or information. In this respect, health policy authorities have begun to identify internet search engines as potential indicators for surveillance and health, such as the GT, a repository of publicly available information on user research patterns and real-time data [17].

In Mangono et al. [18], GT was used to provide insights and potential indicators of important changes in information-seeking patterns during COVID-19 with various pandemic-related terms, such as: coronavirus symptoms, urgent care near me, health insurance, social distancing, and "Chinese virus", among others. They compared the surveys with new monthly Medicaid orders (an application that provides health coverage to Americans), and used principal component analysis to identify research patterns in the GT.

Rovetta and Bhagavathula [19] investigated online search behavior related to the COVID-19 outbreak and the "infodemic nicknames" circulating in Italy using GT. The titles of articles from the most read national newspapers and government websites were surveyed to investigate the extent and attitudes of several related infodemic nicknames for COVID-19. They defined "infodemic nicknames" as substantially erroneous information, which gave rise to misinterpretations, fake news, episodes of racism, or any other form of misleading information that circulated on the internet. They concluded that Google search query data reflects growing regional and population interest in the pandemic. Searches related to disinfectants, face masks, health newsletters, vaccines, and symptoms related to COVID-19 were the main search keywords. However, many infodemic nicknames circulated in Italy. They also conclude that internet research interest in COVID-19, both at the regional and city levels in Italy, was influenced by tradition, electronic newspapers, and printed media coverage.

In Ceron et al. [20], the authors explored news-checking initiatives in Latin America, using a Markov-based computational method to group tweets into topics and identify their diffusion among different datasets about false information related to COVID-19 across regions, comparing if there was a pattern for Argentina, Brazil, Chile, Colombia, Mexico, and Venezuela.

Multivariate statistical methods were used by Custodio et al. [21], where they perceived that health measures led to a significant reduction in air pollution, but on the other hand, the impact of these measures in aquatic environments was poorly analyzed. In this context, multivariate statistical methods were employed to evaluate the water quality of the rivers of the Mantaro River basin and heavy metal contamination indices during the health crisis associated with the COVID-19 pandemic. The techniques employed were principal component analysis and hierarchical cluster analysis according to Spearman's correlation, which generated a dendrogram where the five chemical elements were grouped into two statistically significant groups, observing a significant increase in the critical value of contamination.

Computational and statistical techniques were used to analyze the heterogeneous spread of the pandemic and estimate the death rates from COVID-19 [22–24]. In Silva et al. [22], the authors estimated the effective reproduction rate number for each epidemiological week in Brazil and designed scenarios based on these values, concluding that the only way to flatten the curve is to decrease the reproduction rate. The work of Shafiq et al. [23] was applied to data from Italy and the results revealed that the model of artificial neural networks is an excellent engineering tool to predict survival and mortality rates, presenting more satisfactory and better results than other studies found in the literature.

Multivariate analysis techniques and dendrograms on the pandemic, as well as data from Brazil, are being used for grouping behaviors. James et al. [25] compared and contrasted data from the USA, India, and Brazil, looking at the trajectories of cases, deaths, and mortality rates, which were analyzed state by state. Dendrograms were used and revealed a similar cluster structure between the USA and India. Both countries had a dense majority cluster and a small collection of outliers. Brazil, on the other hand, presented a quite different structure, with two similarly sized clusters that contained most of the elements and then some outliers.

James and Menzies [26] presented a mathematical framework for determining the behavior of the second outbreak of COVID-19 cases in the United States based on a collection of time series. The data were grouped (dendrograms), identifying the different outbreak behaviors, and in the appendix of the work, the authors applied the technique to the data from Brazil and concluded that the second outbreaks were moderate and of comparable severity to the first outbreaks. This is similar to the USA, which also noted significant similarities based on geography.

In the research by Harb et al. [27], a multivariate analysis was performed on a database of COVID-19 infodemic terms in Brazil over 18 months (1 January 2020 to 30 June 2021), including socioeconomic and political variables. The infodemic terms were divided into five groups, and the analysis was performed every 3 months during the evolution of the pandemic. The study concludes that denial about the pandemic and the influence of political leadership can influence the search for infodemic information, contributing to disorganization in the control of the disease and prevention measures.

In this work, using the methodology of Harb et al. [27], we carried out an approach by combining the database of infodemic vaccination terms, extracted from research in the GT, related to variables of vaccination numbers (first and second doses), in the entire year of 2021 for Brazil. Multivariate analysis with dendrograms and the elbow method was performed to group the states into similar patterns of behavior in the two analyses performed.

3. Methodology

The methodology applied in this work can be seen in the flowchart in Figure 1, based on Harb et al. [27] and Braz et al. [28], and the data obtained were obtained from the 27 federative units (26 states and Distrito Federal (DF)) for the year 2021. Python language was used to perform both data manipulation and data clustering steps, and Tableau software (version 2021.3.3, Tableau Software, Mark Nelson, Seattle, WA, USA) was used to reproduce the map of Brazil with the generated clusters.

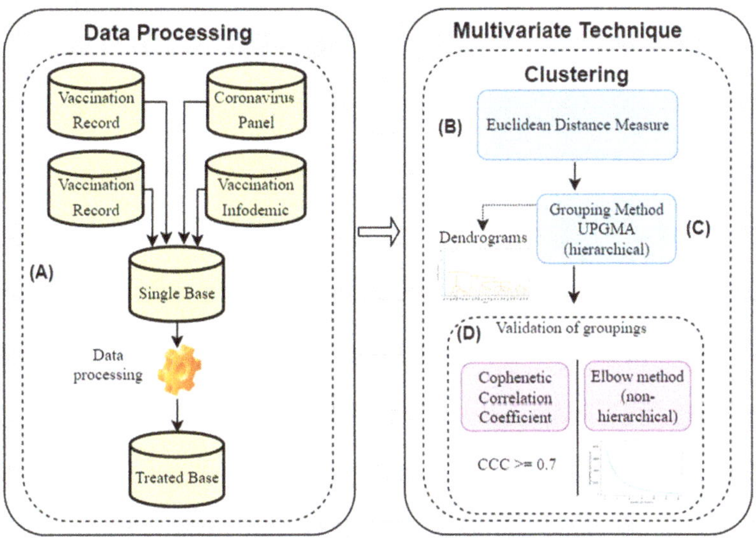

Figure 1. Flowchart of the methodology of the multivariate technique adopted. Processing the data: (**A**) selection and treatment of variables. Multivariate Technique: (**B**) selection of the distance measure; (**C**) selection of the clustering algorithm and (**D**) validation of the clusters.

The two main steps are presented in Figure 1, processing the selected data and clustering. The first step (A) is the extraction and treatment of the variables, generating a single dataset. The subsequent step applies and tests the best parameters for clustering (B) and (C), and at the end, performs cluster validation with two techniques (D).

3.1. Selection and Treatment of Variables for the Database (A)

The first step of the flowchart is the creation of the database, with the selection of variables to be used, the identification of outliers, and data standardization. Four databases available on the internet were used. Three are public databases from the government, and the last database is data extracted from GT. The following steps were considered: data collection, data processing, data mining, data interpretation, and validation. The databases searched were:

1. COVID-19 vaccination records [29]

For each state, three files were made available. After collecting and joining the files, the vaccination fields were selected by age group, counting the people who completed the vaccination cycle (1st and 2nd doses). To achieve a better relationship among the states, the rate per 100,000 inhabitants was calculated for these fields: data divided by the population of each state, multiplied by 100,000.

2. Coronavirus Panel [30]

The fields with information on the number of COVID-19 deaths and number of new cases were selected and the annual average per state was performed.

3. Internet access density [31]

Information on fixed broadband and mobile telephony was selected and the average was determined per each state.

4. Vaccination Infodemic [32]

The infodemic terms about vaccination [27–31,33–36] were selected. A search was performed on GT by the state for each term. The research was carried out using the following filters: geographic, which was used for each Brazilian state; period, which was the chosen year 2021; and categories, which were chosen for all categories and research groups using web searches. The values returned are from 100 (represents a term's peak popularity) to 0 (means there was not enough data about the term). The search was carried out on 1 May 2022, and for each term, the results found in related subjects and related searches were evaluated in order to verify if the term was related to non-infodemic content. As a result, some terms were dropped out.

The infodemic terms selected were: jacare vaccine, turning into jacare, doria vaccine, DNA vaccine, mutated DNA vaccine, vaccine kills, vaccine kills COVID, COVID cancer, COVID cancer vaccine, alcoholic drink COVID vaccine, liquid chip, chip vaccine, COVID hiv vaccine, COVID vaccine Alzheimer, magnetic COVID vaccine, COVID vaccine CoronaVac, COVID fetus vaccine, magnetic COVID vaccine, aluminum coronaVac, vaccine causes autism, vaccine causes impotence, and CoronaVac squint. Some terms are very specific in Brazilian news, such as the term "turning into alligator", researched after the President of Brazil said that after taking the vaccine, a person would become an alligator [37].

After the process of selecting the variables in the databases, outliers were identified, as the clustering technique is sensitive to outliers [38]. However, after analyzing the database, it was decided these values should not be removed. The reason for this is that outliers may form isolated clusters, which, in the case of Brazilian states with more striking characteristics, can happen.

With the database already formed, the standardization of variables was carried out because the use of measurement scales in different magnitudes can distort the analysis, and the most chosen form, among so many techniques, was the standardization z-score, with a mean of zero standard deviation 1 [39,40], shown in Equation (1),

$$Z = \frac{x - \mu}{SD} \tag{1}$$

where x is a data value, μ is the average of the values of the interval, and SD is the standard deviation.

The final database contains information from the 27 federative units for the year 2021. Twelve attributes were selected and treated, as shown in Table 1.

Table 1. Description of the variables selected in the database.

Name Variable	Type	Value
state	text	Brazilian State
number12_17	number	Age range 12 to 17 [1]
number18_64	number	Age range 18 to 64 [1]
number65_69	number	Age range 65 to 69 [1]
number70_74	number	Age range 70 to 74 [1]
number75_79	number	Age range 75 to 79 [1]
number80_	number	Age over 80
medCasosNovos	number	New COVID-19 cases, on average, per day
medObitosNovos	number	COVID-19 deaths, on average, per day
densblf	number	Density of fixed broadband [2]
denstm	number	Density of mobile phone [3]
infodemia	number	Relative volume of research for infodemic terms

[1] Number of people vaccinated with both first and second doses; [2] Number of accesses divided by the number of households; [3] Number of accesses divided by population.

3.2. Selection of the Measure of Distance or Similarity between Each Pair of Objects (B)

After the database was complete, the step was to select the measure of distance or similarity between each pair of objects. According to Metz [41], this approach builds the clusters so that examples belonging to the same cluster have a high similarity and examples belonging to different clusters have a low similarity. The measure chosen was Euclidean distance, the smallest distance between two components. The smaller the distance, the more similar the observations [42], and it is the most used distance metric in cluster analysis [41].

Equation (2) represents the Euclidean distance, where the distance between two observations (i and j) corresponds to the square root of the sum of the squares of the differences between the pairs of observations (i and j) for all p variables:

$$d_{ij} = \sqrt{\sum_{k=1}^{p} \left(x_{ik} - x_{jk}\right)^2} \tag{2}$$

where x_{ik} is the value of the variable k referring to observation i and x_{jk} represents variable k for observation j.

3.3. Selection of the Clustering Algorithm: Hierarchical Method (C)

This step selects the algorithm to maximize the differences between the groups in comparison with the variation within them. They are divided into hierarchical and non-hierarchical methods. According to Berkhin [43], some characteristics must be considered when choosing the clustering algorithm, such as: input types (dissimilarity matrix), attribute types, ability to find groups with different shapes, and outlier tolerance, among others. These characteristics were evaluated, and the hierarchical average method or unweighted pair group method with arithmetic mean (UPGMA) was chosen, which presented the best results in all executions. This method is less sensitive to noise and outliers and is defined by Equation (3):

$$d(i,j) = \frac{1}{|i||j|} \sum_{\substack{x_a \in i \\ x_b \in j}} d(x_a, x_b), \tag{3}$$

where the distance $d(i,j)$ between two groups is given by the average distance between objects of different groups.

One of the main advantages of hierarchical clustering algorithms comes from the representation of their results through dendrograms. A dendrogram is a graphical representation in tree format that presents the hierarchy of the clusters obtained [44]. In our paper, we elaborated on the dendrograms by executing the clustering of the database.

3.4. Validation of the Clusters (D)

At this stage, the quality of the generated clusters was verified through two algorithms. The first is the cophenetic correlation coefficient (*CCC*), which measures the degree of fit between the similarity matrix (phenetic matrix *F*) and the matrix resulting from the simplification provided by the clustering method (cophenetic matrix *C*) [45]. According to the proposal of Rohlf [45], cophenetic correlations >0.7 are admissible for good clusters and it is obtained by Equation (4):

$$CCC = \frac{C\hat{o}v(F,C)}{\sqrt{V(F)V(C)}} \tag{4}$$

The second algorithm is the elbow method, which interprets and validates coherence within cluster analysis, designed to help find the appropriate number of clusters within a dataset (non-hierarchical method). This method allows for evaluation on how the homogeneity or heterogeneity within the clusters varies for the value of each cluster. We can see this "elbow" when plotting its results on a graph, as there is a break in the direction of the curve, possibly informing the number of clusters to be defined [46]. Equation (5) shows the objective function, a squared error function:

$$W(S,C) = \sum_{k=1}^{k} \sum_{X_i \in S_k} |Y_i - C_k|^2 \tag{5}$$

where S is a k-cluster partition of the entity set represented by the vectors Y_i ($i \in I$) in the M-dimensional feature space, consisting of non-empty non-overlapping clusters S_k, each with a centroid C_k (k = 1,2, … K).

4. Computational Results

The database presented for the cluster analysis consisted of 27 observations and 12 indicators (as shown in Table 1). The data were organized in a spreadsheet and standardized as the indicators have different measurement units or scales and can change or alter the grouping structure.

4.1. Analysis with All Age Ranges of Vaccination

The dissimilarity measure used was the Euclidean distance, and for the composition of the clusters, the Average method was used, which presented the best *CCC* result, as can be seen in Table 2. It shows the comparison of the results of the execution of the most used methods for clustering. For each method, the *CCC* value, number of clusters, and the cut-off value in the dendrograms are presented.

Table 2. Comparison of the results of the execution of the most used methods for clustering.

Methods	*CCC*	Cluster Number	Cut Dendrogram
Average	0.887	6	3.1
Centroid	0.884	6	2.5
Complete	0.734	7	3.6
Single	0.808	7	1.6
Ward	0.647	6	5.0

The cut made on the axis of dissimilarity of the dendrogram was at a height of 3.1, which demonstrated the composition of six probable groups, as can be seen in Figure 2. Maranhão (MA) and Pará (PA) are highlighted as having particularities different from the others and greater similarities to each other in terms of the behavior of the indicators studied for the observed period. Observing the generated dendrogram, the first DF and second São Paulo (SP) groups were considered groups with an isolated state.

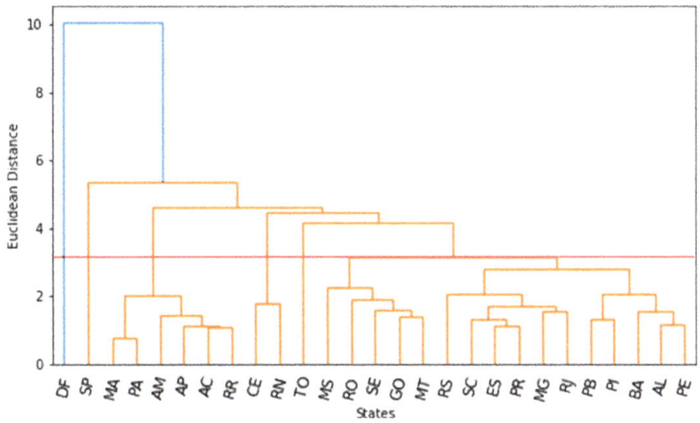

Figure 2. Dendrogram generated for the database with all age groups. The multivariate analysis technique was applied to the 2021 data, grouping Brazil in a different format from the Brazilian regions, which can be observed in the branches of the tree. It presents satisfactory *CCC* (0.887), and six clusters were suggested.

DF is the state with the highest average number of new cases per day of COVID-19 and SP is the largest state in terms of the Brazilian population, and, therefore, they may not have similarities with the other states in the clusters, both with just one state. The third group has six states (MA, PA, Amazonas (AM), Amapá (AP), Acre (AC), and Roraima (RR)), with a representation of 22.22%. The fourth group is composed of two states (Ceará (CE) and Rio Grande do Norte (RN)). The fifth group (Tocantins (TO)), composed of only one state, did not show similarities in behavior with other observations. Finally, the sixth group, the largest in the number of states with sixteen, encompasses states from all Brazilian geographic regions, representing 59.26% of the total states.

To evaluate the generated dendrogram, the *CCC* was employed, presenting a result of 0.887, a value admissible as a good cluster [45]. Thus, the *CCC* result obtained in this research shows an adequate adjustment of the applied clustering method.

The application of the elbow method contributed to defining the value of the number of clusters to be used in the non-hierarchical k-means technique for forming clusters, given that it collaborates in the optimization of clusters [47]. Figure 3 presents the result of the elbow method for the studied database.

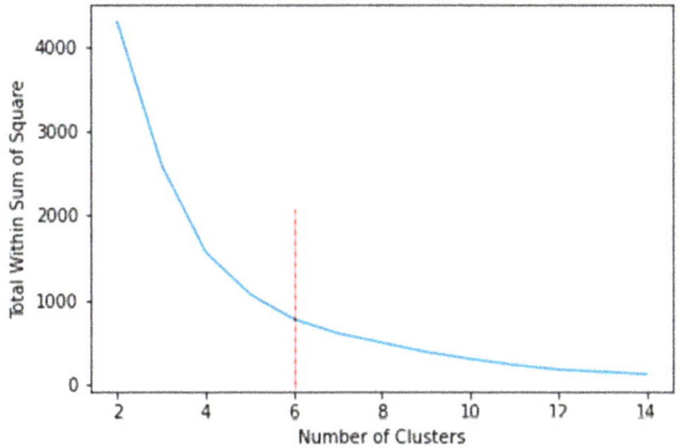

Figure 3. Curve of elbow method generated for the database with all age groups. The number of the cluster was determined by looking at the point position on the "elbow" arm.

The line traces the variation explained as a function of the number of clusters and chooses the elbow of the curve as the number of six clusters to be used.

For better visualization of the dendrogram result, the spatial distribution of the states was performed on the Brazilian map. The map (Figure 4) shows that this technique allowed for the clustering of the states, presenting some dispersed points, possibly because of some local peculiarities distributed within the group. However, there is a predominance of clusters within the same Brazilian region, these being within the same group, according to the method used.

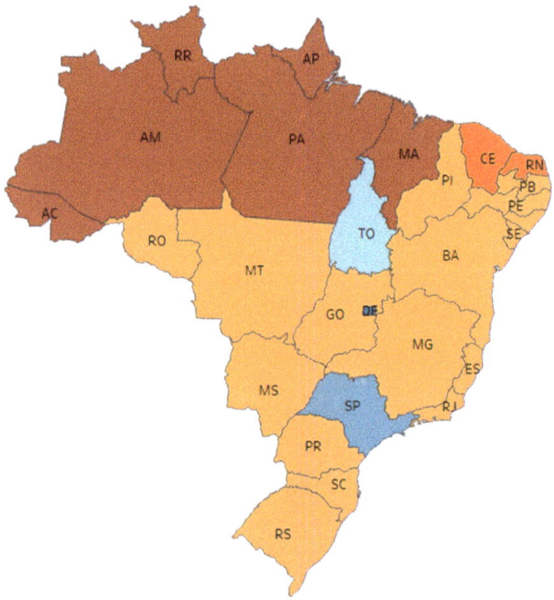

Figure 4. Map of Brazil generated from the dendrogram result with six clusters (with all age groups). Two large clusters contain 80% of the Brazilian states, presenting a great homogeneity in the behavior of the population.

It was noticed that two of the six groups generated present the majority of Brazilian states, indicating similarities in the behavior of the indicators studied. The group with the states MA, PA, AM, AP, AC, and RR are the states with less internet proliferation [48]. In the largest group, with sixteen states, a possible advance in vaccination is observed in Brazil for the year 2021, especially in the South and Southeast regions, which have a high percentage of the immunized population, but areas in the North and Northeast regions still have low immunization rates for COVID-19 [49].

The North and Northeast are places with a lower Human Development Index, younger populations, less educated residents, lower income, and more residents of small towns. In these places, the end of the pandemic seemed farther away than it did for large São Paulo centers (individual clusters), which already have high vaccine coverage with two doses, according to scientists [49], and also showed a lower rate of infodemic searches.

The state of DF, on the other hand, had the highest number of infodemic surveys on vaccination in 2021.

4.2. Analysis with Vaccination of the Elderly 64 Years or Older

The establishment of priority groups for vaccination is an important strategy, based on epidemiological indicators and the characterization of the vulnerability of the groups [50]. Thus, elderly citizens were the first to be vaccinated in Brazil, prioritized by age group.

However, for the second analysis, elderly people aged over 64 years were chosen to represent about 13% of the population [51].

The Euclidean distance was chosen as the dissimilarity measure, and for the composition of the clusters, the average method was used, which also presented better results, with a value of $CCC = 0.889$. The cut on the dendrogram dissimilarity axis was at the height of 1.8, which results in the agglomeration of eight probable groups, as can be seen in Figure 5.

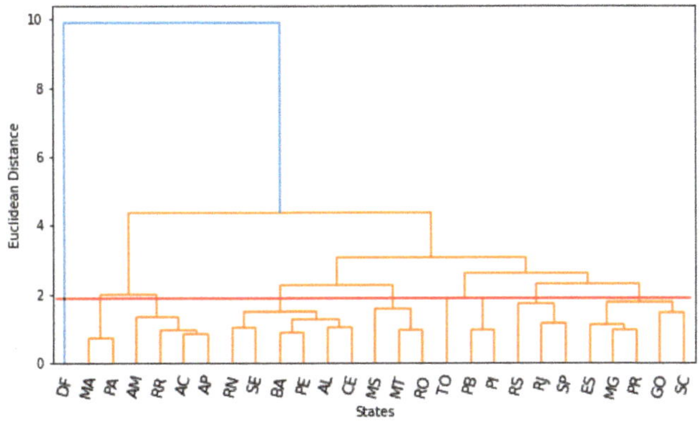

Figure 5. Dendrogram generated with data of the elderly 64 years or older. The multivariate analysis technique was applied to the 2021 data, grouping Brazil in a different format from the Brazilian regions, which can be observed in the branches of the tree. It presents satisfactory *CCC* (0.889), and eight clusters were suggested.

The first cluster, DF, a cluster with a single state, stands out isolated as it is the state with the highest average number of new cases per day (outlier) in just one state. The second group is composed of two states (MA and PA). The next cluster with the states AM, RR, AC, and AP has 14.81% representation, followed by the cluster with the highest number of states, six in all (22.22%) and from the same geographic region (RN, Sergipe (SE), Bahia (BA), Pernambuco (PE), Alagoas (AL) and CE). The fifth, sixth, and seventh clusters, with three states: Mato Grosso do Sul (MS), Mato Grosso (MT), and Rondônia (RO); TO, Paraíba (PB), and Piauí (PI); and Rio Grande do Sul (RS), Rio de Janeiro (RJ), and SP. The last cluster represents 18.53% of the states, with Espírito Santo (ES), Minas Gerais (MG), Paraná (PR), Goiás (GO), and Santa Catarina (SC).

Figure 6 presents the result of the elbow method for the database of elderly people aged 64 and over, suggesting the number of eight clusters.

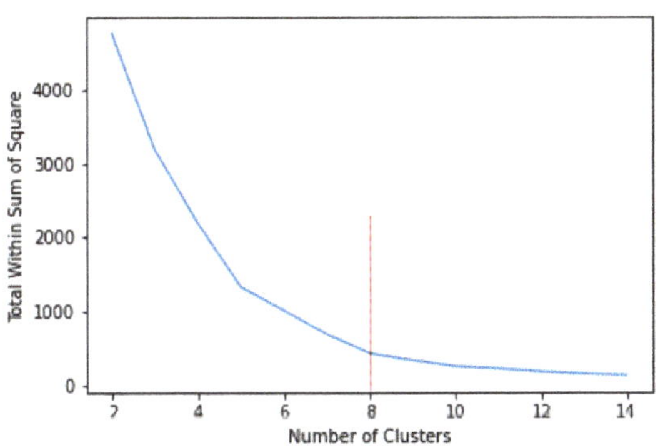

Figure 6. Curve of elbow method generated with data from the elderly 64 years or older. The number of the cluster was determined by looking at the point position on the "elbow" arm.

It can be seen that the second cluster analysis resulted in different groups from the previous analysis, as illustrated in Figure 5, which is better visualized on the Brazilian Map in Figure 7. More clusters were found with similar characteristics among themselves, and differences among the clusters.

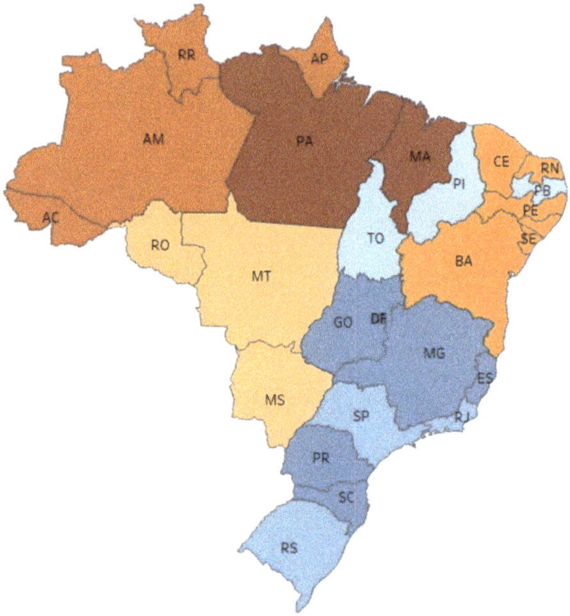

Figure 7. Map of Brazil generated from the result of the dendrogram with eight clusters (age group aged 64 and over). States well distributed in the clusters, showing heterogeneity in the behavior of the population.

This difference in divisions is very marked, in eight clusters tested with data from the elderly (13% of the population) to six clusters (all age groups) and with a different number of states per cluster, possibly due to factors such as:

- vaccination of the elderly was the first in the vaccination calendar (and continued throughout the year 2021) and had no results on the efficacy of vaccines, leading to mistrust [52];
- vaccination campaigns were starting without many disclosures and some calendars did not contain information on dates for each group and where to receive the doses. Each state had the autonomy to prepare the calendar and dissemination;
- vaccines were missing in many states and started applications through the capitals (sometimes not arriving in the interiors). According to [53], there were three errors that led to the lack of vaccines: the government did not anticipate and buy vaccines in 2020, there was a lack of definition on who should be vaccinated first, and a lack of training caused a waste of doses;
- vaccination infodemics were well-exposed and circulated on social media and the Internet, generating fear and vaccine hesitation. People began to discredit science, believe in fake news, and act against science [52].
- in addition, the elderly were identified as the most vulnerable in the dissemination of fake news [54], and they are seven times more likely to spread false news compared to people under 29 years [55]. This generated a pertinent concern because the presence of the elderly as internet users has been growing and this has been shown to be the largest proportional increase among the age groups [55].

The more infodemics shared, the greater the amount of fear and mistrust in the population. In this way, the patterns of behavior among the 27 federative units could be more heterogeneous.

5. Conclusions

The high demand for information corresponding to the growing popularity of COVID-19 vaccination research in news sources highlights the importance of public health officials working with the media to ensure that information is correct. This is because a high number of searches for vaccination infodemics can make it difficult for vaccination campaigns to be productive.

In this sense, the work used infodemic information on COVID-19 vaccination in Brazil, collected from the GT for the year 2021, with information on the number of vaccinated population and other important variables to perform a multivariate analysis of data and employ dendrograms to cluster the Brazilian states.

The use of the clustering technique, for the two analyses performed, resulted in six and eight clusters, respectively. Different results were found in the number of clusters and the aggregated states, composed of states with a high probability of having similar characteristics within each group and differences from the others. The results obtained with the *CCC* indicated a good degree of fit between the dendrograms and the dissimilarity matrices, allowing inferences to be made from the graphic representation. Finally, the UPGMA clustering algorithm was the most efficient, providing the lowest degrees of linkage and the highest *CCC* values.

In the analysis with vaccination data of the elderly aged 64 years or older, more heterogeneity in the patterns is visualized. During this period, the population had distrust and fear of vaccine efficacy, generating more sharing and infodemic research on vaccines. Vaccination in the states followed at different rates.

After analyzing the generated database, six clusters with more clustered states and more homogeneity was observed. The number of vaccinations increased in the age groups as clarifications were made about the importance and efficacy of the vaccine, leading to a significant decrease in both the number of cases and the number of deaths per day, and, of course, more vaccines were applied with improvements in the rate of evolution of pandemic numbers.

Future works should combine more information, expand data to two years of vaccination (2021 and 2022), and add other variables that express social and sociodemographic inequality, such as age and sex, which would provide more potential explanations for the behavior of the Brazilian population. It is intended to use the technique of Bayesian networks that would help policymakers and health managers understand which variables are most relevant.

Author Contributions: Conceptualization, M.d.P.H. and C.R.F.; Methodology, M.d.P.H. and C.R.F.; Software, M.d.P.H.; Validation, M.d.P.H. and M.S.; Formal analysis, M.d.P.H., M.S. and C.R.F.; Investigation, M.d.P.H. and C.R.F.; Data curation, M.d.P.H.; Writing—original draft, M.d.P.H.; Writing—review and editing, M.d.P.H., L.S., T.A., N.V. and C.R.F.; Visualization, L.S., T.A. and N.V.; Supervision, L.S., N.V. and C.R.F. All authors have read and agreed to the published version of the manuscript.

Funding: The authors would like to thank the Coordination for the Improvement of Higher Education (CAPES) and Dean of Research and Graduate Programs of the Federal University of Para (PROPESP/UFPA) for their program supporting qualified publishing (PAPQ)-funding number 02/2022, which facilitated the funding that enabled the payment of publication fees.

Institutional Review Board Statement: Not applicable.

Informed Consent Statement: Not applicable.

Data Availability Statement: Data is available on github, access link: https://github.com/mpenhaharb/InfodemicsVaccine (accessed on 25 July 2022).

Conflicts of Interest: The authors declare no conflict of interest.

Abbreviations

The following abbreviations are used in this manuscript:

AC	Acre
AL	Alagoas
AM	Amazonas
AP	Amapá
BA	Bahia
CCC	Cophenetic Correlation Coefficient
CE	Ceará
DF	Distrito Federal
ES	Espírito Santo
GO	Goiás
GT	Google Trends
MA	Maranhão
MG	Minas Gerais
MS	Mato Grosso do Sul
MT	Mato Grosso
PA	Pará
PB	Paraiba
PE	Pernambuco
PI	Piauí
PR	Paraná
RJ	Rio de Janeiro
RN	Rio Grande do Norte
RO	Rondônia
RR	Roraima
RS	Rio Grande do Sul
SC	Santa Catarina
SD	Standard Deviation
SE	Sergipe
SP	São Paulo
TO	Tocantins
UPGMA	Unweighted Pair Group Method with Arithmetic Mean
WHO	World Health Organization

References

1. Guo, Y.; Cao, Q.; Hong, Z.; Tan, Y.; Chen, S.; Jin, H.; Tan, K.; Wang, D.; Yan, Y. The origin, transmission and clinical therapies on coronavirus disease 2019 (COVID-19) outbreak—An update on the status. *Mil. Med. Res.* **2020**, *7*, 11. [CrossRef] [PubMed]
2. Lu, R.; Zhao, X.; Li, J.; Niu, P.; Yang, P.; Wu, H.; Wang, W.; Song, H.; Huang, B.; Zhu, N.; et al. Genomic characterisation and epidemiology of 2019 novel coronavirus: Implications for virus origins and receptor binding. *Lancet* **2020**, *395*, 565–574. [CrossRef]
3. World Health Organization. Coronavirus Disease (COVID-19-2019) Situation Reports—51. Available online: https://www.who.int/docs/default-source/coronaviruse/situation-reports/20200311-sitrep-51-covid-19.pdf?sfvrsn=1ba62e57_10 (accessed on 3 April 2022).
4. World Health Organization. Coronavirus Disease (COVID-19) Dashboard. Available online: https://covid19.who.int (accessed on 6 June 2022).
5. Kouzy, R.; Jaoude, J.A.; Kraitem, A.; Alam, M.B.E.; Karam, B.; Adib, E.; Zarka, J.; Traboulsi, C.; Akl, E.W.; Baddour, K. Coronavirus Goes Viral: Quantifying the COVID-19 Misinformation Epidemic on Twitter. *Cureus* **2020**, *12*, 3. [CrossRef] [PubMed]
6. Ghebreyesus, T.A. In Proceedings of the Munich Security Conference. Munich, Germany, 15 February 2022; Available online: https://www.who.int/director-general/speeches/detail/munich-security-conference (accessed on 15 March 2022).
7. Department of Global Communications. UN Tackles "Infodemic" of Misinformation and Cybercrime in COVID-19 Crisis. Available online: https://www.un.org/en/un-coronavirus-communications-team/un-tackling-%E2%80%98infodemic%E2%80%99-misinformation-and-cybercrime-COVID-19 (accessed on 15 March 2022).
8. Leem, M.; You, M. Direct and Indirect Associations of Media Use with COVID-19 Vaccine Hesitancy in South Korea: Cross-sectional Web-Based Survey. *J. Med. Internet Res.* **2022**, *24*, e32329. [CrossRef]
9. Batista, S.R.; de Souza, A.S.S.; Nogueira, J.; de Andrade, F.B.; Thumé, E.; Teixeira, D.S.d.C.; Lima-Costa, M.F.; Facchini, L.A.; Nunes, B.P. Comportamentos de proteção contra COVID-19 entre adultos e idosos brasileiros que vivem com multimorbidade: Iniciativa ELSI-COVID-19. *Cad. Saúde Pública* **2020**, *36*, e00196120. [CrossRef]

10. Organização Pan-Americana de Saúde. Dez Ameaças à Saúde Global em 2019. Available online: https://www.paho.org/pt/noticias/17-1-2019-dez-ameacas-saude-que-oms-combatera-em-2019 (accessed on 20 March 2022).
11. Sociedade Brasileira de Imunização. Especialistas se Reúnem para Debater o Fenômeno da Hesitação Vacinal no Brasil. Available online: https://sbim.org.br/noticias/1619-especialistas-se-reunem-para-debater-o-fenomeno-da-hesitacao-vacinal-no-brasil (accessed on 25 March 2022).
12. Google Noticias. Coronavírus (COVID_19). Available online: https://news.google.com/covid19/map?hl=pt-BR&gl=BR&ceid=BR%3Apt-419 (accessed on 5 June 2022).
13. Johns Hopking. Vaccination Progress across the World. Available online: https://coronavirus.jhu.edu/vaccines/international (accessed on 25 June 2022).
14. Nexo Jornal. Como Bolsonaro Atacou e Atrasou a Vacinação na Pandemia. Available online: https://www.nexojornal.com.br/expresso/2021/03/21/Como-Bolsonaro-atacou-e-atrasou-a-vacina%C3%A7%C3%A3o-na-pandemia (accessed on 10 June 2022).
15. Unicamp. Negacionismo na Pandemia: A Virulência da Ignorância. Available online: https://www.unicamp.br/unicamp/noticias/2021/04/14/negacionismo-na-pandemia-virulencia-da-ignorancia (accessed on 10 June 2022).
16. Wilson, K.; Brownstein, J.S. Early detection of disease outbreaks using the Internet. *Can. Méd. Assoc. J.* **2009**, *180*, 829–831. [CrossRef]
17. Arora, V.S.; Mckee, M.; Stuckler, D. Google Trends: Opportunities and limitations in health and health policy research. *Health Policy* **2019**, *123*, 338–341. [CrossRef]
18. Mangono, T.; Smittenaar, P.; Caplan, Y.; Huang, V.H.; Sutermaster, S.; Kemp, H.; Sgaier, S.H. Information-Seeking Patterns During the COVID-19 Pandemic Across the United States: Longitudinal Analysis of Google Trends Data. *J. Med. Internet Res.* **2021**, *23*, e22933. [CrossRef]
19. Rovetta, A.; Bhagavathula, A.S. COVID-19-Related Web search behaviors and infodemic attitudes in Italy: Infodemiological Study. *JMIR Public Health Surveill* **2020**, *6*, e19374. [CrossRef]
20. Ceron, W.; Sanseverino, G.G.; Santos, M.L.; Quiles, M.G. COVID-19 fake news diffusion across Latin America. *Soc. Netw. Anal. Min.* **2021**, *11*, 47. [CrossRef]
21. Custodio, M.; Peñaloza, R.; Alvarado, J.; Chanamé, F.; Maldonado, E. Surface Water Quality in the Mantaro River Watershed Assessed after the Cessation of Anthropogenic Activities Due to the COVID-19 Pandemic. *Pol. J. Environ. Stud.* **2021**, *30*, 3005–3018. [CrossRef]
22. Silva, R.M.; Mendes, C.F.; Manchein, C. Scrutinizing the heterogeneous spreading of COVID-19 outbreak in Brazilian territory. *Phys. Biol.* **2021**, *18*, 025002. [CrossRef]
23. Shafiq, A.; Çolak, A.B.; Sindhu, T.N.; Lone, S.A.; Alsubie, A.; Jarad, F. Comparative study of artificial neural network versus parametric method in COVID-19 data analysis. *Results Phys.* **2022**, *38*, 105613. [CrossRef]
24. Shafiq, A.; Sindhu, T.N.; Alotaibi, N. A novel extended model with versatile shaped failure rate: Statistical inference with F-19 applications. *Results Phys.* **2022**, *36*, 105398. [CrossRef]
25. James, N.; Menzies, M.; Bondell, H. Comparing the dynamics of COVID-19 infection and mortality in the United States, India, and Brazil. *Phys. D Nonlinear Phenom.* **2022**, *432*, 133158. [CrossRef]
26. James, N.; Menzies, M. COVID-19 in the United States: Trajectories and second surge behavior. *Chaos Interdiscip. J. Nonlinear Sci.* **2020**, *30*, 091102. [CrossRef]
27. Harb, M.A.; Silva, L.; Vijaykumar, N.L.; Silva, M.S.; Francês, C.R. An Analysis of the Deleterious Impact of the Infodemic during the COVID-19 Pandemic in Brazil: A Case Study Considering Possible Correlations with Socioeconomic Aspects of Brazilian Demography. *Int. J. Environ. Res. Public Health* **2022**, *19*, 3208. [CrossRef]
28. Braz, A.M.; Oliveira, I.J.; Cavalcanti, L.C.; Almeida, A.C.; Chávez, E.S. Cluster Analysis for Landscape Typology. *Mercator* **2020**, *19*, e19011. [CrossRef]
29. OpenDataSUS. Registros de Vacinação COVID19. Available online: https://opendatasus.saude.gov.br/dataset/covid-19-vacinacao/resource/5093679f-12c3-4d6b-b7bd-07694de54173 (accessed on 15 March 2022).
30. Painel Coronavírus. Dados COVID-19. Available online: https://covid.saude.gov.br/ (accessed on 16 March 2022).
31. Painéis de Dados ANATEL. Banda Larga Fixa. Available online: https://informacoes.anatel.gov.br/paineis/acessos/bandalarga-fixa (accessed on 15 March 2022).
32. Trends. Veja o Que o Mundo está pesquisando. Available online: https://trends.google.com.br/trends/?geo=BR (accessed on 22 March 2022).
33. Rovetta, A.; Bhagavathula, A.S. Global Infodemiology of COVID-19: Analysis of Google Web searches and instagram hashtags. *J. Med. Internet Res.* **2020**, *22*, e20673. [CrossRef]
34. Agência da Hora. Top 5 Fake News Mais Absurdas Sobre a Vacina. Available online: https://www.ufsm.br/midias/experimental/agencia-da-hora/2021/11/11/top-5-fake-news-mais-absurdas-sobre-a-vacina/ (accessed on 1 March 2022).
35. Diaz, L.C. The Lies That Are Told against Vaccines for COVID-19. Available online: https://www.revistaquestaodeciencia.com.br/artigo/2022/01/13/mentiras-que-se-contam-contra-vacinas-para-covid-19 (accessed on 4 March 2022).
36. Brasil de Fato. Você não vai se Transformar em Jacaré: 10 Mentiras Sobre Vacinas que Circulam por aí. Available online: https://www.brasildefato.com.br/2020/12/19/voce-nao-vai-se-transformar-em-jacare-10-mentiras-sobre-vacinas-que-circulam-por-ai (accessed on 2 March 2022).

37. Silva, R. De "Jacaré" a "Vacina do Doria": Relembre Frases de Bolsonaro Sobre Vacinação. Available online: https://www.agazeta.com.br/es/politica/de-jacare-a-vacina-do-doria-relembre-frases-de-bolsonaro-sobre-vacinacao-0121 (accessed on 5 April 2021).
38. Patel, V.R.; Mehta, R.G. Impact of Outlier Removal and Normalization Approach in Modified k-Means Clustering Algorithm. *Int. J. Comput. Sci. Issues* **2011**, *8*, 331. Available online: https://www.ijcsi.org/articles/Impact-of-outlier-removal-and-normalization-approach-in-modified-kmeans-clustering-algorithm.php (accessed on 7 April 2022).
39. De Barros Vilela, G., Jr. Estatística: Teste Z (ou Escore Padronizado). Available online: http://www.cpaqv.org/estatistica/teste_z.pdf (accessed on 1 May 2022).
40. Fávero, L.L.; Belfiore, P.P.; Silva, F.L.; Chan, B.L. *Análise de Dados: Modelagem MULTIVARIADA para Tomada de Decisões*; Elsevier: Amsterdam, The Netherlands, 2009; ISBN 8535230467.
41. Metz, J. Interpretação de Clusters Gerados por Algoritmos de Clustering Hierárquico. Master's Thesis, Instituto de Ciências Matemáticas e de Computação, Universidade de São Paulo (USP), São Carlos, Brazil, 2006. [CrossRef]
42. Machado, R.L. Desenvolvimento de um Algoritmo Imunológico para Agrupamento de Dados. Bachelor's Thesis, Universidade de Caxias do Sul, Caxias do Sul, Brazil, 2011. Available online: https://repositorio.ucs.br/handle/11338/1486 (accessed on 20 April 2022).
43. Berkhin, P. *Survey of Clustering Data Mining Techniques*; Accruel Software: San Jose, CA, USA, 2006; Available online: https://faculty.cc.gatech.edu/~{}isbell/classes/reading/papers/berkhin02survey.pdf (accessed on 1 May 2022).
44. Vicini, L. Análise Multivariada da Teoria à Prática. Available online: http://w3.ufsm.br/adriano/livro/Caderno%20dedatico%20multivariada%20-%20LIVRO%20FINAL%201.pdf (accessed on 20 April 2022).
45. Rohlf, F.J. Adaptative hierarquical clustering schemes. *Syst. Zool.* **1970**, *19*, 58–82. [CrossRef]
46. Kodinariya, T.M.; Makwana, P.R. Review on Determining Number of Cluster in K-Means Clustering. *Int. J. Adv. Res. Comput. Sci. Manag. Stud.* **2013**, *1*, 2321–7782.
47. Syakur, M.A.; Khotimah, B.K.; Rochman, E.M.; Satoto, B.D. Integration K-Means clustering method and elbow method for identification of the best customer profile cluster. *Conf. Ser. Mater. Sci. Eng.* **2018**, *336*, 012017. Available online: https://iopscience.iop.org/article/10.1088/1757-899X/336/1/012017 (accessed on 20 April 2022). [CrossRef]
48. IBGE Educa. Uso de Internet, Televisão e Celular no Brasil. Available online: https://educa.ibge.gov.br/jovens/materias-especiais/20787-uso-de-internet-televisao-e-celular-no-brasil.html (accessed on 10 April 2022).
49. Fundação Oswaldo Cruz. COVID-19: Balanço de dois anos da Pandemia Aponta Vacinação como Prioridade. Available online: https://portal.fiocruz.br/noticia/covid-19-balanco-de-dois-anos-da-pandemia-aponta-vacinacao-como-prioridade (accessed on 11 April 2022).
50. Souto, E.P.; Kabad, J. Hesitação vacinal e os desafios para enfrentamento da pandemia de COVID-19 em idosos no Brasil. *Rev. Bras. Geriatr. Gerontol.* **2020**, *23*, e210032. [CrossRef]
51. IBGE População. Projeção da População do Brasil e das Unidades da Federação. Available online: https://www.ibge.gov.br/apps/populacao/projecao/index.html (accessed on 6 June 2022).
52. Agência da Hora. Por que a Vacinação Contra COVID-19 no Brasil não Segue o Ritmo de Campanhas Anteriores? Available online: https://www.ufsm.br/midias/experimental/agencia-da-hora/2021/05/10/por-que-a-vacinacao-contra-covid-19-no-brasil-nao-segue-o-ritmo-de-campanhas-anteriores/ (accessed on 6 June 2022).
53. BBC News. 3 Erros que Levaram à Falta de Vacinas Contra COVID-19 no Brasil. Available online: https://www.bbc.com/portuguese/brasil-56160026 (accessed on 6 June 2022).
54. Estabel, L.B.; Luce, B.F.; Santini, L.A. Idosos, fake news e letramento informacional. *Rev. Bras. Bibliotecon. Doc.* **2020**, *16*, 1–15. Available online: https://rbbd.febab.org.br/rbbd/article/view/1348/1206 (accessed on 2 June 2022).
55. Yabrude, A.Z.; Souza, A.M.; Campos, C.W.; Bohn, L.; Tiboni, M. Desafios das Fake News com Idosos durante Infodemia sobre COVID-19: Experiência de Estudantes de Medicina. *Rev. Bras. Educ. Med.* **2020**, *44*, e0140. [CrossRef]

computation

MDPI

Article

COVID-19 Vaccines Related User's Response Categorization Using Machine Learning Techniques

Ahmed Shahzad [1], Bushra Zafar [1], Nouman Ali [2], Uzma Jamil [1], Abdulaziz Jarallah Alghadhban [3], Muhammad Assam [4], Nivin A. Ghamry [5] and Elsayed Tag Eldin [6,*]

1 Department of Computer Science, Government College University, Faisalabad 38000, Pakistan
2 Department of Software Engineering, Mirpur University of Science & Technology (MUST), Mirpur 10250, Pakistan
3 Department of Software Engineering, College of Computer Science and Information, King Suad University, Riyadh 11451, Saudi Arabia
4 Department of Software Engineering, University of Science and Technology, Bannu 28100, Pakistan
5 Faculty of Computers and Artificial Intelligence, Cairo University, Giza 3750010, Egypt
6 Faculty of Engineering and Technology, Future University in Egypt New Cairo, New Cairo 11835, Egypt
* Correspondence: elsayed.tageldin@fue.edu.eg

Abstract: Respiratory viruses known as coronaviruses infect people and cause death. The multiple crown-like spikes on the virus's surface give them the name "corona". The pandemic has resulted in a global health crisis and it is expected that every year we will have to fight against different COVID-19 variants. In this critical situation, the existence of COVID-19 vaccinations provides hope for mankind. Despite severe vaccination campaigns and recommendations from health experts and the government, people have perceptions regarding vaccination risks and share their views and experiences on social media platforms. Social attitudes to these types of vaccinations are influenced by their positive and negative effects. The analysis of such opinions can help to determine social trends and formulate policies to increase vaccination acceptance. This study presents a methodology for sentiment analysis of the global perceptions and perspectives related to COVID-19 vaccinations. The research is performed on five vaccinations that include Sinopharm, Pfizer, Moderna, AstraZeneca, and Sinovac on the Twitter platform extracted using Twitter crawling. To effectively perform this research, tweets datasets are categorized into three groups, i.e., positive, negative and natural. For sentiment classification, different machine learning classifiers are used such as Random Forest (RF), Naive Bayes (NB), Decision Tree (DT), Logistic Regression (LR), and Support Vector Machine (SVM). It should be noted that the Decision tree classifier achieves the highest classification performance in all datasets as compared to the other machine learning algorithms. For COVID-19 Vaccine Tweets with Sentiment Annotation (CVSA), the highest accuracy obtained is 93.0%, for the AstraZeneca vaccine dataset 90.94%, for the Pfizer vaccine dataset 91.07%, 88.01% accuracy for the Moderna vaccine dataset, for the Sinovac vaccine dataset 92.8% accuracy, and 93.87% accuracy for the Sinopharm vaccine dataset, respectively. The quantitative comparisons demonstrate that the proposed research achieves better accuracy as compared to state-of-the-art research.

Keywords: COVID-19; vaccines; Twitter; sentiment analysis; classification; machine learning

Citation: Shahzad, A.; Zafar, B.; Ali, N.; Jamil, U.; Alghadhban, A.J.; Assam, M.; Ghamry, N.A.; Eldin, E.T. COVID-19 Vaccines Related User's Response Categorization Using Machine Learning Techniques. *Computation* **2022**, *10*, 141. https://doi.org/10.3390/computation10080141

Academic Editors: Simone Brogi and Vincenzo Calderone

Received: 2 July 2022
Accepted: 12 August 2022
Published: 18 August 2022

1. Introduction

Machine learning and deep learning models are used in various real-time domains such as industrial automation, design of design support systems for medical domains and multimedia analysis [1–5]. Pandemics occur and lead to extensive morbidity and mortality worldwide. In December of 2019, a case of pneumonia of unknown origin was reported in Wuhan, China. From there, the epidemic of the coronavirus swiftly spread to other countries [6–10], leading to the widespread outbreak of COVID-19 on the mainland. The severe acute respiratory syndrome coronavirus is causing a pandemic of coronavirus disease 2019 (COVID-19) all over the globe, and China is one of the countries affected (SARS-COV-2). China was the first country to have an outbreak of the disease. It was also the first country to respond with harsh measures, such as lockdowns and rules about wearing face masks. China was also one of the first countries to get the outbreak under control. The coronavirus (COVID-19) viruses have made their way to many parts of the world. This virus has a high rate of spread and is harmful to humans [11].

Italy was the first European country to experience a significant COVID-19 outbreak, with the detection of the first case on the 21 February 2020 in the province of Lodi in the region of Lombardy. While each province in Italy had confirmed cases of the virus by mid-March 2020, the diffusion of the outbreak in the country was very heterogeneous. The majority of cases were concentrated in Lombardy in the north of the country [12,13].

The World Health Organization (WHO) called the COVID-19 outbreak the sixth public health emergency of international concern (PHEIC) on 30 January 2020. On 11 March 2020, the WHO said that COVID-19 had become a pandemic [14]. This year's new coronavirus killed 85,522 people on 9 April 2020, and the case fatality rate (CFR) was 5.95%. COVID-19 has been classified by the WHO as having a very high global risk. Because lockdowns have been implemented in so many areas, the pandemic scenario has impacted virtually every aspect of society, including the economy [15,16]. Coronavirus disease (COVID-19) is a pandemic and an issue that exists in more than 200 nations throughout the world. Many countries have been badly affected by COVID-19 and lots of people have died in the last two years [16]. The high volume of international travel was the primary factor in the disease's dissemination around the globe; the presence of local contagious links played a secondary role. For example, in 2018, more than 4 billion individuals, or almost six out of every ten persons on the planet, traveled worldwide by means of commercial airplanes [17].

In response to the unusual spread of the illness, there have been concerted attempts on a worldwide scale to collaborate on combating the pandemic. The creation of a vaccine is one of the potential strategies that may be used to combat the COVID-19 pandemic. A chemical that stimulates the development of adaptive immunity in the body and hence assists in the body's fight against various illnesses and diseases is known as a vaccine [17–19]. Many organizations have developed vaccines to avoid and overcome this situation. People have to vaccinate themselves to reduce the threat of this malignant disease [20]. For this, they need some opinion about different types of vaccines available in the market to select the most suitable vaccine for themselves. Social media platforms such as Twitter have proved to be a valuable resource that provides instantaneous access for information tracking and evaluation. In pandemic times, Twitter has been used in various studies as a source of information, e.g., back in 2009 during the HINI outbreak [21]. Twitter has been widely used in various studies for the identification of user's concerns, misinformation spread and sentiment analysis [22]. Twitter users have expressed their opinions regarding COVID-19 vaccination. Only a few research studies have analyzed public sentiments towards COVID-19 vaccination. This research will help them to select their desired vaccines from Sinovac, Pfizer, Moderna, AstraZeneca and Sinopharm. To the best of our knowledge, in previous studies, researchers have tested two to three vaccinations and found accuracy. This research collects and analyzes opinions on five major vaccinations and identifies the most effective machine learning (ML) algorithm to predict the sentiment analysis about five different types of COVID-19 vaccines. In addition to this, the research aims to analyze the sentiments of people towards COVID-19 vaccination on the basis of data obtained from social media. The proposed research will address the following research questions:

- What are people's sentiments toward COVID-19 vaccination on the social media Twitter platform?
- What is the most effective machine learning algorithm to predict the sentiment analysis about five different types of COVID-19 vaccines?

Supervised intelligence enables complex and larger data to be processed and analyzed along with the desired results being achieved. Machine learning offers a novel approach to bringing together the methodologies of fundamental research and technical analysis.

We aim to find better results for the sentiment classification of COVID-19 vaccination by applying ML models. The key contributions of this research are as follows:

- This research presents a methodology for sentiment analysis of the perceptions and perspectives of public tweets related to COVID-19 vaccination. In this regard, a global dataset has been created by extracting tweets related to people's sentiments towards COVID-19 vaccination.
- The TextBlob approach has been applied to determine the polarity of sentiments into positive, negative and neutral. Different supervised machine learning models were applied to the annotated dataset in order to obtain optimal performance.
- In related state-of-the-art research, the researchers have tested two to three vaccinations for sentiment classification. This research will collect opinions on five vaccinations including Sinovac, Pfizer, Moderna, AstraZeneca and Sinopharm and aims to discover which vaccine produces the best results. The proposed research is validated by comparing the performance with the state-of-the-art approaches.

The rest of the article is organized into five sections. Section 2 presents a comprehensive review of the related work. The proposed methodology is discussed in Section 3. Section 4 provides a description of the datasets used for experiments, the metrics used for evaluation and a discussion of the results. Section 5 concludes the research and provides directions for future research.

2. Related Work

This section presents a review of the recent literature on the COVID-19 pandemic which emphasizes the importance of effective vaccination for the whole population.

Machine learning and neural networks have applications in difference domains such as aerial image classification [23–26], face recognition [27], Internet of Things [28,29], healthcare [30–32] and sentiment analysis, etc. Manguri et al. [33] stated that the rise of social data on the internet has accelerated. This leads to study in order to obtain access to the data and information for a variety of academic and commercial purposes. The global COVID-19 sickness has now expanded internationally, and social data on the web includes numerous real-life incidents that happened in everyday life. Many people, including media outlets and government institutions, are disseminating the newest information and viewpoints on the coronavirus. The Twitter data was crawled from Twitter social media through a python programming language, and sentiment analysis was performed using the text blob library in python. The evaluation results of sentiment analysis are shown as a graphical representation based on the data. The information originated from Twitter, where it was discovered via the use of a search for two distinct hashtag keywords: (COVID-19 and coronavirus). In another study [34], the authors argued that a global infrastructure to enable both normal and pandemic/epidemic adult vaccination is urgently needed because of the global connections. Since the number of older persons is continually increasing, the need for a framework to propose vaccinations and establish strong platforms to distribute them was obvious. For older individuals, their families, communities, and nations, adult vaccination as a policy has the potential to protect and improve medical, social, and economic results. COVID-19 vaccinations will soon be available, but it is important to remember that currently, a number of vaccines are available that can keep adults healthy.

Meena et al. [35] pointed out that social media talks about healthcare were an excellent starting point for assessing people's feelings. COVID-19 vaccination was the primary hope of practically every human being on Earth. Many people took to Twitter to express their feelings in response to Russia's first vaccination announcement. Data from tweets were analyzed for the emotions and psychology of the people and the issue of interest they were discussing. The social emotions were disclosed and displayed using computational approaches and algorithms, such as machine-learned and LDA. Sentiment analysis is a technique for recognizing and categorizing views or feelings represented in the source material. A vast amount of data that is rich in sentiment is generated by various types of social media, such as tweets, status updates, blog posts, and so on. The application of sentiment analysis to this user-generated data may be highly helpful in identifying the perspective of the general population. Because of the existence of slang phrases and misspellings, Twitter sentiment analysis is more complex than conventional sentiment analysis. On Twitter, the maximum number of characters permitted is 140. According to authors, there are two methodologies that are employed for interpreting the sentiment gleaned from the text. These are the knowledge-based approach. Alliheibi [36], mentioned that individuals in Saudi Arabia who had received the COVID-19 vaccination were studied via their tweets. People's replies were classified using computational lexical-semantic approaches. The findings show that the majority of Saudi Arabians have an unfavorable view of the government's COVID-19 immunization take-up campaign. According to the findings, the use of data mining applications in government institutions and departments can identify trends that could have an adverse impact on policies and practices, as well as help government institutions make appropriate decisions and adopt reliable and workable policies and procedures.

Yousefinaghani et al. [37] pointed out that COVID-19 vaccinations are the subject of an estimated 4.5 million tweets being analyzed in their investigation. It is possible that Twitter, as it was in this study, may be an effective tool for promoting public health by increasing vaccination uptake and decreasing vaccine resistance. Public health officials might benefit from better knowing vaccine feelings and opinions in order to amplify good postings with supportive language and debunk negative ones with confrontational language that spreads misinformation. Public health organizations may also be able to use Twitter and other media to raise positive messaging and actively minimize negative and opposing messages.

Ezhilan et al. [38] performed a study using a convolutional neural network and a recurrent neural network built for sentiment analysis based on text data related to Twitter data sentiment analysis. CNN and RNN sentiment classifiers performed better than other sentiment classifiers, such as SVM, logistic regression, and Nave Bayes, in terms of accuracy and recall, according to the empirical assessment in this study. Also shown in the study was the performance of general-purpose emotion analyzers such as text blob and Vader. Understanding public opinions regarding coronavirus and COVID-19 helps to detect the rise in dread sentiment and unpleasant feelings, which were important for developing much-needed remedies to stop the rapid spread of the pandemic. The use of exploratory and descriptive text analytics and data visualization methodologies helps to uncover the most basic of ideas. Andrzejczak-Grzadko et al. [39] observed that the Vaccine side effects are widespread, although individuals respond to immunizations in various ways. Manufacturers give a list of their goods' adverse effects. Adverse responses indicate that immunizations are effective and that the immune system is reacting. It compares the AstraZeneca and Pfizer vaccines' side effects. These responses were more prevalent after the first dosage of the AstraZeneca vaccination than after the first and second doses of the Pfizer vaccine, although they were less common after the Pfizer formulation. The survey was made available on the internet. It was performed on patients who had been immunized with Pfizer or AstraZeneca vaccines. The participants were questioned about adverse effects such as injection site discomfort, arm pain, muscle pain, headache, fever, chills, and exhaustion after receiving the first and second doses of the vaccinations. A total of 705 persons responded to the survey. Pfizer had vaccinated 196 of them, whereas AstraZeneca had immunized 509. A total of 96.5% of those who received the first dose of the AstraZeneca vaccine had at least one post-vaccination response. All of the adverse effects mentioned in the survey were reported by 17.1% of respondents. Vaccine responses were recorded by 93.9% of those who received the first Pfizer dosage, while just 2% of those who received the second dose suffered all of the adverse events listed in the survey. Most of the subjects had post-vaccinal reactions after the second dose of the Pfizer vaccine: 54.8% had more adverse reactions, and 15.8% had fewer adverse reactions than after the first dose, and 29.4% had the same side effects after the first and second doses of the Pfizer vaccine.

Saeed et al. [40] stated that some people were reluctant to get their children vaccinated because they were afraid of the unknown. The first and second post-vaccination side effects of the Sinopharm COVID-19 vaccine were shown to be common and moderate, predictable, non-serious, and not life-threatening. For the first time, the Sinopharm vaccine's adverse effects have been evaluated among an age group, and the findings might help lessen public vaccination skepticism. Dubey [41] performed a study to explain. In India, the campaign to prevent COVID-19 began on 16 January 2021. Oxford-Covishield AstraZeneca's and Bharat Biotech's Covaxin were two vaccines employed in this campaign. This initiative has already surpassed 600,000 people in its first four days, and the government has declared that it would be increased in the following days to secure residents' immunity. However, there is still a segment of the population that is skeptical about the COVID-19 vaccine. It was carried out to examine the emotions expressed in India's tweets about these two vaccinations. While the majority of the public has favorable feelings about these vaccinations, the study indicates that there are also negative feelings about them, which are linked to emotions such as fear and wrath. Dumre et al. [42] performed a statistical and sentiment analysis and observed that people in India have begun developing opinions towards them as a result of the impending availability of a vaccine against COVID-19. An investigation of the attitudes and viewpoints of individuals with respect to vaccinations. Out of 200 participants, 32 doctors and 35 participants were vaccinated. The main objectives were to analyze the response to the survey and draw conclusions with the help of data analysis techniques and performed sentiment analysis on participants' responses to identify what stops people from getting vaccinated.

Cotfas et al. [43] described that machine learning-based posture detection was used to analyze the one-month time between the initial announcement of a coronavirus vaccine and the first real immunization procedure outside of the limited clinical trials. The best classifier was selected after a thorough evaluation of the performance of a number of different conventional and deep learning methods. The suggested method was able to classify the tweets into three primary categories, namely in favor, against, and neutral, with an accuracy of 78.94%. The authors in [44] analyzed that the tweets were categorized into four different emotions based on their content: fear, sadness, rage, and joy. A pleasant environment was produced in the healthcare authorities by using phrases such as "thank you", "well", and "good" instead of terms that instill dread in the minds of those who hear them. In light of these findings, local governments have been pushed to impose fact-checkers on social media to combat misleading propaganda. There has been a lack of research on how to verify and categorize tweets, which has led to a rise in the spread of false information. As a result, the authors used Bert, a unique deep-learning model, to obtain better classification accuracy in comparison to standard models of ML. Bert's 89% accuracy outperformed other models including LR, SVM, and LSTM, according to the results. The research results helped to clarify public opinion on pandemics and provided a guideline to medical authorities, public, and private sector employees to overcome unnecessary concern during pandemics.

3. Research Methodology

This research presents a framework for sentiment analysis of COVID-19 vaccines. We have used python as a programming language and several libraries for text mining that will be explained. Figure 1 demonstrates the steps of the proposed methodology framework for sentiment analysis organized in four multiple layers. In the first step, data crawling and pre-processing are performed. The second layer is the learning layer where the pre-processing data will be spilt into training (70% data) and test (30% data) subsets. The training test ratio is chosen in accordance with state-of-the-art research. Empirical studies show that the best results are obtained if we use 20–30% of the data for testing, and the remaining 70–80% of the data for training [17,45]. The training will be used to train five different ML models namely Random Forest (RF), Naive Bayes (NB), Decision Tree (DT), Logistic Regression (LR) and Support Vector Machine (SVM). Again, the run-time behavior of five trained models using model-based testing techniques will be used to check the model's predictions. The third layer is the evaluation layer, the performance of models will be compared on the basis of the evaluation metrics. The task of sentiment classification can usually be seen as a two-class classification (positive and negative). In this research, we add one class namely Neutral, to get the Twitter sentiment. This type of work is essentially a matter of text classification. The fourth layer is the result layer, it presents an analysis and discussion of the results.

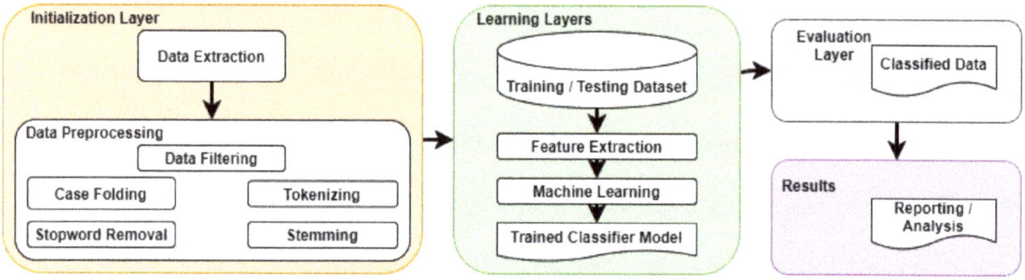

Figure 1. Block diagram of the Proposed Research.

3.1. Data Collection

This study performed sentiment analysis for COVID-19 vaccination. For this purpose, the dataset contains tweets related to the COVID-19 vaccination. To extract the tweets from Twitter with specific keywords, such as COVID-19 vaccine, corona Vaccine, COVID-19 vaccination, and corona vaccination. For this research, we have extracted tweets from tweeter using developer account access keys and Python popular library tweepy. We have searched Twitter using keyword search. Hence, we grabbed about 25,004 tweets containing our search keywords. Then pandas library was used to store the tweets in a data frame and then in a CSV file for further manipulation. In this study, AstraZeneca, Pfizer, Sinovac and Sinopharm employed 5001, 5001, 5001, 5001 and 5000 tweets, respectively.

3.2. Data Pre-Processing

Data pre-processing is applied to pre-process the text when building an ML-based system based on tweet data. Text pre-processing includes the following steps: The text shown before and after applying some of the pre-processing steps is shown in Table 1.

- Case folding is the removal of the case-sensitive text by changing the text to upper or lower case. In this study, the lowercase text was applied.
- Dataset filtering/document filtering removes special characters, mentions, links, URLs, hashtags, single characters, non-ASCII characters, punctuation, number and whitespace.
- Tokenizing means splitting the text into words. The list of tokens is used for further processing.
- Stop Word removal indicates that any words that are considered to be irrelevant or possibly irrelevant are removed.
- Stemming means converting words into prevailing words.

Table 1. Data Pre-processing.

Process	Data
Original Tweet	"China to launch roadmap to 'live with the virus' as two new local vaccines using technologies similar to those of the Novavax and Oxford-AstraZeneca vaccines, are now available in the country, according to Chinese officials and medical experts.
Case Folding	"China to launch roadmap to 'live with the virus' as two new local vaccines using technologies similar to those of the novavax and oxford-astraZeneca vaccines, are now available in the country, according to chinese officials and medical experts.
Document Filtering	china 'to launch roadmap to live with the virus as two new local vaccines using technologies similar to those of the novavax and oxford-astraZeneca vaccines are now available in the country according to chinese officials and medical experts
Tokenizing	'china', 'to', 'launch', 'roadmap', 'to', 'live', 'with' 'the', 'virus' , 'as', 'two', 'new', 'local', 'vaccines' ,'using', 'technologies', 'similar', 'to', 'those', 'of ', 'the novavax' ,'and', 'oxford-astraZeneca', 'vaccines', 'are', 'now', 'available', 'in', 'the' 'country', 'according', 'to', 'chinese', 'officials', 'and', 'medical', 'expert's'.
Stopword Removal	'china', "launch', 'roadmap', 'live', 'virus' , 'local', 'vaccines' ,'using', 'technologies', 'similar', 'the novavax', 'oxford-astraZeneca', 'vaccines', 'available', 'country', 'according', 'chinese', 'officials', 'medical', 'expert's'.

3.3. *Feature Selection/Extraction*

After, the pre-processing stage, the data was processed to select the feature sets. For feature selection, TextBlob, a well-known lexicon-based approach for performing natural language processing (NLP) tasks on the raw text was used. TextBlob is a python package that allows you to manipulate text input using a programming interface. By using TextBlob, one can analyze sentiments in text, create part of speech (pos) tags, extract noun phrases, translate, classify and more. TextBlob package comes with different in-built functions that support the task of language processing. It works for many different languages such as Arabic, Spanish, English, etc. It works in conjunction with NLTK [11].

3.4. *Machine Learning Algorithms*

Machine learning (ML) is a popular use of artificial intelligence since it automates the system and allows it to learn and improve from diverse experiences without being programmed. Computer programs can teach how to learn by giving them access to data and allowing them to utilize it for learning in ML. The learning process in ML begins with seeing the data through examples or instructions that humans offer; these observations enable ML to look for patterns in order to make the best predictions. Five different ML models were used to train the classifier and evaluate classification performance using the test dataset. These are discussed below.

3.4.1. Random Forest

The RF model is an ensemble model that generates high-precision predictions by combining the results obtained from several sub-trees.The supervised ML method known as RF may be used for both classification and regression analysis. The term "forest" refers to a collection of independent Decision Trees that are combined in order to reduce the amount of variance and provide more accurate data forecasts. L. Breiman [46,47] created the random forest algorithm in 2001, and it has been shown to be a very effective tool for classification and regression analysis across a variety of domains. The approach, which combines the predictions from a number of different randomized Decision Trees and then takes the average of those forecasts, has been shown to work well in circumstances in which the number of variables is more than the number of observations. In addition to this, it can be adapted to a wide range of ad hoc learning challenges and it may provide metrics of changing importance, both of which make it suited for use with large-scale problems [47]. An RF can be represented as:

$$RF = mode\{tR_1, tR_2, tR_3, \cdots, tR_n\} \quad (1)$$

$$RF = mode\{\sum_{i=1}^{n} tR_i\} \quad (2)$$

where $tR_1, tR_2, tR_3, \cdots, tR_n$ represent the Decision Trees in RF and n denotes the number of trees.

3.4.2. Naive Bayes

The Bayes Theorem's premise of class conditional independence is used in the NB classification technique. This indicates that the existence of one characteristic in the likelihood of a certain event has no bearing on the presence of another, and each predictor has an equal impact on the outcome. Multinomial NB, Bernoulli NB, and Gaussian NB are the three kinds of NB classifiers. Text categorization, spam detection, and recommendation systems are all applications of this technology. Classifiers are programs that give a class to an object or case based on the values of attributes used to characterize this item or case from a pre-defined list. To do so, NB classifiers employ a probabilistic method, in which they attempt to predict the outcome [48].

3.4.3. Decision Tree

DTs are a technique for non-parametric supervised learning that may be used for classification and regression. DT is a model for ML that may be used for the problem-solving process of regression as well as classification. The purpose of this project is to build a model that can accurately forecast the value of a target variable by gleaning fundamental decision rules from the features of the data. A tree may be thought of as a piecewise constant's approximation [49]. Until the splits become atomic, the model employs the binary technique to split the dataset into n number of subsets. When a data subset cannot be further split, it is said to be atomic. A DT with multiple branches of varying sizes is used in conjunction with partitioning the dataset into an incremental method of construction. The DT was employed in this investigation with a max depth hyper-parameter to minimize complexity and overcome model over-fitting [17].

3.4.4. Logistic Regression

Logistic Regression is a statistical approach to data analysis in which one or more variables are utilized to determine the outcome. When the target variable is categorical, the optimum learning model to utilize is LR, which is the regression model that was used to estimate the likelihood of class members. Linear Regression uses a logistic function to estimate probabilities for the association between the categorical dependent variable and one or more independent variables [50]. Logical regression is used whenever the dependent variables are categorical, such as "true" and "false" or "yes" and "no", rather than continuous, as in the case of Linear Regression, which is employed if the dependent variables are continuous. Although both regression models seek to identify correlations between data inputs, logistic regression is often used when dealing with binary classification challenges such as spam detection since it is more effective at handling these problems. Logistic Regression is a technique that may be used to solve a classification issue. It generates a binomial outcome by stating, in terms of 0 and 1, the probability of an event happening or not occurring, taking into the process.

The prediction of whether a tumor is malignant or benign, for example, or if an Email is spam or not, are both instances of the binomial results that may be obtained by Logistic Regression. There can also be a multinomial result of Logistic Regression, such as predicting the favorite cuisine: Chinese, Italian, Mexican and others. There can also be ordinal outcomes, such as product ratings ranging from 1 to 5, and so on. As a result, Logistic Regression is concerned with the categorical prediction of the target variable. Whereas Linear Regression, on the other hand, is concerned with the prediction of values of continuous variables, such as real estate prices over a three-year period [50].

$$g(x) = \frac{L}{1 + e^{-k(v - v_o)}} \tag{3}$$

The values for the S-shaped curve and the variable v of the LR ranges from $-\infty$ to $+\infty$ for actual numbers. To boost the performance of LR, the hyperparameter "liblinear" was utilized in this study. The hyperparameter 'multi-class' is set as 'multinomial' considering its effectiveness for binary classification problems.

3.4.5. Support Vector Machine

A support vector machine(SVM), which was created by Vladimir Vapnik, is a supervised learning model that can be used to both classify and regress data [51]. On the other hand, the most popular use for it is in the realm of classification problems; in this context, it is used to generate a hyperplane on which the distance between two classes of data points is maximized. The decision boundary is a hyperplane that divides the different categories of data points that are located on each side of the plane (e.g., oranges vs. apples) [51]. SVMs are capable of dealing with problems relating to both classification and regression. This method requires that the hyperplane, which acts as the decision boundary, be defined. A decision plane is necessary whenever there is a need to divide a set of things that belong to different categories. The items may or may not be separated linearly [51].

3.5. Label Prediction

All datasets' tweet data was labeled. The model was chosen in the previous stage was then used to predict the label.

4. Results and Discussion

This section presents the accuracy results of sentiment analysis carried out using five distinct methods applied to two distinct datasets, with the second dataset being further subdivided into five distinct vaccination datasets. The accuracy, precision, recall, F1 score, and support measurement are derived from the Random Forest, Naive Bayes, Decision Tree, Logistic Regression, and Support Vector Machine (SVM).

4.1. Description of Datasets

We have used two datasets for this research, Dataset 1 [52]: COVID-19 Vaccine Tweets with Sentiment Annotation (CVSA) and Dataset 2: COVID-19 vaccines related user's response crawled from Twitter platform to analyze the opinions about vaccines. Dataset 2 is further divided according to five known vaccine datasets, i.e., AstraZeneca, Pfizer, Sinovac, Moderna and Sinopharm, respectively. CVSA has 6000 rows and 3 columns (Tweets id, label, Tweets text). The Sinovac, Pfizer, Moderna, AstraZeneca and the Sinopharm datasets have 5001 rows and 5 columns, respectively (Srno, Datetime, Tweet Id, Text, Username).

4.2. Evaluation Metrics

This section explores the evaluation metrics utilized used for the quantitative evaluation of the proposed research. The metrics used for evaluation of the proposed research are:

I. Confusion matrix: The confusion matrix is often used in ML to analyze or show how models behave in supervised classification contexts [53,54]. It is a square matrix with rows representing the actual class of the examples and columns representing their anticipated class. The confusion matrix defined a comparison between actual and predicted values. The confusion matrix is an N $\times$ N matrix, where N is the number of classes or outputs. For two classes, we obtain a 2 $\times$ 2 matrix. Whereas for three classes or outputs, we obtain a 3 $\times$ 3 confusion matrix. The rows indicate the actual class of the instances. The confusion matrix has four terms to understand: True Positive (Tp), False Positive (Fp), True Negative (Tn), and False Negative (Fn). The datasets used in this research have three outputs or classes, Positive, Neutral, and Negative. In the multi-class classification problem, we won't get Tp, Tn, Fp, and Fn values directly as in the binary classification problem. We need to calculate for each class.

This matrix includes all of the raw information that was created by a classification model when it was applied to a specific data set. This information pertains to the predictions that were produced. It is standard practice to use a testing data set that was not used during the learning phase of a model in order to assess the correctness of the model's ability to generalize its findings. This is performed to see if the model was able to generalize its findings. A confusion matrix may provide the basis for the creation of a great number of artificial, one-dimensional performance metrics. Precision, recall and the F-score, etc. are the performance indicators that can be computed from the confusion matrix. In association with a 2 × 2 cost matrix, a confusion matrix can also be used to compute cost-sensitive performance indicators in cases when different types of errors are not assumed to be equal. The selection of the optimal performance indicator directly relates to the objectives of the learning problem. The confusion matrix is shown in Table 2.

Table 2. Confusion matrix.

		Predicted Case	
		Negative	**Positive**
Actual Case	Negative	Tn = True Negative correct prediction of the negative case	Fp = False Positive incorrect prediction of the positive case
	Positive	Fn = False Negative incorrect prediction of the negative case	Tp = True Positive correct prediction of the positive case

II. Recognition Accuracy (ACC): The classification accuracy (ACC) is the most generally used statistic for evaluating classification performance. It is defined as the total number of instances (TWEETS) correctly classified divided by the number of examples (TWEETS) in the dataset under consideration. It can be stated numerically as:

$$Accuracy = ACC = \frac{Tp + Tn}{Tp + Tn + Fn + Fp} \tag{4}$$

III. Recall: Recall is also used for performance measurement. Recall can be defined as the ratio between tweets classified correctly to the total number of tweets available in the database. Recall in the formula form can be written as:

$$Recall = Sensitivity = \frac{Tp}{Tp + Fn} \tag{5}$$

IV. Precision: It is also known as positive predictive value (PPV), precision is widely used for performance measurement purposes. Precision can be defined as the ratio between tweets classified correctly to the total number of tweets classified. Precision in the formula form can be written as:

$$Precision = \frac{Tp}{Tp + Fp} \tag{6}$$

V. F-measure/F1-Score: The F-score is the harmonic mean of recall and accuracy; a higher value implies better predicting ability. System performance cannot be assessed just on the basis of accuracy or recall. The following formula may be used to determine the F-score:

$$F\text{-}score = \frac{2 \times (Precision \times Recall)}{Precision + Recall} \quad (7)$$

4.3. Results for Dataset 1: COVID-19 Vaccine Tweets with Sentiment Annotation

This section presents a detailed analysis of the results obtained for dataset 1 by applying five ML algorithms. The results are demonstrated using a confusion matrix and bar graphs of different ML classifiers. Figure 2 shows the confusion matrices obtained by applying different ML algorithms. As discussed earlier, Random Forest is an approach to supervised ML that may be flexible and is used for both classification and regression analysis. It can be observed that when the Random Forest algorithm is applied to the dataset, 81.94% accuracy is obtained. Confusion matrix Figure 2a shows the results of precision, recall, F1-score and accuracy obtained by applying the Random Forest algorithm to the dataset. These values are calculated by using Tp, Tn, Fp and Fn parameters. The precision, recall, F1-score and accuracy achieved by applying the random forest ML model are 89.94%, 67.76%, 69.9% and 81.94%, respectively, and are shown in Table 3.

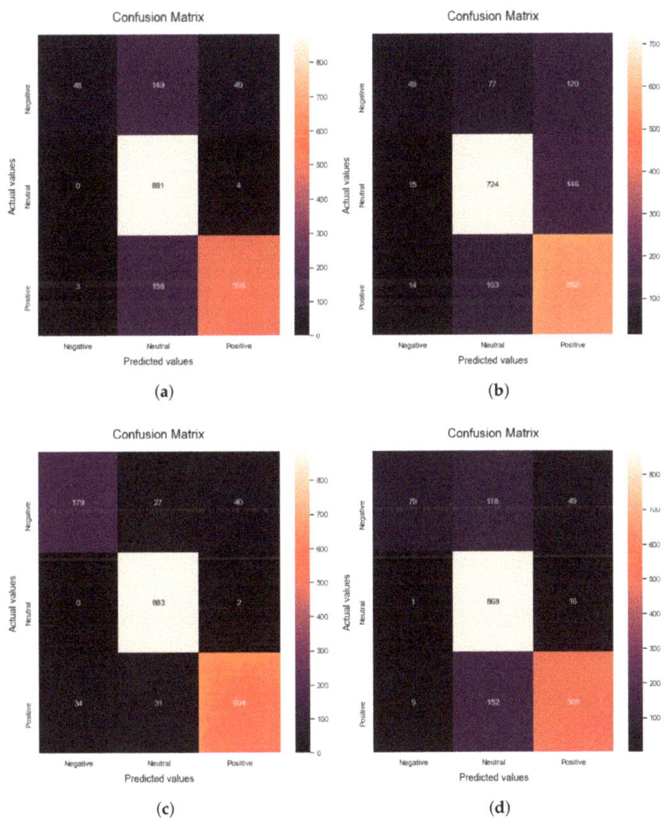

Figure 2. *Cont.*

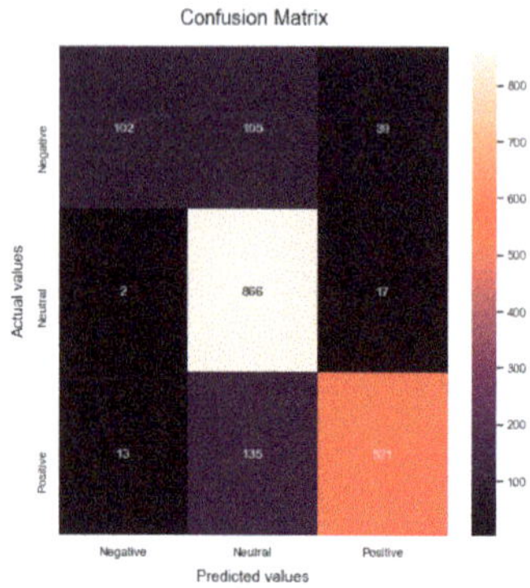

(**e**)

Figure 2. Confusion matrix: (**a**) using Random Forest (**b**) Naive Bayes (**c**) Decision Tree (**d**) Logistic Regression (**e**) SVM.

Table 3. Machine learning Performance on COVID-19 vaccine tweets with sentiment annotation.

Classifier Name	Accuracy%	Precision%	Recall%	F1-Score%
Random Forest	81.94	89.18	67.76	69.9
Naive Bayes	75.67	71.55	63.19	63.2
Decision Tree	93.0	90.43	88.27	89.24
Logistic Regression	82.5	85.35	71.36	74.47
SVM	84.78	87.0	75.05	78.31

The second algorithm used for the evaluation of the proposed research is Naive Bayes. It is a method of classification that is based on the Bayesian concept of conditional independence of class membership. This indicates that the existence of one characteristic does not have an influence on the likelihood of another characteristic being present in a given outcome and that each predictor has an equal impact on the given outcome. This technique is primarily used in text classification, spam identification, and recommendation systems. It can be observed that when the Naive Bayes algorithm is run on the dataset, 75.67% accuracy is obtained. Confusion matrix Figure 2b shows the confusion matrix obtained by applying the Naive Bayes algorithm to the data set. Experimental results demonstrate that the precision, recall, F1-score, and accuracy scores using the NB algorithm are 71.55%, 63.19%, 63.2% and 75.67%, respectively, as shown in Table 3.

The third classifier applied for the evaluation of the proposed research is the Decision Tree. It is a kind of supervised learning that does not rely on parameters and may be used for classification and regression. The goal of this project is to come up with a model that can predict the value of a target variable by finding and using simple decision rules that are based on the data. It can be observed that when the decision tree algorithm is run on the dataset, 93% accuracy is obtained. Figure 2c shows the confusion matrix obtained by applying the Naive Bayes algorithm to the dataset. The precision, recall, F1-score and accuracy scores using the DT algorithm are 90.43%, 88.27%, 89.24% and 93%, respectively. When the dependent variable is categorical—that is, when it has binary outputs such as "true" and "false" or "yes" and "no"—logistic regression is the method of choice to analyze the data. It can be observed that when the Logistic Regression algorithm is run on the dataset, 82.5% accuracy is obtained. Confusion matrix Figure 2d shows the results of precision, recall, F1-score, support and accuracy obtained by applying the Logistic Regression algorithm to the data set. The precision, recall, F1-score and accuracy obtained by applying the logistic regression algorithm are 85.35%, 71.36%, 74.47% and 82.5%, respectively.

The last algorithm used for the evaluation of the proposed research is the support vector machine, which is a popular supervised learning model used for both data classification and regression. It works by creating a hyperplane with the greatest distance between two classes of data points. The decision boundary is a hyperplane that separates the classes of data points on each side of the plane. It can be observed that when the SVM algorithm is run on the dataset, 84.78% accuracy is obtained. Confusion matrix Figure 2e shows the results of precision, recall, F1-score and accuracy obtained by applying the Linear SVM algorithm to the data set. These values are calculated by using Tp, Tn, Fp and Fn parameters. The precision, recall, F1-score and accuracy obtained using the SVM classifier are 87.0%, 75.05%, 78.31% and 84.78%, respectively. It can be evidently seen from Table 3 that the proposed research demonstrates the highest accuracy using the Decision Tree classifier. Figure 3 provides a graphical comparison of the precision, recall, F1-score and accuracy results obtained by applying the different ML classifiers. It can be safely concluded that the DT classifier outperforms the other ML classifiers in terms of classification accuracy for sentiment analysis.

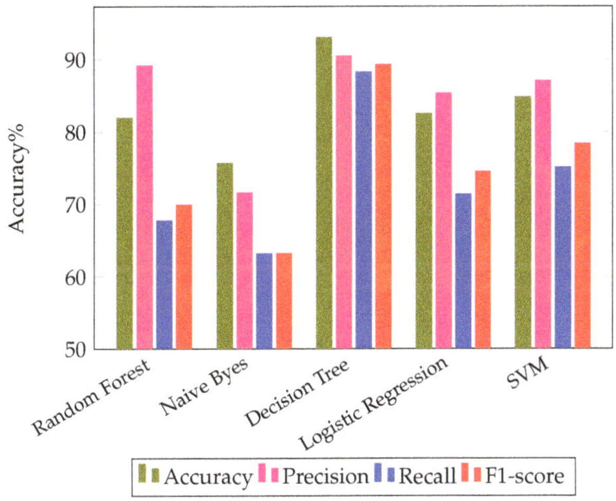

Figure 3. Machine learning Performance on COVID-19 vaccine tweets with sentiment annotation.

4.4. Results for Dataset 2

This section presents a discussion on the experimental results obtained for dataset 2. The dataset is partitioned into five subsets; one representing each vaccine type.

4.4.1. Results for AstraZeneca Dataset

The first classifier applied to the dataset is the Random Forest method, which achieves an accuracy of 81.41%. The Precision, recall, F1-score, support, and accuracy statistics produced by using the Random Forest technique on the dataset are shown in confusion Matrix Figure 4a. The Tp, Tn, Fp, and Fn parameters are used to compute these values. The computed scores of precision, recall, F1-score, and accuracy are 87.27%, 69.32%, 74.19% and 81.81%, respectively, as can be seen in Table 4. The second algorithm used for the evaluation of the proposed research is the Multinomial Naive Bayes algorithm, and it results in 75.28% accuracy. The confusion matrix for the Naive Bayes algorithm is shown in Figure 4b. The precision, recall, F1-score, and accuracy achieved by applying the NB are 70.46%, 70.9%, 69.76% and 75.28%, respectively.

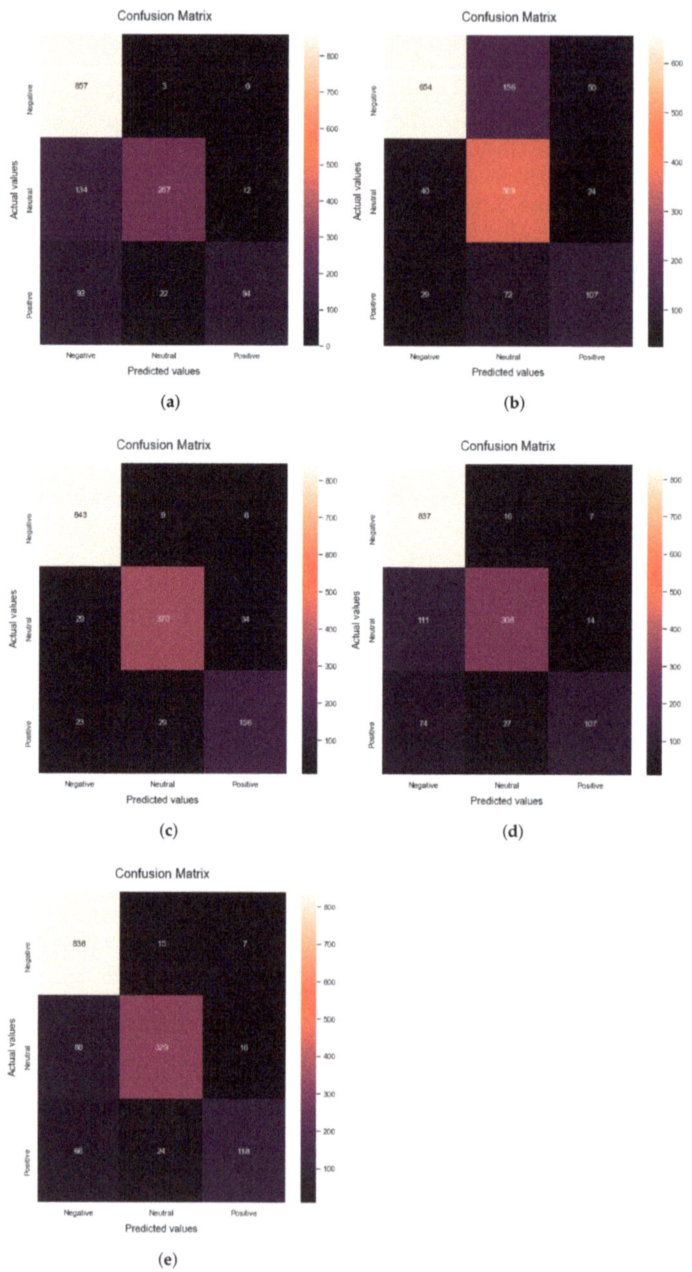

Figure 4. Confusion matrix for AstraZeneca Dataset: (**a**) using Random Forest (**b**) Naive Bayes (**c**) Decision Tree (**d**) Logistic Regression (**e**) SVM.

Table 4. Machine learning performance on AstraZeneca dataset.

Classifier Name	Accuracy%	Precision%	Recall%	F1-Score%
Random Forest	81.81	87.27	69.32	74.19
Naive Bayes	75.28	70.46	70.9	69.76
Decision Tree	90.94	87.33	86.09	86.67
Logistic Regression	83.41	84.41	73.3	77.47
SVM	85.61	85.86	76.72	80.09

When the Decision Tree algorithm is applied to the dataset, the accuracy is found to be 90.94%. The precision, recall, F1-score, support, and accuracy results obtained by using the Naive Bayes algorithm on the dataset are shown in confusion matrix Figure 4c. The TP, TN, FP, and FN parameters are used to calculate these values. Precision, recall, F1-score, and accuracy are 87.33%, 86.09%, 86.67%, and 90.94%, respectively, as can be seen in Table 4. The fourth algorithm used for the evaluation of the proposed research is the Logistic Regression algorithm, which yields an accuracy of 83.41%. Figure 4d shows the Confusion matrix for the LR algorithm. The matrix gives the values of precision, recall, F1-score, and accuracy; as 84.41%, 73.3%, 77.07% and 83.41%, respectively.

The last algorithm used for the evaluation of the proposed research is the SVM, which results in an accuracy of 85.86%. The precision, recall, F1-score, support and accuracy results obtained by using the Linear SVM algorithm on the dataset are shown in confusion Matrix Figure 4e. These values are calculated by using Tp, Tn, Fp, and Fn parameters. The precision, recall, F1-score, support and accuracy scores as obtained using the proposed research are 85.86%, 76.72%, 80.09% and 85.61%, respectively, as shown in Table 4. Figure 5 shows a graphical comparison of the different algorithms for the AstraZeneca vaccine dataset. It can be evidently seen that the highest accuracy is obtained by applying the Decision Tree algorithm.

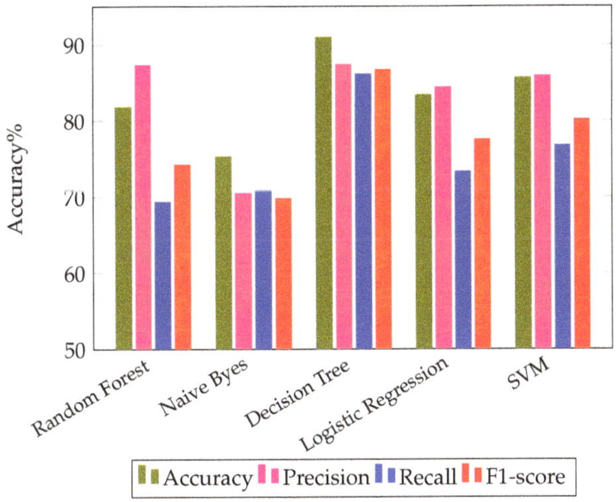

Figure 5. Machine learning Performance on AstraZeneca dataset.

4.4.2. Result of Pfizer Vaccines Dataset

This subsection presents the results obtained for the Pfizer vaccine dataset. When the Random Forest algorithm is applied to the dataset, it achieves an accuracy of 74.42%. The precision, recall, F1-score, support, and accuracy results obtained by using the Random Forest algorithm on the dataset are shown in confusion matrix Figure 6a. The Tp, Tn, Fp, and Fn parameters are used to calculate these values. Precision, recall, F1-score and accuracy are 81.63%, 64.19%, 66.33% and 74.42%, respectively, according to this matrix as can be seen in Table 5. The multinomial Naive Bayes algorithm results in 71.02% accuracy. The precision, recall, F1-score and accuracy results obtained by using the Naive Bayes algorithm on the dataset are shown in confusion Matrix Figure 6b. The values of precision, recall, F1-score and accuracy as computed from the matrix are 67.09%, 65.13%, 65.58% and 71.02%, respectively.

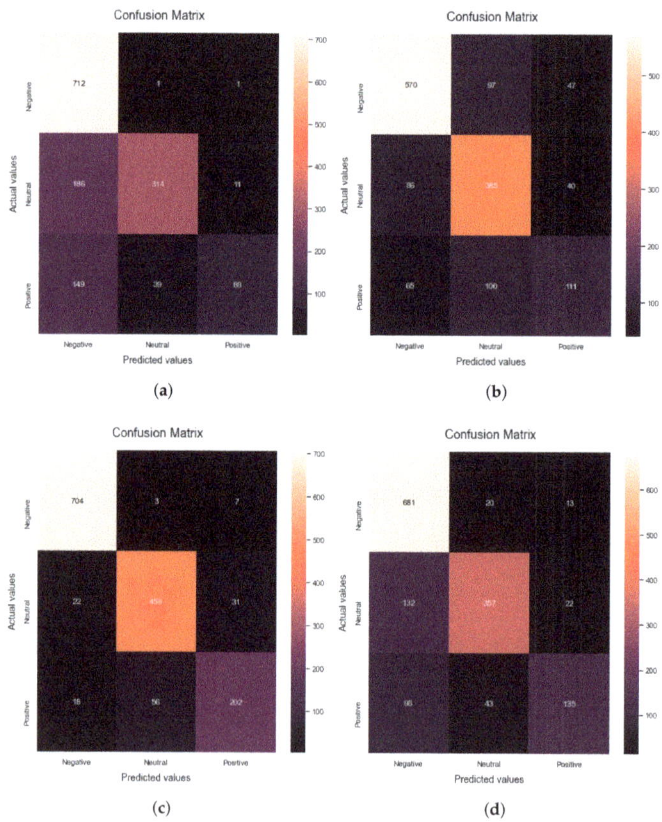

Figure 6. *Cont.*

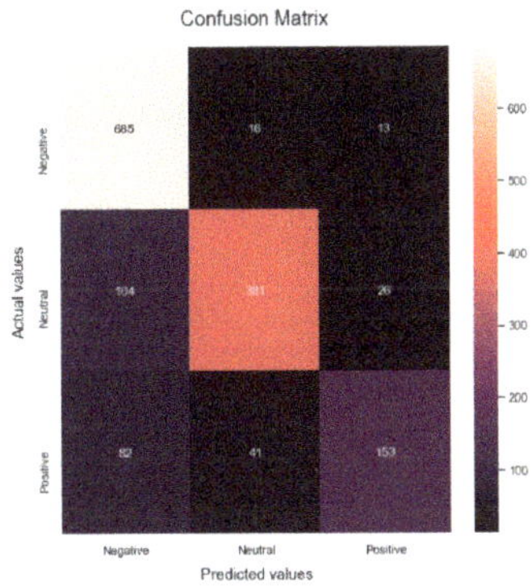

(e)

Figure 6. Confusion matrix for Pfizer Dataset: (**a**) using Random Forest (**b**) Naive Bayes (**c**) Decision Tree (**d**) Logistic Regression (**e**) SVM.

Table 5. Machine learning performance on Pfizer vaccine dataset.

Classifier Name	Accuracy%	Precision%	Recall%	F1-Score%
Random Forest	74,42	81.63	64.19	66.33
Naive Bayes	71.02	67.09	65.13	65.58
Decision Tree	91.07	89.36	87.48	88.3
Logistic Regression	78.72	79.72	71.38	73.68
SVM	81.21	81.77	75.31	77.37

When the Decision Tree algorithm is applied to the dataset, it yields a result of 91.07% accuracy. The precision, recall, F1-score and accuracy results obtained by using the Decision Tree algorithm on the dataset are shown in confusion matrix Figure 6c. The precision, recall, F1-score and accuracy obtained by applying the DT algorithm are 89.36%, 87.48%, 88.3% and 91.07%, respectively, as can be seen in Table 5, respectively.

When the Logistic Regression algorithm is applied to the dataset, the accuracy is found to be 78.72%. The precision, recall, F1-score and accuracy results obtained by using the Logistic Regression algorithm on the dataset are shown in the confusion matrix Figure 6d. The values are determined by means of Tp, Tn, Fp and Fn parameters. This matrix tells the values of precision, recall, F1-score and accuracy as 79.72%, 71.38%, 73.68% and 78.72%, respectively. The last algorithm used for the evaluation of the proposed research is the SVM, which results in an accuracy of 81.21%. The precision, recall, F1-score and accuracy results obtained by using the Linear Support Vector Machine (SVM) algorithm on the dataset are shown in Figure 6e. The precision, recall, F1-score and accuracy scores obtained are 81.77%, 75.31%, 77.37% and 81.21%, respectively, as shown in Table 5. A graphical comparison of different ML algorithms is presented in Figure 7. The Decision Tree outperforms other classifiers and achieves the highest accuracy for sentiment classification.

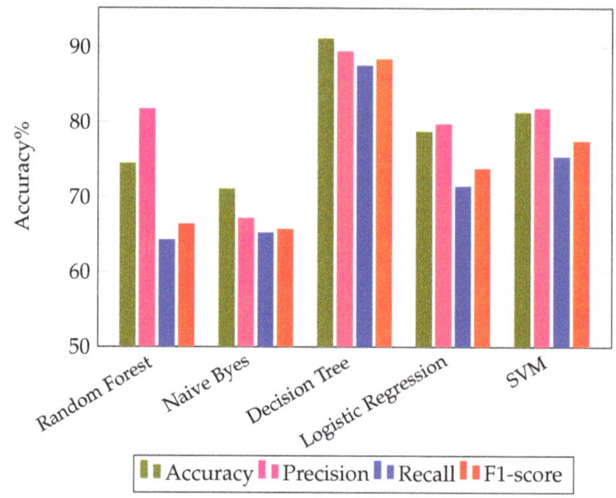

Figure 7. Machine learning Performance on Pfizer dataset.

4.4.3. Results for Sinovac Vaccine Dataset

This subsection presents the results for the Sinovac vaccine dataset. The first algorithm used for evaluation on Sinovac dataset is the Random Forest, and it achieves an accuracy of 79.01%. The precision, recall, F1-score and accuracy results obtained by applying the Random Forest algorithm to the dataset are shown in Figure 8a. The Tp, Tn, Fp and Fn parameters are used to calculate these values. Precision, recall, F1-score, and accuracy obtained for the RF are 85.28%, 67.28%, 70.27% and 79.01%, respectively, as summarized in Table 6. The second algorithm used for evaluation is the Naive Bayes, which results in an accuracy of 72.22%. Confusion matrix Figure 8b shows the results of precision, recall, F1-score and accuracy obtained by applying the multinomial NB algorithm on the dataset. The values of precision, recall, F1-score and accuracy as computed from the matrix are 71.3%, 66.64%, 66.64% and 72.22%, respectively.

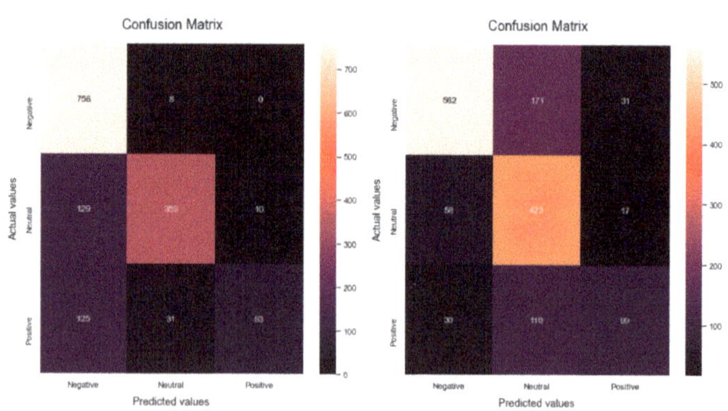

(a) (b)

Figure 8. *Cont.*

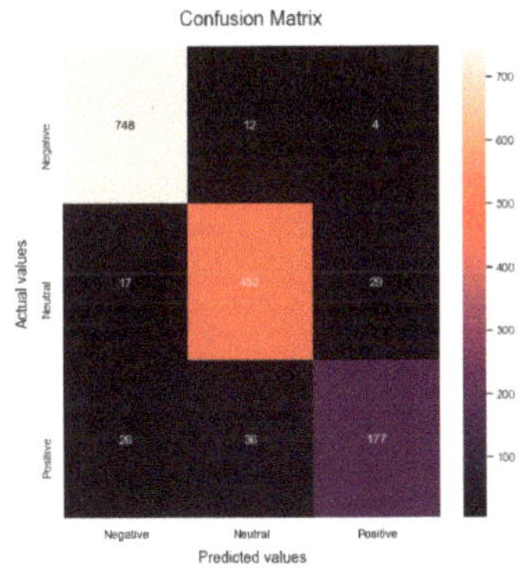

(c)

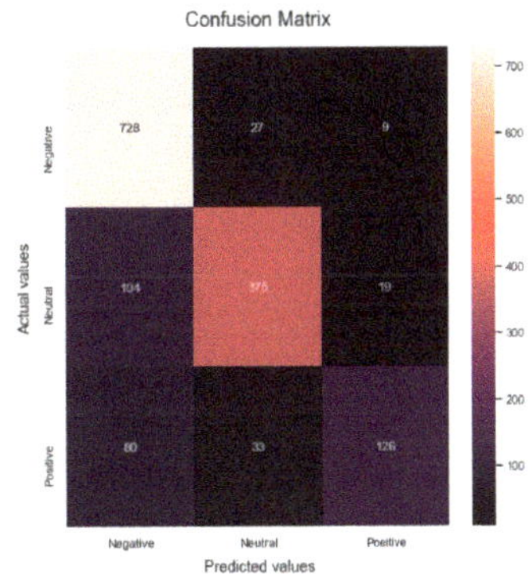

(d)

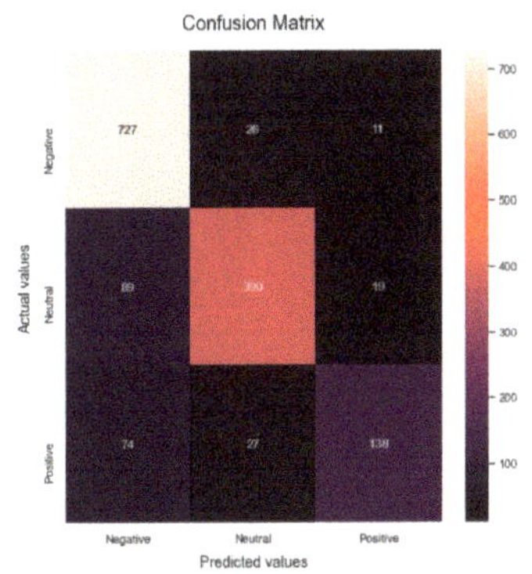

(e)

Figure 8. Confusion matrix for Sinovac Dataset (**a**) using Random Forest (**b**) Naive Bayes (**c**) Decision Tree (**d**) Logistic Regression (**e**) SVM.

Table 6. Machine learning performance on Sinovac vaccine dataset.

Classifier Name	Accuracy%	Precision%	Recall%	F1-Score%
Random Forest	79.01	85.28	67.28	70.27
Naive Bayes	72.22	71.3	66.64	67.06
Decision Tree	92.8	91.55	88.6	89.86
Logistic Regression	81.88	82.62	74.44	77.13
SVM	83.61	83.95	77.07	79.54

When the Decision Tree algorithm is applied to the dataset, it yields a result of 92.8% accuracy. The precision, recall, F1-score and accuracy results obtained by using the Decision Tree algorithm on the dataset are shown in confusion matrix Figure 8c. The Tp, Tn, Fp, and Fn parameters are used to calculate these values. Precision, recall, F1-score and accuracy values are 91.55%, 88.6%, 89.06% and 92.8%, respectively, according to this matrix. When the Logistic Regression algorithm is applied to the dataset, it yields an accuracy of 81.88%. The precision, recall, F1-score and accuracy results obtained by using the LR algorithm on the dataset are shown in Figure 8d. This matrix gives the values of precision, recall, F1-score and accuracy are 82.62%, 74.44%, 77.13% and 81.88%, respectively, as summarized in Table 6.

The last algorithm used for the evaluation of the proposed research is the support vector machine. When SVM is applied to the dataset, it yields an accuracy of 83.61%. The precision, recall, F1-score and accuracy results obtained by using the Linear SVM algorithm on the dataset are shown in confusion matrix Figure 8e. These values are calculated by using Tp, Tn, Fp and Fn parameters. This matrix tells the values of precision, recall, F1-score and accuracy are 83.95%, 77.07%, 79.54%, and 83.61%, respectively, as shown in Table 6. Figure 9 provides a graphical comparison of the performance of different ML classifiers. It can be evidently seen that the highest performance for sentiment classification is obtained using the Decision Tree classifier.

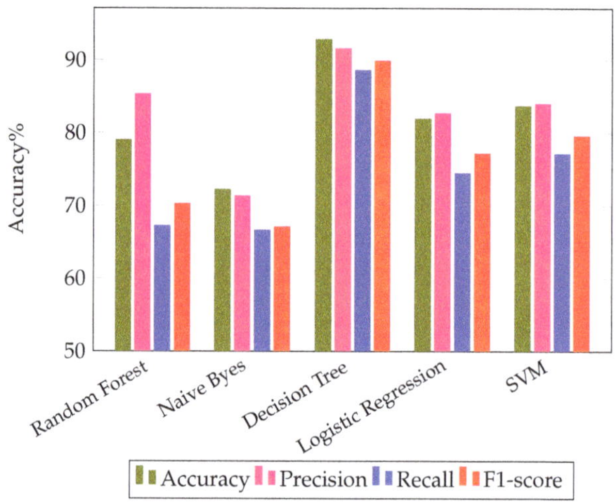

Figure 9. Machine learning Performance on Sinovac dataset.

4.4.4. Results for Moderna Vaccines Dataset

In this subsection, a discussion on the performance of ML algorithms for the Moderna vaccine dataset is presented. It can be seen that when the Random Forest algorithm is applied to the dataset, it achieves an accuracy of 77.75%. The precision, recall, F1-score and accuracy results obtained by using the RF algorithm on the dataset are shown in confusion matrix Figure 10a. The Tp, Tn, Fp, and Fn parameters are used to calculate these values. Precision, recall, F1-score, and accuracy are 85.18%, 64.65%, 67.87% and 77.75%, respectively, as are shown in Table 7.

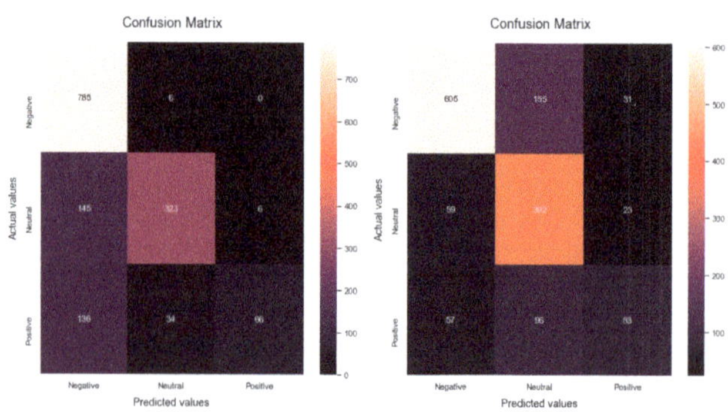

(a) (b)

Figure 10. *Cont.*

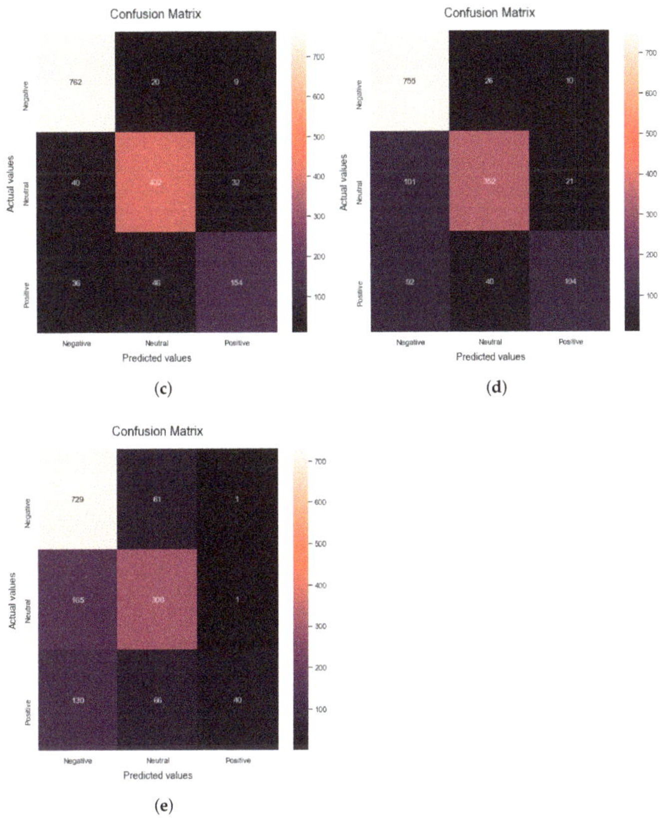

Figure 10. Confusion matrix for Moderna Dataset (**a**) using Random Forest (**b**) Naive Bayes (**c**) Decision Tree (**d**) Logistic Regression (**e**) SVM.

Table 7. Machine learning performance on Moderna dataset.

Classifier Name	Accuracy%	Precision%	Recall%	F1-Score%
Random Forest	77.75	85.18	64.65	67.87
Naive Bayes	71.95	68.49	64.79	64.91
Decision Tree	88.01	85.28	82.4	83.64
Logistic Regression	80.68	80.3	71.26	73.94
SVM	82.81	81.95	74.62	77.12

The second algorithm used for the evaluation of the proposed research is the Naive Bayes algorithm. When NB is applied to the dataset, the accuracy is found to be 71.95%. The precision, recall, F1-score and accuracy results obtained by using the NB algorithm on the dataset are shown in Figure 10b. These values are calculated by using Tp, Tn, Fp and Fn parameters. This matrix gives the values of precision, recall, F1-score and accuracy as 68.49%, 64.79%, 64.91% and 71.95%, respectively. When the Decision Tree algorithm is applied to the dataset, an accuracy of 88.01% is obtained. The precision, recall, F1-score and accuracy results obtained by using the Decision Tree algorithm on the dataset are shown in confusion matrix Figure 10c. The precision, recall, F1-score and accuracy are 85.28%, 82.4%, 83.64%, and 88.01%, respectively, as are summarized in Table 7.

When the Logistic Regression algorithm is applied to the dataset, an accuracy of 80.68% is obtained. Figure 10d shows the confusion matrix for the LR algorithm. The values of precision, recall, F1-score and accuracy as obtained from the confusion matrix are 80.3%, 71.26%, 73.94% and 80.68%, respectively. Finally, the SVM algorithm is applied to the dataset and it achieves an accuracy of 82.81%. The precision, recall, F1-score and accuracy results obtained by using the Linear SVM algorithm on the dataset are shown in confusion matrix Figure 10e. The Tp, Tn, Fp and Fn parameters are used to calculate these values. The computed values of precision, recall, F1-score and accuracy computed from this matrix are 81.95%, 74.62%, 77.12% and 82.81%, respectively, as shown in Table 7. A graphical comparison of different ML classifiers for the Moderna vaccine dataset is presented in Figure 11. As discussed earlier, the highest performance for sentiment classification is obtained with the Decision Tree classifier as compared to the other ML algorithms.

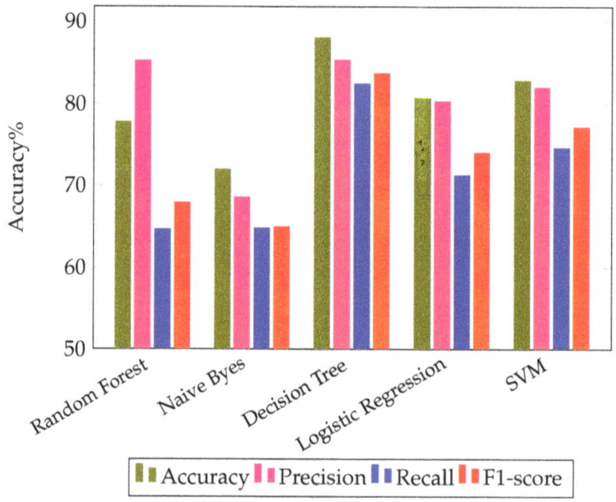

Figure 11. Machine learning Performance on Moderna dataset.

4.5. Results for Sinopharm Vaccines Dataset

This subsection summarizes the results of the sinopharm vaccine dataset. When the Random Forest algorithm is applied to the dataset, an accuracy of 83.61% is obtained. The precision, recall, F1-score and accuracy results obtained by using the RF algorithm on the dataset are shown in confusion matrix Figure 12a. The precision, recall, F1-score and accuracy are 89.14%, 73.32%, 78.09%, and 83.61%, respectively, according to this matrix. When the Naive Bayes algorithm is applied to the dataset, it yields a result of 74.48% accuracy. The precision, recall, F1-score and accuracy results obtained by using the NB algorithm on the dataset are shown in Figure 12b. The values of precision, recall, F1-score and accuracy computed from the matrix are 74.6%, 73.94%, 71.8% and 74.48%, respectively, as shown in Table 8.

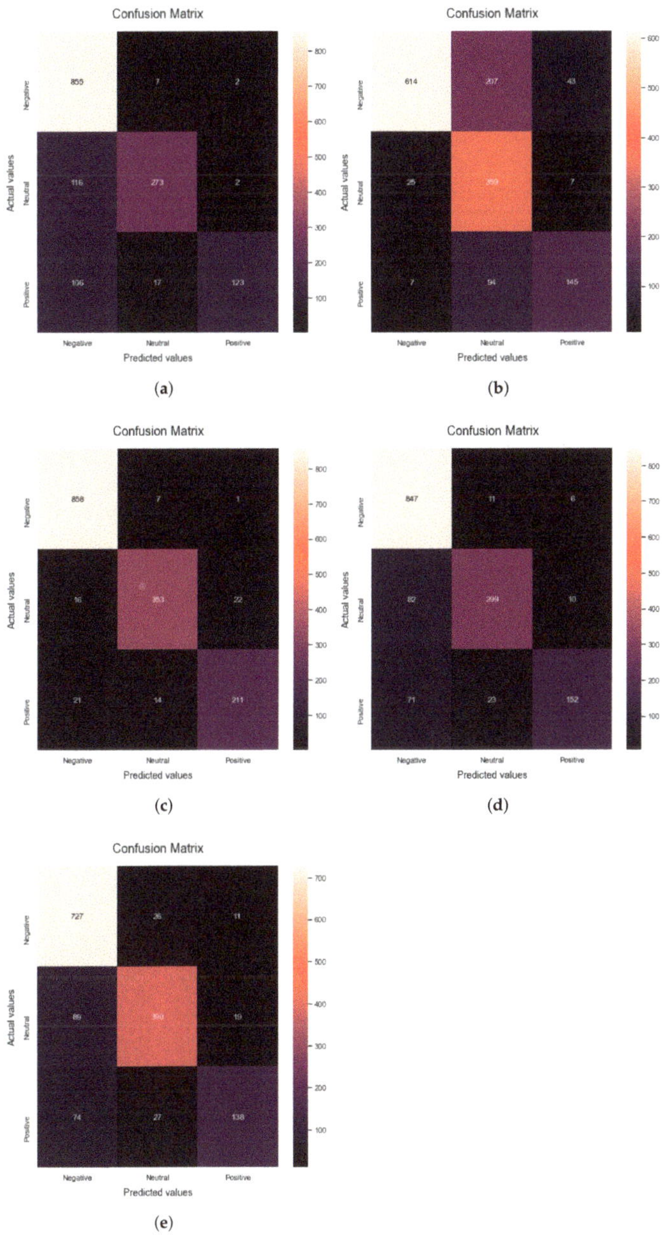

Figure 12. Confusion matrix for Sinopharm Dataset (**a**) using Random Forest (**b**) Naive Bayes (**c**) Decision Tree (**d**) Logistic Regression (**e**) SVM.

Table 8. Machine learning performance on Sinopharm vaccine dataset.

Classifier Name	Accuracy%	Precision%	Recall%	F1-Score%
Random Forest	83.61	89.14	73.32	78.09
Naive Bayes	74.48	74.6	73.94	71.8
Decision Tree	93.87	92.77	90.51	91.57
Logistic Regression	86.48	88.32	78.76	82.3
SVM	87.67	89.0	81.09	84.21

When the Decision Tree algorithm is applied to the dataset, it yields a result of 93.87% accuracy. Figure 12c shows the confusion matrix for the proposed research. The precision, recall, F1-score and accuracy values computed from the matrix are 92.77%, 90.51 %, 91.57% and 93.87%, respectively, as shown in Table 8.

The fourth algorithm used for the evaluation of the proposed research is the Logistic Regression. When the LR algorithm is applied to the dataset, an accuracy of 86.48% is obtained. The precision, recall, F1-score and accuracy results obtained by using the LR algorithm on the dataset are displayed in the confusion matrix Figure 12d. The Tp, Tn, Fp and Fn parameters are used to calculate these values. The obtained values of precision, recall, F1-score and accuracy are 88.32%, 78.76%, 82.3% and 86.48%, respectively. The last algorithm used for the evaluation of the proposed research is the SVM. When the SVM algorithm is applied to the dataset, it yields an accuracy of 87.67%. The precision, recall, F1-score and accuracy results obtained by using the Linear SVM algorithm on the dataset are shown in confusion matrix Figure 12e. The computed values of precision, recall, F1-score and accuracy for the SVM algorithm are 89.0%, 81.09%, 84.21%, and 87.67%, respectively, as shown in Table 8. Figure 13 shows a graphical performance comparison of different ML algorithms for the Sinopharm vaccine dataset. It can be evidently seen that the Decision Tree algorithm outperforms the rest thereby achieving the highest accuracy for sentiment classification.

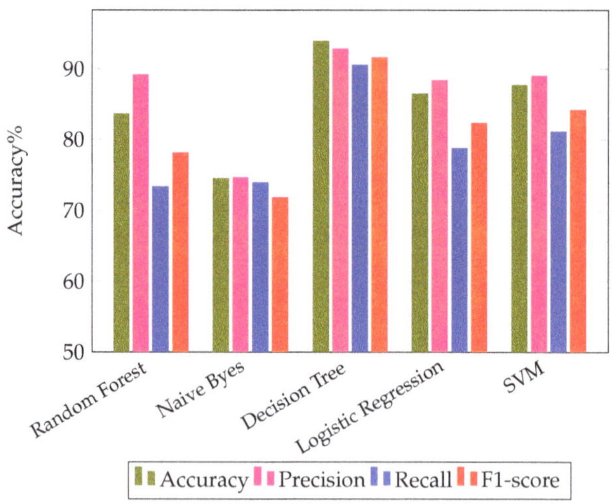

Figure 13. Machine learning Performance on Sinopharm dataset.

4.6. Comparison with State-of-the-Art Research

This article presents a performance comparison of five different ML algorithms for sentiment classification. All of these ML models were deployed using the COVID-19 vaccination tweets dataset that was collected in this study and the annotated sentiment dataset. Training and testing were carried out with the help of the annotated datasets provided by TextBlob. Figure 14 shows a graphical performance comparison of accuracy achieved by different ML algorithms on all datasets used in this research. Each bar illustrates the performance of different ML classifiers for the different tweet datasets. For example, the first bar shows the classification accuracy of different ML classifiers for CVSA dataset. Likewise, the other bars demonstrate the results achieved by various ML algorithms for AstraZeneca, Pfizer, Sinovac, Moderna and Sinopharm tweets datasets, respectively. It can be evidently seen, that the highest performance for all datasets is obtained with the Decision Tree algorithm as compared to the other ML classifiers.

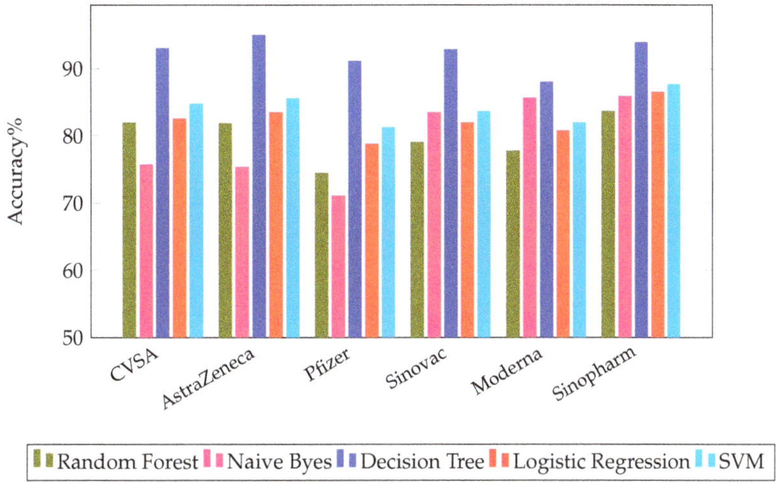

Figure 14. Performance comparison of machine learning algorithms on datasets used in the research.

To represent the significant performance of the proposed research, this subsection presents a comparison in the context of other studies. Table 9 demonstrates the accuracy of results from state-of-the-art related research. As discussed earlier, the DT algorithm achieves the highest performance in the case of all datasets used for this research. Hence, the accuracy achieved using DT is shown in comparison with the state-of-the-art research in Table 9. Results suggest that the proposed approach is significantly better than other studies in terms of accuracy. Despite using the different models in other studies, the proposed research showed superior performance with the Decision Tree classifier and obtained better accuracy for sentiments, which is significantly higher than previous studies. The key findings of this research can be summarized as follows:

- The ratio of positive sentiments is high as compared to the ratio of negative sentiments in tweets related to COVID-19 vaccinations as can be seen in Figure 15. The highest percentage of positive opinions is observed for the Moderna vaccine based on people's sentiments.
- Based on data on people's perceptions, the ratio of sentiments for positive, neutral and negative sentiments may vary. However, on average, it may be concluded that the number of neutral sentiments is higher than the positive and negative sentiments.
- The Decision Tree ML model proved to perform better as compared to the other four models. Tree-based ML models can be a good choice for obtaining higher classification performance when dealing with tweets' textual data.

Table 9. Performance comparison with the state-of-the-art research.

Year	Reference	Model	Dataset	Accuracy%
2021	[43]	SVM	Annotated COVID-19 vaccination	68.88
2021	[43]	CNN	Annotated COVID-19 vaccination	65.71
2021	[43]	BERT	Annotated COVID-19 vaccination	78.94
2021	[11]	SVM	Sinovac vaccine	85
2021	[11]	SVM	Pfizer vaccine	78
2022	This study	Decision Tree	AstraZeneca vaccine	91.07
2022	This study	Decision Tree	Pfizer Vaccine	91.07
2022	This study	Decision Tree	Moderna vaccine	88.01
2022	This study	Decision Tree	Sinovace vaccine	92.8
2022	This study	Decision Tree	Sinopharm vaccine	93.87
2022	This study	Decision Tree	CVSA	93.0

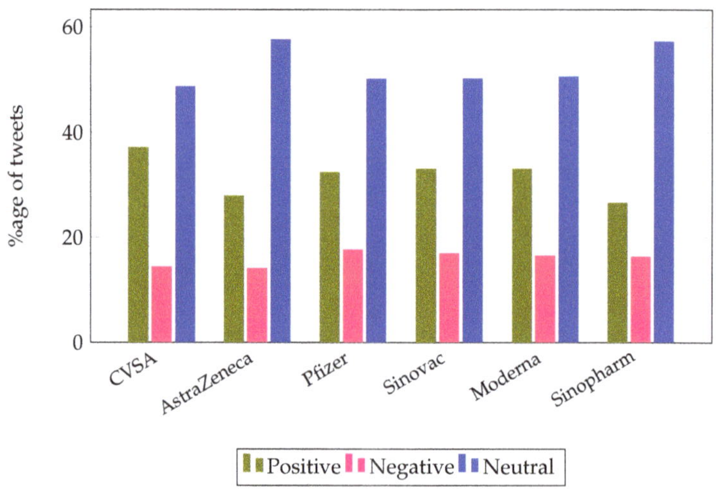

Figure 15. Percentage of positive, negative and neutral tweets in datasets used for this research.

5. Conclusions

The WHO encourages rapid immunization of the whole population to reduce the danger of disease transmission and death. The government authorities, medical experts, and social workers recommend getting the vaccination, however, people have expressed worries and misgivings about the potential for side effects and other medical consequences. Sentiment analysis of social trends can help in effective decision making. This paper presents a framework for analyzing people's worldwide perceptions and attitudes towards Covid-19 vaccines AstraZeneca, Pfizer, Sinovac, Moderna, and Sinopharm, respectively. We have evaluated the performance of five different machine learning classifiers for sentiment analysis. The quantitative comparisons demonstrate that the proposed research achieves better performance as compared to the state-of-the-art research. Based on the experimental results, the highest performance is obtained using the Decision Tree classifier, i.e., 93.0% using CVSA dataset, 93.87% using Sinopharm dataset, Sinovac dataset 92.8%, Pfizer dataset 91.07%, AstraZeneca dataset 90.94%, and for Moderna dataset 88.01%, respectively. In future, we aim to enhance the classification accuracy by applying different pre-processing techniques such as creating a normalization dictionary. Another approach to enhance the performance can be the implication of oversampling or under-sampling techniques such as SMOTE to handle imbalanced data. Additionally, the parameters in each classification model can be fine-tuned to obtain an increase in classifier performance. In future, the performance of deep learning models will be accessed for sentiment classification in order to achieve better accuracy results.

Author Contributions: Conceptualization, A.S., B.Z. and U.J.; Data curation, A.S., B.Z., N.A., A.J.A. and M.A.; Formal analysis, A.S., B.Z., N.A. and U.J.; Investigation, A.S.; Methodology, A.S., B.Z., A.J.A. and U.J.; Project administration, B.Z. , N.A., A.J.A., N.A.G. and E.T.E.; Resources, M.A., N.A.G. and E.T.E.; Software, M.A.; Supervision, B.Z., N.A., N.A.G. and E.T.E.; Validation, U.J.; Writing—original draft, A.S., B.Z., N.A., U.J., A.J.A., M.A. and E.T.E.; Writing—review & editing, N.A.G. All authors have read and agreed to the published version of the manuscript.

Funding: This research received no external funding.

Conflicts of Interest: The authors declare no conflict of interest.

References

1. Khalil, T.; Akram, M.U.; Khalid, S.; Dar, S.H.; Ali, N. A study to identify limitations of existing automated systems to detect glaucoma at initial and curable stage. *Int. J. Imaging Syst. Technol.* **2021**, *31*, 1155–1173. [CrossRef]
2. Aslam, M.A.; Salik, M.N.; Chughtai, F.; Ali, N.; Dar, S.H.; Khalil, T. Image classification based on mid-level feature fusion. In Proceedings of the 2019 15th International Conference on Emerging Technologies (ICET) IEEE, Peshawar, Pakistan, 2–3 December 2019; pp. 1–6.
3. Rasheed, A.; Zafar, B.; Rasheed, A.; Ali, N.; Sajid, M.; Dar, S.H.; Habib, U.; Shehryar, T.; Mahmood, M.T. Fabric defect detection using computer vision techniques: A comprehensive review. *Math. Probl. Eng.* **2020**, *2020*, 8189403. [CrossRef]
4. Riaz, F.; Jabbar, S.; Sajid, M.; Ahmad, M.; Naseer, K.; Ali, N. A collision avoidance scheme for autonomous vehicles inspired by human social norms. *Comput. Electr. Eng.* **2018**, *69*, 690–704. [CrossRef]
5. Ali, N.; Bajwa, K.B.; Sablatnig, R.; Chatzichristofis, S.A.; Iqbal, Z.; Rashid, M.; Habib, H.A. A novel image retrieval based on visual words integration of SIFT and SURF. *PLoS ONE* **2016**, *11*, e0157428. [CrossRef] [PubMed]
6. Wang, S.; Xu, H.; Kotian, R.P.; D'souza, B.; Rao, S.S. A study on psychological implications of COVID-19 on nursing professionals. *Int. J. Healthc. Manag.* **2021**, *14*, 300–305. [CrossRef]
7. Ceylan, Z. Estimation of COVID-19 prevalence in Italy, Spain, and France. *Sci. Total Environ.* **2020**, *729*, 138817. [CrossRef] [PubMed]
8. Wang, L.; Li, J.; Guo, S.; Xie, N.; Yao, L.; Cao, Y.; Day, S.W.; Howard, S.C.; Graff, J.C.; Gu, T.; et al. Real-time estimation and prediction of mortality caused by COVID-19 with patient information based algorithm. *Sci. Total Environ.* **2020**, *727*, 138394. [CrossRef] [PubMed]
9. Zhu, N.; Zhang, D.; Wang, W.; Li, X.; Yang, B.; Song, J.; Zhao, X.; Huang, B.; Shi, W.; Lu, R.; et al. A novel coronavirus from patients with pneumonia in China, 2019. *N. Engl. J. Med.* **2020**, *382*, 727–733. [CrossRef]
10. Du Toit, A. Outbreak of a novel coronavirus. *Nat. Rev. Microbiol.* **2020**, *18*, 123. [CrossRef]
11. Nurdeni, D.A.; Budi, I.; Santoso, A.B. Sentiment analysis on COVID-19 vaccines in Indonesia: From the perspective of Sinovac and Pfizer. In Proceedings of the 2021 3rd East Indonesia Conference on Computer and Information Technology (EIConCIT) IEEE, Surabaya, Indonesia, 9–11 April 2021; pp. 122–127.
12. Odone, A.; Delmonte, D.; Scognamiglio, T.; Signorelli, C. COVID-19 deaths in Lombardy, Italy: Data in context. *Lancet Public Health* **2020**, *5*, e310. [CrossRef]
13. Battiston, P.; Kashyap, R.; Rotondi, V. Reliance on scientists and experts during an epidemic: Evidence from the COVID-19 outbreak in Italy. *SSM-Popul. Health* **2021**, *13*, 100721. [CrossRef] [PubMed]
14. Gomes, G. Naming the coronavirus disease (COVID-19) and the virus that causes it. *Braz. J. Implantol. Health Sci.* **2020**, *2*, 1–3.
15. Abid, K.; Bari, Y.A.; Younas, M.; Tahir Javaid, S.; Imran, A. Progress of COVID-19 Epidemic in Pakistan. *Asia Pac. J. Public Health* **2020**, *32*, 154–156. [CrossRef]
16. Setiati, S.; Azwar, M.K. COVID-19 and Indonesia. *Acta Med. Indones.* **2020**, *52*, 84–89. [PubMed]
17. Reshi, A.A.; Rustam, F.; Aljedaani, W.; Shafi, S.; Alhossan, A.; Alrabiah, Z.; Ahmad, A.; Alsuwailem, H.; Almangour, T.A.; Alshammari, M.A.; et al. COVID-19 Vaccination-Related Sentiments Analysis: A Case Study Using Worldwide Twitter Dataset. *Healthcare* **2022**, *10*, 411. [CrossRef] [PubMed]
18. Hung, I.F.; Poland, G.A. Single-dose Oxford–AstraZeneca COVID-19 vaccine followed by a 12-week booster. *Lancet* **2021**, *397*, 854–855. [CrossRef]
19. Chagla, Z. The BNT162b2 (BioNTech/Pfizer) vaccine had 95% efficacy against COVID-19 7 days after the 2nd dose. *Ann. Intern. Med.* **2021**, *174*, JC15. [CrossRef] [PubMed]

20. Bono, S.A.; Siau, C.S.; Chen, W.S.; Low, W.Y.; Faria de Moura Villela, E.; Pengpid, S.; Hasan, M.T.; Sessou, P.; Ditekemena, J.D.; Amodan, B.O.; et al. Adults' acceptance of COVID-19 vaccine for children in selected lower-and middle-income countries. *Vaccines* **2021**, *10*, 11. [CrossRef]
21. Marcec, R.; Likic, R. Using twitter for sentiment analysis towards AstraZeneca/Oxford, Pfizer/BioNTech and Moderna COVID-19 vaccines. *Postgrad. Med. J.* **2021**, *98*, 544–550. [CrossRef]
22. Chew, C.; Eysenbach, G. Pandemics in the age of Twitter: Content analysis of Tweets during the 2009 H1N1 outbreak. *PLoS ONE* **2010**, *5*, e14118. [CrossRef] [PubMed]
23. Ali, N.; Zafar, B.; Iqbal, M.K.; Sajid, M.; Younis, M.Y.; Dar, S.H.; Mahmood, M.T.; Lee, I.H. Modeling global geometric spatial information for rotation invariant classification of satellite images. *PLoS ONE* **2019**, *14*, e0219833. [CrossRef]
24. Tufail, A.B.; Ullah, I.; Khan, R.; Ali, L.; Yousaf, A.; Rehman, A.U.; Alhakami, W.; Hamam, H.; Cheikhrouhou, O.; Ma, Y.K. Recognition of Ziziphus lotus through Aerial Imaging and Deep Transfer Learning Approach. *Mob. Inf. Syst.* **2021**, *2021*, 4310321. [CrossRef]
25. Zafar, B.; Ashraf, R.; Ali, N.; Iqbal, M.K.; Sajid, M.; Dar, S.H.; Ratyal, N.I. A novel discriminating and relative global spatial image representation with applications in CBIR. *Appl. Sci.* **2018**, *8*, 2242. [CrossRef]
26. Ali, N.; Bajwa, K.B.; Sablatnig, R.; Mehmood, Z. Image retrieval by addition of spatial information based on histograms of triangular regions. *Comput. Electr. Eng.* **2016**, *54*, 539–550. [CrossRef]
27. Sajid, M.; Ali, N.; Dar, S.H.; Zafar, B.; Iqbal, M.K. Short search space and synthesized-reference re-ranking for face image retrieval. *Appl. Soft Comput.* **2021**, *99*, 106871. [CrossRef]
28. Asif, M.; Khan, W.U.; Afzal, H.; Nebhen, J.; Ullah, I.; Rehman, A.U.; Kaabar, M.K. Reduced-complexity LDPC decoding for next-generation IoT networks. *Wirel. Commun. Mob. Comput.* **2021**, *2021*, 2029560. [CrossRef]
29. Fatima, S.; Aslam, N.A.; Tariq, I.; Ali, N. Home security and automation based on internet of things: A comprehensive review. In Proceedings of the IOP Conference Series: Materials Science and Engineering, Topi, Pakistan, 9–10 June 2020; IOP Publishing: Bristl, UK, 2020; Volume 899, p. 012011.
30. Tufail, A.B.; Ullah, I.; Khan, W.U.; Asif, M.; Ahmad, I.; Ma, Y.K.; Khan, R.; Ali, M. Diagnosis of diabetic retinopathy through retinal fundus images and 3D convolutional neural networks with limited number of samples. *Wirel. Commun. Mob. Comput.* **2021**, *2021*, 6013448. [CrossRef]
31. Shabbir, A.; Rasheed, A.; Shehraz, H.; Saleem, A.; Zafar, B.; Sajid, M.; Ali, N.; Dar, S.H.; Shehryar, T. Detection of glaucoma using retinal fundus images: A comprehensive review. *Math. Biosci. Eng.* **2021**, *18*, 2033–2076. [CrossRef]
32. Ahmad, I.; Ullah, I.; Khan, W.U.; Ur Rehman, A.; Adrees, M.S.; Saleem, M.Q.; Cheikhrouhou, O.; Hamam, H.; Shafiq, M. Efficient algorithms for E-healthcare to solve multiobject fuse detection problem. *J. Healthc. Eng.* **2021**, *2021*, 9500304. [CrossRef]
33. Manguri, K.H.; Ramadhan, R.N.; Amin, P.R.M. Twitter sentiment analysis on worldwide COVID-19 outbreaks. *Kurd. J. Appl. Res.* **2020**, *5*, 54–65. [CrossRef]
34. Privor-Dumm, L.A.; Poland, G.A.; Barratt, J.; Durrheim, D.N.; Knoll, M.D.; Vasudevan, P.; Jit, M.; Bonvehí, P.E.; Bonanni, P.; International Council on Adult Immunization. A global agenda for older adult immunization in the COVID-19 era: A roadmap for action. *Vaccine* **2021**, *39*, 5240–5250. [CrossRef] [PubMed]
35. Meena, R.; Thulasi Bai, V. Russia's COVID-19 Vaccine: Social discussion and first emotions. *Res. Sq.* **2020**, 1–13. doi: [CrossRef]
36. Alliheibi, F.M.; Omar, A.; Al-Horais, N. Opinion Mining of Saudi Responses to COVID-19 Vaccines on Twitter. *Int. J. Adv. Comput. Sci. Appl.* **2021**, *12*, 72–78. [CrossRef]
37. Yousefinaghani, S.; Dara, R.; Mubareka, S.; Papadopoulos, A.; Sharif, S. An analysis of COVID-19 vaccine sentiments and opinions on Twitter. *Int. J. Infect. Dis.* **2021**, *108*, 256–262. [CrossRef]
38. Ezhilan, A.; Dheekksha, R.; Anahitaa, R.; Shivani, R. Sentiment analysis and classification of COVID-19 tweets. In Proceedings of the 2021 5th International Conference on Trends in Electronics and Informatics (ICOEI) IEEE, Tirunelveli, India, 3–5 June 2021; pp. 821–828.
39. Andrzejczak-Grzadko, S.; Czudy, Z.; Donderska, M. Side effects after COVID-19 vaccinations among residents of Poland. *Eur. Rev. Med. Pharmacol. Sci.* **2021**, *25*, 4418–4421. [PubMed]
40. Saeed, B.Q.; Al-Shahrabi, R.; Alhaj, S.S.; Alkokhardi, Z.M.; Adrees, A.O. Side effects and perceptions following Sinopharm COVID-19 vaccination. *Int. J. Infect. Dis.* **2021**, *111*, 219–226. [CrossRef] [PubMed]
41. Dubey, A.D. Public Sentiment Analysis of COVID-19 Vaccination Drive in India. *SSRN* **2021**, 3772401. doi: [CrossRef]
42. Dumre, R.; Sharma, K.; Konar, K. Statistical and sentimental analysis on vaccination against COVID-19 in India. In Proceedings of the 2021 International Conference on Communication information and Computing Technology (ICCICT) IEEE, Mumbai, India, 25–27 June 2021; pp. 1–6.
43. Cotfas, L.A.; Delcea, C.; Roxin, I.; Ioanăş, C.; Gherai, D.S.; Tajariol, F. The longest month: Analyzing COVID-19 vaccination opinions dynamics from tweets in the month following the first vaccine announcement. *IEEE Access* **2021**, *9*, 33203–33223. [CrossRef]
44. Chintalapudi, N.; Battineni, G.; Amenta, F. Sentimental analysis of COVID-19 tweets using deep learning models. *Infect. Dis. Rep.* **2021**, *13*, 329–339. [CrossRef] [PubMed]
45. Rustam, F.; Khalid, M.; Aslam, W.; Rupapara, V.; Mehmood, A.; Choi, G.S. A performance comparison of supervised machine learning models for COVID-19 tweets sentiment analysis. *PLoS ONE* **2021**, *16*, e0245909. [CrossRef] [PubMed]
46. Amaratunga, D.; Cabrera, J.; Lee, Y.S. Enriched random forests. *Bioinformatics* **2008**, *24*, 2010–2014. [CrossRef]

47. Biau, G.; Scornet, E. A random forest guided tour. *Test* **2016**, *25*, 197–227. [CrossRef]
48. Borgelt, C.; Gebhardt, J. A naive bayes style possibilistic classifier. In Proceedings of the 7th European Congress on Intelligent Techniques and Soft Computing (EUFIT'99), Aachen, Germany, 13–16 September 1999.
49. Charbuty, B.; Abdulazeez, A. Classification based on decision tree algorithm for machine learning. *J. Appl. Sci. Technol. Trends* **2021**, *2*, 20–28. [CrossRef]
50. Shah, K.; Patel, H.; Sanghvi, D.; Shah, M. A comparative analysis of logistic regression, random forest and KNN models for the text classification. *Augment. Hum. Res.* **2020**, *5*, 12. [CrossRef]
51. Bhagat, A.; Sharma, A.; Chettri, S. Machine learning based sentiment analysis for text messages. *Int. J. Comput. Technol.* **2020**, *7*, 103–109.
52. Gabriel, P. COVID-19 Vaccine Tweets with Sentiment Annotation. Available online: https://www.kaggle.com/datasets/datasciencetool/covid19-vaccine-tweets-with-sentiment-annotation (accessed on 28 March 2022).
53. Latif, A.; Rasheed, A.; Sajid, U.; Ahmed, J.; Ali, N.; Ratyal, N.I.; Zafar, B.; Dar, S.H.; Sajid, M.; Khalil, T. Content-based image retrieval and feature extraction: A comprehensive review. *Math. Probl. Eng.* **2019**, *2019*, 9658350. [CrossRef]
54. Shabbir, A.; Ali, N.; Ahmed, J.; Zafar, B.; Rasheed, A.; Sajid, M.; Ahmed, A.; Dar, S.H. Satellite and scene image classification based on transfer learning and fine tuning of ResNet50. *Math. Probl. Eng.* **2021**, *2021*, 5843816. [CrossRef]

computation

MDPI

Article

Capturing the Complexity of COVID-19 Research: Trend Analysis in the First Two Years of the Pandemic Using a Bayesian Probabilistic Model and Machine Learning Tools

Javier De La Hoz-M [1,*], Susana Mendes [2], María José Fernández-Gómez [3] and Yolanda González Silva [4]

[1] Facultad de Ingeniería, Universidad del Magdalena, Santa Marta 470004, Colombia
[2] MARE/ARNET, School of Tourism and Maritime Technology, Polytechnic of Leiria, 2520-614 Peniche, Portugal
[3] Institute of Biomedical Research of Salamanca, 37008 Salamanca, Spain
[4] Centro de Salud Ponferrada III, Gerencia de Asistencia Sanitaria del Bierzo (GASBI), 24403 Ponferrada, Spain
* Correspondence: jdelahoz@unimagdalena.edu.co

Abstract: Publications about COVID-19 have occurred practically since the first outbreak. Therefore, studying the evolution of the scientific publications on COVID-19 can provide us with information on current research trends and can help researchers and policymakers to form a structured view of the existing evidence base of COVID-19 and provide new research directions. This growth rate was so impressive that the need for updated information and research tools become essential to mitigate the spread of the virus. Therefore, traditional bibliographic research procedures, such as systematic reviews and meta-analyses, become time-consuming and limited in focus. This study aims to study the scientific literature on COVID-19 that has been published since its inception and to map the evolution of research in the time range between February 2020 and January 2022. The search was carried out in PubMed extracting topics using text mining and latent Dirichlet allocation modeling and a trend analysis was performed to analyze the temporal variations in research for each topic. We also study the distribution of these topics between countries and journals. 126,334 peer-reviewed articles and 16 research topics were identified. The countries with the highest number of scientific publications were the United States of America, China, Italy, United Kingdom, and India, respectively. Regarding the distribution of the number of publications by journal, we found that of the 7040 sources *Int. J. Environ. Res. Public Health*, *PLoS ONE*, and *Sci. Rep.*, were the ones that led the publications on COVID-19. We discovered a growing tendency for eight topics (Prevention, Telemedicine, Vaccine immunity, Machine learning, Academic parameters, Risk factors and morbidity and mortality, Information synthesis methods, and Mental health), a falling trend for five of them (Epidemiology, COVID-19 pathology complications, Diagnostic test, Etiopathogenesis, and Political and health factors), and the rest varied throughout time with no discernible patterns (Therapeutics, Pharmacological and therapeutic target, and Repercussion health services).

Keywords: COVID-19; topic modeling; latent Dirichlet allocation; machine learning; text mining

Citation: De La Hoz-M, J.; Mendes, S.; Fernández-Gómez, M.J.; González Silva, Y. Capturing the Complexity of COVID-19 Research: Trend Analysis in the First Two Years of the Pandemic Using a Bayesian Probabilistic Model and Machine Learning Tools. *Computation* **2022**, *10*, 156. https://doi.org/10.3390/computation10090156

Academic Editors: Simone Brogi and Vincenzo Calderone

Received: 28 July 2022
Accepted: 16 August 2022
Published: 8 September 2022

1. Introduction

In March 2020, the World Health Organization declared the coronavirus outbreak a pandemic [1]. Since then, given the novelty of the disease, the scientific community has mobilized rapidly, reaching a considerably high number of scientific publications. As a result of the above, monitoring the rising database in medicine is becoming increasingly difficult, rendering traditional standard procedures such as systematic reviews and meta-analyses inappropriate approaches in an area as dynamic as the novel coronavirus [2]. Given the large number of publications, an approach that is more direct and has a broader reach is required.

Larsen and von Ins [3] stated that the worldwide increase in scientific literature can lead to researchers feeling overwhelmed and, therefore, their ability to carry out a review and follow-up of new research is effectively decimated.

Several comprehensive studies have been published on various aspects of the pandemic, including symptoms, treatments, and comorbidities [4–6]. Bibliometric analysis of studies on the COVID-19 pandemic has also been carried out [7–10]. However, the majority of the research looked at papers that were published during the first months of the COVID-19 pandemic being declared. As a result, several papers released since then have yet to be examined.

Our goals were to analyze the available scientific literature on COVID-19, identify the research topic, and describe the evolution of COVID-19 research to date, using a machine learning-based methodology. The significant worries of society about various facets of the pandemic's effects make scientific knowledge synthesis more vital than ever. Given the growing diversity of research topics related to COVID-19, quantitative studies are needed to better understand and answer the following concerns:

- Question 1 (Q1): What were the key publishing sources and major contributions to COVID-19 research?
- Question 2 (Q2): What are the major research topics in this field?
- Question 3 (Q3): How do these research topics evolve with time?
- Question 4 (Q4): What are the distributions of these topics across countries and journals?

2. Materials and Methods

2.1. Data Collection

Interventional Searching was conducted on 15 February 2022, using PubMed E-utilities using the following query: "COVID-19 (Title/Abstract) AND English (LA) AND Journal Article (PT] AND 2020/02/01 (dp]: 2022/01/31 (dp]". The illness COVID-19, rather than the virus, was the focus of this research. As a result, alternative search phrases or concepts were ignored in this inquiry. For each article, we obtained the title, keywords, abstract, date of publication, list of author affiliations, journal name, and PubMed identification number.

We regarded the country of affiliation of the first author to be the nation of origin of the article. If a nation's name was not contained in the affiliation, we utilized the most recently mentioned geographic entity and manually connected it to a country; for example, "Bogota" was linked to "Colombia".

We used bibliometric analysis to answer Q1. This enables the sample of publications to be used to determine various elements of scientific production [11,12]. In this section of the investigation, data were processed using bibliometrix [11], an open-source software written in the R programming language [13].

2.2. Data Preprocessing

Preprocessing is the first step in text mining techniques and their application, playing a crucial role in the entire procedure [14]. To increase the coherence of the topics, each abstract was tokenized using bigrams which are the combination of consecutive unigrams. Although preprocessing seems trivial, since the text is downloaded to the computer as a readable format, it must be converted to lowercase and punctuation marks, dashes, brackets, numbers, space blanks and other characters removed. In addition, a standard list of words called "stopword" was identified and eliminated, since their main function is to make a sentence grammatically correct (i.e., articles and prepositions).

Data preprocessing was carried out using the web-based tool LDAShiny [15], a package to R programming language [13]. As a result of these operations, a document term matrix was created (dtm).

2.3. Identifying Research Topics

The topic model technique Latent Dirichlet Allocation (LDA) [15] was used to answer Q2, Q3, and Q4. It is based on Bayesian models and is seen as a development of Probabilistic Latent Semantic Analysis [16,17].

A topic may be defined as a multinomial distribution of words in the vocabulary where each word has a different probability within each topic [18]. LDA is one of the unsupervised text mining methods, in which themes or topics of documents can be identified from a larger collection of compiled documents, called corpus. LDA adds a prior sparse Dirichlet distribution on items in a document, using sampling Gibb [19] to generatively assign the probabilities of the topics of each term, and then group the documents into their respective topics, assuming that the documents exhibit a combination of multiple subjects in different proportions. The goal of using LDA is to infer or estimate the latent variables, that is, to compute their conditional distribution documents. Equation (1) shows the statistical assumptions behind the LDA's generative process.

$$p(\beta_K, \theta_D, z_D, w_D) = \prod_{k=1}^{K} p(\beta_K|\eta) \prod_{m=1}^{M} p(\theta_m|\alpha) \prod_{n=1}^{N} p(z_{m,n}|\theta_m) P(w_{m,n}|z_{m,n}, \beta_{m,k}) \quad (1)$$

where M denotes the number of documents, N is number of words in a given document, and each topic k is a multinomial distribution over the vocabulary and comes from a Dirichlet distribution $\beta_k \sim \mathrm{Dir}(\eta)$, the Dirichlet parameter η defines the smoothing of the words within topics, and α is the smoothing of the topics within documents. Every document is represented as a distribution over the topics and comes from a Dirichlet distribution $\theta_m \sim \mathrm{Dir}(\alpha)$. The joint distribution of all the hidden variables, β_K (topics), θ_M (document topic proportions within M), z_M (word topic assignments), and observed variables w_M (words in documents). The per-word topic assignment $z_{m,n}$, and the per-document topic distribution θ_m, are the latent variables and are not observed. Moreover, the word $w_{m,n}$ depends on the per-word topic assignment $z_{m,n}$ and on all the topics β_k (we retrieve the probability of $w_{m,n}$ (row) from $z_{m,n}$ (column) within the $K \times V$ topic matrix). We would have to condition on the only observed variable, that is the words within the documents, to infer the hidden structure with statistical inference. The conditional probability, also known as the posterior, is expressed by Equation (2).

$$p(\beta_K, \theta_M, z_M|w_M) = \frac{p(\beta_K, \theta_M, z_M, w_M)}{p(w_M)} \quad (2)$$

Although the posterior cannot be computed exactly due to the denominator [16], a close enough approximation to the true posterior can be achieved with statistical posterior inference. Mainly two types of inference techniques can be discerned: variational-based algorithms [20] and sampling-based algorithms [21]. An example of a sampling-based algorithm is the Gibbs sampler [22].

A simplified geometric interpretation of LDA is presented in Figure 1 considering only three words (w1,w2,w3) in the V-vocabulary and it is represented as a word simplex (V-dimensional). The word simplex is related to all the probability distribution of words. In addition, it can be seen how the topics, modeled as vocabulary distributions, are located within the simplex word (Figure 1). Figure 1 shows only three topics T, represented as a simplex topic of dimension (T-1). Thus, the documents modeled as distributions on the topics, are points on the simplex topic. For example document 1 would belong to topic 1; document 2 exhibits the same proportion in the three topics; while document 3 does not have proportions of topic 2.

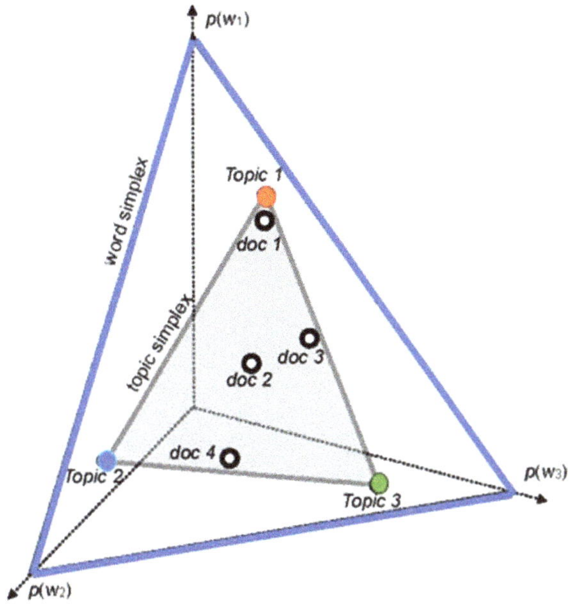

Figure 1. Geometric interpretation of LDA as a (V-1)-dimensional word simplex with V = w1, w2 and w3, with each point representing a discrete distribution of word probabilities. A point that is closer to one of the corners implies that the word has a higher probability mass. (Adapted from [18]).

2.3.1. Creation of LDA Model

LDA was used to extract meaningful information from the discovered articles. We combined the titles and abstracts of each article into a single variable. This variable was then used to serve as the text corpus for the entire data set.

Topic models are document latent variable models that leverage word correlations and latent semantic topics in a collection of texts [20]. This concept presupposes that the predicted number of topics k (i.e., latent variables) must be known in advance. Thus, the selection process of the right number of topics for a given collection of articles is not trivial. Simulations were carried out varying k from 4 to 30. 500 iterations were performed with the inference algorithm called Gibbs sampling [19]. A topic coherence metric [20] was used to estimate the quality of the LDA model. This is a measure of the human interpretability of a model of topics, and is believed to be a better indicator than computational metrics such as perplexity [23].

After determining the number of topics, we evaluated the most likely subject of each article and designated it as the article's primary topic.

2.3.2. Labeling Topics

Because algorithmic analyses are relatively restricted in their capacity to identify latent meanings of human language and the topics are not semantically labeled for the LDA model, manual labeling is regarded as a standard in topic modeling [24]. To provide a semantically correct interpretation, the topic was manually labeled by experienced clinicians and researchers independently using three sources of information: the most frequent word lists (most likely), a sample of the titles, and the abstracts of the five articles classified with the highest probability of belonging to a topic (Supplementary Materials, Table S1).

2.4. Quantitative Indices Used to Analyze the Trend of Topics

It is difficult to comprehend the subjects and trends intuitively due to the vast number of articles and hence the number of words. As a result, we employ certain quantitative indicators given by Xiong et al. [25]. The indexes are described below. The distribution of topics over time is obtained by

$$\theta_k^y = \frac{\sum_{d \in m} \theta_{dk}}{n^m} \tag{3}$$

where $d \in m$ represents the articles published in a given month, θ_{dk} is the proportion of the k-th topic in each item and n^m is the total number of articles published in the month [25].

Topic distribution across journals is defined as the ratio of the k-th topic in the journal

$$\theta_k^j = \frac{\sum_{d \in j} \theta_{dk}}{n^j} \tag{4}$$

where, $d \in j$ represents the articles in a particular journal, θ_{dk} the proportion of the k-th topic on each item, and n^j is the total number of articles published in the journal j.

The proportion of the k-th topic in country c is defined as the topic distribution over countries, that is

$$\theta_k^c = \frac{\sum_{d \in c} \theta_{dk}}{n^c} \tag{5}$$

where $d \in c$ represents the articles in a specific country, θ_{dk} is the proportion of the k-th topic in each article, and n^c is the total number of papers from the country c.

Topic distribution over time within a specific country, is defined as

$$\theta_k^{c,y} = \frac{\sum_{d \in c \cap d \in m} \theta_{dk}}{n^{c,m}} \tag{6}$$

where $d \in c \cap d \in m$ represents documents produced in a certain country over a certain month, θ_{dk} is the proportion of the k-th topic in each document, and $n^{c,m}$ the number of documents from country in month m.

We used simple regression slopes for each topic to facilitate the characterization of the topics in terms of their tendency [22]. The month was a dependent variable, and the proportion of the topics in the corresponding month was the response variable. The slopes derived by regression were positive or negative, and were classed as positive or negative trends, respectively. The statistical significance level was set at 0.01.

3. Results

3.1. Search Results

The initial database containing the documents retrieved after running the search query contained 161,421 documents; this sample was subjected to a filtering process in which repeated and poorly classified documents were eliminated, as well as those that did not contain a summary. There were a total of 126,334 papers in the final sample. Table 1 shows the summary produced, comprising basic statistics on the dataset studied.

A scientific production global map shows that COVID-19 research has been undertaken in all nations (excluding El Salvador, Central African Republic, South Sudan, Eritrea, Somaliland, Turkmenista, and the Democratic Republic of Korea) (Figure 2).

The top ten countries were the United States of America (26,814, 21.22%), China (11,375, 9.0%), Italy (7722, 6.11%) percent), United Kingdom (7522, 5.95 % percent), India (6726, 5.32%), Canada (3591, 2.84%), Spain (3465, 2.74%), Germany (3129, 2.48%), France (3129, 2.48%) and Iran (2843, 2.25%).

The results show that the articles published during the period between February 2020 and January 2022, experienced a compound monthly growth rate close to 34.6% (from 101 to 126,334) (Table 2).

In terms of sources (of the 7040 registered), the International Journal of Environmental Research and Public Health, PLoS ONE, and Scientific Reports have published the largest number of articles on COVID-19, having collectively published close to 5% of all publications on COVID-19 in the study period (Table 3).

Table 1. Main statistics about the COVID-19 collection.

Description		Result
Main information about data	Timespan	February 2020: January 2022
	Sources	7040
	Documents	126,334
	Average years from publication	1.46
Document contents	Keywords plus (id)	13,001
	Author's keywords (de)	112,867
Authors	Authors	440,259
	Author appearances	960,863
	Authors of single-authored Documents	5374
	Authors of multi-authored Documents	434,885
Authors collaboration	Single-authored documents	6698
	Documents per author	0.287
	Authors per document	3.48
	Co-authors per documents	7.61
	Collaboration index	3.64

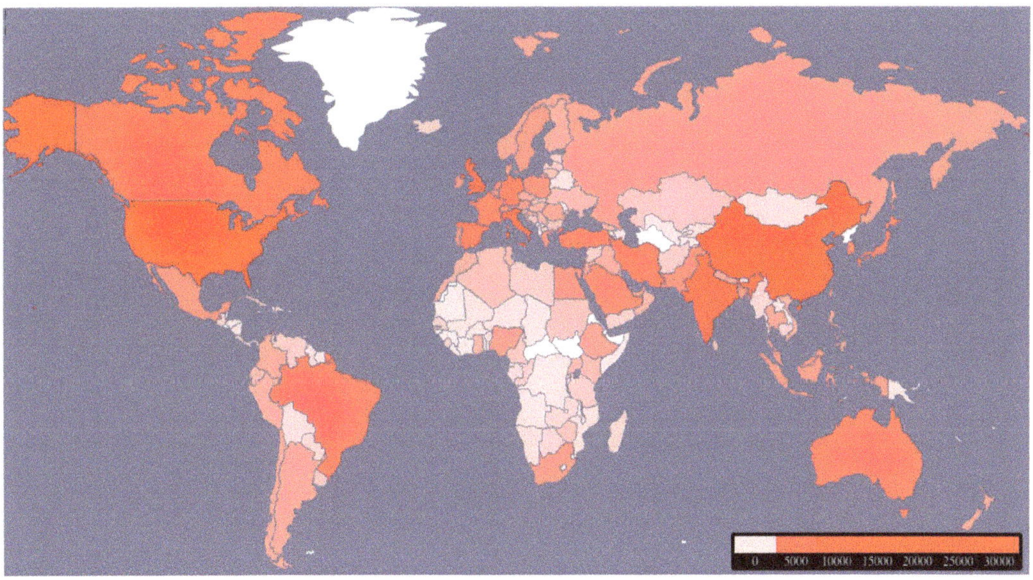

Figure 2. Geographical origin distribution of the 126,334 articles published on COVID-19 analyzed.

3.2. LDA Modeling and Topics

The LDA model with the highest coherence contains 16 topics. Table 4 shows for each of them the 15 most common terms, the label, and the number of published articles referring to them. The topics with the highest number of articles were: t_16 (Political and health factors), t_13 (Mental health), and t_15 (Etiopathogenesis), while the t_9 (Information synthesis methods) had the lowest number of articles.

3.2.1. Trend of Topics

The trend of each of the 16 topics over time was discovered. It can be observed that the probabilities of eight of them gradually increased over time (t_2 Prevention, t_3 Telemedicine, t_4 Vaccine immunity, t_5 Machine learning, t_7 Academic parameters, t_8, Risk factors and morbidity and mortality, t_9 Information synthesis methods, and t_13 Mental health), in five of them the probability decreased (t_6 Epidemiology, t_10 COVID-19 pathology complications, t_12 Diagnostic test, t_15 Etiopathogenesis, and t_16 Political and health factors), while the remainder fluctuated over time (t_1 Therapeutics, t_11 Pharmacological and therapeutic target, and t_14 Repercussion health services), without prominent trends (Figure 3).

Table 2. Main statistics about the COVID-19 collection.

Month	Year	Number	Accumulated
February	2020	101	101
March	2020	558	659
April	2020	2082	2741
May	2020	3476	6217
June	2020	4255	10,472
July	2020	4685	15,157
August	2020	4307	19,464
September	2020	4819	24,283
October	2020	5193	29,476
November	2020	4765	34,241
December	2020	4718	38,959
January	2021	17,640	56,599
February	2021	6345	62,944
March	2021	5984	68,928
April	2021	5421	74,349
May	2021	5578	79,927
June	2021	5803	85,730
July	2021	6121	91,851
August	2021	5529	97,380
September	2021	5736	103,116
October	2021	5880	108,996
November	2021	5546	114,542
December	2021	5593	120,135
January	2022	6199	126,334

Table 3. Top 10 most important sources in terms of number of publications.

Source	Abbreviation	*n*	(%)
International Journal of Environmental Research and Public Health	*Int. J. Environ. Res. Public Health*	3304	2.62
PLoS ONE	*PLoS ONE*	2057	1.63
Scientific Reports	*Sci. Rep.*	1348	1.07
Frontiers in Psychology	*Front. Psychol.*	997	0.79
BMJ Open	*BMJ Open*	923	0.73
Journal of Clinical Medicine	*J. Clin. Med.*	900	0.71
Journal of Medical Virology	*J. Med. Virol.*	865	0.68
Cureus	*Cureus*	817	0.65
Frontiers in Public Health	*Front. Public Health*	813	0.64
International Journal of Infectious Diseases	*Int. J. Infect. Dis.*	786	0.62

3.2.2. Topic Distributions of Various Journals

In Figure 3, we depict the topic distribution of journals as a heatmap, with the intensity of the pixel representing the probability that a given topic is mentioned in a certain journal. Although the content of many of the journals included in our study overlaps to some extent, it is feasible to identify journals that have relatively wide scopes, while others appear to specialize in certain topics. For instance, the journals Frontiers In Psychology and Frontiers In Psychiatry focus on the topic t_13 (Mental health). In addition, we performed a hierarchical cluster analysis on the contents of the selected journals by computing the Euclidean distance between each pair of journals. Dendrogram is shown on the left panel of Figure 4, where journals were classified into seven groups. Two of the 30 journals considered in the analysis formed the isolated cluster 6 (Vaccines) and cluster 7 (BMJ Case Rep.) while the remaining journals can be classified into five groups.

Table 4. 16 topics discovered from 126,334 articles published on COVID-19 in the period February 2020–January 2022. Each topic shows the 15 most likely terms (that is, the words with the highest probability), the label, and the number of published articles belonging to each topic.

Topic	Label	Top_terms	Articles *n* (%)
t_1	Therapeutics	treatment, trial, clinic, group, therapi, control, drug, effect, treat, clinic_trial, dose, efficaci, receiv, improv, random	3671 (2.91)
t_2	Prevention	survei, worker, particip, health, risk, healthcar, associ, prevent, cross, pandem, section, cross_section, factor, behavior, protect	6380 (5.05)
t_3	Telemedicine	servic, women, pandem, clinic, provid, telemedicin, visit, telehealth, health, pregnant, access, deliveri, person, consult, medic	4857 (3.84)
t_4	Vaccine inmunity	vaccin, antibodi, immun, respons, dose, igg, neutral, infect, anti, effect, hesit, mrna, individu, receiv, level	4146 (3.28)
t_5	Machine learning	model, base, predict, method, data, perform, propos, mask, develop, learn, imag, system, valid, time, detect	6781 (5.37)
t_6	Epidemiology	case, infect, countri, data, rate, number, transmiss, model, death, popul, spread, measur, epidem, diseas, outbreak	10,784 (8.54)
t_7	Academic parameters	student, pandem, educ, nurs, onlin, learn, medic, experi, social, train, school, particip, resid, program, media	7693 (6.09)
t_8	Risk factors and morbidity and mortality	mortal, risk, associ, sever, outcom, diseas, hospit, icu, higher, admiss, factor, death, cohort, group, clinic	10,665 (8.44)
t_9	Information synthesis methods	review, systemat, search, analysi, systemat_review, includ, meta, literatur, report, meta_analysi, databas, evid, data, pubm, identifi	1955 (1.55)
t_10	COVID-19 pathology complications	symptom, case, diseas, sever, clinic, report, infect, ct, children, present, group, find, pneumonia, includ, acut	7655 (6.06)
t_11	Pharmacological and therapeutic target	protein, drug, viral, human, viru, cell, target, bind, ac, spike, infect, potenti, activ, genom, variant	7467 (5.91)
t_12	Diagnostic test	test, posit, detect, pcr, sampl, infect, rt, neg, rt_pcr, assai, sensit, viral, diagnost, respiratori, swab	5635 (4.46)
t_13	Mental health	pandem, health, mental, anxieti, mental_health, stress, depress, psycholog, symptom, associ, social, impact, level, particip, increas	12,236 (9.69)
t_14	Repercusion health services	pandem, period, cancer, surgeri, compar, lockdown, impact, increas, emerg, time, surgic, decreas, number, chang, march	6412 (5.08)
t_15	Etiopathogenesis	infect, diseas, sever, respiratori, syndrom, acut, cell, acut_respiratori, immun, respiratori_syndrom, sever_acut, respons, system, inflammatori, associ	12,080 (9.56)
t_16	Political and health factors	health, pandem, public, system, manag, challeng, respons, global, commun, diseas, develop, emerg, public_health, provid, impact	17,917 (14.18)

3.2.3. Topic Distribution over Country

Following the methodology used in the analysis of journals, in Figure 5, we can see a heatmap with a dendrogram in the left panel. We only considered 35 countries for the analysis, of which 30 are considered leaders in the field of scientific research based on their publication volume according to the Nature Index [26]. In general, topics t_9 (Information synthesis methods), t_3 (Telemedicine) and t_1 (Therapeutics) were the ones that generated less interest from the countries evaluated, while t_16 (Political and health factors) was the most prevalent in South Africa, Australia, Ireland, Canada, Singapore, United Kingdom and United States of America (USA).

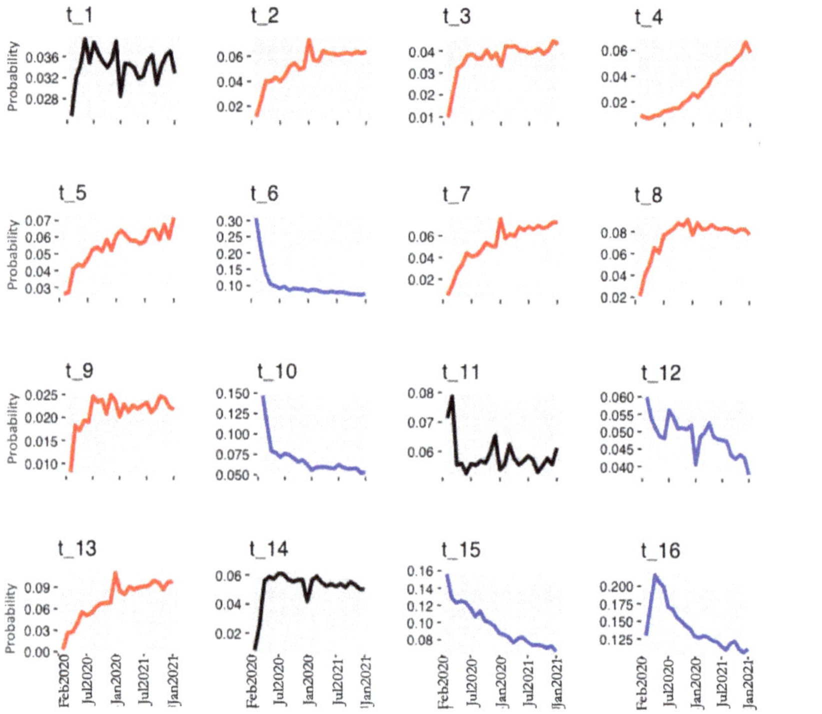

Figure 3. Topic trends research in COVID-19 during February 2020-January 2022.The red color indicates topics with increasing tendency, blue with decreasing tendency, and black fluctuating.

We also investigated the distribution of topics by country over time to determine how topics changed in various countries over time.

In general, t_4 (Vaccine inmunity) was the topic that showed a positive trend in all the countries (except Ireland) considered in the analysis, while t_15 (Etiopathogenesis) and t_16 (Political and health factors) showed a negative or fluctuating trend in the countries analyzed (Table 5).

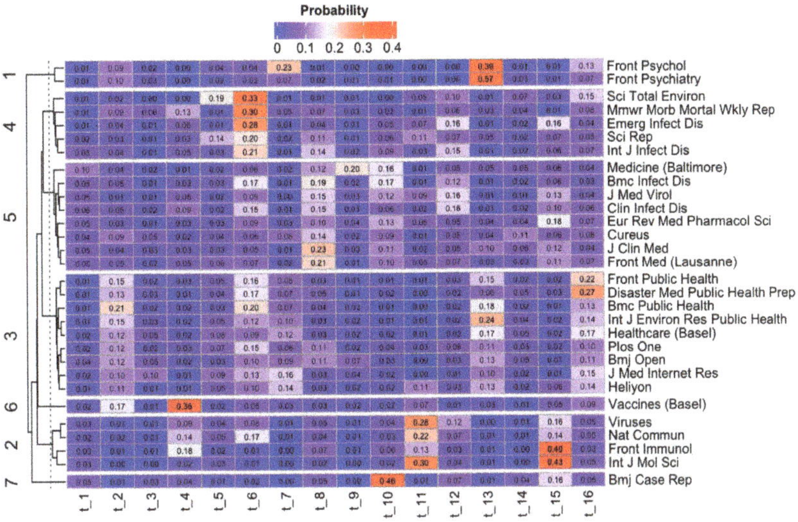

Figure 4. Heatmap overview of the proportional topic in the top-30 analyzed journals. Values are in percentages and row totals sum up to 100%.

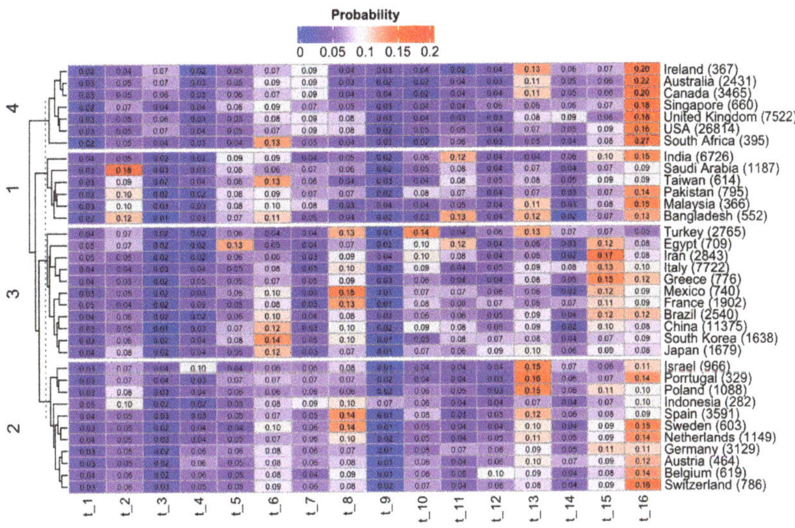

Figure 5. Heatmap overview of the proportional topic in the top-35 analyzed countries. Values are in percentages and row totals sum up to 100%. In parentheses, the number of articles is shown.

Table 5. Topic trends research in COVID-19 during February 2020–January 2022. Red color indicates increasing tendency, blue decreasing tendency, and white fluctuating or no prominent trends.

Country	t_1	t_2	t_3	t_4	t_5	t_6	t_7	t_8	t_9	t_10	t_11	t_12	t_13	t_14	t_15	t_16
USA																
China																
Italy																
United Kingdom																
India																
Spain																
Canada																
Germany																
Iran																
Turkey																
Brazil																
Australia																
France																
Japan																
South Korea																
Saudi Arabia																
Netherlands																
Poland																
Israel																
Pakistan																
Switzerland																
Greece																
Mexico																
Egypt																
Singapore																
Belgium																
Taiwan																
Sweden																
Bangladesh																
Austria																
South Africa																
Ireland																
Malaysia																
Portugal																
Indonesia																

4. Discussion

The rapid increase in publications related to COVID-19 is unprecedented in the scientific literature, even compared to the Zika virus outbreak in Latin America (January 2016), when the WHO declared a health emergency of international concern [27]. In this case, there were only 644 publications on PubMed for the first six months after the declaration, which highlights the big difference with the 15,557 publications on COVID-19 between February and July 2020. Another pertinent comparison can be made with the global pandemic caused by influenza A (H1N1), first detected in North America in 2009 [28]. In fact, while the first publication of clinical trials on COVID-19 was made 44 days after the declaration of a pandemic by the WHO [29], for H1N1, this occurred 190 days after the declaration [28]. However, not only the number of articles published was exceptional, but also the period of time between data collection and publication of the articles was surprising. This faster publication procedure was largely made possible by a shorter peer review process. Horbach [30] evidenced this in his study with the peer-review process of 14 medical publications. In fact, journal processing time was lowered by 49%. Researchers on topics related to COVID 19 have worked beyond their means, both researching and reviewing the literature, while it seems reasonable that journals might find it difficult to attract reviewers with relevant experience, as they are likely to be active scientists, it seems that journals are finding enough reviewers willing to review articles related to the coronavirus in a very short time [31].

Wanting to share information quickly has often led to a decrease in the evaluation time of articles and more lax reviews, having accepted articles of lower quality, prioritizing immediacy in information over quality [32]

The above confirms the growing public and scientific interest, given the fact that the disease represents a major threat to public health worldwide, but also to the economic and social consequences associated with it. Therefore, it is not surprising that COVID-19 research has seen an unprecedented increase since the beginning of the pandemic [33].

The results also suggest that the USA exceeds countries such as China, Italy, the United Kingdom, India, Canada, Spain, Germany, France, and Iran in number of articles published. This fact is not surprising given the amount of USA government funds that were invested in COVID-19 research [33]. Publications from other geographic areas are substantially less abundant, with gaps particularly visible in Africa, Latin America, Eastern Europe, and Central Asia.

Some systematic reviews have been published on COVID-19, these require a lot of research time and have generally focused on specific aspects of the pandemic [4–6]. Those works also analyzed reports on COVID-19 in the media [34], social networks such as Twitter [35], and Sina-Weibo (a Twitter system used in China) [36].

Unlike the aforementioned reviews, this study did not focus on specific aspects of the pandemic, but instead reviewed all the scientific literature related to COVID-19 during the two years after the pandemic was declared. In particular, LDA allowed for the evaluation of the variation of the research in the medium term. This technique also offers the possibility to conduct a more in-depth analysis on a particular topic identified. We identified 16 topics (namely, t_1 Therapeutics, t_2 Prevention; t_3 Telemedicine, t_4 Vaccine immunity, t_5 Machine learning t_6 Epidemiology, t_7 Academic parameters, t_8 Risk factors and morbidity and mortality, t_9 Information synthesis methods, t_10 COVID-19 pathology complications, t_11 Pharmacological and therapeutic target, t_12 Diagnostic test, t_13, Mental health, t_14 Repercusion health services, t_15 Etiopathogenesis and t_16 Political and health factors) it was possible to categorize the scientific papers on COVID-19 that were published during the first two years of the pandemic.

Älgå et al. [2] explored the scientific literature on COVID-19 (16,670 articles, using PubMed as in our study) in the time period between February and June 2020 using LDA. In this case, 14 topics were identified (namely, Therapies and vaccines, Risk factors, Health care response, Epidemiology, Disease transmission, Impact on health care practices, Radiology, Epidemiological modeling, Clinical manifestations, Protective measures, Immunology, Pregnancy, and Psychological impact). Therefore, it was observed that some of the topics coincide with some labeled in this study. However, there were differences regarding the most prevalent topics. While [2] reported that the most prevalent topics were health care response, clinical manifestations, and psychological impact, in our case they were t_16 Political and health factors, t_13, Mental health, and t_15 Etiopathogenesis. These differences can be explained by the time period evaluated. Furthermore, since the COVID-19 epidemic is still ongoing, the topics of study will most likely continue to change over time.

Among the rising academic attempts to address COVID-19 problems, a large portion of the research has naturally concentrated on elements relating to Political and health factors, Epidemiology and Risk factors, morbidity and mortality, Mental health, and, Etiopathogenesis.

It should be noted that the study was constrained by the exclusion of grey literature, books, book chapters, reviews, and reports. The data was acquired entirely from the PubMed database and only scientific articles were considered. Academics may opt to conduct future research using other databases, such as Scopus and Web of Science, which include non-indexed journals not included in PubMed. In this sense, future research might compare the findings of this study to those obtained from other databases.

In sum, the findings of this study may be used to illustrate how the medical research community reacts and what issues are prioritized. On the other hand, it was easy to identify how research efforts were distributed globally and how they changed over time.

5. Conclusions

Scientific research and data play a very important role in the early control and prevention of disease outbreaks and epidemics. It is of great interest to quickly share all information with the public, researchers, government organizations, and institutes, both nationally and internationally. An example of this was the surprising amount of studies on COVID-19 that have been published since the novel coronavirus was originally identified.

In this work, the variations in the COVID-19 study that were available over the first two years of the pandemic were highlighted. Therefore, this study demonstrated that the United States of America, China, and Italy have leading roles in COVID-19 research. In addition, through LDA modeling, a list of 16 topics was obtained and important temporal trends could be identified.

In sum, the outcomes can provide new study guidelines, as well as aid in understanding research trends, in the context of worldwide occurrences, useful for academics and policymakers. Furthermore, the results achieved showed that topic modeling is a quick and efficient way to evaluate the progress of a huge and quickly developing a research topic, such as COVID-19. Additionally, and perhaps even more importantly, the methodology used has the potential to identify topics for future research, not only in studies on pandemics but also as a tool for the identification and review of scientific literature in other fields which may be of great public interest.

Supplementary Materials: The following supporting information can be downloaded at: https://www.mdpi.com/article/10.3390/computation10090156/s1, Table S1: Five-top papers for each estimated topic, based on topic-document probability matrix.

Author Contributions: Conceptualization J.D.L.H.-M., S.M. and M.J.F.-G.; methodology, J.D.L.H.-M. and S.M.; software J.D.L.H.-M.; validation, Y.G.S., and M.J.F.-G.; formal analysis, J.D.L.H.-M., S.M. and M.J.F.-G.; investigation, J.D.L.H.-M. and Y.G.S.; data curation, J.D.L.H.-M.; writing—original draft preparation, J.D.L.H.-M.; writing—review and editing, S.M., Y.G.S. and M.J.F.-G.; visualization, J.D.L.H.-M.; supervision, M.J.F.-G. and S.M.; project administration, S.M. and M.J.F.-G. All authors have read and agreed to the published version of the manuscript.

Funding: This research received no external funding.

Institutional Review Board Statement: Not applicable.

Informed Consent Statement: Not applicable.

Data Availability Statement: Not applicable.

Acknowledgments: Regarding Susana Mendes, this work was funded by national funds through FCT - Fundação para a Ciência e a Tecnologia, I.P., under the project MARE (UIDB/04292/2020 and UIDP/04292/2020) and the project LA/P/0069/2020 granted to the Associate Laboratory ARNET.

Conflicts of Interest: The authors declare no conflict of interest.

References

1. *WHO Announces COVID-19 Outbreak a Pandemic;* Word Health Organization-Regional Office for Europe: Copenhagen, Denmark, 2020; Available online: http://www.euro.who.int/en/health-topics/health-emergencies/coronavirus-COVIDcovid-19/news/news/2020/3/who-announces-COVID-19-outbreak-a-pandemic (accessed on 1 June 2020).
2. Älgå, A.; Eriksson, O.; Nordberg, M. Analysis of Scientific Publications during the Early Phase of the COVID-19 Pandemic: Topic Modeling Study. *J. Med. Internet Res.* **2020**, *22*, e21559. [CrossRef] [PubMed]
3. Larsen, P.O.; von Ins, M. The rate of growth in scientific publication and the decline in coverage provided by science citation index. *Scientometrics* **2010**, *84*, 575–603. [CrossRef] [PubMed]
4. Li, X.; Guan, B.; Su, T.; Liu, W.; Chen, M.; Waleed, K.B.; Zhu, Z. Impact of cardiovascular disease and cardiac injury on in-hospital mortality in patients with COVID-19: A systematic review and meta-analysis. *Heart* **2020**, *106*, 1142–1147. [CrossRef] [PubMed]

5. Parasa, S.; Desai, M.; Chandrasekar, V.T.; Patel, H.K.; Kennedy, K.F.; Roesch, T.; Sharma, P. Prevalence of gastrointestinal symptoms and fecal viral shedding in patients with coronavirus disease 2019: A systematic review and meta-analysis. *JAMA Netw. Open.* **2020**, *3*, e2011335. [CrossRef]
6. Cortegiani, A.; Ingoglia, G.; Ippolito, M.; Giarratano, A.; Einav, S. A systematic review on the efficacy and safety of chloroquine for the treatment of COVID-19. *J. Crit. Care* **2020**, *57*, 279–283. [CrossRef]
7. Aristovnik, A.; Ravšelj, D.; Umek, L. A bibliometric analysis of COVID-19 across science and social science research landscape. *Sustainability* **2020**, *12*, 9132. [CrossRef]
8. Haghani, M.; Bliemer, M.C.; Goerlandt, F.; Li, J. The scientific literature on Coronaviruses, COVID-19 and its associated safety-related research dimensions: A scientometric analysis and scoping review. *Saf Sci.* **2020**, *129*, 104806. [CrossRef]
9. Doanvo, A.; Qian, X.; Ramjee, D.; Piontkivska, H.; Desai, A.; Majumder, M. Machine learning maps research needs in COVID-19 literature. *Patterns* **2020**, *1*, 100123. [CrossRef]
10. Mao, X.; Guo, L.; Fu, P.; Xiang, C. The status and trends of coronavirus research: A global bibliometric and visualized analysis. *Medicine* **2020**, *99*, e20137. [CrossRef]
11. Aria, M.; Cuccurullo, C. Bibliometrix: An R-toolfor comprehensive science mapping analysis. *J. Informetr.* **2017**, *11*, 959–975. [CrossRef]
12. Cobo, M.J.; López-Herrera, A.G.; Herrera-Viedma, E.; Herrera, F. Science mapping software tools: Review, analysis, and cooperative study among tools. *J. Assoc. Inf. Sci. Technol.* **2011**, *62*, 1382–1402. [CrossRef]
13. R Core Team. R: A Language and Environment for Statistical Computing. 2019. Available online: https://www.r-project.org (accessed on 1 May 2021).
14. Vijayarani, S.; Ilamathi, M.J.; Nithya, M. Preprocessing techniques for text mining-an overview. *Int. J. Comput. Sci. Commun. Netw.* **2015**, *5*, 7–16.
15. De La Hoz-M, J.; Fernández-Gómez, M.J.; Mendes, S. LDAShiny: An R package for exploratory review of scientific literature based on a Bayesian probabilistic model and machine learning tools. *Mathematics* **2021**, *9*, 1671. [CrossRef]
16. Blei, D.M.; Ng, A.Y.; Jordan, M.I. Latent Dirichlet allocation. *J. Mach. Learn Res.* **2003**, *3*, 93–1022.
17. Hornik, K.; Grün, B. Topicmodels: An R package for fitting topic models. *J. Stat. Softw.* **2011**, *40*, 1–30. [CrossRef]
18. Syed, S.; Borit, M.; Spruit, M. Using Machine Learning to Uncover Latent Research Topics in Fishery Models. *Rev. Fish. Sci. Aquac.* **2018**, *26*, 319–336. [CrossRef]
19. Geman, S.; Geman, D. Stochastic relaxation, Gibbs distributions, and the Bayesian restoration of images. *IEEE Trans. Pattern Anal. Mach. Intell.* **1984**, *6*, 721–741. [CrossRef]
20. Blei, D.M.; Lafferty, J.D. A correlated topic model of science. *Ann. Appl. Stat.* **2007**, *1*, 17–35. [CrossRef]
21. Roder, M.; Both, A.; Hinneburg, A. Exploring the space of topic coherence measures. In Proceedings of the Eighth ACM International Conference on Web Search and Data Mining, New York, NY, USA, 31 January–6 February 2015; pp. 399–408.
22. Porteous, I.; Newman, D.; Ihler, A.; Asuncion, A.; Smyth, P.; Welling, M. Fast Collapsed Gibbs Sampling for Latent Dirichlet Allocation. In Proceedings of the 14th ACM SIGKDD International Conference on Knowledge Discovery and Data Mining (KDD '08), Las Vegas, NV, USA, 24–27 August 2008; ACM Press: New York, NY, USA, 2008; pp. 569–577. [CrossRef]
23. Griffths, T.L.; Steyvers, M. Finding scientific topics. *Proc. Natl. Acad. Sci. USA* **2004**, *101* (Suppl. S1), 5228–5235. [CrossRef]
24. Chang, J.; Gerrish, S.; Wang, C.; Boyd-Graber, J.L.; Blei, D.M. Reading tea leaves: How humans interpret topic models. In *Advances in Neural Information Processing Systems*; MIT Press: Cambridge, MA, USA, 2009; pp. 288–296.
25. Xiong, H.; Cheng, Y.; Zhao, W.; Liu, J. Analyzing scientific research topics in manufacturing field using a topic model. *Comput. Ind. Eng.* **2019**, *135*, 333–347. [CrossRef]
26. Nature Index. Available online: https://www.natureindex.com/annual-tables/2021/country/all/all (accessed on 19 March 2022).
27. Wilder-Smith, A.; Osman, S. Public health emergencies of international concern: A historic overview. *J. Travel Med.* **2020**, *27*, taaa227. [CrossRef] [PubMed]
28. Greenberg, M.E.; Lai, M.H.; Hartel, G.F.; Wichems, C.H.; Gittleson, C.; Bennet, J.; Basser, R.L. Response to a monovalent 2009 influenza A (H1N1) vaccine. *New Eng. J. Med.* **2009**, *361*, 2405–2413. [CrossRef] [PubMed]
29. Borba, M.G.S.; Val, F.F.A.; Sampaio, V.S.; Alexandre, M.A.A.; Melo, G.C.; Brito, M.; Lacerda, M.V.G. Effect of high vs low doses of chloroquine diphosphate as adjunctive therapy for patients hospitalized with severe acute respiratory syndrome coronavirus 2 (SARS-CoV-2) infection: A randomized clinical trial. *JAMA Netw. Open* **2020**, *3*, e208857. [CrossRef] [PubMed]
30. Horbach, S.P. Pandemic publishing: Medical journals strongly speed up their publication process for COVID-19. *Quant. Sci. Stud.* **2020**, *1*, 1056–1067. [CrossRef]
31. Dewan, H.; Nishan, M.; Sainudeen, S.; Sanskriti Jha, K.; Mahobia, A.; Tiwari, R.V.C. COVID 19 Scoping: A Systematic Review and Meta-Analyisis. *J. Pham. Bioallied Sci.* **2021**, *13*, S938. [CrossRef]
32. Darsono, D.; Rohmana, J.A.; Busro, B. Against COVID-19 Pandemic: Bibliometric Assessment of World Scholars' International Publications related to COVID-19. *J. Komun. Ikat. Sarj. Komun. Indones.* **2020**, *5*, 75–89. [CrossRef]
33. Funding Opportunities Specific to COVID-19. National Institutes of Health. U.S. Department of Health and Human Services. Available online: https://grants.nih.gov/grants/guide/COVID-Related.cfm (accessed on 10 April 2022).
34. Liu, Q.; Zheng, Z.; Zheng, J.; Chen, Q.; Liu, G.; Chen, S.; Ming, W.K. Health communication through news media during the early stage of the COVID-19 outbreak in China: Digital topic modeling approach. *J. Med. Internet Res.* **2020**, *22*, e19118. [CrossRef]

35. Abd-Alrazaq, A.; Alhuwail, D.; Househ, M.; Hamdi, M.; Shah, Z. Top concerns of tweeters during the COVID-19 pandemic: Infoveillance study. *J. Med. Internet Res.* **2020**, *22*, e19016. [CrossRef]
36. Han, X.; Wang, J.; Zhang, M.; Wang, X. Using social media to mine and analyze public opinion related to COVID-19 in China. *Int. J. Environ. Res. Public Health* **2020**, *17*, 2788. [CrossRef]

computation

MDPI

Article

Evaluation of the Effectiveness of Community Activities Restriction in Containing the Spread of COVID-19 in West Java, Indonesia Using Time-Series Clustering

Dhika Surya Pangestu [1], Sukono [2] and Nursanti Anggriani [2]

[1] Master's Program of Mathematics, Faculty of Mathematics and Natural Sciences, Universitas Padjadjaran, Sumedang, Jatinangor 45363, West Java, Indonesia
[2] Department of Mathematics, Faculty of Mathematics and Natural Sciences, Universitas Padjadjaran, Sumedang, Jatinangor 45363, West Java, Indonesia
* Correspondence: sukono@unpad.ac.id

Abstract: The purpose of this research is to classify time-series data on the number of daily COVID-19 cases based on the dynamics. This research aims to evaluate the effectiveness of community activity restrictions in suppressing the number of new cases of COVID-19 in cities and regencies in West Java. We performed time-series clustering on daily positive case data for COVID-19 in 27 cities and regencies in West Java Province, Indonesia for this study. The k-medoids clustering algorithm was used for clustering, with shape-based lock step measures, specifically, the cross correlation-based distance. We used daily new infected cases data for COVID-19 in 27 cities and regencies in West Java Province during the worst situation. We used data from 1 July 2021 to 31 September 2021 and from 1 January 2022 to 31 May 2022, during the Emergency Community Activity Restriction period (PPKM). According to our findings, the optimal number of clusters that could be formed from the data we had was 4 clusters for the first period and 2 clusters for the second period, with silhouette value of 0.2633 and 0.6363, respectively. For the first period, we discovered that PPKM was successful in clusters 1 and 2, namely in 25 cities/districts in West Java, except for Bogor and Depok, while for the second period, we found PPKM to be effective in reducing the number of COVID-19 cases throughout cities and regencies in West Java. This shows there is an improvement from the implementation of PPKM in the first period. We also found that the cluster that was formed was not only influenced by the effectiveness of the PPKM, but also by geography. The closer a city is to a hotspot region for the spread of COVID-19, the earlier the increase in the number of new COVID-19 cases will occur.

Keywords: COVID-19 cases; West Java Province; k-medoids clustering algorithm; shape-based lock step measures; cross the correlation-based distance

Citation: Pangestu, D.S.; S.; Anggriani, N. Evaluation of the Effectiveness of Community Activities Restriction in Containing the Spread of COVID-19 in West Java, Indonesia Using Time-Series Clustering. *Computation* **2022**, *10*, 153. https://doi.org/10.3390/computation10090153

Academic Editors: Simone Brogi and Vincenzo Calderone

Received: 14 August 2022
Accepted: 1 September 2022
Published: 4 September 2022

1. Introduction

COVID-19 was verified to have been first seen on 2 March 2020, in Indonesia. At the time, two persons had been exposed to COVID-19 through interaction with Japanese residents. This was discovered when a Japanese citizen was diagnosed with the coronavirus after leaving Indonesia and landing in West Java [1]. Since the virus's first appearance, the number of COVID-19 cases in Indonesia has steadily increased, with 6,054,973 persons affected as of 31 May 2021 [2]. According to the Worldometer, Indonesia is ranked 14th in the world and 4th in Asia for COVID-19 positive cases [3]. West Java is one of Indonesia's provinces. According to the 2020 population census, West Java has the greatest population in Indonesia, totaling 48,274,162 persons [4]. West Java, being the province with the highest population, is one of the provinces that provide a significant portion of the total number of COVID-19 cases in Indonesia. West Java reported that there have been 1,107,911 confirmed cases of COVID-19 as of 31 October 2021 [5], with a total of 216 active cases.

Figure 1 shows the dynamic of the COVID-19 daily new case in West Java from March 2020 to May 2022. According to Figure 1, the worst of the COVID-19 outbreak in West Java happened between February and March of 2022. At the time, the number of daily COVID-19 cases in West Java Province reached 16,251 cases per day, on 17 February 2022; this kind of significant increase in daily COVID-19 cases happened twice in West Java. Previously, it happened between July and August of 2021. At the time, the number of daily COVID-19 cases in West Java Province reached 11,101 cases per day on 13 July 2021. The government introduced emergency PPKM on 3–25 July 2021, followed by PPKM 4 levels on 26 July–2 August 2021 [6], to reduce the number of daily instances of COVID-19, which climbed rapidly in the period July–August 2021. However, assessing the success of this intervention would be difficult without an analysis that describes how the COVID-19 pandemic will behave. The COVID-19 pandemic has wreaked havoc on infrastructure, the economy, and, most crucially, human lives. Furthermore, the impact of policies implemented may differ in each city and district in West Java Province, depending on how the community views it, the number of first instances when the intervention is implemented, and so on. As a result, it is crucial to conduct a study of the impact of policies enacted by West Java's cities and regencies. This may be accomplished by grouping cities and regencies with comparable dynamics of daily COVID-19 instances. Cluster analysis can be used to do this.

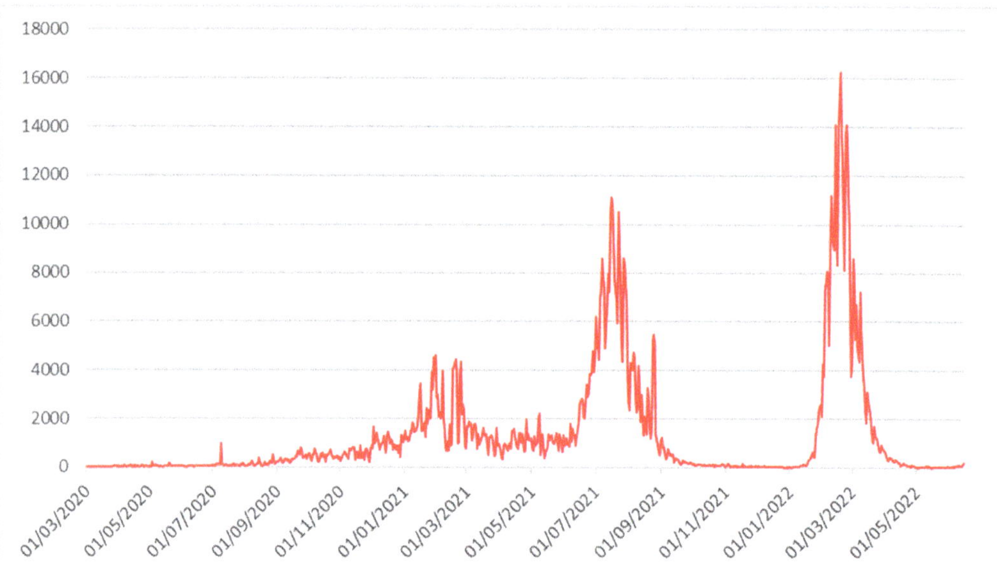

Figure 1. Daily positive cases of COVID-19 in West Java.

Cluster analysis is a technique for identifying groups in a data collection in order to gather data in one group that is relatively similar to other groups while having apparent distinctions. Cluster analysis may be applied to time series data, which has somewhat different grouping techniques and algorithms than cross-sectional data. Clustering on time-series data is commonly used to uncover intriguing patterns in a collection of time-series data [7,8]. The clustering of time-series data is classified into two categories: The first group is used to detect patterns that emerge often in the dataset [9]. The second group is a strategy for detecting patterns that appear unexpectedly in a data collection, or patterns that are significantly different from other data in the same dataset [10]. There have been several studies regarding cluster analysis on COVID-19 data.

In 2020, Zarikas et al. [11] conducted time-series clustering on data on COVID-19 cases with country data. Hierarchical analysis was used with the Euclidean distance measure. The variables used are active cases, active cases per population, and active cases per population and per area. Zarikas et al. found that the surface area of each country is a parameter influencing the criticality of the situation, i.e., geography matters. Later in the same year in 2020, Alvarez et al. [12] proposed a clustering method for identifying groups of countries with a similar spread of the coronavirus. The variable of interest is the number of daily infections per country. The method used is a non-parametric method, namely Hierarchical Trees (HT) and the Minimum Spanning Trees (MST). Alvarez et al. found that there were groups of countries with differentiated contagion dynamics, both in the number of contagions and at the time of the greatest transmission of the disease. It is concluded that the actions taken by the countries, the speed at which they were taken, and the number of tests carried out may explain part of the differences in the dynamics of contagion. Abdullah et al. [13] in 2021 conducted a study on time series clustering on Indonesian COVID-19 case data. He used confirmed, death, and recovered cases data of COVID-19 provinces in Indonesia, with the method utilized being K-means clustering. Abdullah et al. found that there were three provincial clusters in Indonesia based on the spread of COVID-19 that occurred. Elsi et al. [14] in 2020 conducted a mapping of Indonesia's national food security during the COVID-19 pandemic. The method used was K-medoids clustering with the variable of interest, monthly per capita expenditure in urban and rural areas by province, and groups of goods consisting of 33 data records (2011–2018). Elsi et al. found that 42% of Indonesia still has low food security as evidenced by the fulfillment of higher food needs than non-food.

Based on our literature review, we did not find any time-series clustering that specifically discusses the effectiveness of policies taken in an area. Generally, research on time-series clustering in COVID-19 only pays attention to location and compares the dynamics of cases between regions. To evaluate the containment policy for the spread of COVID-19, containment policy needs to be carried out simultaneously between regions, and the type of containment policy used in each area is relatively similar. Moreover, this requires special attention to the type of distance measures used in the clustering process.

In this study, we propose a method to evaluate the effectiveness of COVID-19 containment policies applied to an area. We propose a clustering method using shape-based lock-step distance measures, namely cross-correlation. Cross-correlation is a measure of distance that shows the similarity between datasets. This makes the similarity between time-series data in one cluster maximum and minimizes the similarity between different clusters. In addition, the nature of cross-correlation-based distance, which is a lock-step distance measure, makes the clustering carried out to compare the raw values of the data at the same time frame. The novelty of this research is the usage of shape-based lock-step distance measures in the clustering process. In the context of evaluating policies on handling COVID-19, our proposed method provides a series of clusters that are generated from data on COVID-19 new cases, on the exact date of the case of the same incident across different places. Thus, the resulting cluster can be more accurate in describing the development of COVID-19 cases in each formed cluster because the comparisons were made on the same date for each region.

This study aims to evaluate the effectiveness of implementing community activity restrictions (PPKM) in suppressing the spread of COVID-19 in 27 cities and regencies in West Java. As a result, in this study, we used a cluster analysis with lock step distance measures to determine the impact of the government's policies on the cities and regencies in West Java. This is a strategic thing to do, because we will be able to create numerous clusters based on the peculiarities of the dynamics of COVID-19. The results of this study can be used by the government to evaluate the effectiveness of containment policies taken in each region and identify structural similarities in the dynamics of COVID-19 that occur in each region. Thus, we hoped that the government will be able to evaluate the effectiveness of the policies taken more objectively and able to formulate better policies in dealing with the spread of COVID-19 with this information.

2. Overview

2.1. Coronavirus Disease 2019 (COVID-19)

SARS-CoV-2 is a virus that infects the respiratory tract and produces the infectious illness COVID-19. The World Health Organization (WHO) initially learned about this new virus on 31 December 2019, in Wuhan, China [15]. Coronavirus is a virus that spreads from animal to animal and can infect people. This virus's native hosts are bats, although numerous other animal species have been discovered as potential contributors. MERS-CoV can be transferred to humans through camels, but SARS-CoV-1 can be transmitted through civets [16]. A person who tests positive for COVID-19 might experience a wide range of symptoms, from minor aches and pains to major sickness. Symptoms might emerge anywhere between 2 and 14 days after being exposed to the virus. The most common symptoms of COVID-19 infection are fever and cough, however, there are other signs and symptoms to consider [17]. On 11 March 2020, WHO declared COVID-19 a pandemic [15]. According to statistics from China at the time [18], adults, particularly those with congenital defects, have a higher risk of getting infected by severe COVID-19 cases and a higher fatality rate than younger persons. According to data from the European Economic Area/European Union (for countries where data are available), roughly 20–30% of confirmed COVID-19 patients are hospitalized and 2% have severe disease. People with more severe symptoms, on the other hand, are more likely than those with less severe symptoms to get tested. As a result, the real proportion of persons who need to be hospitalized as a percentage of the overall number of infected people is lower than the figures reflect. Those aged 60 and up, as well as those with a congenital illness, are more likely to be hospitalized [19].

2.2. Time-Series Clustering

Clustering is a technique for identifying groups in a data collection to obtain data that are relatively similar in one group and have distinct distinctions between them [20,21]. Time-series clustering is a unique sort of clustering. A temporal sequence is made up of a series of nominal symbols from a certain alphabet, while a time series is made up of a continuous series of real value elements [22]. Because the feature values of time-series data vary with time, they are categorized as dynamic data. This implies that the value of each time-series point is one or more observations made chronologically. Time-series data are a sort of temporal data that contain a lot of dimensions and a lot of spaces [23,24]. Clustering on time-series data is commonly used to uncover intriguing patterns in a collection of time-series data [7,8]. The clustering of time-series data is classified into two categories: The first group is used to detect patterns that emerge often in the dataset [9]. The second group is a strategy for detecting patterns that appear unexpectedly in a data collection, or patterns that are significantly different from other data in the same dataset [10].

In brief, locating clusters of time-series data may help solve real-world issues in a variety of domains, such as identifying dynamic changes in time series and detecting connections across time series [25]. It may be used to locate firms with comparable stock price movements in a financial database, for example. Predictions and recommendations: a hybrid method that combines clustering and per-cluster function approximation can assist users in making predictions and making suggestions [26–28]. For example, in scientific databases, this can address difficulties such as predicting today's patterns by locating the solar magnetic wind. Pattern discovery: searching the database for intriguing patterns. Different daily sales trends of particular items at a store, for example, can be identified in a marketing database. In the phenomenon of COVID-19, time-series clustering has been carried out for creating the home dwell time clusters [29], estimation of the dynamics of COVID-19 in states [30], and also for the COVID-19 pandemic evolution [31]. This study utilizes time-series clustering to locate cities and regencies with comparable dynamics of daily positive COVID-19 instances, as well as how the daily case patterns are with the introduction of community activity restrictions (PPKM).

3. Materials and Methods

3.1. Materials

The study's research object is daily COVID-19 infected case data from 27 cities and regencies in West Java Province. The data were collected from 1 July 2021 to 31 September 2021 and from 1 January 2022 to 31 May 2022, during the Emergency Community Activity Restriction period (PPKM). Pikobar-West Java COVID-19 Information and Coordination Center [32] was the source of the data used in this study. For data management and visualization, we used R software version 4.1.2 [33]. R is a programming language for statistical computing and graphics created by statisticians Ross Ihaka and Robert Gentleman. Currently, R is supported by the R Core Team and the R Foundation for Statistical Computing based in Vienna, Austria. For data visualization and transformation, we use ggplot2 [34] and reshape [35] packages. As for the time-series clustering process, we use TSdist [36], factoextra [37], and NbClust packages [38].

3.2. Methods

3.2.1. Clustering Daily Positive Case Data Using K-Medoids with Cross-Correlation Based Distance

K-medoids is comparable to clustering or partitioning around medoids (PAM). The k-medoids approach is based on identifying k representative items among the data set's objects. A representative object, known as a centroid, is used in clustering. The representative object in k-Medoids is also known as the group medoid. K-medoid can be used on objects with very big values that vary from the data distribution to address the difficulty of utilizing k-means. This approach is preferable to most non-hierarchical clustering algorithms based on the minimal value of the sum of the squared estimate of error because it is more resilient (SSE) [39].

3.2.2. Calculating Cross-Correlation Based Distance

Calculating the distance metric utilized in k-medoids is the first step. We employ a distance measure with shape-based lock step distance features in this study. We use this metric to create clusters from raw data values and compare them with the same latency. We employ a distance metric called time-series distance that is based on the cross-correlation between two numerical time series. The distance between two numerical time series based on cross-correlation is determined as follows [40],

$$d_{i,j} = \sqrt{\frac{1 - \rho^2_{i,j,0}}{\sum_{k=1}^{max} \rho^2_{i,j,k}}}, \quad (1)$$

where $\rho^2_{i,j,k}$ shows the cross-correlation between the two-time series x_i and y_j at lag k and max is the maximum lag.

3.2.3. Determining the Number of Optimal Clusters with Elbow Methods

The elbow method was then used to calculate the number of clusters in this investigation. This approach is useful for calculating the number of clusters that should be used. The user searches for changes in slope to discover the ideal number of clusters using the elbow technique, in which the number of squares in each number of clusters is computed and graphed, and the user looks for changes in slope to determine the optimal number of clusters. The following formula is used to determine the SSE of the elbow method,

$$SSE = \sum_{k=1}^{k} \sum_{x_i \in S_k} ||x_i - c_k||^2, \quad (2)$$

where k is the number of groups in the algorithm used, x_i is the number of data, and c_k is the number of cluster members in the k-th cluster. The elbow method examines the proportion of variance expressed as a function of the number of clusters [41]. The basic concept is to pick a point when the increased cost is no longer worth the declining return [42]. This is a visual method, starting with $k = 2$, and growing at each step by 1 step, while calculating the clusters and the costs associated with increasing the number of clusters. At some values for k, the cost drops drastically, and after that it starts to slope, and this is the optimal value of k. The reason is that after the k, when the value of k is increased or the number of clusters is increased, the new clusters that are formed no longer have a significant difference with those that have been created previously, or the new clusters created will be very similar to some of the existing clusters [43].

3.2.4. Clustering Daily Positive Case Data Using K-Medoids

The last step is clustering using k-medoids. The steps for clustering using the k-medoids method are as follows:

(a) Calculate the distance of each object using cross correlation-based distance with Equation (1).
(b) Calculate v_j for each object j with $d_i = \sum_{j=1}^{n} d_{ij}$

$$v_j = \sum_{i=1}^{n} \frac{d_{ij}}{d_i},\ j = 1, \ldots, n, \tag{3}$$

where

d_{ij} : Cross correlation distance matrix elements
v_j: Standardize the number of rows for each column j

(c) Sort v_j from smallest to largest. Choose k clusters that have the first smallest v_j as the center (medoid).
(d) Allocate non-medoid objects to the nearest medoid based on the cross correlation-based distance.
(e) Calculate the total distance from the non-medoid cluster to the center.
(f) Define a new medoid for each cluster which is an object that minimizes the total distance to other objects in the cluster. Update the existing medoid in each cluster by replacing it with a new medoid obtained from the existing cluster.
(g) Allocate non-medoid objects to the nearest medoid based on the cross correlation-based distance.
(h) Calculate the total distance from the non-medoid cluster to the center.
(i) If the number of new centers differs from the total distance of the cluster centers in the first iteration, change the center (medoid). Otherwise, the iteration is stopped and the result becomes the final cluster.

The number of groups (k) in k-medoids is selected based on the elbow method.

3.3. *Cluster Internal Validation*

The intrinsic information in the data is utilized to assess the quality of the clustering that has been done in the internal validation of the dataset, which uses the cluster partition as input. The size of the cluster division that represents its compactness, connectivity, and separation is chosen for internal validation [44]. Connectivity is a metric that represents closeness [45]. Separation quantifies the distance between cluster centroids, whereas compactness examines the homogeneity of the clusters produced by looking at intra-cluster variation. Compactness and separation have a trend that shows the opposite trend, so the method that is widely used is to combine the two sizes into one integrated size.

The homogeneity of the clusters created was measured using silhouette width in this study. Silhouette width is a metric that takes into account both compactness and non-linear separation [46]. The average silhouette width for each observation is the average of silhouette value $S(i)$ for every i objects that belong to a certain cluster. The silhouette value indicates the amount of confidence in the dataset's placement in a cluster of specific observations, with a value near to 1 indicating a good cluster and a value close to −1 indicating a bad cluster, with the *i*-th observation, defined as

$$S(i) = \frac{b_i - ai}{\max(b_i, a_i)}, \tag{4}$$

where *ai* is the average distance between *i* and all other observations in the same cluster, and b_i is the average distance between *i* and observations in the nearest neighboring cluster, i.e.,

$$a_i = \frac{1}{n(C(i))} \sum_{j \in C(i)} dist(i, j), \tag{5}$$

$$b_i = \underset{C_k \in C \backslash C(i)}{min} \sum_{j \in C(i)} \frac{dist(i, j)}{n(C_k)}, \tag{6}$$

where $C(i)$ is a cluster containing observations *i*, $dist(I,\ j)$ is a measure of the distance used between observations *i* and *j*, and $n(C)$ is the cardinality of cluster *C*. After the cluster is validated, then the cluster formed will be investigated and interpreted in accordance with the phenomenon of the implementation of restrictions on community activities (PPKM) that occurred in West Java. The data and syntax used in this research can be accessed at https://github.com/DhikaSuryaP/COVID-19-Clustering-in-West-Java (accessed on 15 July 2022).

4. Results

The k-medoids clustering approach was employed in the cluster analysis of daily positive COVID-19 cases in 27 cities and regencies in West Java, with the distance measure utilized being cross-correlation-based distance. Clustering was carried out in two PPKM periods with the worst COVID-19 case dynamics in West Java.

- The first clustering period is 1 July 2021–30 September 2021.
- The second clustering period is 1 January 2022–31 May 2022.

4.1. Optimal Cluster Number Selection

This study used daily COVID-19 positive case data from 27 cities and regencies in West Java Province. The ideal number of clusters for daily positive case data for COVID-19 in 27 cities and regencies in West Java province was determined using the elbow method. Figure 2 depicts the ideal number of clusters for daily COVID-19 positive cases in West Java cities and regencies.

Figure 2 shows that the ideal number of clusters to generate in this study is four clusters for the first clustering period and two clusters for the second clustering period. This is proven by the total change within the sum of the square that occurs begin to dampen, implying that the difference between the total within the sum of the square in the cluster is no longer significant, or that the cluster that is formed no longer has a significant difference after the numbers of clusters are enlarged by more than the ideal number. As a result, the number of clusters produced in this study will be four for the first clustering period and two for the second clustering period.

4.2. Clusters Internal Validation

Table 1 depicts clusters of daily positive COVID-19 cases in 27 West Java cities and regencies.

After getting the temporary cluster in Table 1, we iterated the process three times. The cluster's outcome did not change until the third iteration. After data on the number of daily COVID-19 cases in 27 cities and regencies in West Java were clustered, the clusters were internally validated. This is done by looking at the silhouette width value to guarantee that the produced cluster is homogeneous. Table 2 displays the silhouette width values together with the number of clusters that have attempted to form.

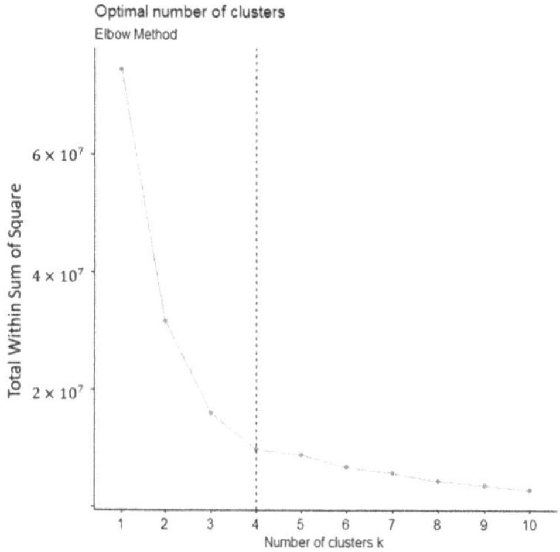

(**a**)

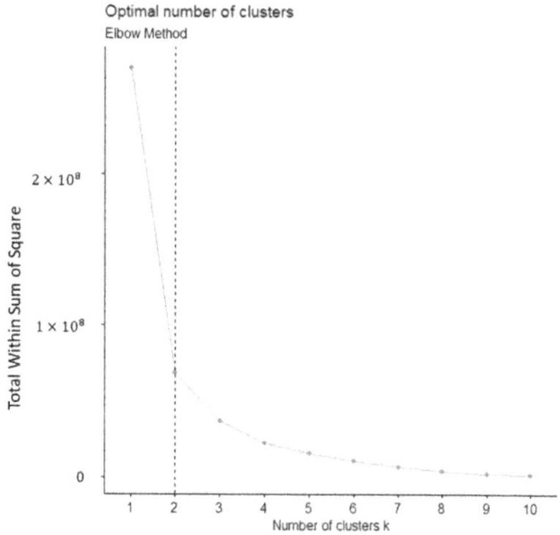

(**b**)

Figure 2. The optimal number of clusters based on the elbow method: (**a**) Clustering period 1 July 2021–30 September 2021, (**b**) clustering period 1 January 2022–31 May 2022.

Table 1. Clusters of daily COVID-19 cases in cities and districts in West Java.

Cluster	Periods	Cities/Districts
1	First Period	KAB. BANDUNG, KAB. BANDUNG BARAT, KAB. CIAMIS, KAB. CIANJUR, KAB. INDRAMAYU, KAB. KARAWANG, KAB. PANGANDARAN, KAB. SUBANG, KAB. SUKABUMI, KAB. TASIKMALAYA, KOTA BANDUNG, KOTA BANJAR, KOTA SUKABUMI, KOTA TASIKMALAYA
2		KAB. BEKASI, KAB. BOGOR, KAB. CIREBON, KAB. GARUT, KAB. KUNINGAN, KAB. MAJALENGKA, KAB. PURWAKARTA, KAB. SUMEDANG, KOTA BEKASI, KOTA CIMAHI, KOTA CIREBON
3		KOTA BOGOR
4		KOTA DEPOK
1	Second Period	KAB. BANDUNG, KAB. BANDUNG BARAT, KAB. CIAMIS, KAB. CIANJUR, KAB. CIREBON, KAB. GARUT, KAB. INDRAMAYU, KAB. KARAWANG, KAB. KUNINGAN, KAB. MAJALENGKA, KAB. PANGANDARAN, KAB. PURWAKARTA, KAB. SUBANG, KAB. SUKABUMI, KAB. SUMEDANG, KAB. TASIKMALAYA, KOTA BANDUNG, KOTA BANJAR, KOTA CIMAHI, KOTA CIREBON, KOTA SUKABUMI, KOTA TASIKMALAYA
2		KAB. BEKASI, KAB. BOGOR, KOTA BEKASI, KOTA BOGOR, KOTA DEPOK

Table 2. Silhouette width values in the clusters tested in this study.

Number of Clusters	2	3	4	5	6	7	8	9	10
First Period	0.2514	**0.2605**	0.2633	0.1952	0.1915	0.1720	0.1765	0.1247	0.1027
Second Period	**0.6363**	0.3056	0.3258	0.3339	0.3073	0.3154	0.3281	0.3016	0.2992

The highest silhouette width value achieved by the first-period cluster was 0.2633 and the second-period cluster was 0.6363, according to Table 2. This demonstrates that the development of four clusters for the first period and two clusters for the second period in this study was suitable since it demonstrated that the highest degree of confidence in cluster member placement was attained, as in the graph given in Figure 3.

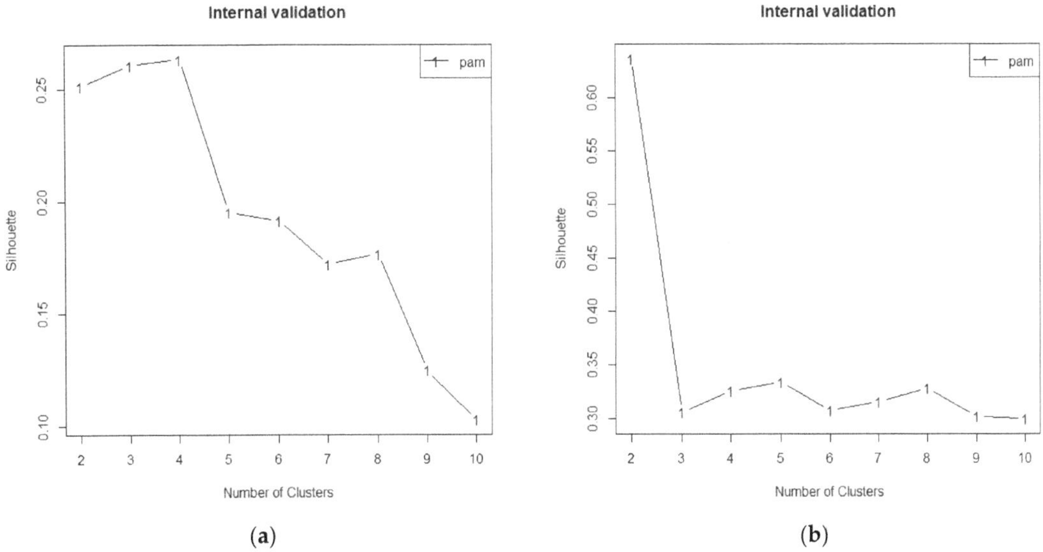

Figure 3. Silhouette width plot for each number clusters: (**a**) Clustering period 1 July 2021–30 September 2021, (**b**) clustering period 1 January 2022–31 May 2022.

From Figure 3, it can be seen that the highest silhouette value was achieved when the number of clusters formed was four clusters for the first period and two clusters for the second period. After three iterations and internal validation, it was found that Table 1 shows the final cluster for the number of active COVID-19 cases in 27 cities and regencies in West Java, meaning that for clusters that have been formed, further analysis can be done.

4.3. First Period (1 July 2021–30 September 2021) Clustering Results

Figure 4 shows the development of the number of daily positive cases of COVID-19 in Cluster 1 for the first period.

Figure 4 shows the development of the number of daily positive cases of COVID-19 in Cluster 1 in the period 1 July 2021–30 September 2021. This cluster is the cluster with the most members, namely 14 cities/regencies. Characteristics that can be observed through the graph in this cluster: it can be seen that this cluster is a cluster in which, when the emergency PPKM was implemented, namely on 3 July 2021, the number of daily positive cases of COVID-19 that occurred was increasing. Then, 14 days after the first time the emergency PPKM was carried out, namely on 17 July 2021, in this cluster there were signs of a decrease in daily positive cases of COVID-19. When PPKM 4 Level was implemented, namely 26 July 2021–2 August 2021, in this cluster, almost all cluster members began to experience a significant decrease in cases when compared to the worst conditions that had been experienced before. The last characteristic of this cluster is the steady decline in cases, which continued until 30 September 2021.

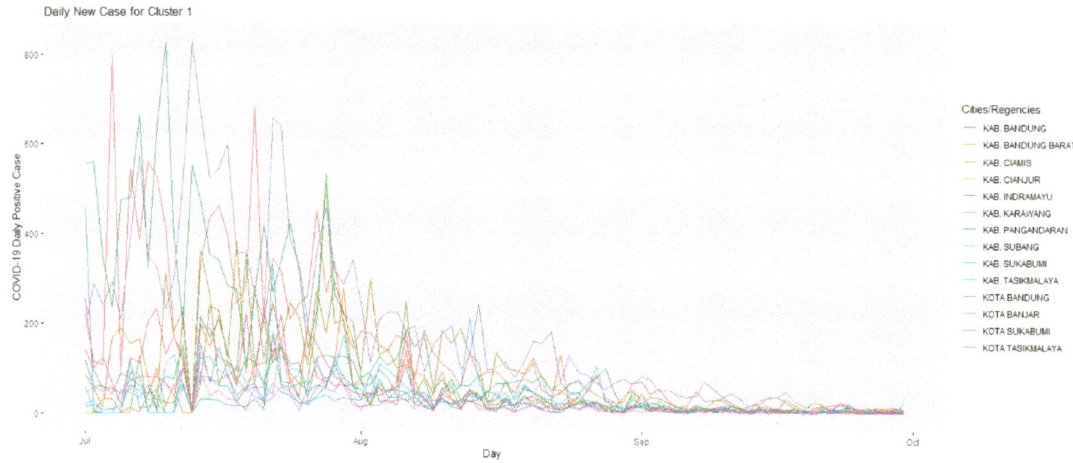

Figure 4. Daily positive cases of COVID-19 in cluster 1 for the first period.

Figure 5 shows the development of daily positive cases of COVID-19 in Cluster 2. This cluster consists of 11 cities/regencies for the period 1 July 2021–30 September 2021. Characteristics that can be observed through the graph in this cluster: it can be seen that this cluster is a cluster that, when PPKM starting to be implemented, namely on 3 July 2021, was not experiencing an increase in daily positive cases, different from Cluster 1. This cluster began to experience an increase which began on 13 July 2021, and this lasted until 25 July 2021, then began to experience a steady decline after the Level 4 PPKM began to be implemented, namely on 26 July 2021. In the period August 2021–September 2021, cities and towns and cities districts that are members of Cluster 2 no longer experienced a significant increase in cases, and were steadily decreasing. This indicates that the policies taken were appropriate to reduce the number of daily positive cases of COVID-19 that occurred in Cluster 2, given in Figure 6.

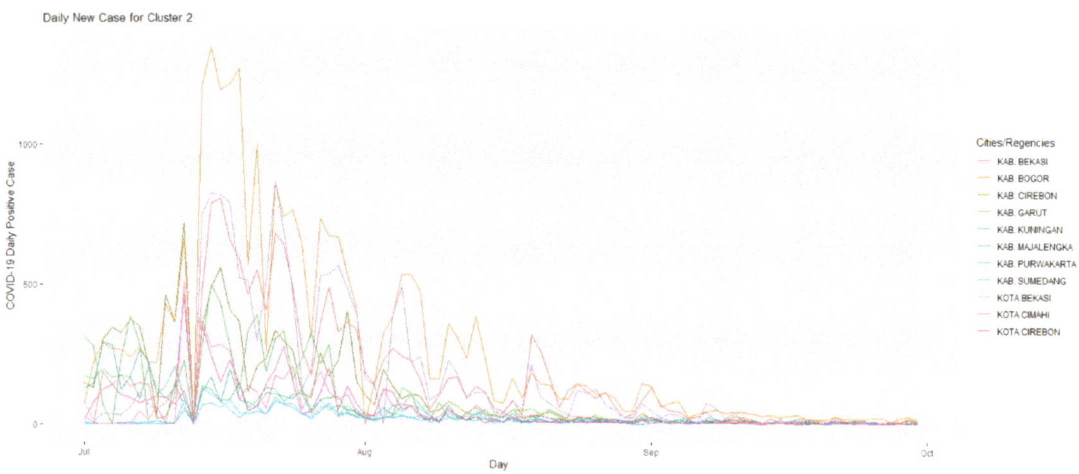

Figure 5. Daily positive cases of COVID-19 in Cluster 2 for the first period.

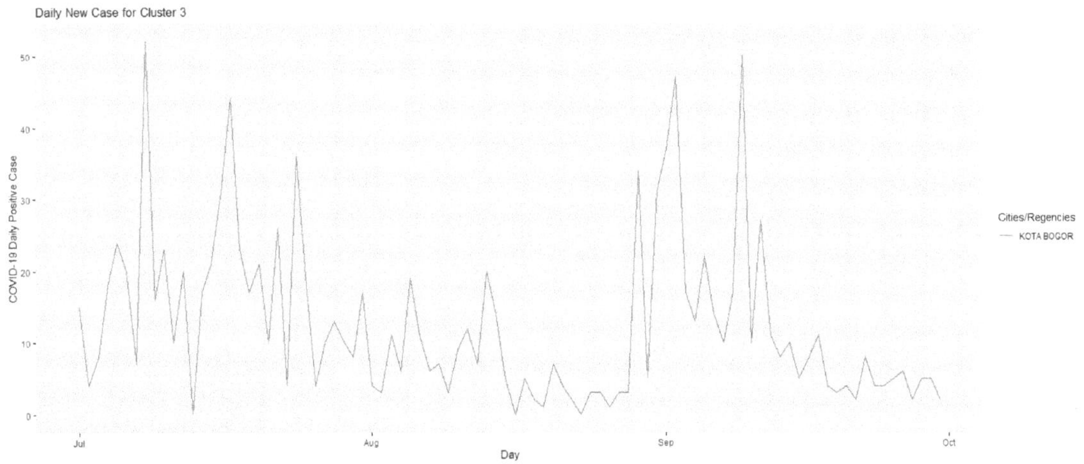

Figure 6. Daily positive cases of COVID-19 in Cluster 3 for the first period.

Figure 6 shows the development of daily positive cases of COVID-19 in cluster 3. This cluster is unique because it only consists of one city, namely Bogor City. The unique characteristic that can be observed through the graph in this cluster is that it experienced two peaks, namely when the emergency PPKM was implemented, on 3 July 2021 and in the period 26 August 2021–16 September 2021. This cluster experienced a decrease when the Level 4 PPKM was implemented, namely on 26 July 2021, and continued to decline until 23 August 2021, but what distinguishes this cluster from other clusters is that in this cluster, there was an increase again on 26 August 2021–11 September 2021; then, they again experienced a decrease in cases that continued to occur until 30 September 2021. This cluster is a cluster that experienced a second wave since the emergency PPKM and Level 4 PPKM were implemented. This indicates an ineffective implementation of PPKM in this city, as a given in Figure 6.

Figure 7 shows the development of daily positive cases of COVID-19 in Cluster 4. This cluster is unique because it only consists of one city, the same as Cluster 3, namely Depok City. The unique characteristic that can be observed through the graph in this cluster is that it can be seen that this cluster experiences a peak or worst-case scenario in different periods when compared to other clusters. This cluster peaked on 21 August 2021–26 August 2021, and this happened suddenly. In contrast to other clusters, which in the same period actually decreased, when the emergency PPKM and Level 4 PPKM were implemented, this cluster experienced a fluctuating number of daily positive cases of COVID-19. This cluster experienced a significant decline on 27 August 2021. At that time, the number of daily cases that occurred was 110 cases, very different from the previous day where the number of daily positive cases was 3341 cases. Furthermore, during 28 August 2021–30 September 2021, this cluster experienced a steady decline. This cluster is unique with peaks that occurred at different periods than other clusters, and it occurred very suddenly, which indicates an ineffective implementation of PPKM in this city.

4.4. Second Period (1 January 2022–31 May 2022) Clustering Results

In the second period, the number of clusters formed was less than in the previous period. During this period, two clusters were formed based on the daily number of new COVID-19 cases in West Java. Figure 8 shows the development of the number of daily positive cases of COVID-19 in Cluster 1 for the second period.

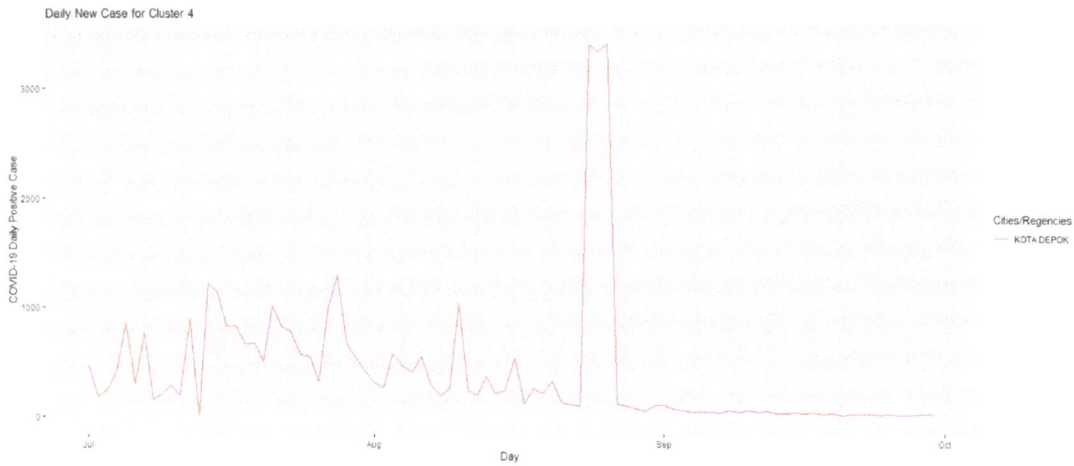

Figure 7. Daily positive cases of COVID-19 in Cluster 4 for the first period.

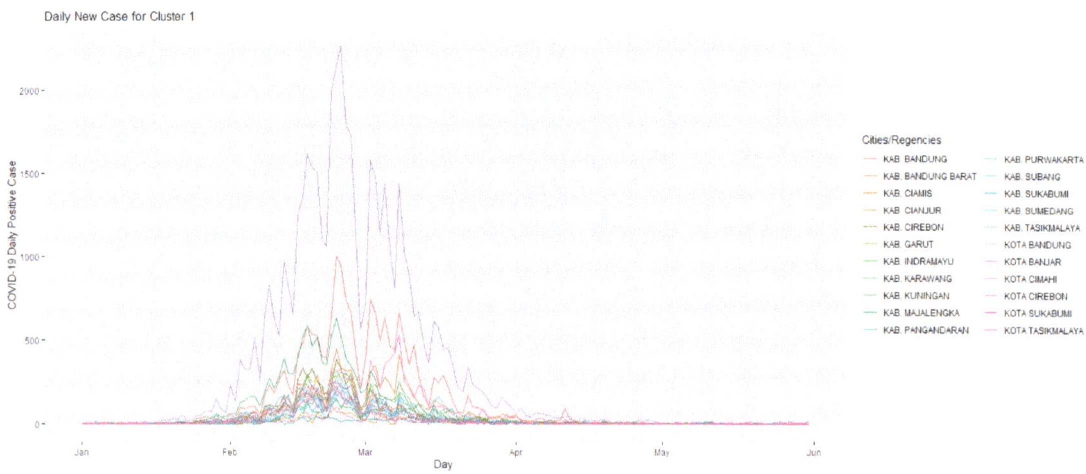

Figure 8. Daily positive cases of COVID-19 in Cluster 1 for the second period.

Figure 8 shows the development of the number of daily positive cases of COVID-19 in Cluster 1 in the period 1 January 2022–31 May 2022. In the clustering conducted in Period 2, this cluster is the cluster with the most members, namely 22 cities/regencies. In this cluster, PPKM began to be implemented on 4 January 2022. When PPKM was first implemented, the number of cases that occurred was still relatively low. This is different from Period 1 where PPKM began to be implemented when COVID-19 cases began to experience a significant increase. In this cluster, the increase in the number of COVID-19 cases began to occur at the end of January 2022, and the peak occurred at the end of February 2022. The daily number of COVID-19 cases began to decline in early March 2022, until finally on 4 April 2022, PPKM was relieved by the government. Since then, the number of daily COVID-19 cases has continued to decline until May 2022.

Figure 9 shows the development of the daily number of positive COVID-19 cases in Cluster 2 for the period 1 July 2021–30 September 2021. In the clustering conducted in Period 2, this cluster only consists of five cities and regencies. In this cluster, the start time of PPKM is still the same as the previous cluster, namely on 4 January 2022. However, this cluster has differences from the previous Cluster 1; namely, in this cluster, it can be seen that the increase in the number of daily cases of COVID-19 started earlier, that is, from mid-January. In addition, in this cluster, the peak that occurred was earlier than Cluster 1, which occurred in mid-February. The decrease in the number of daily cases in this cluster occurred at the end of February and continued until 31 May 2022.

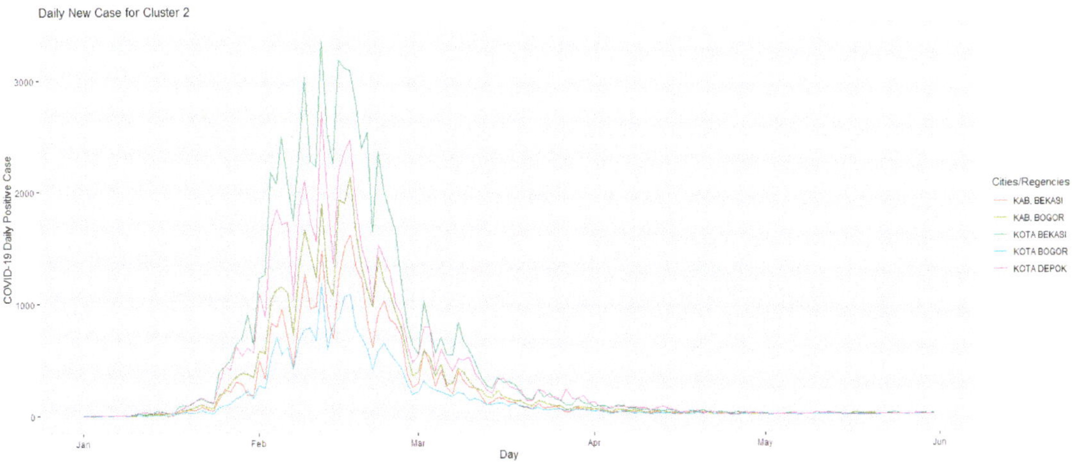

Figure 9. Daily positive cases of COVID-19 in Cluster 2 for the second period.

The clusters formed in this study have unique characteristics and features and can be used as an initial evaluation of the effectiveness of PPKM. This is indicated by the differences that can be extracted from the clusters that were formed. Further discussion of clustering results and their implications for the effectiveness of PPKM implementation is included in the next section.

5. Discussion

In this study, we succeeded in forming clusters for daily COVID-19 cases in 27 cities and regencies in West Java Province. We formed them based on the results of the elbow method and validated the clustering results using silhouette width. The validation results show that for the first and second period, the formation of four and two clusters provides the highest level of confidence, respectively. Based on the cluster, it was found that the use of time-series clustering was effective in extracting insight from data on the number of COVID-19 cases that occurred during PPKM.

For the first period clustering (1 July 2021–30 September 2021). We found some insights that could be useful for immediate use or further study. Based on the four clusters that we formed, we found that the implementation of the emergency community activity restriction (PPKM), and the Level 4 PPKM had succeeded in reducing the number of cases. daily COVID-19 in 25 cities and regencies in West Java. This is reflected in Figures 4 and 5, namely the graphs for Clusters 1 and 2. The two graphs show that there has been a decline in cases since 2 August 2021. In these two clusters, almost all cluster members began to experience a significant decrease in cases when compared to the worst conditions that had been experienced previously. Another characteristic of this cluster is the steady decline in cases, which continued until 31 October 2021. We also found that two cities had different daily developments of COVID-19 positive cases compared to other cities and regencies in West Java Province during the period implementation of restrictions on community activities (PPKM). The cities of Bogor and Depok had different developments in the number of daily COVID-19 cases compared to other cities and regencies in West Java during PPKM. Bogor City experienced two significant spikes in cases, namely when

PPKM was implemented and Level 4 PPKM was implemented, while Depok City faced its worst situation 19 days after PPKM was implemented. This means that the cities of Bogor and Depok have different daily COVID-19 case dynamics compared to other cities and regencies, which indicates the need for different treatment for the two cities. In the city of Bogor, it was found that the peak of cases occurred twice, namely during PPKM and after PPKM. This indicates that the implementation of PPKM is not effective in that city, so that a more stringent PPKM execution accompanied by an extension of its duration is an option that can be considered for the City of Bogor. As for the City of Depok, the highest number of cases actually occurred after the PPKM period, and during the PPKM period, the number of COVID-19 cases in this city tended to be less. This indicates that the implementation of PPKM in this city is inconsistent, especially in the mid-to-late PPKM period. This means that supervision on the implementation of PPKM in this city need to be tightened, especially during the period leading up to the end of PPKM when the number of daily cases tends to decrease.

For the second period (1 January 2022–31 May 2022), based on the two clusters formed, we found that there was a delay between the two clusters. In the first cluster, the increase in COVID cases began at the end of January, while in the second cluster, the increase in the number of cases occurred earlier, namely in mid-January. This difference turns out to have an impact on when the peak number of cases occurs. This is shown in Cluster 1, where the peak of new cases occurred at the end of February, while in Cluster 2 the peak occurred in mid-February. There is a gap of one month between the start of the increase in cases and the peak of cases that occur. One of the reasons for the difference in the start of improvement between these two clusters is location. Members of Cluster 2 are cities and regencies that are part of the JABODETABEK (Jakarta–Bogor–Depok–Tangerang–Bekasi) area. At that time, Jakarta was the first location to experience an increase in the number of COVID-19 cases in Indonesia. This is a logical reason for an earlier increase in the number of cases in Cluster 2. This also confirms the findings of Zarikas et al. [11] which stated that geography had an effect on the dynamics of COVID-19 cases. For the effectiveness of the implementation of PPKM in Period 2, there was no significant difference between Clusters 1 and 2. This is because the time since the peak occurred until the number of new cases was relatively low, more or less the same for these two clusters, which was about one month and there was no significant increase in cases until May 2022. In Cluster 1, new cases began to stabilize at a low level in mid-April, while in Cluster 2, new cases began to stabilize at a low level in early April. This shows that in this second period, all cities and districts in West Java implemented PPKM effectively, and showed an improvement from the implementation of PPKM in the first period.

After discussing the clusters in the first and second periods, we know that in each of these periods, the number of clusters was different. In the first period, four clusters were formed, while in the second period, two clusters were formed. Why does this happen? If we try to observe the dynamics of new cases of COVID-19 that occur, in the first period, the City of Depok and the City of Bogor had such unique dynamics that the two cities formed their own respective clusters. The ineffective implementation of PPKM in these two cities led to the formation of these new clusters. However, in the second period, the implementation of PPKM in both cities improved. Thus, in the second period, the dynamics of cases that occurred in the cities of Depok and Bogor were more or less the same as other cities and regencies that were members of Cluster 2. This shows that the variability of the effectiveness of containment policies affects the number of clusters formed. In the first period, the implementation of PPKM in Clusters 1 and 2 proved effective in reducing the number of new cases of COVID-19, while in Clusters 3 and 4, the implementation of PPKM was not effective, and the two clusters had very different case dynamics compared to other cities. In the second period, the effectiveness of the implementation of PPKM was more or less the same, or PPKM is effectively applied to all cities and districts in West Java. This resulted in fewer clusters being formed in the second period than in the first period, which was only two clusters. The thing that distinguishes Clusters 1 and 2 in this second period is only geography. The results of this study indicate that the proposed clustering method is successful in classifying cities and regencies based on the dynamics of new COVID-19 cases that occur. The proposed method is sensitive to the effectiveness of containment policies and also to the geography of each city and regency.

In this study, a method was proposed to evaluate the COVID-19 spread containment policy. The proposed time-series method succeeded in providing an insightful cluster for evaluating the effectiveness of the policy. However, it should be noted that to apply this method to new data, there are two conditions that need to be met: (1) The policies applied, whether the types of policies applied are relatively similar between regions, and (2) the policies need to implemented simultaneously across regions. The proposed method is suitable for use in the same policy conditions and carried out simultaneously in various regions. This is evident from the proposed classification method which succeeded in capturing features of the effectiveness of implementing PPKM in 27 cities and regencies in West Java Province. If the two conditions mentioned above are not met, the clustering method proposed in this study is not suitable for use.

6. Conclusions

In this study, we clustered the time-series data of daily COVID-19 cases in 27 cities and regencies in West Java Province. We did clustering during PPKM implementation with the worst number of new COVID-19 cases, namely 1 July 2021–30 September 2021, and 1 January 2022–31 May 2022. The distance measure that we used for time-series clustering in this study was a type of shape-based lock-step distance measures, namely cross-correlation distance and to determine the optimal number of clusters, we used the elbow method. After the cluster was formed, we did internal validation for the cluster using the silhouette width. The results of our study found that the optimal number of clusters that could be formed from the data we had was four clusters for the first period and two clusters for the second period. For the first period, we found that from the 27 cities and regencies that we studied, there were 25 cities/districts that belong to Cluster 1 and 2, and they showed that the implementation of the emergency PPKM and the Level 4 PPKM was effective for the number of daily positive cases of COVID-19. This indicates that for the majority of cities and regencies in West Java, PPKM is the right policy to implement. In addition, there are two cities that have unique patterns when compared to other cities and regencies, namely the cities of Bogor and Depok. The City of Depok showed an increasing trend in the number of new COVID-19 cases during PPKM, while the City of Bogor experienced an increasing trend after PPKM was implemented. This shows that the implementation of PPKM did not succeed in reducing the number of daily new cases of COVID-19 in the two cities. For the effectiveness of the implementation of PPKM in Period 2, there was no significant difference between Clusters 1 and 2. This is because the time from the peak occurred until the number of new cases was relatively low for these two clusters, about one month, and there was no significant increase in cases until May 2022. In Cluster 1, new cases began to stabilize at a low level in mid-April, while in Cluster 2 new cases began to stabilize at a low level in early April. This shows that in this second period, all cities and districts in West Java implemented PPKM well and showed an improvement from the implementation of PPKM in the first period.

We recommend that the government pay more attention to areas that are close to the hotspot regions for the spread of COVID-19. In this case, the most common is the state capital or provincial capital. If the effectiveness of the PPKM determines when the spread starts to stabilize at a low level, then the geographic location of the city and district determines when the spread begins to accelerate. This means that if it is identified that an area is starting to experience an acceleration in the spread of COVID-19, then other adjacent areas need to be the first areas to implement a containment policy. This is an effort that can be made by the government to contain the rate of COVID-19 so that the spread does not reach a dangerous level. We hope that the government can continue to control the spread of COVID-19 at a safe level with this information.

The clusters show that time-series clustering can be used to evaluate policies in different areas within the same time frame. Using the right distance measures can provide insightful clusters for COVID-19 policy and management. While it is certainly useful, the clusters produced in this study only differentiate between regions that have successfully implemented PPKM and those that have not. In fact, information about the success rate of each region can be different, and this information is strategic information to obtain because this can be a reference for the government for the ideal implementation of PPKM. Thus, in future research, it is very good to consider the hierarchy of the success rates of PPKM in each region in the cluster.

Author Contributions: Conceptualization, D.S.P. and S.; methodology, N.A.; software, D.S.P.; validation, D.S.P., S. and N.A.; formal analysis, S.; investigation, N.A.; resources, N.A.; data curation, D.S.P.; writing—original draft preparation, D.S.P.; writing—review and editing, S. and N.A.; visualization, D.S.P.; supervision, N.A.; project administration, S.; funding acquisition, S. All authors have read and agreed to the published version of the manuscript.

Funding: This research was funded by the Padjadjaran Postgraduate Excellence Scholarship (BUPP) grant number 1595/UN.3.1/PT.00/2021.

Data Availability Statement: Publicly available datasets were analyzed in this study. This data can be found here: https://github.com/DhikaSuryaP/COVID-19-Clustering-in-West-Java, accessed on 13 August 2022.

Acknowledgments: The authors would like to thank the Dean of the Faculty of Mathematics and Natural Sciences, Universitas Padjadjaran, and the Directorate of Research and Community Service (DRPM), who has given a grant: Pendidikan Magister Menuju Doktor Bagi Sarjana Unggul (PMDSU/BUPP).

Conflicts of Interest: The authors declare no conflict of interest.

References

1. Nuraini, T.N. Kronologi Munculnya Covid-19 Di Indonesia Hingga Terbit Keppres Darurat Kesehatan. Available online: https://www.merdeka.com/trending/kronologi-munculnya-covid-19-di-indonesia-hingga-terbit-keppres-darurat-kesehatan-kln.html (accessed on 15 July 2022).
2. COVID-19 Response Acceleration Task Force; Peta Sebaran. 2022. Available online: https://covid19.go.id/peta-sebaran (accessed on 15 July 2022).
3. Worldometer; COVID-19 Cases by Country. 2022. Available online: https://www.worldometers.info/coronavirus/ (accessed on 15 July 2022).
4. West Java Central Agency on Statistics (BPS). *Hasil Sensus Penduduk 2020 Di Provinsi Jawa Barat*; West Java Central Agency on Statistics (BPS): Bandung, Indonesia, 2020.
5. Regional Government of West Java Province West Java Province COVID-19 Case Statistics Dashboard. Available online: https://dashboard.jabarprov.go.id/id/dashboard-pikobar/trace/statistik (accessed on 15 July 2022).
6. Kompas Kebijakan Covid-19 Dari PSBB Hingga PPKM Empat Level. Kompaspedia 2021. Available online: https://kompaspedia.kompas.id/baca/infografik/kronologi/kebijakan-covid-19-dari-psbb-hingga-ppkm-empat-level (accessed on 15 July 2022).
7. Wang, H.; Wang, W.; Yang, J.; Yu, P.S. Clustering by Pattern Similarity in Large Data Sets. *Proc. ACM SIGMOD Int. Conf. Manag. Data* **2002**, *2*, 394–405. [CrossRef]
8. Das, G.; Lin, K.-I.; Mannila, H.; Renganathan, G.; Smyth, P. Rule Discovery from Time Series. *KDD* **1998**, *1*, 16–22.
9. Fu, T.C.; Chung, F.L.; Ng, V.; Luk, R. Pattern Discovery from Stock Time Series Using Self-Organizing Maps. *Work. Notes KDD2001 Work. Temporal Data Min.* **2001**, *1*, 26–29.
10. Keogh, E.; Lonardi, S.; Chiu, B.Y.C. Finding Surprising Patterns in a Time Series Database in Linear Time and Space. *Proc. ACM SIGKDD Int. Conf. Knowl. Discov. Data Min.* **2002**, 550–556. [CrossRef]
11. Zarikas, V.; Poulopoulos, S.G.; Gareiou, Z.; Zervas, E. Clustering Analysis of Countries Using the COVID-19 Cases Dataset. *Data Br.* **2020**, *31*, 105787. [CrossRef] [PubMed]
12. Alvarez, E.; Brida, J.G.; Limas, E. Comparisons of COVID-19 Dynamics in the Different Countries of the World Using Time-Series Clustering. *medRxiv* **2020**. [CrossRef]
13. Abdullah, D.; Susilo, S.; Ahmar, A.S.; Rusli, R.; Hidayat, R. The Application of K-Means Clustering for Province Clustering in Indonesia of the Risk of the COVID-19 Pandemic Based on COVID-19 Data. *Qual. Quant.* **2022**, *56*, 1283–1291. [CrossRef] [PubMed]
14. Elsi, Z.R.S.; Pratiwi, H.; Efendi, Y.; Rusdina, R.; Alfah, R.; Windarto, A.P.; Wiza, F. Utilization of Data Mining Techniques in National Food Security during the Covid-19 Pandemic in Indonesia. *J. Phys. Conf. Ser.* **2020**, *1594*, 012007. [CrossRef]

15. World Health Organization. Coronavirus disease 2019 (COVID-19): Situation Report, 51. 2020. Available online: https://apps.who.int/iris/handle/10665/331475 (accessed on 14 June 2022).
16. ECDC COVID-19 Situation Update for the EU/EEA and the UK. Available online: www.ecdc.europa.eu (accessed on 26 June 2022).
17. CDC CDC FAQ on COVID-19. Available online: https://www.cdc.gov/coronavirus/2019ncov/faq.html#fSymptoms-&-Emergency-Warning-Signs (accessed on 14 June 2022).
18. World Health Organization. Coronavirus Disease 2019 (COVID-19): Situation Report, 57. 2020. Available online: https://cdn.who.int/media/docs/default-source/searo/indonesia/covid19/external-situation-report-57_2-june-2021.pdf?sfvrsn=cb275259_5 (accessed on 14 June 2022).
19. ECDC Q & A on COVID-19: Basic Facts. Available online: www.ecdc.europa.eu (accessed on 15 July 2022).
20. Kaufman, L.; Rousseeuw, P.J. *Finding Groups in Data: An Introduction to Cluster Analysis*; John Wiley: New York, NY, USA, 1990.
21. Omran, M.G.H.; Engelbrecht, A.P.; Salman, A. An Overview of Clustering Methods. *Intell. Data Anal.* **2007**, *11*, 583–605. [CrossRef]
22. Shahnawaz, M.; Ranjan, A.; Danish, M. Temporal Data Mining: An Overview. *Int. J. Eng. Adv. Technol. IJEAT Oct.* **2011**, *1*, 2249–8958.
23. Warren Liao, T. Clustering of Time Series Data—A Survey. *Pattern Recognit.* **2005**, *38*, 1857–1874. [CrossRef]
24. Lin, J.; Vlachos, M.; Keogh, E.; Gunopulos, D. Ierative Incremental Clustering of Time Series. *Lect. Notes Comput. Sci.* **2004**, *2992*, 106–122. [CrossRef]
25. He, W.; Feng, G.; Wu, Q.; He, T.; Wan, S.; Chou, J. A New Method for Abrupt Dynamic Change Detection of Correlated Time Series. *Int. J. Climatol.* **2012**, *32*, 1604–1614. [CrossRef]
26. Pavlidis, N.G.; Plagianakos, V.P.; Tasoulis, D.K.; Vrahatis, M.N. Financial Forecasting through Unsupervised Clustering and Neural Networks. *Oper. Res.* **2006**, *6*, 103–127. [CrossRef]
27. Sfetsos, A.; Siriopoulos, C. Time Series Forecasting with a Hybrid Clustering Scheme and Pattern Recognition. *IEEE Trans. Syst. Man Cybern. Part A Syst. Hum.* **2004**, *34*, 399–405. [CrossRef]
28. Mirri, S.; Delnevo, G.; Roccetti, M. Is a COVID-19 second wave possible in Emilia-Romagna (Italy)? Fore-casting a future outbreak with particulate pollution and machine learning. *Computation* **2020**, *8*, 74. [CrossRef]
29. Huang, X.; Li, Z.; Lu, J.; Wang, S.; Wei, H.; Chen, B. Time-Series Clustering for Home Dwell Time during COVID-19: What Can We Learn from It? *ISPRS Int. J. Geo-Inf.* **2020**, *9*, 675. [CrossRef]
30. Rojas-Valenzuela, I.; Valenzuela, O.; Delgado-Marquez, E.; Rojas, F. Estimation of COVID-19 Dynamics in the Different States of the United States during the First Months of the Pandemic. *Eng. Proc.* **2021**, *3*, 53. [CrossRef]
31. Brida, J.G.; Alvarez, E.; Limas, E. Clustering of Time Series for the Analysis of the COVID-19 Pandemic Evolution. *Econ. Bull.* **2021**, *41*, 1082–1096.
32. Pikobar Statistik Kasus COVID-19 Provinsi Jawa Barat. Available online: https://dashboard.jabarprov.go.id/id/dashboard-pikobar/trace/statistik (accessed on 16 June 2022).
33. R Core Team. *R: A Language and Environment for Statistical Computing*; R Foundation for Statistical Computing: Vienna, Austria, 2021.
34. Wickham, H. *Ggplot2: Elegant Graphics for Data Analysis*; Springer: New York, NY, USA, 2016; ISBN 978-3-319-24277-4.
35. Wickham, H. Package 'Reshape'. 2015. Available online: https://cran.rproject.org/web/packages/reshape2/reshape2.pdf (accessed on 15 July 2022).
36. Mori, U.; Mendiburu, A.; Lozano, J.A. Distance Measures for Time Series in r: The TSdist Package. *R J.* **2016**, *8*, 455–463. [CrossRef]
37. Kassambara, A.; Mundt, F. Factoextra: Extract and Visualize the Results of Multivariate Data Analyses. 2020. Available online: https://CRAN.R-project.org/package=factoextra (accessed on 15 July 2022).
38. Charrad, M.; Ghazzali, N.; Boiteau, V.; Niknafs, A. NbClust: An R Package for Determining the Relevant Number of Clusters in a Data Set. *J. Stat. Softw.* **2014**, *61*, 1–36. [CrossRef]
39. Park, H.S.; Jun, C.H. A simple and fast algorithm for K-medoids clustering. *Expert Syst. Appl.* **2009**, *36*, 3336–3341. [CrossRef]
40. Davis, R.A.; Matsui, M.; Mikosch, T.; Wan, P. Applications of distance correlation to time series. *Bernoulli* **2018**, *24*, 3087–3116. [CrossRef]
41. Bholowalia, P.; Kumar, A. EBK-Means: A Clustering Technique Based on Elbow Method and K-Means in WSN. *Int. J. Comput. Appl.* **2014**, *105*, 975–8887.
42. Thorndike, R.L. Who Belongs in the Family? *Psychometrika* **1953**, *18*, 267–276. [CrossRef]
43. Kodinariya, T.M.; Makwana, P.R. Review on Determining of Cluster in K-Means. *Int. J. Adv. Res. Comput. Sci. Manag. Stud.* **2013**, *1*, 90–95.
44. Brock, G.; Pihur, V.; Datta, S.; Datta, S. ClValid: An R Package for Cluster Validation. *Solid State Commun.* **2008**, *25*, 1–22. [CrossRef]
45. Handl, J.; Knowles, J.; Kell, D.B. Computational Cluster Validation in Post-Genomic Data Analysis. *Bioinformatics* **2005**, *21*, 3201–3212. [CrossRef]
46. Rousseeuw, P.J. Silhouettes: A Graphical Aid to the Interpretation and Validation of Cluster Analysis. *J. Comput. Appl. Math.* **1987**, *20*, 53–65. [CrossRef]

computation

Article

Assessing Traffic Congestion Hazard Period due to Commuters' Home-to-Shopping Center Departures after COVID-19 Curfew Timings

Majed Alinizzi, Husnain Haider * and Mohammad Alresheedi

Department of Civil Engineering, College of Engineering, Qassim University, Buraydah 51452, Qassim, Saudi Arabia; mfanzy@qu.edu.sa (M.A.); m.alresheedi@qu.edu.sa (M.A.)
* Correspondence: husnain@qec.edu.sa or h.hussain@qu.edu.sa

Abstract: In addition to a wide range of socio-economic impacts, traffic congestion during the era of the COVID-19 pandemic has been identified as a critical issue to be addressed. In urban neighborhoods, the timespan of traffic congestion hazard (H_{TC}) after the curfew lift is subjected to the commuters' decisions about home-to-shopping center departures. The decision for departing early or late for shopping depends on both the internal (commuter related) and external (shopping center related) factors. The present study developed a practical methodology to assess the H_{TC} period after the curfew timings. An online questionnaire survey was conducted to appraise the commuters' perception about departure time and to assess the impact of eight internal (family size, involvement in other activities, nature of job, education level, age, number of vehicles, number of children, and availability of personal driver) and three external (availability of shopping center of choice in near vicinity, distance to shopping center, and size of the city) factors on their decision. With an acceptable 20% response rate, Chi-square and Cramer's V tests ascertained family size and involvement in other activities as the most significant internal factors and availability of shopping center of choice as the primary external factor. Age, number of children, and size of the city influenced to some extent the commuters' decisions about early or delayed departure. Large associations were found for most of the factors, except education level and availability of drivers in a household. Fuzzy synthetic evaluation (FSE) first segregated the commuters' responses over a four level-rating system: no delay (0), short delay (1), moderate delay (3), and long delay (5). Subsequently, the hierarchical bottom-up aggregation effectively determined the period of highest traffic congestion. Logical study findings revealed that most (about 65%) of the commuters depart for shopping within 15 min after the curfew lift, so H_{TC} in the early part (the first one hour) of the no curfew period needs attention. The traffic regulatory agencies can use the proposed approach with basic socio-demographic data of an urban neighborhood's residents to identify the H_{TC} period and implement effective traffic management strategies accordingly.

Keywords: traffic congestion; departure delay; COVID-19; traffic delay; commuter perception; chi-square test; fuzzy synthetic evaluation (FSE)

Citation: Alinizzi, M.; Haider, H.; Alresheedi, M. Assessing Traffic Congestion Hazard Period due to Commuters' Home-to-Shopping Center Departures after COVID-19 Curfew Timings. *Computation* **2022**, *10*, 132. https://doi.org/10.3390/computation10080132

Academic Editors: Simone Brogi and Vincenzo Calderone

Received: 26 June 2022
Accepted: 28 July 2022
Published: 2 August 2022

1. Introduction

On December 2019, the authorities in Wuhan City (Hubei Province of China), reported the novel coronavirus disease COVID-19 epidemic. This disease is resulted from acute respiratory syndrome SARS-CoV-2 [1]. Since then, the COVID-19 has attracted global attention because of the affirmed risk of human-to-human transmission [2]. On 11 March 2019, the World Health Organization (WHO) announced COVID-19 as a global pandemic [3]. As of January 2022, over 352 million people have been affected by COVID-19 (SARS-CoV-2), and around 5.6 million affected have died. The United States, with over 7 million cases and 889,000 deaths, is on the top of the global counts, followed by India with more than 39 million cases (490,000 deaths) and Brazil with over 24 million cases (623,000 deaths) [4].

The COVID-19 pandemic not only infected people and took lives; it significantly disrupted all types of socioeconomic activities around the globe [5]. The fear of the virus itself, governmental restrictions, and curfew timings have restricted the movement of citizens for business, education, recreation, and religious activities.

The COVID-19 pandemic situation posed new inquiries to traffic engineers and planners. The pandemic situation demands contemporary transportation planning approaches to deal with changing mobility and activity habits [6]. The situation also demands regional specific planning, keeping in view the exclusive socio-economic and environmental characteristics of the region [7]. Particularly in the early stage, many countries adopted varying strategies to attenuate the socioeconomic and health impacts of the COVID-19 pandemic on their communities. The Kingdom of Saudi Arabia (KSA) among many other countries also implemented precautionary measures to prevent the transmission of the COVID-19 infection to protect the health of citizens and residents of the Kingdom. One of these measures is forcing partial and total public curfews on certain cities. A complete public curfew was enforced on large cities with high transmission rates of COVID-19 such as Riyadh, while a partial public curfew was imposed on the relatively smaller regions with low spread rate of COVID-19 such as the Qassim Region.

Despite the fact the public curfew policies are important in the age of COVID-19, studying the impact of such policies on transportation systems has yet to be considered. Large cities with overcrowded populations such as Riyadh City, KSA, continue to suffer from congested roads. In the efforts to mitigate day-to day traffic congestion, different traffic management strategies have been developed and implemented. For example, during rush hours, the traffic department in Riyadh City controls the entrances and the exits on main roads to regulate traffic flows and thus mitigate possible congestion. Such policies may not be adequate during public curfews. Immediately after lifting a curfew, people rush to the roadways to arrange their necessities on one side. On the other hand, some people delay their shopping trips to avoid that early congestion and contribute to traffic congestion in the last part prior to curfew. Recently, certain cities in Saudi Arabia have adopted a partial curfew policy. The question of interest is whether there would be a sudden traffic congestion after a public curfew is lifted or higher congestion subject to the later part of the no-curfew duration. Information on possible traffic congestion in response to different curfew policies will help the decision-makers in evaluating the potential congestion hazard, revising and modifying the curfew durations, and anticipating the congested road sections so that the public may be advised to avoid such roads during certain times.

Traffic congestion occurs when too many vehicles use common main streets and service points with limited capacity, which can potentially lead to an increase in traffic flow and impact commuter's travel time [8]. The adoption of imposing partial public curfews in cities resulted in high traffic congestion on the main streets, particularly around shopping areas, right after a curfew is lifted. For instance, the population of Riyadh City is approximately under 5 million with about 985,000 vehicles flooding the city main streets daily [9]. Since post-curfew traffic may adversely impact main streets, it is of great importance to measure the impacts of different curfew policies on traffic flows. To anticipate the effect of the public curfew policies on the urban traffic system in Saudi Arabia, it is important to understand different factors that affect the commuter decision about early or delayed post-curfew departure to shopping areas. This would be of high importance in supporting decision-making and giving more useful insights on the implementation of suitable public curfew policies.

Since the beginning of the pandemic, several studies have been conducted on the impacts of COVID-19 on urban traffic and transportation systems. Bucsky [10] investigated the changes in traffic behavior in Hungary during the COVID-19 pandemic and found significant changes in travel mode and decline in public transit users by almost 80%, while car usage increased to 65% in response to the spread of the disease. This was supported by Jenelius and Cebecauer [11], who found almost a 60% decrease in public transit users in Sweden during the COVID-19 outbreak. Abdullah et al. [12] developed a binary logistic model to assess commuters' travel mode choices (public and private) during the pandemic and identified gender, income, job type, education level, vehicle ownership, and safety precautions as important influencing factors. Other researchers reported the impact of COVID-19 on consumer travel behavior. Consumer grocery trips right after curfew (which was referred to as panic buying) was observed to increase during the COVD-19 pandemic compared to other types of shopping and leisure activities [13–17]. The type of shopping was reported to be dependent on the curfew duration and restrictions. The impact of the COVID-19 pandemic on both traffic volume and car accidents was studied by Katrakazas et al. [15] using tracking technologies in Greece and Saudi Arabia. Their study showed a reduction in travel volume with an almost 41% reduction in car accidents. Similar findings were observed by Saladie et al. [16], who found a 63% reduction in travel volume along with 74% decline in car accidents in Spain during the year of 2020. In some of the European countries such as Sweden, Germany, and Austria, the public preferred travelling for short distances compared to long distances. Furthermore, the number of people travelling to commercial areas and city centers decreased in response to the lockdowns and COVID-19 restrictions in those countries [18,19].

While the above studies focused on the impact of the COVID-19 pandemic on changes in public travel modes and transportation safety, limited studies exist on the COVID-19 impact on traffic congestion. Huang et al. [20] conducted a data-driven analysis of travel behavior during the pandemic in China. Various factors were found to influence the travel behavior and traffic congestion, such as means of transportation, distance, and location. Muley et al. [21] studied the impact of staged and sequential COVID-19 preventive measures on traffic mobility in several intersections in Qatar. Their study found that although the volumes were significantly reduced to almost 30%, traffic patterns were similar before and after the implementation of the measures. Moreover, traffic violations and accidents showed a drop of 73% and 37%, respectively, 56t as a response to the preventive measures. Recently, Xu et al. [22] investigated the changes in traffic patterns before and during the COVID-19 pandemic in Shanghai, China. Their study found that the central areas were more affected by the travel restrictions during the pandemic compared to suburban areas, in which a decrease in the traffic congestion was observed. An analytical framework was proposed based on the traffic characteristics and areas to help with policy decision-making of urban road transportation systems during the pandemic. Loo and Huang [23] studied the changes in traffic congestion patterns due to the enforced curfew in Hong Kong by calculating a congestion index. Their study showed that under the curfew law, morning peak-hour congestion was reduced with a significant drop in congestion index in the central areas and urban cores.

Like other countries, after the nationwide spread of the COVID-19 outbreak in Saudi Arabia, partial and total public curfew policies were adopted to mitigate the transmission rate of the virus. From the review of the literature, while the majorities of the studies focused on the impact of COVID-19 on travel modes and car accidents, none of the past studies have identified the factors that affect the commuter decision about early or delayed post-curfew departure to shopping areas and city centers. Hence, it is difficult for the cities' transportation ministries to identify the high congestion periods during the time of no curfew and subsequently plan and implement traffic management strategies.

To the best of our knowledge, no methodology in the literature assesses traffic congestion during the no curfew period based on socio-demographic data of an urban neighborhood's residents. To avoid traffic congestion, the present study primarily aimed to investigate the commuter's decision of home-to-shopping center departure (HSD) after a public curfew is lifted. Primary objectives of the study are to: (i) assess the impact of different internal (commuter) and external (shopping center) factors affecting traffic congestion through interview-based surveys, (ii) perform statistical analysis to establish the significance of the factors on the departure delay decision, and (iii) develop a multicriteria analysis-based approach to come up with the traffic congestion hazard period within the no curfew timing. The results of this study are intended to support decision makers in anticipating possible traffic congestion due to different curfew strategies and developing or reviewing appropriate traffic management strategies accordingly.

2. Materials and Methods

2.1. Traffic Congestion Hazard Evaluation Framework

Figure 1 presents the traffic congestion hazard evaluation framework developed in the present study. First, potential internal (commuter related) and external (shopping center related) departure delay factors (D_F) were identified through published literature and expert opinion in brainstorming sessions. Internal factors cover commuters' household and socioeconomic characteristics (e.g., age, education, and family size), while external factors are associated with the type (commuter's choice), location (distance from commuter's residence) of the shopping area, and the size of the commuter's city. Subsequently, a questionnaire survey was developed that encompassed all the factors and the departure delay duration (D_D) in the form of dichotomous or multiple-choice questions. After securing ethics approval from the funding agency, the online version of the questionnaire survey was distributed to around 300 participants. The sample size selection process considered participants with different age groups, family sizes, job sectors, and education levels, residing in cities of different sizes (small, medium, and large). Details for city size classification are given in the subsequent sections. Third, the responses received were statistically analyzed using Chi-square and Cramer's V tests to establish the association between the D_F and D_D. Fuzzy set theory has been recognized in dealing with imprecise and subjective judgment [24,25] and was found as an appropriate approach for the linguistic nature of available data in the present study. Hence, spatial variations of traffic congestion during the no-curfew interval were assessed using fuzzy synthetic evaluation (FSE).

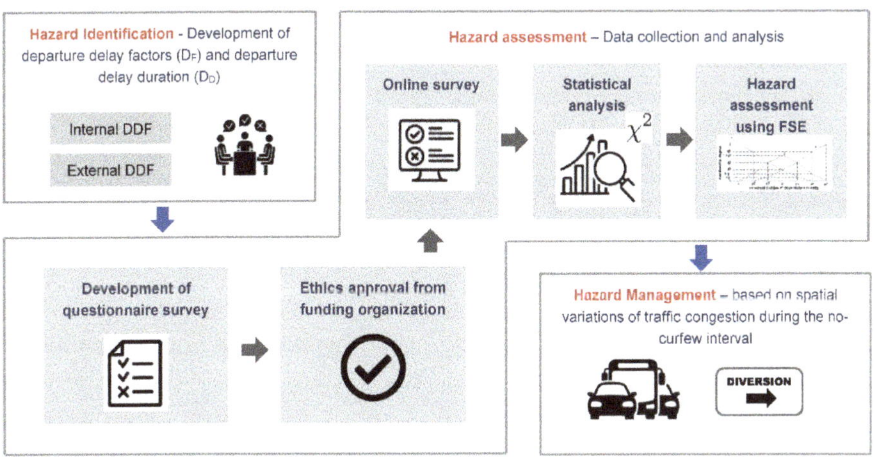

Figure 1. Vignette of Traffic Congestion Hazard Evaluation Framework.

2.2. Development of Departure Delay Factors and Questionnaire Survey

After the implementation of curfew, traffic congestion was recognized in the central cities of almost all the provinces across KSA. Adopting a partial curfew policy may exacerbate the problem of congested roads in large cities such as Riyadh City, capital of Saudi Arabia. Nevertheless, shopping areas in small- to medium-sized cities also face traffic congestion prone to post-curfew traffic. Although large cities like Riyadh, with a population of around 7.4 million, are more prone to congestion, capital cities of other provinces also face the similar issues during the no-curfew period. The cities of Buraydah, with a population of around 0.7 million, and Madinah, with 1.52 million, were classified as medium-sized cities, while Hail City, with a population of 407,000, was classified as a small sized-city [26,27].

To determine the post-curfew traffic congestion problem in shopping areas, the response of commuters' departure times after curfew is lifted needs to be known. In order to understand the commuter perception of delaying the post-curfew departure time, different internal and external DDF and expected D_D were identified. Subsequently, both the D_F and D_D were translated into a survey format, which was approved by the ethics department at Qassim University, Saudi Arabia. Table 1 presents the D_F, D_D, and the corresponding questions asked from the commuters. The online questionnaire survey was sent to varying people living in Riyadh, Buraydah, Madinah, and Hail cities using an online questionnaire. Each observation in the travel survey represents the perception of one commuter of how long he (or she) delayed their departure based on the internal factors. The survey targeted individual households commuting from their houses between 6:00–10:00 am, the no-curfew period.

Table 1. Internal and external factors for evaluating commuter home-to-shopping center departure delay after COVID-19 curfew timings.

No	Factors	Units	LD [1]	MD [2]	SD [3]	Questions Asked
			Departure delay factors (D_F)			
1.	Internal Factors					
1.1	Family size	No	>10	5–10	<5	How many persons live in your house?
1.2	Involvement in other personal activities	Y/N	Yes	-	No	If sometimes delay, are you involved in some personal activities?
1.3	Nature of job	-	PS [4]	GS [5]	NPO [6]	What is your job sector?
1.4	Education level	-	HS [7]	D/G [8]	HE [9]	What is your education level?
1.5	Age	years	>50	35–50	≤35	Which of the following age group you belong to?
1.6	Number of vehicles	No	>2	2	1	What is the number of cars in your household?
1.7	Number of children	No	>2	1–2	0	How many children are there in your house?
1.8	Availability of driver	Y/N	Yes	-	No	Do you have a driver in your household?
2.	External Factors					
2.1	Availability of shopping center of choice	Y/N	No	-	Yes	What is the size of the nearest shopping center to your residence? Which types of shopping center do you prefer to shop from?
2.2	Distance to Shopping center	Km	>4	2–4	<2	What is the approximate distance from your residence to the nearest shopping center?
2.3	Size of the city	Population	Large	Medium	Small	What is the region of your residency?
			Departure delay duration (D_D)			
3.1	Departure delay	Y/N	Yes	-	No	Do you ever delay home-to-shop centers departure time to avoid traffic congestion after a public curfew is lifted?
3.2	Delay time	minutes	>30	15–30	<15	If sometimes delay, on average how many minutes do you delay?

[1] Long Delay (LD), [2] Moderate Delay (MD), [3] Short Delay (SD), [4] Private Sector (PS), [5] Government Sector (GS), [6] Non-profit Organization (NPO), [7] High School (HS), [8] Diploma/Graduation (D/G), [9] Higher Education (HE).

A random sampling approach was used to select a representative sample. First, common characteristics (internal factors) were identified such as family size (1.1 in Table 1), job type (1.3), age (1.4), and education level (1.5). Particular to the COVID-19 situation, these characteristics were used in a study on the COVID-19 impact on transportation mode selection [12]. Second, characteristics (D_F) specific to the present study, such as involvement in activities (1.2), number of vehicles (1.6), and number of children (1.7), and availability of driver (1.8) were identified using expert opinion and personal observations.

Finally, recent studies were consulted to identify the characteristics (external factors) affecting traffic congestion in urban areas, such as size of city [28] and distance and type of shopping center [29].

In Table 1, we tried to capture the commuter's personal household related aspects through the internal D_F. The last column of the table presents the related questions asked from the commuter. For example, the households with many family members may decide for a longer delay to their home-to-shopping center departure (HSD). Similarly, older commuters or the families with more children may also delay their departures for grocery shopping. Generally, people working in the private sector are busier and may delay their trips. Availability of more than one car and a driver can also relax an individual, which delays HSD. We also thought that external factors, such as shopping center of choice (e.g., convenient store, medium-sized market, and supermarket) and its availability in the near vicinity, can also influence the commuter's decision and thus included them in the questionnaire. Finally, the information regarding the departure delay duration was also gathered through the questions mentioned in the last column.

2.3. Statistical Analysis

The data collected through the questionnaire provided the information regarding both the D_F and D_D. Statistical analysis of the collected data was performed to estimate the percentage frequencies of all the D_F for different D_D (long, moderate, and short delay). Before using this data to the proposed model for identifying the peak congestion period, the Chi-square independence test established the level of association (weak, moderate, or strong) between the D_F and D_D. An example of the null and alternative hypothesis for family size is given in the following.

H_0: ***The null hypothesis:*** *Family size of the commuter is a perfectly independent factor and does not affect the home-to-shopping center departure delay.*

H_a: ***The alternative hypothesis:*** *Family size of the commuter is a dependent factor and somehow affects the home-to-shopping center departure delay.*

Similar hypotheses were applicable for the remaining D_D. The Chi-square method is based on expected frequencies at which the null hypothesis holds. The expected frequencies for all the D_F against the given D_D were calculated using the following relationship [30]:

$$e_{ij} = \frac{o_i \times o_j}{N} \tag{1}$$

where e_{ij} denotes the expected frequency, o_i and o_j presents the marginal column and row frequencies respectively, and N is the total number of responses.

As the o_i and o_j differ, the residuals were estimated as:

$$r_{ij} = o_{ij} - e_{ij} \tag{2}$$

A larger r_{ij} value (absolute) denotes that there is a large difference between the observed responses and the null hypothesis. Subsequently, all the residuals were added to estimate the chi-square (χ^2) test statistic as:

$$\chi^2 = \sum \frac{(o_{ij} - e_{ij})^2}{e_{ij}} \tag{3}$$

In the next step, the independence of the variables in the given population, in terms of p-value, was estimated for a given χ^2 and degree of freedom using the following equation.

$$d_f = (i-1) \times (j-1) \tag{4}$$

where i and j are the number of rows and column (categories) in the contingency table.

To reject the null hypothesis of independence, the calculated chi-squared values were compared with the critical values from the chi-squared distribution at $p < 0.05$. The critical values are 3.84, 5.99, 7.82, 9.49, 11.07, and 12.59 with corresponding d_f of 1, 2, 3, 4, 5, and 6. The chi-square values higher than the critical values reject the null hypothesis of independence.

As the performance of the chi-square test depends on an adequately large sample size, the significance estimated by this test does not inform the degree of effect. Therefore, the effect size can give the magnitude of effect. The strength of association between the D_F and D_D was estimated using the effect size (ES) of the chi-square test for each D_F using Cramér's V, which essentially is a kind of Pearson correlation for categorical variables as used in the present study. It was determined by:

$$V = \sqrt{\frac{\chi^2}{n \cdot d_f}} \quad (5)$$

where n is the total number of responses, dividing χ^2 by the number, and taking the square root. Cohen [31] presented the interpretation of effect size using the Cramér's V method. For the df of 5 or higher, the fields have small association if ES ≤ 0.04, medium association if $0.04 <$ ES ≤ 0.13, and large association if ES > 0.22 [31].

2.4. *Traffic Congestion Hazard Period Assessment*

To identify the time segment with the highest congestion during the no-curfew interval, the following assumptions were established through the brainstorming sessions supported by practical observations in Saudi Arabia and news reporting by electronic and print media during the era of COVID-19 curfews around the globe:

- Commuters with immediate (no delay) departures contributed to traffic congestion in the earliest time segment of the no-curfew period.
- Commuters with shortly (<15 min) delayed departures contributed to traffic congestion in the early-middle time segment of the no-curfew period.
- Commuter with moderately (15–30 min) delayed departures contributed to traffic congestion in the middle time segment of the no-curfew period.
- Commuters with long (>30 min) delayed departures contributed to traffic congestion in the last time segment of the no-curfew period.

The statistical analysis in the previous section established the linkage between the D_F and D_D. The information obtained from survey responses was aggregated using the multicriteria evaluation model to identify the time segment with the highest traffic congestion hazard. The hierarchical-based Fuzzy Synthetic Evaluation (FSE) approach presented in Figure 2 aggregated the information in the form of D_F. The step-by-step procedure given in the following was modified from Zhao et al. [25]:

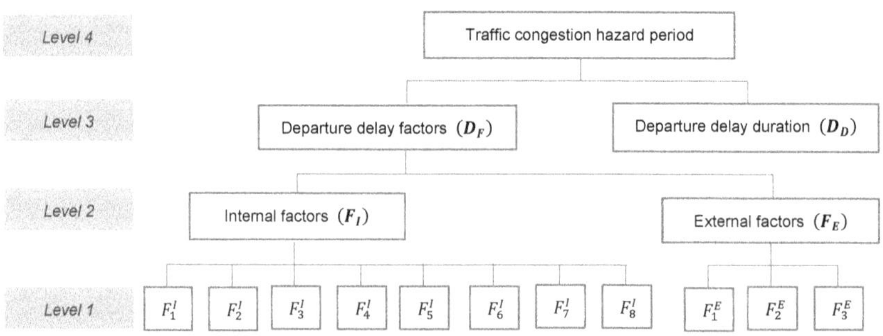

Figure 2. Hierarchical-based Fuzzy Synthetic Evaluation approach for assessing traffic congestion period. Details of internal and external factors shown at level 1 are given in Table 1.

Levels 1 and 2: Estimate the impact of internal and external factors on departure delay

The impact for each internal factor (F_I) was obtained from the questionnaire survey given in Table 1. The UoD linguistically defines a four level-rating (S_j = 0, 1, 3, 5) to evaluate the impact as no delay (0), short delay (1), moderate delay (3), and long delay (5). The term f_{i0}^I essentially describes the degree of association of each factor to these levels. The following equation describes this step in the matrix form:

$$\left(I_i^I\right)_{1*4} = \left(f_{i0,}^I f_{i1,}^I f_{i3,}^I f_{i5}^I\right) \quad (6)$$

where I_i^I represents the internal factor (i = 1, 2, ... , n) and n is the total number of internal factors.

The impact of each internal factor (F_i^I) was calculated by the following equation:

$$F_i^I = \sum_{i=1}^{4} \left(S_j * f_{ij}^I\right) \quad (7)$$

To calculate the overall impact of internal factors at level 2 of Figure 2, the importance weight of each internal factor was estimated using the following equation:

$$w_{i,}^I = F_i^I / \sum_{i=1}^{k} F_i^I \quad (8)$$

The FSE method aggregates the weighted matrix given in Equation (7) and the evaluation matrix given in Equation (6) and presents the results in the form of the following equation to estimate the membership functions for each internal factor, where i = 1, 2, ... , t:

$$d_{tj}^I = \sum_{i=1}^{k} w_i^I * f_{ij}^I,\ k = 8 \quad (9)$$

$$\left(D_t^I\right)_{1*4} = \left(W_i^I\right)_{1\,*\,8} * \left(F_i^I\right)_{8\,*\,4} = \left(d_{t0,}^I\ d_{t1,}^I\ d_{t3,}^I\ d_{t5}^I\right) \quad (10)$$

Knowing the membership functions of t number of factors' groups at level 2, the overall impact of internal factors (F_I) can be estimated as:

$$F_I = \sum_{i=1}^{4} \left(S_j * d_{tj}^I\right) \quad (11)$$

Similarly, the overall impact of external factors (F_E) was estimated as:

$$F_E = \sum_{i=1}^{4} \left(S_j * d_{tj}^E\right) \quad (12)$$

Levels 3 and 4: Calculate the overall impact of departure delay factors and departure delay duration on the traffic congestion hazard period

To estimate the impact of D_F and D_D on traffic congestion hazard period, their respective importance weights were estimated using the following equation:

$$w_{Gt}^{DF} = \left(\sum_{i=1}^{k} F_i\right) / \sum_{t=1}^{q} \left(\sum_{i=1}^{k} F_i\right)_t,\ q = 2 \quad (13)$$

where w_{Gt}^{DF} are the importance weights of the two sub-groups (F_I and F_E) of departure delay factors, t represents the number of groups (q = 2), and i denotes the number of factors under F_I (k = 8) and F_E (k = 3).

Furthermore,

$$d_{Allj}^{DF} = \sum_{t=1}^{q} w_{Gt}^{DF} * d_{tj}^{DF},\ q = 2 \quad (14)$$

$$\left(D_{All}^{DF}\right)_{1*4} = \left(W_{Gt}^{DF}\right)_{1*2} \times \left(D_G^{DF}\right)_{2*4} = \left(d_{t0,}^{DF} d_{t1,}^{DF} d_{t3,}^{DF} d_{t5}^{DF}\right) \quad (15)$$

$$D_F = \sum_{i=1}^{2} \left(S_j * d_{tj}^{DF}\right) \quad (16)$$

Similarly,

$$\left(D_i^{DD}\right)_{1*4} = \left(f_{i0,}^{DD} f_{i1,}^{DD} f_{i3,}^{DD} f_{i5}^{DD}\right) \tag{17}$$

and the impact of departure delay duration (D_D) at level 3 can be estimated as:

$$D_D = \sum_{i=1}^{4} \left(S_j * f_{ij}^{DD}\right) \tag{18}$$

Finally, the impact of D_F was aggregated with the D_D at level 4 to estimate the traffic congestion hazard period (H_{TC}) as:

$$H_{TC} = \sqrt{D_F \times D_D} \tag{19}$$

3. Results

3.1. Survey Responses

The online survey was sent to around 250 respondents in June 2020. After the asked time to return the questionnaire, a satisfactory response rate of 20% was received from 50 participants residing in the four provinces of the country [32]. As the survey responses provided the opinion of the participants about both the departure delay and delay duration, statistical analysis estimated the percent frequencies for all the DF against different DD. Figure 3a–k presents the stacked bar charts for all the internal and external departure delay factors as given in Table 1. Figure 3a shows that overall, the small-sized families with less than 10 persons had lesser tendency to delay the HSD, while 80% of the large-sized families with more than 10 persons delayed their HSD for over 30 min. As per Figure 3b, the respondents busy in some work at home mostly delayed their departure, i.e., 40% of the busy persons delayed their departure for over 30 min. It was also found that people who work in the private sector (usually busy schedule of possible online working) delay their shopping visits (Figure 3c). The results revealed that 80% of the respondents working in non-profit organizations did not delay their HSD, while only 43% of public sector employees departed immediately after the curfew lift.

Figure 3d illustrates that highly educated people have a general tendency to go early for shopping after lifting of curfew period. Around 80% of the highly educated respondents reported their departure within 15 min after the curfew lift. It can be seen in Figure 3e that around 55% of people older than 35 years of age delay their HSD more than 15 min in comparison to 30% of younger respondents. In Figure 3f, commuters having more than one vehicle have more leisure to delay their shopping visits. Around 60% of people with one car immediately depart after the curfew timing, while around 70% of owners with two vehicles leave after 15 min. Larger families with more children are generally busy and delay their HSD (see Figure 3g). For instance, around 90% families with more than two children do not immediately depart to shop. Availability of driver is another factor that can delay the home to shopping area departure. Figure 3h shows that 40% of households with drivers delayed their trips for more than 15 min in comparison to 26% of households with no drivers.

Figure 3i shows that the commuters (65%) who do not reside nearby the shopping center of their choice usually delay their HSD, because by delaying their trip they might face less traffic. Likewise, longer distance to the shopping center of commuters' choice also leads to HSD delay, as illustrated in Figure 3j. Figure 3k reveals that the citizens of smaller cities leave for shopping within 30 min after the lift of curfew period. Finally, Figure 3l displays that 44% of the commuters did not delay their shopping visits, while 18% of them delayed their trips for more than 30 min. The distribution of departure delay shown in Figure 3l and the impacts of different factors described in Figure 3a–k demand a more detailed methodology to identify the potential period with highest traffic congestion.

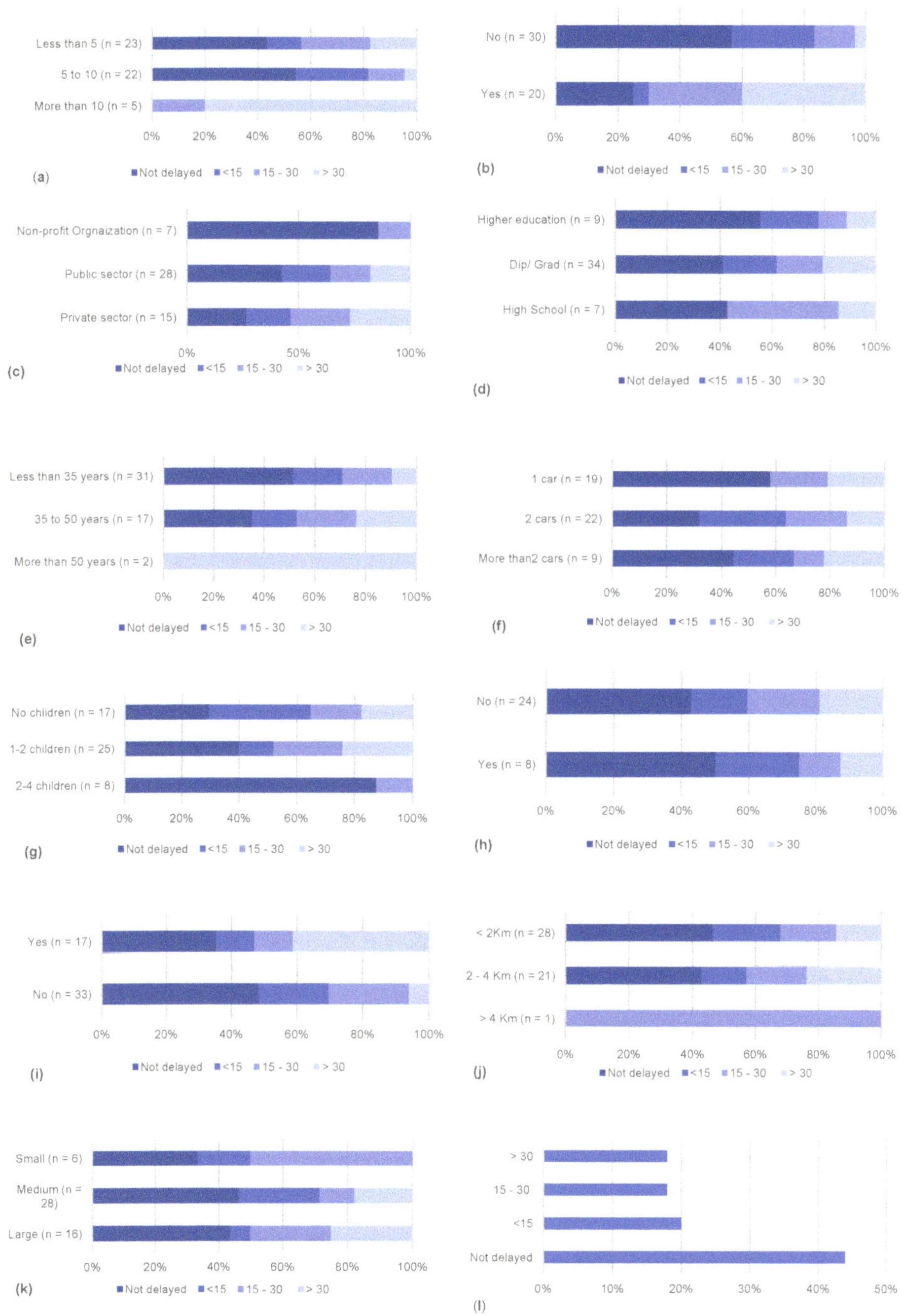

Figure 3. Stacked bar charts showing percentage frequencies for each departure delay factor for each departure delay duration: (**a**) family size, (**b**) involvement in other activities, (**c**) nature of job, (**d**) education level, (**e**) age, (**f**) number of vehicles, (**g**) number of children, (**h**) availability of driver, (**i**) availability of shopping center of choice, (**j**) distance to shopping center, and (**k**) size of the city; (**l**) percentage distribution of departure delay.

3.2. Statistical Analysis

Prior to applying the survey findings to assess the traffic congestion hazard period (HTC) using the hierarchical framework presented in Figure 2, the level of association between the D_F and D_D was established using the Chi-square independence test. The null and alternate hypothesis were established between all the internal and external departure delay factors and departure delay duration. An example for family size is given in Section 2.3. Table 2 presents the rationale to develop the null and alternative hypothesis for all the departure delay factors. The table also presents the results of Chi-square tests, which are described in the following section.

Table 2. Hypothesis and the level of association between D_F and D_D.

No	Factors	Rationale of the Hypothesis	Chi-Square (χ^2)	Significance at $p < 0.05$	Cramer's V	Association
1.	Internal Factors					
1.1	Family size	Commuters with large family size delay home-to-shopping center departure (HSD) for longer duration.	18.8	Significant	0.43	Large
1.2	Involvement in other personal activities	Commuters involved in personal activities delay HSD for longer duration.	16.5	Significant	0.60	Large
1.3	Nature of job	Commuters working in the private sector are busier and delay HSD for longer duration.	7.7	Not significant	0.28	Large
1.4	Education level	Less educated commuters' activities delay HSD for longer duration.	4.5	Not significant	0.21	Medium
1.5	Age	Older commuters delay HSD for longer duration.	11.4	Not significant	0.34	Large
1.6	Number of vehicles	Commuters having more than one vehicle delay HSD for longer duration.	8.3	Not significant	0.29	Large
1.7	Number of children	Households with more children delay HSD for longer duration.	11.5	Not significant	0.34	Large
1.8	Availability of driver	Presence of a driver in a household delay HSD for longer duration.	0.8	Not significant	0.12	Medium
2.	External Factors					
2.1	Availability of shopping center of choice	Non-availability of shopping center of commuter's choice delays HSD for longer duration.	9.5	Significant	0.44	Large
2.2	Distance to shopping center	Longer distance from shopping center of commuter choice delays HSD for longer duration.	5.1	Not significant	0.23	Large
2.3	Size of the city	Citizens of larger cities delay HSD for longer duration.	7.8	Not significant	0.28	Large

Due to space limitations, the detailed calculations for establishing the association between family size and DD are not given here. First, the obtained data was populated in the contingency table that shows the observed frequency of responses. Table 3 displays the information as percentages for better understanding and was used to develop Figure 3. There are three family size categories and four departure delay categories. The data clearly shows that 44% of the respondents did not delay HSD, while 38% delayed their shopping trips for more than 15 min. The table also shows that 80% of the commuters with family size larger than 10 delayed their HSD for more than 30 min.

The subsequent step establishes the expected frequencies, which are essentially the frequencies we expect in the observed data if the null hypothesis holds. The expected frequencies calculated using Equation (1) are given in Table 4. Next, the residuals as the difference between the observed and expected frequencies were determined using Equation (2). Table 5 shows the Chi-square values for each cell of the contingency table determined from Equation (3). The final χ^2 value (sum of the columns' total of Table 5) was found to be 18.8. Equation (4) estimated the degree of freedom ($d_f = 6$), which informs the probability of finding $\chi^2 \geq 18.8 \approx 0.0188$. Finally, the scale of association between the groups was found using Equation (5). Chi-square values for df = 6 and p-value < 0.05 showed that only two internal factors (family size and involvement in other activities) and one external factor (availability of shopping center of choice) significantly depend on departure delay duration with Chi-squared values higher than the critical values. However, with the small number of responses (n = 50), Cramer's V test established the magnitude of effect for all the internal and external factors. Table 2 presents that all the factors have a large magnitude of effect, except education level and presence of driver in a household. Hence, all the selected factors were used for assessing the H_{TC} with the help of the FSE methodology described in Section 2.4.

Table 3. Observed percentage frequencies for family size (n = 50).

Departure Delay	Family Size			Total
	More Than 10 (n = 5)	**5 to 10 (n = 22)**	**Less Than 5 (n = 23)**	
Not delayed	0%	55%	43%	44%
<15	0%	27%	13%	18%
15–30	20%	14%	26%	20%
>30	80%	5%	17%	18%
Total	100%	100%	100%	100%

Table 4. Expected frequencies for perfectly independent variables.

Departure Delay	Family Size (No of Persons)			Total
	>10	**5–10**	**<5**	
Not delayed	2.2	9.68	10.12	22
<15	0.9	3.96	4.14	9
15–30	1	4.4	4.6	10
>30	0.9	3.96	4.14	9
Total	5	22	23	50

Table 5. Contingency table showing calculated Chi-square values.

Departure Delay	Family Size (No of Persons)		
	>10	**5–10**	**<5**
Not delayed	2.2	0.6	0.0
<15	0.9	1.1	0.3
15–30	0.0	0.4	0.4
>30	10.7	2.2	0.0
Total	13.8	4.3	0.7

3.3. *Assessment of Traffic Congestion Hazard Period*

Equation (6) calculated the degree of association of each factor (internal of external) to the four-level-rating (S_j = 0, 1, 3, 5) as described in Section 2.4. The term I_i^I for family size was calculated as:

$$\left(I_1^I\right)_{1*4} = \left(f_{i0,}^I f_{i1,}^I f_{i3,}^I f_{i5}^I\right) = (0.44, 0.26, 0.20, 0.10)$$

The impact of family size F_1^I was estimated using Equation (7) as:

$$F_i^I = \sum_{i=1}^{4}\left(S_j * f_{ij}^I\right) = 0 \times 0.44 + 1 \times 0.26 + 3 \times 0.20 + 5 \times 0.1 = 1.36$$

Similarly, F_i^I for all the internal factors were calculated.

Importance weights of all the internal factor were estimated to calculate the overall impact of internal factors at level 2 of Figure 2, using Equation (8):

$$\begin{aligned} w_1^I &= F_i^I / \textstyle\sum_{i=1}^{k} F_i^I \\ &= 1.36/(1.36 + 1.76 + 2.08 + 1.68 + 1.16 + 1.56 + 1.24 + 0.88) = 1.36/11.72 = 0.116 \end{aligned}$$

Similarly, the weights of all the internal and external factors were calculated.

Subsequently, to apply FSE, Equation (10) estimated the membership functions for internal factors:

$$\begin{aligned} \left(D_1^I\right)_{1*4} &= \left(W_1^I\right)_{1*8} * \left(F_1^I\right)_{8*4} \\ &= [0.116\ \ 0.150\ \ 0.177\ \ 0.143\ \ 0.099\ \ 0.133\ \ 0.106\ \ 0.075] \\ &\quad \times \begin{bmatrix} 0.44 & 0.26 & 0.2 & 0.1 \\ 0.44 & 0.26 & 0.0 & 0.3 \\ 0.44 & 0.02 & 0.32 & 0.22 \\ 0.44 & 0.08 & 0.40 & 0.08 \\ 0.44 & 0.30 & 0.22 & 0.04 \\ 0.44 & 0.16 & 0.30 & 0.1 \\ 0.44 & 0.24 & 0.30 & 0.02 \\ 0.44 & 0.48 & 0.0 & 0.08 \end{bmatrix} \end{aligned}$$

$$D_1^I = \begin{bmatrix} 0.44 & 0.197 & 0.230 & 0.132 \end{bmatrix}$$

Similarly, the membership functions for external factors were estimated as:

$$\left(D_2^E\right)_{1*4} = \left(W_2^E\right)_{1*8} * \left(F_2^E\right)_{8*4}$$

$$= \begin{bmatrix} 0.39 & 0.228 & 0.382 \end{bmatrix} \times \begin{bmatrix} 0.44 & 0.22 & 0.0 & 0.34 \\ 0.44 & 0.3 & 0.24 & 0.02 \\ 0.44 & 0.08 & 0.3 & 0.18 \end{bmatrix}$$

$$D_2^E = \begin{bmatrix} 0.44 & 0.185 & 0.169 & 0.206 \end{bmatrix}$$

Then, with known membership functions of t number of factors' groups at level 2, the overall impact of each group (F_I and F_E) was estimated using Equations (11) and (12):

$$F_I = \textstyle\sum_{i=1}^{4}\left(S_j * d_{tj}^I\right) = 0 \times 0.44 + 1 \times 0.197 + 3 \times 0.230 + 5 \times 0.132 = 1.547$$

$$F_E = \textstyle\sum_{i=1}^{4}\left(S_j * d_{tj}^E\right) = 0 \times 0.44 + 1 \times 0.185 + 3 \times 0.169 + 5 \times 0.206 = 1.722$$

At level 4, the impacts of D_F and D_D on H_{TC} were aggregated by estimating their respective importance weights using Equation (13):

$$w_{F_I}^{DF} = \frac{1.547}{1.547 + 1.722} = 0.473$$

$$w_{F_E}^{DF} = \frac{1.722}{1.547 + 1.722} = 0.527$$

and

$$\left(D_{All}^{DF}\right)_{1*4} = \begin{bmatrix} 0.473 & 0.527 \end{bmatrix} \times \begin{bmatrix} 0.44 & 0.197 & 0.23 & 0.132 \\ 0.44 & 0.185 & 0.169 & 0.206 \end{bmatrix}$$
$$= \begin{bmatrix} 0.44 & 0.191 & 0.198 & 0.171 \end{bmatrix}$$

and

$$D_F = 1.64$$

Similarly,

$$\left(D_i^{DD}\right)_{1*4} = \begin{bmatrix} 0.0 & 0.20 & 0.54 & 0.90 \end{bmatrix}$$
$$D_D = 1.64$$

Finally, the impacts of D_F and D_D were aggregated to estimate the traffic congestion hazard period (H_{TC}) using Equation (19) as:

$$H_{TC} = \sqrt{D_F \times D_D} = 1.64$$

4. Discussion

The methodology developed in the present research contains two primary phases. The first phase identifies internal and external factors encompassing socio-economic variables of the commuters residing in small, medium, and large cities which can affect departure (early or late) decision. All the internal and external factors were identified through literature and expert opinion during brain storming sessions. The results illustrated in Section 3.1 highlights some important characteristics of the commuters during pandemic restrictions. Man-Keun et al. [33] investigated the impact of family size on grocery shopping and found that large-sized households prefer large discount stores even if not located in the near vicinity. Preparing for grocery shopping for large families also takes much longer in comparison to small families in order to find several missing items to meet the needs of family members [34]. Age, education level, and job type play an important role in decisions to commute after curfew lift. Their findings are in line with a recent past study on selection of traffic modes during COVID-19 by Abdullah et al. [12]. While evaluating the impact of job sector on commuter departure decision, it was also reported that private sector employees remained busier than public sector ones due to a faster transition and technical support in the private sector during the COVID-19 pandemic [35].

The study considered size of the city as a factor contributing to traffic congestion after curfew lift. High congestion has always been associated with larger cities [36]; nevertheless, ever-increasing population, inadequate capacity of streets, and mixed land uses pose diverse impacts of traffic in smaller cities as well, such as high accident frequencies in urban centers [28]. The findings of the present study revealed that departure patterns in small- and medium-sized cities are almost consistent with those in large cities during the COVID-19 period. Another reason for such findings is that the small- and medium-sized cities in this study are also capital cities of their respective provinces, with all the types of commercial, public, and residential land uses as large cities. Statistical analysis established that almost all the factors have large magnitude of effect, except education level and presence of driver in a household.

The past studies effectively employed FSE for risk assessment based on human perception and uncertain expert judgment. Akter et al. [37] used FSE and Intergovernmental Panel on Climate Change (IPCC) methods to develop risk assessment maps. They found that FSE eliminated the uncertainties associated with expert judgment in different IPCC methods and generated one risk map for a known hazard domain. Zhao et al. [28] used the FSE approach for risk assessment of green building projects in Singapore. Their approach aggregated the likelihood of occurrence and risk criticality of risk factors. They used a questionnaire survey method to interview experienced (over 10 years) project managers to ascertain the risks involved in green building projects. Their approach was modified for traffic hazard assessment in the present research. Same values of D_F and D_D affirm the computational accuracy of the proposed approach. Based on the UoD defined prior to conducting the commuter survey, the calculated H_{TC} value of 1.64 ascertains the highest congestion after 20 min of curfew lift, where "0" corresponds to immediate home-to-shopping center departure right after the curfew lift; "1" to a departure within 15 min; "3" to a delay between 15 and 30 min; and "5" to a departure delay of more than 30 min.

Figure 4 illustrates a theoretical display of the highest traffic congestion hazard period due to home-to-shopping center departure delay by commuters. Based on the questionnaire survey, the findings of this study show that 44% of the people depart as soon as the curfew lifts. Without much delay, an additional 20% of commuters leave their homes (within 15 min of the lift) for shopping that further increase the traffic congestion on the urban roads and streets. In the next 15 min (between 15 and 30 min after the lift), almost 82% of the total commuters have left their homes for shopping. The remaining 18% of commuters, who leave their homes after half an hour, are assumed to have a mindset of using the last part of the no curfew period. The behavior of the commuters illustrated in Figure 4 directs to the traffic congestion hazard in the early part of the no curfew period. The figure also shows that around 60% to 80% of the residents in an area occupy the capacity of urban streets and parking areas during his period. As per Figure 4, the highest possible congestion can be expected after half an hour of curfew lift to the end of the first hour. Responsible traffic agencies can adopt appropriate traffic management measures during this part of the day during a pandemic. The measures may include the following actions by a traffic regulatory agency [38]: reducing the need and length of a trip, promoting nonmotorized and public transport, promoting carpooling, shifting peak-hour travel, and diverting travel from congested locations. For the specific scenario of traffic in Saudi Arabia, the last two measures seem more practical.

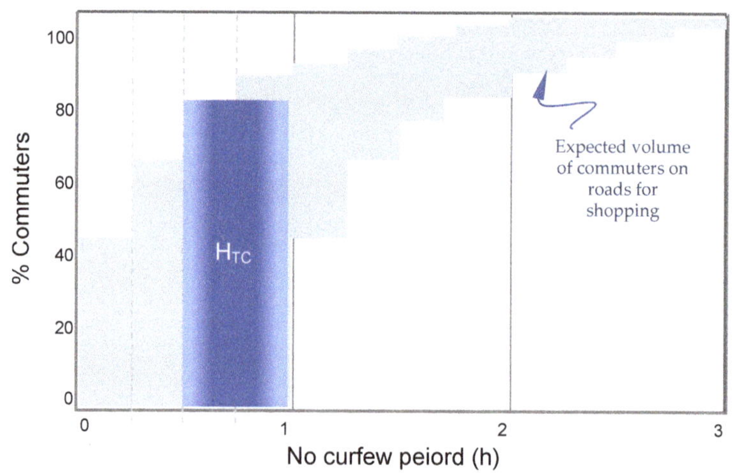

Figure 4. A theoretical illustration of highest traffic congestion period after COVID-19 curfew timings. Assumed no curfew period is 3 h based on 2021 no curfew timings in Saudi Arabia.

Modern transport system models (TSM) integrated with traditional approaches can evaluate the potential effects of traffic demand variations during COVID-19. For instance, floating car data can be used in TSM supported with big data for travel demand analysis [39]. Using the proposed methodology, the regulators can use socio-demographic data to identify the traffic congestion hazard period after the curfew timing. The data regarding the internal or external factors with higher significance and large effect size should be given importance. Interestingly, such internal D_F need simple and easily available data from the General Authority of Statistics in Saudi Arabia, such as family size, nature of job, age of the commuter, and number of children in a household. The only important external factor is availability of shopping center of choice in the near vicinity. As most of the cities in Saudi Arabia possess similar cultural practices, layouts of urban streets, and types of shopping centers, the percentages found in the present research can be used to assess these factors. Accordingly, the traffic regulators can identify the highest congestion period in an urban setting during the pandemic era.

The subjective organization of the proposed traffic congestion hazard assessment framework during pandemics' curfew periods is a limitation of the present study. Future studies following a similar approach using larger data sets can establish the credibility of the objective source basis and a practical reference value of the methods used.

5. Conclusions

Traffic congestion is evident after curfew lifting during the era of the COVID-19 pandemic. In an urban neighborhood, a congestion episode depends on the percentage of commuters leaving for home-to-shopping centers over the span of the no curfew period. The study found that departing early or delaying the shopping trip depends on certain internal (commuter related) and external (shopping related) factors. Among internal factors, family size and business (involvement in other activities) were found to be the most significant factors affecting the departure delay, while availability of shopping center of choice significantly affected the decision amongst the external factors. Age, number of children, and size of the city also influenced the commuters' decision about delaying the departure. Commuters' departure patterns in the small- and medium-sized cities (capitals of respective provinces) were found to be consistent with large cities during the COVID-19 period, primarily due to similar commercial, public, and residential activities.

Chi-square and Cramer's V tests established the statistical significance of the association between the departure delay factors and the departure delay duration. Chi-square values widely ranged between 0.8 and 18.8 for internal factors and from 5.1 to 9.5 for external factors. Cramers' V established large associations for most of the factors, except education level and availability of driver in a household. Fuzzy synthetic evaluation (FSE) can effectively ascertain the period of highest traffic congestion based on the commuters' responses. The study revealed that traffic congestion hazard in the early part (precisely the second half of the first hour) after the curfew lift needs particular attention of the traffic regulatory agencies. Future studies can validate the findings of the present research by implementing the proposed approach in different areas and conducting traffic monitoring studies after the curfew lift during pandemics.

Author Contributions: M.A. (Majed Alinizzi), funding, supervision, conceptualization, and writing—review and editing; H.H., conceptualization, methodology, analysis, validation, and writing-original draft preparation, M.A. (Mohammad Alresheedi), data collection, methodology, and writing—review and editing. All authors have read and agreed to the published version of the manuscript.

Funding: This research was funded by the Deanship of Scientific Research, Qassim University, grant No. 10014-qec-2020-1-1-L.

Institutional Review Board Statement: Not applicable.

Informed Consent Statement: Not applicable.

Data Availability Statement: All the shareable data is given in the main text.

Acknowledgments: The authors gratefully acknowledge Qassim University, represented by the Deanship of Scientific Research, for the financial support of this research under the number (10014-qec-2020-1-1-L) during the academic year 1440 AH/2020 AD.

Conflicts of Interest: The authors declare no conflict of interest.

References

1. Wu, F.; Zhao, S.; Yu, B.; Chen, Y.; Wang, W.; Song, Z.; Hu, Y.; Tao, Z.; Tian, J.; Pei, Y.; et al. A new coronavirus associated with human respiratory disease in China. *Nature* **2020**, *579*, 265–269. [CrossRef] [PubMed]

2. Wang, M.; Jiang, A.; Gong, L.; Luo, L.; Guo, W.; Li, C.; Zheng, J.; Li, C.; Yang, B.; Zeng, J. Temperature significant change COVID-19 transmission in 429 cities. *medRxiv* **2020**. [CrossRef]
3. WHO. *WHO Characterizes COVID-19 as a Pandemic*; WHO: Geneva, Switzerland, 2020.
4. Worldometer. Available online: https://www.worldometers.info/coronavirus/#countriesm (accessed on 24 January 2022).
5. Kim, K. Impacts of COVID-19 on transportation: Summary and synthesis of interdisciplinary research. *Trans. Res. Interdiscip. Perspect.* **2021**, *9*, 100305. [CrossRef] [PubMed]
6. Cartenì, A.; Henke, I. Transportation Planning, Mobility Habits and Sustainable Development in the Era of COVID-19 Pandemic. *Sustainability* **2022**, *14*, 2968. [CrossRef]
7. Russo, F.; Corrado, R. Regional transport plans: From direction role denied to common rules identified. *Sustainability* **2021**, *13*, 9052. [CrossRef]
8. Khawagi, W. The problem of traffic congestion in Saudi Arabia. *Int. J. Sci. Eng. Res.* **2017**, *8*, 1632–1638.
9. Majhad, H.; Bramantoro, A.; Syamsuddin, I.; Yunianta, A.; Basori, A.; Prabuwono, A.; Barukab, O. A traffic congestion framework for smart Riyadh City based on IoT services. *Int. J. Adv. Comput. Sci. Appl.* **2018**, *9*, 2018. [CrossRef]
10. Bucsky, P. Modal share changes due to COVID-19: The case of Budapest. *Transport. Res. Interdiscip. Perspect.* **2020**, *8*, 100141. [CrossRef] [PubMed]
11. Jenelius, E.; Cebecauer, M. Impacts of COVID-19 on public transport ridership in Sweden: Analysis of ticket validations, sales and passenger counts. *Transp. Res. Interdiscip. Perspect.* **2020**, *8*, 100242. [PubMed]
12. Abdullah, M.; Ali, N.; Javid, M.A.; Dias, C.; Campisi, T. Public transport versus solo travel mode choices during the COVID-19 pandemic: Self-reported evidence from a developing country. *Transp. Eng.* **2021**, *5*, 100078. [CrossRef]
13. Parady, G.; Tanaguchi, A.; Takami, K. Travel behavior changes during the COVID-19 pandemic in Japan: Analyzing the effects of risk perception and social influence on going-out self-restriction. *Transp. Res. Interdiscip. Perspect.* **2020**, *7*, 100181. [CrossRef]
14. Abdullah, M.; Dias, C.; Muley, D.; Shahin, M.d. Exploring the impacts of COVID-19 on travel behavior and mode preferences. *Transp. Res. Interdiscip. Perspect.* **2020**, *8*, 100255. [CrossRef] [PubMed]
15. Katrakazas, C.; Michelaraki, E.; Sekadakis, M.; Yannis, G. A descriptive analysis of the effect of the COVID-19 pandemic on driving behavior and road safety. *Transp. Res. Interdiscip. Perspect.* **2020**, *7*, 100186. [CrossRef] [PubMed]
16. Saladie, O.; Bustamante, E.; Gutierrez, A. COVID-19 lockdown and reduction of traffic accidents in Tarragona province, Spain. *Transp. Res. Interdiscip. Perspect.* **2020**, *8*, 1000218. [CrossRef] [PubMed]
17. Loske, D. The impact of COVID-19 on transport volume and freight capacity dynamics: An empirical analysis in German food retail logistics. *Transp. Res. Interdiscip. Perspect.* **2020**, *6*, 100165. [CrossRef]
18. Dahlberg, M.; Edin, P.-A.; Grönqvist, E.; Lyhagen, J.; Östh, J.; Siretskiy, A.; Toger, M. Effects of the COVID-19 Pandemic on Population Mobility under Mild Policies: Causal Evidence from Sweden. *arXiv* **2020**, arXiv:2004.09087. [CrossRef]
19. Heiler, G.; Reisch, T.; Hurt, J.; Forghani, M.; Omani, A.; Hanbury, A.; Karimipour, F. Country-wide mobility changes observed using mobile phone data during COVID-19 pandemic. In Proceedings of the 2020 IEEE International Conference on Big Data, Atlanta, GA, USA, 10–13 December 2020.
20. Huang, J.; Wang, H.; Fan, M.; Zhuo, A.; Sun, Y.; Li, Y. Understanding the impact of the COVID-19 pandemic on transportation-related behaviors with human mobility data. In Proceedings of the 26th ACM SIGKDD International Conference on Knowledge Discovery & Data Mining, Virtual Event, CA, USA, 6–10 July 2020; pp. 3443–3450.
21. Muley, D.; Ghanim, M.; Mohammad, A.; Kharbeche, M. Quantifying the impact of COVID-19 preventive measures on traffic in the State of Qatar. *Transp. Policy* **2021**, *103*, 45–59. [CrossRef]
22. Xu, P.; Li, W.; Hu, X.; Wu, H.; Li, J. Spatiotemporal analysis of urban road congestion during and post COVID-19 pandemic in Shanghai, China. *Transp. Res. Interdiscip. Perspect.* **2022**, *13*, 100555. [CrossRef]
23. Loo, B.; Huang, Z. Spatio-temporal variations of traffic congestion under work from home (WFH) arrangements: Lessons learned from COVID-19. *Cities* **2022**, *124*, 103610. [CrossRef] [PubMed]
24. Alinizzi, M.; Haider, H.; Almoshaogeh, M.; Alharbi, F.; Alogla, S.M.; Al-Saadi, G.A. Sustainability Assessment of Construction Technologies for Large Pipelines on Urban Highways: Scenario Analysis using Fuzzy QFD. *Sustainability* **2020**, *12*, 2648. [CrossRef]
25. Zhao, X.; Hwang, B.G.; Gao, Y. A fuzzy synthetic evaluation approach for risk assessment: A case of Singapore's green projects. *J. Clean. Product.* **2016**, *115*, 203–213. [CrossRef]
26. Saudi Ministry of Health. Available online: https://www.moh.gov.sa (accessed on 5 January 2022).
27. Saudi Ministry of Interior. Available online: https://www.moi.gov.sa (accessed on 5 January 2022).
28. Wang, M.; Debbage, N. Urban morphology and traffic congestion: Longitudinal evidence from US cities. *Comput. Environ. Urban Syst.* **2021**, *89*, 101676. [CrossRef]
29. Hawkins-Mofokeng, R.; Tlapana, T.; Ssemugooma, D.K. Effects of Traffic Congestion on Shopping Location Choice in the Greater eThekwini Region. *J. Bus. Manag. Rev.* **2022**, *3*, 372–386. [CrossRef]
30. SPSS Tutorials. Available online: https://www.spss-tutorials.com/chi-square-independence-test/ (accessed on 10 February 2022).
31. Cohen, J. *Statistical Power and Analysis for the Behavioral Sciences*, 2nd ed.; Hisdale, N.J., Ed.; Lawrence Erlbaum Associates: Mahwah, NJ, USA, 1988; pp. 79–80.
32. Visser, P.S.; Krosnick, J.A.; Marquette, J.; Curtin, M. Mail surveys for election forecasting? An evaluation of the Columbus Dispatch poll. *Public Opin. Q.* **1996**, *60*, 181–227. [CrossRef]
33. ManKeun, K.; Lee, Y.; Kim, S. The influence of household types on food and grocery store choices. *J. Rural Dev.* **2018**, *41*, 89–121.

34. Thomson, E. *Family Size Preferences, International Encyclopedia of the Social & Behavioral Sciences*, 2nd ed.; Elsevier: Amsterdam, The Netherlands, 2015; pp. 805–808.
35. Deloitte, Perspectives-Working from Home during the Coronavirus Crisis Is Far Less Common among Public Authorities than in the Private Sector. Available online: https://www2.deloitte.com/ch/en/pages/public-sector/articles/working-from-home-during-coronavirus-less-common-among-public-authorities.html (accessed on 27 July 2022).
36. Chang, Y.S.; Lee, Y.J.; Choi, S.S.B. Is there more traffic congestion in larger cities?—Scaling analysis of the 101 largest US urban centers. *Transp. Policy* **2017**, *59*, 54–63. [CrossRef]
37. Akter, M.; Momtaz, J.; Rubaiya, K.; Dewan, S.K.; Anisul, H.; Munsur, R.; Mashfiqus, S. Risk assessment based on fuzzy synthetic evaluation method. *Sci. Total Environ.* **2019**, *658*, 818–829. [CrossRef] [PubMed]
38. Strickland, S.G.; Berman, W. *Congestion Control and Demand Management, Public Road-Winter*; Federal Highway Administration: Washington, DC, USA, 1995; Volume 58.
39. Croce, A.; Musolino, G.; Rindone, C.; Vitetta, A. Estimation of travel demand models with limited information: Floating car data for parameters' calibration. *Sustainability* **2021**, *16*, 8838. [CrossRef]

computation

Article

A Prospective Method for Generating COVID-19 Dynamics

Kamal Khairudin Sukandar [1], Andy Leonardo Louismono [1], Metra Volisa [1], Rudy Kusdiantara [1,2], Muhammad Fakhruddin [3,*], Nuning Nuraini [1,2] and Edy Soewono [1,2]

1 Faculty of Mathematics and Natural Sciences, Institut Teknologi Bandung, Bandung 40132, Indonesia; kamalkhairudin@students.itb.ac.id (K.K.S.); andyleonardo@students.itb.ac.id (A.L.L.); metravolisa@students.itb.ac.id (M.V.); rudy@math.itb.ac.id (R.K.); nuning@math.itb.ac.id (N.N.); esoewono@itb.ac.id (E.S.)
2 Center of Mathematical Modeling and Simulation, Institut Teknologi Bandung, Bandung 40132, Indonesia
3 Department of Mathematics, Faculty of Military Mathematics and Natural Sciences, The Republic of Indonesia Defense University, IPSC Area, Sentul, Bogor 16810, Indonesia
* Correspondence: muhammad.fakhruddin@idu.ac.id

Abstract: Generating dynamic operators are constructed here from the cumulative case function to recover all state dynamics of a Susceptible–Exposed–Infectious–Recovered (SEIR) model for COVID-19 transmission. In this study, recorded and unrecorded EIRs and a time-dependent infection rate are taken into account to accommodate immeasurable control and intervention processes. Generating dynamic operators are built and implemented on the cumulative cases. All infection processes, which are hidden in this cumulative function, can be recovered entirely by implementing the generating operators. Direct implementation of the operators on the cumulative function gives all recorded state dynamics. Further, the unrecorded daily infection rate is estimated from the ratio between IFR and CFR. The remaining dynamics of unrecorded states are directly obtained from the generating operators. The simulations are conducted using infection data provided by Worldometers from ten selected countries. It is shown that the higher number of daily PCR tests contributed directly to reducing the effective reproduction ratio. The simulations of all state dynamics, infection rates, and effective reproduction ratios for several countries in the first and second waves of transmissions are presented. This method directly measures daily transmission indicators, which can be effectively used for the day-to-day control of the epidemic.

Keywords: COVID-19; SEIR models; dynamics generator; unrecorded infections; Richard's curve

Citation: Sukandar, K.K.; Louismono, A.L.; Volisa, M.; Kusdiantara, R.; Fakhruddin, M.; Nuraini, N.; Soewono, E. A Prospective Method for Generating COVID-19 Dynamics. *Computation* **2022**, *10*, 107. https://doi.org/10.3390/computation10070107

Academic Editors: Simone Brogi and Vincenzo Calderone

Received: 11 May 2022
Accepted: 17 June 2022
Published: 24 June 2022

1. Introduction

The COVID-19 pandemic is an ongoing global disease caused by the severe acute respiratory syndrome coronavirus 2 (SARS-CoV-2). The virus was reported to be first identified in December 2019 in Wuhan, China [1]. It was suspected that the original infection came from animals in the Wuhan Animal Market [2]. The number of cases grew exponentially, with more evidence of newly infected people who had never been visiting the market. This fact provided evidence that human-to-human transmission was the primary source of the fast transmission [3]. Immediately, the cases spread throughout all provinces in the country and even passed the borders through the neighboring countries. At the end of January 2020, the government of Wuhan imposed a total lockdown, preventing people from entering and leaving the city of Wuhan. The strict lockdown was also extended in response to the rapid spread of the virus [4].

Effort to predict the progress of COVID-19 transmission was made using the early data to obtain insight into infection characteristics. Zhang et al. used the stochastic model of the SEIR (Susceptible–Exposed–Infected–Recovery) model and provided the forecasts on the number of cases in several provinces in China, i.e., Shanghai, Beijing, Guangdong, Zhejiang, Chongqing, and Hunan using [5]. It was estimated that the virus transmission would be significantly disappearing in those regions in March 2020. The compartmental model-based predictions on COVID-19 figures were also conducted by Bertozzi et al. [6], who used the generic SIR (Susceptible–Infected–Recovery) model. They studied the COVID-19 spread in California and predicted the end of the first outbreak in August 2020. Not limited to that, other work utilized the Richard's Curve to yield the extrapolated figure of the infection trajectories. The work conducted by Nuraini et al. predicted that the spread would reach the peak in late March 2020 and soon decrease significantly until totally vanishing in April 2020 [7].

Learning from the experience of many countries during the first wave of transmission, the detection of infected persons was a crucial aspect. The availability of a sufficient amount of diagnostic tests was necessary. In the early phase of the pandemic, many developing countries were struggling to provide the proper amount of specimens to detect COVID-19. During the first wave transmission, as a non-manufacturer of Polymerase Chain Reaction (PCR) Reagents for real-time COVID-19 detection and due to the limitations of world supply, health authorities in Indonesia could not fulfill the daily PCR testing target as was recommended by the WHO ([8,9]). The lack of testing capacity certainly implies the low recorded cases as compared to total infections. Consequently, as many countries were already able to contain the disease within two months, other countries, including Indonesia, were still facing the outbreaks for a more extended period.

The complication of COVID-19 transmission is mainly related to the inability of the authorities to record all infected people and people's behavior toward the disease. It is a challenge for epidemiologists to construct simple models that can accommodate the most important phenomena. Compartmental models are very widely used in the construction of the disease transmission [10]. The simplest compartmental model for direct transmission is known as the SIR model, which contains susceptible (S), infectious (I), and recovered (R) compartments. Ross already applied this model in the early 20th century [11]. SIR-type models for COVID-19 transmission were used extensively in the early phase of the pandemic. Typical observations in the early transmission focused on predicting the outbreak's peak and the disappearance of the disease by exploiting the daily COVID-19 data. Yang et al. predicted the epidemic's future using the modified SEIR model linked with artificial intelligence. For daily progress, Susanto et al. estimated the effective reproduction ratio using the transmission data in Italy [12]. We have constructed, in Section 2, the basic formulation of the generating operators for the simple SEIR dynamics, which are then generalized to accommodate both recorded and unrecorded cases, as given in Section 3. Using the cumulative infections data provided by Worldometers, the simulations are conducted for country comparisons, i.e., Brazil, China, Germany, India, Indonesia, Islamic Republic of Iran, Italy, Japan, Republic of Korea, and Singapore.

2. Generating Operator in a Simple SEIR Model

During the early transmission of COVID-19, there was pressure in each affecting country to measure the daily reproduction ratio and predict the time when the outbreak was slowing and disappearing. With limited data and information, the simple SEIR model was used extensively. We formulate the concept of a generating operator to extract all states, which was first introduced in [13].

2.1. Model Formulation

We start the SEIR transmission model of COVID-19 with susceptible compartment S, exposed compartment under incubation period E, infected and infectious compartment I, and recovered compartment R. The overall process of infections is shown in Figure 1.

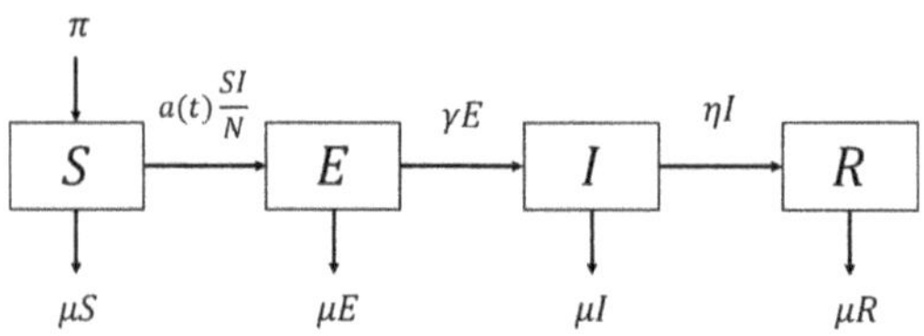

Figure 1. Flow diagram of the simple SEIR model.

The governing equations of the simple SEIR model are formulated as:

$$\begin{aligned}
\frac{dS}{dt} &= \pi - a(t)\frac{SI}{N} - \mu S, \\
\frac{dE}{dt} &= a(t)\frac{SI}{N} - \gamma E - \mu E, \\
\frac{dI}{dt} &= \gamma E - \eta I - \mu I, \\
\frac{dR}{dt} &= \eta I - \mu R,
\end{aligned} \tag{1}$$

where the parameters π, μ, γ, η, are the recruitment rate, the natural death rate, the inverse of the incubation period, and the inverse of the recovery period, respectively. The infection rate is given as a time-dependent parameter $a(t)$ to accommodate the intervention process, which is not measurable in the field. In this model, the total population is assumed constant, i.e., $N = S + E + I + R = \frac{\pi}{\mu}$. When it comes to the interaction process, we assume that the population is well-mixed, which can be physically analogous to 'well-stirred' individuals that force infected and susceptible to all-to-all interaction at all times. This assumption simplifies the mathematics evolutionary processes, which makes the analytical solutions possible [14].

A detailed description of parameters introduced in system (1) is given in Table 1.

Table 1. Descriptions of parameters given in system (1) and (24).

Parameters	Definition	Value	Source
N	Number of overall population	adjusted	[15]
π	Natural recruitment rate	adjusted	[16]
μ	Natural death rate	$\frac{1}{70\times365}$	[16]
$a(t)$	Infection rate	estimated	-
$\omega(t)$	Transition rate	adjusted	-
γ^{-1}	Incubation period of COVID-19	$\frac{1}{6}$	[17]
η^{-1}	Infection period of COVID-19	$\frac{1}{14}$	[18]

Note that those being labeled with 'estimated' in Table 1 will be evaluated using the generating operator, while the others will be adjusted to specific regions. The transition rate $\omega(t)$ exists in the generalized model, the values of which will be further explained in the next section.

In the case of constant infection rate $a(t) = a$, the basic reproduction ratio R_0, representing the average number of secondary infections caused by one infected person in the early pandemic [19], is given as

$$R_0 = \sqrt{\frac{a\gamma}{(\gamma+\mu)(\eta+\mu)}}. \tag{2}$$

As time evolves, the basic reproduction ratio is no longer appropriate to measure the progress of the transmission. The corresponding effective reproduction ratio, denoted with R_{0E}, is intended for tracking in the progress of transmission, which is given as:

$$R_{0E} = \sqrt{\frac{a(t)\gamma}{(\gamma+\mu)(\eta+\mu)}\frac{S(t)}{N}}. \tag{3}$$

The effective reproduction ratio is basically a basic reproduction ratio but with the time-dependent transmission rate and additional term of $S(t)/N$. This formula is obtained by implementing the NGM method without substituting the Disease Free Equilibrium [12].

In the following subsection, the state generating operator and a method for estimating the transmission rate $a(t)$ will be constructed. This construction gives a more adaptable estimate to track the progress of transmission involving intervention in the field.

2.2. *Cumulative Case Data for Constructing the Generating Operator*

The inability of the timely and accurate collection of COVID-19 data in daily case reports occurs in many countries. Discrepancies of confirmed official COVID-19 data were reported from many countries, such as Bangladesh [20], India [21], and the USA ([22,23]). The quality of the COVID-19 data certainly contributes to the consistency of the model and the accuracy of prediction.

The fluctuation of the daily cases also contributes to the prediction bias due to errors in data fitting. The choice of cumulative data for generating strategic indicators is mainly due to the smooth profile of the data to allow accurate fitting. Detail transmission behavior is kept within the cumulative case data, which can be recovered by identifying the proper generating operator. The S-curve shape of the cumulative data is best fitted with (one of them) Richard's curve.

We start with data fitting of cumulative cases using the Generalized Linear Growth Model (GLGM), widely known as Richard's Curve ([24,25]). The model comprises four parameters, denoted by $C_i, i \in 1,2,3,4$. The value of C_1 acts as the final epidemic size, with $\lim_{t\to\infty} \mathcal{K}(t) = C_1$, whereas C_3 represents the intrinsic growth rate. The higher this value, the steeper the curve at the early outbreak. The other two values are C_2 and C_4, which both act as the adjuster. While the former adjusts the lag phase of the curve, the latter is strongly related to the adjustment of the initial value at $t = 0$ [26]. The general form of Richard's Curve is given by Equation (4) as follows:

$$\mathcal{K}(t) = \frac{C_1}{(1 + C_2\exp(-C_3(t - C_4)))^{\frac{1}{C_2}}}. \tag{4}$$

All the parameters that exist in the explicit formula of Richard's Curve are extracted by applying the optimization scheme to obtain the minimum deviation between the data $\hat{\mathcal{K}}(t)$ and the fitted formula $\mathcal{K}(t)$. The optimization problem on parameter estimation can be written as

$$\min_{C_i\in\mathcal{D}} \sum_{i=1}^{N} \left(\mathcal{K}(t_i) - \hat{\mathcal{K}}(t_i)\right)^2, \tag{5}$$

where $\mathcal{D}$ is the search domain of the parameters and N represents the length of cumulative data.

The construction of the generating operator starts with the definition of the additional compartment $K(t)$, representing the cumulative cases at time t, which is given as follows.

$$K(t) - I(t) + R(t) \tag{6}$$

Take the first derivative of K, then we have

$$\begin{aligned}\frac{dK(t)}{dt} &= \frac{dI(t)}{dt}+\frac{dR(t)}{dt}\\ &= \gamma E(t)-(\eta+\mu)I(t)-(\eta I(t)-\mu R(t))\\ &= \gamma E(t)-\mu K(t).\end{aligned} \tag{7}$$

Solving for $E(t)$, then we have

$$\gamma E(t)=\frac{dK(t)}{dt}+\mu K(t). \tag{8}$$

From the result above, we can express $E(t)$ as a function of $K(t)$ as follows

$$E(t)=\frac{1}{\gamma}\left(\frac{dK(t)}{dt}+\mu K(t)\right). \tag{9}$$

Further, by taking the derivative of $E(t)$, the we obtain the daily new exposed

$$a(t)\frac{SI}{N}=\frac{dE(t)}{dt}+(\gamma+\mu)E=\frac{1}{\gamma}\left(\frac{d^2K(t)}{dt^2}+(\gamma+2\mu)\frac{dK(t)}{dt}+\mu(\gamma+\mu)K(t)\right). \tag{10}$$

Let $\mathcal{X}(t)=\begin{bmatrix}I(t)\\R(t)\end{bmatrix}$ depicting the dynamics of active infections and total recovery simultaneously. Thus, the third and fourth equation in system (1) can be rewritten as:

$$\mathcal{X}'(t)+\mathcal{A}\mathcal{X}(t)=\mathcal{F}(t), \tag{11}$$

where $\mathcal{A}=\begin{bmatrix}\eta+\mu & 0\\ -\eta & \mu\end{bmatrix}$ and $\mathcal{F}(t)=\begin{bmatrix}\gamma E(t)\\0\end{bmatrix}$. With initial value $\mathcal{X}(0)=\begin{bmatrix}I(0)\\R(0)\end{bmatrix}$, the solution of a system can be obtained by applied the integration factor.

$$\mathcal{X}(t)=e^{-At}\mathcal{X}(0)+e^{-At}\int_0^t e^{A\tau}\mathcal{F}(s)d\tau. \tag{12}$$

Then, we have the solution for $I(t)$ and $R(t)$ as follows

$$I(t) = I(0)e^{-(\eta+\mu)t}+e^{-(\eta+\mu)t}\int_0^t\left(\frac{dK(\tau)}{d\tau}+\mu K(\tau)\right)e^{(\eta+\mu)\tau}d\tau, \tag{13}$$

$$R(t) = R(0)e^{-\mu t}+\eta e^{-\mu t}\int_0^t I(\tau)e^{\mu\tau}d\tau, \tag{14}$$

where $I(0)$ and $R(0)$ is given by the data of initial active cases and total recovery. Substituting Equation (13) to Equation (14), the $K(t)$-related formula of $R(t)$ is given by

$$\begin{aligned}R(t) &= (R(0)+I(0)(1-e^{-\eta t}))e^{-\mu t}+\\ &\quad \eta e^{-\mu t}\int_0^t e^{-\eta\tau}\int_0^\tau\left(\frac{dK(\sigma)}{d\sigma}+\mu K(\sigma)\right)e^{(\eta+\mu)\sigma}d\sigma d\tau\end{aligned} \tag{15}$$

Assuming that the number of population, N, is constant, we have the dynamics of susceptible individuals written as follows.

$$S(t)=N-E(t)-I(t)-R(t). \tag{16}$$

Now, consider an equation of E in (1). We can find $a(t)$ by manipulating the equation. We have that

$$a(t)=\frac{\left(\pi-\mu S-\frac{dS}{dt}\right)N}{SI}, \tag{17}$$

where S and I can be written as Equations (13) and (16), respectively. By that, the estimation of all states, as well as the transmission rate, can be generated using the information of cumulative infections. Summarizing the above construction, we formulate the generating operator

$$\mathcal{T} = \begin{bmatrix} \mathcal{T}_1 \\ \mathcal{T}_2 \end{bmatrix} : C^1[0,\infty] \times (C^1[0,\infty])^2 \to C^1[0,\infty] \times (C^1[0,\infty])^2, \tag{18}$$

$\mathcal{T}_i, i = 1,2$ for the SEIR dynamics is given as follows

$$\mathcal{T}_1 = \frac{1}{\gamma}\left(\frac{d}{dt} + \mu\right) \tag{19}$$

$$\mathcal{T}_2 = \int_0^t e^{As}\bar{F}(\mathcal{T}_1)ds, \tag{20}$$

where

$$\bar{F} = (\gamma\mathcal{T}_1, 0)^T \tag{21}$$

and A is given in (31). Hence, we have

$$\mathcal{T}_1(K(t)) = E(t) \tag{22}$$

$$e^{-At}(\mathcal{X}_0 + \mathcal{T}_2(K(t))) = \mathcal{X}(t) = (I(t), R(t))^T. \tag{23}$$

Figure 2 this illustrates the flow of how this dynamics generator works on the estimation of all state dynamics, including the time-dependent rate of transmission by means of the empirical fit to Richard's Curve.

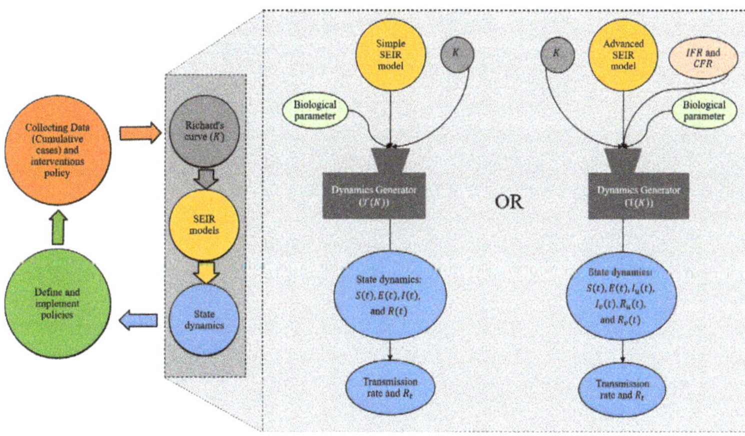

Figure 2. Diagram of the approach for estimating all state dynamics of the SEIR models using the dynamics generator.

3. Generalized SEIR for Second Wave Transmission of COVID-19

Many countries suffered badly during the second wave of COVID-19, which came unexpectedly after the period of relaxation at the end of the first wave ([27–29]). The phenomena of hospitals filling up, beds becoming scarce, and death rates exploding became constant daily news. Most countries implemented massive PCR testing as recommended by the WHO to isolate the positive cases in the population. Naturally, the simple SEIR model will not be realistic in representing the transmission.

3.1. Model Construction

We generalize the model (1) to accommodate for the intervention effect of COVID-19. The consequence of the COVID-19 testing capacity is also taken into account by distinguishing the recorded and unrecorded infections. It is assumed that the recorded infected people will have the self-awareness to lessen the contact with others than those who are not recorded. The mathematical model of the advanced SEIR is given as follows

$$\begin{aligned}\frac{dS}{dt} &= \pi - a(t)\frac{SI_n}{N} - \mu S,\\ \frac{dE}{dt} &= a(t)\frac{SI_n}{N} - \gamma E - \mu E,\\ \frac{dI_n}{dt} &= (1-\omega(t))\gamma E - \eta I_n - \mu I_n,\\ \frac{dI_s}{dt} &= \omega(t)\gamma E - \eta I_s - \mu I_s,\\ \frac{dR_n}{dt} &= \eta I_n - \mu R_n,\\ \frac{dR_s}{dt} &= \eta I_s - \mu R_s.\end{aligned} \tag{24}$$

with the assumption of the constant total population, we have

$$\frac{dN}{dt} = \pi - \mu N = 0, \tag{25}$$

and $N = \dfrac{\pi}{\mu}$.

The two compartments I and R are split into two, with indexes n and s, which stand for unrecorded and recorded (and isolated for treatment), respectively. In the previous assumptions, people in the I_s compartment do not have a chance to infect the susceptible individuals due to the isolation and hospitalization. In addition, people in the I_n can cause infections by making contact with people in the S compartment. Depicted in Figure 3, people will be either identified as an unrecorded or recorded infected person once they leave the E compartment. Infected individuals will recover after a period of time and become immune to the virus. No difference is assumed in the infection period, which implies the same value for the recovery rate for both recorded and unrecorded infections. More details about the parameters of the generalized model are given in Table 1.

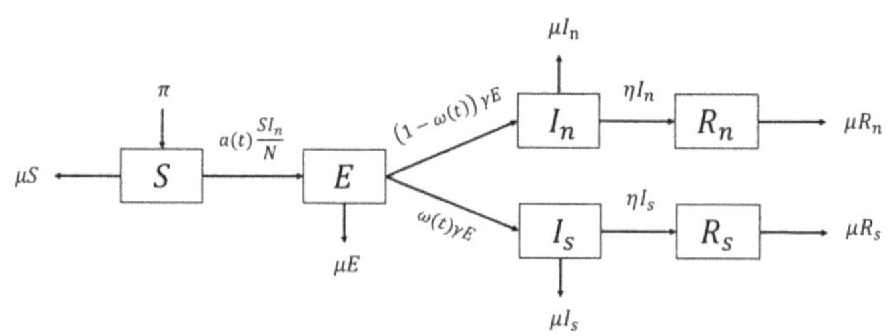

Figure 3. Flow diagram of the generalized SEIR Model.

In the case of constant infection rate $a(t) = a$ and transition rate $\omega(t) = \omega$, we have the formulation of basic reproduction number given, as follows

$$R_1 = \sqrt{\frac{a(1-\omega)\gamma}{(\gamma+\mu)(\eta+\mu)}}. \tag{26}$$

for zero transition rate, $\omega = 0$, R_1 reduces to R_0. Following the same derivation in the previous section, we construct the effective reproduction ratio

$$R_{1E} = \sqrt{\frac{a(t)(1-\omega(t))\gamma}{(\eta+\mu)(\gamma+\mu)}\frac{S(t)}{N}}. \tag{27}$$

Following the similar construction in Section 2, we define a new cumulative compartment as in (6)

$$K(t) = I_s(t) + R_s(t) \tag{28}$$

from the first derivative of K,

$$\begin{aligned} \frac{dK(t)}{dt} &= \frac{dI_s(t)}{dt} + \frac{dR_s(t)}{dt} \\ &= \omega(t)\gamma E(t) - \mu K(t). \end{aligned} \tag{29}$$

we obtain the daily new recorded cases $\omega(t)E(t)$, in the form

$$\omega(t)E(t) = \frac{1}{\gamma}\left(\frac{dK(t)}{dt} + \mu K(t)\right). \tag{30}$$

Let $\mathbf{X}(t) = \begin{bmatrix} I_s(t) \\ R_s(t) \\ I_n(t) \\ R_n(t) \end{bmatrix}$ depict the dynamics of active infections and total recovery simultaneously. Thus, the fourth and sixth equations in system (24) can be rewritten as

$$\mathbf{X}'(t) + \mathbf{A}\mathbf{X}(t) = \mathbf{F}(t), \tag{31}$$

where $\mathbf{A} = \begin{bmatrix} \eta+\mu & 0 & 0 & 0 \\ -\eta & \mu & 0 & 0 \\ 0 & 0 & \eta+\mu & 0 \\ 0 & 0 & -\eta & \mu \end{bmatrix}$ and $\mathbf{F}(t) = \begin{bmatrix} \omega(t)\gamma E(t) \\ 0 \\ (1-\omega(t))\gamma E(t) \\ 0 \end{bmatrix}$. With the initial value $\mathbf{X}(0) = \begin{bmatrix} I_s(0) \\ R_s(0) \\ I_n(0) \\ R_n(0) \end{bmatrix}$, the solution of a system can be obtained by applying the integration factor as follows

$$\mathbf{X}(t) = e^{-\mathbf{A}t}\mathbf{X}(0) + e^{-\mathbf{A}t}\int_0^t e^{\mathbf{A}\tau}\mathbf{F}(\tau)d\tau. \tag{32}$$

The explicit form of I_s, R_s, I_n, and R_n can be given as:

$$I_s(t) = I_s(0)e^{-(\mu+\eta)t} + e^{-(\mu+\eta)t}\int_0^t \left(\frac{dK(\tau)}{d\tau} + \mu K(\tau)\right)e^{(\mu+\eta)\tau}d\tau \tag{33}$$

and

$$R_s(t) = R_s(0)e^{-\mu t} + \eta e^{-\mu t}\int_0^t I_s(\tau)e^{\mu\tau}d\tau. \tag{34}$$

$$I_n(t) = I_n(0)e^{-(\mu+\eta)t} + e^{-(\mu+\eta)t}\int_0^t \left(\frac{1-\omega(t)}{\omega(t)}\right)\left(\frac{dK(\tau)}{d\tau} + \mu K(\tau)\right)e^{(\mu+\eta)\tau}d\tau \tag{35}$$

and

$$R_n(t) = R_n(0)e^{-\mu t} + \eta e^{-\mu t}\int_0^t I_n(\tau)e^{\mu\tau}d\tau. \quad (36)$$

Assuming the number of population, N, is constant, then

$$S(t) = N - E(t) - I_n(t) - I_s(t) - R_n(t) - R_s(t). \quad (37)$$

from the first equation of S in (24), we can find $a(t)$

$$a(t) = \frac{\left(\pi - \mu S - \frac{dS}{dt}\right)N}{SI_n}, \quad (38)$$

with S and I_n given as in the previous derivation.

Summarizing from the above derivation, we generalize the construction (18)

$$\mathbf{T} = \begin{bmatrix} \mathbf{T}_1 \\ \mathbf{T}_2 \end{bmatrix} : C^1[0,\infty] \times (C^1[0,\infty])^4 \to C^1[0,\infty] \times (C^1[0,\infty])^4, \quad (39)$$

with $\mathbf{T}_i, i = 1,2$ for the generalized SEIR dynamics as follows

$$\mathbf{T}_1 = \frac{1}{\gamma}\left(\frac{d}{dt} + \mu\right) \quad (40)$$

$$\mathbf{T}_2 = \int_0^t e^{As}\bar{G}(\mathbf{T}_1)ds, \quad (41)$$

where

$$\bar{G} = \left(\mathbf{T}_1, 0, \frac{1-\omega}{\omega}\mathbf{T}_1, 0\right)^T \quad (42)$$

and A is given in (31). The five states $\omega(t)E(t), I_s, R_s, I_n$, and R_n are produced by these operators as follows

$$\mathbf{T}_1(K(t)) = \omega(t)E(t) \quad (43)$$

$$e^{-At}(\mathbf{X}_0 + \mathbf{T}_2(K(t))) = \mathbf{X}(t) = (I_s(t), R_s(t), I_n(t), R_n(t))^T. \quad (44)$$

3.2. Estimation of $\omega(t)$

Referring to Figure 3, new infections are separated into recorded and unrecorded cases. While the former will be immediately quarantined and treated and hence unable to infect susceptible individuals, the latter remains unidentified and then will spread the virus. From the daily new infected persons, $\gamma E(t)$, the portion $\omega\gamma E(t)$ is recorded through testing, which will enter the I_s compartment. The rest of the portion, $(1-\omega)\gamma E(t)$ will remain unrecorded and enter the unrecorded $I_n(t)$. By that, ω, $0 \le \omega \le 1$ depicts the share of the recorded newly infected population against its total. This parameter represents the ability of the "random" selection of the test target to capture the positive cases.

The capacity of some countries to cover all infections is strongly related to their ability to provide the testing kits [30]. In the early pandemic, many countries struggled to fulfill the demand for COVID-19 testing kits, resulting in the low value of ω. Nevertheless, in early 2021, several countries were able to conduct more massive daily testings [31], making the figure of testing capacity change dramatically. In response, it is reasonable to set the value of ω to vary over time (time-dependent) and hence denoted with $\omega(t)$. The dynamics of $\omega(t)$ will be estimated using the Infection Fatality Ratio (IFR) and Case Fatality Ratio (CFR).

In epidemiology, a CFR is the proportion of deaths from a certain disease compared to the total number of people diagnosed/confirmed with the disease for a particular period [32]. Similarly, the IFR also applies to infectious disease transmission to represent the proportion of deaths among all infected individuals, including all recorded and unrecorded subjects [33]. This quantity is closely related to the CFR but with additional accounts for unapparent infections among healthy people. The observed CFR in time t is defined by the total number of deaths, $D(t)$, divided by the total number of confirmed cases at time t, $K(t)$, i.e., $CFR(t) = \frac{D(t)}{K(t)}$, whereas the IFR is defined based on the total number of infections. Formally, $IFR = \frac{\tilde{D}}{\tilde{\mathcal{I}}}$, where $\tilde{D}$ and $\tilde{\mathcal{I}}$ denote the median of total deaths and estimated total infections considered from the early pandemic until a certain specified time. The $\tilde{\mathcal{I}}$ will be estimated by involving the data of total tests. The total number of infections is estimated by assuming that each person is only tested once, and the distribution of infections among the entire population is equal. The total infections, $\tilde{\mathcal{I}}$, for each country follow the definition introduced in [34], which is defined by dividing the total confirmed cases with the total tests conducted and multiplying it with its population size, i.e.,

$$\tilde{\mathcal{I}} = \left(\frac{\tilde{K}}{\tilde{T}}\right) \cdot N \tag{45}$$

where $\tilde{T}$ is the total tests performed until a certain specified time. Note that this method estimates the constant value of IFR. This argument should confirm that this parameter is a virus-related parameter, which assumes that no significant mutation affecting the virulence will lead to a constant value of IFR [35]. The estimated constant CFR for related diseases in some countries can be seen in [36].

Dividing the estimated IFR by the observed time-dependent CFR depicts the share of infected individuals that were recorded. By that, the time-dependent reporting rate $\omega(t)$ is defined as follows

$$\omega(t) = \frac{\hat{IFR}}{CFR(t)}. \tag{46}$$

the value of IFR is always less than that of CFR, resulting in the values $0 < \omega(t) \leq 1$. The greater the value $\omega(t)$, the more the infectious persons were isolated.

4. Numerical Simulations

4.1. Simple SEIR Model

In this section, the numerical simulations for Model (1) that resulted from implementing the generating operator on the fitted cumulative function $\mathcal{K}$ are shown. COVID-19 data are selected from ten countries representing different population sizes: Brazil, China, Germany, India, Indonesia, Islamic Republic of Iran, Italy, Japan, Singapore, and South Korea. The COVID-19 data are taken from the official website of Worldometer [37], consisting of the daily number of active cases and total recovery. The data were taken during the early transmission period ranges from late February until September 2021. In these simulations, only the first 60 days after the initial transmission will be used and analyzed. The interval for each country may vary depending on the initial transmission.

All biological parameters for the selected countries are chosen as the same. The natural death rate, denoted by μ, was assumed to be $\mu = \frac{1}{70\times365}$, referring to the average human life expectancy. As of December 2020, Our World in Data claimed that the life expectancy of humans was about 70 years [16]. The remaining biological parameters are listed in Table 1.

4.1.1. Fitted Cumulative Data

The simulation began by estimating the closest GLGM dynamics to the provided data. All the parameters were obtained by solving the optimization problem (5). The calculation was conducted numerically using the built-in function in MATLAB. Notice that the global minimizer is difficult to obtain using the numerical method. Thus, the initial guess was varied following the Sobol sequence in 4-D so that the result would be close to the global minimizer. The estimated parameters for the ten observed countries are given in Table 2. The second and third columns in the Table indicate the time interval of the data used in the calculation. The interval varies among all countries depending on the initial transmission of the virus. The last four columns provide the estimated C_i for each country. These values depict the characteristics of the virus spread in each country and, hence, may differ from one country to another, though it was the same virus that spreads. For instance, the data fitting suggests that the value of C_3 for India is significantly higher compared to that for Indonesia. This result indicates that the spread of the virus in India is more significant compared to that in Indonesia. This fact can be seen in Figure 4, where the graph for India is much steeper than that for Indonesia.

Table 2. Parameter estimation of the cumulative dynamics using GLGM and the early pandemic data.

Country	Start Date	End Date	C_1	C_2	C_3	C_4
Brazil	25 February 2020	25 April 2020	52,934	0.4448	0.1005	58.1919
China	22 January 2020	22 March 2020	77,469	1.4611	0.2683	17.9661
Germany	15 February 2020	15 April 2020	150,171	0.3334	0.1204	43.4199
India	15 February 2020	15 April 2020	29,061	0.4595	0.1104	60.8589
Indonesia	2 March 2020	1 May 2020	21,032	0.1043	0.0445	50.9905
Iran	19 February 2020	19 April 2020	80,453	1.3952	0.1428	45.6531
Italy	15 February 2020	15 April 2020	174,575	0.0264	0.0743	38.7238
Japan	15 February 2020	15 April 2020	12,102	4.1299	0.3758	52.4032
Singapore	15 February 2020	15 April 2020	9846	0.0001	0.1358	14.9104
South-Korea	15 February 2020	15 April 2020	10,298	12.0304	1.0719	70.0849

Figure 4 illustrates the estimated models of cumulative infections together with the data for the first 60 days after the initial transmission. Overall, the general behavior of the data was well-fitted by the rendered S-curve Richard's model. For instance, in China and South Korea, the cumulative infections started to ramp up rapidly in the early pandemic yet began to decline within the first 60 days of transmission, giving us the perfect S-curve models. The countries China, Germany, Iran, Italy, Japan, and South Korea underwent a sharp increase in transmission and start to bend down before the end of the 60-day period. For the rest of the countries, the total infections were stagnantly increasing and there was no sign of sloping down within the first 60 days.

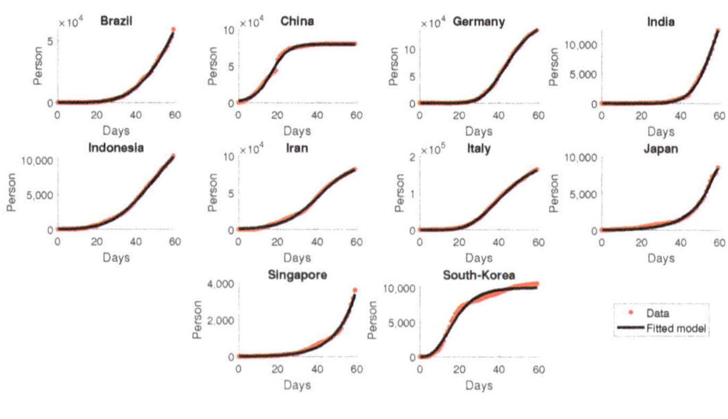

Figure 4. Fitting results $\bar{\mathcal{C}}$ of cumulative data for several countries in the early pandemic. The red dots represent the actual data, and the solid black lines represent the fitted model $\bar{\mathcal{C}}$.

With the direct substitution of $\bar{\mathcal{C}}$ on the right-hand side of (8), we obtain the estimate of the daily new cases $\gamma E(t)$. Figure 5 illustrates the estimation of daily new cases compared to the actual data for the ten observed countries. In general, the estimated dynamics resemble the real data of daily new cases. The model can also identify the peak time of the daily new cases.

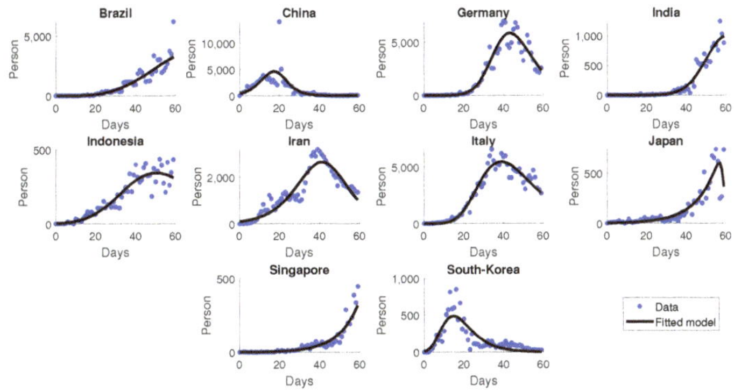

Figure 5. Simulation results of daily new cases $\gamma\mathcal{T}_1(\hat{\mathcal{K}})$ for several countries in the early pandemic. Blue dots represent the actual data, and the solid black lines represent the simulation.

4.1.2. Simulation of SEIR Dynamics

Simulation of E, I and R are obtained directly from substituting $\hat{\mathcal{K}}$ into $\mathcal{T}_1$ and $\mathcal{T}_2$, respectively. For the ten observed countries, the dynamics of EIR (Exposed–Infected–Recovery) are given in Figure 6, omitting the S compartment form visualization due to its scale problem.

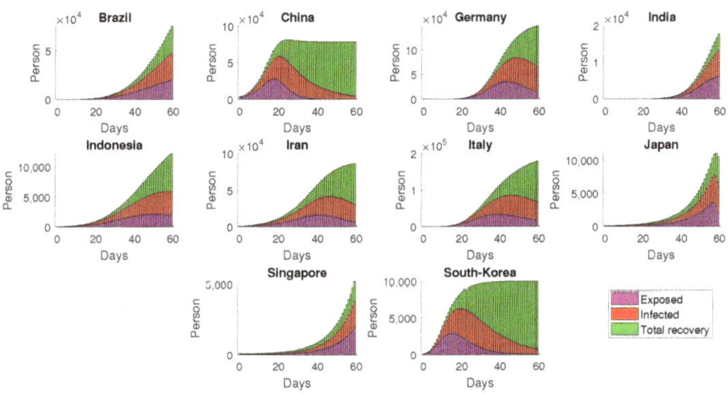

Figure 6. Simulation of estimated SEIR in several countries by implementing the generating operator.

Given in Figure 6, the number of exposed cases dropped significantly in China and South Korea within the first 60 days, leaving no exposed cases in the late simulation. The success of the two countries in controlling the disease was the result of mobility restriction across the country [38], as well as the public participation in the implementation of COVID-19 protocols [39]. In other countries, such as Brazil and Singapore, the virus seems to not be rapidly spreading. However, the exposed cases gradually increased and had no sign of significant decrease within two months. Although Model (1) does not explicitly accommodate various interventions in the field, the time-dependent infection rate $u(t)$ could represent the daily measure of infection due to the unmeasurable intervention and control.

4.1.3. Dynamics of the Effective Reproduction Number

The basic reproduction numbers of the observed countries are given in Table 3 using Equation (2). Since the basic reproduction number calculation only applies to the autonomous system, we drop the time dependency of the transmission rate. Thus, we use the average number of the 60-day transmission rate. In comparison, the time-dependent effective reproduction ratio is depicted in Figure 7 as a measure of the daily performance of virus transmission. It is shown in Figure 7 that, except for China and South Korea, other countries took much longer to reduce the effective reproduction ratio to below one.

Table 3. Estimated basic reproduction number R_0 for the simple model.

Country	R_0	Country	R_0
Brazil	3.79	Iran	3.63
China	2.65	Italy	3.39
Germany	1.22	Japan	1.28
India	3.81	Singapore	0.74
Indonesia	3.46	South-Korea	2.74

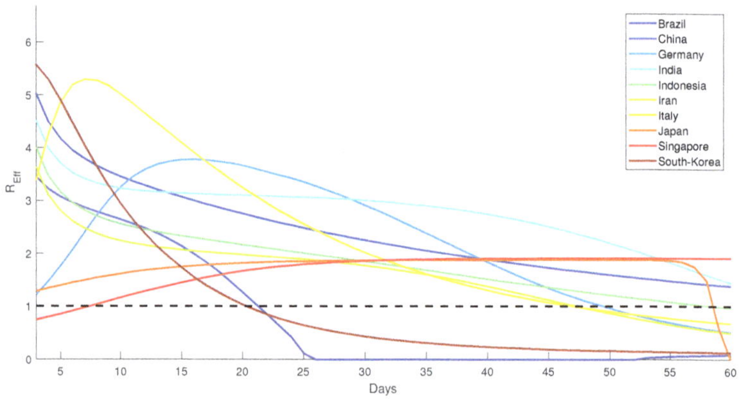

Figure 7. Dynamics of the effective reproduction number (R_{0E}) for several countries.

In general, the estimated models depict the behavior of the pandemic over time. Furthermore, the estimated parameters can be used to conduct the prediction of how the pandemic behaves in each country. The short-term prediction can be an option since parameters would change over time as the new data are retrieved. Moreover, extension beyond the period of observation could not be expected for the forecast [40].

4.2. Generalized SEIR Model

This section emphasizes the implementation of the operators described in Section 3 for the generalized SEIR model, which involves the recorded and unrecorded infections. The cumulative data were taken from [37] during the 60-day second wave period of transmission of each of the ten selected countries. The same parameters in Table 1 were used for simulations, and the values of $\omega(t)$ were estimated using the information of IFR and CFR.

4.2.1. Fitted Cumulative Data

The same construction of cumulative dynamics using GLGM as performed in Section 2.2 is used for the second wave period. Table 4 shows the selected time interval for each country in which the second wave transmission is believed to occur, together with its estimated parameters for the GLGM. Given in Figure 8, the figures for cumulative infections were significantly increasing in the observed time intervals, which indicate the resurgence of the pandemic. In addition, it is shown that the rendered model fits the provided data well.

Table 4. Parameter estimation of the cumulative dynamics using GLGM for the second transmission.

Country	Start Date	End Date	C_1	C_2	C_3	C_4
Brazil	19 February 2021	19 April 2021	5,303,359	0.0001	0.0407	29.5592
China	1 January 2021	1 March 2021	2803	0.3080	0.1394	16.0797
Germany	25 March 2021	25 May 2021	1,033,461	0.0001	0.0559	21.1068
India	1 April 2021	31 May 2021	18,662,604	0.1584	0.0615	29.2508
Indonesia	15 June 2021	14 August 2021	2,356,958	0.1620	0.0559	32.1398
Iran	26 March 2021	24 May 2021	1,144,360	0.0001	0.0572	23.8976
Italy	1 November 2021	31 December 2020	1,440,383	0.0001	0.0616	15.6811
Japan	23 July 2021	21 September 2021	880,483	0.5768	0.0899	27.9857
Singapore	7 July 2020	4 September 2020	11,938	0.4967	0.1125	18.7243
South-Korea	24 November 2020	23 January 2021	48,828	0.2316	0.0649	26.5325

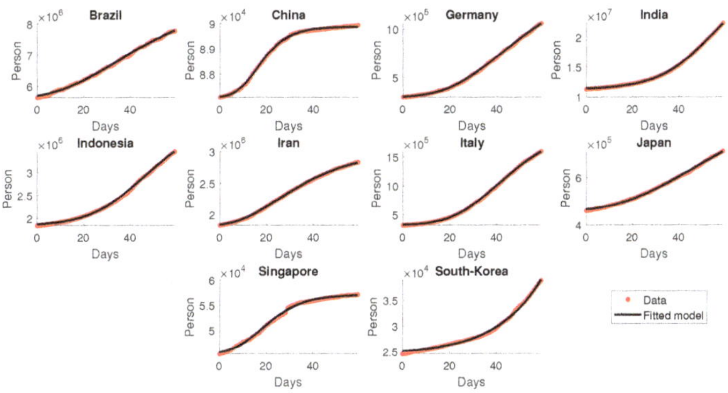

Figure 8. Fitting results $\bar{\mathcal{K}}$ of cumulative data for the second wave transmission.

Figure 9 shows good agreement between the simulations and the data of daily new cases. All the depicted figures are considered to be the resurgence of cases after the first hit ends, e.g., China [41], Germany [42], Italy [43], and India [44].

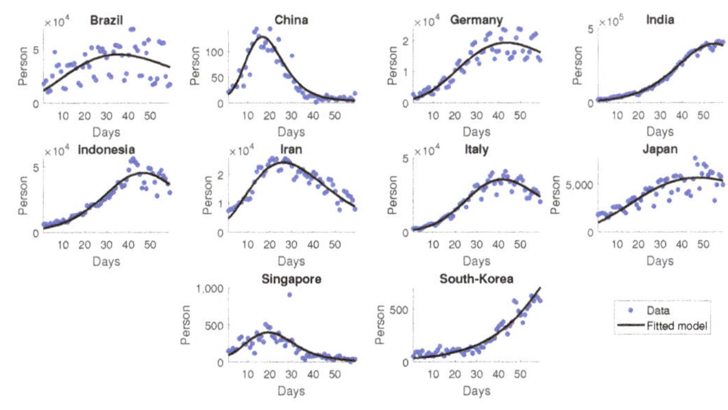

Figure 9. Simulation results of daily new cases $\gamma \mathbf{T}_1(\bar{\mathcal{K}})$ for the second wave transmission.

4.2.2. Estimated $\omega(t)$

The estimation of $\omega(t)$ starts with the estimation of both IFR and $CFR(t)$. As stated earlier in Section 3.2, the estimated value of IFR is assumed to be constant over time, yet the CFR depends on time. The observed time-dependent CFR, which is defined as $CFR(t) = \frac{D(t)}{K(t)}$, is evaluated by utilizing the data retrieved from Worldometer [37]. On the other hand, the estimation of IFR is obtained by first estimating the number of total infections using Equation (45), implementing the data of total tests performed by each country that is publicly provided by OurWorldinData [31].

Figure 10 shows the estimation of IFR for the ten countries. Italy and Germany have higher IFR values than other regions, while Singapore is considered the lowest. This result shows that even though the simple formula was claimed to underestimate the true IFR [45], the general pattern for the observed ten countries resulted in consistent results [34].

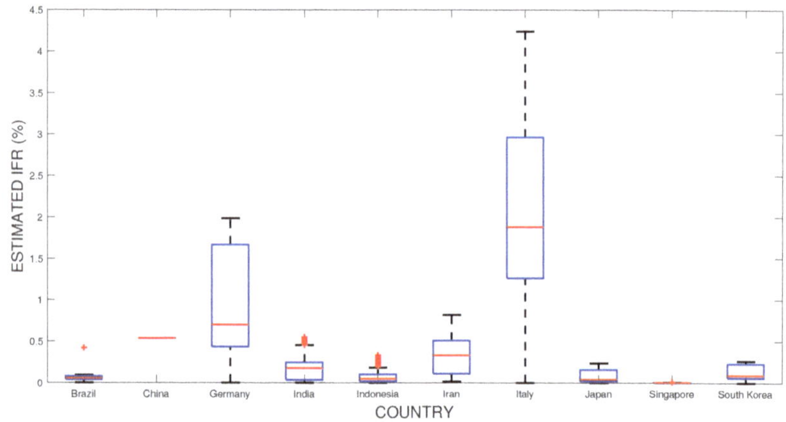

Figure 10. Estimated IFR for the ten observed countries. The blue boxes represents the range from the lower to upper quartiles, with the median (red line) was chosen to represent the single point estimated IFR. The black add signs depicts its maximum and minimum values of the estimated IFR.

Utilizing the observed $CFR(t)$ and estimated IFR given in Figure 10, the dynamics of $\omega(t)$ in each country's time interval are evaluated by dividing the observed CFR by the estimated IFR. The dynamics of $\omega(t)$ for each country are depicted in Figure 11.

Overall, other than Indonesia and Japan, the share of those being reported against the total number of infections increased. According to ([46,47]), the number of deaths in Indonesia was rapidly increasing from April to May 2021, making the observed CFR increase as well [48]. Since the estimated IFR remains constant, the dynamics $\omega(t)$ were significantly declining. The increasing CFR indicates that the total deaths increase more significantly than the total recorded infections. Assuming those total deaths are linearly dependent on the total infections, then the increasing CFR also indicates the number of total unapparent infected individuals. The unrecorded infections should be increasing once the daily test decreases. Indonesia experienced a significant decline in daily testing in August 2021, from about 160,000 to only 100,000 specimens a day, confirming the substantial drop in transition rate $\omega(t)$ in this country. The same argument holds for explaining the slight decline in the transition rate for Japan in the late simulation. Figure 11a,b shows how the daily test in both countries experienced a significant decline in mid-August 2021 and May 2021, respectively. For the rest of the countries, the daily test delivered to the population is relatively increasing, making no significant increase in the unapparent infected individuals and decreasing the observed CFR.

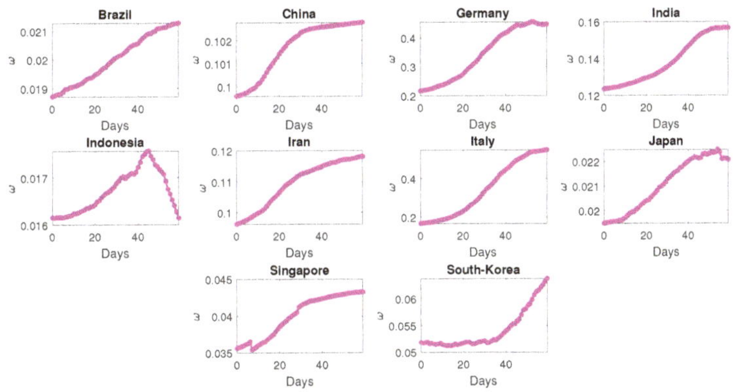

(a)

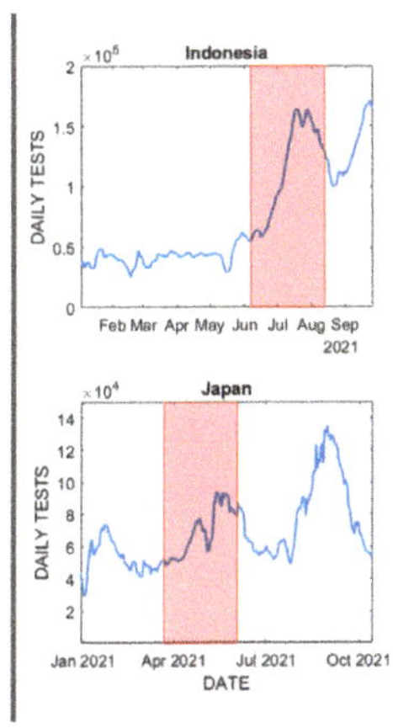

(b)

Figure 11. (**a**) Dynamics of estimated $\omega(t)$, 60 days after the initial second wave emerged; (**b**) the daily test experienced a significant drop in Indonesia and Japan during the simulation interval (in the red-shaded area).

4.2.3. Dynamics of the Generalized SEIR Model

Given the estimation of $\omega(t)$ and the provided cumulative data, the number of exposed populations at time-t can be estimated by means of the generating operators for the advanced model, namely $\mathbf{T_1}$. In addition, the other unobservable compartments are obtained by implementing $\mathbf{T_2}$, resulting in an estimation of unrecorded active cases and the total recovered. The numerical simulation comparing the figures of infected people being recorded or not is depicted in Figure 12. In general, the number of unapparent cases is estimated to be way higher compared to that of being identified. These results are strongly related to the calculated values of $\omega(t)$ for each country, which are identified as very low (around 10%), on average. On the other hand, the opposite results are found in Germany and Italy, which is a result of the relatively higher share of recorded infections.

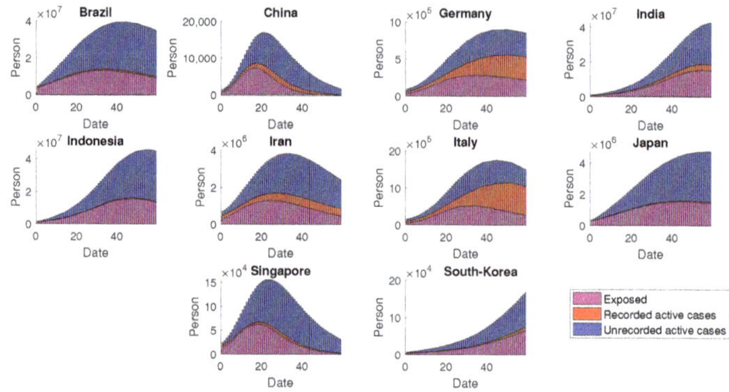

Figure 12. Dynamics of exposed and infected individuals: both recorded and unrecorded cases. Some recorded cases may not be visible due to the higher figures of exposed and unrecorded infected individuals.

The performance of $\omega(t)$ is related to the ratio between the recorded and unrecorded recovery. Figure 13 shows that other than China and Germany, the proportions of unrecorded recovery are much higher than the recorded recovery. This finding is consistent with the fact that the lower the values of $\omega(t)$ for certain countries, the higher the share of unrecorded total recovery.

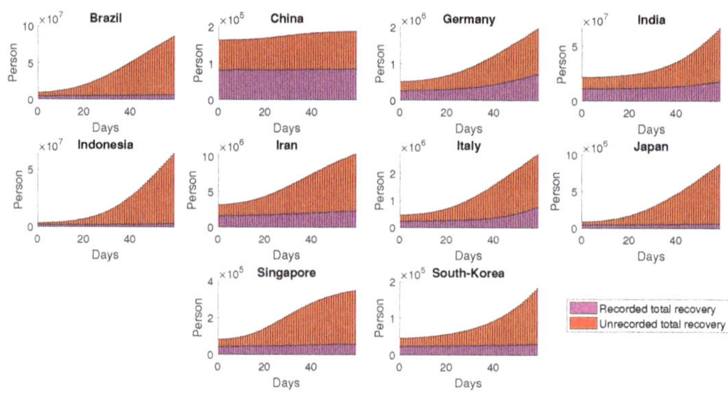

Figure 13. Estimated dynamics of total recovery resulted from both recorded and unrecorded cases.

Finally, the estimated effective reproductive ratio is shown in Figure 14. It is shown that in the case of second-wave transmission, the effective reproductive ratios decreased to a level one much faster than those in the first-wave transmission. This evidence justifies that the role of massive testing played a significant role in controlling the transmission. Since early 2020, the evaluation of the effective reproductive ratio played a vital role in regulating proper interventions related to COVID-19. Germany, Italy and other European countries have been using the calculation of R_{Eff} since the early pandemic [49], which was also followed by other countries such as Indonesia [50] and India [51].

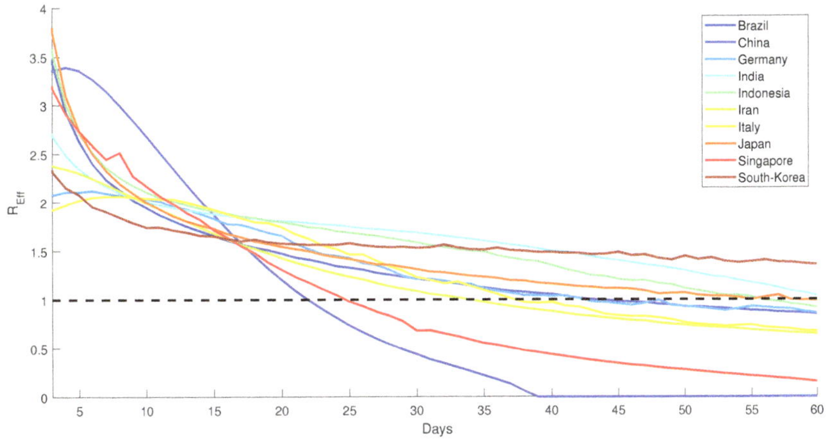

Figure 14. Effective reproduction number of countries, 60 days from the initial second wave of COVID-19 first identified.

4.3. More about the Effective Reproduction Number

Specifically, in the second transmission, it is intriguing to learn how the countries' intervention-related parameters, i.e., $\omega(t)$ and $a(t)$, evolve over time. Depicted in Figure 15 is the countries' profile situated in the contour plot, which is representing the effective reproduction number but omitting the term $S(t)/N$, i.e., $\sqrt{\dfrac{a(1-\omega)\gamma}{(\mu+\gamma)(\mu+\eta)}}$. This formula is nothing but the effective reproduction number, which has not taken the dynamics of the susceptible populations into account. On day 1 of the second transmission, it is indicated that all countries experienced significant transmission of COVID-19 with a relatively low transition rate ω. As time evolved, all countries simultaneously moved to the left with lower reproduction numbers, and we ended up with five countries that were assigned with reproduction numbers higher than one at day 60. To be compared with that depicted in Figure 14, there were only two countries that had effective reproduction numbers higher than one at day 60. Since the reproduction numbers depicted in Figure 15 are omitting the role of $S(t)/N$, there are three countries that had a significant effect of susceptible population size on suppressing the effective reproduction number, i.e., Indonesia, Japan, and Brazil. In other words, the three mentioned countries have passed below one in regards to R_{1E} because of the significant deviation of $S(t)$ compared to N. Since the dynamics of S at every time point are dependent on all other variables, this indicates that the unrecorded infections and recovery have had a significant effect on these countries. Acknowledging that these three are densely populated countries, the high estimated number of unrecorded infections and recovery has resulted in an R_{1E} of below one at day 60.

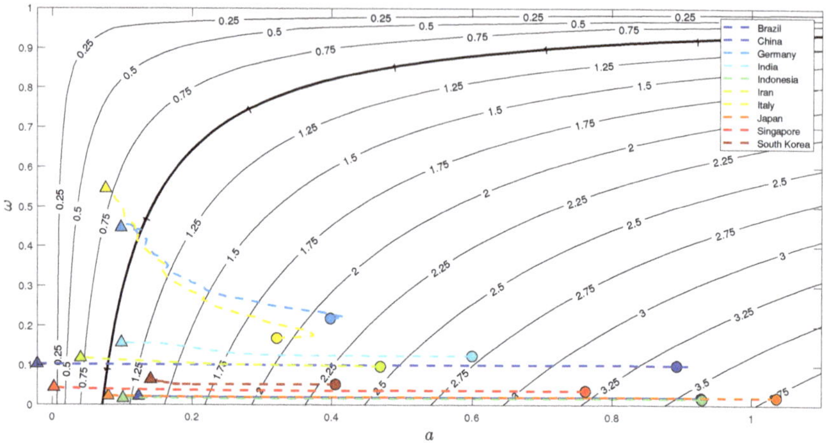

Figure 15. Evolution of the reproduction number in the 60-day simulation of the second transmission. The solid black lines are the level sets of R_0 taken from (27) for constant ω and a. The evolution starts from circle-shaped graphs and ends with triangular-shaped graphs.

Lastly, the depiction of transmission given in Figure 15 also gives us insight into how countries handle the second wave. Germany and Italy are two European countries with high total tests. As time evolves, the snippets move to the left with a higher transition rate ω. On the 60th day, it is clear that these two countries are separated from the rest of the countries due to their high testing capacity. The snippets move to the left for China, Singapore, and Iran, with the transition rate leveling off. These results are expected to confirm the fact that the number of tests conducted was not high, but the large-scale intervention could be more effective [52]. The rest of the countries are dominated by the densely populated countries that were not really strict with lockdown and COVID-19 testing [53]. However, the fact that the estimated unrecorded infections and recovery are relatively high causes the R_{1E} to pass below one even though the transition rate remains low or the transmission rate remains high.

5. Conclusions

This study proposes a new approach to obtain the explicit solutions for each state's dynamics in the SEIR models, or the so-called dynamics generator. There are three crucial components in the construction of the dynamics generator; cumulative data, Richard's Curve, and the proposed compartmental models. The idea of this approach is to fit the cumulative empirical data to Richard's Curve (K) and then define the relations between K and other state dynamics in the SEIR models. Using basic knowledge of linear algebra and calculus, the generator can be constructed to generate all state dynamics in terms of K. In other words, we have constructed a method that generates all state dynamics by means of the empirical data of cumulative cases. Cumulative recorded data was chosen due to its monotonic profile, which has the advantage of choosing a satisfactory fitted cumulative function.

In terms of the compartmental models, we have demonstrated the derivation of the dynamics generator for both simple and advanced models. The constructed dynamics generator produces all state dynamics of the SEIR model, including the figures of the hidden infections using the advanced model. One of the perks of using this approach is to also evaluate the time-dependent rate of transmission, which summarizes all individuals or governmental interventions.

Specifically for the advanced model, we estimated the rate of unrecorded cases using the Case Fatality Rate (CFR) and the estimated Infections Fatality Rate (IFR), which is constructed from the daily Polymerase Chain Reaction (PCR) test. The remaining unrecorded states are then generated directly from the dynamics generator. It is shown that the increase in the number of daily PCR tests significantly reduces the effective reproduction ratio and quickly lowers the ratio to a controllable level. This method gives an important indicator that could be used for daily control of the epidemic, even though it is hard to measure the effect of specific interventions such as mask covering.

Eventually, we have seen that the approach is well-used to generate all state dynamics of the SEIR models, given the cumulative data in a particular period that follows the general S-curve. Once the data does not follow the general S-curve, such as a double S-curved-like data, the standard Richard's Curve will no longer be relevant. Hence, this study highlights room for improvement by considering other explicit functions other than Richard's Curve that can be relevant for the non-S-curved empirical data.

Author Contributions: Conceptualization, E.S. and N.N.; methodology, E.S., K.K.S., A.L.L., M.V. and M.F.; software, K.K.S., A.L.L., M.V., R.K. and M.F.; validation, E.S., N.N., K.K.S. and M.F.; formal analysis, K.K.S., A.L.L. and M.V.; investigation, K.K.S., A.L.L. and M.V.; resources, E.S. and N.N.; data curation, K.K.S., A.L.L., M.V. and R.K.; writing—original draft preparation, K.K.S., A.L.L. and M.V.; writing—review and editing, E.S., M.F., R.K. and N.N.; visualization, K.K.S., A.L.L. and M.V.; supervision, E.S. and N.N.; project administration, E.S., N.N. and M.F.; funding acquisition, E.S. All authors have read and agreed to the published version of the manuscript.

Funding: Part of the research by E.S. is funded by the Indonesian RistekBrin Competitive Grant No. 120I/IT1.C02/TA.00/2021.

Institutional Review Board Statement: Not applicable.

Informed Consent Statement: Not applicable.

Data Availability Statement: The initial data were openly accessible at: https://www.worldometers.info/coronavirus/#countries (accessed on 1 June 2021).

Conflicts of Interest: The authors declare there is no conflict of interest.

References

1. Novel Coronavirus (2019-nCoV)—Situation Report 1. Available online: https://www.who.int/docs/default-source/coronaviruse/situation-reports/20200121-sitrep-1-2019-ncov.pdf (accessed on 5 April 2021).
2. Andersen, K.G.; Rambaut, A.; Lipkin, W.I.; Holmes, E.C.; Garry, R.F. The proximal origin of SARS-CoV-2. *Nat. Med.* **2020**, *26*, 450–452. [CrossRef]
3. Huang, C.; Wang, Y.; Li, X.; Ren, L.; Zhao, J.; Hu, Y.; Zhang, L.; Fan, G.; Xu, J.; Gu, X.; et al. Clinical features of patients infected with 2019 novel coronavirus in Wuhan, China. *Lancet* **2020**, *395*, 497–506. [CrossRef]
4. Singhal, T. A review of coronavirus disease-2019 (COVID-19). *Indian J. Pediatr.* **2020**, *87*, 281–286. [CrossRef]
5. Zhang, Y.; You, C.; Cai, Z.; Sun, J.; Hu, W.; Zhou, X.H. Prediction of the COVID-19 outbreak in China based on a new stochastic dynamic model. *Sci. Rep.* **2020**, *10*, 21522 . [CrossRef]
6. Bertozzi, A.; Franco, E.; Mohler, G.; Short, M.; Sledge, D. The challenges of modeling and forecasting the spread of COVID-19. *Proc. Natl. Acad. Sci. USA* **2020**, *117*, 16732–26738. [CrossRef]
7. Nuraini, N.; Khairudin, K.; Apri, M. Modeling simulation of COVID-19 in Indonesia based on early endemic data. *Commun. Biomath. Sci.* **2020**, *3*, 1–8. [CrossRef]
8. Sucahya, P.K. Barriers to COVID-19 RT-PCR Testing in Indonesia: A Health Policy Perspective. *J. Indones. Health Policy Adm.* **2020**, *5*, 36–42 . [CrossRef]
9. Indonesia's Lab Problems Persist, Testing Rate Remains Below 1%. Available online: https://www.thejakartapost.com/news/2020/10/22/ri-lab-problems-persist-testing-rate-remains-below-1.html (accessed on 2 May 2021).
10. Yang, Z.; Zeng, Z.; Wang, K.; Wong, S.S.; Liang, W.; Zanin, M.; Liu, P.; Cao, X.; Gao, Z.; Mai, Z.; et al. Modified SEIR and AI prediction of the epidemics trend of COVID-19 in China under public health interventions. *J. Thorac. Dis.* **2020**, *12*, 165. [CrossRef]
11. Ross, R. An application of the theory of probabilities to the study of a priori pathometry—Part I. *Proc. R. Soc. Lond. Ser. A Contain. Pap. A Math. Phys. Character* **1916**, *92*, 204–230.
12. Susanto, H.; Tjahjono, V.; Hasan, A.; Kasim, M.; Nuraini, N.; Putri, E.; Kusdiantara, R.; Kurniawan, H. How many can you infect? Simple (and naive) methods of estimating the reproduction number. *Commun. Biomath. Sci.* **2020**, *3*, 28–36. [CrossRef]
13. Soewono, E. On the analysis of COVID-19 transmission in Wuhan, Diamond Princess and Jakarta-cluster. *Commun. Biomath. Sci.* **2020**, *3*, 9–18. [CrossRef]
14. Azque-Herrerias, F.; Munuzuri-Perez, V.; Galla, T. Stirring does not make populations well mixed. *Sci. Rep.* **2018**, *8*, 4068. [CrossRef]

15. World Population by Countries. Available online: https://www.worldometers.info/world-population/#density (accessed on 9 July 2021).
16. Human Life Expectancy. Available online: https://www.worldometers.info/world-population/indonesia-population/ (accessed on 7 March 2021).
17. Lauer, S.A.; Grantz, K.H.; Bi, Q.; Jones, F.K.; Zheng, Q.; Meredith, H.R.; Azman, A.S.; Reich, N.G.; Lessler, J. The incubation period of coronavirus disease 2019 (COVID-19) from publicly reported confirmed cases: Estimation and application. *Ann. Intern. Med.* **2020**, *172*, 577–582. [CrossRef]
18. Transmission of SARS-CoV-2: Implications for Infection Prevention Precautions. Available online: https://www.who.int/news-room/commentaries/detail/transmission-of-sars-cov-2-implications-for-infection-prevention-precautions (accessed on 6 April 2021).
19. Diekmann, O.; Heesterbeek, J.; Roberts, M.G. The construction of next-generation matrices for compartmental epidemic models. *J. R. Soc. Interface* **2010**, *7*, 873–885. [CrossRef]
20. Ahamad, M.G.; Tanin, F.; Talukder, B.; Ahmed, M.U. Officially Confirmed COVID-19 and Unreported COVID-19—Like Illness Death Counts: An Assessment of Reporting Discrepancy in Bangladesh. *Am. J. Trop. Med. Hyg.* **2021**, *104*, 546. [CrossRef]
21. Vasudevan, V.; Gnanasekaran, A.; Sankar, V.; Vasudevan, S.A.; Zou, J. Disparity in the quality of COVID-19 data reporting across India. *BMC Public Health* **2021**, *21*, 1211. [CrossRef]
22. Woolf, S.H.; Chapman, D.A.; Sabo, R.T.; Weinberger, D.M.; Hill, L. Excess deaths from COVID-19 and other causes, March-April 2020. *J. Am. Med. Assoc.* **2020**, *324*, 510–513. [CrossRef]
23. Ioannidis, J.P.; Cripps, S.; Tanner, M.A. Forecasting for COVID-19 has failed. *Int. J. Forecast.* **2020**, *38*, 423–438 . [CrossRef]
24. Richards, F. A flexible growth function for empirical use. *J. Exp. Bot.* **1959**, *10*, 290–301. [CrossRef]
25. Lei, Y.; Zhang, S. Features and partial derivatives of Bertalanffy-Richards growth model in forestry. *Nonlinear Anal. Model. Control* **2004**, *9*, 65–73. [CrossRef]
26. Lee, S.Y.; Lei, B.; Mallick, B. Estimation of COVID-19 spread curves integrating global data and borrowing information. *PLoS ONE* **2020**, *15*, e0236860. [CrossRef]
27. Germany Is Poised to Tighten Lockdown as COVID-19 Cases Surge Again. Available online: https://www.wsj.com/livecoverage/covid-2021-03-22/card/h7nDLKXUj1H0yZJsgeoI (accessed on 2 August 2021).
28. S. Korea Sees Rise in Cases after Relaxing Social Distancing Rules. Available online: http://www.koreaherald.com/view.php?ud=20201102000137 (accessed on 4 July 2021).
29. 'The Perfect Storm': Lax Social Distancing Fuelled a Coronavirus Variant's Brazilian Surge. Available online: https://www.nature.com/articles/d41586-021-01480-3 (accessed on 4 July 2021).
30. Fiore, V.G.; DeFelice, N.; Glicksberg, B.S.; Perl, O.; Shuster, A.; Kulkarni, K.; O'Brien, M.; Pisauro, M.A.; Chung, D.; Gu, X. Containment of COVID-19: Simulating the impact of different policies and testing capacities for contact tracing, testing, and isolation. *PLoS ONE* **2021**, *16*, e0247614. [CrossRef]
31. Daily COVID-19 Tests. Available online: https://ourworldindata.org/grapher/daily-COVID-19-tests-smoothed-7-day (accessed on 7 September 2021).
32. Case Fatality Rate. Available online: https://www.britannica.com/science/case-fatality-rate (accessed on 23 August 2021).
33. Streeck, H.; Schulte, B.; Kümmerer, B.M.; Richter, E.; Höller, T.; Fuhrmann, C.; Bartok, E.; Dolscheid-Pommerich, R.; Berger, M.; Wessendorf, L.; et al. Infection fatality rate of SARS-CoV2 in a super-spreading event in Germany. *Nat. Commun.* **2020**, *11*, 5829. [CrossRef]
34. Teppone, M. One Year of COVID-19 Pandemic: Case Fatality Ratio and Infection Fatality Ratio. A Systematic Analysis of 219 Countries and Territories. *Preprints* **2021**, 1–13. [CrossRef]
35. Ioannidis, J.P. Infection fatality rate of COVID-19 inferred from seroprevalence data. *Bull. World Health Organ.* **2021**, *99*, 19. [CrossRef]
36. Mallapaty, S. How deadly is the coronavirus? Scientists are close to an answer. *Nature* **2020**, *582*, 467–469. [CrossRef]
37. Reported Cases and Deaths by Country or Territory. Available online: https://www.worldometers.info/coronavirus/#countries (accessed on 21 March 2021).
38. Kraemer, M.U.; Yang, C.H.; Gutierrez, B.; Wu, C.H.; Klein, B.; Pigott, D.M.; Open COVID-19 Data Working Group; du Plessis, L.; Faria, N.R.; Li, R.; et al. The effect of human mobility and control measures on the COVID-19 epidemic in China. *Science* **2020**, *368*, 493–497. [CrossRef]
39. Choi, S.; Ki, M. Estimating the reproductive number and the outbreak size of COVID-19 in Korea. *Epidemiol. Health* **2020**, *42*, e2020011. [CrossRef]
40. Chowell, G. Fitting dynamic models to epidemic outbreaks with quantified uncertainty: A primer for parameter uncertainty, identifiability, and forecasts. *Infect. Dis. Model.* **2017**, *2*, 379–398. [CrossRef]
41. China Reports Most New COVID-19 Cases Since January amid Delta Surge. Available online: https://www.reuters.com/world/china/china-reports-96-new-coronavirus-cases-aug-3-vs-90-day-ago-2021-08-04/ (accessed on 5 August 2021).
42. Coronavirus Digest: Germany Cases Surge to New Record. Available online: https://www.dw.com/en/coronavirus-digest-germany-cases-surge-to-new-record/a-55292392 (accessed on 5 August 2021).
43. Lai, A.; Bergna, A.; Menzo, S.; Zehender, G.; Caucci, S.; Ghisetti, V.; Rizzo, F.; Maggi, F.; Cerutti, F.; Giurato, G.; et al. Circulating SARS-CoV-2 Variants in Italy, October 2020–March 2021. *Virol. J.* **2021**, *18*, 168. [CrossRef]

44. Visaria, A.; Dharamdasani, T. The complex causes of India's 2021 COVID-19 surge. *Lancet* **2021**, *397*, 2464. [CrossRef]
45. Luo, G.; Zhang, X.; Zheng, H.; He, D. Infection fatality ratio and case fatality ratio of COVID-19. *Int. J. Infect. Dis.* **2021**, *113*, 43–46. [CrossRef] [PubMed]
46. Epidemiologist Urges Evaluation as Indonesia's COVID-19 Deaths Increase. Available online: https://www.ugm.ac.id/en/news/21140-epidemiologist-urges-evaluation-as-indonesia-s-COVID-19-deaths-increase (accessed on 5 August 2021).
47. COVID-19 in Southeast Asia: All Eyes on Indonesia. Available online: https://theconversation.com/COVID-19-in-southeast-asia-all-eyes-on-indonesia-164244 (accessed on 10 August 2021).
48. Novel Coronavirus (2019-nCoV)—Situation Report 60. Available online: https://cdn.who.int/media/docs/default-source/searo/indonesia/covid19/external-situation-report-60_23-june-2021.pdf?sfvrsn=15d6c3ad_5 (accessed on 1 July 2021).
49. Germany: Infection R-Rate Still Above 1, but Restrictions Still Lifted. Available online: https://www.dw.com/en/germany-infection-r-rate-still-above-1-but-restrictions-still-lifted/a-53383279 (accessed on 14 June 2022).
50. Indonesia's R0, Explained. Available online: https://www.thejakartapost.com/news/2020/06/01/indonesias-r0-explained.html (accessed on 14 June 2022).
51. India's Omicron Surge Explained: Reproduction Number up, Doubling Time down. Available online: https://www.business-standard.com/article/current-affairs/india-s-omicron-surge-explained-reproduction-number-up-doubling-time-down-122010900082_1.html (accessed on 14 June 2022).
52. Liu, S.; Ermolieva, T.; Cao, G.; Chen, G.; Zheng, X. Analyzing the effectiveness of COVID-19 lockdown policies using the time-dependent reproduction number and the regression discontinuity framework: Comparison between countries. *Eng. Proc.* **2021**, *5*, 8.
53. Andriani, H. Effectiveness of large-scale social restrictions (PSBB) toward the new normal era during COVID-19 outbreak: A mini policy review. *J. Indones. Health Policy Adm.* **2020**, *5*, 61–65.

computation

MDPI

Article

Explainable Artificial Intelligence Approach for the Early Prediction of Ventilator Support and Mortality in COVID-19 Patients

Nida Aslam

Department of Computer Science, College of Computer Science and Information Technology, Imam Abdulrahman Bin Faisal University, P.O. Box 1982, Dammam 31441, Saudi Arabia; naslam@iau.edu.sa

Abstract: Early prediction of mortality and risk of deterioration in COVID-19 patients can reduce mortality and increase the opportunity for better and more timely treatment. In the current study, the DL model and explainable artificial intelligence (EAI) were combined to identify the impact of certain attributes on the prediction of mortality and ventilatory support in COVID-19 patients. Nevertheless, the DL model does not suffer from the curse of dimensionality, but in order to identify significant attributes, the EAI feature importance method was used. The DL model produced significant results; however, it lacks interpretability. The study was performed using COVID-19-hospitalized patients in King Abdulaziz Medical City, Riyadh. The dataset contains the patients' demographic information, laboratory investigations, and chest X-ray (CXR) findings. The dataset used suffers from an imbalance; therefore, balanced accuracy, sensitivity, specificity, Youden index, and AUC measures were used to investigate the effectiveness of the proposed model. Furthermore, the experiments were conducted using original and SMOTE (over and under sampled) datasets. The proposed model outperforms the baseline study, with a balanced accuracy of 0.98 and an AUC of 0.998 for predicting mortality using the full-feature set. Meanwhile, for predicting ventilator support a highest balanced accuracy of 0.979 and an AUC of 0.981 was achieved. The proposed explainable prediction model will assist doctors in the early prediction of COVID-19 patients that are at risk of mortality or ventilatory support and improve the management of hospital resources.

Keywords: deep learning; explainable artificial intelligence; machine learning; mortality; prediction; ventilator support

Citation: Aslam, N. Explainable Artificial Intelligence Approach for the Early Prediction of Ventilator Support and Mortality in COVID-19 Patients. *Computation* **2022**, *10*, 36. https://doi.org/10.3390/computation10030036

Academic Editors: Simone Brogi and Vincenzo Calderone

Received: 21 January 2022
Accepted: 23 February 2022
Published: 28 February 2022

1. Introduction

Severe acute respiratory syndrome coronavirus 2 (SARS-COV-2) also known as COVID-19, was first diagnosed in China in late 2019. Since then, it has infected around 222 countries worldwide and as of 7 January 2022, the total number of cases is approximately 301,121,144, including 92,107 patients in critical condition [1]. COVID-19 patients can be symptomatic or asymptomatic. Symptomatic patients' stages can be mild, moderate, or severe. Severe cases may lead to failure of the respiratory system or mortality at the same time. Although the probability of severe cases in patients is not high, sometimes a moderate-stage patient may quickly experience serious complications and need immediate hospitalization and intensive care. Because of this uncertainty, hospitals are sometimes confronted with a huge number of COVID-19-critical patients requiring ventilator support. Similarly, due to the unpredictable nature of COVID-19 [2], it is very crucial to develop an early warning system to predict which patients will deteriorate.

Several artificial intelligence (AI)-based systems have been developed for early diagnosis using clinical data, chest X-rays (CXR) [3], CT scans [4], hybrid data [5], mortality predictions, ventilatory support identification, contact tracing, drug discovery, perceptions of people based on social media data analysis, etc. Studies have proven the significance of AI techniques for combating the COVID-19 pandemic [6,7]. Therefore, early identification of patients who need ventilator support is very crucial for treating patients in a timely manner, as well as for hospitals to manage their resources effectively. Furthermore, it will also assist hospital systems to prioritize their patients. Several studies have been conducted for the early identification of hospitalized patients who are vulnerable to deterioration and require ventilator support, using machine learning (ML) and deep learning (DL).

However, the integration of ML and DL techniques has led to remarkable outcomes in healthcare. These techniques have enhanced the decision-making process, but due to the complexity of the models they engender a lack of interpretability. ML and DL techniques are opaque and represent a form of black-box technique, and fail to provide justification for their respective predictions or decisions [8]. Intrinsically there is a trade-off between predictive power and interpretability; for example, DL models have high prediction accuracy, but the complexity of the model increases the model's opacity. Similarly, ML algorithms such as the decision tree (DT) have good interpretability but sometimes produce low prediction results compared to DL models [9]. The innate opaqueness of the model has raised the need for transparent and interpretable systems that can assist healthcare professionals in making decisions.

Nevertheless, ML and DL have high predictive power but lack interpretability [10]. To deal with the challenges of ML and DL, recent trends have evolved towards explainable artificial intelligence (EAI) techniques. Although EAI is not a new field, it only integrates interpretability and transparency into the ML and DL models [11]. EAI systems lead to better, more trustworthy, and more interpretable decisions.

Recently, research trends have moved towards EAI [12]. Accordingly, from a healthcare perspective, EAI needs to consider different data modalities to achieve the required result [13]. This necessitates that healthcare professionals should be able to understand the rationale behind the how and why of a particular decision. Therefore, in the current study, EAI techniques are used to predict mortality and identify those COVID-19 patients whose condition is rapidly deteriorating and who may require ventilator support.

Contribution

The objective of the study is twofold, i.e., to predict mortality and ventilatory support. The main contributions of the study are:

- An attempt to propose a model with better predictive power and interpretability compared to the benchmark study, which can help healthcare professionals to make better and more retraceable decisions.
- The proposal of an evidence-based and interpretable decision-making system using EAI techniques for the early prediction of mortality and susceptibility to ventilator support.
- Identification of highly significant risk factors for the early prediction of ventilatory support and mortality.
- To the best of the author's knowledge, very few studies have investigated EAI for the early identification of the COVID-19 patients who are at risk for mortality and deterioration.
- Overall, the proposed model outperformed the baseline study.

The rest of this article is organized as follows: Section 2 presents the literature review; Section 3 shows the material and methods used in the study; Section 4 discusses the experimental results; Section 5 contains the discussion; Section 6 concludes the paper.

2. Related Studies

Due to the integration of technology in the healthcare and electronic health records, a huge number of studies have been conducted to combat COVID-19 from different perspectives such as diagnosis, triage, prognosis, epidemiology, contact tracing, drug efficacy, genome structure analysis, etc. [14]. Furthermore, different types of data such as chest X-rays, CT scans, and clinical data have been extensively used for the diagnosis of COVID-19 patients, early prediction of mortality, identification of patients requiring ventilator support, and remote triaging of COVID-19 patients. The following section discusses some of the recent studies on early mortality prediction and ventilator support.

2.1. *AI-Based Studies for Early Identification of COVID-19 Patients for Ventilator Support*

Early prediction of COVID-19 patients who are at risk of deterioration can reduce the risk of mortality and help hospitals to manage their resources. Varzaneh et al. [15] proposed an AI-based model for early prediction of intubation in COVID-19-hospitalized patients, using several ML algorithms such as the decision tree (DT), support-vector machine (SVM), k-nearest neighbor (KNN), and multilayer perceptron (MLP) for the classification. However, the feature selection was performed using a bioinspired technique, i.e., the horse herd optimization algorithm (HOA). The study achieved an accuracy of 0.938 by integrating DT and HOA. However, the dataset contains only 13% of patient samples for the intubation class. Furthermore, the study has identified that in the current dataset some significant paraclinical attributes are missing.

However, Zhang et al. [16] proposed a DL-based model to identify the at-risk COVID-19-hospitalized patients for the mechanical ventilator (MV) after 24 h. The significance of the study is that all relevant patient data such as laboratory results, medications, demographic information, signs and symptoms, and all clinical procedures were used. For the data imputation, the attention method was used. Furthermore, a comparison was made among the ML and DL models, and it was found that the proposed DBNet outperformed with an AUC (area under the curve) of 0.80 and an F1 score of 0.798. However, the study produced a good outcome and utilized patient samples from multiple hospitals, but conversely, needed huge attributes.

Similarly, Aljouie et al. [17] utilized different ML techniques to predict mortality and identify ventilator requirement for COVID-19-hospitalized patients. The study used patients' clinical data, laboratory results, comorbidity, and CXR findings. Feature selection was performed with ReliefF, while extreme gradient boosting (XGB), random forest (RF), SVM and logistic regression (LR) were used for classification. Similar to the other studies, the dataset suffers from a class imbalance; therefore, several data-augmentation techniques were applied. The study found that CXR data are more significant in predicting ventilatory support as compared to comorbidity, lab results, and other demographic features. A highest AUC of 0.87 and a balanced accuracy of 0.81 were achieved using the features selected from ReliefF, classification with SVM, and data augmentation with random undersampling. Similar to Aljouie, Bae et al. [18] also performed a study that predicted ventilator support and mortality of COVID-19-hospitalized patients using CXR and radiomic features. However, the study only utilized radiomic data. Furthermore, the number of patient samples in the study was few compared to Aljouie et al.'s study. Latent discriminant analysis (LDA), RF, quadrant discriminant analysis (QDA), and the DL model were used for the classification. They found that the DL model with radiomic features produced an AUC of 0.79, a sensitivity of 0.71 and a specificity of 0.71, respectively. The study was multicenter, but the number of samples was lower.

Furthermore, Balbi et al. [19] used LR for identifying patients that required ventilator support using CXR, demographic, clinical, and laboratory data. The objective was to identify the significance of the features and found that along with CXR, patients' medical history and other vitals have significantly enhanced the prediction. Conversely, another study [20] utilized CXR for predicting MV support using the DenseNet121 model. The significance of the study is that the patients that require MV were identified 3 days before the event. The model achieved an accuracy of 0.90, and a sensitivity and specificity of 0.86 and 0.84, respectively. The study was only limited to CXR for the prediction. Integration of other clinical data at admission might improve the performance of the model.

2.2. *AI-Based Studies to Predict Early Mortality in COVID-19 Patients*

Similar to the proposed study, Aljouie et al. [17] and Bae et al. [18], proposed a model for two tasks, i.e., mortality and ventilator support prediction for COVID-19-hospitalized patients. In the first study, the author utilized different categories of data and found that comorbidity alone can help in predicting the mortality of COVID-19 patients. However, in the second study, CXR and radiomic data were used and found that the integration of radiomic data improved the early prediction of mortality. Aljouie et al. [15] achieved an AUC of 0.83 and a balanced accuracy of 0.80 using RF. Meanwhile, in Bae et al.'s study [18], the DL model achieved an AUC of 0.83, sensitivity of 0.79, and specificity of 0.74. As previously discussed, the number of patient samples in Aljouie et al. was larger than that of the Bae et al. study; furthermore, the study also utilized different types of patient data such as clinical, comorbidity, demographic, and CXR.

Pourhomayoun and Shakibi [21] developed different ML models such as SVM, ANN, DT, RF, KNN, and LR for the prediction of mortality in COVID-19 patients. The study utilized a dataset from demographically different countries and a huge number of samples as compared to the previous studies mentioned in the literature review. However, the dataset contained huge number of missing values. They achieved an accuracy of 0.89 using ANN. In addition, Khan et al. [22] made a comparison between the ML and DL algorithms to predict mortality in COVID-19 patients using the dataset proposed in Pourhomayoun and Shakibi [21]. They found the significance of the DL method in the early prediction of mortality as compared to the ML algorithms. The DL model achieved enhanced results, with an accuracy of 0.97, a sensitivity of 0.97, and a specificity of 1. Furthermore, the number of features used to train the model was also reduced in the study [22].

Moreover, a study was performed to predict mortality among COVID-19 patients who are in severe condition [23]. The study aimed to identify the risk factors based on different categories of patient data such as clinical, demographic, comorbidity, laboratory tests, radiological data, etc. The study was performed using a small sample size of 150 patients in Romania. The LR model was used for the classification. The study found that D-dimmer, C-reactive protein (CRP), and high heartbeat are the most significant mortality predictors for COVID-19 patients. Furthermore, a correlation was found between these features and patients that needed ICU addition and ventilation. The study identified the limitation that some of the patients in the dataset were in severe condition due to late hospitalization.

Similarly, Pezoulas et al. [24] utilized the EAI concept to identify patients at risk of ICU and mortality among COVID-19 patients, using several ML models such as the gradient boosting (GB) algorithm for classification. Experiments were conducted using 214 patients and used clinical, demographic, comorbidity, and lab results. They achieved an accuracy of 0.79 and 0.81 in predicting mortality among COVID-19 patients in ICU.

Recently, a study was made by Moulaei et al. [25] to compare the performance of different ML techniques to predict mortality using data at the time of admission to hospital. The dataset contained the patients' clinical, demographic, and laboratory results. A total of 54 features were initially selected from the dataset. Using the genetic algorithm (GA), 38 features were selected. Several ML models such as XGB, RF, MLP, KNN, NB, reinforcement learning, and J48 were used for classification. They were found to outperform, with an accuracy of 0.95, sensitivity of 0.90, and specificity of 0.95. Nevertheless, the study has achieved significant results and used a dataset with a sample of 1500 patients. However, the dataset suffers from a huge imbalance due to the fact that the mortality probability in COVID-19 patients is not high. Furthermore, the dataset was collected from one center.

Nevertheless, most of the previously mentioned studies have successfully utilized ML and DL for the early identification of mortality and ventilation support in COVID-19 patients. However, some of the studies have utilized a very limited dataset size. The largest open-source dataset that can be utilized for the prediction of mortality and ventilation support was proposed by Aljouie et al. [17]. Another benefit is that it does not only contain radiological data but also contains patients' clinical, demographic, and laboratory results and comorbidities. Therefore, in the current study, we utilize the data introduced in Aljouie et al.'s study. Most of the previous studies utilized ML and DL learning; however, in the current study we will use the concept of EAI to identify the risk factors that contribute to the identification of COVID-19 patients at risk of mortality and ventilator support. To the best of the authors' knowledge, there is so far only one study (Pezoulas et al. [24]) that has implemented EAI for the prediction of mortality. In the proposed study, we will aim to produce a model with a better outcome compared with the baseline study. Furthermore, we will also attempt to produce the outcome in a more interpretable format using EAI.

3. Materials and Methods

The study was performed using Python ver. 3.9.7. The libraries used during the implementation were Tensor Flow Keras ver. 2.5.2, Dalex ver. 1.4.1, Matplotlib ver. 3.4.3, Sklearn ver. 0.24.2, Pandas ver. 1.3.4 and NumPy ver. 1.19.5. Several sets of experiments were conducted to determine the significance of different categories of attributes in the early prediction of mortality and identification of at-risk patients who require ventilator support. The study mainly consisted of two objectives: to predict the patients who will need ventilatory support and to predict the mortality of the patients. Therefore, we carried out experiments with three cases defined in Table 1. The first case predicted mortality, while cases 2 and 3 predicted ventilator support for COVID-19-hospitalized patients. For each case, three sets of experiments were performed using the full-feature set, with selected features using the EAI feature importance method for all three cases. Meanwhile, the third experiment for case 1 was performed using only the comorbidity feature and for cases 2 and 3 only CXR features were used. These features were used to further investigate the findings made by Aljouie et al. [17]. They found that mortality in COVID-19 patients could be predicted using the comorbidity feature, while CXR functions could be used to predict ventilatory support.

Table 1. Distribution of patient samples per category for cases 1, 2, and 3.

Class Attribute	Cases	Target Class	No of Samples	Selected Features	Total Number of Samples
Vital Status	Case 1	Deceased	136	Platelet_count, age, Hgb, WBC, CXR_zone_11, gender, CXR_zone_12, MCV, CXR_zone_9, MCHC, CXR_zone_10, CXR_zone_5, CXR_zone_8, T2D, CXR_zone_1, CXR_zone_2, CXR_zone_6, CXR_zone_4, CKD, Asthma	1513
		Alive	1377		
Ventilator Support Status	Case 2	Mechanical Ventilator (MV)	184	Platelet_count, age, Hgb, WBC, MCV, MCHC, CXR_zone_6, CXR_zone_1, gender, asthma, CXR_zone_3, T2D, CXR_zone_11, CXR_zone_4, HTN, CXR_zone_12, CXR_zone_9, CAD, CXR_zone_10, CXR_zone_5	1508
		Noninvasive Ventilation (NIV)	111		
		No Ventilatory Support (NVS)	1213		
	Case 3	Ventilatory Support (VS)	295	Platelet_count, WBC, CXR_zone_9, CXR_zone_11, age, Hgb, gender, CXR_zone_12, MCHC, CXR_zone_1, MPV, MCH, CXR_zone_5, CKD, CXR_zone_4, CXR_zone_10, CXR_zone_8, MCV, HTN, T2D	
		No Ventilatory Support (NVS)	1213		

3.1. Exploratory Dataset Analysis

The study was conducted using retrospective data from COVID-19-hospitalized patients in the Kingdom of Saudi Arabia (KSA). The dataset was introduced and used in the study by Aljouie et al. [17]. The dataset consists of 5739 patient demographics, clinical and laboratory investigations, and CXR findings. Moreover, the dataset includes two target attributes, namely patient outcome (deceased or alive) and ventilatory support. The inclusion criteria for the patient sample included in the current study correspond to those of Aljouie et al. to find the mortality and ventilatory support dataset. Table 1 shows the number of samples per category for case 1 (target class vital status (deceased, alive)), case 2 (ventilatory support status (mechanical ventilator (MV), noninvasive ventilation (NIV) and no ventilator support (NVS)) and case 3 (ventilation support status (ventilation support (VS) and no ventilation support (NVS)). Five values were missing in the ventilator support status attribute, so they were removed in cases 2 and 3.

The dataset contains demographic features (gender, age), laboratory results from complete blood count CBC (hematocrit, hemoglobin, mean corpuscular hemoglobin concentration (MCHC), mean corpuscular hemoglobin (MCH), mean corpuscular volume (MCV), mean platelet volume (MPV), red blood cells (RBC), Platelet count, red cell distribution width (RDW), white blood cells (WBC), and radiological findings and comorbidity (cancer, coronary artery disease (CAD), hypertension (HTN), asthma, chronic obstructive pulmonary disease (COPD), type II diabetes mellitus (T2D), liver cirrhosis (LC), chronic hepatitis B (CHB), chronic hepatitis C (HCV) and chronic kidney disease (CKD)). Age and all CBC attributes are numeric, while the remaining attributes are categorical. The CXRs are annotated in twelve zones, as shown in Figure 1. Initially the CXR is divided in two upper (A) and lower zone (B) and also the junction (C). Then these zones are further divided into twelve zones which indicate the points where the radiologist assign severity levels. The zone attributes consist of three possible values (0–2) indicating the severity of ground glass opacity (GGO). Zero indicates the absence of GGO. Ultimately, the dataset contains 35 predictors and 2 class attributes.

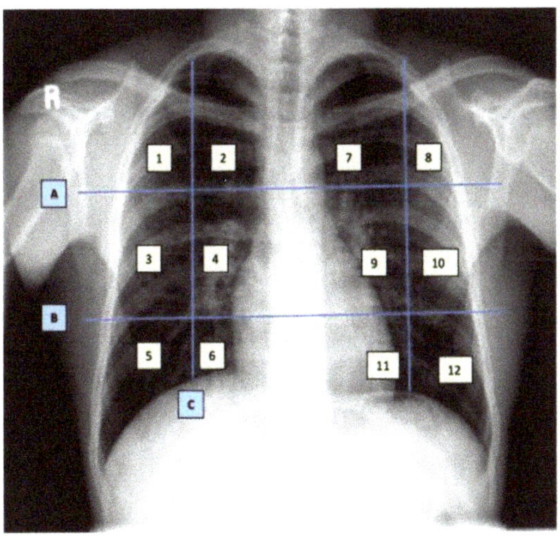

Figure 1. Chest X-ray zone segmentation and annotation [17].

Furthermore, for the exploratory analysis of the dataset, the age attribute was discretized by applying equal-width binning. The minimum patient age in the dataset was 19 and the maximum age was 107. A bin length of 10 was used, with the first bin range being [19–29), [29–39) and so on. The number of bins was 9. Figures 2 and 3 represent the age distribution of the patients according to their ventilation status and vital status. As seen in

Figure 2, the huge number of patients that need MV support were in the range of 50–59. However, for most of the age ranges, the number of patients for the other two categories, i.e., noninvasive ventilator (NV) and no ventilator (NO) was similar. Correspondingly, for the mortality prediction, the maximum number of survived patients was in the range of [49–59). However, the number of deceased patients was high, in the range of [69–79), as shown in Figure 3. The mean age of the patients in the dataset was 54.83. Similarly, Figure 4 indicates the distribution of comorbidity in patients according to their ventilator support status, while Figure 5 indicates the distribution of comorbidity according to the patient's outcome, i.e., deceased or alive. The dataset contains the sample of the hospitalized COVID-19 patients, and it can be seen from Figures 4 and 5 that the most common chronic diseases are hypertension and Type II diabetes. The dataset contains a huge number of male samples as compared to female samples. Furthermore, Figure 6 represents the correlation of CBC attributes.

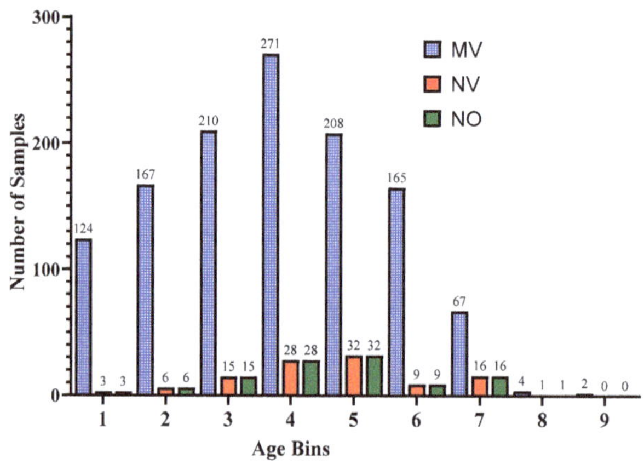

Figure 2. Distribution of patient age range according to ventilator support cases.

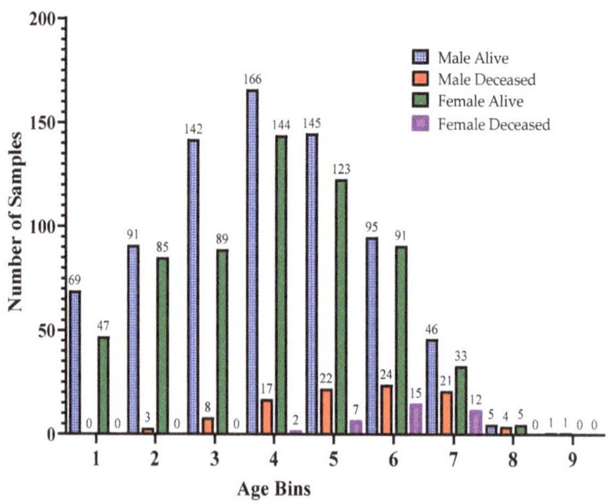

Figure 3. Distribution of patient age range according to patients' vital status.

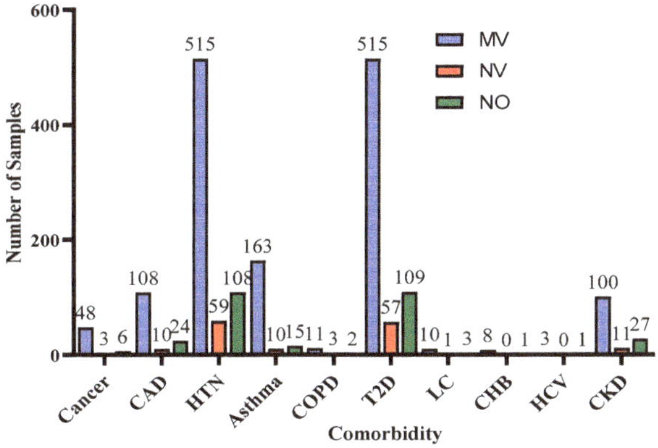

Figure 4. Distribution of comorbidity according to ventilator support cases.

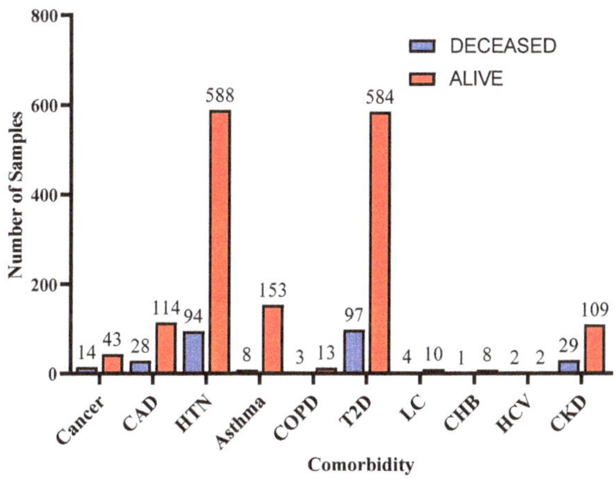

Figure 5. Distribution of comorbidity according to patients' vital status.

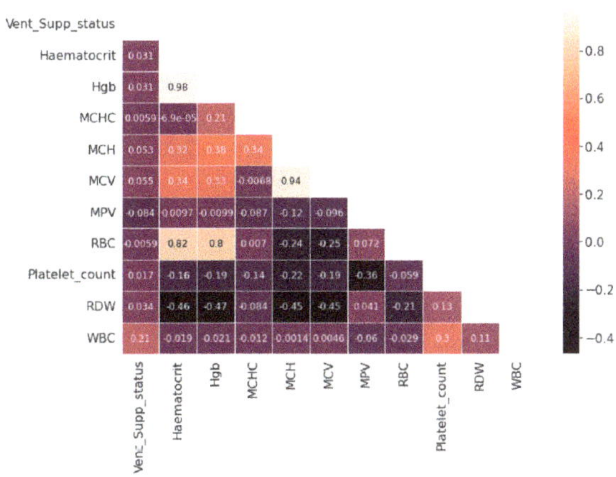

Figure 6. Correlation map for the Complete Blood Count attribute.

3.2. Deep Learning Model

In the last decades, DL models have been extensively used and investigated for various prediction tasks. Instead of using handcrafted features and then applying the traditional ML technique, DL models are better able to understand and learn complex patterns from the data. These models are feedforward and contain three main layers, i.e., the input layer, hidden layers, and the output layer. The input layer is used to obtain data from the source and provide it to the model for further processing, the hidden layers are mainly used to collect the complex pattern from the data, and the output layer is used to classify the data. The backpropagation technique is used to update the weights of the model using a gradient descent algorithm.

Gradient descent is a first-order derivative function used for optimization in DL. The function measures the effect of parameter values on model performance. The gradient descent equation is shown below:

$$y = x - \gamma \nabla f(x) \tag{1}$$

where y represents the current outcome, x represents the true values, and $f(x)$ represents the predicted outcomes. The negative sign indicates the reduction in GD, and γ represents the gradient factor, also known as the learning rate. The GD function aims to reduce the cost function, i.e., $f(w, y)$ and achieve the local minima. It is an iterative function and is represented as

$$\nabla f(x) = \frac{\vartheta}{\vartheta \theta_x} x(\theta_0, \theta_1) \tag{2}$$

Based on the aim of the study, we performed three set of experiments. In the proposed study, three deep learning models were developed with slight variations in the input and output layers, based on the number of features for the input and the number of class labels. We used three sets of data as input to the models (full features, selected features, and comorbidity features). The full-feature set size was 34, the input layer was defined with 34 neurons, and the selected-feature set size was 9, so the input layer was designed with 9 neurons, and the comorbidity-feature set size was 10, so the input layer contained 10 neurons. In addition to this output class, we have 2 output requirements, binary and multiclass i.e., 3 classes. Therefore, the structure of the output layer was modified accordingly: for binary classification, we used 1 neuron in the output layer, while for 3 classes, we used 3 neurons in the output layer.

The structure of the model includes 13 layers. The 12 layers were hidden layers with 1024, 1024, 512, 512, 256, 256, 128, 128, 64, 64, 32, and 32 neurons. Rectified linear unit (ReLU) was used as the activation function for the hidden layers, while dropout layers with a rate of 20% were added after two consecutive hidden layers in order to avoid model overfitting. Sigmoid and softmax activation functions were used at the output layer to perform binary and multiclass classification, respectively.

The equation for the ReLU is

$$\mathrm{ReLU}(\mathrm{x}) = \max(0, \mathrm{x}) \tag{3}$$

ReLU activation function is used to deal with negative values. If the input is negative, the output is zero, and thus the neuron does not participate in model processing for that particular epoch. This makes the neural network sparser and more efficient. Meanwhile, the sigmoid equation is mentioned below:

$$\mathrm{Sigmoid}(\mathrm{x}) = \frac{1}{1 + \mathrm{e}^{-\mathrm{x}}} \tag{4}$$

The sigmoid function provides the model with a value that is between 0 and 1. This is useful because we can use the resulting value as a probability for a particular class.

$$\text{softmax}\left(\vec{z}\right)_i = \frac{e^{z_i}}{\sum_{j=1}^{K} e^{z_i}} \quad (5)$$

where $\vec{z}$ is the input vector for the softmax function, z_i is elements of the input vectors, e^{z_i} is the standard exponential function applied to each element, and K represents the number of classes. The softmax function turns a vector of K into values between 0 and 1. These values are considered as probabilities.

The DL model was optimized using the Adam optimizer [26]. Model configurations include the Adam optimization algorithm with a learning rate of 0.001. Moreover, the loss was calculated using binary and categorical cross entropy, and the accuracy metric was used to evaluate the model's accuracy. To train the model, we used 200 epochs with a batch size of 128. The structure of the model is shown in Figure 7.

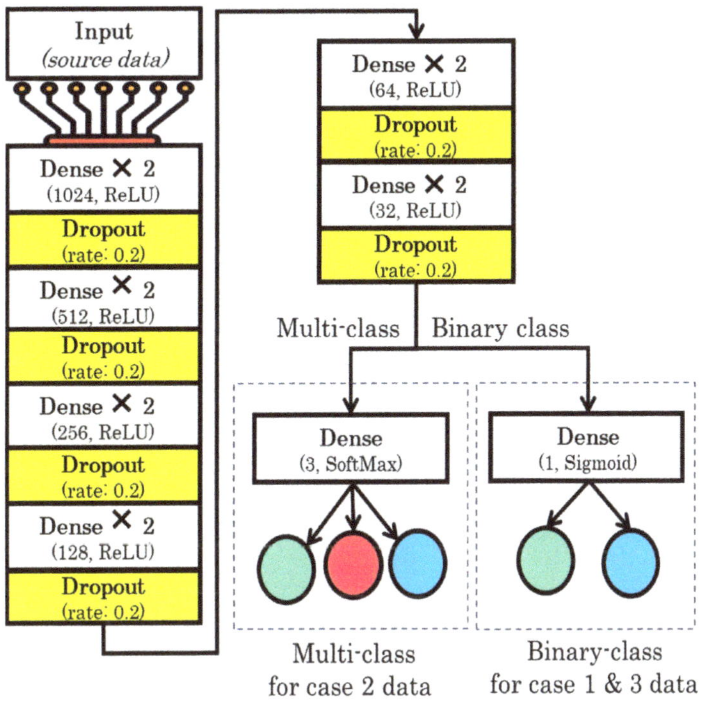

Figure 7. Structure of the proposed model.

3.3. Explainability of the Proposed Model

An Agnostic modeling approach was used to incorporate the interpretability of the proposed model without compromising the model's performance. This method is not model-specific; it interprets the model's behavior without considering the internal logic of the model [27]. In the current study, Shapley was used to find the average impact of the attributes on model performance. Figure 8 shows the mean (SHAP) value for mortality prediction; Figure 9 shows the mean (SHAP) value for predicting patients requiring ventilatory support (case 2 multi-class); Figure 10 indicates the mean (SHAP) value for predicting patients who require ventilator support (case 3 binary class).

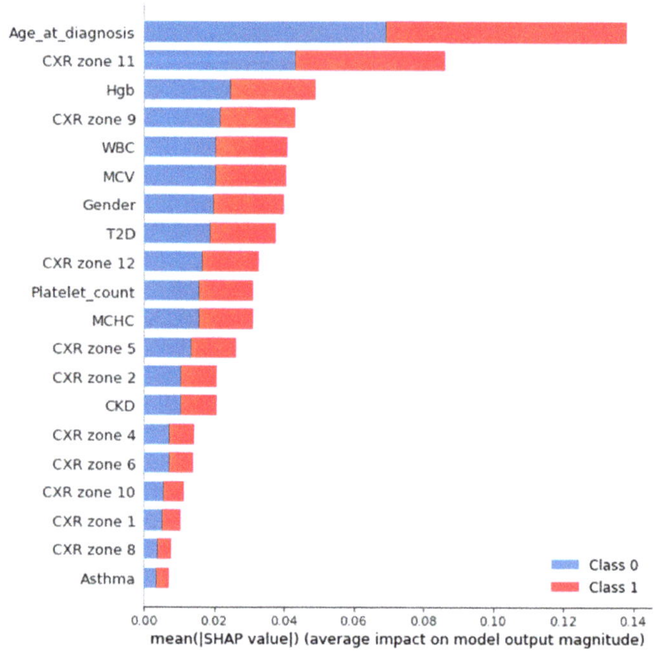

Figure 8. Mean (SHAP value) for the prediction of mortality (Case 1).

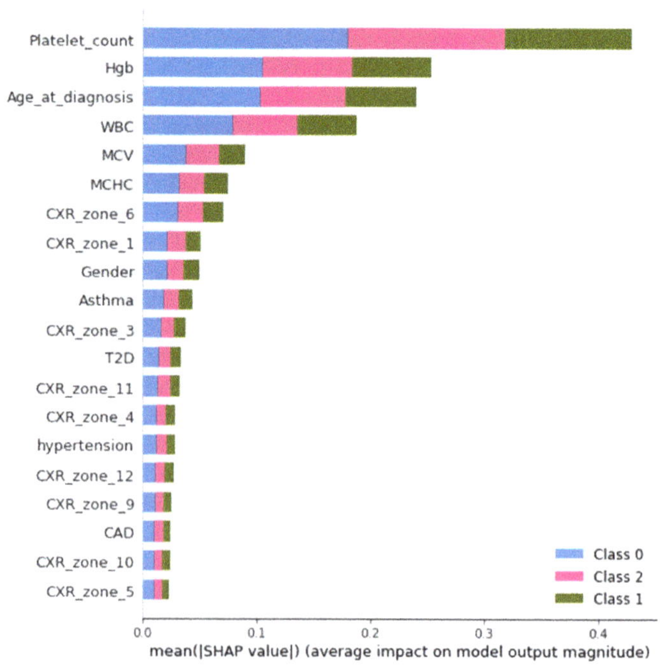

Figure 9. Mean (SHAP value) for the prediction of ventilatory support patients (Case 2).

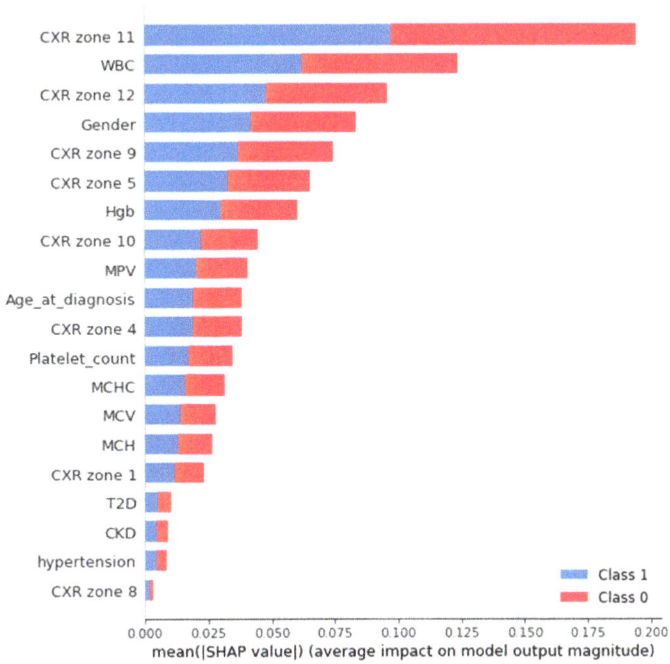

Figure 10. Mean (SHAP value) for the prediction of ventilatory support patients (Case 3).

Furthermore, the interpretability was further enhanced by using the induced decision tree for case 1, case 2, and case 3. Figures 11–13 present the decision tree for all three cases. Experiments were also conducted for all the three cases using DT and is included in Tables A1–A3 in Appendix A. The performance of DT is investigated because it is an interpretable model. But sometimes due to the tradeoff among the interpretability and model performance, as seen in the appendix the result of DT is not significant.

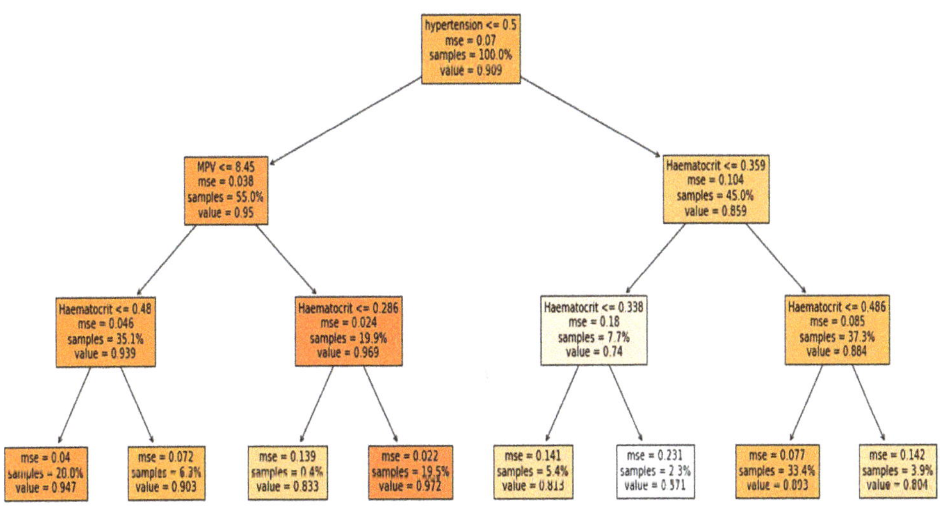

Figure 11. Case 1: Mortality rate with binary classification.

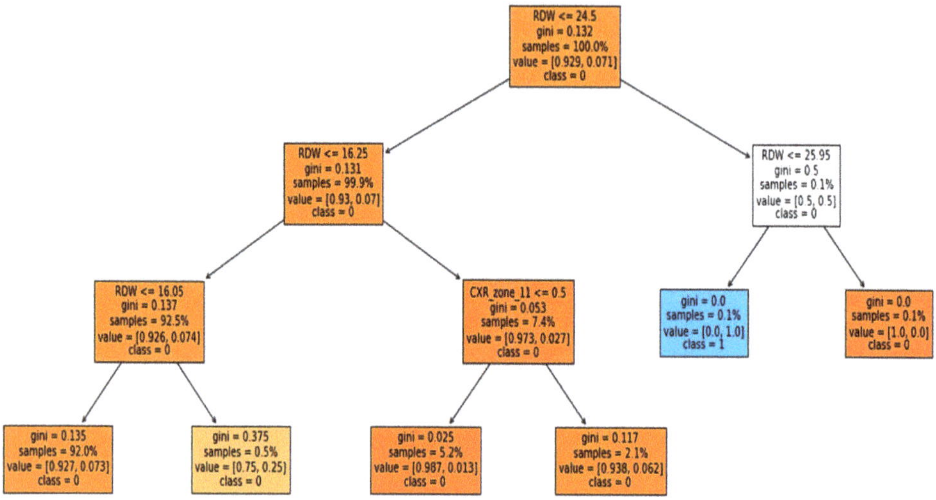

Figure 12. Case 2: Ventilator with multiclassification.

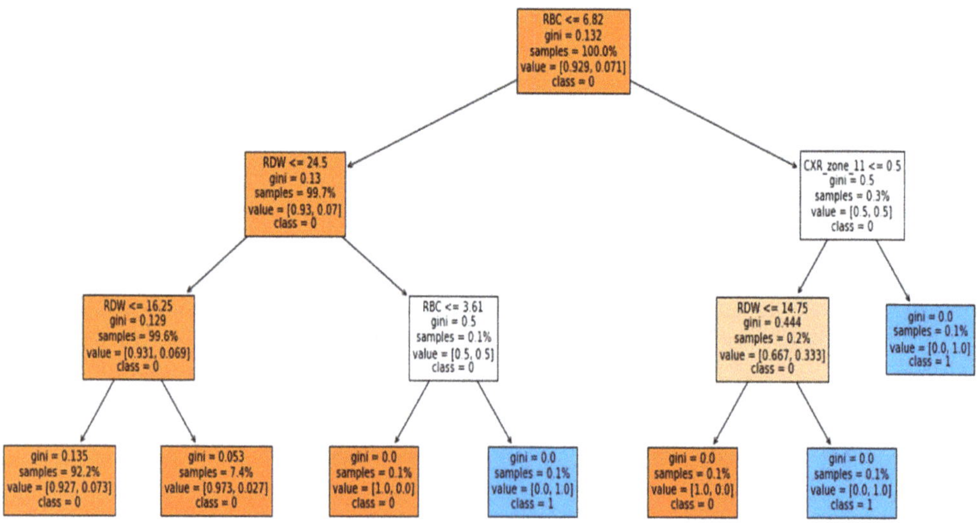

Figure 13. Case 3: Ventilator with binary classification.

3.4. Evaluation Measures

The performance of the proposed model was compared using balanced accuracy, sensitivity, specificity, Youden index, and area under the curve (AUC). There are several evaluation measures that can be used for investigating the performance of the classification algorithms. Among these measures, some of the measures are highly influenced by class distribution, such as accuracy, precision, and recall [28]. Therefore, the unbalanced class problem uses measures such as balanced accuracy, AUC, and Youden index. Correspondingly, balanced accuracy and AUC are also used in the baseline study.

Sensitivity (SN) represents the true positive rate of the model and is calculated using the following equation:

$$SN = \frac{\text{sum(correctly predicted positive class)}}{\text{sum(positive class samples in the dataset)}} \quad (6)$$

Specificity represents the true negative rate and is calculated using the equation below:

$$SP = \frac{\text{sum(correctly predicted Negative class samples)}}{\text{sum(negative class samples in the dataset)}} \quad (7)$$

As can be seen from Equations (6) and (7), the above measures are not affected by class distribution.

Similarly, balanced accuracy is the mean of the sensitivity and the specificity:

$$\text{Balanced Accuracy} = \frac{SP + SN}{2} \quad (8)$$

Likewise, the Youden index (YI) is one of the measures used specifically to determine the effectiveness of the diagnostic test. It evaluates the discriminative power of the diagnostic test. It is computed using the equation below:

$$\text{Youden Index} = SP + SN - 1 \quad (9)$$

The YI values range from 0–1. A smaller value indicates poor diagnostic capability, while a value closer to 1 indicates the significance of the test.

Furthermore, the model discriminative power is further validated using AUC. As is the case with the Youden index, the value of AUC also ranges from 0–1, with a value closer to zero indicating poor performance and a value closer to 1 indicating significant performance of the model.

4. Results

This section presents the results for all three cases that predict mortality and ventilator support in COVID-19 patients. Furthermore, the dataset suffers from an imbalance, which is why data-sampling techniques were applied, such as SMOTE with oversampling and SMOTE with undersampling. A k-fold cross-validation technique with a value of k equal to 10 was used to partition the data for all experiments. In the k-fold cross-validation, the dataset is initially divided into k-segments, where $(k - 1)$ segments are used to train the model and one segment to test in each iteration. Furthermore, the training segments were then divided into training and validation sets. The validation set was used for parameter tuning. Table 2 represents the testing result of the proposed model for case 1, i.e., to predict the patient's vital status as surviving or deceased. Similar sensitivity is achieved using the full- and selected-feature sets with the original dataset without any data-sampling technique. The highest sensitivity was achieved using the full-feature set and dataset after SMOTE undersampling. However, for other measures such as specificity, balanced accuracy, Youden index, and AUC, the highest results were obtained using the full-feature set after SMOTE with oversampling. A similar AUC was achieved with full and selected features for SMOTE oversampling with the full-feature set and original dataset with selected features. Meanwhile, the best overall AUC was obtained using selected features and with SMOTE oversampling. In the baseline study by Aljouie et al. [17], the best balanced accuracy of 0.78 and AUC of 0.85 was achieved using LR and oversampling data with the comorbidity attribute. However, in the proposed study, the highest AUC achieved was 0.875 and the balanced accuracy was 0.904 using the comorbidity attributes and the original dataset. We found that after oversampling the dataset, a similar AUC was achieved using comorbidity attributes.

Table 2. Result of the proposed model for mortality prediction (Case 1).

Feature Set	Technique	SN	SP	Bal-Acc	YI	AUC
Full Features	Original Dataset	**0.990**	0.920	0.955	0.910	0.990
	SMOTE with Oversampling	0.985	**0.988**	**0.986**	**0.973**	**0.998**
	SMOTE with Undersampling	**0.994**	0.933	0.963	0.927	**0.991**
Selected Features	Original Dataset	**0.990**	0.937	0.964	0.927	**0.998**
	SMOTE with Oversampling	0.979	0.953	0.966	0.933	**0.999**
	SMOTE with Undersampling	0.977	0.924	0.950	0.901	0.974
Comorbidity Features	Original Dataset	0.810	0.940	0.875	0.750	0.904
	SMOTE with Oversampling	0.837	0.819	0.828	0.657	0.903
	SMOTE with Undersampling	0.830	0.890	0.860	0.720	0.899

Furthermore, experiments were also conducted to identify patients who are at risk of the ventilator support. Initially the experiments were carried out for the multiclass, i.e., to predict patients requiring a mechanical ventilator (MV), a noninvasive ventilator (NV), or no ventilator (NV) using all features, selected features, and CXR features. CXR features were used because in the baseline study the author found that CXR features can be used to predict the ventilatory support of COVID-19 patients. Table 3 presents the performance of the proposed DL model using different-feature sets. The table contains the testing results. Analogous to case 1, the model produced the best performance with the full-feature set in this case. It indicates the significance of all the features in predicting patients for ventilator support. However, in this case, SMOTE with the undersampling dataset set achieved the best results for all evaluation measures. For mortality prediction (case 1), a similar AUC was achieved using the full- and selected-feature sets. However, in this case there was a significant difference in the model AUC using full and selected features. Furthermore, the baseline study using the CXR feature achieved the highest results, i.e., balanced accuracy of 0.52 and AUC of 0.76 using XGB and oversampling data. However, the proposed study outperformed the baseline with a balanced accuracy of 0.838 and an AUC of 0.842 using SMOTE oversampling. We also found that the oversampling technique enhanced performance with CXR features compared to the original and undersampled datasets.

Table 3. Result of the proposed model for ventilator-support prediction (Multiclass) (Case 2).

Feature Set	Technique	SN	SP	Bal-Acc	YI	AUC
Full Features	Original Dataset	0.815	0.802	0.8085	0.617	0.835
	SMOTE with Oversampling	0.837	0.937	0.887	0.775	0.907
	SMOTE with Undersampling	**0.969**	**0.989**	**0.979**	**0.958**	**0.981**
Selected Features	Original Dataset	0.834	0.821	0.8275	0.655	0.837
	SMOTE with Oversampling	0.858	0.915	0.8865	0.773	0.905
	SMOTE with Undersampling	0.935	0.912	0.9235	0.847	0.932
CXR Features	Original Dataset	0.712	0.878	0.795	0.592	0.823
	SMOTE with Oversampling	0.814	0.862	0.838	0.676	0.842
	SMOTE with Undersampling	0.683	0.856	0.7695	0.539	0.773

Lastly, the experiments conducted for the binary class (ventilator support vs. no support) were performed to predict which patients would need ventilation. The result of the proposed model is shown in Table 4 using the test set. Comparable to case 2 (multiclass), the binary class with the undersampled data and with full features also achieved the best results. After converting the multiclass to binary class, the performance of the model was slightly improved. However, there was a significant difference between the full-feature and different datasets' results, i.e., original and over- and undersampling. Similarly, a baseline study also achieved the highest results using undersampling data with a balanced accuracy of 0.79 and an AUC of 0.82. They found that using XGB with the undersampled dataset gave the best result to predict whether or not the patient was at the risk of needing ventilator support. Meanwhile, the proposed study outperformed the baseline study with an AUC of 0.904 and a balanced accuracy of 0.875 using the original dataset. All results demonstrate the significance of the proposed model for all three cases.

Table 4. Result of the proposed model for ventilator-support prediction (Binary class) (Case 3).

Feature Set	Technique	SN	SP	Bal-Acc	YI	AUC
Full Features	Original Dataset	0.863	0.806	0.835	0.670	0.959
	SMOTE with Oversampling	0.907	0.936	0.921	0.843	0.983
	SMOTE with Undersampling	**0.972**	**0.996**	**0.984**	**0.968**	**0.990**
Selected Features	Original Dataset	0.936	0.975	0.955	0.911	0.982
	SMOTE with Oversampling	0.914	0.963	0.938	0.876	0.958
	SMOTE with Undersampling	0.948	0.985	0.966	0.933	0.984
CXR Features	Original Dataset	0.810	0.940	0.875	0.750	0.904
	SMOTE with Oversampling	0.837	0.819	0.828	0.657	0.903
	SMOTE with Undersampling	0.830	0.890	0.860	0.720	0.899

5. Discussion

Owing to the dynamic clinical indications of COVID-19 and sometimes a sudden deterioration in the condition of moderate-stage patients, it is crucial to develop an automated model that can preemptively predict which patients are at risk for ventilator support and mortality. Furthermore, there is a need to provide a model that can provide a reliable explanation to healthcare professionals. Therefore, in the proposed study, the DL model was used along with the EAI to predict mortality and ventilator support. Several studies have investigated the use of demographic features, lab tests, signs and symptoms, and radiological findings for the prediction. Consequently, demographic, clinical features, comorbidity, and CXR zone features were used in the proposed study.

Among the demographic features, age and gender were found to be significant feature. Similarly, ref. [15–18,23] found age as one of the key features for predicting intubation in COVID-19 patients. However, Bae et al. [18] used radiomics features and two demographic features (age and gender) to predict mortality and ventilator support. The radiomic scores were assigned by experienced radiologists, who found that radiomic features greatly enhanced the performance of the algorithms. Furthermore, Zhang et al. [16] included information on medication and found that patients taking medication for respiratory disease and pneumonia were more likely to end up on a ventilator. Conversely, Balbi et al. [19] found that some of the patients that were diagnosed as COVID-19-negative with the RT-PCR test, while the CXR analysis of the patients revealed pneumonia. Similarly, some of the patients had no significant signs on the CXR but were predicted to be positive using the RT-PCR test. Nevertheless, they found that CXR attributes can only be used for the prediction if other features such as SpO_2, PaO_2, and some other clinical features are available. Likewise, Kulkarni et al. [20] used the CXR for early detection of COVID-19 patients requiring ventilator support using the DL model, and found that CXR features can be used to perform the prediction 3 days in advance.

Conversely, [21,22] predict mortality based on demographic, comorbidity, and symptoms. Notwithstanding this, the studies provided significant results; however, the studies lacked some of the significant lab tests and CXR attributes from the dataset. Correspondingly, Aljouie et al. [17] also found that comorbidity alone can predict mortality in COVID-19 patients. In addition, Khan et al. [22] examined three comorbidities (cardiac problems, diabetes, hypertension) as significant features. However, Pezoulas et al. [24] found that some lab tests are a significant attribute in predicting mortality, while Moulaei et al. [25] discovered that shortness of breath and extra oxygen therapy are among the top features to predict mortality.

Nevertheless, the current study has produced significant results; however, there is always room for further improvement. The study was conducted with a dataset from a single center and country. Furthermore, some of the clinical attributes identified as significant such as CPR, D-dimmer, heartbeat, SpO_2, and PaO_2, etc. are missing from the dataset. In order to further validate the performance of the proposed model, it needs to be experimented with the multicenter dataset, and other features identified as significant in the previous literature also need to be considered. Correspondingly, the dataset also suffers from an imbalance due to the low mortality rate among COVID-19 patients. Therefore, the measures unaffected by class distribution were used in the proposed study. Additionally, the impact of COVID-19 vaccination must also be considered.

6. Conclusions

To sum up, the current study investigated the application of the DL model to predict mortality and the need for ventilator support in COVID-19 patients. The dataset includes COVID-19 patients' demographic information, laboratory results, comorbidity, and CXR. To alleviate the data-imbalance issue, the SMOTE data-sampling technique was applied to both under- and oversampling. Features were selected using the EAI feature importance technique. The optimization of the DL model was performed using the Adam optimizer. Several sets of experiments were performed using full features, selected features, and comorbidity features only to predict mortality, and CXR findings to predict ventilator support. Experimental results showed that the proposed study outperformed the baseline study, with a balanced accuracy of 0.98 and an AUC of 0.998 for predicting mortality. When identifying patients on ventilator support, the model achieved a balanced accuracy of 0.979 and an AUC of 0.981. Furthermore, EAI is used to incorporate interpretability into the proposed DL model and to identify the impact of attributes on the proposed model's performance. Shapley was used to compute the influence of attributes, and an induced decision tree was used to extract the rules from the model. In particular, the proposed model can be used as a tool that can assist doctors to predict at-risk patients and aid hospitals to manage and plan their resources effectively. Conversely, this study can potentially be extended to examine performance using the multicenter and multicountry dataset. In addition, some of the significant lab investigation results and COVID-19 vaccinations must be considered.

Funding: This research received no external funding.

Institutional Review Board Statement: Not applicable.

Informed Consent Statement: Not applicable.

Data Availability Statement: The study used an open-source dataset accessible from the weblink https://data.mendeley.com/datasets/r6t9tmzzmz/3 (accessed on 7 December 2021).

Conflicts of Interest: The authors declare no conflict of interest.

Appendix A. Decision Tree Results

Table A1. Case 1—Decision Tree (Binary-Class, Deceased:0, Alive:1).

Feature Set	Technique	SN	SP	Bal-Acc	YI	AUC
Full Features	Original Dataset	0.884	0.250	0.567	0.134	0.567
	SMOTE with Oversampling	0.817	0.752	0.784	0.569	0.784
	SMOTE with Undersampling	0.905	0.184	0.544	0.089	0.545
Selected Features	Original Dataset	0.429	0.837	0.633	0.266	0.632
	SMOTE with Oversampling	0.812	0.779	0.796	0.592	0.796
	SMOTE with Undersampling	0.406	0.831	0.618	0.237	0.619
Comorbidity Features	Original Dataset	0.458	0.858	0.658	0.317	0.658
	SMOTE with Oversampling	0.793	0.706	0.749	0.499	0.749
	SMOTE with Undersampling	0.444	0.864	0.654	0.309	0.654

Table A2. Case 2—Decision Tree (Multiclass, No_Ventilator:0, NIV:1, MV:2).

Feature Set	Technique	SN	SP	Bal-Acc	YI	AUC
Full Features	Original Dataset	0.415	0.748	0.582	0.163	0.582
	SMOTE with Oversampling	0.747	0.874	0.811	0.622	0.811
	SMOTE with Undersampling	0.404	0.749	0.577	0.153	0.577
Selected Features	Original Dataset	0.380	0.760	0.570	0.141	0.570
	SMOTE with Oversampling	0.769	0.885	0.827	0.655	0.827
	SMOTE with Undersampling	0.426	0.744	0.585	0.171	0.586
Comorbidity Features	Original Dataset	0.344	0.653	0.498	-0.003	0.498
	SMOTE with Oversampling	0.737	0.868	0.802	0.605	0.803
	SMOTE with Undersampling	0.397	0.721	0.558	0.117	0.558

Table A3. Case 3—Decision Tree (Multiclass, No_Ventilator:0, NIV:1, MV:1).

Feature Set	Technique	SN	SP	Bal-Acc	YI	AUC
Full Features	Original Dataset	0.402	0.863	0.633	0.266	0.632
	SMOTE with Oversampling	0.758	0.784	0.771	0.543	0.772
	SMOTE with Undersampling	0.542	0.833	0.687	0.375	0.688
Selected Features	Original Dataset	0.333	0.846	0.590	0.180	0.591
	SMOTE with Oversampling	0.774	0.748	0.761	0.522	0.763
	SMOTE with Undersampling	0.581	0.843	0.712	0.423	0.713
Comorbidity Features	Original Dataset	0.458	0.779	0.618	0.237	0.619
	SMOTE with Oversampling	0.731	0.744	0.738	0.476	0.738
	SMOTE with Undersampling	0.357	0.792	0.574	0.149	0.575

References

1. Worldometer-COVID-19. Available online: https://www.worldometers.info/coronavirus/#countries (accessed on 7 January 2022).
2. Li, X.; Liao, H.; Wen, Z. A consensus model to manage the non-cooperative behaviors of individuals in uncertain group decision making problems during the COVID-19 outbreak. *Appl. Soft Comput.* **2021**, *99*, 106789. [CrossRef]
3. Khan, I.U.; Aslam, N. A deep-learning-based framework for automated diagnosis of COVID-19 using X-ray images. *Information* **2020**, *11*, 419. [CrossRef]

4. Jin, C.; Chen, W.; Cao, Y.; Xu, Z.; Tan, Z.; Zhang, X.; Deng, L.; Zheng, C.; Zhou, J.; Shi, H.; et al. Development and evaluation of an artificial intelligence system for COVID-19 diagnosis. *Nat. Commun.* **2020**, *11*, 5088. [CrossRef] [PubMed]
5. Khan, I.U.; Aslam, N.; Anwar, T.; Alsaif, H.S.; Chrouf, S.M.; Alzahrani, N.A.; Alamoudi, F.A.; Kamaleldin, M.M.; Awary, K.B. Using a Deep Learning Model to Explore the Impact of Clinical Data on COVID-19 Diagnosis Using Chest X-ray. *Sensors* **2022**, *22*, 669. [CrossRef]
6. El-Rashidy, N.; Abdelrazik, S.; Abuhmed, T.; Amer, E.; Ali, F.; Hu, J.W.; El-Sappagh, S. Comprehensive Survey of Using Machine Learning in the COVID-19 Pandemic. *Diagnostics* **2021**, *11*, 1155. [CrossRef]
7. Xu, Z.; Su, C.; Xiao, Y.; Wang, F. Artificial intelligence for COVID-19: Battling the pandemic with computational intelligence. *Intell. Med.* **2021**, *2*, 13–29. [CrossRef]
8. Doran, D.; Schulz, S.; Besold, T.R. What does explainable AI really mean? A new conceptualization of perspectives. *arXiv* **2017**, arXiv:1710.00794.
9. Bologna, G.; Hayashi, Y. Characterization of symbolic rules embedded in deep dimlp networks: A challenge to transparency of deep learning. *J. Artif. Intell. Soft Comput. Res.* **2017**, *7*, 265–286. [CrossRef]
10. Carvalho, D.V.; Pereira, E.M.; Cardoso, J.S. Machine Learning Interpretability: A Survey on Methods and Metrics. *Electronics* **2019**, *8*, 832. [CrossRef]
11. Goebel, R.; Chander, A.; Holzinger, K.; Lecue, F.; Akata, Z.; Stumpf, S.; Kieseberg, P.; Holzinger, A. Explainable AI: The New 42? In *Machine Learning and Knowledge Extraction*; Springer: Cham, Switzerland, 2018; pp. 295–303. [CrossRef]
12. Holzinger, A.; Kieseberg, P.; Weippl, E.; Tjoa, A.M. Current Advances, Trends and Challenges of Machine Learning and Knowledge Extraction: From Machine Learning to Explainable AI. In *Machine Learning and Knowledge Extraction*; Springer: Cham, Switzerland, 2018; pp. 1–8. [CrossRef]
13. Lötsch, J.; Kringel, D. Explainable Artificial Intelligence (XAI) in Biomedicine: Making AI Decisions Trustworthy for Physicians and Patients. *Biomedinformatics* **2022**, *2*, 1–17. [CrossRef]
14. Pham, Q.V.; Nguyen, D.C.; Huynh-The, T.; Hwang, W.J.; Pathirana, P.N. Artificial Intelligence (AI) and Big Data for Coronavirus (COVID-19) Pandemic: A Survey on the State-of-the-Arts. *IEEE Access* **2020**, *8*, 130820–130839. [CrossRef] [PubMed]
15. Varzaneh, Z.A.; Orooji, A.; Erfannia, L.; Shanbehzadeh, M. A new COVID-19 intubation prediction strategy using an intelligent feature selection and K-NN method. *Inform. Med. Unlocked* **2022**, *28*, 100825. [CrossRef] [PubMed]
16. Zhang, K.; Jiang, X.; Madadi, M.; Chen, L.; Savitz, S.; Shams, S. DBNet: A novel deep learning framework for mechanical ventilation prediction using electronic health records. In Proceedings of the 12th ACM Conference on Bioinformatics, Computational Biology, and Health Informatics, Gainesville, FL, USA, 1–4 August 2021.
17. Aljouie, A.F.; Almazroa, A.; Bokhari, Y.; Alawad, M.; Mahmoud, E.; Alawad, E.; Alsehawi, A.; Rashid, M.; Alomair, L.; Almozaai, S.; et al. Early Prediction of COVID-19 Ventilation Requirement and Mortality from Routinely Collected Baseline Chest Radiographs, Laboratory, and Clinical Data with Machine Learning. *J. Multidiscip. Healthc.* **2021**, *14*, 2017–2033. [CrossRef] [PubMed]
18. Bae, J.; Kapse, S.; Singh, G.; Gattu, R.; Ali, S.; Shah, N.; Marshall, C.; Pierce, J.; Phatak, T.; Gupta, A.; et al. Predicting Mechanical Ventilation and Mortality in COVID-19 Using Radiomics and Deep Learning on Chest Radiographs: A Multi-Institutional Study. *Diagnostics* **2021**, *11*, 1812. [CrossRef]
19. Balbi, M.; Caroli, A.; Corsi, A.; Milanese, G.; Surace, A.; Di Marco, F.; Novelli, L.; Silva, M.; Lorini, F.L.; Duca, A.; et al. Chest X-ray for predicting mortality and the need for ventilatory support in COVID-19 patients presenting to the emergency department. *Eur. Radiol.* **2021**, *31*, 1999–2012. [CrossRef]
20. Kulkarni, A.R.; Athavale, A.M.; Sahni, A.; Sukhal, S.; Saini, A.; Itteera, M.; Zhukovsky, S.; Vernik, J.; Abraham, M.; Joshi, A.; et al. Deep learning model to predict the need for mechanical ventilation using chest X ray images in hospitalised patients with COVID-19. *BMJ Innov.* **2021**, *7*, 261–270. [CrossRef]
21. Pourhomayoun, M.; Shakibi, M. Predicting mortality risk in patients with COVID-19 using machine learning to help medical decision-making. *Smart Health* **2021**, *20*, 100178. [CrossRef]
22. Khan, I.U.; Aslam, N.; Aljabri, M.; Aljameel, S.S.; Kamaleldin, M.M.; Alshamrani, F.M.; Chrouf, S.M. Computational Intelligence-Based Model for Mortality Rate Prediction in COVID-19 Patients. *Int. J. Environ. Res. Public Health* **2021**, *18*, 6429. [CrossRef]
23. Timpau, A.S.; Miftode, R.S.; Petris, A.O.; Costache, I.I.; Miftode, I.L.; Rosu, F.M.; Anton-Paduraru, D.T.; Leca, D.; Miftode, E.G. Mortality Predictors in Severe COVID-19 Patients from an East European Tertiary Center: A Never-Ending Challenge for a No Happy Ending Pandemic. *J. Clin. Med.* **2022**, *11*, 58. [CrossRef]
24. Pezoulas, V.C.; Kourou, K.D.; Papaloukas, C.; Triantafyllia, V.; Lampropoulou, V.; Siouti, E.; Papadaki, M.; Salagianni, M.; Koukaki, E.; Rovina, N.; et al. A Multimodal Approach for the Risk Prediction of Intensive Care and Mortality in Patients with COVID-19. *Diagnostics* **2022**, *12*, 56. [CrossRef]
25. Moulaei, K.; Shanbehzadeh, M.; Mohammadi-Taghiabad, Z.; Kazemi-Arpanahi, H. Comparing machine learning algorithms for predicting COVID-19 mortality. *BMC Med. Inform. Decis. Mak.* **2022**, *22*, 2. [CrossRef] [PubMed]
26. Kingma, D.P.; Ba, J. Adam: A method for stochastic optimization. In Proceedings of the 3rd International Conference for Learning Representations, San Diego, CA, USA, 7–9 May 2015.
27. Sewwandi, R.H. Local Model-Agnostic Explanations for Machine Learning and Time-Series Forecasting Models. Ph.D. Thesis, Monash University, Subang Jaya, Malysia, 2022.
28. Tharwat, A. Classification assessment methods. *Appl. Comput. Inform.* **2021**, *17*, 168–192. [CrossRef]

computation

MDPI

Article

Estimation of Daily Reproduction Numbers during the COVID-19 Outbreak

Jacques Demongeot [1,*], Kayode Oshinubi [1], Mustapha Rachdi [1], Hervé Seligmann [1,2], Florence Thuderoz [1] and Jules Waku [3]

1 Laboratory AGEIS EA 7407, Team Tools for e-Gnosis Medical & Labcom CNRS/UGA/OrangeLabs Telecom4Health, Faculty of Medicine, University Grenoble Alpes (UGA), 38700 La Tronche, France; Kayode.Oshinubi@univ-grenoble-alpes.fr (K.O.); Mustapha.Rachdi@univ-grenoble-alpes.fr (M.R.); varanuseremius@gmail.com (H.S.); florence.thuderoz@gmail.com (F.T.)
2 The National Natural History Collections, The Hebrew University of Jerusalem, Jerusalem 91404, Israel
3 UMMISCO UMI IRD 209 & LIRIMA, University of Yaoundé I, P.O. Box 337, Yaoundé 999108, Cameroon; jules.waku@gmail.com
* Correspondence: Jacques.Demongeot@univ-grenoble-alpes.fr

Abstract: (1) Background: The estimation of daily reproduction numbers throughout the contagiousness period is rarely considered, and only their sum R_0 is calculated to quantify the contagiousness level of an infectious disease. (2) Methods: We provide the equation of the discrete dynamics of the epidemic's growth and obtain an estimation of the daily reproduction numbers by using a deconvolution technique on a series of new COVID-19 cases. (3) Results: We provide both simulation results and estimations for several countries and waves of the COVID-19 outbreak. (4) Discussion: We discuss the role of noise on the stability of the epidemic's dynamics. (5) Conclusions: We consider the possibility of improving the estimation of the distribution of daily reproduction numbers during the contagiousness period by taking into account the heterogeneity due to several host age classes.

Keywords: daily reproduction number; COVID-19 outbreak; discrete epidemic growth equation; discrete deconvolution; COVID-19 in several countries

Citation: Demongeot, J.; Oshinubi, K.; Rachdi, M.; Seligmann, H.; Thuderoz, F.; Waku, J. Estimation of Daily Reproduction Numbers during the COVID-19 Outbreak. *Computation* **2021**, *9*, 109. https://doi.org/10.3390/computation9100109

Academic Editor: Simone Brogi

Received: 22 September 2021
Accepted: 8 October 2021
Published: 18 October 2021

1. Introduction

1.1. Overview and Literature Review

Following the severe acute respiratory syndrome outbreak caused by coronavirus SARS CoV-1 in 2002 [1] and the Middle East Respiratory Syndrome outbreak caused by coronavirus MERS-CoV in 2012 [2], the COVID-19 disease caused by coronavirus SARS CoV-2 is the third coronavirus outbreak to occur in the past two decades. Human coronaviruses, including 229E, OC43, NL63 and HKU1, are a group of viruses that cause a significant percentage of all common colds in humans [3]. SARS CoV-2 can be transmitted from person to person by respiratory droplets and through contact and fomites. Therefore, the severity of disease symptoms, such as cough and sputum, and their viral load, are often the most important factors in the virus's ability to spread, and these factors can change rapidly within only a few days during an individual's period of contagiousness. This ability to spread is quantified by the basic reproduction number R_0 (also called the average reproductive rate), a classical epidemiologic parameter that describes the transmissibility of an infectious disease and is equal to the number of susceptible individuals that an infectious individual can transmit the disease to during his contagiousness period. For contagious diseases, the transmissibility is not a biological constant: it is affected by numerous factors, including endogenous factors, such as the concentration of the virus in aerosols emitted by the patient (variable during his contagiousness period), and exogenous factors, such as geo-climatic, demographic, socio-behavioral and economic factors governing pathogen transmission (variable during the outbreak's history) [4–8].

Due to these exogenous factors, R_0 might change seasonally, but these factor variations are not significant if a very short period of time is considered. R_0 depends also on endogenous factors such as the viral load of the infectious individuals during their contagiousness period, and the variations in this viral load [9–15] must be considered in both theoretical and applied studies on the COVID-19 outbreak, in which the authors estimate a unique reproduction number R_0 linked to the Malthusian growth parameter of the exponential phase of the epidemic, during which R_0 is greater than 1 (Figure 1). The corresponding model has been examined in depth, because it is useful and important for various applications, but the distribution of the daily reproduction number R_j at day j of an individual's contagiousness period is rarely considered within a stochastic framework [16–20].

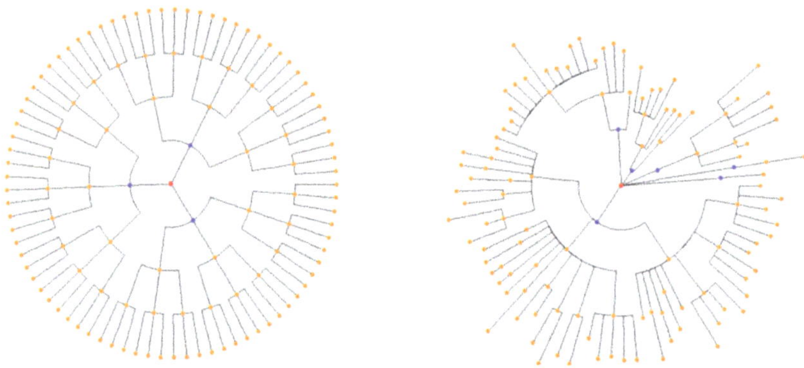

Figure 1. Spread of an epidemic disease from the first infectious "patient zero" (in red), located at the center of its influence sphere comprising the successive generations of infected individuals, for the same value of the reproduction number $R_0 = 3$, with a deterministic dynamic (**left**) and a stochastic one (**right**), with standard deviation σ of the uniform distribution on an interval centered on R_0 and with a random variable time interval i between infectious generations (after [16]).

We therefore defined a partial reproduction number for each day of an individual's contagiousness period, and, assuming initially that this number was the same for all individuals, we obtained the evolution equation for the number of new daily cases in a population. Assuming that the distribution of partial reproduction numbers (referred to as daily for simplicity) was subject to fluctuations, we calculated the consequences for their estimation, and we estimated them for a large number of countries, taking a duration of contagiousness of 3 followed by 7 days.

When this distribution is considered, it is possible to calculate its entropy as a parameter quantifying its uniformity and to simulate the dynamics of the infectious disease either using a Markovian model such as that defined in Delbrück's approach [17] or a classical discrete or ODE SIR deterministic model. In the Markovian case, R_0 can be calculated from the evolutionary entropy defined by L. Demetrius as the Kolmogorov–Sinaï entropy of the corresponding random process [18], which measures the stability of the invariant measure, dividing the population into the subpopulations S (individuals susceptible to but not yet infected with the disease), I (infectious individuals) and R (individuals who have recovered from the disease and now have immunity to it). In the deterministic case, R_0 corresponds to the Malthusian parameter quantifying its exponential growth, and the stability of the asymptotic steady state depends on the subdominant eigenvalue [19,20].

1.2. Calculation of R_0

In epidemiology, there are essentially two broad ways to calculate R_0, which correspond to the individual-level modeling and to the population-level modeling. At the individual level, if we suppose the susceptible population size constant (hypothesis valid during the exponential phase of an epidemic), the daily reproduction rates of an individual are typically non-constant over his contagiousness period, and the calculations we present in the following define a new method for estimating R_0, as the sum of the daily reproduction rates. This new approach allows us to have a clearer view on the respective influence on the transmission rate by endogenous factors (depending on the level of immunologic defenses of an individual) and exogenous factors (depending on environmental conditions).

2. Materials and Methods

The methodology chosen starts from an attempt to reconstruct an epidemic dynamic from the knowledge of the number R_{ikj} of people infected at day j by a given infectious individual i during the kth day of his period of contagiousness of length r. By summing up the number of new infectious individuals X_{j-k} present on day $j - k$ where started their contagiousness, we find that the number of new infected people on day j is equal to:

$$X_j = \Sigma_{k=1,r}\, \Sigma_{i=1}\, X_{j-k}\, R_{ikj} \quad (1)$$

We will assume in the following that R_{ikj} is the same, equal to R_k, for all individuals I and day j, then depends only on day k. Then, we have:

$$X_j = \Sigma_{k=1,r}\, R_k\, X_{j-k} \quad (2)$$

The convolution Equation (2) is the basis of our modelling of the epidemic dynamics.

2.1. The Contagion Mechanism from a First Infectious Case Zero

Let us suppose that the secondary infected individuals are recruited from the centre of the sphere of influence of an infectious case zero and that the next infected individuals remain on a sphere centred on case 0, by just widening its radius on day 2. Therefore, the susceptible individuals C(j), which each infectious on day $j - 1$ can recruit, are on a part of the sphere of influence of case 0 reached at day j (rectangles on Figure 2).

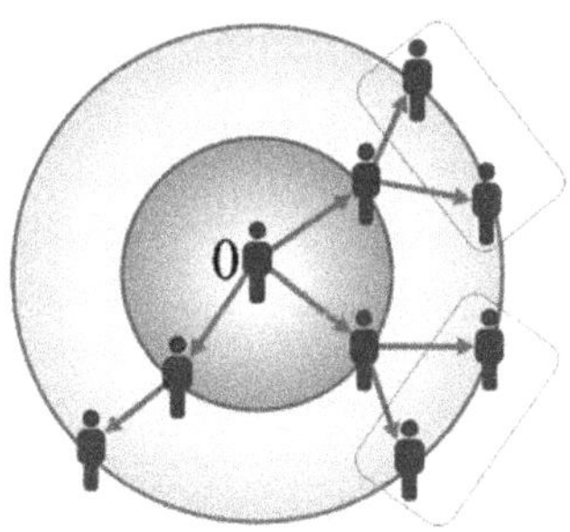

Figure 2. Spread of an epidemic disease from a first infectious case 0 (located at its influence sphere centre) progressively infecting its neighbours in some regions (rectangles) on successive spheres.

2.2. The Biphasic Pattern of the Virulence Curve of Coronaviruses

Mostly, the clinical course of patients with seasonal influenza shows a biphasic occurrence of symptoms with two distinct peaks. Patients have a classic influenza disease followed by an improvement period and a recurrence of the symptoms [11]. The influenza RNA virus shedding (the time during which a person might be contagious to another person) increases sharply one half to one day after infection, peaks on day 2 and persists for an average total duration of 4.5 days, between 3 and 6 days, which explains why we will choose in the following contagiousness duration these extreme values, i.e., either 3 or 6 days, depending on the positivity of the estimated daily reproduction numbers. It is common to consider this biphasic evolution of influenza clinically: after incubation of one day, there is a high fever (39–40 °C), then a drop in temperature before rising, hence the term "V" fever. The other symptoms, such as coughing, often also have this improvement on the second day of the flu attack: after a first feverish rise (39–39.5 °C), the temperature drops to 38 °C on the second day, then rises before disappearing on the 5th day, the fever being accompanied by respiratory signs (coughing, sneezing, clear rhinorrhea, etc.). By looking at the shape of virulence curves observed in coronavirus patients [12–16], we often see this biphasic pattern.

2.3. *Relationships between Markovian and ODE SIR Approaches*

In the following, we suppose that the susceptible population size remains constant, which constitutes a hypothesis valid during the exponential phase of epidemic waves. The Markovian stochastic and ODE deterministic approaches are linked by a common background consisting of the birth and death process approach used in the kinetics of molecular reactions by Delbrück [17], then in dynamical systems theory by numerous authors [18–23], namely in modelling of the epidemic spread in exponential growth. In the ODE approach, the Malthusian parameter is the dominant eigenvalue, and the equivalent in the Markovian approach is the Kolmogorov–Sinai entropy (called evolutionary entropy in [24–26]).

2.3.1. First Method for Obtaining the SIR Equation from a Deterministic Discrete Mechanism

Let us suppose the model is deterministic and denote by X_j the number of new infected cases at day j ($j \geq 1$), and R_k ($k = 1, \ldots, r$) the daily reproduction number at day k of the contagiousness period of length r for all infectious individuals. Then, we have obtained Equation (2) by supposing that the contagiousness behaviour is the same for all the infectious individuals:

$$X_j = \sum_{k=1,r} R_k \, X_{j-k},$$

which says that the X_{j-k} new infected at day $j - k$ give $R_k\, X_{j-k}$ new infected on day j, throughout a period of contagiousness of r days, the R_k's being possibly different or zero. For example, if $r = 3$, for the number X_5 of new cases at day 5, equation $X_5 = R_1X_4 + R_2X_3 + R_3X_2$ means that new cases at day 4 have contributed to new cases at day 5 with the term R_1X_4, R_1 being the reproduction number at first day of contagiousness of new infected individuals at day 4.

In matrix form, we obtain:

$$X = MR, \tag{3}$$

where $X = (X_j, \ldots, X_{j-r-1})$ and $R = (R_1, \ldots, R_r)$ are r-dimensional vectors and M is the following r-r matrix:

$$M = \begin{bmatrix} X_{j-1}, & X_{j-2}, \ldots, & X_{j-r} \\ X_{j-k-1}, & X_{j-k-2}, \ldots, & X_{j-k-r} \\ X_{j-r} & X_{j-r-1}, \ldots, & X_{j-2r+1} \end{bmatrix} \tag{4}$$

It is easy to show that, if $X_0 = 1$ and $r = 5$ (estimated length of the contagiousness period for COVID-19 [12–21]), we obtain:

$$X_5 = {R_1}^5 + 4{R_1}^3R_2 + 3{R_1}^2R_3 + 3R_1{R_2}^2 + 2R_2R_3 + 2R_1R_4 + R_5 \tag{5}$$

The length r of the contagiousness period can be estimated from the ARIMA series of the stationary random variables Y_j's, equal to the X_j's without their trend, by considering the length of the interval on which the auto-correlation function remains more than a certain threshold, e.g., 0.1 [4]. For example, by assuming $r = 3$, if $R_1 = a$, $R_2 = b$ and $R_3 = c$, we obtain:

$$\begin{gathered} X_0 = 1,\ X_1 = a,\ X_2 = a^2 + b + c,\ X_3 = a^3 + 2ab,\ X_4 = a^4 + 3a^2b + b^2 + 2ac, \\ X_5 = a^5 + 4a^3b + 3ab^2 + 3a^2c + 2bc,\ X_6 = a^6 + 5a^4b + 4a^3c + 6a^2b^2 + 6abc + b^3 + c^2, \\ X_7 = a^7 + 6a^5b + 5a^4c + 10a^3b^2 + 12a^2bc + 4ab^3 + 3b^2c + 3ac^2 \end{gathered} \tag{6}$$

If R_1 and R_2 are equal, respectively, to a and b, and if a = b = R/2, c = 0, then, X_5 behaves like:

$$X_5 = R^5/32 + R^4/4 + 3R^3/8 \tag{7}$$

If R = 2, $\{X_j\}_{i=1,\infty}$ is the Fibonacci sequence, and more generally, for R > 0, the generalized Fibonacci sequence. Let us suppose now that b = c = 0 and a depends on the day j: a_j = > C(j), where C(j) represents the number of susceptible individuals, which can be met by one contagious individual at day j. If infected individuals (supposed to all be contagious) at day j are denoted by I_j, we have:

$$X_j = \Delta I_j/\Delta j = (I_{j+1} - I_j)/(j + 1 - j) = \nu C(j)I_j \tag{8}$$

Let us suppose, as in Section 2.1, that the first infectious individual 0 recruits from the centre of its sphere of influence secondary infected individuals remaining in this sphere, and that the susceptible individuals recruited by the I_j infectious individuals present at day j are located on a part of the sphere of centered on the first infectious 0 obtained by widening its radius (Figure 2). Then, we can consider that the function C(j) increases, then saturates due to the fact that an infectious individual can meet only a limited number of susceptible individuals as the sphere grows. We can propose for C(j) the functional form C(j) = S(j)/(c + S(j)), where S(j) is the number of susceptible individuals at day j. Then, we can write the following equation, taking into account the mortality rate μ:

$$X_j = \Delta I_j/\Delta j = \nu C(j)I_j - \mu I_j = \nu I_j\, S(j)/(c + S(j)) - \mu I_j \tag{9}$$

This discrete version of epidemic modeling is used much less than the classic continuous version, corresponding to the ODE SIR model, with which we will show a natural link. Indeed, the discrete Equation (9) is close to SIR Equation (10), if the value of c is greater than that of S:

$$dI/dt = \nu IS/(c + S) - \mu I \tag{10}$$

2.3.2. Second Method for Obtaining the SIR Equation from a Stochastic Discrete Mechanism

Another way to derive the SIR equation is the probabilistic approach, which comes from the microscopic equation of molecular shocks by Delbrück [17] and corresponds to a classical birth-and-death process: if at least one event (with rates of contact ν, birth f, death μ or recovering ρ) occurs in the interval (t, t + dt), and by supposing that births compensate deaths, leaving constant the total size N of the population, we have:

$$\begin{aligned}\text{Probability}(\{S(t + dt) = k, I(t + dt) = N - k\}) &= P(S(t) = k, I(t) = N - k)\,[1 - [\mu k + \nu k(N - k) - fk - \rho(N - k)]dt] \\ &+ P(S(t) = k - 1, I(t) = N - k + 1)\,[f(k - 1) + \rho(N - k + 1)]dt \\ &- P(S(t) = k+1, I(t) = N - k - 1)\,[\mu(k + 1) + \nu(k + 1)\,(N - k - 1)]dt\end{aligned} \tag{11}$$

Hence, we have, if $P_k(t)$ denotes Probability($\{S(t) = k, I(t) = N - k\}$):

$$\begin{aligned}dP_k(t)/d &= [P(S(t + dt) = k, I(t + dt) = N - k) - P(S(t) = k, I(t) = N - k)]/dt \\ &= - P(S(t) = k, I(t) = N - k)\,[\mu k + \nu k\,(N - k) - fk\text{-}\rho(N - k)] \\ &+ P(S(t) = k - 1, I(t) = N - k + 1)\,[f(k - 1) + \rho(N - k + 1)] \\ &- P(S(t) = k + 1, I(t) = N - k - 1)\,[\mu(k + 1) + \nu(k + 1)(N - k - 1)],\end{aligned}$$

and we obtain:

$$dP_k(t)/dt = -[\mu k + \nu k(N - k) - fk - \rho(N - k)]P_k(t) + [f(k - 1) + \rho(N - k + 1)]P_{k-1}(t) - [\mu(k + 1) + \nu(k + 1)(N - k1)]P_{k+1}(t)$$

Then, by multiplying by s^k and summing over k, we obtain the characteristic function of the random variable S. If births do not compensate deaths, we have:

$$\begin{aligned} \text{Probability}\ (\{S(t+dt)=k, I(t+dt)=j\}) &= P(S(t)=k, I(t)=j)\,(1-[\mu k+\nu kj-fk-\rho j]dt) \\ &+ P(S(t)=k-1, I(t)=j+1)\,[f(k-1)+\rho(j+1)]dt \\ &- P(S(t)=k+1, I(t)=j-1)\,[\mu(k+1)+\nu(k+1)(j-1)]dt \end{aligned} \tag{12}$$

If S and I are supposed to be independent and if the coefficients ν, f, μ and ρ are sufficiently small, S and I are Poisson random variables [27], whose expectations E(S) and E(I) verify:

$$\begin{aligned} dE(S)/dt &= fE(S) - \nu E(SI) - \mu E(S) + \rho E(I) \\ \text{or, if } f = \mu,\ dE(S)/dt &\approx E(I)\,[-\nu E(S) + \rho], \end{aligned} \tag{13}$$

leading to the SIR equation for the variables S, I and R considered as deterministic:

$$dS/dt = -\nu SI + \rho R,\ dI/dt = \nu SI - kI - \mu I,\ dR/dt = kI - \rho R \tag{14}$$

3. Results

3.1. Distribution of the Daily Reproduction Numbers R_j's along the Contagiousness Period of an Individual. A Theoretical Example Where They Are Supposed to Be Constant during the Epidemics

If R_0 denotes the basic reproduction number (or average transmission rate) in a given-population, we can estimate the distribution V (whose coefficients are denoted $V_j = R_j/R_o$) of the daily reproduction numbers R_j along the contagious period of an individual, by remarking that the number X_j of new infectious cases at day j is equal to $X_j = I_j - I_{j-1}$, where I_j is the cumulated number of infectious at day j, and verifies the convolution equation (equivalent to Equation (2)):

$$X_j = \sum_{k=1,r} R_k X_{j-k},\ \text{giving in continuous time}: \ X(t) = \int_1^r R(s)X(t-s)ds, \tag{15}$$

where r is the duration of the contagion period, estimated by $1/(\rho + \mu)$, ρ being the recovering rate and μ the death rate in SIR Equation (14). r and S can be considered as constant during the exponential phases of the pandemic, and we can assume that the distribution V is also constant; then, V can be estimated by solving the linear system (equivalent to Equation (3)):

$$R = M^{-1}X \tag{16}$$

where M is given by Equation (4). Equation (16) can be solved numerically, if the pandemic is observed during a time greater than $1/(\rho + \mu)$. We will first demonstrate an example of how the matrix M can be repeatedly calculated for consecutive periods of length equal to that of the contagiousness period (supposed to be constant during the outbreak), giving matrix series M_1, M_2, … Following Equation (4), we put the values of X_i's in the two matrices below, with r = 3 for two periods, the first from day 1 to day 3 and the second from day 4 to day 6.

$$M_1 = \begin{bmatrix} X_4 & X_3 & X_2 \\ X_3 & X_2 & X_1 \\ X_2 & X_1 & X_o \end{bmatrix},\ M_2 = \begin{bmatrix} X_6 & X_5 & X_4 \\ X_5 & X_4 & X_3 \\ X_4 & X_3 & X_2 \end{bmatrix}, \ldots,$$

where, after Equation (6), M_1 and M_2 can be calculated from the R_j's as:

$$M_1 = \begin{bmatrix} R_1^4 + 3R_1^2R_2 + 2R_1R_3 + R_2^2 & R_1^3 + 2R_1R_2 + R_3 & R_1^2 + R_2 \\ R_1^3 + 2R_1R_2 + R_3 & R_1^2 + R_2 & R_1 \\ R_1^2 + R_2 & R_1 & 1 \end{bmatrix},$$

and M_2 is given by:

$$\begin{bmatrix} R_1^6+5R_1^4R_2+4R_1^3R_3+6R_1R_2R_3+6R_1^2R_2^2+R_2^3+R_3^2 & R_1^5+4R_1^3R_2+3R_1^2R_2+2R_2R_3+3R_3R_1^2 & R_1^4+3R_1^2R_2+2R_1R_3+R_2^2 \\ R_1^5+4R_1^3R_2+3R_1^2R_2+2R_2R_3+3R_3R_1^2 & R_1^4+3R_1^2R_2+2R_1R_3+R_2^2 & R_1^3+2R_1R_2+R_3 \\ R_1^4+3R_1^2R_2+2R_1R_3+R_2^2 & R_1^3+2R_1R_2+R_3 & R_1^2+R_2 \end{bmatrix}$$

Additionally, from Equation (2), if, for instance, j = 8 and r = 3, then we have the expression below, which means that the new cases on the 8th day depend on the new cases detected on the previous days 7, 6 and 5, supposed to be in a period of contagiousness of 3 days:

$$X_8 = \sum_{k=1,3} R_k X_{8-k} = R_1X_7 + R_2X_6 + R_3X_5 \quad (17)$$

Let us suppose now that the initial R_j's on a contagiousness period of 3 days, are equal to:
$\begin{bmatrix} R_1 \\ R_2 \\ R_3 \end{bmatrix} = \begin{bmatrix} 2 \\ 1 \\ 2 \end{bmatrix}$, then matrix M defined by $M_{ij} = X_{7-(i+j)}$ gives the R_j's from Equation (16), hence allows the calculation of $X_j = \Sigma_{k=1,3} R_k X_{j-k}$.

The inverse of M is denoted by M^{-1} and verifies: $R = M^{-1}X$, where $X = (X_6, X_5, X_4)$, with $X_1 = 1$, $X_2 = 2$, $X_3 = 5$, $X_4 = 14$, $X_5 = 37$, $X_6 = 98$ and we obtain:

$$M_1^{-1} = \begin{bmatrix} 37 & 14 & 5 \\ 14 & 5 & 2 \\ 5 & 2 & 1 \end{bmatrix}^{-1} = \begin{bmatrix} -1/4 & 1 & -3/4 \\ 1 & -3 & 1 \\ -3/4 & 1 & 11/4 \end{bmatrix},$$

and a deconvolution gives the resulting R_j's:
$\begin{bmatrix} -1/4 & 1 & -3/4 \\ 1 & -3 & 1 \\ -3/4 & 1 & 11/4 \end{bmatrix} \begin{bmatrix} 98 \\ 37 \\ 14 \end{bmatrix} = \begin{bmatrix} 2 \\ 1 \\ 2 \end{bmatrix} = \begin{bmatrix} R_1 \\ R_2 \\ R_3 \end{bmatrix}$, thanks to the following calculation:

$$R_1 = -49/2 + 37 - 21/2 = 2$$

$$R_2 = 98 - 111 + 14 = 1$$

$$R_3 = -147/2 + 37 + 77 = 2$$

We obtain for the resulting distribution of daily reproduction numbers the exact replica of the initial distribution. We obtain the same result by replacing M_1 by the matrix M_2.

3.2. Distribution of the Daily Reproduction Numbers R_j's When They Are Supposed to Be Random

Let us consider a stochastic version of the deterministic toy model corresponding to Equation (17), by introducing an increasing noise on the R_j's, e.g., by randomly choosing their values following a uniform distribution on the three intervals: [2 − a, 2 + a], [1 − a/2, 1 + a/2] and [2 − a, 2 + a] (for having a U-shape behavior), with increasing values of a, from 0.1 to 1, in order to see when the deconvolution would give negative resulting R_j's, with conservation of the average of their sum R_0, if the random choice of the values of the R_j's at each generation is repeated, following the stochastic version of Equation (2): $X_j = \Sigma_{k=1,r} (R_k + \varepsilon_k) X_{j-k}$, where r is the contagiousness period duration and ε_k is a noise perturbing R_k, whose distribution is chosen uniform on the interval [0, 2a] for k = 1,3, and [0, a] for k = 2. This choice is arbitrary, and the main reason of the randomization is to show that the deconvolution can give negative results for R_k's, as those observed for increasing values of a, from 0.1 to 1, with explicit calculations for three consecutive periods, from day 1 to day 3, from day 4 to day 6, and from day 7 to day 9.

For each random choice of the values of the daily reproduction numbers R_j's, we can calculate a matrix M_1 corresponding to Equation (3). Its inversion into the matrix ${M_1}^{-1}$ makes it possible to solve the problem of deconvolution of Equation (2)—that is to say, to obtain new R_j's as a function of the observed X_k's. We can then calculate a new matrix M_2 from these new Rj's and thus continue during an epidemic the estimation of the daily reproduction numbers R_j's from the successive matrices $M_1, M_2, \ldots$, and observed X_k's.

1. For a = 0.1, let us randomly and uniformly choose the initial distribution of the daily reproduction numbers R_1 in the interval [1.9, 2.1], R_2 in [0.95, 1.05] and R_3 in [1.9, 2.1] as $R_1 = 2.1$, $R_2 = 0.95$, $R_3 = 2.1$. Then, the transition matrix M_1 is equal to:

$$M_1 = \begin{bmatrix} 41.7391 & 15.351 & 5.36 \\ 15.351 & 5.36 & 2.1 \\ 5.36 & 2.1 & 1 \end{bmatrix} \text{ and we have:}$$

$$M_1^{-1} = \begin{bmatrix} -0.2154195 & 0.92857143 & -0.7953515 \\ 0.92857143 & -2.95 & 1.2178571 \\ -0.7953515 & 1.2178571 & 2.705584 \end{bmatrix}$$

From $X_6 = 113.491$, $X_5 = 41.7391$, $X_4 = 15.351$, resulting R_j's are: $R_1 = 2.1$, $R_2 = 0.95$, $R_3 = 2.1$.

The next initial R_j's are chosen as: $R_1 = 2$, $R_2 = 0.95$, $R_3 = 1.9$ and we have:

$$X_7 = 2X_6 + 0.95X_5 + 1.9X_4 = 226.982 + 39.652 + 29.17 = 295.8$$

$$X_8 = 2X_7 + 0.95X_6 + 1.9X_5 = 591.6 + 107.816 + 79.304 = 778.72$$

Then, we obtain the matrices M_2 and ${M_2}^{-1}$:

$$M_2 = \begin{bmatrix} 295.8 & 113.491 & 41.7391 \\ 113.491 & 41.7391 & 15.351 \\ 41.7391 & 15.351 & 5.36 \end{bmatrix}$$

$$M_2^{-1} = \begin{bmatrix} -0.07779371 & 0.20964295 & 0.00524305 \\ 0.20964295 & -1.0123552 & 1.26721348 \\ 0.00524305 & 1.26721348 & -3.48354228 \end{bmatrix}$$

Then, the resulting R_j's equal: $R_1 = 2.0279$, $R_2 = 7.6158$, $R_3 = -16.426$.
The next initial R_j's are: $R_1 = 2$, $R_2 = 1.05$, $R_3 = 1.9$ and we have:

$$X_9 = 2X_8 + 1.05X_7 + 1.9X_6 = 1557.44 + 310.59 + 215.63 = 2083.66$$

$$X_{10} = 2X_9 + 1.05X_8 + 1.9X_7 = 4167.32 + 817.656 + 562.02 = 5546.996$$

From these values of X_9 and X_{10}, we obtain the matrices M_3 and ${M_3}^{-1}$:

$$M_3 = \begin{bmatrix} 2083.66 & 778.72 & 295.8 \\ 778.72 & 295.8 & 113.491 \\ 295.8 & 113.491 & 41.7391 \end{bmatrix}$$

$$M_3^{-1} = \begin{bmatrix} 0.02596375 & -0.05192766 & -0.04280771 \\ -0.05192766 & 0.0256605 & 0.29823273 \\ -0.04280771 & 0.29823273 & -0.48358035 \end{bmatrix}$$

Then, the resulting R_j's equal: $R_1 = 2.486$, $R_2 = -2.33$, $R_3 = 7.38769$.

2. For a = 1, let us choose the initial R_1 in [1, 3], R_2 in [0.5, 1.5] and R_3 in [1, 3], e.g., $R_1 = 1$, $R_2 = 1.355$ and $R_3 = 1.1$. Then, the transition matrix M_1 is equal to:

$$M_1 = \begin{bmatrix} 9.101 & 4.81 & 2.355 \\ 4.81 & 2.355 & 1 \\ 2.355 & 1 & 1 \end{bmatrix} \text{ and its inverse is given by:}$$

$$M_1^{-1} = \begin{bmatrix} -1.11983471 & 2.02892562 & 0.60828512 \\ 2.02892562 & -2.93801653 & -1.84010331 \\ 0.60828512 & -1.84010331 & 1.40759184 \end{bmatrix}$$

New cases are: $X_6 = 18.209$, $X_5 = 9.101$, $X_4 = 4.81$, $X_3 = 2.355$, $X_2 = 1$, $X_1 = 1$, and by deconvoluting, we obtain the resulting R_j's equal to: $R_1 = 1$, $R_2 = 1.355$, $R_3 = 1.1$, i.e., the exact initial distribution.

Let us now consider new initial R_j's: $R_1 = 1$, $R_2 = 1$, $R_3 = 1$. That gives a new matrix M_2, with new X_7 and X_8 calculated from the new initial R_j's, by using the former values of $X_6, \ldots, X_2$:

$$X_7 = X_6 + X_5 + X_4 = 18.209 + 9.101 + 4.81 = 32.12$$

$$X_8 = X_7 + X_6 + X_5 = 32.12 + 18.209 + 9.101 = 59.43$$

Hence, we obtain:

$$M_2 = \begin{bmatrix} 32.12 & 18.209 & 9.101 \\ 18.209 & 9.101 & 4.81 \\ 9.101 & 4.81 & 2.36 \end{bmatrix} \text{ and}$$

$$M_2^{-1} = \begin{bmatrix} -0.35061537 & 0.1839519 & 0.97925345 \\ 0.1839519 & -1.47916605 & 2.31025157 \\ 0.97925345 & 2.31025157 & -8.0783421 \end{bmatrix}$$

and the resulting R_j's equal: $R_1 = 2.90$, $R_2 = 5.4888$, $R_3 = -14.696$.

We calculate X_9 and X_{10} using new initial R_j's: $R_1 = 3.0$, $R_2 = 0.5$, $R_3 = 2.9$:

$$X_9 = 3X_8 + 0.5X_7 + 2.9X_6 = 178.29 + 16.06 + 52.81 = 247.16$$

$$X_{10} = 3X_9 + 0.5X_8 + 2.9X_7 = 741.48 + 29.715 + 93.148 = 864.343$$

Hence, we obtain:

$$M_3 = \begin{bmatrix} 247.16 & 59.43 & 32.12 \\ 59.43 & 32.12 & 18.209 \\ 32.12 & 18.209 & 9.101 \end{bmatrix} \text{ and}$$

$$M_3^{-1} = \begin{bmatrix} 0.00718287 & -0.00805357 & -0.00923703 \\ -0.00805357 & -0.22288084 & 0.47435642 \\ -0.00923703 & 0.47435642 & -0.80659958 \end{bmatrix}$$

and the resulting R_j's equal: $R_1 = 3.66898$, $R_2 = -33.857$, $R_3 = 61.32$.

More precise simulation results are given in Table 1, which summarizes computations made for random choices of R_j's distributions, for a = 0.1 and a = 1 and until time 20. These simulations show a great sensitivity to noise, but a qualitative conservation of their U-shaped distribution along the contagiousness period of individuals. More precisely, because of the presence of noise on the Rj's, we cannot always obtain positive values from the data for the Rj's by applying the deconvolution, which explains the presence of negative values in empirical examples, as in the theoretical noised examples. A way to solve this problem could be to suppose that noise is stationary during all of the growth period of a wave, then calculate the Rj's for all running time windows of length equal to the contagiousness duration and then obtain the mean of the Rj's corresponding to these windows. As this stationary hypothesis is not widely accepted, we prefer to keep negative values and focus on the shape of the distribution of the Rj's.

Table 1. Simulation results obtained for extreme noises a = 0.1 and a = 1, showing great variations of deconvoluted distribution of daily reproduction numbers X_j's and a qualitative conservation of their U-shaped distribution along contagiousness period.

a	Initial R_j's	t	X_t	X_{t+1}	X_{t+2}	Resulting R'_js	R_0	Distribution Shape, Sign R_0
0.1	2.1; 0.95; 2.1	4	15.35	31.74	113.5	2.1; 0.95; 2.1	5.15	U-shape, positive
	2; 0.95; 1.9	6	113.5	295.8	778.7	2.03; 7.6; −16.4	−6.77	Inverted U-shape, negative
	2; 1.06; 1.9	8	778.7	2083.7	5547	2.49; −2.33; 7.39	7.55	U-shape, positive
	1.9; 1.05; 1.9	10	5547	14,207	36,776	2.69; −16.7; 43.8	29.8	U-shape, positive
	1.9; 0.95; 1.9	12	36,776	93,910	240,359	2.92; 1.68; −6.7	−2.1	Decreased shape, negative
	1.9; 1; 1.9	14	240,359	622,149	1,605,227	2.3; −4.83; 14.3	11.8	U-shape, positive
	2; 1.05; 1.9	16	1,605,227	4,331,630	11,561,153	2.76; 27; −70	−40.2	Inverted U-shape, negative
	1.9; 1; 1.95	18	11,561,153	29,558,395	76,502,587	2.5; −6.48; 17.9	13.9	U-shape, positive
	2; 1; 2.1	20	76,502,587	2,076,519	556,226,772	2.67; −7.6; 19.7	14.8	U-shape, positive
1	1; 1.355; 1.1	4	4.81	9.1	18.21	1; 1.355; 1.1	3.455	Inverted U-shape, positive
	1; 1; 1	6	18.21	32.12	59.43	2.9; 5.49; −14.7	−6.31	Inverted U-shape, negative
	3; 0.5; 2.9	8	59.43	247.16	864.34	3.7; −33.9; 61.3	31.1	U-shape, positive
	2.6; 0.7; 2.6	10	864.34	2574.82	7942	3; −1.79; 7.14	8.35	U-shape, positive
	2.5; 0.75; 1.5	12	7942.2	23,083.1	67,526.6	3.35; 2.54; −11.6	−5.71	Decreased shape, negative
	2.4; 0.8; 2.4	14	67,526.6	199,590	588,437	2.58; −0.5; 4.8	6.88	U-shape, positive
	2; 1; 2	16	588,437	1,511,517	4,010,652	2.72; −1.08; 3.19	4.83	U-shape, positive
	2.3; 1.15; 2.3	18	4,010,652	12,316,150	36,415,885	2.88; −7.9; 21.7	16.7	U-shape, positive
	2.8; 0.6; 2	20	36,415,885	117,375,471	375,133,150	3.7; 4.1; −17	−9.2	Inverted U-shape, negative

3.3. Distribution of the Daily Reproduction Numbers R_j's. The Real Example of France

Figure 3 gives the effective transmission rates R_e calculated between 20–25 October 2020 just before the second lockdown in France [28,29]. As the second wave of the epidemic is still in its exponential phase, it is more convenient (i) to consider the distribution of the marginal daily reproduction numbers and (ii) to calculate its entropy and simulate the epidemic dynamics using a Markovian model [4]. By using the daily new infected cases given in [30], we can calculate, as in Section 3.1, the inverse matrix M^{-1} for the period from 20 to 25 October 2020 (exponential phase of the second wave), by choosing 3 days for the duration of contagiousness period and the following raw data for new infected cases: 20,468 for 20 October, then 26,676, 41,622, 42,032, 45,422 and 52,010 for 25 October. Then, for France between 15 February and 27 October 2020, we obtain the daily reproduction numbers given in Figure 3 with a U-shape as observed for influenza viruses.

We have:

$$M^{-1} = \begin{bmatrix} 45,422 & 42,032 & 41,622 \\ 42,032 & 41,622 & 26,676 \\ 41,622 & 26,676 & 20,468 \end{bmatrix}^{-1} = \begin{bmatrix} -0.0000163989812 & -0.0000292188776 & 0.00007142863 \\ -0.0000292188776 & 0.0000938161392 & -0.0000628537817 \\ 0.00007142863 & -0.0000628537817 & -0.00001447698 \end{bmatrix}$$

Hence, we can deduce the daily R_j's, i.e., the vector (R_1, R_2, R_3):

$$\begin{bmatrix} -0.0000163989812 & -0.0000292188776 & 0.00007142863 \\ -0.0000292188776 & 0.0000938161392 & -0.0000628537817 \\ 0.00007142863 & -0.0000628537817 & -0.00001447698 \end{bmatrix} \begin{bmatrix} 52,010 \\ 45,422 \\ 42,032 \end{bmatrix} = \begin{bmatrix} -0.852911911949567 & -1.32717986039119 & 3.00228812555347 \\ -1.51967382631645 & 4.26131667592337 & -2.64187015405365 \\ 3.71500298367996 & -2.85494447414886 & -0.60849658654673 \end{bmatrix}$$

$$= \begin{bmatrix} 0.82219725466 \\ 0.0997726955533 \\ 0.2515619229844 \end{bmatrix} = \begin{bmatrix} R_1 \\ R_2 \\ R_3 \end{bmatrix}$$

The effective reproduction number is equal to $R_0 \approx 1.174$, a value close to that calculated directly (Figure 3), giving V = (0.7, 0.085, 0.215), with a maximal daily reproduction number the first day of the contagiousness period. The entropy H of V is equal to:

$$H = -\Sigma_{k=1,r} V_k \, Log(V_k) = 0.25 + 0.21 + 0.33 = 0.79.$$

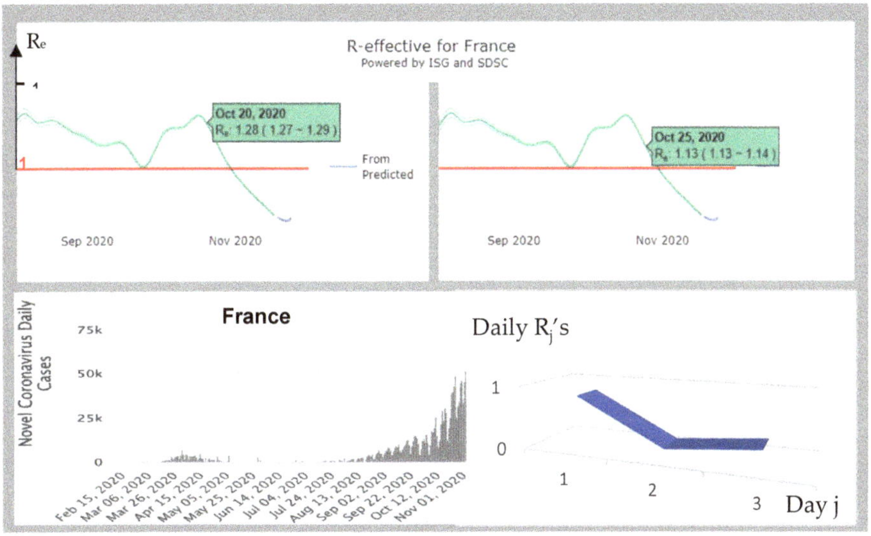

Figure 3. Top: estimation of the effective reproduction number R_e's for 20 October and the 25 October 2020 (in green, with their 95% confidence interval) [28,29]. **Bottom left**: daily new cases in France between 15 February and 27 October [30]. **Bottom right**: U-shape of the evolution of the daily R_j's along the 3-day contagiousness period of an individual.

3.4. Calculation of the Rj's for Different Countries

3.4.1. Chile

By using the daily new infected cases given in [30], we can calculate M^{-1} for the period from 1 to 12 November 2020 (endemic phase), by choosing 6 days for the duration of the contagiousness period and the following 7-day moving average data for the new infected cases (Figure 4): 1400 for 1 November, then 1370, 1382, 1359, 1362, 1405, 1389, 1385, 1384, 1387, 1394 and 1408 for 12 November.

We have:

$$M^{-1} = \begin{bmatrix} 1394 & 1387 & 1384 & 1385 & 1389 & 1405 \\ 1387 & 1384 & 1385 & 1389 & 1405 & 1362 \\ 1384 & 1385 & 1389 & 1405 & 1362 & 1359 \\ 1385 & 1389 & 1405 & 1362 & 1359 & 1382 \\ 1389 & 1405 & 1362 & 1359 & 1382 & 1370 \\ 1405 & 1362 & 1359 & 1382 & 1370 & 1400 \end{bmatrix}^{-1} =$$

$$\begin{bmatrix} -0.05714222 & 0.01016059 & -0.00901664 & 0.01474588 & 0.00640175 & 0.03539322 \\ 0.01016059 & -0.01827291 & 0.0106261 & -0.00763363 & 0.02139586 & -0.01613675 \\ -0.00901664 & 0.0106261 & -0.00544051 & 0.02150289 & -0.01468484 & -0.00286391 \\ 0.01474588 & -0.00763363 & 0.02150289 & -0.01796266 & -0.00553414 & -0.00509801 \\ 0.00640175 & 0.02139586 & -0.01468484 & -0.00553414 & -0.00305831 & -0.00452917 \\ 0.03539322 & -0.01613675 & -0.00286391 & -0.00509801 & -0.00452917 & -0.00686198 \end{bmatrix}$$

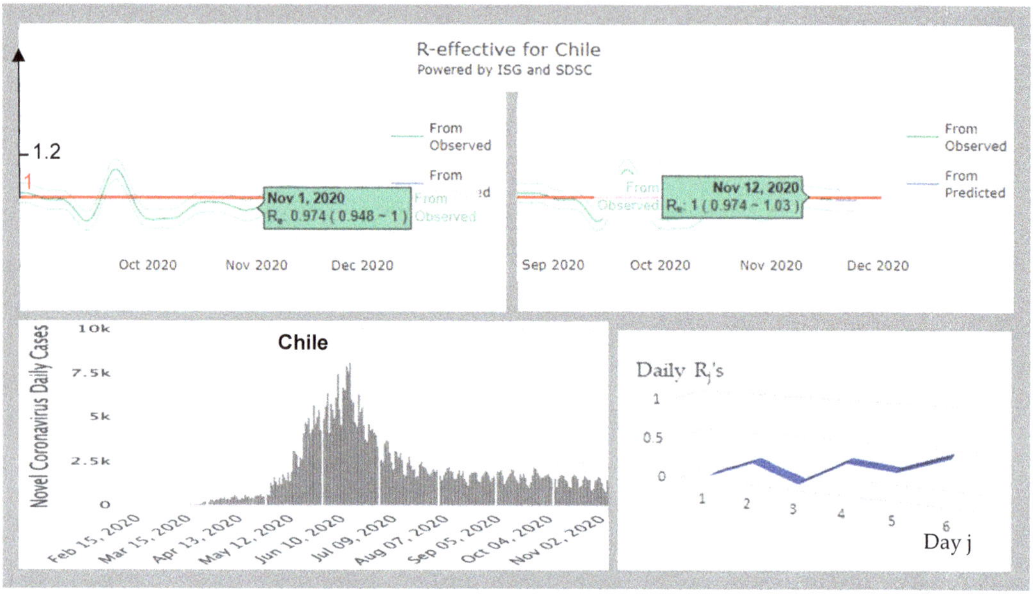

Figure 4. Top: estimation of the effective reproduction number R_e's for the 1 November and the 12 November 2020 (in green, with their 95% confidence interval) [28,29]. **Bottom left**: Daily new cases in Chile between 1 November and 12 November [30]. **Bottom right**: U-shape of the evolution of the daily R_j's along the infectious 6-day period of an individual.

Hence, after deconvolution, we obtain:

$$R = \begin{bmatrix} -0.36256122 \\ 0.22645436 \\ 0.01488726 \\ 0.33918287 \\ 0.28557502 \\ 0.50696243 \end{bmatrix}$$

The effective reproduction number is equal to $R_0 \approx 1.011$, a value close to that calculated directly, with a maximal daily reproduction number the last day of the contagiousness period. Due to the negativity of R_1, we cannot derive the distribution V and therefore calculate its entropy. As entropy is an indicator of non-uniformity, an alternative could be to calculate it by shifting values of Rj's upwards by the value of their minimum.

The quasi-endemic situation in Chile since the end of August, which corresponds to the increase of temperature and drought at this period of the year [4], gives a cyclicity of the new cases occurrence whose period equals the length of the contagiousness period of about 6 days, analogue to the cyclic phenomenon observed in simulated stochastic data of Section 3.2. with a similar U-shaped distribution of the R_j's.

3.4.2. Russia

By using the daily new infected cases given in [30], we can calculate M^{-1} for the period from 30 September to 5 October 2020 (exponential phase of the second wave), by choosing 3 days for the duration of the contagiousness period and the following raw data for new infected cases (Figure 5): 7721 for 30 September, then 8056, 8371, 8704, 9081, 9473 for 5 October.

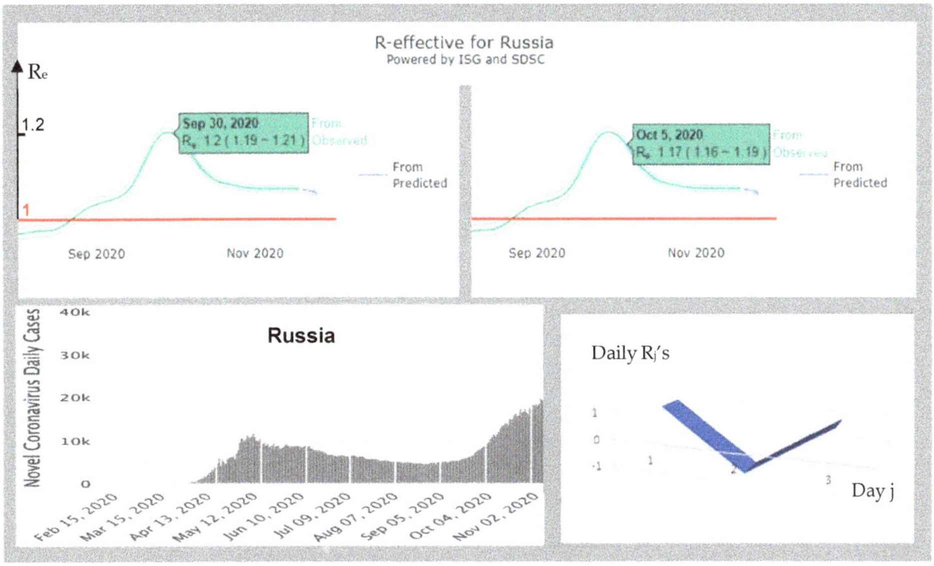

Figure 5. Top: estimation of the effective reproduction number R_e's for 30 September and the 5 October 2020 (in green, with their 95% confidence interval) [28,29]. **Bottom left**: Daily new cases in Russia between 15 February and 21 November [30]. **Bottom right**: U-shape of the evolution of the daily R_j's along the 3-day contagiousness period.

We have:

$$M^{-1} = \begin{bmatrix} 9081 & 8704 & 8371 \\ 8704 & 8371 & 8056 \\ 8371 & 8056 & 7721 \end{bmatrix}^{-1} \text{ and}$$

$$\begin{bmatrix} 0.031553440566948 & -0.027594779248393 & -0.005417732076268 \\ -0.027594779248393 & -0.00482333528665 & 0.034950483895551 \\ -0.005417732076268 & 0.034950483895551 & -0.030463575061795 \end{bmatrix} \begin{bmatrix} 9473 \\ 9081 \\ 8704 \end{bmatrix} = \begin{bmatrix} R_1 \\ R_2 \\ R_3 \end{bmatrix},$$

where:

$$R_1 = 298.905742490698404 - 250.588190354656833 - 47.155939991836672 = 1.161612144205$$

$$R_2 = -261.405343820026889 - 43.80070773806865 + 304.209011826875904 = -0.997039731220$$

$$R_3 = -51.322175958486764 + 317.385344255498631 - 265.15495733786368 = 0.90821095914$$

The effective reproduction number is equal to $R_0 \approx 1.073$, a value close to that calculated directly, with a maximal daily reproduction number the first day of the contagiousness period. Due to the negativity of R_2, we cannot derive the distribution V and therefore calculate its entropy. The period studied corresponds to a local slow increase of new infected cases at the start of the second wave in Russia, which looks like a staircase succession of slightly inclined 4-day plateaus followed by a step: at the beginning of October, in Russia, new tightened restrictions (but avoiding lockdown) appeared [31], which could explain the change of the value of the slope observed in the new daily cases [30].

3.4.3. Nigeria

By using the daily new infected cases given in [30], we can calculate M^{-1} for the period from 5 November to 10 November (endemic phase), by choosing 3 days for the duration of the contagiousness period and the following raw data for the new infected cases (Figure 6): 141 for 5 November, then 149, 133, 161, 164, and 166 for 10 November.

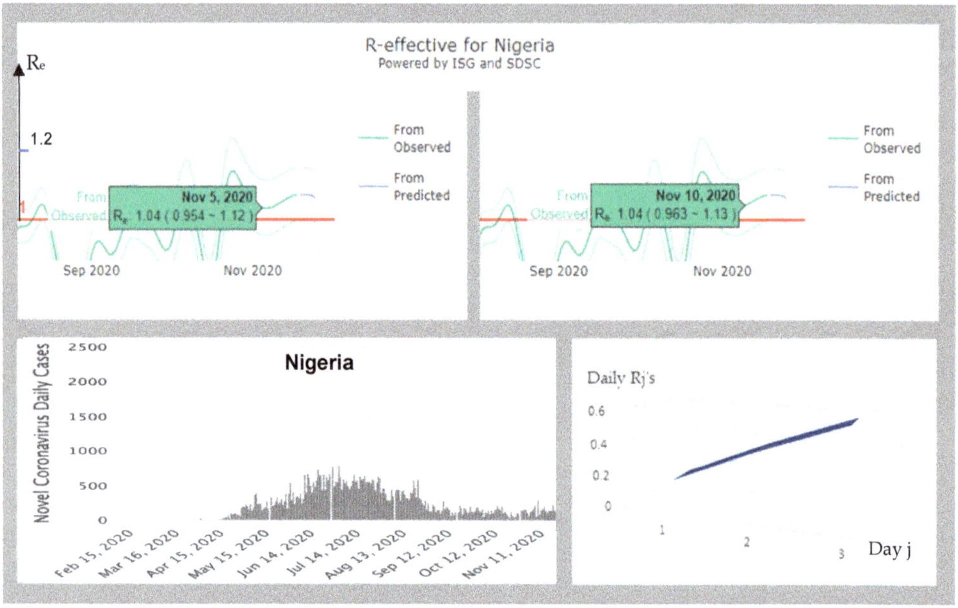

Figure 6. Top: estimation of the effective reproduction number R_e's for 5 November and 10 November 2020 (in green, with their 95% confidence interval) [28,29]. **Bottom left**: Daily new cases in Nigeria between 15 February and 21 November [30]. **Bottom right**: increasing evolution of the daily R_j's along the 3-day contagiousness period of an individual.

We have:

$$M^{-1} = \begin{bmatrix} 164 & 161 & 131 \\ 161 & 131 & 149 \\ 131 & 149 & 141 \end{bmatrix}^{-1} = \begin{bmatrix} 0.01796807 & 0.01502897 & -0.03283028 \\ 0.01502897 & -0.02832263 & 0.01575332 \\ -0.03283028 & 0.01575332 & 0.02141264 \end{bmatrix}$$

After deconvolution, we obtain:

$$R = \begin{bmatrix} 0.16177513 \\ 0.38618314 \\ 0.58115333 \end{bmatrix}$$

The effective reproduction number is equal to $R_0 \approx 1.129$, value close to that calculated directly, with a maximal daily reproduction number the last day of the contagiousness period. The distribution V equals (0.143, 0.342, 0.515) and its entropy H is equal to:

$$H = -\Sigma_{k=1,r} V_k \, \mathrm{Log}(V_k) = 0.29 + 0.37 + 0.34 = 1.$$

In Appendix C, Table A1 gives the shape of the R_j's distribution for 194 countries.

3.5. Weekly Patterns in Daily Infected Cases

Daily new infected cases are highly affected by weekdays, such that case numbers are lowest at the start of the week and increase afterwards. This pattern is observed at the world level, as well as at the level of almost every single country or USA state. Hence, in order to estimate biologically meaningful reproduction numbers, clean of weekly patterns due to administrative constraints, analyses have to be restricted to specific periods shorter than a week, or at rare occasions when patterns escape the administrative constraints. This weekly phenomenon occurs during exponential increase as well as decrease phases of the pandemic and during endemic periods in numbers of daily cases (Figure 6). In addition, the daily new infected case record is discontinuous for many countries/regions, which frequently publish, on Monday or Tuesday, a cumulative count for that day and the weekend days. For example, Sweden typically publishes only four numbers over one week, the one on Tuesday cumulating cases for Saturday, Sunday and the two first weekdays. Discontinuity in records further limits the availability of data enabling detailed analyses of daily reproduction numbers and can be considered as extreme weekday effects on new case records due to various administrative constraints.

We calculated Pearson correlation coefficients r between a running window of daily new case numbers of 20 consecutive days and a running window of identical duration with different intervals between the two running windows. These Pearson correlation coefficients r typically peak with a lag of seven days between the two running windows.

The mean of these correlations are for each day of the week from Tuesday (data making up for the weekend underestimation) to Monday: 0.571, 0.514 (0.081), 0.383 (0.00008), 0.347 (0.000003), 0.381 (0.000006), 0.468 (0.000444) and 0.558 (0.03916), with, in parentheses, the p-value of the one-tailed paired t-test showing that the correlation observed with running windows starting Tuesday are more than the others (see also supplementary material). This could reflect a biological phenomenon of seven infection days. However, examination of the frequency distributions of lags for r maxima reveals, besides the median lag at 7 days, local maxima for multiples of 7 (14, 21, 28, 35, etc.). About 50 percent of all local maxima in r involve lags that are multiples of seven (seven included).

This excludes a biological causation, except if data periodicity comes from an entrainment by the weekly "Zeitgeber" of census, near the duration of the contagiousness interval. We tried to control for weekdays using two methods, and combinations thereof. For the first method, we calculated z-scores for each weekday, considering the mean number of cases for each weekday, and subtracted that mean from the observed number for a day (Figure 7). This delta was then divided by the standard deviation of the number of cases for that weekday. The mean and standard variation are calculated across the whole period of study for each weekday.

The second method implies data smoothing using a running window of 5 consecutive days, where the mean number of new cases calculated across the five days is subtracted from the number of new cases observed for the third day. Hence, data for a given day are compared to a mean including two previous, and two later days (Figure 8).

We constructed two further datasets, where z-scores are applied in the first to data after smoothing from the second method and are applied in the second data after smoothing from the first method (not shown) (Figures 9 and 10).

These four datasets from daily new cases database [30] transformed according to different methods and combinations thereof designed to control for weekday were analysed using the running window method. Despite attempts at controlling for weekday effects, the median lag was always seven days across all four transformed datasets, and local maxima in lag distributions were multiples of seven. After data transformations, about 50 percent of all local maxima were lags that are multiples of seven, seven included.

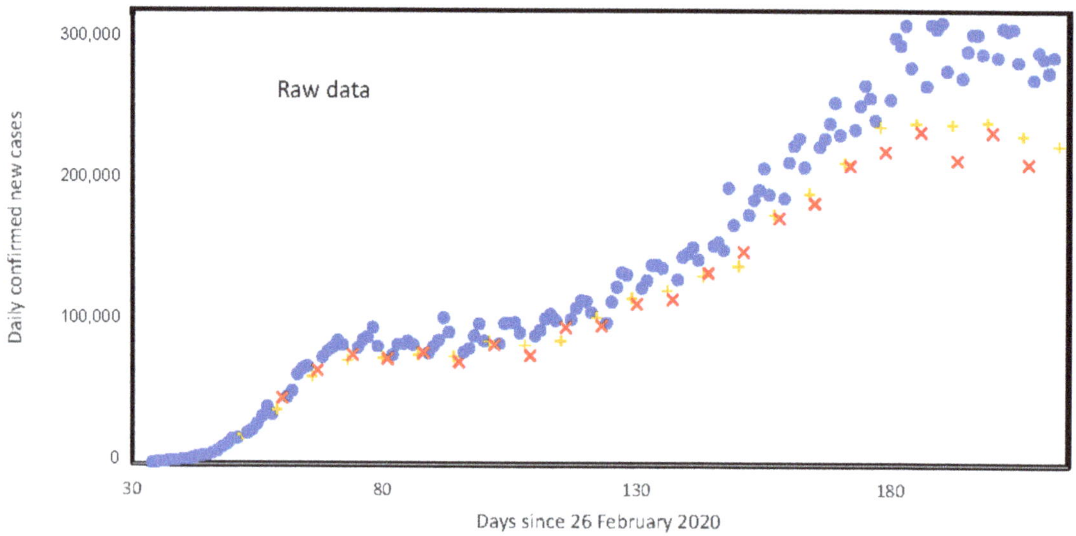

Figure 7. Confirmed world daily new cases (from [30]) as a function of days since 26 February until 23 August 2020 + indicates Sundays, X indicates Mondays.

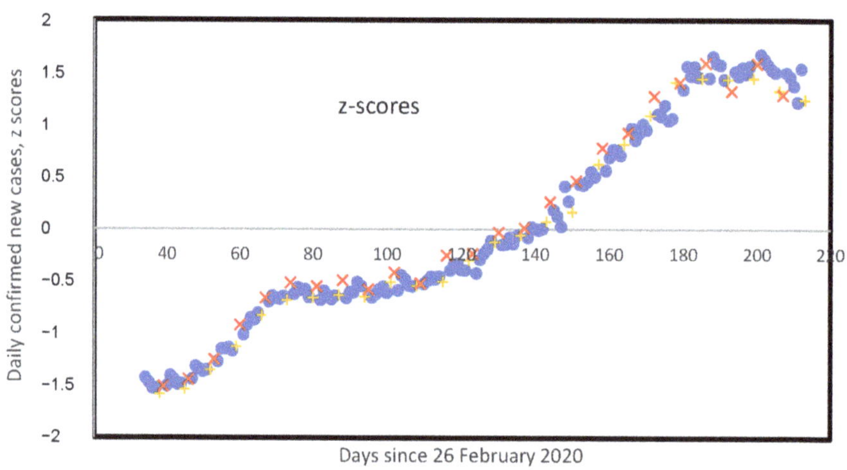

Figure 8. Z-transformed scores of confirmed world daily new cases [30], from Figure 6, as a function of days since 26 February 2020 until 23 August 2020 + indicates Sundays, X indicates Mondays. Z-transformations are specific to each weekday.

Visual inspection of plots of these transformed data versus time for daily new infected cases from the whole world shows systematic local biases in daily new infected cases (after transformation) on Sundays and Mondays, for all four transformed datasets, with Sundays and/or Mondays as local minima and/or local maxima, according to which method or combination thereof was applied to the data. Hence, the methods we used failed to neutralize the weekly patterns in daily new cases due to administrative constraints. This issue highly limits the data available for detailed analyses of daily new cases aimed at estimating biologically relevant estimates of reproduction numbers at the level of short temporal scales.

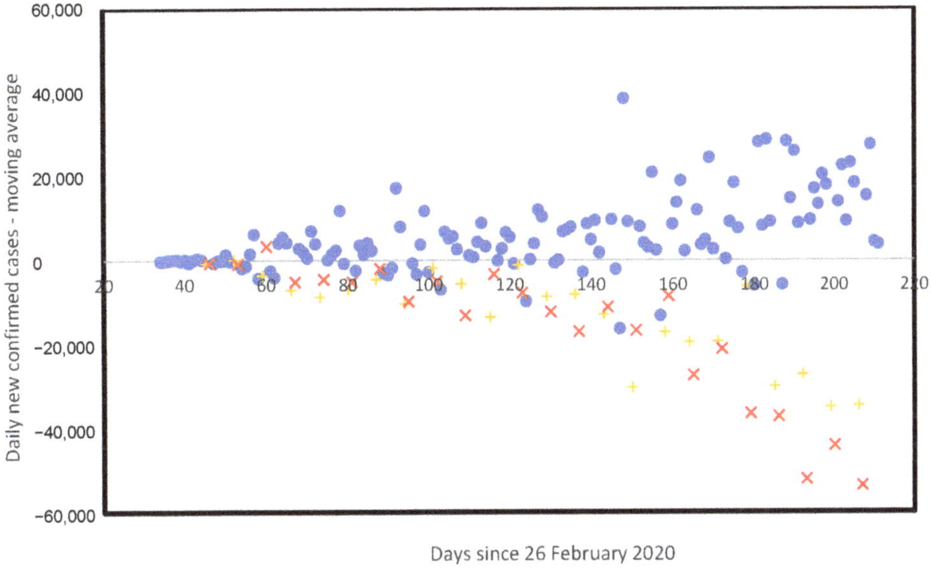

Figure 9. Smoothed confirmed world daily new cases [30], from Figure 7, as a function of days since 26 February 2020 until 23 August 2020 + indicates Sundays, X indicates Mondays. For each specific day j, the mean number of confirmed daily new cases calculated for days j − 1, j − 2, j, j + 1 and j + 2 is subtracted from the number for day j.

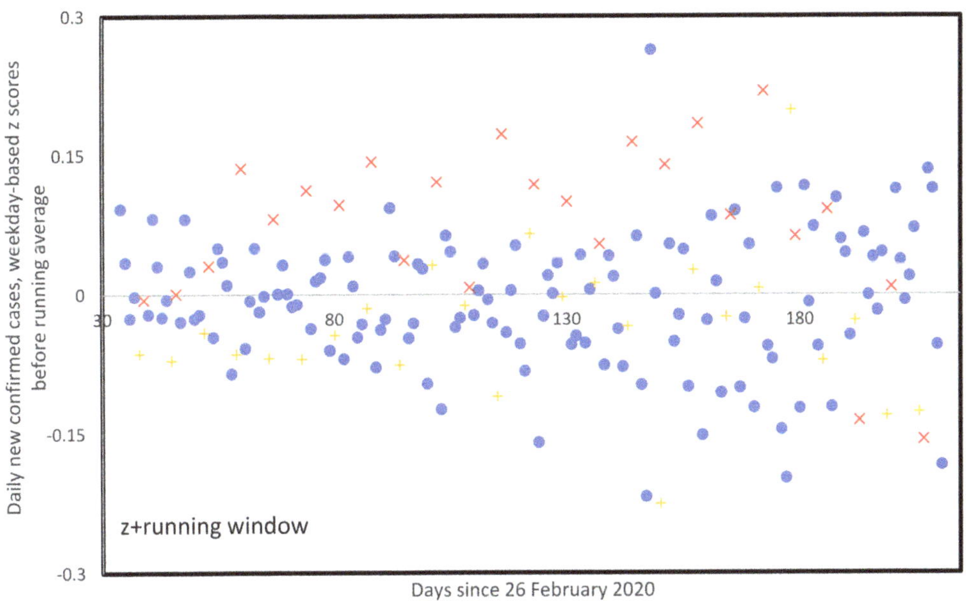

Figure 10. Smoothed confirmed world daily new cases [30] applied to z-scores from Figure 8, as a function of days since 26 February 2020 until 23 August 2020 + indicates Sundays, X indicates Mondays. Z-transformations are specific to each weekday. For specific day j, the mean number of confirmed new cases calculated for days j − 1, j − 2, j, j + 1, j + 2 is subtracted from the number for day j.

By smoothing on five consecutive days of raw data (confirmed world daily new infected cases [24]) and applying the z-transformation, we obtain a better result in Figure 11 than in Figure 10 in order to neutralize the weekly pattern. The reason is that the smoothing largely eliminates the counting defect during weekends due either to fewer hospital admissions and/or less systematic PCR tests or to a lack of staff at the end of the week to perform the counts.

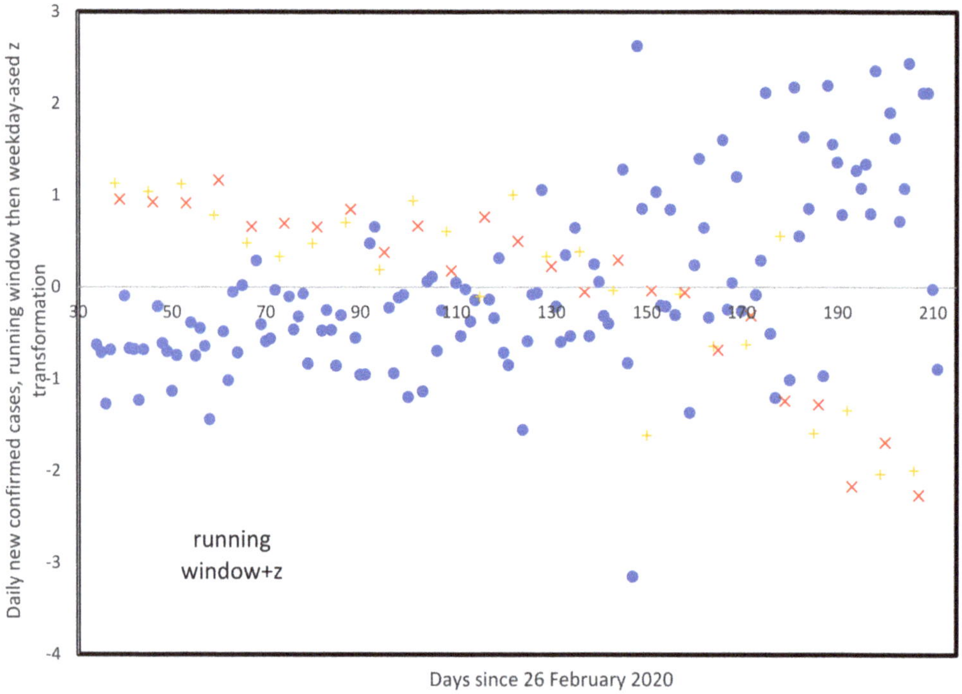

Figure 11. Z-transformed scores of smoothed confirmed world daily new cases [30] smoothed data from Figure 9, as a function of days since 26 February 2020 until 23 August 2020. + indicates Sundays, X indicates Mondays. Z-transformations are specific to each weekday.

4. Discussion

The duration of the contagiousness period, as well as the daily virulence, are not constant over time. Three main factors, which are not constant during a pandemic, can explain this:

- In the virus transmitter, the transition between the mechanisms of innate (the first defense barrier) and adaptive (the second barrier) immunity may explain a transient decrease in the emission of the pathogenic agent during the phase of contagiousness [15],
- In the environmental transmission channel, many geophysical factors that vary over time can influence the transmission of the virus (temperature, humidity, altitude, etc.) [4–8],
- In the recipient of the virus, individual or public policies of prevention, protection, eviction or vaccination, which evolve according to the epidemic severity and the awareness of individuals and socio-political forces, can change the sensitivity of the susceptible individuals [32].

It is therefore very important to seek to estimate the average duration of the period of contagiousness of individuals and the variations, during this phase of contagiousness, of the associated daily reproduction numbers [33–39]. If the duration of the contagiousness phase is more than 3–5 days, for example ±7 days, the periodicity of seven days observed for the new daily cases could result of an entrainment of the dynamics of new cases driven by the social "Zeitgeber" represented by the counting of new cases, less precise during the weekend (probably underestimated in many countries not working at this time). That questions the deconvolution over 3 and 5 days, giving some negative R_j. In a future work, we will compare our results with those obtained by deconvolutions on contagiousness durations between 3 and 12 days in order to obtain possibly more realistic values for the R_j's, and hence, have perhaps a double explanation for the 7 days periodicity, both sociological and biological. Before this future work, we have extended our study using a duration r = 3 of contagiousness to r = 7. The results are given in Appendix B: they show the same existence of identical variations of U-shape type but they specify the values of R_j's, more often positive and of more realistic magnitude, while keeping a sum approximately equal to R_0.

Rhodes and Demetrius have pointed out the interest of the distribution of the daily reproduction numbers [24] with respect to the classical unique R_0, even time-dependent [25]. In particular, they found that this distribution was generally not uniform, which we have confirmed here by showing many cases where we observe the biphasic form of the virulence already observed in respiratory viruses, such as influenza. The entropy of the distribution makes it possible to evaluate the intensity of its corresponding U-shape. This entropy is high if the daily reproduction numbers are uniform, and it is low if the contagiousness is concentrated over one or two days. If some Rj are negative, it is still possible to calculate this uniformity index, by shifting their distribution by a translation equal to the inverse of the negative minimum value.

We have neglected in the present study the natural birth and death rates by supposing them identical, but we could have taken into account the mortality due to the COVID-19. The discrete dynamics of new cases can be considered as Leslie dynamics governed by the matrix equation:

$$X_j = L\, X_{j-1},$$

where X_j is the vector of the new cases living at day j and L is the Leslie matrix given by:

$$L = \begin{bmatrix} R_1 & R_2 & R_3 & \cdots & \cdots & R_r \\ b_1 & 0 & 0 & \cdots & \cdots & 0 \\ 0 & b_2 & 0 & \cdots & \cdots & 0 \\ \vdots & \vdots & \ddots & \cdots & \cdots & \vdots \\ \vdots & \vdots & \vdots & \ddots & \cdots & \vdots \\ 0 & 0 & 0 & \cdots & b_{r-1} & 0 \end{bmatrix} \text{ and } X_{j-1} = \begin{bmatrix} X_{j-1} \\ X_{j-2} \\ X_{j-3} \\ \vdots \\ \vdots \\ X_{j-r} \end{bmatrix},$$

where $b_j = 1 - \mu_j \leq 1, \forall\, i = 1, \ldots, r$, is the recovering probability between days j and j + 1.

The dynamical stability for L^2 distance to the stationary infection age pyramid $P = \lim_j X_j/\Sigma_{i=j,j-r+1}X_i$ is related to $|\lambda - \lambda'|$, the modulus of the difference between the dominant and sub-dominant eigenvalues of L, namely $\lambda = e^R$ and λ', where R is the Malthusian growth rate and P is the left eigenvector of L corresponding to λ. The dynamical stability for the distance (or symmetrized divergence) of Kullback–Leibler to P considered as stationary distribution is related to the population entropy H [26–32], which is defined if $l_j = \prod_{i=1,j-1} b_i$ and $p_j = l_j R_j/\lambda^j$, as follows:

$$H = -\Sigma_{j=1,r} p_j \,\mathrm{Log}(p_j)/\Sigma_{j=1,r}\, j p_j \tag{18}$$

The mathematical characterization by the population entropy defined in Equation (16) of the stochastic stability of the dynamics described by Equation (16) has its origin in the theory of large deviations [40–42]. This notion of stability pertains to the rate at which the system returns to its steady state after a random exogenous and/or endogenous perturbation and it could be useful to quantify further the variations of the distribution of the daily reproduction numbers observed for many countries [43–53].

In summary, the main limitations of the present study are:

- The hypothesis of spatio-temporal stationarity of the daily reproduction numbers is no longer valid in the case of rapid geo-climatic changes, such as sudden temperature rises, which decrease the virulence of SARS CoV-2 [4], or mutations affecting its transmissibility.
- The still approximate knowledge of the duration r of the period of contagiousness necessitates a more in-depth study at variable durations, by retaining the value of r, which makes all of the daily reproduction numbers positive.
- The choice of uniform random fluctuations of the daily reproduction numbers is based on arguments of simplicity. A more precise study would undoubtedly lead to a unimodal law varying throughout the contagious period, the average of which following a U-shaped curve, of the type observed in the literature on a few real patients [10,54–58].

5. Conclusions and Perspectives

Concerning contagious diseases, public health physicians are constantly faced with four challenges. The first concerns the estimation of the basic reproduction number R_0. The systematic use of R_0 simplifies the decision-making process by policymakers, advised by public health authorities, but it is too much of a caricature to account for the biology behind the viral spread. We have observed in the COVID-19 outbreak that it was non-constant during an epidemic wave due to exogenous and endogenous factors influencing both the duration of the contagiousness period and the daily transmission rate during this phase [54–56]. Then, the first challenge concerns the estimation of the mean duration of the infectious period for infected patients. As for the transmission rate, realistic assumptions made it possible to obtain an upper limit to this duration [45], mainly due to the lack of viral load data in large patient cohorts (see Figure A1 in Appendix A from [57–59]), in order to better guide the individual quarantine measures decided by the authorities in charge of public health. This upper bound also makes it possible to obtain a lower bound for the percentage of unreported infected patients, which gives an idea of the quality of the census of cases of infected patients, which is the second challenge facing specialists of contagious diseases. The third challenge is the estimation of the daily reproduction number over the contagiousness period, which was precisely the topic of the present paper. A fourth interesting challenge for this community is the extension of the methods developed in the present paper to the contagious non-infectious diseases (i.e., without causal infectious agent), such as social contagious diseases [59–61], the best example being that of the pandemic linked to obesity, for which many concepts and modelling methods remain available.

Eventually, our approach using marginal daily reproduction numbers involving a certain level of noise in the dynamics of new daily infected cases defines a stochastic framework which describes phenomenologically the exponential phase as our results show for countries such as France, Russia, Sweden, etc. This stochastic modelling allows a better understanding of the role of the contagiousness period length and of the heterogeneity (e.g., the U-shape) of its daily reproduction number distribution in the COVID-19 outbreak dynamics [62–65]. On the medical level, the important message about the U-shape is that COVID-19 is similar to other viral diseases, such as influenza, with two successive reactions from the two immune defense barriers, innate cellular immunity first, which is not sufficient if symptoms persist, then adaptive immunity (cellular and humoral), which results in a transient decrease in contagiousness between the two phases. The medical recommendations are, in this case, never to take a transient improvement for a permanent disappearance of the symptoms. One could indeed, for a public health use, be satisfied after estimating the sum of the R_j's, that is to say, R_0 or the effective R_e. For an individual health use, it is important to know the existence of a minimum of the R_j's, which generally corresponds to a temporary clinical improvement, after the partial success of the innate immune defenses. This makes it possible to prevent the patient from continuing to respect absolute isolation and therapeutic measures, even if a transient improvement occurs; otherwise, they risk, as in the flu, a bacterial pulmonary superinfection (a frequent cause of death in the case of COVID-19). On the theoretical level, the interest of the proposed method is its generic character: it can be applied to all contagious diseases, within the very general framework of Equation (1), which makes no assumption about the spatial heterogeneity or the longitudinal constancy of the daily reproduction numbers. The deconvolution of Equation (1) poses a new theoretical problem when it is offered in this context, and our future research will propose new avenues of research in this field.

Supplementary Materials: The following are available online at https://www.mdpi.com/article/10.3390/computation9100109/s1. Table S1. Presentation of the Pearson correlation coefficients between 20 numbers of world daily new cases observed between the days 34 to 53 after the 24 January 2020 (date of the start of the Covid-19 outbreak with confirmed cases in Europe) and series of 20 numbers of world daily new cases observed in running windows of length 20 days until day 213.

Author Contributions: Conceptualization, J.D. and J.W.; methodology, J.D., K.O., M.R., H.S. and J.W.; K.O. and F.T. have performed the calculations and Figures. All authors have equally participated to the other steps of the article elaboration. All authors have read and agreed to the published version of the manuscript.

Funding: This research received no external funding.

Institutional Review Board Statement: Not applicable.

Informed Consent Statement: Not applicable.

Data Availability Statement: Data are available on public databases https://renkulab.shinyapps.io/COVID-19-Epidemic-Forecasting/_w_e213563a/?tab=ecdc_pred&country=France, (accessed on 22 November 2020). and https://www.worldometers.info/coronavirus/, (accessed on 2 November 2020).

Acknowledgments: The authors hereby give their thanks to the framework of the University of Excellence Concept "Research University in Helmholtz Association I Living the Change".

Conflicts of Interest: The authors declare no conflict of interest.

Appendix A

Figure A1 shows a U-shaped evolution for the viral load in real [57] and in simulated [58] COVID-19 patients, and in real influenza-infected animals for the viral load and the body temperature [59].

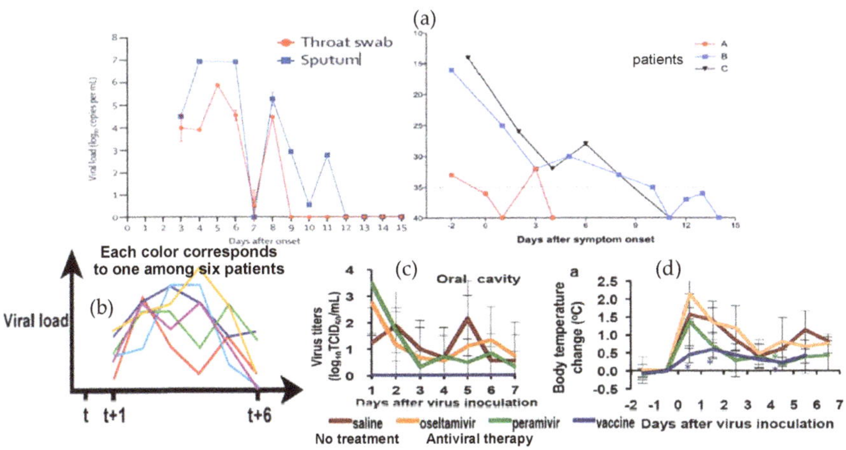

Figure A1. (**a**) Viral load in real COVID-19 patients [10], (**b**) in influenza-simulated patients [57] and (**c**) in real influenza-infected animals (red curve [58]), and (**d**) body temperature in real influenza-infected animals (red curve [58]).

Appendix B

1. Beginning of the pandemic in France from 21 February 2020 to 9 March 2020

   The numbers of new cases are:
   21 February 2, 4, 19, 18, 39, 27, 56, 20, 67, 126, 209, 269, 236, 185 **9 March**
   Then, the matrix M is defined by:

$$M = \begin{bmatrix} 236 & 269 & 209 & 126 & 67 & 20 & 56 \\ 269 & 209 & 126 & 67 & 20 & 56 & 27 \\ 209 & 126 & 67 & 20 & 56 & 27 & 39 \\ 126 & 67 & 20 & 56 & 27 & 39 & 18 \\ 67 & 20 & 56 & 27 & 39 & 18 & 19 \\ 20 & 56 & 27 & 39 & 18 & 19 & 4 \\ 56 & 27 & 39 & 18 & 19 & 4 & 2 \end{bmatrix}$$

and we have:

$$M^{-1} = \begin{bmatrix} -5.884 \times 10^{-5} & 5.399 \times 10^{-5} & -1.555 \times 10^{-4} & 7.241 \times 10^{-3} & -5.146 \times 10^{-3} & -1.255 \times 10^{-2} & -1.277 \times 10^{-2} \\ 5.399 \times 10^{-5} & -1.714 \times 10^{-4} & 7.324 \times 10^{-3} & -6.862 \times 10^{-3} & -1.139 \times 10^{-2} & 1560 \times 10^{-2} & -3.242 \times 10^{-3} \\ -1.555 \times 10^{-4} & 7.324 \times 10^{-3} & -6.862 \times 10^{-3} & -1.177 \times 10^{-2} & -1.592 \times 10^{-2} & -2.441 \times 10^{-3} & -4.780 \times 10^{-4} \\ 7.241 \times 10^{-3} & -6.862 \times 10^{-3} & -1.177 \times 10^{-2} & -2.164 \times 10^{-2} & -6.654 \times 10^{-3} & -1.0780 \times 10^{-2} & -9.514 \times 10^{-3} \\ -5.146 \times 10^{-3} & -1.139 \times 10^{-2} & 1.592 \times 10^{-2} & -6.654 \times 10^{-3} & -3.692 \times 10^{-3} & 2.797 \times 10^{-2} & 2.637 \times 10^{-2} \\ 1.255 \times 10^{-2} & 1.560 \times 10^{-2} & -2.441 \times 10^{-3} & -1.078 \times 10^{-2} & 2.797 \times 10^{-2} & 2.555 \times 10^{-2} & -3.125 \times 10^{-2} \\ 1.277 \times 10^{-2} & -3.242 \times 10^{-3} & -4.780 \times 10^{-4} & 9.514 \times 10^{-3} & 2.637 \times 10^{-2} & -3.125 \times 10^{-2} & -7.828 \times 10^{-4} \end{bmatrix}$$

Because, $X = \begin{bmatrix} 185 \\ 236 \\ 269 \\ 209 \\ 126 \\ 67 \\ 20 \end{bmatrix}$, hence $R = M^{-1}\,X = \begin{bmatrix} 0.239 \\ 0.052 \\ -0.783 \\ -0.295 \\ 1.189 \\ 3.060 \\ 3.122 \end{bmatrix}$ and we can represent the evolution of X_j's on Figure A2.

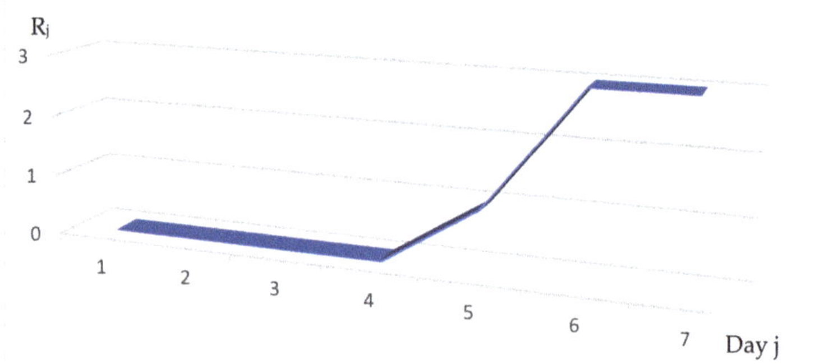

Figure A2. Values of the daily reproduction numbers R_j along the period of contagiousness of length 7 days.

The evolution of the Xj's along the period of contagiousness shows at day 4 a sharp increase and a saturation.

2. Exponential phase in France from 25 October 2020 to 7 November 2020

   The numbers of new cases are:

7 November 83,334, 58,581, 56,292, 39,880, 35,912, 51,104, 45,258, 33,447, 46,185, 44,705, 34,194, 31,360, 25,123, 48,808 25 **October**

Then, the matrix M is defined by:

$$M = \begin{bmatrix} 58,581 & 56,292 & 39,880 & 35,912 & 51,104 & 45,258 & 33,447 \\ 56,292 & 39,880 & 35,912 & 51,104 & 45,258 & 33,447 & 46,185 \\ 39,880 & 35,912 & 51,104 & 45,258 & 33,447 & 46,185 & 44,705 \\ 35,912 & 51,104 & 45,258 & 33,447 & 46,185 & 44,705 & 34,194 \\ 51,104 & 45,258 & 33,447 & 46,185 & 144,705 & 34,194 & 31,360 \\ 45,258 & 33,447 & 46,185 & 44,705 & 34,194 & 31,360 & 25,123 \\ 33,447 & 46,185 & 44,705 & 34,194 & 31,360 & 25,123 & 48,808 \end{bmatrix}$$

and we obtain

$$R = \begin{bmatrix} 2.867 \\ -1.231 \\ 1.351 \\ -2.705 \\ -0.155 \\ 0.223 \\ 0.769 \end{bmatrix}$$

The Figure A3 shows an evolution of the Xj's with a U-shape on the three first days along the period of contagiousness with a sum of R_j's equal to 1.11, close to the effective reproduction number R_e = 1.13 [28].

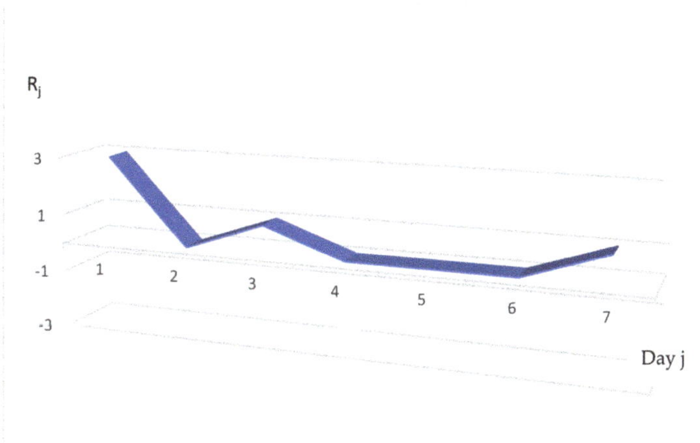

Figure A3. Values of the daily reproduction numbers R_j along the period of contagiousness of length 7 days.

3. Beginning of the pandemic in the USA from 21 February 2020 to 5 March 2020

The number of new cases are:

21 February 20, 0, 0, 18, 4, 3, 0, 3, 5, 7, 25, 24, 34, 63 **5 March**

Then, we have:

$$R = \begin{bmatrix} 0.466 \\ 0.584 \\ 1.547 \\ -1.044 \\ 0.174 \\ 0.297 \\ 0.692 \end{bmatrix}$$

The evolution of the Xj's shows in Figure A4 a U-shape on day 4 with a sum of R_j's equal to 2.72, less than the effective reproduction number R_e = 3.27 [28].

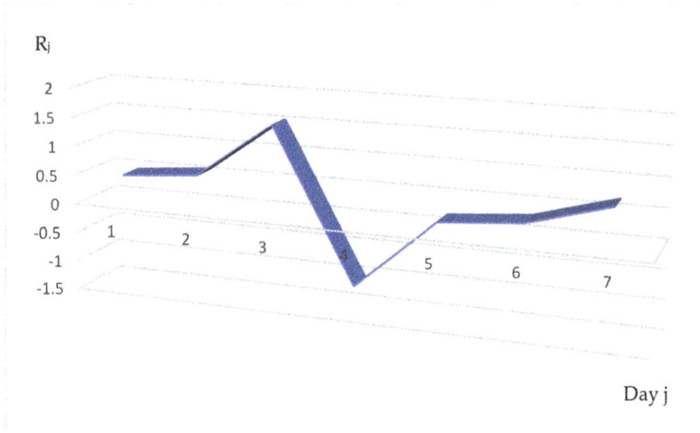

Figure A4. Values of the daily reproduction numbers R_j along the period of contagiousness of length 7 days.

4. USA exponential phase from 1 November 2020 to 4 November 2020

The numbers of new cases are:

N 14 163,961, 183,792, 167,665, 150,535, 159,565, 120,924, 108,248, 135,385, 136,292, 129,663, 113,709, 105,745, 86,030, 75,285 **N 1**

Then, we have:

$$R = \begin{bmatrix} 0.020 \\ -0.439 \\ 0.583 \\ -0.367 \\ 0.497 \\ -0.056 \\ 1.113 \end{bmatrix}$$

The evolution of the Xj's shows in Figure A5 a U-shape on the four last days with a sum of R_j's equal to 1.35, close to the effective reproduction number R_e = 1.24 [28].

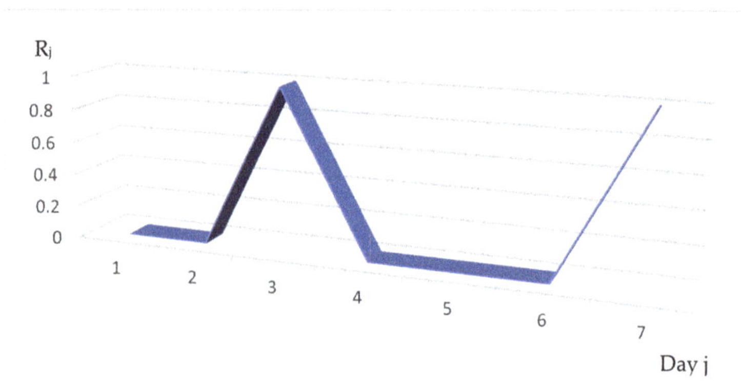

Figure A5. Values of the daily reproduction numbers R_j along the period of contagiousness of length 7 days.

5. Beginning of the pandemic in the UK from 23 February 2020 to 7 March 2020

The number of new cases are:
23 February 4, 0, 0, 0, 3, 4, 3, 12, 3, 11, 33, 26, 43, 41 **7 March**
Then, we have:

$$R = \begin{bmatrix} -0.388 \\ -1.189 \\ 1.334 \\ 1.960 \\ 4.862 \\ -0.170 \\ 3.479 \end{bmatrix}$$

Figure A6 shows an evolution of the Xj's with a U-shape on the three last days along the period of contagiousness with a sum of R_j's equal to 9.88, higher than the effective reproduction number $R_e = 2.95$ [28].

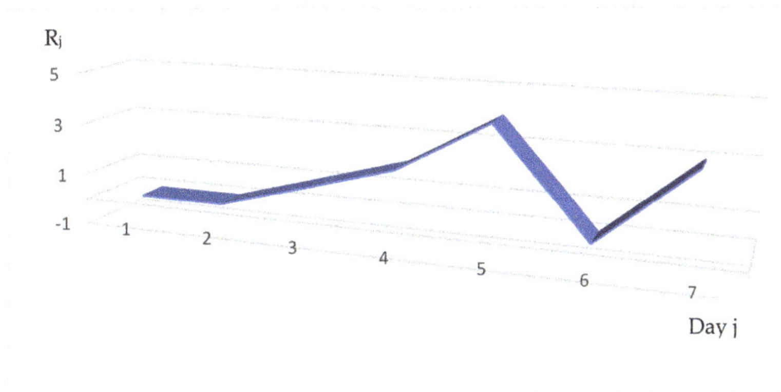

Figure A6. Values of the daily reproduction numbers R_j along the period of contagiousness of length 7 days.

6. UK exponential phase from 17 October 2020 to 30 October 2020

The numbers of new cases are:
30 October 24,350, 23,014, 24,646, 22,833, 20,843, 19,746, 22,961, 20,484, 21,195, 26,624, 21,282, 18,761, 16,943, 16,133 **17 October**
Then, we have:

$$R = \begin{bmatrix} 0.020 \\ 0.334 \\ 0.462 \\ -0.098 \\ -0.134 \\ -0.043 \\ 0.526 \end{bmatrix}$$

Figure A7 shows an evolution of the Xj's with a U-shape on the five last days along the period of contagiousness with a sum of R_j's equal to 1.07, close to the effective reproduction number $R_e = 1.06$ [28].

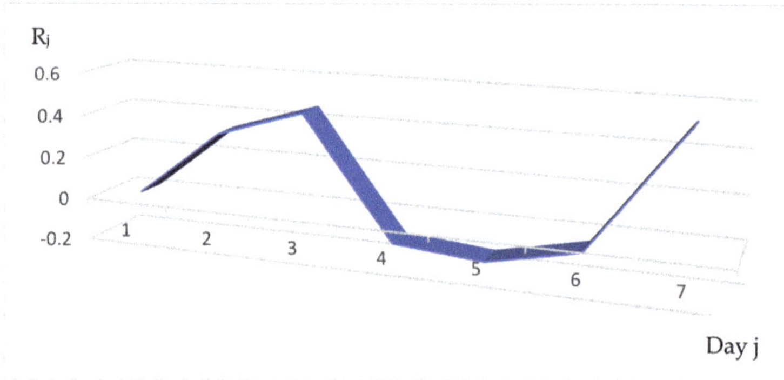

Figure A7. Values of the daily reproduction numbers R_j along the period of contagiousness of length 7 days.

Appendix C

Table A1 is built from new COVID-19 cases at the start of the first and second waves for 194 countries; it shows 42 among these 194 countries having a U-shape evolution of their daily R_j's twice, for 12.12 ± 6 expected with 0.95 confidence ($p < 10^{-12}$), and 189 times, a U-shape evolution for all countries and waves (397), for 99.3 ± 9 expected with 0.95 confidence ($p < 10^{-24}$). Hence, the U-shape is the most frequent evolution of daily R_j's, which confirms the comparison with the behavior of seasonal influenza (see Section 2.2).

Table A1. Calculation of the daily R_j's and shape of their distribution for 194 countries and for the two first waves.

No	All Countries Country Name	R_0	First Wave R_j's	U-Shape	R_0	Second Wave R_j's	U-Shape
1	AFGHANISTAN	0.65	0.17; 0.09; 0.39	YES	0.04	−1.38; −0.36; 1.78	INCR
2	ALGERIA	1.25	3.93; −6.21; 3.53	YES	0.91	1.28; −1.06; 0.69	YES
3	ARUBA	5.46	10.31; −39.32; 34.47	YES	1.10	1.54; −1.60; 1.16	YES
4	ANDORRA	1.36	1.00; 0.79; −0.43	DECR	0.12	4.34; −1.63; −2.59	DECR
5	ANGOLA	0.63	0.33; 1.42; −1.12	INV	1.70	9.22; −1.58; −5.94	DECR
6	ANTIGUA	1.92	0.00; 1.25; 0.67	INV	2.13	−0.40; 1.33; 1.20	INV
7	ALBANIA	0.96	0.48; 0.50; −0.02	INV	0.66	1.98; −0.56; −0.76	DECR
8	ARGENTINA	0.73	0.57; −1.28; 1.44	YES	0.36	1.27; 0.75; −1.66	DECR
9	ARMENIA	4.43	17.99; −36.99; 23.43	YES	0.86	1.41; −0.97; 0.42	YES
10	AUSTRALIA	2.79	−1.02; 3.47; 0.34	YES	1.50	−0.88; 0.68; 1.70	INCR
11	AUSTRIA	1.17	−1.78; −0.05; 3.00	INCR	2.08	0.62; −3.55; 5.01	YES
12	AZERBAIJAN	1.16	1.23; −1.32; 1.25	YES	0.37	10.36; −6.45; −3.54	YES
13	BAHAMAS	0.57	−0.13; −0.98; 1.68	YES	1.22	0.22; −0.86; 1.86	YES
14	BAHRAIN	1.10	−0.74; 0.28; 1.56	INCR	1.14	1.98; −2.69; 1.85	YES
15	BANGLADESH	1.04	2.37; −2.97; 1.64	YES	0.99	0.86; −0.69; 0.82	YES
16	BARBADOS	1.86	0.86; −0.64; 1.64	YES	1.14	0.22; −0.81; 1.73	YES
17	BELARUS	1.57	−2.37; −4.58; 8.52	YES	1.07	−0.33; 0.24; 1.16	INCR
18	BELGIUM	0.43	11.66; −15.63; 4.41	YES	2.23	1.17; −2.39; 3.45	YES
19	BELIZE	0.99	0.80; 0.42; −0.23	DECR	0.51	1.77; −0.21; −1.05	DECR
20	BENIN	0.85	0.81; 0.47; −0.43	DECR	0.85	1.17; 0.22; −0.54	DECR
21	BHUTAN	15.00	14.00; 15.00; −14.00	INV	1.08	0.80; 0.57; −0.29	DECR
22	BOLIVIA	2.17	8.47; −1.17; −5.13	DECR	1.61	0.96; −0.30; 0.95	YES
23	BOSNIA	0.09	−1.06; −1.05; 2.20	INCR	1.56	−0.57; −0.51; 2.64	INCR
24	BOTSWANA	28.47	0.22; 0.00; 28.25	YES	28.43	0.22; −0.05; 28.26	YES
25	BRAZIL	0.77	0.31; 1.08; −0.62	INV	0.46	1.21; 0.16; −0.91	DECR
26	BRUNEI	1.08	0.10; −0.15; 1.13	YES	1.00	1.00; −1.00; 1.00	YES
27	BULGARIA	5.06	14.73; −66.02; 56.35	YES	0.75	1.34; −0.98; 0.39	YES
28	BURKINA FASO	1.08	0.72; −0.34; 0.70	YES	0.94	0.31; 0.24; 0.39	YES
29	BURUNDI	1.33	1.33; −0.67; 0.67	YES	2.18	0.53; 1.80; −0.15	INV
30	CABO VERDE	0.82	−0.08; −0.26; 1.16	YES	0.19	0.56; 1.37; −1.74	INV
31	CAMBODIA	0.34	0.08; 0.25; 0.01	INV	0.27	0.06; 0.15; 0.06	INV
32	CAMEROON	2.17	2.36; 1.25; −1.44	DECR	2.48	0.50; −0.25; 2.23	YES
33	CANADA	1.10	−0.55; −0.72; 2.37	YES	0.44	2.36; −0.44; −1.48	DECR
34	CAR	1.66	−0.07; 0.64; 1.09	INCR	0.33	0.44; −0.22; 0.11	YES
35	CHAD	1.19	0.77; −1.15; 1.57	YES	0.77	1.19; 0.25; −0.67	DECR
36	CHILE	1.00	0.72; 0.17; 0.11	DECR	1.64	0.37; −4.45; 5.72	YES
37	CHINA	1.10	0.90; −0.49; 0.69	YES	0.87	1.16; 0.60; −0.89	DECR

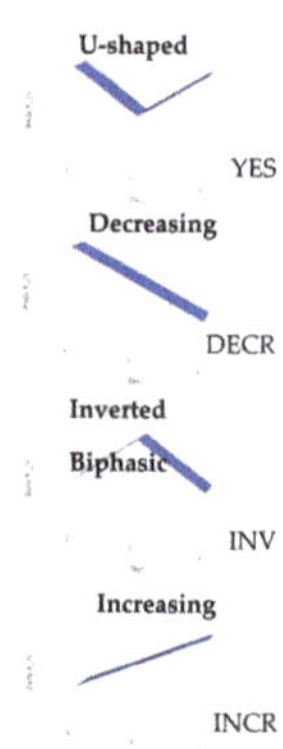

Table A1. *Cont.*

No	All Countries Country Name	First Wave R_0	First Wave R_j's	First Wave U-Shape	Second Wave R_0	Second Wave R_j's	Second Wave U-Shape
38	COLUMBIA	1.00	1.75; −0.86; 0.11	YES	1.47	−1.14; 3.08; −0.47	INV
39	COMOROS	3.75	0.00; −2.75; 6.5	YES	1.65	−0.58; 1.24; 0.99	INV
40	CONGO DEM	0.03	−0.37; −0.39; 0.79	YES	0.88	0.66; 0.74; −0.52	INV
41	CONGO REP	0.92	0.92; 0.92; −0.92	DECR	0.39	−0.12; 0.19; 0.32	INCR
42	COSTA RICA	0.50	−2.79; −3.84; 7.13	YES	1.26	1.21; −0.85; 0.90	YES
43	COTE D'VOIRE	1.18	−0.49; −0.63; 2.30	YES	2.09	4.32; −7.09; 4.86	YES
44	CROTIA	0.75	0.53; 0.79; −0.57	INV	0.57	0.68; −0.64; 0.53	YES
45	CUBA	0.48	−37.25; 16.17; 21.56	INCR	0.78	0.34; −0.73; 1.17	YES
46	CURACAO	0.50	3.00; −1.00; −1.50	DECR	4.19	1.93; −4.01; 6.27	YES
47	CYPRUS	0.69	0.27; 2.49; −2.07	INV	0.45	−0.42; 1.76; −0.89	INV
48	CZECH	0.16	−0.16; 3.88; −3.56	INV	0.88	1.88; −1.41; 0.41	YES
49	DENMARK	0.80	−0.11; 0.41; 0.50	INCR	0.64	−0.03; 4.65; −3.98	INV
50	DJIBOUTI	0.17	1.23; 0.24; −1.30	DECR	0.36	0.64; 0.41; −0.69	DECR
51	DOMINICAN	1.02	1.05; −0.31; 0.28	YES	1.57	0.32; −0.06; 1.31	YES
52	DOMINICA	7.75	2.00; −4.00; 9.75	YES	0.67	−0.36; 0.72; 0.31	INV
53	ECUADOR	1.46	−0.47; 1.06; 0.87	INV	1.14	0.73; −0.14; 0.55	YES
54	EGYPT	0.84	0.30; 0.37; 0.17	INV	0.51	11.99; −3.76; −7.72	DECR
55	EL SALVADOR	1.70	−0.20; 0.59; 1.31	INCR	0.66	−0.76; −14.49; 15.91	YES
56	EQUITORIAL G.	0.38	0.85; −0.20; −0.27	DECR	1.48	0.81; −0.66; 1.33	YES
57	ERITREA	1.18	1.44; −0.05; −0.21	DECR	0.80	1.02; 0.20; −0.42	DECR
58	ESTONIA	0.87	1.96; 0.82; −1.91	DECR	3.04	−0.70; −1.80; 5.54	YES
59	ESWATINI	0.94	1.41; −1.42; 0.95	YES	0.71	−0.02; 1.52; −0.79	INV
60	ETHIOPIA	0.80	−0.56; −1.45; 2.81	YES	1.24	0.34; 0.13; 0.77	YES
61	FIJI	2.00	0.00; 1.00; 1.00	INCR	0.50	0.75; −0.50; 0.25	YES
62	FINLAND	1.14	0.91; −0.42; 0.65	YES	2.41	0.56; −2.38; 4.23	YES
63	FRANCE	1.17	0.82; 0.10; 0.25	YES	2.17	0.88; −0.86; 2.15	YES
64	GABON	0.97	0.20; 0.47; 0.30	INV	0.19	−0.51; 0.00; 0.70	INCR
65	GAMBIA	0.83	−0.25; 0.43; 0.65	INCR	0.37	−0.38; 0.00; 0.75	INCR
66	GEORGIA	1.23	0.16; 0.43; 0.64	INCR	0.79	1.52; −0.49; −0.24	YES
67	GERMANY	0.73	0.15; −1.04; 1.62	YES	0.79	1.15; −0.56; 0.20	YES
68	GHANA	1.48	0.55; 0.70; 0.23	INV	0.62	0.13; −0.81; 1.30	YES
69	GREECE	0.71	0.33; −0.27; 0.65	YES	0.71	0.95; 0.28; −0.52	DECR
70	GRENADA	14.00	−5.00; 3.00; 16.00	INCR	0.10	−0.15; 0.00; 0.25	INCR
71	GUADELOUPE	1.35	0.00; 0.76; 0.59	INV	1.35	0.00; 0.76; 0.59	YES
72	GUATEMALA	0.25	2.01; −0.70; −1.06	YES	0.27	1.19; −0.11; −0.81	DECR
73	GUIANA FRENCH	0.88	1.30; −0.38; −0.04	YES	0.43	0.99; 0.27; −0.83	DECR
74	GUINEA	0.46	0.65; −0.56; 0.37	YES	1.68	0.21; 0.68; 0.79	INCR
75	GUINEA BISSAU	1.14	0.06; 1.59; −0.51	INV	4.20	−0.11; 0.04; 4.27	INCR
76	GUYANA	2.38	−3.45; −0.20; 6.03	INCR	4.23	−0.53; 0.58; 4.18	INCR
77	HAITI	0.60	0.30; −0.13; 0.43	YES	0.61	0.32; 0.42; −0.13	INV
78	HONDURAS	0.57	−2.94; 3.12; 0.39	INV	1.64	0.13; 0.54; 0.97	INCR
79	HONGKONG	0.04	0.95; −0.69; −0.22	YES	0.24	2.50; −8.79; 6.53	YES
80	HUNGARY	0.90	0.66; −0.12; 0.36	YES	1.93	1.91; −2.72; 2.74	YES
81	ICELAND	2.28	−0.85; 3.93; −0.80	INV	0.66	0.84; 0.22; −0.40	NO
82	INDIA	0.98	1.82; 0.53; −1.37	DECR	0.96	1.08; −0.57; 0.45	YES
83	INDONESIA	0.95	0.67; 0.88; −0.60	INV	0.99	1.06; −0.03; −0.03	YES
84	IRAN	1.04	1.73; −0.67; −0.02	YES	0.90	6.62; −6.62; 0.90	YES
85	IRAQ	0.77	0.15; −0.35; 0.96	YES	0.96	0.77; −0.40; 0.59	YES
86	IRELAND	2.16	−2.83; −5.64; 10.63	YES	1.12	1.12; −0.39; 0.39	YES
87	ISRAEL	0.21	−1.39; 1.08; 0.52	INV	1.16	−0.16; 0.44; 0.88	INCR
88	ITALY	1.04	2.24; −1.85; 0.65	YES	3.69	1.65; −7.89; 9.93	YES
89	JAMAICA	0.43	0.13; 0.06; 0.24	YES	2.47	−0.34; 2.06; 0.75	INV
90	JAPAN	1.02	0.69; 0.88; −0.55	INV	1.16	0.61; 0.42; 0.13	DECR
91	JORDAN	2.53	10.82; −18.20; 9.91	YES	0.93	1.28; 0.57; −0.92	DECR
92	KAZAKHSTAN	0.60	0.53; −5.45; 5.52	YES	2.06	−0.05; 2.37; −1.26	INV
93	KENYA	1.14	0.05; 0.65; 0.44	INV	1.18	0.47; 1.34; −0.63	INV
94	KOREA REP.	1.00	0.12; 0.87; 0.01	INV	1.04	0.60; −0.03; 0.47	YES
95	KOSOVO	1.02	1.00; 1.02; −1.00	INV	0.99	1.31; −0.29; −0.03	YES
96	KUWAIT	0.88	0.5; −0.34; 0.67	YES	1.10	0.58; −0.84; 1.36	YES
97	KYRGYZSTAN	0.17	−0.73; 0.26; 1.64	INCR	1.05	0.28; −0.32; 1.09	YES
98	LAO PDR	0.50	0.50; 0.50; −0.50	DECR	0.15	0.33; 0.74; −0.92	INV
99	LATVIA	0.74	1.97; −0.76; −0.47	YES	0.50	0.40; −0.22; 0.32	YES
100	LEBANON	1.03	0.57; 0.12; 0.34	YES	0.90	0.23; 0.06; 0.61	YES
101	LESOTHO	7.08	−2.86; 7.22; 2.72	INV	1.42	0.37; 1.51; −0.46	INV
102	LIBERIA	0.31	0.18; −0.04; 0.17	YES	4.56	0.14; 4.61; −0.19	INV
103	LIBYA	0.96	0.19; −0.71; 1.48	YES	0.79	−0.42; 0.56; 0.65	INCR
104	LITHUANIA	0.83	0.56; 0.11; 0.16	YES	2.49	−0.90; −0.52; 3.91	INCR
105	LUXEMBOURG	0.24	−8.55; −3.75; 12.54	INCR	1.48	1.16; −0.91; 1.23	YES

Table A1. *Cont.*

No	All Countries Country Name	R_0	First Wave R_j's	U-Shape	R_0	Second Wave R_j's	U-Shape
106	MACAO	0.29	1.14; 2.29; −3.14	INV	-	-	-
107	MADAGASCAR	0.94	0.61; −0.16; 0.49	YES	0.75	0.38; −1.54; 1.91	YES
108	MALAWI	1.12	−0.23; 0.53; 0.82	INCR	6.46	−0.41; 0.99; 5.88	INCR
109	MALAYSIA	1.25	0.38; 2.79; −1.92	INV	1.30	−0.57; 1.82; 0.05	INV
110	MALDIVES	0.83	0.60; −0.53; 0.76	YES	1.05	−0.27; 0.70; 0.62	INV
111	MALI	0.64	0.59; 0.42; −0.37	DECR	7.78	−2.64; −4.96; 15.38	YES
112	MALTA	1.06	1.15; 0.24; −0.33	DECR	0.99	−0.73; 1.81; −0.09	INV
113	MAURITANIA	1.76	−0.94; 0.29; 2.41	INCR	1.14	0.73; −0.41; 0.82	YES
114	MAURITIUS	4.49	−4.05; 0.36; 8.18	INCR	0.35	1.41; 0.53; −1.59	DECR
115	MAYOTTE	5.46	−9.46; −2.50; 17.42	INCR	1.05	0.72; −0.17; 0.50	YES
116	MEXICO	0.86	−1.39; 3.07; −0.82	INV	2.53	−0.55; 0.10; 2.98	INCR
117	MOLDOVA	1.03	2.73; −0.67; −1.03	DECR	0.36	1.27; 0.66; −1.57	DECR
118	MONACO	3.15	0.52; −1.93; 4.56	YES	0.54	1.02; −0.12; −0.36	DECR
119	MONGOLIA	10.25	1.25; 19.25; −10.25	INV	0.68	0.91; 0.25; −0.48	DECR
120	MONTENEGRO	1.37	2.94; −3.90; 2.33	YES	0.66	2.36; 0.26; −1.96	DECR
121	MOROCCO	0.90	0.36; 1.41; −0.87	INV	0.95	0.95; −0.15; 0.15	YES
122	MOZAMBIQUE	0.72	0.92; 0.001; −0.20	DECR	0.70	2.46; −2.45; 0.69	YES
123	MYANMAR	1.12	−0.75; 1.07; 0.80	INV	1.15	−1.36; −2.17; 4.68	YES
124	NAMIBIA	0.68	1.37; −1.82; 1.13	YES	1.22	−0.26; 0.95; 0.53	INV
125	NEPAL	0.74	0.35; 0.76; −0.37	INV	0.78	0.11; 0.58; 0.09	INV
126	NETHERLAND	1.19	0.11; 0.11; 0.97	YES	1.04	1.05; −0.99; 0.98	YES
127	NEW CALEDONIA	5.00	−2.00; 2.00; 5.00	YES	1.00	1.00; −1.00; 1.00	YES
128	NEW ZEALAND	0.74	2.30; −3.40; 1.84	YES	0.72	−0.52; 0.43; 0.81	INCR
129	NICARAGUA	0.97	−0.03; 0.97; 0.03	INV	1.02	0.86; 0.14; 0.02	DECR
130	NIGER	0.63	0.28; −0.12; 0.47	YES	2.21	−0.14; 0.39; 1.96	INCR
131	NIGERIA	1.13	0.16; 0.39; 0.58	INCR	1.02	1.38; −0.65; 0.29	YES
132	MACEDONIA	0.74	1.83; −1.16; 0.07	YES	0.74	1.26; −0.10; −0.42	DECR
133	NORWAY	0.77	−0.19; −0.61; 1.57	YES	2.13	6.02; −10.80; 6.91	YES
134	OMAN	3.70	0.39; 0.12; 3.19	YES	9.80	−16.87; 39.41; −12.74	INV
135	PAKISTAN	1.22	−0.61; 1.07; 0.76	INV	1.19	0.55; −0.11; 0.75	YES
136	PALESTINE	0.96	−0.18; −0.23; 1.37	YES	1.06	−0.21; 0.18; 1.09	INCR
137	PANAMA	0.96	0.16; 0.56; 0.24	INV	0.79	1.22; −0.16; −0.27	DECR
138	PAPAU NEW G.	0.49	0.35; −1.96; 2.10	YES	0.88	−0.39; 0.04; 1.23	INCR
139	PARAGUAY	0.59	−1.52; 1.90; 0.21	INV	1.20	−3.20;3.06; 1.34	INV
140	PERU	0.89	8.30; −2.47; −4.94	DECR	0.53	3.98; −4.72; 1.27	YES
141	PHILLIPPINES	1.15	0.89; −0.08; 0.34	YES	1.54	0.07; 2.84; −1.37	INV
142	POLAND	0.92	2.32; −1.89; 0.49	YES	1.31	1.71; −1.63; 1.23	YES
143	POLYNESIA	0.66	0.22; 0.20; 0.24	YES	0.21	−1.05; 1.09; 0.17	INV
144	PORTUGAL	1.56	−1.34; −8.29; 11.19	YES	3.89	1.13; −4.00; 6.76	YES
145	QATAR	0.80	−0.84; −1.99; 3.63	YES	1.03	0.62; 0.61; −0.20	INV
146	ROMANIA	0.88	0.90; 0.06; −0.08	DECR	0.95	1.23; −0.48; 0.20	YES
147	RUSSIA	1.07	1.16; −1.00; 0.91	YES	0.87	0.83; −5.77; 5.81	YES
148	RWANDA	1.80	3.20; 2.20; −3.60	DECR	0.14	3.93; −2.75; −1.04	YES
149	SAO TOME	1.44	0.44; 0.64; 0.36	INV	2.67	2.25; −3.45; 3.87	YES
150	SAN MARINO	5.10	0.28; 1.14;3.68	INCR	0.26	−0.05; 2.32; −2.01	INV
151	SAUDI ARABIA	0.90	−1.70; 2.94; −0.34	INV	0.98	−1.05; 0.54; 1.49	INCR
152	SENEGAL	0.72	−0.19; 1.48; −0.57	INV	1.59	0.73; 0.23; 0.63	YES
153	SERBIA	1.62	−0.40; 0.47; 1.55	INCR	0.82	2.02; −0.94; −0.26	YES
154	SEYCHELLES	0.48	0.30; 0.51; −0.33	INV	0.54	0.38; −0.19; 0.35	YES
155	SIERRA LEONE	2.23	−2.93; −0.80; 5.96	INCR	1.37	0.95; −1.25; 1.67	YES
156	SINGAPORE	1.33	1.15; 0.51; −0.33	DECR	2.83	1.61; −2.44; 3.66	YES
157	SLOVAK	0.99	−2.67; 1.90; 1.76	INV	0.74	0.97; −0.73; 0.50	YES
158	SLOVENIA	0.75	1.56; −0.71; −0.10	DECR	0.64	1.47; −0.47; −0.36	YES
159	SOMALIA	1.18	−0.16; 1.51; −0.17	INV	0.29	0.86; 0.57; −1.14	DECR
160	SOUTH AFRICA	0.87	0.22; 0.73; −0.08	INV	1.49	0.20; −0.04; 1.33	YES
161	SOUTH SUDAN	0.58	0.10; 0.16; 0.32	INCR	1.72	0.63; −0.63; 1.72	YES
162	SPAIN	0.38	−0.18; 0.27; 0.29	INCR	0.51	1.21; −0.86; 0.16	YES
163	SRI LANKA	2.13	2.73; −0.75; 0.15	YES	0.79	0.42; 1.00; −0.63	INV
164	ST KITTS NEVIS	2.00	0.00; 1.00; 1.00	INCR	1.07	0.25; 0.18; 0.64	YES
165	ST LUCIA	1.13	−0.53; −0.04; 1.70	INCR	1.00	1.00; −1.00; 1.00	YES
166	ST VINCENT	0.04	−0.29; 0.24; 0.10	INV	0.69	−0.24; 0.35; 0.58	INCR
167	SUDAN	0.36	−1.46; 2.34; −0.52	INV	2.00	0.00; 2.00; 0.00	INV
168	SURINAME	10.34	2.70; 18.77; −11.13	INV	1.63	2.95; −1.25; −0.07	YES
169	SWEDEN	0.56	0.58; −1.20; 1.18	YES	1.21	0.67; −0.91; 1.45	YES
170	SWITZERLAND	1.21	1.25; 0.13; −0.17	DECR	0.28	0.89; 1.18; −1.79	INV
171	SYRIA	1.43	1.39; 4.13; −4.09	INV	0.18	0.31; −0.68; 0.55	YES
172	TAIWAN	1.88	−0.13; 1.38; 0.63	INV	0.66	−5.21; 13.83; −7.96	INV
173	TAJIKISTAN	1.02	0.71; −0.60; 0.91	YES	1.49	1.83; −0.17; −0.17	YES
174	TANZANIA	0.91	−1.50; 0.18; 2.23	INCR	1.89	3.42; 14.26; −15.79	INV

Table A1. *Cont.*

No	All Countries Country Name	First Wave R_0	R_j's	U-Shape	Second Wave R_0	R_j's	U-Shape
175	THAILAND	0.69	0.42; 0.07; 0.20	YES	2.71	−1.77; −0.75; 5.23	INCR
176	TIMOR LESTE	5.00	1.00; 0.00; 4.00	YES	1.33	0.00; 1.00; 0.33	INV
177	TOGO	0.08	6.05; −6.18; 0.21	YES	1.14	0.18; 0.09; 0.87	YES
178	TRINIDAD	0.32	−0.26; 1.46; −0.88	INV	0.55	0.26; 0.03; 0.26	YES
179	TUNISIA	1.53	0.77; −0.04; 0.80	YES	2.77	−3.21; −2.41; 8.39	INCR
180	TURKEY	1.15	−1.50; −1.13; 3.78	INCR	2.21	19.82; −47.90; 30.29	YES
181	UAE	0.97	2.07; −1.11; 0.01	YES	1.15	1.25; −0.64; 0.54	YES
182	UGANDA	0.95	0.87; −0.37; 0.45	YES	0.64	0.44; −0.06; 0.26	YES
183	UKRAINE	0.96	1.35; −1.04; 0.65	YES	0.30	3.10; 1.07; −1.73	DECR
184	UK	0.76	−0.02; −0.76; 1.54	YES	1.03	0.43; 0.82; −0.22	INV
185	USA	8.42	31.42; −99.18; 76.18	YES	0.49	3.32; −0.38; −2.45	DECR
186	URUGUAY	0.63	0.71; 0.31; −0.39	DECR	1.03	−0.23; 0.35; 0.91	INCR
187	UZBEKISTAN	0.95	0.04; 0.10; 0.81	INCR	0.90	−0.03; −0.39; 1.32	YES
188	VENEZUELA	1.54	1.65; 2.95; −3.06	INV	0.82	1.09; −2.53; 2.26	YES
189	VIETNAM	3.29	−0.84; −0.39; 4.52	YES	1.43	0.76; −0.11; 0.78	YES
190	VIRGIN ISLANDS	0.51	0.01; −0.06; 0.56	YES	0.33	0.44; −0.22; 0.11	YES
191	WEST GAZA	1.00	−1.00; −2.00; 4.00	YES	0.98	0.59; −0.11; 0.50	YES
192	YEMEN	0.70	−0.34; 0.17; 0.86	INCR	1.50	1.00; 0.00; 0.50	YES
193	ZAMBIA	0.75	0.25; −0.13; 0.63	YES	1.12	1.11; −0.44; 0.45	YES
194	ZIMBABWE	1.44	0.24; 0.60; 0.60	INCR	1.62	1.08; −1.12; 1.66	YES

References

1. Yu, I.T.S.; Li, Y.; Wong, T.W.; Tam, W.; Chan, A.T.; Lee, J.H.W.; Leung, D.Y.C.; Ho, T. Evidence of airborne transmission of the severe acute respiratory syndrome virus. *N. Engl. J. Med.* **2004**, *350*, 1731–1739. [CrossRef]
2. Assiri, A.; McGeer, A.; Perl, T.M.; Price, C.S.; Al Rabeeah, A.A.; Cummings, D.A.T.; Alabdullatif, Z.N.; Assad, M.; Almulhim, A.; Makhdoom, H.; et al. Hospital Outbreak of Middle East Respiratory Syndrome Coronavirus. *N. Engl. J. Med.* **2013**, *369*, 407–416. [CrossRef] [PubMed]
3. Gaunt, E.R.; Hardie, A.; Claas, E.C.J.; Simmonds, P. Epidemiology and Clinical Presentations of the Four Human Coronaviruses 229E, HKU1, NL63, and OC43 Detected over 3 Years Using a Novel Multiplex Real-Time PCR Method. *J. Clin. Microbiol.* **2010**, *48*, 2940–2947. [CrossRef]
4. Demongeot, J.; Flet-Berliac, Y.; Seligmann, H. Temperature decreases spread parameters of the new covid-19 cases dynamics. *Biology* **2020**, *9*, 94. [CrossRef] [PubMed]
5. Ahmed, H.M.; Elbarkouky, R.A.; Omar, O.A.M.; Ragusa, M.A. Models for COVID-19 Daily confirmed cases in different countries. *Mathematics* **2021**, *9*, 659. [CrossRef]
6. Barlow, J.; Vodenska, I. Socio-Economic Impact of the Covid-19 Pandemic in the US. *Entropy* **2021**, *23*, 673. [CrossRef] [PubMed]
7. Seligmann, H.; Iggui, S.; Rachdi, M.; Vuillerme, N.; Demongeot, J. Inverted covariate effects for mutated 2nd vs 1st wave Covid-19: High temperature spread biased for young. *Biology* **2020**, *9*, 226. [CrossRef]
8. Seligmann, H.; Vuillerme, N.; Demongeot, J. Summer COVID-19 Third Wave: Faster High Altitude Spread Suggests High UV Adaptation. Available online: https://www.medrxiv.org/content/10.1101/2020.08.17.20176628v1 (accessed on 22 September 2021).
9. Carrat, F.; Vergu, E.; Ferguson, N.M.; Lemaitre, M.; Cauchemez, S.; Leach, S.; Valleron, A.J. Time Lines of Infection and Disease in Human Influenza: A Review of Volunteer. *Am. J. Epidemiol.* **2008**, *167*, 775–785. [CrossRef]
10. Pan, Y.; Zhang, D.; Yang, P.; Poon, L.L.M.; Wang, Q. Viral load of SARS-CoV-2 in clinical samples. *Lancet Infect. Dis.* **2020**, *20*, 411–412. [CrossRef]
11. Wölfel, R.; Corman, V.M.; Guggemos, W.; Seilmaier, M.; Zange, S.; Müller, M.A.; Niemeyer, D.; Jones, T.C.; Vollma, P.; Rothe, C.; et al. Virological assessment of hospitalized patients with COVID-2019. *Nature* **2020**, *581*, 465–469. [CrossRef]
12. Liu, W.D.; Chang, S.Y.; Wang, J.T.; Tsai, M.J.; Hung, C.C.; Hsu, C.L.; Chang, S.C. Prolonged virus shedding even after seroconversion in a patient with COVID-19. *J. Infect.* **2020**, *81*, 318–356. [CrossRef]
13. Ferretti, L.; Wymant, C.; Kendall, M.; Zhao, L.; Nurtay, A.; Abeler-Dörner, L.; Parker, M.; Bonsall, D.; Fraser, C. Quantifying SARS-CoV-2 transmission suggests epidemic control with digital contact tracing. *Science* **2020**, *368*, eabb6936. [CrossRef]
14. Cheng, H.Y.; Jian, S.W.; Liu, D.P.; Ng, T.C.; Huang, W.T.; Lin, H.H. Contact Tracing Assessment of COVID-19 Transmission Dynamics in Taiwan and Risk at Different Exposure Periods Before and After Symptom Onset. *JAMA Intern. Med.* **2020**, *180*, 1156–1163. [CrossRef]
15. He, X.; Lau, E.H.Y.; Wu, P.; Deng, X.; Wang, J.; Hao, X.; Lau, Y.C.; Wong, J.Y.; Guan, Y., Tan, X.; et al. Temporal dynamics in viral shedding and transmissibility of COVID-19. *Nat. Med.* **2020**, *26*, 672–675. [CrossRef] [PubMed]
16. Lacoude, P. Covid-19: Le Début de la Fin? Available online: https://www.contrepoints.org/2020/07/22/376624-covid-19-lx10-debut-dx10-la-fin-1 (accessed on 22 November 2020).

17. Delbrück, M. Statistical fluctuations in autocatalytic reactions. *J. Chem. Phys.* **1940**, *8*, 120–124. [CrossRef]
18. De Jesús Rubio, J. Stability Analysis of the Modified Levenberg-Marquardt Algorithm for the Artificial Neural Network Training. *IEEE Trans. Neural Netw. Learn. Syst.* **2021**, *32*, 3510–3524. [CrossRef] [PubMed]
19. de Jesús Rubio, J. Adapting H-Infinity Controller for the Desired Reference Tracking of the Sphere Position in the Maglev Process. *Inf. Sci.* **2021**, *569*, 669–686. [CrossRef]
20. Chiang, H.S.; Chen, M.Y.; Huang, Y.J. Wavelet-Based EEG Processing for Epilepsy Detection Using Fuzzy Entropy and Associative Petri Net. *IEEE Access* **2019**, *7*, 103255–103262. [CrossRef]
21. De Jesús Rubio, J.; Pan, Y.; Pieper, J.; Chen, M.Y.; Sossa Azuela, J.H. Advances in Robots Trajectories Learning via Fast Neural Networks. *Front. Neurorobot.* **2021**, *15*, 1–3.
22. Vargas, D.M. Superpixels extraction by an Intuitionistic fuzzy clustering algorithm. *JART* **2021**, *19*, 140–152. [CrossRef]
23. Soriano, L.A.; Zamora, E.; Vazquez-Nicolas, J.M.; Hernandez, G.; Barraza Madrigal, J.A.; Balderas, D. PD Control Compensation Based on a Cascade Neural Network Applied to a Robot Manipulator. *Front. Neurorobot.* **2020**, *14*, 577749. [CrossRef] [PubMed]
24. Demetrius, L. Boltzmann, Darwin and the directionality theory. *Phys. Rep.* **2013**, *530*, 1–86. [CrossRef]
25. Rhodes, C.J.; Demetrius, L. Evolutionary Entropy Determines Invasion Success in Emergent Epidemics. *PLoS ONE* **2010**, *5*, e12951. [CrossRef] [PubMed]
26. Demongeot, J.; Demetrius, L. Complexity and Stability in Biological Systems. *Int. J. Bifurc. Chaos* **2015**, *25*, 40013. [CrossRef]
27. Garcia, N. Birth and death processes as projections of higher-dimensional Poisson processes. *Adv. Appl. Probab.* **1995**, *4*, 911–930. [CrossRef]
28. Renkulab. COVID-19 Daily Epidemic Forecasting. Available online: https://renkulab.shinyapps.io/COVID-19-Epidemic-Forecasting/_w_e213563a/?tab=ecdc_pred&%20country=France (accessed on 22 November 2020).
29. Scire, J.; Nadeau, S.A.; Vaughan, T.; Gavin, B.; Fuchs, S.; Sommer, J.; Koch, K.N.; Misteli, R.; Mundorff, L.; Götz, T.; et al. Reproductive number of the COVID-19 epidemic in Switzerland with a focus on the Cantons of Basel-Stadt and Basel-Landschaft. *Swiss Med. Wkly.* **2020**, *150*, w20271. [CrossRef]
30. Worldometer. Reported Cases and Deaths by Country or Territory. Available online: https://www.worldometers.info/coronavirus/ (accessed on 2 November 2020).
31. DW. Coronavirus: Russia Tightens Restrictions, but Avoids Lockdown. 2021. Available online: https://www.dw.com/en/coronavirus-russia-restrictions-pandemic-lockdown/a-55301714 (accessed on 14 October 2021).
32. Seligmann, H.; Vuillerme, N.; Demongeot, J. Unpredictable, Counter-Intuitive Geoclimatic and Demographic Correlations of COVID-19 Spread Rates. *Biology* **2021**, *10*, 623. [CrossRef]
33. Breban, R.; Vardavas, R.; Blower, S. Theory versus data: How to calculate R_0? *PLoS ONE* **2007**, *2*, e282. [CrossRef]
34. Demetrius, L. Demographic parameters and natural selection. *Proc. Natl. Acad. Sci. USA* **1974**, *71*, 4645–4647. [CrossRef]
35. Demetrius, L. Statistical mechanics and population biology. *J. Stat. Phys.* **1983**, *30*, 709–750. [CrossRef]
36. Demongeot, J.; Demetrius, L. La dérive démographique et la sélection naturelle: Etude empirique de la France (1850–1965). *Population* **1989**, *2*, 231–248.
37. Demongeot, J. Biological boundaries and biological age. *Acta Biotheor.* **2009**, *57*, 397–419. [CrossRef]
38. Gaudart, J.; Ghassani, M.; Mintsa, J.; Rachdi, M.; Waku, J.; Demongeot, J. Demography and Diffusion in epidemics: Malaria and Black Death spread. *Acta Biotheor.* **2010**, *58*, 277–305. [CrossRef]
39. Demongeot, J.; Hansen, O.; Hessami, H.; Jannot, A.S.; Mintsa, J.; Rachdi, M.; Taramasco, C. Random modelling of contagious diseases. *Acta Biotheor.* **2013**, *61*, 141–172. [CrossRef]
40. Wentzell, A.D.; Freidlin, M.I. On small random perturbations of dynamical systems. *Russ. Math. Surv.* **1970**, *25*, 1–55.
41. Donsker, M.D.; Varadhan, S.R.S. Asymptotic evaluation of certain Markov process expectations for large time. I. *Comm. Pure Appl. Math.* **1975**, *28*, 1–47. [CrossRef]
42. Freidlin, M.I.; Wentzell, A.D. *Random Perturbations of Dynamical Systems*; Springer: New York, NY, USA, 1984.
43. Scarpino, S.V.; Petri, G. On the predictability of infectious disease outbreaks. *Nat. Commun.* **2019**, *10*, 898. [CrossRef] [PubMed]
44. Liu, Z.; Magal, P.; Seydi, O.; Webb, G. Understanding Unreported Cases in the covid-19 Epidemic Outbreak in Wuhan, China, and Importance of Major Public Health Interventions. *Biology* **2020**, *9*, 50. [CrossRef] [PubMed]
45. Demongeot, J.; Griette, Q.; Magal, P. Computations of the transmission rates in SI epidemic model applied to COVID-19 data in mainland China. *R. Soc. Open Sci.* **2020**, *7*, 201878. [CrossRef]
46. Adam, D.C.; Wu, P.; Wong, J.Y.; Lau, E.H.Y.; Tsang, T.K.; Cauchemez, S.; Leung, G.M.; Cowling, B.J. Clustering and superspreading potential of SARS-CoV-2 infections in Hong Kong. *Nat. Med.* **2020**, *26*, 1714–1719. [CrossRef]
47. Nishiura, H.; Lintona, N.M.; Akhmetzhanov, A.R. Serial interval of novel coronavirus (COVID-19) infections. *Int. J. Infect. Dis.* **2020**, *93*, 284–286. [CrossRef] [PubMed]
48. Gaudart, J.; Landier, J.; Huiart, L.; Legendre, E.; Lehot, L.; Bendiane, M.K.; Chiche, L.; Petitjean, A.; Mosnier, E.; Kirakoya-Samadoulougou, F.; et al. Factors associated with spatial heterogeneity of Covid-19 in France: A nationwide ecological study. *Lancet Public Health* **2021**, *6*, e222–e231. [CrossRef]
49. Bakhta, A.; Boiveau, T.; Maday, Y.; Mula, O. Epidemiological Forecasting with Model Reduction of Compartmental Models. Application to the COVID-19 Pandemic. *Biology* **2021**, *10*, 22. [CrossRef]
50. Roques, L.; Bonnefon, O.; Baudrot, V.; Soubeyrand, S.; Berestycki, H. A parsimonious approach for spatial transmissionand heterogeneity in the COVID-19 propagation. *R. Soc. Open Sci.* **2020**, *7*, 201382. [CrossRef]

51. Griette, Q.; Demongeot, J.; Magal, P. A robust phenomenological approach to investigate COVID-19 data for France. *Math. Appl. Sci. Eng.* **2021**, *3*, 149–160.
52. Griette, Q.; Magal, P. Clarifying predictions for COVID-19 from testing data: The example of New-York State. *Infect. Dis. Model.* **2021**, *6*, 273–283.
53. Oshinubi, K.; Rachdi, M.; Demongeot, J. Analysis of Daily Reproduction Rates of COVID-19 Using Current Health Expenditure as Gross Domestic Product Percentage (CHE/GDP) across Countries. Available online: https://www.medrxiv.org/content/10.1101/2021.08.27.21262737v1 (accessed on 22 September 2021).
54. Kawasuji, H.; Takegoshi, Y.; Kaneda, M.; Ueno, A.; Miyajima, Y.; Kawago, K.; Fukui, Y.; Yoshida, Y.; Kimura, M.; Yamada, H.; et al. Transmissibility of COVID-19 depends on the viral load around onset in adult and symptomatic patients. *PLoS ONE* **2020**, *15*, e0243597. [CrossRef]
55. Kim, S.E.; Jeong, H.S.; Yu, Y.; Shin, S.U.; Kim, S.; Oh, T.H.; Kim, U.J.; Kang, S.J.; Jang, H.C.; Jung, S.I.; et al. Viral kinetics of SARS-CoV-2 in asymptomatic carriers and presymptomatic patients. *Int. J. Infect. Dis.* **2020**, *95*, 441–443. [CrossRef]
56. Murphy, B.R.; Rennels, M.B.; Douglas, R.G., Jr.; Betts, R.F.; Couch, R.B.; Cate, T.R., Jr.; Chanock, R.M.; Kendal, A.P.; Maassab, H.F.; Suwanagool, S.; et al. Evaluation of influenza A/Hong Kong/123/77 (H1N1) ts-1A2 and cold-adapted recombinant viruses in seronegative adult volunteers. *Infect. Immun.* **1980**, *29*, 348–355. [CrossRef]
57. Chao, D.L.; Halloran, M.E.; Obenchain, V.J.; Longini, I.M., Jr. FluTE, a Publicly Available Stochastic Influenza Epidemic Simulation Model. *PLoS Comput. Biol.* **2010**, *6*, e1000656. [CrossRef] [PubMed]
58. Itoh, Y.; Shichinohe, S.; Nakayama, M.; Igarashi, M.; Ishii, A.; Ishigaki, H.; Ishida, H.; Kitagawa, N.; Sasamura, T.; Shiohara, M.; et al. Emergence of H7N9 Influenza A Virus Resistant to Neuraminidase Inhibitors in Nonhuman Primates. *Antimicrob. Agents Chemother.* **2015**, *59*, 4962–4973. [CrossRef]
59. Demongeot, J.; Taramasco, C. Evolution of social networks: The example of obesity. *Biogerontology* **2014**, *15*, 611–626. [CrossRef] [PubMed]
60. Demongeot, J.; Hansen, O.; Taramasco, C. Complex systems and contagious social diseases: Example of obesity. *Virulence* **2015**, *7*, 129–140. [CrossRef]
61. Demongeot, J.; Jelassi, M.; Taramasco, C. From Susceptibility to Frailty in social networks: The case of obesity. *Math. Pop. Stud.* **2017**, *24*, 219–245. [CrossRef]
62. Oshinubi, K.; Rachdi, M.; Demongeot, J. Analysis of reproduction number R_0 of COVID-19 using Current Health Expenditure as Gross Domestic Product percentage (CHE/GDP) across countries. *Healthcare* **2021**, *9*, 1247. [CrossRef]
63. Oshinubi, K.; Rachdi, M.; Demongeot, J. Functional Data Analysis: Transition from Daily Observation of COVID-19 Prevalence in France to Functional Curves. Available online: https://www.medrxiv.org/content/10.1101/2021.09.25.21264106v1 (accessed on 22 September 2021).
64. Oshinubi, K.; Ibrahim, F.; Rachdi, M.; Demongeot, J. Modelling of COVID-19 Pandemic vis-à-vis Some Socio-Economic Factors. Available online: https://www.medrxiv.org/content/10.1101/2021.09.30.21264356v1 (accessed on 22 September 2021).
65. Oshinubi, K.; Rachdi, M.; Demongeot, J. The application of ARIMA model to analyse incidence pattern in several countries. *J. Math. Comput. Sci.* **2021**, *26*, 41–57.

computation

MDPI

Article

Least-Squares Finite Element Method for a Meso-Scale Model of the Spread of COVID-19

Fleurianne Bertrand * and Emilie Pirch

Department of Computational Mathematics, Humboldt-Universität zu Berlin, 12489 Berlin, Germany; pirchemi@math.hu-berlin.de
* Correspondence: fb@math.hu-berlin.de

Abstract: This paper investigates numerical properties of a flux-based finite element method for the discretization of a SEIQRD (susceptible-exposed-infected-quarantined-recovered-deceased) model for the spread of COVID-19. The model is largely based on the SEIRD (susceptible-exposed-infected-recovered-deceased) models developed in recent works, with additional extension by a quarantined compartment of the living population and the resulting first-order system of coupled PDEs is solved by a Least-Squares meso-scale method. We incorporate several data on political measures for the containment of the spread gathered during the course of the year 2020 and develop an indicator that influences the predictions calculated by the method. The numerical experiments conducted show a promising accuracy of predictions of the space-time behavior of the virus compared to the real disease spreading data.

Keywords: COVID-19; least-squares finite element method; susceptible-exposed-infected-quarantined-recovered-deceased (SEIQRD)

Citation: Bertrand, F.; Pirch, E. Least-Squares Finite Element Method for a Meso-Scale Model of the Spread of COVID-19. *Computation* **2021**, *9*, 18. https://doi.org/10.3390/computation9020018

Academic Editor: Simone Brogi

Received: 15 December 2020
Accepted: 28 January 2021
Published: 5 February 2021

1. Introduction

The outbreak of the global pandemic caused by the novel virus responsible for COVID-19 had, and still has, a great impact on the life of the global human population. Human lives are threatened greatly by this highly infectious virus with higher probability of death and long-term damages to individuals of higher age or with a compromised immune system. Due to this delicate situation of global influence, various political measures have to be taken to prevent the virus from spreading as much as possible before an effective vaccine can be developed and distributed among the population to ensure immunity of a substantial part of the population that eventually causes the virus to die out. The most prominent question in the meantime, however, is that of the measures to be taken to 'flatten the curve' of new infections as the virus seems to spread exponentially if exposure is not regulated in any way. Among the measures already taken by the governments are curfews, lockdowns of whole cities and countries, quarantines of people exposed to the virus or that recently have been to areas with a high impact, travel restrictions, and—most commonly propagated measure on social media—social-distancing. But, to this point, there does not seem to be a general (political) consensus about the safest plan to slow the spread of the virus and which measures are the most effective, imposed on the people in exactly which level of strictness. This calls for a scientific modeling of the epidemiological behavior of this virus to form a plausible foundation for regulations. Such a model needs to extract some patterns thereof from the scattered data collected during the time of first notice in late 2019 until the latest developments today and to convert them into functions that can effectively predict new developments in the future. Regulating factors, such as exposure and mortality rates, can hopefully be witnessed and then, in turn, used to optimize the political measures accordingly.

The well-known epidemiological models of SIR type (susceptible-infected-recovered) have been extensively analyzed, and we refer to Reference [1] for an overview. This model works with a separation of the general population that needs to be studied into certain compartments (S, I, R) that have different roles in the spread of and affection by the virus and have a different use in the respective models. This compartment list can be extended to account for the further specifications of the disease, and we refer to Reference [2] for an overview. For example, the SEIR model includes an exposed group and the SEIRD (susceptible-exposed-infected-recovered-deceased) model separates truly recovered and deceased. The choice of these compartments for COVID-19 modeling has been the subject of many recent publications. The experts in modeling seem to agree that a COVID-19 model should account for asymptomatic transmission and that a quarantined group might be relevant (see Reference [3–9]).

A further challenge is the modeling of the spatial spread of the epidemic diseases in geographical regions. Several works, therefore, coupled the classical SIR model with inter-city networks, as in Reference [10,11]. To this aim, the classical epidemiological models of SIR type have been recast in the variational setting of analytical mechanics in Reference [12] with continuum partial differential equation models with diffusion terms describing the spatial variation in epidemics. First, mechanical and mathematical investigations in this direction were pursued in Reference [13,14] and seem very promising. A derivation of such a coupled system of PDEs without particular reference to an established SIR model has been conducted in Reference [15], where the authors have shown how the epidemiological dynamics can be expressed in PDEs step-by-step. For a mathematical analysis of a similar SIR model, we refer to Reference [16]. Another link can be drawn to the field of machine learning, as neural network predictors have proven themselves recently in similar fields, such as traffic and social modeling. Deep learning structures have been used to develop predictors for the COVID-19 virus spread. The techniques of using training data to be fed to the neural network that automatically computes a possible prediction are a great advantage in comparison to classical FEM methods that need a detailed model and a system of PDEs thought-out beforehand. A work on this forecast of the regional spread and intensity of the virus prevalence is presented in Reference [17]. Limitations, however, are exactly these training data, or the lack thereof, as at the beginning of the pandemic there might have not been a big enough variety of data to train the algorithm properly, and this can be linked to a choice of which data to use to make a most fitting prediction, until newer case numbers and their distribution are known.

In this work, we opted for a continuum partial differential equation model as in Reference [13,14] but added the quarantined compartment. Moreover, instead of a classical variational formulation, an approximation of the solution is obtained with a mixed formulation involving the fluxes of the variable accounting for the number of individuals in each group. This variational formulation is chosen to be of Least-Squares type, such that the linearization is relatively straightforward, the solving procedure involves a positive definite matrix, and we can use the inherent error estimator for adaptive strategies. We will map out the country of Germany with respect to accumulating regions and incorporate the ideas of travel restrictions and contact limitations imposed on the population. A further advantage of this approach is that it will give us the possibility to account for the political interventions made by the government in a hope to contain the spread of the virus in affected areas. To give an analysis of the spread of the virus under the already existing imposed political measures, data on restrictions, such as travel and contact reduction or bans, have been studied in the example of Germany. During early stages of the virus development in Europe, the case counts in this country have been significantly smaller than the ones of the neighboring countries. As respective measures of regulation have been taken early-on in March and April with rising numbers and a successful containment of the spread was achieved due to fast decreasing new daily infection rates, this serves as an indicator that the political decisions taken could have been effective. Another aspect is the division of the country in individual states, similar to the USA, with their own respective government that could more or less individually regulate the graveness of the measures, while the state intervened with German-wide restrictions only a few times during the time period of March until November. Such federal "infection containment acts" have been imposed, for example, during the lockdown in March with rather strict almost-curfew measures and then the permission to the individual states of relaxation of these acts, e.g., of the contact restrictions from single-household contacts to two-household rules or small groups and then successive enlargement of the number of people allowed at public gatherings or festivities. Eventually, the "lockdown light" has been re-inforced following the alarming high numbers of new daily infections. The indication as "light" is a terminology chosen by the government to contrast the "regular" lockdown in March that had stricter regulations imposed on businesses and catering that caused the economy to recess slightly.

We aimed at presenting a solution technique to the system of PDEs constructed by the SEIQRD (susceptible-exposed-infected-quarantined-recovered-deceased) model using a Least-Squares Method to predict the regional spread of COVID-19 in the country of Germany. We rely on data gathered by the Federal Statistical Office of Germany on actual numbers of infections, the reduction of incoming and outgoing flights, and contact restrictions as political reactions to contain the spread. These data serve to develop an indicator that is a key part in our calculations and shows at which time containment regulations gripped and give rise to a likely decrease (or increase) in subsequent new numbers of infections and their regional spread. Interpolation is used to fit and avoid losses of data and the resulting predicted versus real-life data will be presented in order to show the applicability of our Least-Squares solution method. To this end, this paper is structured subsequently in 5 more sections. In Section 2, the SEIQRD model is stated, and, in Section 3, the Least-Squares Method and the resulting first-order system to be solved are discussed. Following this, we develop the special discretization of the system in Section 4 and focus on the explanation of the parameters and their fitting using our indicator in Section 5. The numerical results are presented and analyzed in Section 6.

2. Model

We opt to change the usual SEIRD (susceptible-exposed-infected-recovered-deceased) model for epidemiological studies to a SEIQRD model that also takes into account a *quarantined* compartment of the population infected with the virus.

This model assumes that the living population is divided into five compartments: the *susceptible* population $S(\mathbf{x},t)$, the *exposed* population $E(\mathbf{x},t)$, the *infected* population $I(\mathbf{x},t)$, the *recovered* population $R(\mathbf{x},t)$, the *quarantined* population $Q(\mathbf{x},t)$, and *deceased* population $D(\mathbf{x})$. As in the works of Reference [13,14], we do not consider the birth rate nor the general (non-COVID-19) mortality rate and denote with $n(\mathbf{x})$ the sum of the living population, i.e.,

$$n(\mathbf{x}) = \sum_{i \in \{S,E,I,Q,R,D\}} \phi_i(\mathbf{x},t)\,, \tag{1}$$

with the functions ϕ_i representing the respective compartments for convenience of formulating the coupled PDE model. Note that, since we consider the compartment D of the deceased population, n does not vary over the time.

We distinguish between recovery rates γ_i, contact rates β_i, the inverse of the incubation period σ, a backflow η, and the quarantining rate δ.

Following Reference [13], we denote by γ_E the asymptomatic recovery rate and recall that it is the proportion of change in the exposed group that never enters the *infected* group (as they stay undetected) towards the recovered group. In the sense of the subsequent notations, that means that there is a decrease in the number of exposed people and an increase of recovered people.

$$E \xrightarrow{\gamma_E} R \qquad \frac{\partial}{\partial t}\phi_E \mathrel{-}= \gamma_E E, \quad \frac{\partial}{\partial t}\phi_R \mathrel{+}= \gamma_E E\,. \tag{2}$$

Similarly, γ_R denotes the infected recovery rate, i.e., the infected people that do show symptoms and, therefore, enter the regulated process of quarantine as an intermediate step (see below) before entering the recovered population.

$$I \xrightarrow{\gamma_R} R \qquad \frac{\partial}{\partial t}\phi_I \mathrel{-}= \gamma_R I, \quad \frac{\partial}{\partial t}\phi_R \mathrel{+}= \gamma_R I\,. \tag{3}$$

σ is the inverse of the incubation period that indicates how fast exposed individuals change to infected individuals after known exposure to the virus.

$$E \xrightarrow{\sigma} I \qquad \frac{\partial}{\partial t}\phi_E \mathrel{-}= \sigma E, \quad \frac{\partial}{\partial t}\phi_I \mathrel{+}= \sigma E. \tag{4}$$

One particularity of the new virus is that as of now the status of immunity of recovered patients is unclear. Therefore, we opt for a model that assumes that not all recovered patients are immune; thus, the backflow $\eta R(\mathbf{x}, t)$ is included that carries the proportion of recovered patients that are not immune back to the susceptible individuals with rate η.

$$R \xrightarrow{\eta} S \qquad \frac{\partial}{\partial t}\phi_R \mathrel{-}= \eta R, \quad \frac{\partial}{\partial t}\phi_S \mathrel{+}= \eta R. \tag{5}$$

We now want to consider the additional effect of the quarantine and choose a quarantine scheme connected to the infected, exposed and recovered, as a natural way to symbolize that quarantined people can be both in a state of yet non-discovered infection, being asymptomatic, healthy, or symptomatic (which means visibly showing symptoms that a possible infection with the virus might be accounted for). This quarantine rate should change with time and based on political decisions, as it has been mandatory for returnees from highly affected areas to undergo self-quarantine for several days while waiting for the result of the test that indicates the infection status. Quarantined individuals can recover or decease, as seen below.

$$Q \xrightarrow{\gamma_Q} R \qquad \frac{\partial}{\partial t}\phi_Q \mathrel{-}= \gamma_Q Q, \quad \frac{\partial}{\partial t}\phi_R \mathrel{+}= \gamma_Q Q, \tag{6}$$

$$I \xrightarrow{\delta} Q \qquad \frac{\partial}{\partial t}\phi_I \mathrel{-}= \delta I, \quad \frac{\partial}{\partial t}\phi_Q \mathrel{+}= \delta I. \tag{7}$$

Moreover, we follow the thoughts of Reference [10] and make the deceased linearly dependant on the quarantine, as the death of these individuals is connected to a visible infection that needs treatment in medical facilities that impose a strict quarantine on these patients. Thus, we get

$$Q \xrightarrow{\gamma_D} D \qquad \frac{\partial}{\partial t}\phi_Q \mathrel{-}= \gamma_D Q, \quad \frac{\partial}{\partial t}\phi_D \mathrel{+}= \gamma_D Q, \tag{8}$$

with the fatality rate γ_D.

In order to model the tendency of outbreaks to cluster towards large population centers, we follow the idea of Reference [13] and consider the Allee effect, which, in a sense, defines a correlation between the density of a population and the fitness of its individuals, with constant parameter α. We, therefore, need to consider the partial derivatives in space and introduce the space of weak derivatives $H^1(\Omega)$ on a simply connected geographical domain $\Omega \subset \mathbb{R}^2$. For ϕ_i sufficiently smooth, the Allee effect now reads

$$\frac{\partial}{\partial t}\phi_S(\mathbf{x},t) = -f(\phi_S,\phi_E,\phi_I,n(\mathbf{x})), \tag{9}$$

with

$$f(\phi_S,\phi_E,\phi_I,n(\mathbf{x})) = \left(1-\frac{\alpha}{n(\mathbf{x})}\right)(\beta_I\phi_S(\mathbf{x},t)\phi_I(\mathbf{x},t)+\beta_E\phi_S(\mathbf{x},t)\phi_E(\mathbf{x},t)), \tag{10}$$

where β_E is the contact rate at which the exposed asymptomatic patients transmit the virus to susceptible individuals, and β_I is the symptomatic contact rate.

Note that, in order to simplify the notation, we skipped the time dependence in the notation of the coefficients. However, those coefficients are supposed to change over time, as we will see in Section 5.

Assuming the population fields are sufficiently smooth, the model consists of the following system of nonlinear coupled partial differential equations over $\Omega \times [0, T]$:

$$\begin{aligned}\frac{\partial}{\partial t}\phi_S(\mathbf{x},t) =& \eta\phi_R(\mathbf{x},t)+\nabla\cdot(n(\mathbf{x})\,\nu_S\nabla\phi_S(\mathbf{x},t))\\ &-\left(1-\frac{\alpha}{n(\mathbf{x})}\right)(\beta_I\phi_S(\mathbf{x},t)\phi_I(\mathbf{x},t)+\beta_E\phi_S(\mathbf{x},t)\phi_E(\mathbf{x},t))'\end{aligned} \tag{11a}$$

$$\begin{aligned}\frac{\partial}{\partial t}\phi_E(\mathbf{x},t) =& \left(1-\frac{\alpha}{n(\mathbf{x})}\right)(\beta_I\phi_S(\mathbf{x},t)\phi_I(\mathbf{x},t)+\beta_E\phi_S(\mathbf{x},t)\phi_E(\mathbf{x},t))\\ &-\sigma\phi_E(\mathbf{x},t)-\gamma_E\phi_E(\mathbf{x},t)+\nabla\cdot(n(\mathbf{x})\,\nu_E\nabla\phi_E(\mathbf{x},t))\end{aligned}, \tag{11b}$$

$$\frac{\partial}{\partial t}\phi_I(\mathbf{x},t) = \sigma\phi_E(\mathbf{x},t)-\delta\phi_I(\mathbf{x},t)-\gamma_R\phi_I(\mathbf{x},t)+\nabla\cdot(n(\mathbf{x})\,\nu_I\nabla\phi_I(\mathbf{x},t)), \tag{11c}$$

$$\frac{\partial}{\partial t}\phi_Q(\mathbf{x},t) = \delta\phi_I(\mathbf{x},t)-\gamma_D\phi_Q(\mathbf{x},t)-\gamma_Q\phi_Q(\mathbf{x},t)+\nabla\cdot(n(\mathbf{x})\,\nu_Q\nabla\phi_Q(\mathbf{x},t)), \tag{11d}$$

(11e)

$$\frac{\partial}{\partial t}\phi_R(\mathbf{x},t) = \gamma_R\phi_I(\mathbf{x},t)+\gamma_E\phi_E(\mathbf{x},t)+\gamma_Q\phi_Q(\mathbf{x},t)-\eta\phi_R(\mathbf{x},t)+\nabla\cdot(n(\mathbf{x})\,\nu_R\nabla\phi_R(\mathbf{x},t)) \tag{11f}$$

(11g)

$$\frac{\partial}{\partial t}\phi_D(\mathbf{x},t) = \gamma_D\phi_Q(\mathbf{x},t), \tag{11h}$$

where the coefficients $\nu_S,\nu_E,\nu_I,\nu_Q,\nu_R,\nu_D$ account for the diffusion aspect; confer with Reference [18–21]. The model is summarized in Figure 1.

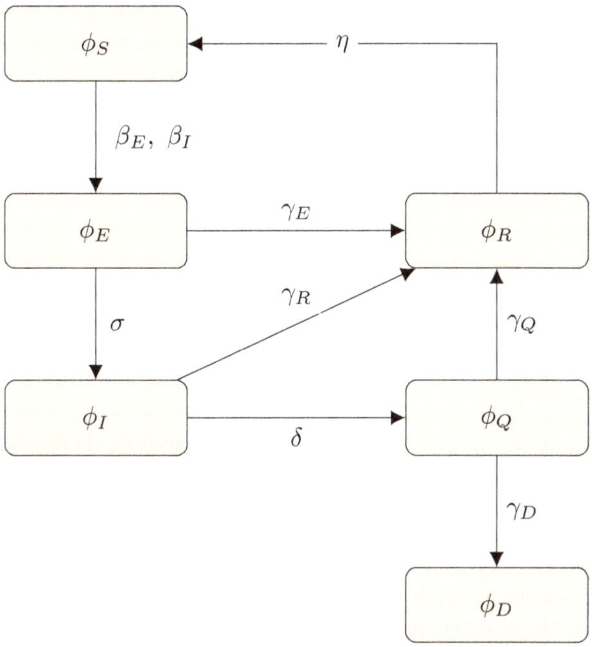

Figure 1. Flow chart depicting the regulating functions $\gamma_k, \sigma, \eta, \beta_i$ $(k = E, R, Q, D, i = E, I)$ and δ for the respective compartments of the population ϕ_j $(j = S, E, I, R, Q, D)$.

3. The Least-Squares Method

The class of Least-Squares Finite Element Methods is based on the idea of the residual minimization of a variational problem and as these methods rely on inner-product projections, they tend to be particularly robust and stable. While traditional finite element methods are usually developed from a variational setting that comes almost directly from the problem to solve at hand, Least-Squares Methods work exactly the other way round by fixing a variational framework before and then fitting the problem into this framework. For an introduction to this class of numerical methods, we refer the reader to Reference [22].

With the notation $\boldsymbol{\phi} = (\phi_S, \phi_E, \phi_I, \phi_Q, \phi_R, \phi_D)^\top$, $\boldsymbol{\nu} = (\nu_S, \nu_E, \nu_I, \nu_Q, \nu_R, \nu_D)^\top$, $A(\mathbf{x}) = n(\mathbf{x}) diag(\nu)$, $\mathbf{f}(\boldsymbol{\phi}) = (-f(\boldsymbol{\phi}), f(\boldsymbol{\phi}), 0, 0, 0, 0)^\top$, as well as

$$B = \begin{pmatrix} 0 & 0 & 0 & 0 & \eta & 0 \\ 0 & -\sigma - \gamma_E & 0 & 0 & 0 & 0 \\ 0 & \sigma & -\delta - \gamma_R & 0 & 0 & 0 \\ 0 & 0 & \delta & -\gamma_Q - \gamma_D & 0 & 0 \\ 0 & \gamma_E & \gamma_R & \gamma_Q & -\eta & 0 \\ 0 & 0 & 0 & \gamma_D & 0 & 0 \end{pmatrix}, \tag{12}$$

the system can be written in a vector form as

$$\frac{\partial}{\partial t}\boldsymbol{\phi} = B\boldsymbol{\phi} + \mathbf{f}(\boldsymbol{\phi}) + \nabla \cdot (A \nabla \boldsymbol{\phi}) \tag{13}$$

for $\boldsymbol{\phi} \in V = L^2(0, T, H^1(\Omega)))^6$ and with $[0, T]$ our time interval of interest. Defining $\boldsymbol{\sigma} = A \nabla \boldsymbol{\phi}$ leads to

$$\frac{\partial}{\partial t}\boldsymbol{\phi} = B\boldsymbol{\phi} + \mathbf{f}(\boldsymbol{\phi}) + \nabla \cdot \boldsymbol{\sigma}\,. \tag{14}$$

The components of σ then belong to the space of integrable divergence, i.e.,

$$\sigma \in \Sigma := L^2(0,T,(H_g(\mathrm{div},\Omega))^6) \text{ with } H_g(\mathrm{div},\Omega) = \{\boldsymbol{\tau} \in H(\mathrm{div},\Omega) : \boldsymbol{\tau}\cdot\mathbf{n} = g \text{ on } \partial\Omega\}, \tag{15}$$

where a Neumann boundary condition g on the boundary $\Gamma = \partial\Omega$ of Ω is prescribed in the space. With $f(\phi_S,\phi_E,\phi_I,n(\mathbf{x})) = \left(1-\frac{\alpha}{n(\mathbf{x})}\right)(\beta_I\phi_S(\mathbf{x},t)\phi_I(\mathbf{x},t) + \beta_E\phi_S(\mathbf{x},t)\phi_E(\mathbf{x},t))$ and the matrix

$$K = \left(1-\frac{\alpha}{n}\right)\begin{pmatrix} 0 & \beta_E & \beta_I & 0 & \dots & 0 \\ \vdots & 0 & 0 & \vdots & & \vdots \\ \vdots & \vdots & \vdots & \vdots & & \vdots \\ 0 & 0 & 0 & 0 & \dots & 0 \end{pmatrix}, \tag{16}$$

we obtain

$$\mathbf{f}(\boldsymbol{\phi}) = \left(-\boldsymbol{\phi}^\top K\boldsymbol{\phi}, \boldsymbol{\phi}^\top K\boldsymbol{\phi}, 0,0,0,0\right)^\top. \tag{17}$$

Using an implicit Euler time discretization, the first-order system reads

$$\mathcal{R}(\boldsymbol{\phi},\sigma;\boldsymbol{\phi}^{\mathrm{old}},\sigma^{\mathrm{old}}) = \begin{pmatrix} \boldsymbol{\phi} - \boldsymbol{\phi}^{\mathrm{old}} - \tau(B\boldsymbol{\phi} + \mathbf{f}(\boldsymbol{\phi}) + \nabla\cdot\sigma) \\ \sigma - A(\boldsymbol{\phi})\nabla\boldsymbol{\phi} \end{pmatrix} = 0\,. \tag{18}$$

Our Least-Squares Finite Element method consists of the least squares minimization of $\mathcal{R}(\boldsymbol{\phi},\sigma;\boldsymbol{\phi}^{\mathrm{old}},\sigma^{\mathrm{old}})$ in $V\times\Sigma$, which means we search $(\boldsymbol{\phi},\sigma) \in V\times\Sigma$

$$\left\|\mathcal{R}(\boldsymbol{\phi},\sigma;\boldsymbol{\phi}^{\mathrm{old}},\sigma^{\mathrm{old}})\right\|_{0,\Omega}^2 \leqslant \left\|\mathcal{R}(\boldsymbol{\psi},\boldsymbol{\tau};\boldsymbol{\phi}^{\mathrm{old}},\sigma^{\mathrm{old}})\right\|_{0,\Omega}^2 \tag{19}$$

for all $(\boldsymbol{\psi},\boldsymbol{\tau}) \in V\times\Sigma$. As the function f is a nonlinear function of $\boldsymbol{\phi}$, we will solve with the Gauss–Newton Multilevel Method proposed in Reference [23]. In fact, the main theorem states that if an iterative method is used which converges uniformly with respect to h, then a stopping criterion of the form

$$\mathrm{res}(\boldsymbol{\phi}_h^{(k)},\sigma_h^{(k)}) \leqslant \lambda h\left\|\mathcal{R}\left(\boldsymbol{\phi}_h^{(k)},\sigma_h^{(k)}\right)\right\|_{0,\Omega}, \tag{20}$$

based on a particular residual is useful with λ independent of h. Here, this residual is defined as the scalar product

$$\mathrm{res}(\boldsymbol{\phi}_h^{(k)},\sigma_h^{(k)}) = \left(\mathcal{R}\left(\boldsymbol{\phi}_h^{(k)},\sigma_h^{(k)}\right), \mathcal{J}(\boldsymbol{\phi}_h^{(k)},\sigma_h^{(k)})[\boldsymbol{\psi}_h,\sigma_h]\right)_{0,\Omega}, \tag{21}$$

with $\mathcal{J}$ the Fréchet derivative of $\mathcal{R}$ (omitting the notation of dependence on the data of the previous step) in the direction $[\boldsymbol{\psi}_h,\sigma_h] \in V_h\times\Sigma_h$ in the discretization space (to be defined in Section 4 below) that we calculate in the following. As the nonlinearity is concentrated in the term $\mathbf{f}(\boldsymbol{\phi})$, we introduce

$$\mathcal{R}_0(\boldsymbol{\phi},\sigma;\boldsymbol{\phi}^{\mathrm{old}},\sigma^{\mathrm{old}}) = \mathcal{R}(\boldsymbol{\phi},\sigma;\boldsymbol{\phi}^{\mathrm{old}},\sigma^{\mathrm{old}}) - \tau(\mathbf{f}(\boldsymbol{\phi}),0)^\top \tag{22}$$

in order to simplify the notation. The variable τ is not to be confused with $\boldsymbol{\tau} \in \Sigma$, as $t = t^{old} + \tau$ indicates the time step performed by the Euler discretization in the Gauss-Newton Multilevel Method in Reference [23].

For the derivative associated with the variable σ, we obtain

$$\frac{\partial}{\partial\theta}\mathcal{R}(\boldsymbol{\phi},\sigma+\theta\boldsymbol{\tau};\boldsymbol{\phi}^{\mathrm{old}},\sigma^{\mathrm{old}})\Big|_{\theta=0} = \begin{pmatrix} \tau\nabla\cdot\boldsymbol{\tau} \\ \boldsymbol{\tau} \end{pmatrix}, \tag{23}$$

and, for the linear part associated with the variable $\boldsymbol{\phi}$, we have

$$\frac{\partial}{\partial \theta}\mathcal{R}_0(\boldsymbol{\phi}+\theta\boldsymbol{\psi},\boldsymbol{\tau};\boldsymbol{\phi}^{\mathrm{old}},\boldsymbol{\sigma}^{\mathrm{old}})\Big|_{\theta=0}=\begin{pmatrix}\boldsymbol{\psi}-\tau B\boldsymbol{\psi}\\ -A\nabla\boldsymbol{\psi}\end{pmatrix}. \tag{24}$$

For the directional derivatives of the function f, we first state

$$\frac{\partial}{\partial \theta}f(\phi_S+\theta\psi_S,\phi_E,\phi_I,n)\Big|_{\theta=0}=\left(1-\frac{\alpha}{n}\right)(\beta_I\psi_S\phi_I+\beta_E\psi_S\phi_E), \tag{25}$$

$$\frac{\partial}{\partial \theta}f(\phi_S,\phi_E+\theta\psi_E,\phi_I,n)\Big|_{\theta=0}=\left(1-\frac{\alpha}{n}\right)(\beta_E\phi_S\psi_E), \tag{26}$$

$$\frac{\partial}{\partial \theta}f(\phi_S,\phi_E,\phi_I+\theta\psi_I,n)\Big|_{\theta=0}=\left(1-\frac{\alpha}{n}\right)(\beta_I\phi_S\psi_I), \tag{27}$$

such that

$$\frac{\partial}{\partial \theta}f(\boldsymbol{\phi}+\theta\boldsymbol{\psi})\Big|_{\theta=0}=\left(1-\frac{\alpha}{n}\right)(\beta_I(\phi_S\psi_I+\psi_S\phi_I)+\beta_E(\phi_S\psi_E+\psi_S\phi_E)), \tag{28}$$

and, with the matrix K and the notation from before, we obtain

$$\frac{\partial}{\partial \theta}\mathbf{f}(\boldsymbol{\phi}+\theta\boldsymbol{\psi})\Big|_{\theta=0}=\left(1-\frac{\alpha}{n}\right)\begin{pmatrix}-\beta_I(\phi_S\psi_I+\psi_S\phi_I)-\beta_E(\phi_S\psi_E+\psi_S\phi_E)\\ \beta_I(\phi_S\psi_I+\psi_S\phi_I)+\beta_E(\phi_S\psi_E+\psi_S\phi_E)\\ 0\\ 0\\ 0\\ 0\end{pmatrix}, \tag{29}$$

$$=\left(-(\boldsymbol{\phi}^T K\boldsymbol{\psi}+\boldsymbol{\psi}^T K\boldsymbol{\phi}),\boldsymbol{\phi}^T K\boldsymbol{\psi}+\boldsymbol{\psi}^T K\boldsymbol{\phi},0,0,0,0\right), \tag{30}$$

$$=(\boldsymbol{\phi}^T K\boldsymbol{\psi}+\boldsymbol{\psi}^T K\boldsymbol{\phi})(-1,1,0,0,0,0)^\top. \tag{31}$$

The Fréchet derivative is now the sum of (23), (24), and (29).

$$\mathcal{J}(\boldsymbol{\phi},\boldsymbol{\sigma})[\boldsymbol{\psi},\boldsymbol{\tau}]=\begin{pmatrix}\tau\nabla\cdot\boldsymbol{\tau}+\boldsymbol{\psi}-\tau(B\boldsymbol{\psi})-\tau(\boldsymbol{\phi}^T K\boldsymbol{\psi}+\boldsymbol{\psi}^T K\boldsymbol{\phi})\\ \boldsymbol{\tau}-A\nabla\boldsymbol{\psi}\end{pmatrix}. \tag{32}$$

4. Finite Element Discretization

In this work, we considered a fixed time step τ, while space-time adaptivity will be considered in a follow-up paper. Therefore, in each time-step, the finite element discretization of the Least-Squares Finite Element Method consists of considering the minimization problem (19) in a finite-dimensional subspace $V_h\times\Sigma_h\subseteq H^1(\Omega)^6\times H_g(\mathrm{div},\Omega)^6$, based on a triangulation $\mathcal{T}_h$ of Ω, i.e., we search $(\boldsymbol{\phi}_h,\boldsymbol{\sigma}_h)$ in $V_h\times\Sigma_h$, satisfying

$$\left\|\mathcal{R}(\boldsymbol{\phi}_h,\boldsymbol{\sigma}_h;\boldsymbol{\phi}_h^{\mathrm{old}},\boldsymbol{\sigma}_h^{\mathrm{old}})\right\|_{0,\Omega}^2\leqslant\left\|\mathcal{R}(\boldsymbol{\psi}_h,\boldsymbol{\tau}_h;\boldsymbol{\phi}_h^{\mathrm{old}},\boldsymbol{\sigma}_h^{\mathrm{old}})\right\|_{0,\Omega}^2 \tag{33}$$

for all $(\boldsymbol{\psi}_h,\boldsymbol{\tau}_h)\in V_h\times\Sigma_h$. As the Least-Squares Method does not require any compatibility of the finite element spaces, we choose $V_h=P^1(\mathcal{T}_h)^6$ as the standard Lagrange element and $\Sigma_h=RT^0(\mathcal{T}_h)^6\cap H_g(\mathrm{div},\Omega)^6$ the Raviart-Thomas element space accounting for the Neumann boundary condition prescribed by the function g. The Raviart-Thomas spaces for arbitrary degree k and dimension n of the $\Omega\subset\mathbb{R}^n$ are defined as

$$RT^k(\mathcal{T}_h)=P^k(\mathcal{T}_h)^n+xP^k(\mathcal{T}_h), \tag{34}$$

where $P^k(T)$ is the space of local polynomials of degree at most k on a triangle $T \in \mathcal{T}_h$. For the case $k = 0$, $n = 2$, this gives

$$RT^0(\mathcal{T}_h) := \{q \in P^1(T) : \forall T \in \mathcal{T}_h \; \exists a \in \mathbb{R}^2 \; \exists b \in \mathbb{R} \; \forall x \in T, \; q(x) = a + bx \\ \text{and } \forall E \in \mathcal{E}_\Omega, \; [q]_E \cdot n_E = 0\} . \tag{35}$$

The local degrees of freedom of the combination $P^1(\mathcal{T}_h) \times RT^0(\mathcal{T}_h)$ are pictured in Figure 2.

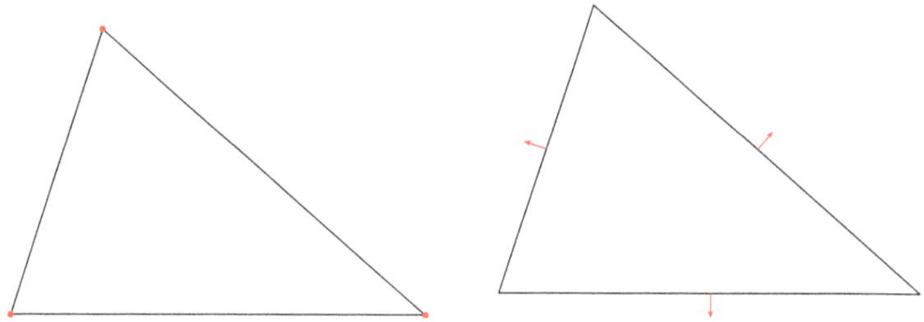

Figure 2. Local degrees of freedom by using P^1- and RT^0 bases in the discretization of the first-order system to be solved with the Least-Squares Method.

The inner basis functions of RT^0 can be defined on the edge-path $\omega_E = T_E^+ \cup T_E^-$, where T_E^+ and T_E^- are the adjacent triangles of the edge E by the following formula:

$$\psi_E(x) := \begin{cases} \pm\frac{1}{2|T|}(x - P_E^{\pm}) & \text{for } x \in T^{\pm} \\ 0 & \text{else,} \end{cases} . \tag{36}$$

Such a basis function is shown in Figure 3. With our computations of the Fréchet derivative, the nonlinear least-squares problem (33) is equivalent to the variational problem

$$\left(\mathcal{R}(\boldsymbol{\phi}_h, \sigma_h), \mathcal{J}(\boldsymbol{\phi}_h, \sigma_h)[\boldsymbol{\psi}_h, \boldsymbol{\tau}_h]\right)_{0,\Omega} = 0 \tag{37}$$

for all $(\boldsymbol{\psi}_h, \boldsymbol{\tau}_h) \in V_h \times \Sigma_h$.

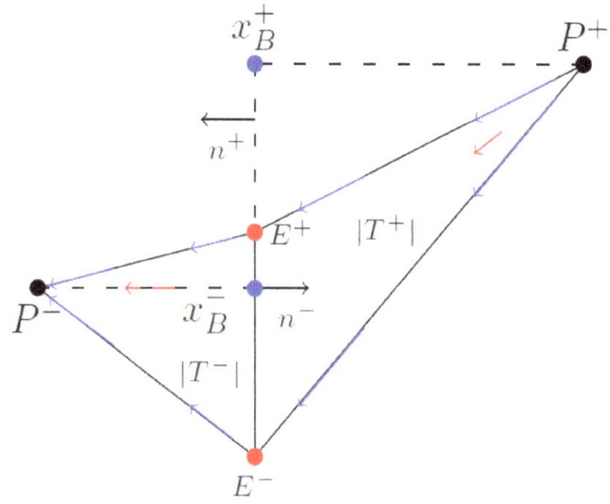

Figure 3. RT^0-basis functions on a triangle patch ω_T.

This is a nonlinear algebraic least-squares problem which we solved using an inexact Gauss-Newton method similar to the one presented in Reference [23]. Successive approximations to the nonlinear least-squares problem are, therefore, obtained by minimizing the linear least-squares problem

$$\mathcal{F}_{\text{lin}}(\delta_{\boldsymbol{\phi}_h}, \delta_{\sigma_h}; \boldsymbol{\phi}_h{}^{(k)}, \sigma_h{}^{(k)}) = \left\| \mathcal{R}\left(\boldsymbol{\phi}_h{}^{(k)}, \sigma_h{}^{(k)}\right) + \mathcal{J}\left(\boldsymbol{\phi}_h{}^{(k)}, \sigma_h{}^{(k)}\right)\begin{pmatrix} \delta_{\boldsymbol{\phi}_h} \\ \delta_{\sigma_h} \end{pmatrix} \right\|_{0,\Omega}^2. \tag{38}$$

Recall that minimizing F_{lin} in $V_h \times \Sigma_h$ is equivalent to the variational formulation

$$\left(\mathcal{R}\left(\boldsymbol{\phi}_h{}^{(k)}, \sigma_h{}^{(k)}\right) + \mathcal{J}\left(\boldsymbol{\phi}_h{}^{(k)}, \sigma_h{}^{(k)}\right)\begin{pmatrix} \delta_{\boldsymbol{\phi}_h} \\ \delta_{\sigma_h} \end{pmatrix}, \mathcal{J}\left(\boldsymbol{\phi}_h{}^{(k)}, \sigma_h{}^{(k)}\right)\begin{pmatrix} \boldsymbol{\psi}_h \\ \tau_h \end{pmatrix}\right)_{0,\Omega} = 0 \tag{39}$$

for all $(\boldsymbol{\psi}_h, \tau_h) \in V_h \times \Sigma_h$. Following the suggestion of the authors, we use

$$\text{res}(\boldsymbol{\phi}_h^{(k)}, \sigma_h^{(k)}) = \left(\mathcal{R}\left(\boldsymbol{\phi}_h^{(k)}, \sigma_h^{(k)}; \boldsymbol{\phi}_h^{\text{old}}, \sigma_h^{\text{old}}\right), \mathcal{J}\left(\boldsymbol{\phi}_h^{(k)}, \sigma_h^{(k)}\right)[\boldsymbol{\psi}_h, \tau_h]\right)_{0,\Omega} \tag{40}$$

as stopping criterion, i.e., the Gauss-Newton iteration is stopped as soon as the nonlinear residual satisfies (20), where we choose $\lambda = 0.2$. The steps are summarized in Algorithm 1.

Algorithm 1: Gauss-Newton for minimization of the nonlinear functional.

Input: solution of the last time step $(\boldsymbol{\phi}_h^{\text{old}}, \sigma_h^{\text{old}})$, parameter λ
$k = 0$
$\boldsymbol{\phi}_h^{(k)} = \boldsymbol{\phi}_h^{\text{old}}$
$\sigma_h^{(k)} = \sigma_h^{\text{old}}$
while $res(\boldsymbol{\phi}_h^{(k)}, \sigma_h^{(k)}) \leqslant \lambda h \left\| \mathcal{R}\left(\boldsymbol{\phi}_h^{(k)}, \sigma_h^{(k)}\right)\right\|_{0,\Omega}$ **do**
  Solve (39)
  $\boldsymbol{\phi}_h^{(k+1)} = \boldsymbol{\phi}_h^{(k)} + \delta_{\boldsymbol{\phi}_h}$
  $\sigma_h^{(k+1)} = \sigma_h^{(k)} + \delta_{\sigma_h}$
  $k = k + 1$
end
Result: $\boldsymbol{\phi}_h^{(k)}, \sigma_h^{(k)}$

---

5. Parameter Fitting

This section is devoted to the description of the parameters $\beta_{E,I}, \sigma, \gamma_{E,R,Q,D}, \delta, \eta$ that are used in the PDEs (11a)–(11f). The key idea is that we assume $\alpha, \beta_E, \beta_I, \delta$ is linearly dependent on some indicator $\theta(\mathbf{x}, t)$ taking into account the political measures. Surely, the linear dependency is an important restriction and nonlinear functions will be considered in a follow-up paper. On the other side, the SIR-type models are based on a linear incidence rate such that this ansatz is expected to give first adequate results. We also let γ_D vary over the time, taking into account that the health system had to learn and to increase the capacities. γ_D does not vary in space. We started with an ansatz corresponding to a polynomial of degree 5, and it turned out that a polynomial of degree 3 is sufficient.

The other parameters are assumed not to be dependent on the political restriction and, therefore, are constant in time.

For the design of this indicator, we took inspiration from the flight data found in Reference [24] for the comparison to the numbers of the COVID-19 not-yet inflicted year 2019 in Germany and the flight reduction in the year 2020 taken from Reference [25]. This data has been collected by the *Statistisches Bundesamt* (Federal Statistical Office of Germany) and is publicly accessible.

The indicator follows the data gathered for the reduction of the number of outgoing and incoming flights, as well as the contact reduction measures imposed by the government, over the time period of the outbreak of COVID-19 in Germany dating from January (or March, as the contact restraints haven been imposed later) until September 2020. The assumption that justifies this indicator is a correlation of the measures and the intensity of virus prevalence within the population. Our model is fed by two aspects, the first being the reduction of flights. This is based on the fact that following the growing international numbers in January, the government took measures of reducing flights to contain the risk of the residential population to be infected by travelling individuals that might come back from a high-risk area. This also gives rise to the question of reasonable initial values for the indicator and draws a connection between these of the compartments presented earlier in Section 2. Figure 4 shows how drastically the number in flights decreases up to April and then slowly increases again but stagnates in August.

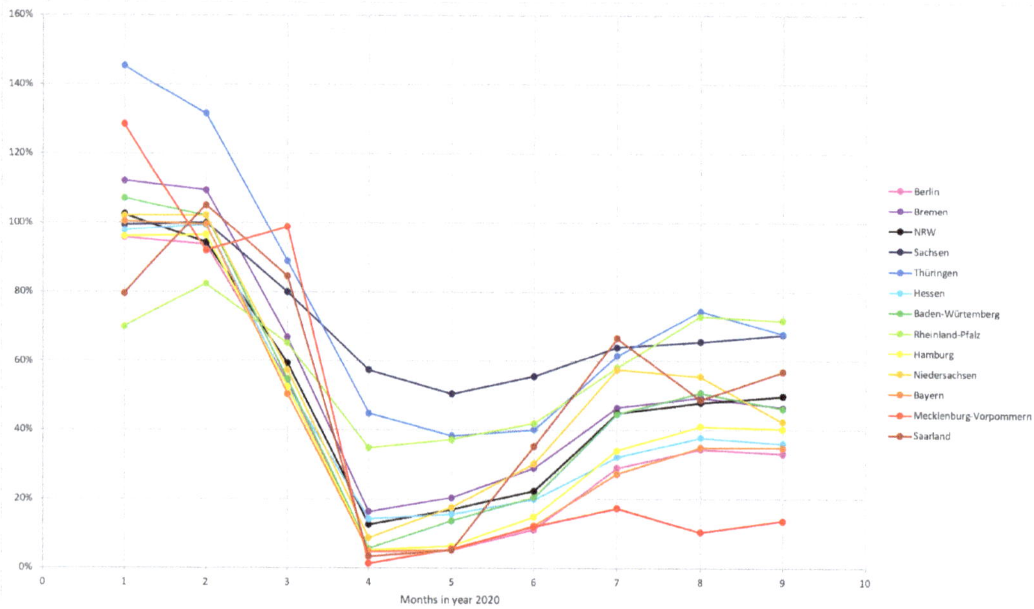

Figure 4. Flight data collected by the Federal Statistical Office of Germany in Reference [25]. A value of 100% is assigned if the number of outgoing and incoming flights in Germany for the respective region is the same as in the year 2019.

This can be linked to our second class of data, the contact restrictions. As the numbers in infections surged in March, a lockdown was announced across Germany with the same regulation imposed in every federal state: Only people belonging from their own household could be met and maximum one other person in public. Big gatherings have been forbidden completely and even travelling restrictions across the federal states (within the country!) have been imposed via bans on touristic stays at hotels. A model that takes these travel restrictions into account has also been considered in Reference [26]. These restrictions have been successively loosened on a private and a public level over the course of May and June and in July, August, and September the situation has been lead towards further normalization by permissions for public gatherings with growing numbers of participants of 100, 200, 350, etc. This tendency is reflected in the flight numbers, as they have been increasing from the depression in April, while being still far away from pre-pandemic numbers. The differences in the states can be seen while studying the respective "infection containment acts" and press releases (given that the numbers are reflected correctly). Not all states, however, have completely discarded the contact restraints in June and July (like Brandenburg and Mecklenburg-Vorpommern) but stayed with a moderate permission to meet an arbitrary number of people belonging to two households or a group of maximum 10 people from different households (like in Bavaria). These, however, are regulations for public meetings, but private gatherings have frequently not been observed, or no regulations have been imposed on private premises whatsoever (Bavaria, since June).

Drawing together these two classes of data we developed an indicator, the numbers of which can be seen in Table 1 and Figure 5. The indicator combines the contact limitations and the travel restrictions in terms of flights to create a weighting in the sense that the spread of the virus in already existing infections stays more close-region bound and the number of new infections is *predicted* to stay lower than an uncontrollable development without any restrictions. Thus, a value of 0.8, for example, indicates that due to travel and contact restrictions active at that time, a reduction of the transmission rates of the virus in our model towards 80% is used in the calculations compared to the uncontrolled case. At the beginning of 2020, restrictions for flights from China were already in place, as well as limitations of large events. Therefore, we chose to set this indicator to 0.8 for January in all federal states. Depending on how fast the government of the respective state were in implementing the measures, we let this indicator decrease until April. Note, for instance, that Bavaria had the strictest regulation in April and has, therefore, the smaller indicator in April. Similarly, the regulations were decreasing in July but remain very strong, and this is the reason why this state has, again, the smallest indicator from July to September.

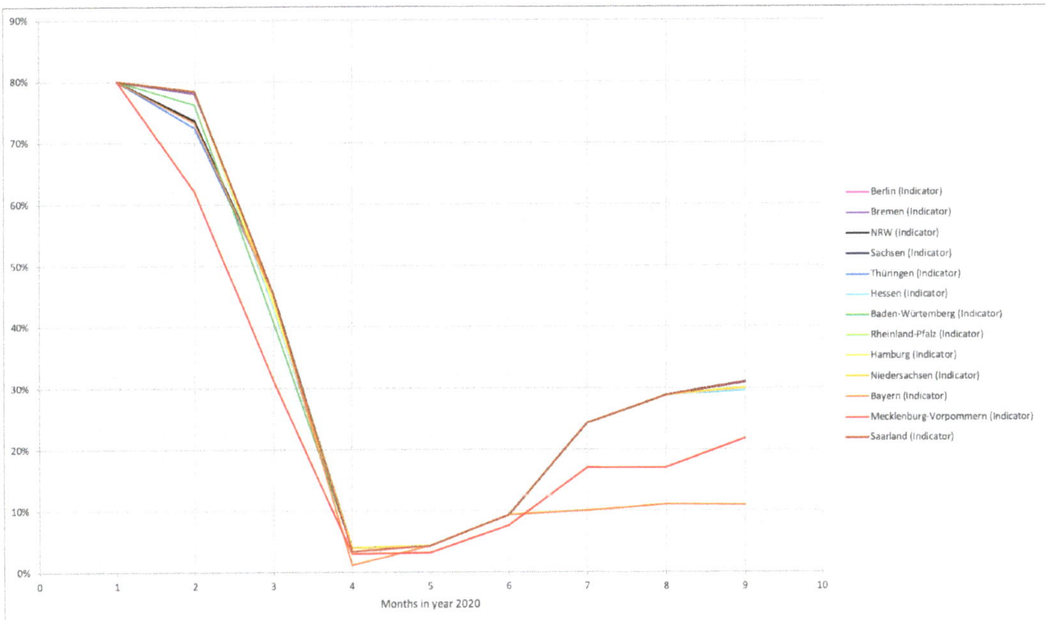

Figure 5. Indicator fitted to the collected data on contact restrictions and flights for comparison.

Table 1. Value of the indicator per month and state as shown in Figure 5.

	1 (Jan.)	2 (Feb.)	3 (Mar.)	4 (Apr.)	5 (May)	6 (Jun.)	7 (Jul.)	8 (Aug.)	9 (Sept.)
Berlin	0.8	0.78	0.45	0.04	0.04	0.09	0.24	0.29	0.31
Bremen	0.8	0.78	0.45	0.04	0.04	0.09	0.24	0.29	0.31
NRW	0.8	0.74	0.45	0.04	0.04	0.09	0.24	0.29	0.31
Sachsen	0.8	0.78	0.45	0.04	0.04	0.09	0.24	0.29	0.31
Thüringen	0.8	0.72	0.45	0.04	0.04	0.09	0.24	0.29	0.31
Hessen	0.8	0.78	0.43	0.04	0.04	0.09	0.24	0.29	0.3
Baden-Würt.	0.8	0.76	0.41	0.04	0.04	0.09	0.24	0.29	0.3
Rheinland-Pfalz	0.8	0.78	0.45	0.04	0.04	0.09	0.24	0.29	0.3
Hamburg	0.8	0.78	0.44	0.04	0.04	0.09	0.24	0.29	0.3
Niedersachsen	0.8	0.78	0.45	0.04	0.04	0.09	0.24	0.29	0.3
Bayern	0.8	0.73	0.45	0.01	0.04	0.09	0.1	0.11	0.11
Mecklenburg-V.	0.8	0.62	0.31	0.03	0.03	0.08	0.17	0.17	0.22
Saarland	0.8	0.78	0.45	0.03	0.04	0.09	0.24	0.29	0.31

6. Numerical Experiment

We performed the numerical experiment with the open source FENICS (see, e.g., Reference [27]). We use a finite-element spatial discretization of Germany, consisting of an unstructured mesh containing 1773 elements. Further results with finer meshes and adaptive mesh refinement strategies will be presented in a follow-up paper. In this project, we restricted ourselves to the time step $\tau = 0.1$ day due to the fact the the coupled PDE had to be solved many times. The initial conditions are the data from the "COVID-19 Dashboard" [28] of the *Robert Koch-Institut*, the leading epidemiological research institute in Germany concerned with data gathering at this time, of February 15th in which evaluations are based on the reporting data transmitted from the health authorities according to IfSG (infection protection acts). Data can be individually chosen for the respective states and regions. On the coast part of the German border, zero Neumann boundary conditions are set, while, on the remaining part, the data from an SRI model without diffusion (nor quarantine) are used.

The data from 15th February to 1st June was used for the calibration for the constant-in-time parameters, i.e., $\sigma, \gamma_{E,R}, \eta$. In order to investigate the sensibility of these coefficients, we also reproduced the calibration using less data, always starting from 15th February. For each *Bundesland* (federal state), we show the results in Figure 6. For the parameter depending on the indicator, the results of this analysis are shown in Figure 7. Figure 8 shows the evolving spatial pattern of the COVID-19 outbreak in Germany. A comparison of the prediction and the data from RKI is shown in Figure 9.

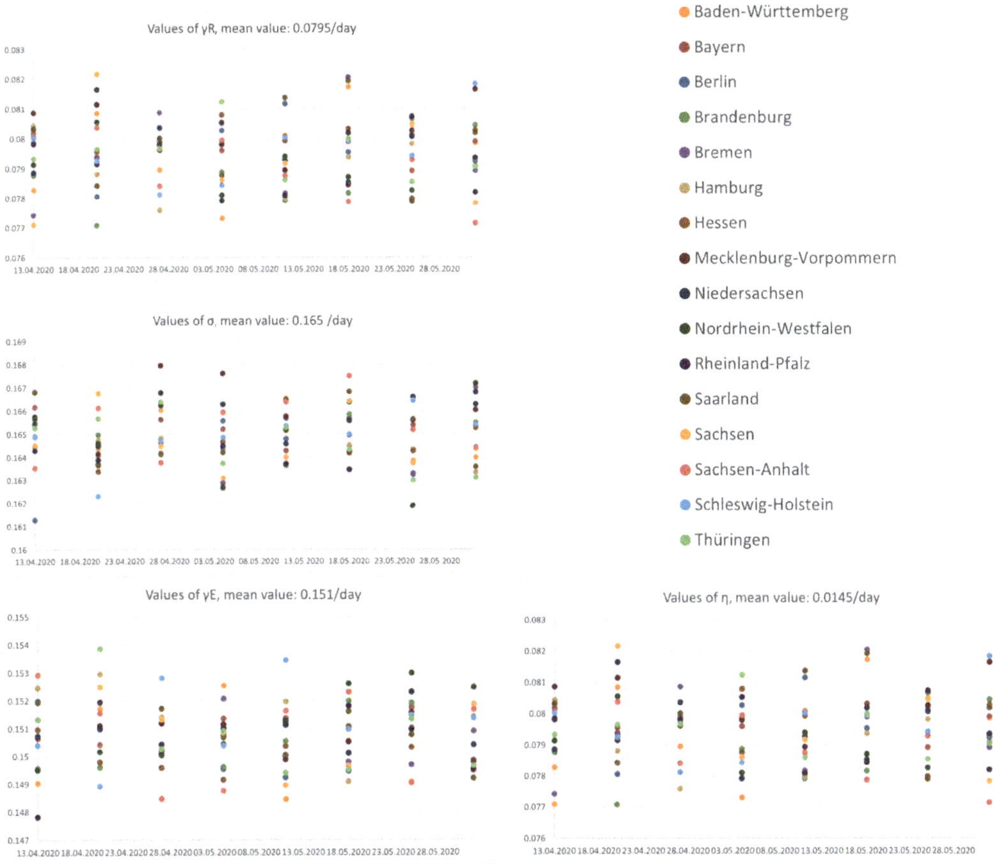

Figure 6. Values of the parameters $\sigma, \gamma_{E,R}, \eta$ with different time period fitting for the respective federal states.

Figure 7. Values of $\gamma_Q(\theta), \beta_{E,I}(\theta), \delta(\theta)$ with different time period fitting for the respective federal states.

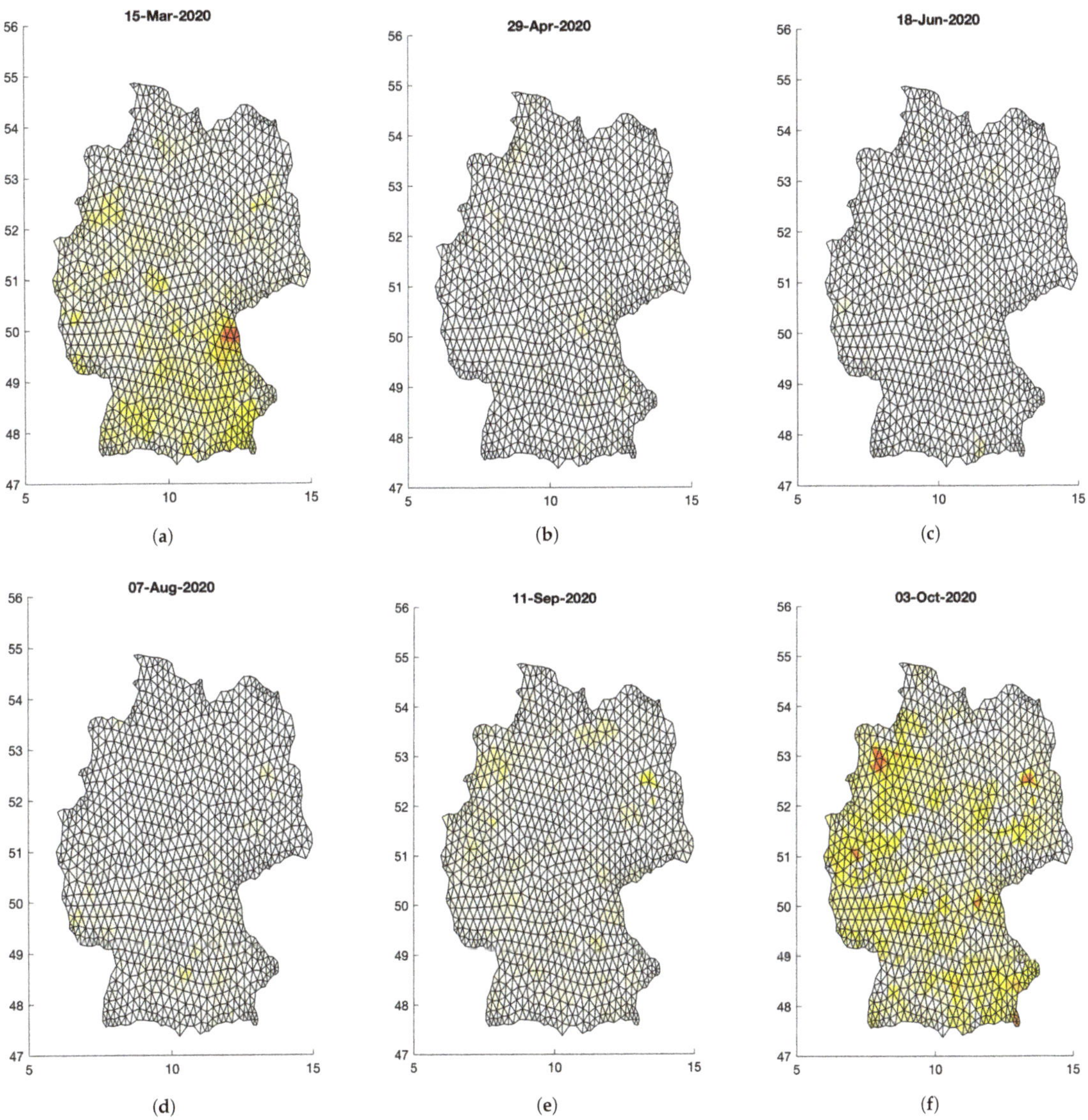

Figure 8. Regional spread of the virus at different time stages after initial outbreak on day 1 (D1). (**a**) D55, (**b**) D100, (**c**) D150, (**d**) D200, (**e**) D235, (**f**) D257.

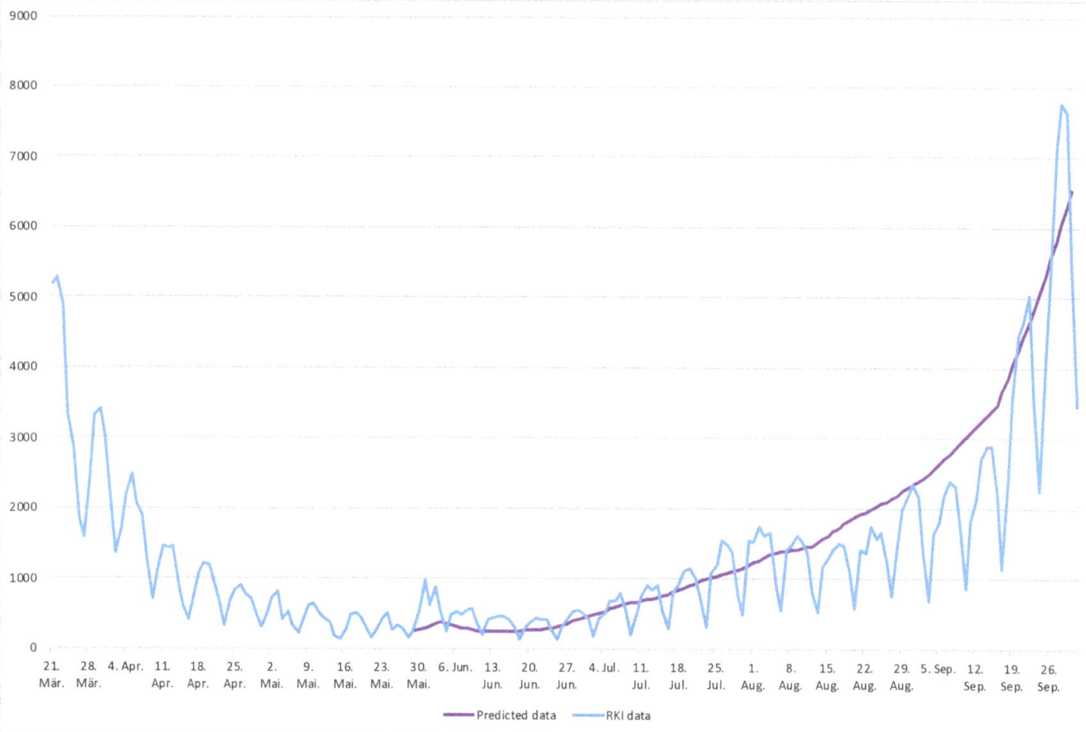

Figure 9. Predicted number of infections in Germany versus real data from RKI.

In order to present an evaluation of the accuracy of the prediction we start by considering the error as the forecast minus the real RKI data. Unfortunately, the RKI data are not monotone due to infrastructural and organizational reasons. For instance, reported new infections are linked to the days of the week in a sense that the public health departments are frequently closed over the weekends and have only started to register new cases also during the weekends after the situation has been severely more tense. Thus, Monday reports contain more new cases than the other days of the week up until Friday, as it can also contain the cases to be accounted towards Saturday and Sunday.

From the RKI data, we, therefore, constructed a piecewise linear interpolation $\mathcal{I}_{RKI,d}$ with

$$\mathcal{I}_{RKI,d}(x) = RKI(x + d - (x \equiv 7))\left(1 - \frac{(x-d) \equiv 7}{7}\right) + RKI(x + 7 + d - (x \equiv 7))\frac{((x-d) \equiv 7)}{7} \quad (41)$$

between each weekday, as well as the average $\mathcal{I}_{RKI,avg7}$ of the last seven days. The difference between the RKI data and these interpolations, as well as the difference between the RKI data and the prediction, are shown in Figure 10. We see that the prediction overshoots the Thursday line, such that the error $N_{predicted} - N_{\mathcal{I}_{RKI,thursday}}$ is positive.The forecast undershoots none of the other lines over the whole prediction time. We note that, until the beginning of August, the forecast undershoots the *avg* line and the error $N_{predicted} - N_{RKI,avg7}$ is negative. After this time, the forecast overshoots all the RKI interpolations until the end of September. The different errors are shown in Figure 11. We remark that the error oscillates taking into account that the RKI data oscillates. Overall, the error remains smaller than the error due to the piecewise linear interpolation of the data.

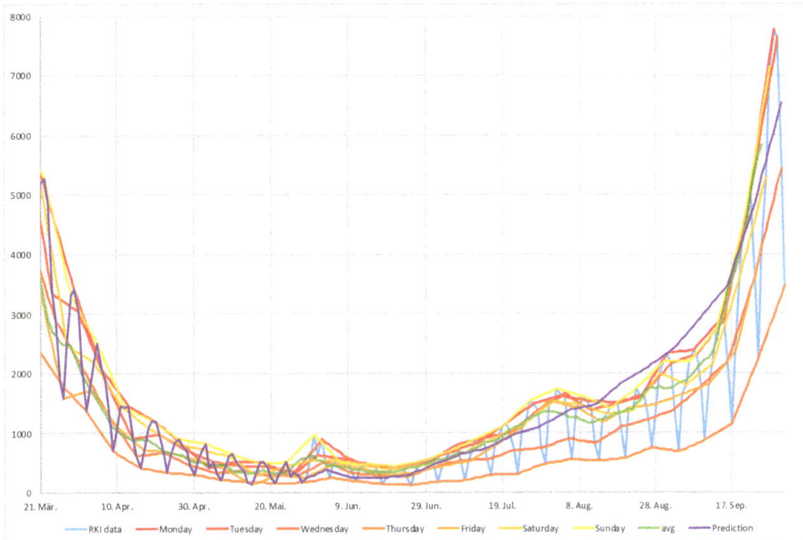

Figure 10. Predicted number of infections in Germany, the interpolation made for each day of the week compared, and the observed data from the RKI.

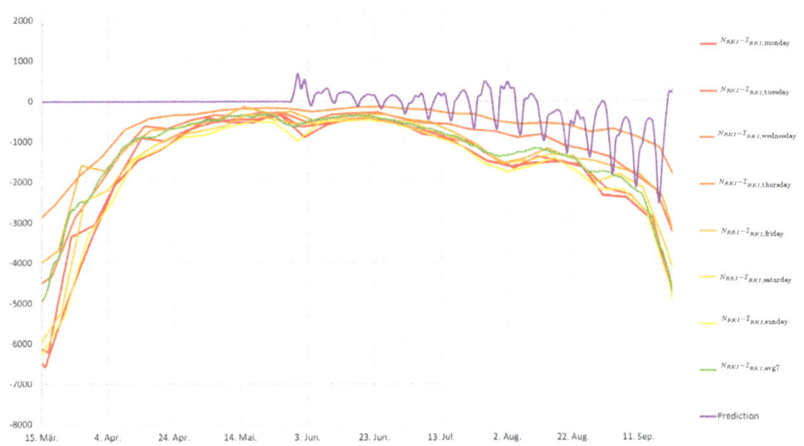

Figure 11. Error curves for the respective day of interpolation and their average marked in green; the calculated prediction is marked in violet.

In order to deal with these discrepancies, we computed the mean absolute percentage error (MAPE) and root-mean-square error (RMSE) for each of the previously mentioned interpolations of the RKI data. These quantities, obtained with

$$MAPE: \frac{1}{n_N}\sum\frac{|N_{predicted} - \mathcal{I}_{RKI,d}|}{\mathcal{I}_{RKI,d}} \quad RMSE: \sqrt{\frac{1}{n_N}\sum(N_{predicted} - \mathcal{I}_{RKI,d})^2}, \tag{42}$$

are given in Table 2. These can be compared to the numbers in the work [10] to find a similar accuracy of the prognostic. We remark, again, that the interpolation has a larger effect:

In order to study how the model is sensitive to the indicator, we perturbed the indicator up to 10%. The results in Figure 12 indicate that small variations are acceptable, as the resulting data stay all in a close proximity, even so still in a reasonable range in the second half of the timeline.

While the results of our numerical experiments look very promising, this is definitely to be accounted to some of the specific decisions we took for tailoring our calculations. For the indicator, we had to set a suitable initial value, for example, which represents the percentage of non-restrictions (100% means no restrictions) at some point. In addition, while the data we collected are a lot, only certain moments where incorporated and it is also always unknown beforehand whether the contact restrictions, for instance, will always be followed directly after press announcement. In this sense, the human choice is a big factor that cannot always be considered accordingly. (We refer to the most recent developments, as a "hard lockdown" has been imposed at the beginning of November that is still active, but the count of new infection cases per day have not decreased to a "satisfactory level" since. One of the reasons could be the dissatisfaction of large parts of the population with the deemed too drastic and restrictive measures, calls for demonstrations and large (and also private) gatherings without proper regard of the distancing measures, the loosening of the rules during Christmas-time, and the like.)

Our employed model is largely based on the works of Reference [10,13,14] and our Least-Squares solution technique shows a consistency with the numerical results presented in these works. However, some adjustments have been made in order to fit the computational work more tightly to the real-life data, thus producing more promising predictions. In Reference [13], the model successively forecasts exposed and infected cases which at this point are of high importance to the public health institutions. Similarly to our interpolation technique, a comparison of an "optimistic" and a "pessimistic" case can be witnessed, with the actual real-life data lying in between. Like the authors of this work, we come to the conclusion that this particular system of PDEs successively models the local virus dynamics on a meso-scale level.

The question of interest for practical relevance of our work remains: Can the predictions be used to influence and support political decisions in terms of virus containment? The answer is yes, but the transmission dynamics have to be investigated more closely in order to limit grave effects (like lockdowns) on the whole of the population. It could be more favorable to single out so-called virus hubs and rather focus on containment strategies in these areas while maintaining a tolerable, moderate policy for the remaining areas. To this end, the authors of Reference [10] present a detailed work on inter-state transmission that can be accounted to the use of the GLEAM network that serves to analyze the dynamics more closely in heavily-affected regions due to tourism and high traffic density. In addition, concrete rates for specific contact restrictions (that also include, for example, school closings, which could be one of the new aspects that we could include, as well in future work) have been used in the model, while we rely on the indicator for parameter fitting. It has to be noted though that the problem of limited testing and the related dark figures arises, that introduces a uncertainty in the data that is used for parameter calibration. Nevertheless, the use of such a network in our model could lead to even more closely fitted spacial predictions of spread and, thus, more detailed timelines, like in Figure 8, where, in part (f), some suggested virus hubs are noticeable of the type that the authors of Reference [10] can predict very accurately with the fine-tuning of the GLEAM network. Like in our approach, the predictions never undershot the actual observed numbers (in the most relevant cases), which indicated a high potential in practical use.

In Reference [14], another approach is shown that uses a machine learning technique to simulate the spread of the virus. A Bayesian learning in OPAL (Occam Plausibility Algorithm) is presented, where the simulation process in terms of more automatically computing spatio-temporal evolving can be seen. Comparing the resulting correlation and Pearson coefficients, our results show a similar accuracy, presenting two solution techniques to such systems of PDEs. Reference [10] presents a mixture of these two suitable techniques via a meso-scale approach, like ours, and refinement via a machine learning technique, the GLEAM network.

Table 2. Mean absolute percentage error (MAPE), root-mean-square error (RMSE), and Pearson coefficients for the different days, based on the interpolation of the data for their 7-day average.

Day	MAPE	RMSE/Max(RKI)	Pearson Coeffcient
no interpolation	0.396	46.253	0.843
Monday	2.623	1620.667	0.884
Tuesday	146.343	1564.583	0.865
Wednesday	99.048	1101.250	0.851
Thursday	59.527	731.861	0.816
Friday	2.256	1202.249	0.875
Saturday	39.708	1405.030	0.883
Sunday	99.898	651.748	0.901
Avg	2.450	11,497.504	0.864

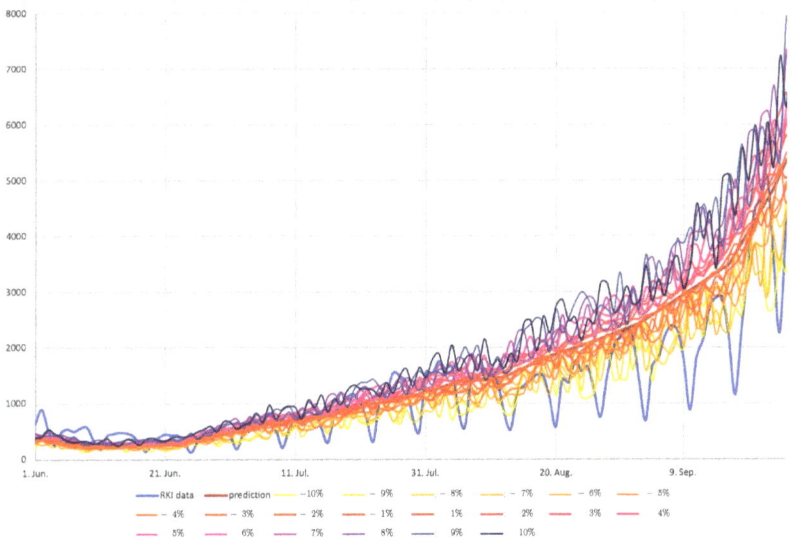

Figure 12. Sensitivity of the model towards the perturbations of the indicator.

Overall, we observe that our sensitivity analysis suggests that our indicator serves as a good tool to tune our predictions taking into account political measures that are taken. These predictions can in turn be used to help politicians and public health offices to take according measures in terms of contact restrictions and medical, as well as supply resource re-evaluation, to limit the virus spread to a tolerable amount and to anticipate spreads in particularly affected areas due to, for example, touristic location.

For future work, we are considering a more refined tailoring of our discretization method. A more technically challenging task due to its complexity and amount of data it produces is to implement the successive solution of the system with more than one Euler time step in one solution procedure. For further theoretical work, we will try to develop more modifications to classical models in the literature to test the limits of the accuracy of our discretization method. Works in actual simulation will be aimed at, as well.

Supplementary Materials: The following is available at https://www.mdpi.com/2079-3197/9/2/18/s1, Table S1: RKI data.

Author Contributions: Both authors contributed equally. Both authors have read and agreed to the published version of the manuscript.

Funding: This research received no external funding.

Data Availability Statement: The data presented in this study are available in the supplementary material provided.

Conflicts of Interest: The authors declare no conflict of interest.

References

1. Hethcote, H.W. The Mathematics of Infectious Diseases. *Siam Rev.* **2000**, *42*, 599–653. [CrossRef]
2. Pastor-Satorras, R.; Castellano, C.; Mieghem, P.V.; Vespignani, A. Epidemic processes in complex networks. *Rev. Mod. Phys.* **2015**, *87*, 925. [CrossRef]
3. He, S.; Peng, Y.; Sun, K. SEIR modeling of the COVID-19 and its dynamics. *Nonlinear Dyn.* **2020**, *101*, 1667–1680. [CrossRef] [PubMed]
4. Maier, B.; Brockmann, D. Effective containment explains subexponential growth in recent confirmed COVID-19 cases in China. *Science* **2020**, *368*, 742–746. [CrossRef] [PubMed]
5. Pedersen, M.; Meneghini, M.G. Quantifying undetected COVID-19 cases and effects of containment measures in Italy. ResearchGate Preprint. Available online: https://www.researchgate.net/profile/Morten_Pedersen2/publication/339915690_Quantifying_undetected_COVID-19_cases_and_effects_of_containment_measures_in_Italy_Predicting_phase_2_dynamics/links/5e76433ea6fdcccd62159b49/Quantifying-undetected-COVID-19-cases-and-effects-of-containment-measures-in-Italy-Predicting-phase-2-dynamics.pdf (accessed on 21 March 2020).
6. Peirlinck, M.; Costabal, F.; Linka, K.; Kuhl, E. Outbreak dynamics of COVID-19 in China and the United States. *Biomech. Model. Mechanobiol.* **2020**, *27*, 1–15. [CrossRef]
7. Kucharski, A.J.; Russell, T.W.; Diamond, C.; Liu, Y.; Edmunds, J.; Funk, S.; Flasche, S. Early dynamics of transmission and control of COVID- 19: A mathematical modelling study. *Lancet Infect. Dis.* **2020**, *20*, 553–558. [CrossRef]
8. Jia, J.; Ding, J.; Liu, S.; Liao, G.; Li, J.; Duan, B.; Wang, G.; Zhang, R. Modeling the Control of COVID-19: Impact of Policy Interventions and Meteorological Factors. *arXiv* **2020**, arXiv:2003.02985.
9. Yongzhen, P.; Shaoying, L.; Shujing, G.; Shuping, L.; Changguo, L. A delayed SEIQR epidemic model with pulse vaccination and the quarantine measure. *Comput. Math. Appl.* **2009**, *58*, 135–145. [CrossRef]
10. Kergaßner, A.; Burkhardt, C.; Lippold, D.; Kergaßner, M.; Pflug, L.; Budday, D.; Steinmann, P.; Budday, S. Memory-based meso-scale modeling of Covid-19-County-resolved timelines in Germany. *Comput. Mech.* **2020**, *66*, 1069–1079. [CrossRef]
11. Lu, Z.; Yu, Y.; Chen, Y.; Ren, G.; Xu, C.; Wang, S.; Yin, Z. A fractional-order SEIHDR model for COVID-19 with inter-city networked coupling effects. *Nonlinear Dyn.* **2020**, *101*, 1717–1730. [CrossRef]
12. Steinmann, P. Analytical Mechanics Allows Novel Vistas on Mathematical Epidemic Dynamics Modelling. *arXiv* **2020**, arXiv:2006.03961.
13. Viguerie, A.; Lorenzo, G.; Auricchio, F.; Baroli, D.; Hughes, T.J.; Patton, A.; Reali, A.; Yankeelov, T.E.; Veneziani, A. Simulating the spread of COVID-19 via a spatially-resolved susceptible–exposed–infected–recovered–deceased (SEIRD) model with heterogeneous diffusion. *Appl. Math. Lett.* **2020**, *111*, 106617. [CrossRef] [PubMed]
14. Jha, P.K.; Cao, L.; Oden, J.T. Bayesian-based predictions of COVID-19 evolution in Texas using multispecies mixture-theoretic continuum models. *Comput. Mech.* **2020**, *66*, 1055–1068. [CrossRef]
15. Cherniha, R.; Davydovych, V. A Mathematical Model for the COVID-19 Outbreak and Its Applications. *Symmetry* **2020**, *12*, 990. [CrossRef]
16. Magal, P.; Noussar, A. Modeling epidemic outbreaks in geographical regions: seasonal influenza in Puerto Rico. *Discret. Contin. Dyn. Syst. Ser. S* **2020**, *13*, 3535. [CrossRef]
17. Wieczorek, M.; Siłka, J.; Połap, D.; Woźniak, M.; Damaševičius, R. Real-time neural network based predictor for cov19 virus spread. *PLoS ONE* **2020**, *15*, e0243189. [CrossRef]
18. Kim, M. Galerkin methods for a model of population dynamics with nonlinear diffusion. *Numer. Methods Partial Differ. Equ.* **1996**, *12*, 59–73. [CrossRef]
19. Kim, M. A numerical method for spatial diffusion in age-structured populations. *Numer. Methods Partial Differ. Equ.* **2010**, *12*, 253–273.
20. Keller, J.; Gerardo-Giorda, L.; Veneziani, A. Numerical simulation of a suceptible-exposed-infectious space-continuous model for the spread of rabies in raccoons across a realistic landscape. *J. Biol. Dyn.* **2013**, *7*, 31–46. [CrossRef]
21. Holmes, E.; Lewis, M.; Banks, J.; Veit, R. Partial differential equations in ecology: Spatial interactions and population dynamics. *Ecology* **1994**, *75*, 17–29. [CrossRef]
22. Bochev, P.; Gunzburger, M. *Least-Squares Finite Element Methods*; Applied Mathematical Sciences; Springer: New York, NY, USA, 2009; Volume 166.
23. Starke, G. Gauss–Newton Multilevel Methods for Least-Squares Finite Element Computations of Variably Saturated Subsurface Flow. *Computing* **1999**, *64*, 323–338. [CrossRef]
24. Statistisches Bundesamt. Luftverkehr auf Hauptverkehrsflughäfen. 2019. Available online: https://www.destatis.de/DE/Themen/Branchen-Unternehmen/Transport-Verkehr/Personenverkehr/Publikationen/Downloads-Luftverkehr/luftverkehr-ausgewaehlte-flugplaetze-2080610197004.pdf?__blob=publicationFile (accessed on 1 October 2020).

25. Statistisches Bundesamt. Luftverkehr. 2020. Available online: https://www.destatis.de/DE/Themen/Branchen-Unternehmen/Transport-Verkehr/Personenverkehr/Publikationen/Downloads-Luftverkehr/luftverkehr-2080600201094.pdf?__blob=publicationFile (accessed on 1 October 2020).
26. Kuhl, E. Data-driven modeling of COVID-19—Lessons learned. *Extrem. Mech. Lett.* **2020**, *40*, 100921. [CrossRef] [PubMed]
27. Alnæs, M.S.; Blechta, J.; Hake, J.; Johansson, A.; Kehlet, B.; Logg, A.; Richardson, C.; Ring, J.; Rognes, M.E.; Wells, G.N. The FEniCS Project Version 1.5. *Arch. Numer. Softw.* **2015**, *3*. [CrossRef]
28. Robert Koch-Institut Germany. COVID-19 Dashboard. 2020. Available online: https://experience.arcgis.com/experience/478220a4c454480e823b17327b2bf1d4/page/page_0/ (accessed on 1 October 2020).

computation

Article

On the Application of Advanced Machine Learning Methods to Analyze Enhanced, Multimodal Data from Persons Infected with COVID-19

Wenhuan Zeng [1,*,†], Anupam Gautam [1,2,†] and Daniel H. Huson [1]

1 Institute for Bioinformatics and Medical Informatics, University of Tübingen, Sand 14, 72076 Tübingen, Germany; anupam.gautam@uni-tuebingen.de (A.G.); daniel.huson@uni-tuebingen.de (D.H.H.)
2 International Max Planck Research School 'From Molecules to Organisms', Max Planck Institute for Developmental Biology, Max-Planck-Ring 5, 72076 Tübingen, Germany
* Correspondence: wenhuan.zeng@uni-tuebingen.de
† These authors contributed equally to this work.

Abstract: The current COVID-19 pandemic, caused by the rapid worldwide spread of the SARS-CoV-2 virus, is having severe consequences for human health and the world economy. The virus affects different individuals differently, with many infected patients showing only mild symptoms, and others showing critical illness. To lessen the impact of the epidemic, one problem is to determine which factors play an important role in a patient's progression of the disease. Here, we construct an enhanced COVID-19 structured dataset from more than one source, using natural language processing to add local weather conditions and country-specific research sentiment. The enhanced structured dataset contains 301,363 samples and 43 features, and we applied both machine learning algorithms and deep learning algorithms on it so as to forecast patient's survival probability. In addition, we import alignment sequence data to improve the performance of the model. Application of Extreme Gradient Boosting (XGBoost) on the enhanced structured dataset achieves 97% accuracy in predicting patient's survival; with climatic factors, and then age, showing the most importance. Similarly, the application of a Multi-Layer Perceptron (MLP) achieves 98% accuracy. This work suggests that enhancing the available data, mostly basic information on patients, so as to include additional, potentially important features, such as weather conditions, is useful. The explored models suggest that textual weather descriptions can improve outcome forecast.

Keywords: COVID-19; machine learning; deep learning; NLP; weather; sentiment analysis

Citation: Zeng, W.; Gautam, A.; Huson, D.H. On the Application of Advanced Machine Learning Methods to Analyze Enhanced, Multimodal Data from Persons Infected with COVID-19. *Computation* **2021**, *9*, 4. https://doi.org/10.3390/computation9010004

Received: 8 December 2020
Accepted: 1 January 2021
Published: 7 January 2021

1. Introduction

The current COVID-19 pandemic, caused by the rapid worldwide spread of the SARS-CoV-2 virus, is affecting many aspects of society, in particular human health (at the time of writing, over 66 million diagnosed cases and 1.5 million deaths [1]), but also social issues [2,3], mental health, and the economy [4]. Researchers from different scientific fields, including immunology, genetics, and bioinformatics, are studying the pandemic to find ways to slow its progression.

Machine learning approaches are also part of this endeavor [5–9]. For example, Shahid et al. [10] use several models, including ARIMA, SVR, LSTM, and Bi-LSTM, for time series prediction of confirmed cases, deaths, and recoveries in ten major countries affected by COVID-19. Shreshth et al. [11] present a machine learning model to predict how the number of cases of COVID-19 will develop, and to forecast when a specific country can expect to see an end of the pandemic, using the FogBus framework. Other researchers have built machine learning models for the classification and diagnosis of COVID-19 that are based on medical images [12,13]. Further, Yan et al. [14] provide an interpretable mortality model that is based on a database of blood samples from 485 infected patients in the region of Wuhan,

China. To date, most machine learning and deep learning research [15,16] on COVID-19 build a classification model on various types of data to investigate which might be the important features to predict a specific outcome. One potential difficulty when running such approaches on publicly available dataset is that the features are originally collected so as to fulfill the needs of the data provider, which then can be a source of bias, when the data is used to address other questions. In particular, features that have high predictive value for the outcome for an infected patient might be missing. Generally speaking, the presence or absence of features will impact the accuracy of a model.

The COVID-19 data provided by Xu et al. [17] contain a large number of samples, but limited features that mainly provide basic information on patients. Here, we seek to improve the usefulness of this data by adding a number of features that might help to increase the accuracy of a predictive model.

Research indicates that local climate plays a roles in pandemic outbreaks [18]. Lowen et al. [19] demonstrated that aerosol spread of the influenza virus is dependent upon both ambient relative humidity and temperature, using guinea pig as a model host. Tan et al. [20] investigated the effect of weather in four cities in China and concluded that SARS outbreaks were significantly associated with the temperature and its variations. For the SARS-CoV-2 virus, there are some contradicting findings. Initial studies suggested a negative correlation between temperature and COVID-19 infection [21], or temperature-independence [22], while other research detected a positive relation between temperature and COVID-19 cases at temperatures below 3 °C [23], and also relates temperature to decrease in spread parameters of the case dynamics [24]. Therefore, local weather factors should be taken into consideration.

Infection and mortality rates differ between countries, as does the response to the pandemic. A study on news platforms and social media indicates that more than half (52%) of all news headlines evoked negative sentiments [25], on the one hand, whereas public positive tweets outweighed negative tweets on the other hand [26]. Application of machine learning algorithms on such data indicates a growth in fear and negative sentiment [27]. To explore this further, in this study we assume that a researchers attitude toward COVID-19, optimistic or pessimistic, will reflect the situation in their country, to some extent, and might be detectable in their publications on the pandemic.

While most previous work focuses on a single data type, in this study, we combine multiple data types. While a number of papers focus on country-wise pandemic prediction [28–30], here we develop a classification model that is based on worldwide data.

We first built an initial structured dataset on patients that tested positive for the virus, based on the work in [17]. We then constructed an enhanced structured dataset by adding new features based on (1) the local weather conditions when the patient was probably infected, and (2) the average weighted average polarity score for research abstracts on the pandemic, per country.

Another reasonable hypothesis is that the specific genome sequence of the virus that affected a given patient may help predict the outcome for the patient. There is research that associates genomic variations with mortality rate of COVID-19 [31], and further research [32] shows that the SARS-CoV-2 virus carries 7.23 mutations per sample compared to the reference, on average. There is work that attempts to predict outcome using machine learning and deep learning methods [33,34]. Both NCBI [35] and GISAID [36,37] provide genomic data for the virus.

Ideally, we would have liked to further enhance the initial dataset by adding virus genome sequences to each sample. Unfortunately, these sequences are not available. So, to explore the use of genomic sequences, we created an additional sequence dataset that consists of unknown patients and their virus sequence, obtained from GISAID.

In this paper, we investigated the application of two algorithms—XGBoost and MLP—to build models both on the initial structured dataset and also on the enhanced structured dataset. In addition, we built a Bi-LSTM model on the sequence dataset. The applied analysis pipelines are summarized in Figure 1.

Based on the initial dataset, we confirm that age is one of the most important factors for predicting survival. When considering the enhanced structured dataset, we find that the weather textual description, followed by local temperature, humidity, and age, arise as the most important features. On the enhanced data, we found that the Extreme Gradient Boosting (XGBoost) method achieved 97% accuracy in predicting a patient's survival. We describe how to predict patient's outcome using a combination of a Multi-Layer Perceptron (MLP) and Bidirectional Long Short-Term Memory (Bi-LSTM), using both the enhanced structured dataset, and the sequence dataset, respectively.

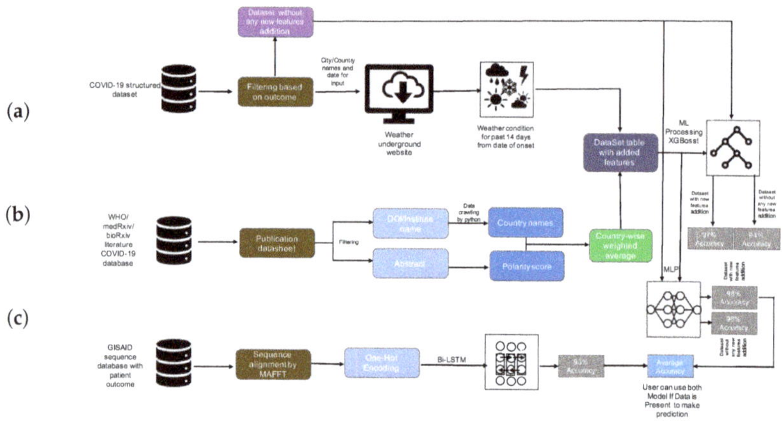

Figure 1. Analysis summary. (**a**) The initial COVID-19 structured dataset was filtered for patients for which the outcome has been recorded, and then, for these items, the weather was determined using the Weather Underground website [38]. (**b**) The WHO, medRxiv, and bioRxiv COVID-19 literature database were filtered and preprocessed to extract author institute/address/country, and these were postprocessed so as to obtain a country-wise research sentiment polarity score. XGBoost and Multi-Layer Perceptron (MLP) were trained on both the initial and the enhanced structured data, and the accuracy of survival prediction was shown to be 94% and 97% (using XGBoost), and 98% and 98% (using MLP), respectively. (**c**) Bidirectional Long Short-Term Memory (Bi-LSTM) was used to train a classification model on the sequence dataset, the accuracy was 93%. Finally, the MLP model and Bi-LSTM models were stacked to jointly predict outcome.

2. Materials

2.1. Data Collection

Data were collected from a number of sources.

2.1.1. COVID-19 Structured Dataset

We downloaded COVID-19 patient data provided by Xu et al. [17] from Github [39], on 21 August 2020 (file latestdata.csv). The dataset includes patient's basic information features, including ID, age, sex, city, province, country, etc. All rows that do not contain a value in the outcome column were dropped, resulting in 307,382 patient data rows out of 2,676,311. The final dataset contained 301,363 patients from 46 countries. All further processing was performed on this dataset.

2.1.2. WHO, medRxiv, and bioRxiv COVID-19 Literature Database

We downloaded a database of literature on COVID-19 from the World Health Organization (WHO) website [40] on 13 April 2020. Of the 5354 downloaded entries, we kept only those whose Journal Name and DOI fields were not blank, which resulted in 4683 publications in 590 journals. This list was extended with COVID-19 SARS-CoV-2 preprints published on medRxiv [41] and bioRxiv [42]. For this we used the bioRxiv API [43] to download the paper information; a total of 8076 entries were downloaded on 27 August 2020. We then analyzed these publications to determine the authors' institute and country; when no country was explicitly given, we used Google Maps [44] and Wikipedia [45] to determine the country in which the author's institute is located. This gave rise to 9577 (1501 of 4683 WHO, 8076 of 8076 medRxiv and bioRxiv) entries. Finally, we merged the two datasets and removed all duplicates, obtaining 9542 (1484 of 1501 WHO, 8058 of 8076 medRxiv and bioRxiv, Additional File 1) entries in total.

2.1.3. GISAID CoV-19 Sequences Dataset

The GISAID sequence repository contains more than 244,000 genomic sequences for SARS-CoV-2. We downloaded all that were labeled as complete, with high coverage, and were found in a human host on 25 August 2020. This resulted in 4957 genome sequences (with metadata). Further, we included the reference SARS-CoV-2 Wuhan genome (NCBI Accession MN908947.3 [46]) to the dataset and collected the patient information from the publication [47]. Finally, we removed all those sequences that did not have a patient status in the metadata file. Our final dataset contained 4720 sequences (Additional File 2).

2.2. COVID-19-Enhanced Structured Dataset

In this paper, we present an enhanced COVID-19 structured dataset, which is based on the above described initial COVID-19 structured dataset. These data were enhanced by adding features that reflect the weather situation in the location of the infected person, and the research sentiment in units of country, as described in the following.

2.3. Addition Feature Construction

It has been demonstrated that there is a link between environmental factors and the development of COVID-19 [48]. It is reasonable to assume that weather plays a role in disease progression. Therefore, we collected temperature, humidity, and textual description of the weather for the city where the patient lives from the Weather Underground website [38]. Assuming that the incubation period of the virus is approximately 14 days, we collected weather data from 14 days before the patient exhibited relevant symptoms (as recorded in the initial structured dataset).

We also wanted to explore the assumption that researchers' attitudes toward COVID-19, either optimistic or pessimistic, reflect the situation in each country, to some extent, and might be detectable in their publications on the pandemic. Therefore, we collected journal publications from the WHO and from the medRxiv and bioRxiv COVID-19 literature database. For each abstract, we determined the author's institution with the help of the paper's DOI and address by institute name. We applied sentiment analysis to obtain a polarity score on each abstract, and then calculated an weighted average polarity score for each country. Figure 2 displays the weighted average polarity score inferred for different countries.

The weather and sentiment features were added to the initial structured dataset so as to produce the enhanced structured dataset, as outlined in Figure 1.

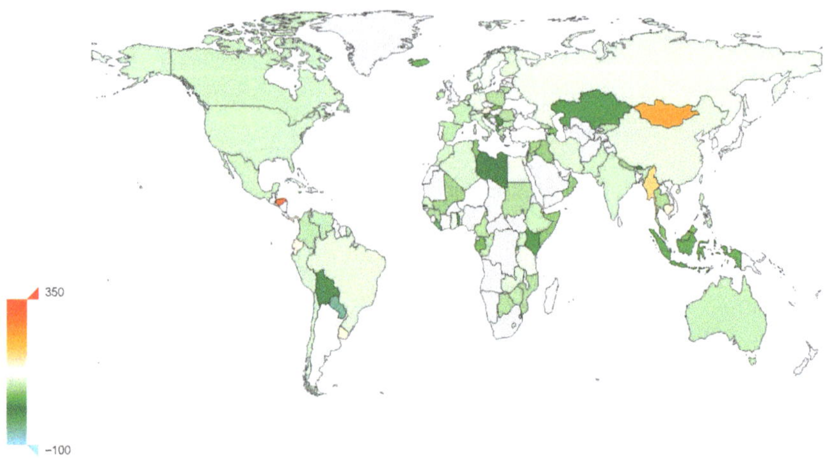

Figure 2. Sentiment polarity score. Average research sentiment polarity score of research, for different countries. Based on a sentiment analysis of abstracts of papers published on COVID-19. One-thousand times the real value.

2.4. Data Processing

2.4.1. Structured Data

The features present in the initial COVID-19 structured dataset include both categorical variables and discrete variables. Each sample in the dataset contains the variables sex, age, the time interval between the patient's onset date, confirmed infected date and admission date, symptoms description, presence of chronic disease, and outcome.

To this initial data, we then added local weather variables (temperature, humidity, and climate description) and the weighted polarity score of the country's scientific research sentiment. The result of this is called the enhanced structured dataset.

To prepare the datasets for building classification models using both XGBoost and MLP (as discussed below), we performed the following steps. We encoded all multi-value text features, such as symptom description (values such as fever, cough, and sputum) or climate description (values such as fair, light rain shower, and cloudy) into three-dimensional embedding vectors, using label encoding on categorical variables such as sex and history of chronic disease (Additional File 3).

We assigned the constant −999 to all missing values. After filtering for samples that have a valid outcome value and city record, we obtained 301,363 samples. Additionally, when we ran MLP, we treated sex and binary chronic disease as categorical features and all others as numerical features, and we normalized all numerical features.

2.4.2. Sequence Data

We performed multiple sequence alignment of the sequence dataset using MAFFT [49], run as follows.

```
mafft --retree 2 --maxiterate 1000 --thread 48 DeathAndAliveForMafft.fasta
  >DeathAndAliveForMafftAlignment1000Iterate.fasta
```

The program required 589 walk-clock minutes to align the 4720 virus genome sequences. The resulting alignment length was 32,015 (Additional File 4).

Furthermore, we applied character-level one-hot encoding on each sequence, mapping each position to a six-dimensional vector (one dimension for each of the four nucleotides, one for the gap character, and one for all ambiguity codes). Each sequence was padded to a fixed length of 33,100 (a multiple of 100), so as to allow us to use 100 time steps in the model described below.

2.5. Data Statistics

We built both a XGBoost model and an MLP model on both the initial structured dataset and on the enhanced structured dataset, respectively.

To evaluate the methods, we split each dataset into a training set and test set in proportion 8:2. Further, to prevent overfitting, we used cross-validation on our training datasets, instead of splitting additional validation sets from the original dataset. As shown in Table 1, the original dataset is typically imbalanced. To address this, we applied the Synthetic Minority Oversampling Technique (SMOTE) [50] to the minority group of each training set, attaining a ratio of positive to negative samples of 10:1. Note that here positive samples refer to patients that survive.

Table 1. Sampling statistics. For the enhanced structured dataset, we report the number of positive and negative samples both in the training set and test set, both before and after oversampling, respectively.

	Enhanced Data		After Oversampling	
	Training Set	**Test Set**	**Training Set**	**Test Set**
Positive samples	236,483	59,117	236,483	59,117
Negative samples	4607	1156	23,648	1156
Total	241,090	60,273	260,131	60,273

3. Methods and Experiment

3.1. Sentiment Analysis

A number of papers have studied the forecasting of pandemics using natural language processing on data obtained from various social media [51–53]. Along these lines, we performed sentiment analysis on the abstracts of research papers (associated with COVID-19) using the Python package Textblob [54], which operates by analyzing text content and assigning emotional values to words based on matches to a built-in dictionary.

3.2. Machine Learning Algorithm

Our focus was on the performance of prediction of survival of the infection, based on either the initial or the enhanced structured dataset.

Here, we use the Extreme Gradient Boosting (XGBoost) [55] method to build a prediction model. XGBoost is a powerful member of the gradient boosting family, which is designed to perform well on sparse features, and is known to perform well on Kaggle tasks. This approach avoids overfitting using its built-in L_1 and L_2 regularization on the target function:

$$Obj = \sum_{i=1}^{n} l(y_i, \hat{y}_i) + \sum_{i=1}^{t} \Omega(f_i). \tag{1}$$

As an additive model, XGBoost consists of k base models, and in most cases we choose the tree model as its base model. Suppose that, for the k-th of t iterations, we train the tree model $f_k(x)$, then

$$\hat{y}_i^t = \sum_{k=1}^{t} f_k(x_i) = \hat{y}_{i-1}^t + f_t(x_i) \tag{2}$$

is the estimated result for the ith sample after t iterations. During construction of each tree, XGBoost minimizes the objective function, with the regularization term show in

Equation (1) in the split phase of each node. In each tree, we calculate the *Gain* of the feature and choose the tree that has the biggest value as the leaf node to be split:

$$Gain = \frac{G_L^2}{H_L + \lambda} + \frac{G_R^2}{H_R + \lambda} - \frac{(G_L + G_R)^2}{H_L + H_R + \lambda} - \lambda. \quad (3)$$

3.3. Deep Learning Algorithms

To broaden our research and to allow a comparison of methods, we also built deep learning models on both the initial and enhanced structured datasets, together with the sequence dataset, respectively (Figure 3).

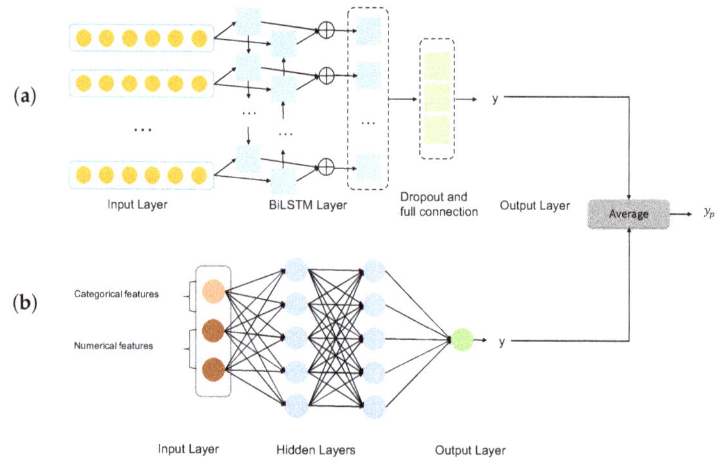

Figure 3. Ensemble deep learning model. (**a**) The MLP is trained on the structured dataset. (**b**) The Bi-LSTM model is trained on the sequence dataset. The two models are stacked in the prediction step.

3.3.1. Multi-Layer Perceptron

As indicated in Figure 3b, we use a simple Multi-Layer Perceptron (MLP) as neural network structure, which has an input layer, hidden layer, and output layer, to build a classification model on the structured dataset.

3.3.2. Bidirectional Long Short-Term Memory

Each sample in our sequence dataset has length 33,100 after alignment and data processing. We can interpret each sequence $X = (x_1, x_2, \cdots, x_n)$ as a time-series, where x_t is the data associated with the tth time point. Recurrent neural networks (RNN) proposed by Elman [56] are commonly used for time series; however, they are not suitable for our task due to the length of the alignments. Long short-term memory (LSTM) [57] is a special variant of RNN. It uses a gate structure in the hidden layer of each time step to protect and control the cell state.

An LSTM cell employs three gates, namely, a forget gate, an input gate, and an output gate, operating as shown in Figure 4. An LSTM learns to memorize and forget specific information during the training step. It provides the ability to capture long-term dependency relationships.

Each gate employs a sigmoid function that aims at producing output values of 0 or 1, defined as

$$\sigma(t) = \frac{1}{1 + e^{-t}} \quad (4)$$

An LSTM does not encode the information in inverse order, so it does not capture the impact of later words on previous words. A bidirectional long short-term memory (Bi-LSTM) overcomes this problem by combining a forward LSTM with a backward LSTM in each time step. This design addresses the issue of bidirectional semantic dependency during model building.

Therefore, we use a Bi-LSTM on our sequence data. Assume we are given a sequence $X = (x_1, x_2, \cdots, x_n)$, where x_t reflects the one-hot encoding. The hidden state of each time point is

$$h_t = [\overrightarrow{h_t}, \overleftarrow{h_t}] \tag{5}$$

In summary, this allows us to consider the impact of the virus sequence information on the patient's condition.

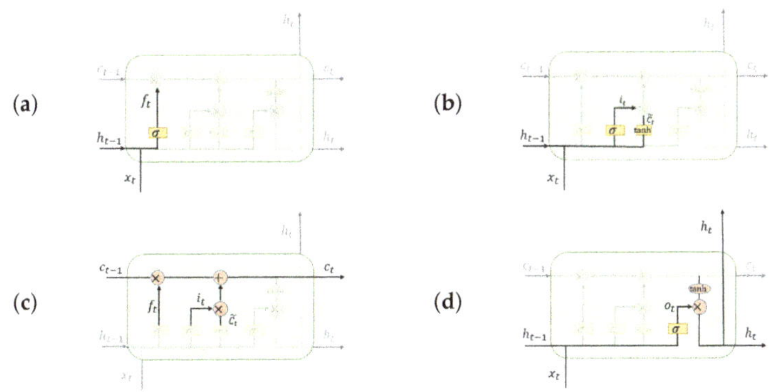

Figure 4. Operation of gates in an LSTM cell. The LSTM determines the hidden state and cell state at the present sequence location as follows. (**a**) A forget gate f_t controls the input of the $(t-1)$th hidden state, (**b**) an input gate i_t controls the input of x_t, (**c**) a transitional phase calculates the tth positions cell state, and then, finally, (**d**) an output gate O_t returns the tth position's hidden state h_t.

Finally, we stacked the MLP and Bi-LSTM deep learning classification models to jointly predict whether the infected patient will survive.

3.4. Implementation

3.4.1. Machine Learning Algorithms

In this study, we ran the XGBoost algorithm both on the initial structured dataset and also on the enhanced structured dataset, the latter additionally containing local weather and research sentiment. To determine the model parameters with the best capacity for prediction, we used GridSearchCV (a function of sklean) to systematically traverse multiple parameter combinations and determine the best parameters through cross-validation. Each subtree in our model is a complicated tree whose maximum depth is 10. Based on the result of model tuning, we set the learning rate to 0.05 and eta to 0.2. Further, we used 1500 estimators, and gamma, alpha, and lambda equal to 0.01, 0.5, and 0.8, respectively.

Each tree was trained on half of the features and half of the samples, chosen at random.

3.4.2. Deep Learning Algorithms

In Figure 3a we show the architecture of the model that accepts aligned sequences. It is a single Bi-LSTM with 128 hidden units and 100 time steps. After randomly dropping 1% of neurons, we use a fully connected layer and ReLU (rectified linear unit) activation function. Output is passed through a sigmoid function.

To model datasets that include both categorical features and normalized numerical features (Figure 3b), we used a 2-layer full connected neural network with 256 hidden units for each layer. To prevent model overfitting, we dropped a neuron with 5% probability during the forward propagation. A sigmoid function was used to determine output.

During training of both models, we split validation set from training set as proportion 1:3, and to moderate bias created by imbalanced data distribution, we set the class weight ratio between positive samples and negative samples to 1:10. After training as described above, we stacked the two models together so as to obtained average probability, passed through a sigmoid function (Figure 3).

4. Results

We evaluated the algorithms' performance using multiple metrics (Table 2).

Table 2. Performance measures. We report accuracy (Acc.), area under the curve (AUC), F1 score, recall, and precision (Prec.) for the named models and datasets. To compare the performance of the models using the initial or enhanced structured datasets, superior values are shown in bold. (for confusion matrices see Additional file 5).

Model	Dataset	Acc.	AUC	F1 Score	Recall	Prec.
XGBoost	Initial structured dataset	0.94	0.61	0.97	0.96	0.98
	Enhanced structured dataset	**0.97**	**0.77**	**0.99**	**0.99**	0.98
MLP	Initial structured dataset	0.98	0.56	0.99	1.0	0.98
	Enhanced structured dataset	0.98	**0.59**	0.99	1.0	0.98
Bi-LSTM	Sequence dataset	0.93	0.73	0.96	1.0	0.93

4.1. Machine Learning Model

The accuracy of the model created by using the initial structured dataset (no added features) is 94%, whereas using the enhanced structured dataset (with added features), the model's accuracy is 97%. As accuracy on an imbalanced dataset is limited, we display the receiver operating characteristic (ROC) curve of both datasets in Figure 5 to provide a further comparison. The enhanced structured dataset has significantly higher area under the ROC curve (AUC) scores than the model built on the initial structured dataset. There also exist tiny differences between the F1 score, recall, and precision of the two models. The method we chose to evaluate the importance score of feature is based on counting the number of times that a feature occurred in a tree. The feature importance for both datasets is shown in Figure 6. For the initial structured dataset, age plays a more important role than other features. For the model based on the enhanced structured dataset, the weather description, temperature, and humidity are more important than age; moreover, the level of importance of weather is higher than that of age. We visualized the frequency of the textual weather description on survivors and non-survivors, respectively (Figure 7). The weighted average research sentiment polarity score does not have an exceptional f score.

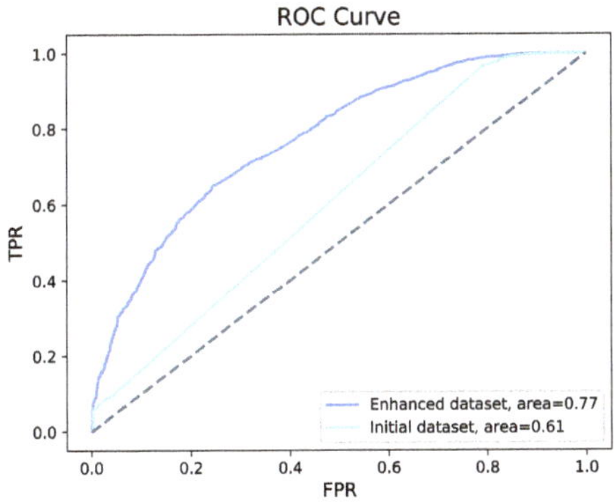

Figure 5. ROC of XGBoost. XGBoost shows an the accuracy of 94% on the initial structured dataset and an accuracy 97% on the enhanced structured dataset, with an increase of the area under the curve from 61% to 77%.

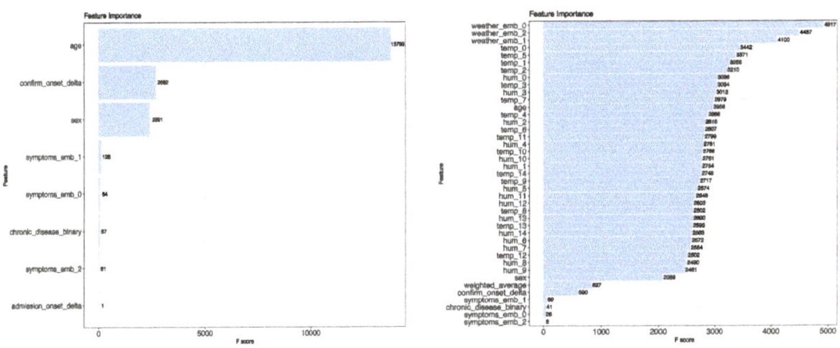

(a) (b)

Figure 6. Feature scores on the enhanced structured dataset. (**a**) XGBoost processing of the initial structured dataset identified age as an important feature. (**b**) XGBoost processing of the enhanced structured dataset identified in the weather as an important feature.

(a) (b)

Figure 7. Textual weather description. (**a**) Word cloud visualization of the frequency of textual weather description for survivors. (**b**) Word cloud visualization of the frequency of textual weather description for non-survivors.

4.2. Deep Learning Model

As shown in Table 2, on both the initial and enhanced structured datasets, the MLP method demonstrated higher accuracy than the XGBoost method. For both datasets, the accuracy using MLP is 98%. However, the ROC curve (Figure 8) indicates that the model shows a better classification ability on the enhanced structured dataset.

Taking sequence data into account, we obtained 93% accuracy and the area under the ROC curve is 0.73, as shown in Figure 9. Among all the models we built, the AUC score was highest when using a Bi-LSTM on the sequence data.

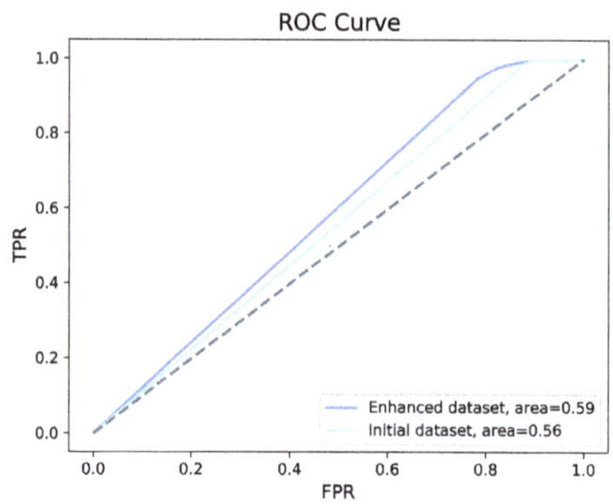

Figure 8. ROC of MLP. MLP shows an accuracy of 98% on both the initial and the enhanced structured dataset, with an increase in area under the curve from 56% to 59%.

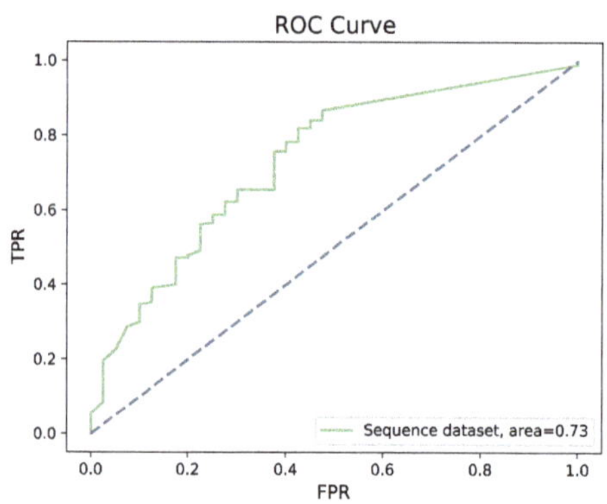

Figure 9. ROC of Bi-LSTM. Bi-LSTM shows an accuracy of 93% on sequence dataset, with an area under the curve of 0.73.

5. Discussion and Conclusions

The performance of machine learning and deep learning methods depends on the amount and quality of available features. Our analysis illustrates that current publicly available data can be enhanced, so as to increase the accuracy of survival prediction by 3% along with positive changes in other model validating metrics, such as AUC (16%), F1 score (2%), and Recall (3%) in case of XGBoost. For MLP the accuracy, F1 score, Recall, and Precision remained the same both for the initial and enhanced structured dataset, but the AUC increased by 3%.

To further evaluate the capability of the proposed models, we repeated the construction of all models on the same datasets, however, with the roles of positive and negative samples reversed, that is, this time considering patients who did *not* survive as positive samples. We observed that for XGBoost and MLP, the models based on the enhanced structured dataset perform better than those based on initial structured dataset in all aspects except recall (see Table 3). Further, it can be observed that even the best model has really poor performances in detecting patients who did not survive, as witnessed by the F1 score of 0.20.

Table 3. Performance measures (predicting death). Considering patients that die as positive samples, we report performance as in the previous Table (for confusion matrices see Additional file 5).

Model	Dataset	Acc.	AUC	F1 Score	Recall	Prec.
XGBoost	Initial structured dataset	0.96	0.60	0.15	**0.19**	0.12
	Enhanced structured dataset	**0.98**	**0.77**	**0.20**	0.13	**0.50**
MLP	Initial structured dataset	0.98	0.55	**0.15**	0.11	**0.21**
	Enhanced structured dataset	0.98	**0.59**	0.13	**0.21**	0.10
Bi-LSTM	Sequence dataset	0.93	0.64	0.21	0.35	0.14

Our study shows how one might enhance a dataset by adding informative features that are not available in the original dataset. Here we demonstrated this for local weather and country-wise research sentiment. Local weather conditions has been implicated as an important feature previous studies.

Our analysis also shows that age is an important factor for survival of COVID-19 as well. However, in the data considered here, the total number of deaths above age 60 were 793 and 2887 survived or were still alive, while in the age group between 40 and 60 there were 421 deaths and 10,346 alive or survived. Therefore, linking mortality to a particular age group is not appropriate based on the current data.

While this analysis suggests that elderly have a higher risk of death, which has already been observed [58,59], saying that mortality is associated with old age is probably generally true for any infectious disease. Age is one of the confounding factors that could be responsible for an increased COVID-19 mortality rate [60,61].

For the model based on the enhanced structured dataset, the weather textual description, followed by local temperature, humidity, and age, appear as the most important features and account for the increase in the accuracy of the model. The most apparent difference in the weather attributes for survivors and non-survivors (Figure 7) is "smoke". This suggests that environmental conditions, in particular air pollution, may play a role in determining the outcome of the disease.

In contrast, in our investigation, the research sentiment score did not show the importance that we had suspected. The values of this feature are never particular high or low, and the highest value of this feature is only 0.35, and thus the difference between the highest score and lowest score is also small. We assume that one of the reasons for this is that academic writing aims for a neutral tone.

The model that we developed on the virus genome dataset failed to provide added predictive power. We suspect that virus genome data would be much more useful, if it were available for the large, structured dataset. However, our study may provide a starting point for further work.

Further, this analysis confirms that enhancing a dataset, rather than just analyzing the originally given features, might lead to a better prediction of a particular outcome. Along with some of the features which should be paid more attention while collecting the data.

There are a number of possible directions for future work. As more viral genomes become available, more powerful Deep Learning methods can be applied to them to help predict patient survival. Additional features such as patient health status, weight, height, medical history should also be integrated. The effect of climate on patient survival warrants more investigation. Finally, methods such as a Recurrent Neural Network-based LSTM might help to study how mutations influence the transmissibility of the virus [62].

Supplementary Materials: Additional files, datasets and models analyzed during our study along with the supplementary materials (like scripts) can be accessed at https://github.com/husonlab/covid19paper. Additional file 1: Merge and processed publication data downloaded from WHO, medRxiv and bioRxiv COVID-19 literature database; Additional file 2: Sequences used for MAFFT alignment downloaded from GISAID and NCBI; Additional file 3: COVID-19 Enhanced structured dataset; Additional file 4: Aligned sequence used for Bi-LSTM; Additional file 5: Confusion matrices for all the build models.

Author Contributions: D.H.H. proposed and guided the project. D.H.H., W.Z., and A.G. wrote the manuscript. W.Z. and A.G. designed the work, wrote the code for data generation, conducted processing, and carried out the analysis. W.Z. carried out the machine learning and deep learning model development and analysis. All authors read and approved the final manuscript.

Funding: This work was supported by the BMBF-funded de.NBI Cloud within the German Network for Bioinformatics Infrastructure (de.NBI) (031A537B, 031A533A, 031A538A,031A533B, 031A535A, 031A537C, 031A534A, 031A532B). Furthermore, we acknowledge support by the Open Access Publishing Fund of University of Tübingen.

Institutional Review Board Statement: Not applicable.

Informed Consent Statement: Not applicable.

Data Availability Statement: Publicly available datasets were analyzed in this study. These datasets can be found here: Xu et al. dataset. [https://doi.org/10.1038/s41597-020-0448-0], preprints from medRxiv and bioRxiv were accessed by using API: [https://api.biorxiv.org/covid19/help], sequence dataset was download from GISAID: [https://www.gisaid.org/] and NCBI: [https://www.ncbi.nlm.nih.gov/search/all/?term=MN908947] and remaining processed and generated dataset can be downloaded from [https://github.com/husonlab/covid19paper].

Acknowledgments: We would like to thank Caner Bağči for helpful discussions on sequence analysis.

Conflicts of Interest: The authors declare no conflicts of interest.

Abbreviations

The following abbreviations are used in this manuscript:

COVID-19	coronavirus disease
SARS-CoV-2	severe acute respiratory syndrome coronavirus 2
ARIMA	Autoregressive Integrated Moving Average model
Bi-LSTM	Bidirectional Long Short-Term Memory
LSTM	Long Short-Term Memory
MLP	Multi-Layer Perceptron
NLP	Natural Language Processing
WHO	World Health Organization
NCBI	National Center for Biotechnology Information
GISAID	Global initiative on sharing all influenza data
RNN	Recurrent Neural Network
SVR	Support Vector Regression
XGBoost	Extreme Gradient Boosting

References

1. WHO Coronavirus Disease (COVID-19) Dashboard. Available online: https://covid19.who.int/ (accessed on 5 January 2021)
2. Torales, J.; O'Higgins, M.; Castaldelli-Maia, J.M.; Ventriglio, A. The outbreak of COVID-19 coronavirus and its impact on global mental health. *Int. J. Soc. Psychiatry* **2020**, *31*, 0020764020915212. [CrossRef]
3. Singh, J.; Singh, J. COVID-19 and its impact on society. *Electron. Res. J. Soc. Sci. Humanit.* **2020**, *2*, 102–105.
4. Holmes, E.A.; O'Connor, R.C.; Perry, V.H.; Tracey, I.; Wessely, S.; Arseneault, L.; Ballard, C.; Christensen, H.; Silver, R.C.; Everall, I.; et al. Multidisciplinary research priorities for the COVID-19 pandemic: A call for action for mental health science. *Lancet Psychiatry* **2020**, *7*, 547–560. [CrossRef]
5. Lalmuanawma, S.; Hussain, J.; Chhakchhuak, L. Applications of machine learning and artificial intelligence for Covid-19 (SARS-CoV-2) pandemic: A review. *Chaos Solitons Fractals* **2020**, *139*, 110059. [CrossRef] [PubMed]

6. Ramchandani, A.; Fan, C.; Mostafavi, A. Deepcovidnet: An interpretable deep learning model for predictive surveillance of covid-19 using heterogeneous features and their interactions. *IEEE Access* **2020**, *8*, 159915–159930. [CrossRef]
7. Wang, P.; Zheng, X.; Li, J.; Zhu, B. Prediction of epidemic trends in COVID-19 with logistic model and machine learning technics. *Chaos Solitons Fractals* **2020**, *139*, 110058. [CrossRef] [PubMed]
8. Mirri, S.; Delnevo, G.; Roccetti, M. Is a COVID-19 Second Wave Possible in Emilia-Romagna (Italy)? Forecasting a Future Outbreak with Particulate Pollution and Machine Learning. *Computation* **2020**, *8*, 74. [CrossRef]
9. Alakus, T.B.; Turkoglu, I. Comparison of deep learning approaches to predict COVID-19 infection. *Chaos Solitons Fractals* **2020**, *140*, 110120. [CrossRef]
10. Shahid, F.; Zameer, A.; Muneeb, M. Predictions for COVID-19 with deep learning models of LSTM, GRU and Bi-LSTM. *Chaos Solitons Fractals* **2020**, *140*, 110212. [CrossRef]
11. Tuli, S.; Tuli, S.; Tuli, R.; Gill, S.S. Predicting the Growth and Trend of COVID-19 Pandemic using Machine Learning and Cloud Computing. *Internet Things* **2020**, *11*, 100222. [CrossRef]
12. Elaziz, M.A.; Hosny, K.M.; Salah, A.; Darwish, M.M.; Lu, S.; Sahlol, A.T. New machine learning method for image-based diagnosis of COVID-19. *PLoS ONE* **2020**, *15*, e0235187. [CrossRef] [PubMed]
13. Barstugan, M.; Ozkaya, U.; Ozturk, S. Coronavirus (Covid-19) classification using ct images by machine learning methods. *arXiv* **2020**, arXiv:2003.09424.
14. Yan, L.; Zhang, H.-T.; Goncalves, J.; Xiao, Y.; Wang, M.; Guo, Y.; Sun, C.; Tang, X.; Jing, L.; Zhang, M. An interpretable mortality prediction model for COVID-19 patients. *Nat. Mach. Intell.* **2020**, *2*, 283–288. [CrossRef]
15. Narin, A.; Kaya, C.; Pamuk, Z. Automatic detection of coronavirus disease (COVID-19) using X-ray images and deep convolutional neural networks. *arXiv* **2020**, arXiv:2003.10849.
16. Magar, R.; Yadav, P.; Farimani, A.B. Potential neutralizing antibodies discovered for novel corona virus using machine learning. *arXiv* **2020**, arXiv:22003.08447.
17. Xu, B.; Gutierrez, B.; Mekaru, S.; Sewalk, K.; Goodwin, L.; Loskill, A.; Cohn, E.L.; Hswen, Y.; Hill, S.C.; Cobo, M.M.; et al. Epidemiological data from the COVID-19 outbreak, real-time case information. *Sci. Data* **2020**, *7*. [CrossRef]
18. Lin, K.; Fong, D.Y.T.; Zhu, B.; Karlberg, J. Environmental factors on the SARS epidemic: Air temperature, passage of time and multiplicative effect of hospital infection. *Epidemiol. Infect.* **2006**, *134*, 223–230. [CrossRef] [PubMed]
19. Lowen, A.C.; Mubareka, S.; Steel, J.; Palese, P. Influenza virus transmission is dependent on relative humidity and temperature. *PLoS Pathog.* **2007**, *3*, 151. [CrossRef]
20. Tan, J.; Mu, L.; Huang, J.; Yu, S.; Chen, B.; Yin, J. An initial investigation of the association between the SARS outbreak and weather: With the view of the environmental temperature and its variation. *J. Epidemiol. Community Health* **2005**, *59*, 186–192. [CrossRef]
21. Prata, D.N.; Rodrigues, W.; Bermejo, P.H. Temperature significantly changes COVID-19 transmission in (sub)tropical cities of brazil. *Sci. Total. Environ.* **2020**, *729*, 138862. [CrossRef]
22. Jamil, T.; Alam, I.; Gojobori, T.; Duarte, C.M. No evidence for temperature-dependence of the COVID-19 epidemic. *Front. Public Health* **2020**, *8*, 436. [CrossRef] [PubMed]
23. Xie, J.; Zhu, Y. Association between ambient temperature and COVID-19 infection in 122 cities from China. *Sci. Total. Environ.* **2020**, *724*, 138201. [CrossRef] [PubMed]
24. Demongeot, J.; Flet-Berliac, Y.; Seligmann, H. Temperature decreases spread parameters of the new COVID-19 case dynamics. association between ambient temperature and COVID-19 infection in 122 cities from China. *Biology* **2020**, *9*, 94. [CrossRef]
25. Aslam, F.; Awan, T.M.; Syed, J.H.; Kashif, A.; Parveen, M. Sentiments and emotions evoked bynews headlines of coronavirus disease (covid-19) outbreak. *Humanit. Soc. Sci. Commun.* **2020**, *7*, 1–9. [CrossRef]
26. Hung, M.; Lauren, E.; Hon, E.S.; Birmingham, W.C.; Xu, J.; Su, S.; Hon, S.D.; Park, J.; Dang, P.; Lipsky, M.S. Social network analysis of covid-19 sentiments: Application of artificial intelligence. *J. Med. Internet Res.* **2020**, *22*, e22590. [CrossRef]
27. Samuel, J.; Ali, G.G.; Rahman, M.; Esawi, E.; Samuel, Y. Covid-19 public sentiment insights and machine learning for tweets classification. *Information* **2020**, *11*, 314. [CrossRef]
28. Souza, F.S.H.; Hojo-Souza, N.S.; Santos, E.B.; Silva, C.M.; Guidoni, D.L. Predicting the disease outcome in COVID-19 positive patients through Machine Learning: A retrospective cohort study with Brazilian data. *medRxiv* **2020**. [CrossRef]
29. Pinter, G.; Felde, I.; Mosavi, A.; Ghamisi, P.; Gloaguen, R. COVID-19 Pandemic Prediction for Hungary: A Hybrid Machine Learning Approach. *Mathematics* **2020**, *8*, 890. [CrossRef]
30. Arora, P.; Kumar, H.; Panigrahi, B.K. Prediction and analysis of COVID-19 positive cases using deep learning models: A descriptive case study of India. *Chaos Solitons Fractals* **2020**, *139*, 110017. [CrossRef]
31. Toyoshima, Y.; Nemoto, K.; Matsumoto, S.; Nakamura, Y.; Kiyotani, K. SARS-CoV-2 genomic variations associated with mortality rate of COVID-19. *J. Hum. Genet.* **2020**, *65*, 1075–1082. [CrossRef]
32. Mercatelli, D.; Giorgi, F.M. Geographic and genomic distribution of SARS-CoV-2 mutations. *Front. Microbiol.* **2020**. [CrossRef] [PubMed]
33. Bhonde, S.; Bhati, M.; Prasad, J. Predictive Analytics to Combat with COVID-19 Using Genome Sequencing. 2020. Available online: https://ssrn.com/abstract=3580692 (accessed on 5 January 2021).
34. Machine Learning for Biology: How Will COVID-19 Mutate Next? Available online: https://towardsdatascience.com/machine-learning-for-biology-how-will-covid-19-mutate-next-4df93cfaf544 (accessed on 5 January 2021).

35. National Center for Biotechnology Information. Available online: https://www.ncbi.nlm.nih.gov/ (accessed on 5 January 2021).
36. Elbe, S.; Buckland-Merrett, G. Data, disease and diplomacy: GISAID's innovative contribution to global health. *Glob. Challenges* **2017**, *1*, 33–46. [CrossRef] [PubMed]
37. Shu, Y.; McCauley, J. GISAID: Global initiative on sharing all influenza data–from vision to reality. *Eurosurveillance* **2017**, *22*, 13. [CrossRef] [PubMed]
38. Weather Underground. Available online: https://www.wunderground.com/ (accessed on 5 January 2021).
39. nCoV2019. Available online: https://github.com/beoutbreakprepared/nCoV2019/tree/master/latest_data (accessed on 5 January 2021).
40. Global Research on Coronavirus Disease (COVID-19). Available online: https://www.who.int/emergencies/diseases/novel-coronavirus-2019/global-research-on-novel-coronavirus-2019-ncov (accessed on 5 January 2021).
41. medRxiv. Available online: https://www.medrxiv.org/ (accessed on 5 January 2021).
42. bioRxiv. Available online: https://www.biorxiv.org/ (accessed on 5 January 2021).
43. API Summary for the Collection of COVID-19 SARS-CoV-2 Preprints from medRxiv and bioRxiv. Available online: https://api.biorxiv.org/covid19/help (accessed on 5 January 2021).
44. Google Map. Available online: https://www.google.com/maps/ (accessed on 5 January 2021).
45. WIKIPEDIA. Available online: https://www.wikipedia.org/ (accessed on 5 January 2021).
46. NCBI Accession MN908947.3. Available online: https://www.ncbi.nlm.nih.gov/search/all/?term=MN908947 (accessed on 5 January 2021).
47. Wu, F.; Zhao, S.; Yu, B.; Chen, Y.M.; Wang, W.; Song, Z.G.; Hu, Y.; Tao, Z.W.; Tian, J.H.; Pei, Y.Y.; et al. A new coronavirus associated with human respiratory disease in China. *Nature* **2020**, *579*, 265269. [CrossRef]
48. Triplett, M. Evidence that higher temperatures are associated with lower incidence of COVID-19 in pandemic state, cumulative cases reported up to March 27, 2020. *medRxiv* **2020**. [CrossRef]
49. Katoh, K.; Standley, D.M. MAFFT multiple sequence alignment software version 7: Improvements in performance and usability. *Mol. Biol. Evol.* **2013**, *30*, 772–780. [CrossRef]
50. Chawla, N.V.; Bowyer, K.W.; Hall, L.O.; Kegelmeyer, W.P. SMOTE: Synthetic minority over-sampling technique. *J. Artif. Intell. Res.* **2020**, *16*, 321–357. [CrossRef]
51. Alessa, A.; Faezipour, M. A review of influenza detection and prediction through social networking sites. *Theor. Biol. Med Model.* **2018**, *15*, 2. [CrossRef]
52. Lee, K.; Agrawal, A.; Choudhary, A. Forecasting influenza levels using real-time social media streams. In Proceedings of the 2017 IEEE International Conference on Healthcare Informatics (ICHI), Park City, UT, USA, 23–26 August 2017; pp. 409–414.
53. Wang, Y.; Xu, K.; Kang, Y.; Wang, H.; Wang, F.; Avram, A. Regional influenza prediction with sampling Twitter data and PDE model. *Int. J. Environ. Res. Public Health* **2020**, *17*, 678. [CrossRef]
54. TextBlob. Available online: https://github.com/sloria/TextBlob (accessed on 5 January 2021).
55. Chen, T.; Guestrin, C. Xgboost: A scalable tree boosting system. In *Proceedings of the 22nd ACM Sigkdd International Conference on Knowledge Discovery and Data Mining*; ACM: New York, NY, USA, 2016; pp. 785–794.
56. Elman, J.L. Finding structure in time. *Cogn. Sci.* **1990**, *14*, 179–211. [CrossRef]
57. Hochreiter, S.; Schmidhuber, J. Long short-term memory. *Neural Comput.* **1997**, *9*, 1735–1780.
58. Verity, R.; Okell, L.C.; Dorigatti, I.; Winskill, P.; Whittaker, C.; Imai, N.; Cuomo-Dannenburg, G.; Thompson, H.; Walker, P.G.; Fu, H.; et al. Estimates of the severity of coronavirus disease 2019: A model-based analysis. *Lancet Infect. Dis.* **2020**. [CrossRef]
59. Glynn, J.R. Protecting workers aged 60–69 years from COVID-19. *Lancet Infect. Dis.* **2020**. [CrossRef]
60. Wang, H.; Li, T.; Barbarino, P.; Gauthier, S.; Brodaty, H.; Molinuevo, J.L.; Xie, H.; Sun, Y.; Yu, E. Dementia care during COVID-19. *Lancet* **2020**, *395*, 1190–1191.
61. Armitage, R.; Nellums, L.B. COVID-19 and the consequences of isolating the elderly. *Lancet Public Health* **2020**, *5*, e256.
62. Korber, B.; Fischer, W.M.; Gnanakaran, S.; Yoon, H.; Theiler, J.; Abfalterer, W.; Hengartner, N.; Giorgi, E.E.; Bhattacharya, T.; Foley, B.; et al. Tracking changes in SARS-CoV-2 Spike: Evidence that D614G increases infectivity of the COVID-19 virus. *Cell* **2020**, *182*, 812–827.

computation

Article

An Accuracy vs. Complexity Comparison of Deep Learning Architectures for the Detection of COVID-19 Disease

Sima Sarv Ahrabi, Michele Scarpiniti *, Enzo Baccarelli and Alireza Momenzadeh

Department of Information Engineering, Electronics and Telecommunications (DIET), Sapienza University of Rome, Via Eudossiana 18, 00184 Rome, Italy; sima.sarvahrabi@uniroma1.it (S.S.A.); enzo.baccarelli@uniroma1.it (E.B.); alireza.momenzadeh@uniroma1.it (A.M.)
* Correspondence: michele.scarpiniti@uniroma1.it; Tel.:+39-06-44585869

Abstract: In parallel with the vast medical research on clinical treatment of COVID-19, an important action to have the disease completely under control is to carefully monitor the patients. What the detection of COVID-19 relies on most is the viral tests, however, the study of X-rays is helpful due to the ease of availability. There are various studies that employ Deep Learning (DL) paradigms, aiming at reinforcing the radiography-based recognition of lung infection by COVID-19. In this regard, we make a comparison of the noteworthy approaches devoted to the binary classification of infected images by using DL techniques, then we also propose a variant of a convolutional neural network (CNN) with optimized parameters, which performs very well on a recent dataset of COVID-19. The proposed model's effectiveness is demonstrated to be of considerable importance due to its uncomplicated design, in contrast to other presented models. In our approach, we randomly put several images of the utilized dataset aside as a hold out set; the model detects most of the COVID-19 X-rays correctly, with an excellent overall accuracy of 99.8%. In addition, the significance of the results obtained by testing different datasets of diverse characteristics (which, more specifically, are not used in the training process) demonstrates the effectiveness of the proposed approach in terms of an accuracy up to 93%.

Keywords: COVID-19; chest X-ray; convolutional neural network; classification; deep learning

Citation: Sarv Ahrabi, S.; Scarpiniti, M.; Baccarelli, E.; Momenzadeh, A. An Accuracy vs. Complexity Comparison of Deep Learning Architectures for the Detection of COVID-19 Disease. *Computation* **2021**, *9*, 3. https://doi.org/10.3390/computation9010003

Received: 2 December 2020
Accepted: 1 January 2021
Published: 6 January 2021

1. Introduction

The coronavirus that appeared in 2019—the severe acute respiratory syndrome (SARS-CoV-2)—has become a matter of considerable public concern. COVID-19 leads to complications such as acute respiratory disorder, heart problems, and secondary infections in a rather high proportion of patients, with an appallingly high mortality rate. Identifying the infected individuals is naturally a matter of the utmost importance not only to undergo treatment, also to be kept away from others to prevent the disease from spreading [1]. Surveillance programs, which are widely implemented, commonly employ serological tests. However, the limited number of test kits available to many countries can be considered as one of the issues regarding the identification of patients [2], where the identification of patients becomes a tough challenge. Furthermore, these tests take between a few hours and a day for the results to be provided. While some countries may lack the means to adequately perform antibody or viral tests, in addition, these types of medical examinations might be error-prone. Consequently, in this critical situation, what requires addressing is a viable alternative to these tests.

The radiology gives a decisive advantage when we monitor how the disease runs its course, and it represents a common approach due to its availability [3–5]. Hence, along with biomedical methods—like polymerase chain reaction (PCR) that allows the diagnosis of infectious diseases—the study of lung X-rays could prove highly beneficial, specifically to countries with less access to biomedical facilities.

Considering that deep learning (DL) architectures have been successfully applied to various fields, including medical image analysis, it could even further enhance our ability to cope with the difficulty of diagnosing the disease [6]. In fact, the capabilities and influences of these state-of-the-art techniques are growing constantly [7–9]. At the present time, the development of deep neural networks (DNNs), capable of detecting COVID-19 symptoms in an accurate (and simultaneously fast) way, has become a matter of concern to many researchers [10]. A set of studies show the ability of DNNs, specifically the convolutional neural networks (CNNs) [11] to efficiently detect the symptoms of COVID-19 in X-rays [12]. A series of recent studies focus on undertaking a comparative study of pretrained DL models applied to the classification of COVID-19, by using Computed Tomography (CT) Scans or X-rays in specific datasets [13–20]. However, the state-of-the-art research contributions adopt, for the most part, "Transfer Learning" [21–24] as the approach to automatic detection of COVID-19 symptoms. These contributions pursue the goal of developing novel methods, however, they possess their own disadvantages. Regarding these methods, in general, it must be stated the key issue is that only at the cost of high complexity does the accuracy of these designed models improve. In other words, a great accuracy is achieved, provided that the systems complexities increase. Otherwise, the results would not be acceptable. The well-known networks, for instance, AlexNet [25], variants of ResNet [26], VGG [27], GoogLeNet [28], EfficientNet [29], and DensNet [30] prove themselves to be powerful in many applications; however, a major drawback to them is that they usually require significant training time, causing a high cost in real-world applications [31].

In our opinion, DL techniques represent a powerful tool for reinforcing the process of automated (binary) classification of X-rays into normal and infected by COVID-19 images. In this regard, the general framework of deep learning applied to the automatic detection of COVID-19 symptoms can be named as COVID-in-Depth (CoDe).

However, since DNNs are computationally demanding and memory-hungry, a number of techniques have been introduced to tackle this issue [32]. A first approach that can be taken is to minimize the size of DNNs, and simultaneously try to maintain the resulting accuracy at a reasonable level. Another technique addresses this issue by reducing the whole number of parameters [33]. A simple model helps to prevent the overfitting when datasets are limited in size [34]. In this paper, we have tried to minimize the size of the neural network and number of parameters. The main advantage of this network, compared to other models, is its simplicity and low complexity, which leads to a major reduction in computational cost, while maintaining the accuracy at a high level. Consequently, the model is perfectly capable of running fast on low-performance computers with high accuracy.

Fresh Contribution of the Paper

Motivated by these considerations, in this paper, we pursue two main goals:

1. We first draw a comparison of the state-of-the-art approaches that work towards the goal of classifying X-rays into normal and COVID-19 categories. This provides an overview on how the state-of-the-art approaches behave on different dataset commonly used in the literature;
2. We also propose a variant of CNNs—a custom-designed architecture with optimized parameters—that performs very well on a recent dataset. In our contribution, we concentrate our concerted efforts, specifically, on reducing the network complexities, whilst simultaneously achieving the accuracy of a superbly high level. To accomplish our goal, we have optimized our model for an excellent performance and a straightforward design. Moreover, in order to assess the proposed architecture and demonstrate its effectiveness, we test it on some additional datasets, not used in the training phase. In similar works, it is rarely observed that a model is evaluated by referring to large external datasets.

Our proposed CNN-based classifier is trained from scratch, which is different from other contributions that adopt a transfer learning approach. Specifically, the main features of the proposed architecture are:

- A considerably high accuracy of COVID-19 identification;
- A highly reduced system complexity, compared to other state-of-the-art models;
- The usability of the model in resource-limited execution environments;
- The assessment of the proposed model by using external datasets not involved in training process.

The rest of this article is organized as follows. Section 2 is devoted to the related work on binary COVID-19 classification from X-rays, by using CNNs. Section 3 presents the related information on data type, preprocessing, and data augmentation. A detailed description of the proposed architecture is demonstrated in Section 4, while details about experimental setup along with the results, and performance evaluation are discussed in Section 5. In Section 6, the capability of the model will be challenged by the act of classifying X-rays of external datasets. A comparison between state-of-the-art contributions is drawn in Section 7. Finally, the conclusion and possible future research directions are outlined in Section 8.

2. Related Work

The majority of previous research contributions has applied pretrained frameworks to classification of COVID-19 infected patients. In [35], the authors utilize the AlexNet architecture as a feature extractor, where the most efficient features are selected using the Relief algorithm and then in the final stage, the classification of the effective features is conducted, by using the support-vector machines (SVM) classifier. The test results demonstrate an accuracy score of 99.18%. However, finding the optimal parameters for the SVM, and also optimal values for the Relief algorithm, can be considered as the limitations of this study.

ResNet-50 CNN, with conventional transfer learning scheme from ImageNet database, has been used in [36–39]. The validation accuracy of these networks have not exceeded 98%, and some of them present a dramatically low degree of accuracy. Moreover, ResNet-50 is utilized as the feature extractor, and the SVM as the classifier in [40]. This work is not an end-to-end network and the low number of COVID-19 X-rays in the dataset (25 images) causes the result not to be so valuable, while the overall accuracy of the study is 95.38%. With modified ResNet-18, [41] develops a deep convolutional generative adversarial network to produce synthetic data, but is not rather able to produce unique synthetic data, since the proposed network is trained separately for each class. The test accuracy for detection of COVID-19 is reported to be 82.91%. A Deep Convolutional Autoencoder approach, COVIDomaly, is proposed by [42]. After performing 3-fold cross-validation, a pooled ROC–AUC of 0.6902 is obtained for the binary classification.

In [43], the authors perform multi-dilation convolutional layers, where the group convlution uses several dilation rates. The training convergence of the model is very erratic, where it fluctuates a lot after 45 epochs, and the accuracy of 97.4% is achieved for COVID/Normal cases. The ability of capsule networks, in order to classify COVID-19 X-rays is examined in study by [44]. The proposed method, CapsNet, achieves an accuracy of 97.24% in binary classification. In [45], the authors investigate a set of different approaches, in which AlexNet, GoogLeNet, and RestNet-18 are used for multi-classification, where the GoogLeNet is adopted as the main deep transfer model for classification of COVID-19 and normal images. Although, the work achieves 99.9% in the validation accuracy, the use of a very small dataset for the training (69 image of COVID-19 without augmentation) causes a low-level reliability. The EfficientNet [29], based on transfer learning, shows a valuable accuracy on several datasets. However, the authors of [46] employ a network for COVID-19 classification, obtaining a validation accuracy that does not exceed 93.9%. Among the various applied deep transfer learning approaches, [47] achieves a high validation accuracy, by using Xception network (99.52%) for the training, however, the results are not efficient enough in the test analysis (97.40%), compared to the validation accuracy.

A recent study by [48] concludes that the validity of the usual testing protocols in most papers dealing with the automatic diagnosis of COVID-19 might be biased and learn to predict features that predict features that are more dependent on the source dataset than relevant medical information. The attempt, made in [49] based on a modified version of AlexNet, results in the accuracy of 98%, while VGG-19 and DenseNet-201 [50] are not capable of achieving higher overall accuracy than 90%. The authors of [51] utilize the standard version of DenseNet-169 and reach a resulting accuracy of 95.72%. The standard version of VGG-16 with synthetic data augmentation technique, for classifying COVID-19, results in the validation accuracy rate of 95% [52]. A model, based on the combination of a CNN and long short-term memory (LSTM), is developed by [53] to diagnose COVID-19 automatically from X-rays. This network is composed of 21 layers and achieves an accuracy of 99.4%, with a long training time of more than 5 hours. However, their operations take advantage of running at high speeds.

The research by [54] focuses only on the screening stage. The synthetic data, which are generated by a conditional deep convolutional generative adversarial network (conditional DC-GAN), is used to augment the training dataset for COVID-19 classification. The proposed method attains a mean accuracy of 96.97%. In addition, the transfer learning method is used to train four CNNs, including ResNet18, ResNet50, SqueezeNet and DenseNet-121, to identify COVID-19 symptoms in the analyzed chest X-ray images, and three of these networks do not exceed a sensitivity rate of 98%, while the results of the other one are not considerable at all [55]. The VGG-19 and the MobileNet-V2 are employed by the authors of [56] and they confirm that these two networks are not capable of classifying the COVID-19 X-ray images. The ResNet-50 and VGG-16 produce comparatively better results than VGG-19 and MobileNet-V2. The AUC scores of ResNet-50 and VGG-16 are evaluated to be 0.6578 and 0.7264, respectively. The Inception-V3 produces better results than other pre-trained networks, however, the highest AUC score in transfer learning experiments is obtained by DenseNet-121 (0.9648). In [57], the authors proposed nCOVnet, by using neural network-based method on VGG-16, to achieve the overall accuracy of 88.10%. However, for the most part, the obtained results are biased due to the small amount of COVID-19 X-rays [43]. It should be considered that the proposed schemes provide performance in different combinations of classification with balanced sets of data. Moreover, the larger number of non-COVID X-rays are properly utilized for the initial training phase that is effectively transferred for diagnosing COVID-19 in the final transfer learning phase.

A critical question, here, would be whether or not an automatic COVID-19 identification, with a correlation of high accuracy and low system complexity is achievable?

3. Data Type and Preprocessing

3.1. X-ray Images DataSet

The data used in this work has been collected from "The Cancer Imaging Archive (TCIA)" [58]—a collection of X-rays and CT images—related to patients with positive COVID-19 tests [59]. We have separated 253 X-rays from the CTs (last modification in September 15, 2020). The related information has been collected from 105 patients, both females and males, with a minimum age of 19 and maximum age of 91 years old. The mortality was of 10 out of 105 patients. The images, provided in the dataset, have been of very low resolutions, and in addition, each image has been presented in a different resolution to other ones. The original images have been in Digital Imaging and Communications in Medicine (DICOM) format. Normal X-rays are selected randomly from the data collection of Mendeley [60] in the 'jpeg' format. Age, sex and any other information regarding the patients of this dataset is not provided due to privacy concerns. The difference of image sizes and the need to manipulate them for some random samples are shown in Figure 1.

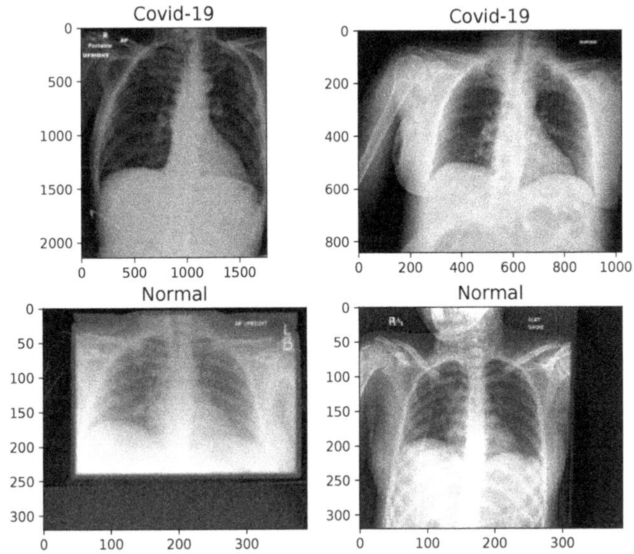

Figure 1. X-ray image samples of two categories before image manipulation.

3.2. Augmentation of Images

The key problem with small and unbalanced datasets is that models trained with them may not be generalized. Hence, these models suffer the issue of overfitting. Data augmentation is an approach to reducing overfitting, by which we are able to increase the amount of data, with only using available data [61]. In this paper, 100 images, with an equal proportion of classes, is held out for testing the model, and the number of samples of each class is balanced by increasing the number of COVID-19 chest X-rays to 500 images. The augmented images are randomly selected from the original X-rays. In total, 1000 (500 normal + 500 COVID-19) images are employed for the training phase. On the other hand, image preprocessing is needed, because the images are not of the same size, as shown in Figure 1, and therefore are converted into the same size for training. In Table 1, we present the manipulations that are applied to the X-rays, before proceeding with the augmentation.

Table 1. Manipulation of the X-ray images before augmentation.

Rotating	Rotate 30% of the images to right and left in maximum 10 degrees.
Shearing	Shear in the x and y axis randomly in 20 degrees.
Cropping	Crop marginal parts of the periphery X-rays.
Resizing	Resize all the images in both width and height equal to 240 pixels.
Elastic distortion	Carried out for 20% of images by setting the grid width and height equal to 2.

For this purpose, we have used the "Augmentor"—a Python package designed to aid the augmentation and artificial generation of image data for machine learning tasks. It is primarily a data augmentation tool, but also incorporate basic image preprocessing functionality. It has an emphasis on providing operations that are typically used in the generation of image data for machine learning problems. The visualization of X-ray images of each class, after manipulation, is shown in Figure 2.

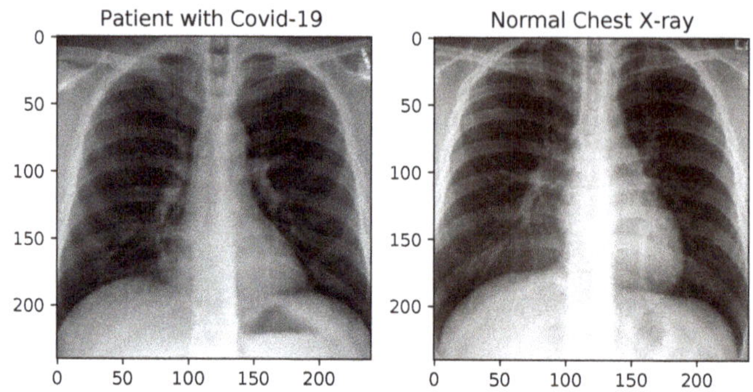

Figure 2. X-ray image samples of two categories after image manipulation.

4. Proposed Method

The proposed method, framed inside the CoDe vision, is explained by discussing the network topology and development parameters. The architecture of the model is explained with the help of diagrams, leading to the CNN model, and its operation and evaluation metrics will be explained later.

4.1. Architecture of the Model

The proposed model deploys Keras functional API, and its overall architecture is presented according to the execution graph of Tensorboard in Figure 3. The learning phase flag in the execution graph, is a 'Boolean tensor' (0 = test, 1 = train) to be passed as input to any Keras function that uses a different behavior at train and test time.

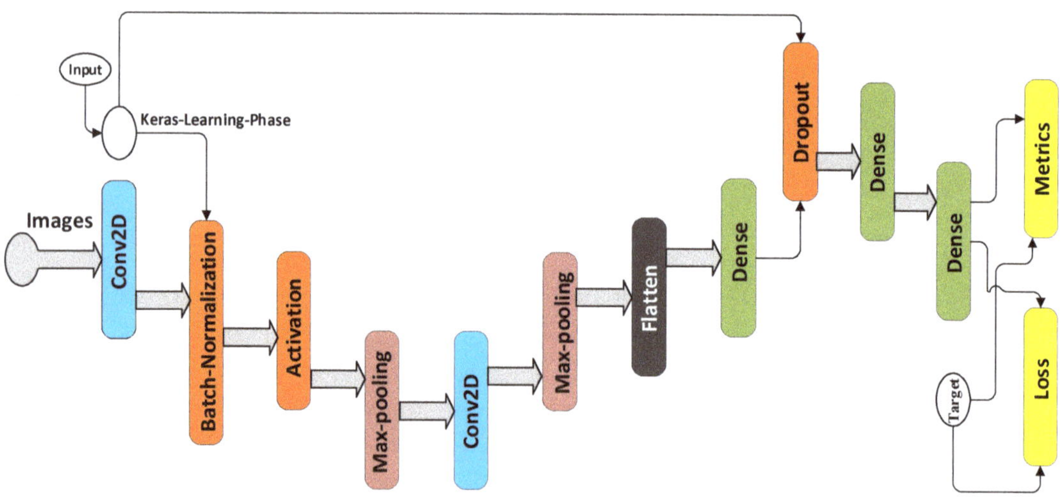

Figure 3. Architecture of the proposed model.

There are 12 layers in this network including convolution, max-pooling, batch normalization, dropout, activation, and fully-connected layers, whose details are summarized in Figure 2. The kernel size of the two convolution layers, i.e., layers 2 and 6 are equal to 3. After each convolution layer, a max-pooling operation is applied to the feature maps, but before the first max-pooling, batch normalization and then the rectified linear unit (ReLU) activation are employed. $ReLU(x) = max(0, x)$ is the element-wise maximum of 0 and the input tensor. The purpose of batch normalization is to normalize the activation of the previous layer at each batch, i.e., applying a transformation that maintains the mean activation close to 0 and the activation standard deviation close to 1 .

The normalization operation is computed by using the following Equations (1) and (2) [62]. Considering the intermediate activation x of a mini-batch of size, we can compute the mean and variance of the batch:

$$\mu_B = \frac{1}{m}\sum_{i=1}^{m} x_i, \qquad \sigma_B^2 = \frac{1}{m}\sum_{i=1}^{m}(x_i - \mu_B)^2, \tag{1}$$

and then compute a normalized version of x, including a small factor ϵ for numerical stability:

$$\widetilde{x}_i = \frac{x_i - \mu_B}{\sqrt{\sigma_B^2 + \varepsilon}}, \tag{2}$$

and finally, $\widetilde{x}_i$ is linearly transformed by γ and β, which are two learned parameters: $y_i = \gamma\widetilde{x}_i + \beta$.

Max-pooling is for reducing the size of the feature map. The parameters for the kernel (filter) size in this work is obtained through brute force technique, and moreover, the stride for convolution and max-pooling operation is set at 2. A 'valid' padding is adopted. To avoid overfitting, we try to keep the model as simple as possible, additionally, we use a dropout layer to make a regularized network section for the inferences. We set the dropout rate to $p = 0.5$, which yields the maximum regularization. Otherwise, in the networks, if all the weights are learned together, normally some of the connections will attain more predictive capability than the others. In such a scenario, as the network is trained iteratively, these powerful connections are learned more, while the weaker ones are ignored [11]. Over many iterations, only a fraction of the node connections is trained and the rest stop participating. In the other words dropout works by probabilistically removing a neuron from designated layers during training or by dropping certain connections [63]. It is worth pointing out that 'Non-trainable parameters', as displayed in Table 2, refer to the number of weights that are not updated during training with back propagation which perform like statistics in the batch normalization layer. They are updated with mean and variance, but they are not "trained with back propagation".

Table 2. Summary of the proposed model.

Layer	Type	Kernel Size	Stride	Output Shape
1	Input layer	-	-	240 × 240 × 3
2	Conv2D	3 × 3	2	119 × 119 × 32
3	Batch Normalization	-	-	119 × 119 × 32
4	Activation	-	-	119 × 119 × 32
5	Max-Pooling 2D	2 × 2	1	59 × 59 × 32
6	Conv2D	3 × 3	1	29 × 29 × 32
7	Max-Pooling 2D	2 × 2	1	14 × 14 × 32
8	Flatten	-	-	6272
9	Dense	-	-	256
10	Dropout	-	-	256
11	Dense	-	-	128
12	Output	-	-	1

Total params: 1,649,185; Trainable params: 1,649,121; Non-trainable params: 64.

4.2. Evaluation Metrics

The performance of the proposed model is evaluated with 5-fold cross-validation. The dataset is divided into two parts, i.e., training and hold out. The held out part is aimed at testing the model, at the end, while the train set is divided into 5 parts. Figure 4 shows the segmentation applied to the dataset.

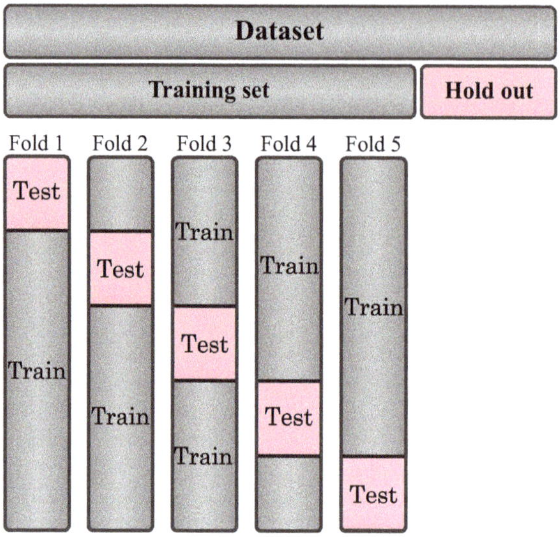

Figure 4. Schematic overview of the 5-fold cross-validation and hold out set.

The training process is carried out 5 times, and the performance of the method is calculated by taking the average of trainings. The evaluation metrics selected in this work, which are commonly used to measure the quality of the model, are: confusion matrix, accuracy, specificity, sensitivity, F1-score, the area under the (Receiver Operating Characteristic) curve (AUC), and the Matthews correlation coefficient (MCC). The calculated parameters of these metrics are based on true positive (TP), true negative (TN), false positive (FP), false negative (FN) rates, as shown in Table 3.

Table 3. The performance metrics for the evaluation of the model.

Performance Metrics	Formula
Sensitivity	$TP/(TP+FN)$
Specificity	$TN/(TN+FP)$
Precision	$TP/(TP+FP)$
F-Score	$2TP/(2TP+FP+FN)$
Accuracy	$(TP+TN)/(TP+FN+FP+TN)$
MCC	$\frac{(TP \times TN - FP \times FN)}{\sqrt{(TP+FP)(TP+FN)(TN+FP)(TN+FN)}}$

5. Experimental Phase

5.1. Primary Preparation

In the experimental setup, first, the dataset is randomly divided into two parts: 653 and 100 images of the dataset belong to the training and hold out respectively. The hold out part is separated with the goal of testing the model after the training is performed. The process is illustrated in Figure 5. Thereafter we perform the augmentation to raw images in training set. All the input images are resized to 240 × 240 pixel size. In the meantime, the dataset is shuffled to overcome the negative effect of the overfitting. The train set is split with respect to 5-fold cross-validation. Afterwards, 25% of each training set is considered as a validation set, in order to use the early stopping strategy. The early stopping and ReduceLROnPlateau methods are employed to monitor the improvement of validation loss, and in the case that no improvement is verified for a 'patience' number of five iterations, the learning rate (lr) is reduced at the $factor$ of 0.1 ($lr_{new} = rl \times factor$).

On the other hand, if the validation loss does not improve for a 'patience' number of 10 iterations, the training will be stopped automatically. The process is now well prepared for the Gradient descent Optimization (SGD) of the Keras. Setting the 'restore_best_weights' to 'True', model weights are restored from the epoch with the best value of the monitored quantity. The training is conducted for 100 epochs, with a batch size of 32. After we reach the best weights, we implement the testing phase by the unseen hold out dataset. The diagram of the process is illustrated in Figure 5 that highlights the division of dataset during the processing.

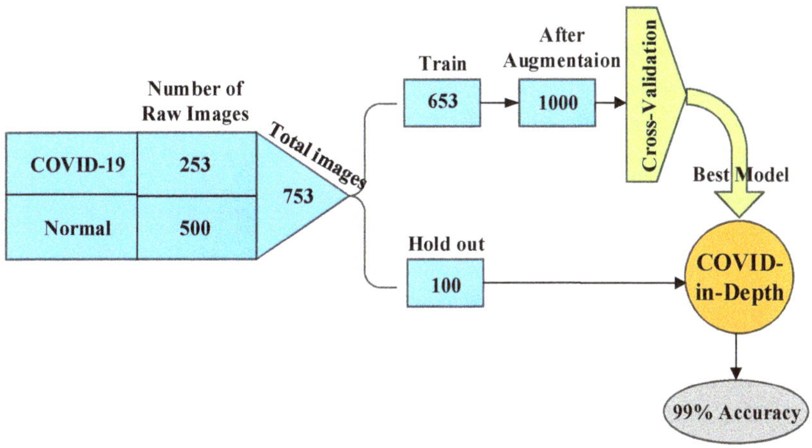

Figure 5. Schematic diagram of the process.

The experimental work is carried out by using Python programming language. Some information about configuration of the model can be found in Table 4.

Table 4. Main information of configuring the model for training.

Programming Language	API	Optimization	Loss	Batch Size	Learning Rate	Epoch
Python	Keras	Gradient Descent (SGD)	Binary-crossentropy	32	0.01	100

5.2. Numerical Results

When the epoch number of training iterations increases, the loss value does not change considerably, suggesting that the model converges well on a relatively optimal state without distinct overfitting or underfitting. The training curves of the loss value and the accuracy for the last fold of training are shown in Figures 6 and 7. The system can be considered a fit model, since the validation error is low, while slightly higher than the training error.

The validity of implementation, associated with the performance metrics, such as precision, recall and F1-score, is shown in Table 5. All the estimations are made through the 'scikit-learn' API [64]. Approximately, for about 400 iterations in 5 times training, the proposed model achieves almost the accuracy of 99.80%, for correct identification of infected cases. Moreover, 100% of not infected people are correctly identified as being healthy. The resultant precision, recall, F1-score, and AUC are all competent enough, to validate the high efficiency of the proposed model. All the measures' values are the average of the 5-fold cross-validation.

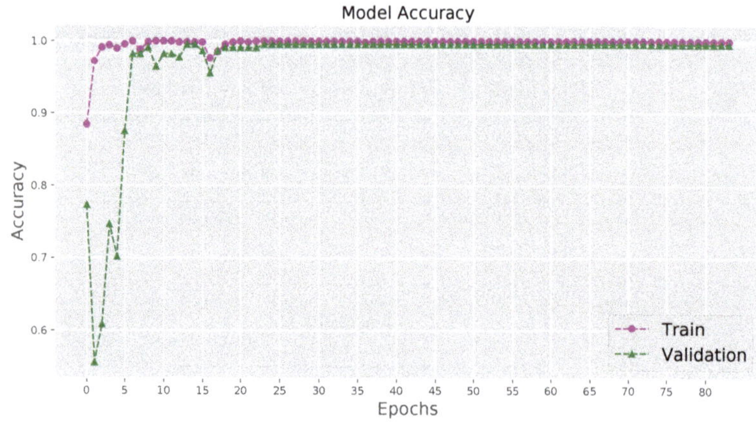

Figure 6. Visualizing the accuracy for the last fold cross-validation training.

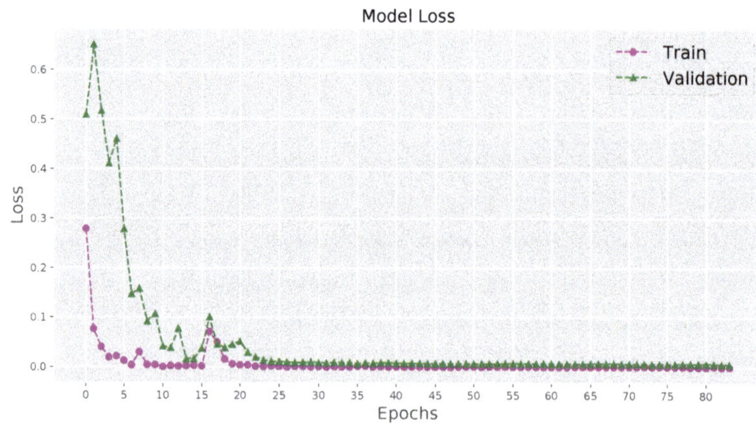

Figure 7. Visualizing the loss for the last fold cross-validation training.

Table 5. Performance of the model in average of 5-fold cross-validation for both classes.

	Precision	Recall	F1-Score	AUC	MCC	Support (Image)
Not infected	0.9980	1.0000	0.9990			
Infected	1.0000	0.9980	0.9990			
Average of two classes	0.9990	0.9990	0.9990	0.9990	0.9980	200

Confusion matrices for all the 5-fold cross-validation training process are presented in Figure 8.

The performance of the network for detecting normal and COVID-19 X-rays, corresponding to the average of the 5-fold cross-validation, is shown in Figure 9. We know that in binary classification, the recall of the positive class is also known as 'sensitivity', and the recall of the negative class is 'specificity'. Therefore, as we can see in the graph of Figure 9, recall of the 'not-infected' is specificity or the true negative rate of the result.

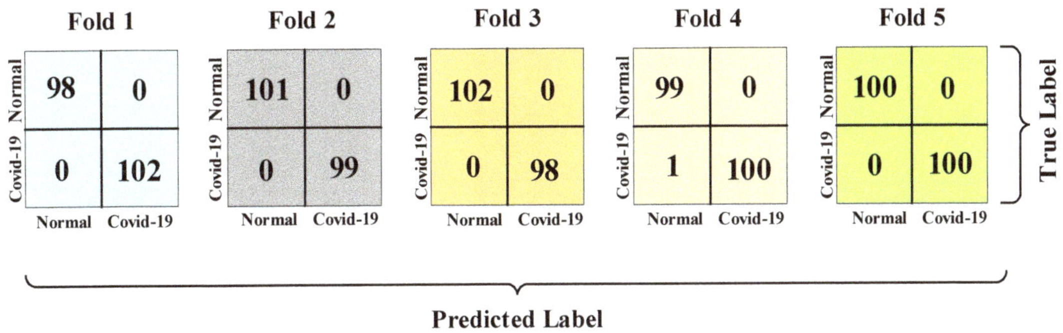

Figure 8. Confusion matrices for the 5-fold cross-validation training process.

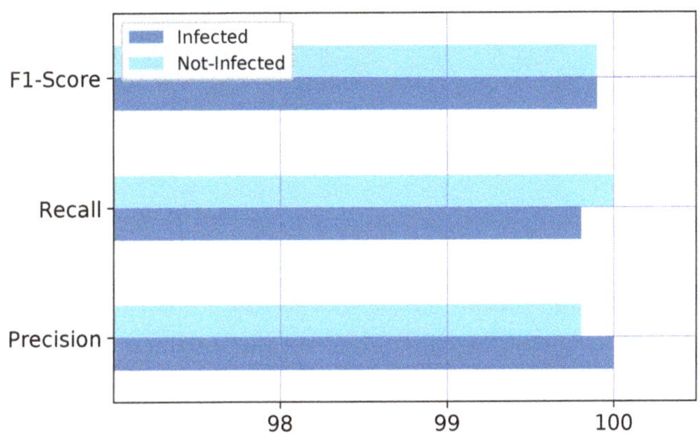

Figure 9. The graph of model performance in average of 5-fold cross-validation for both classes.

5.3. Hold Out Test Dataset

To ensure how well the model works, we employ 100 images belonging to the dataset, which are randomly held out with the same original proportion. The hold out dataset is not involved in the process of training. In total, 66 normal images and 34 images with COVID-19 symptoms evaluate the model performance. The network is capable of identifying all the infected patients correctly and only one misdiagnosing in normal X-rays, with the accuracy of 99%. The confusion matrix for demonstrating the performance is shown in Figure 10, and Table 6 represents the highly considerable performance of the model.

The area under the receiver operating characteristic curve, which is known as AUC, is equal to 0.99.

Table 6. Model performance using hold out test data.

	Precision	Recall	F1-Score	AUC	MCC	Support (Image)
Hold out test data	0.99	0.99	0.99	0.99	0.99	100

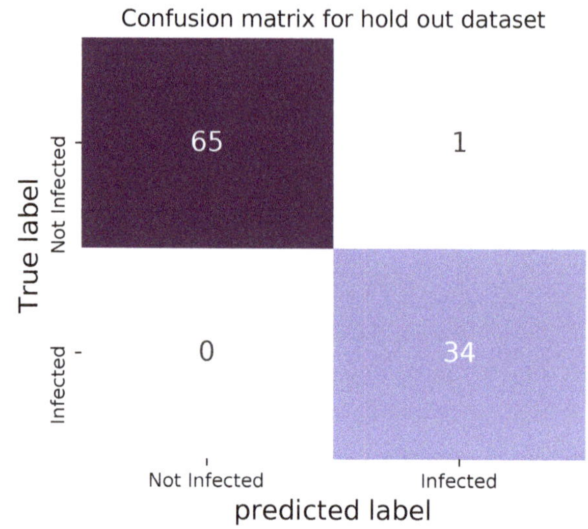

Figure 10. Confusion matrix of testing for hold out X-rays.

5.4. Comparison between Results Obtained by Using Similar Dataset

Although the access to public datasets of COVID-19 X-rays are highly limited, on the other hand, a wide variety of datasets including normal X-rays are reachable. The information about Uniform Resource Locator (URL) of datasets, including the amount of X-rays, which have been employed by other similar classification networks, are presented in Table 7.

Table 7. Available datasets and number of images.

Number	Dataset URL	COVID-19 X-rays	Healthy X-rays
1	https://data.mendeley.com/datasets/rscbjbr9sj/2	–	1583
2	https://github.com/ieee8023/COVID-chestxray-dataset	930	–
3	https://github.com/agchung/Figure1-COVID-chestxray-dataset	55	–
4	https://github.com/shervinmin/DeepCovid	184	1898
5	https://nihcc.app.box.com/v/ChestXray-NIHCC	–	84,312
6	https://www.kaggle.com/andrewmvd/convid19-x-rays	79	–
7	https://www.kaggle.com/paultimothymooney/chest-xray-pneumonia	–	1583
8	https://www.kaggle.com/tawsifurrahman/covid19-radiography-database	1143	1341
9	https://wiki.cancerimagingarchive.net/display/Public/COVID-19	253	–

The COVID-19 dataset [59] used in this work (number 9 in Table 7) is derived from a rather new source that has not been employed by similar works yet, but the normal X-rays are selected through a popular dataset [60] (number 1 in Table 7), and the related article could be observed in [65]. Table 8 presents a comparison between results obtained by several other studies that employ the same dataset of normal X-rays (with different amounts of images) as we have utilized.

As Table 8 shows, the results obtained by the proposed model's performance is considerable compared to other contributions. Although it appears that the accuracy reported by GoogLeNet in [45,66] is equal or slightly higher, however, this item could be easily justified by considering the fact that the training process is performed based on using a quite smaller dataset than ours, and importantly enough, without augmentation.

In addition, the computational cost of the approaches, adopted by the authors (GoogLeNet and VGG-16), are higher.

Table 8. Comparative analysis of binary classification of COVID-19 for the similar utilized dataset.

Network Utilized by Ref.	Accuracy	Sensitivity	Specificity	F1-Score	Precision	AUC	Cross-Validation	Number of Images	
								COVID-19	Normal
CapsNet [44]	0.8919	0.8422	0.9179	0.8421	0.9706	–	10-fold	231	500
CovXNet [43]	0.9740	0.9780	0.9470	0.9710	0.9630	0.9690	5-fold	305	1583
DenseNet121 [56]	0.9839	0.9392	0.9904	–	–	0.9648	8-fold	225	1583
GoogLeNet [45]	0.9990	–	–	–	–	–	–	69	79
VGG-16 [66]	1.0000	1.0000	1.0000	1.0000	1.0000	1.0000	–	16	20
Inception-V3 [56]	0.8803	0.8794	0.9397	0.8788	0.8830	–	–	231	1583
MobileNet-V2 [56]	0.8547	0.8555	0.9273	0.8545	0.8540	–	–	231	1583
ResNet50 [67]	0.9934	0.9100	0.9900	0.9000	0.9000	0.9900	10-fold	239	–
VGG-19 [68]	0.9678	0.9866	0.9646	–	–	–	10-fold	224	504
Proposed	0.9990	0.9980	1.0000	0.9990	0.9990	0.9990	5-fold	500	500

6. Model Evaluation by External Dataset

In order to assess the proposed architecture and demonstrate its effectiveness, we aim at testing it on some datasets not used in the training phase. It could be rarely observed in similar works that a model is evaluated by referring to a considerably large external dataset. However, in this contribution, with the aim of further performance evaluation and ensuring no occurrence of overfitting, we employ other independent datasets of normal and COVID-19 images, which have not been used in the training phase of the proposed model and selected from those available in Table 7. The best weights, achieved by the proposed model, are employed to measure the accuracy of the model for these unseen datasets.

6.1. External Dataset 1

As the first trial, we select two datasets that contain 744 and 1341 COVID-19 and normal images, respectively, for a total of 2085 images. Normal chest X-rays are imported from Kaggle repository "Chest X-Ray Images (Pneumonia)" [69] (number 7 in Table 7) and COVID-19 images from [70] (number 2 in Table 7), both of which are vastly utilized for binary classification, for instance, in [19,23,40,57,71]. The confusion matrix obtained by testing our trained architecture on such datasets is presented in Figure 11.

The obtained results in terms of precision, recall and F1-score, are presented in Table 9. The overall accuracy of the model, in the presence of the independent datasets for the unseen 2085 images, is 92.95%.

Table 9. Performance of the model in external dataset-1 for both classes.

	Precision	Recall	F1-Score	Support (Image)
Not infected	0.9123	0.9851	0.9473	1341
Infected	0.9686	0.8293	0.8936	744
Average of two classes	0.9404	0.9072	0.9204	2085

It is very interesting that the proposed architecture, trained on a different dataset, is capable of achieving very good accuracy, comparable to many state-of-the-art approaches, like those shown in Table 8.

According to the results presented in Figure 11, it is observed that the model could identify the normal X-rays far better than the COVID-19 images. This point can be explained by the fact that the unseen COVID-19 dataset may contain a lot of images that are related to patients at their early stages of the disease, and therefore, cannot be identified as COVID-19 X-rays.

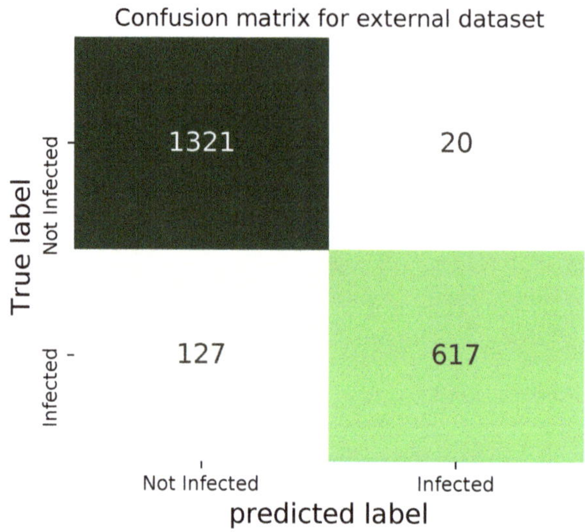

Figure 11. Confusion matrix for the external dataset-1.

6.2. External Dataset 2

Two of the datasets, presented in Table 7, are selected here for the second trial: dataset number 6 and 8 are COVID-19 and normal cases, respectively. In total, 1538 images are employed in the testing process. The confusion matrix obtained by testing our trained architecture on such datasets is shown in Figure 12.

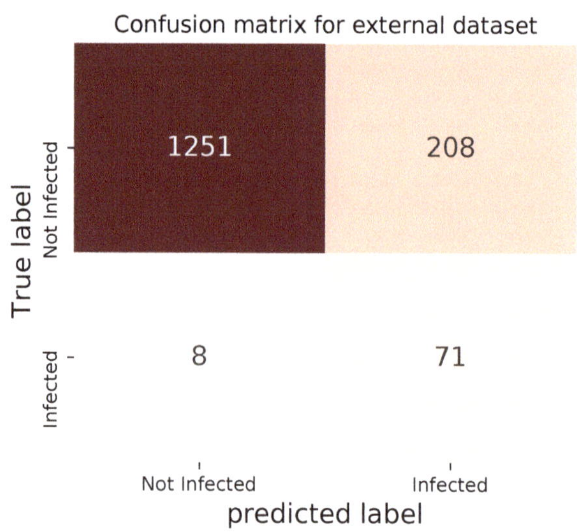

Figure 12. Confusion matrix for the external dataset-2.

The obtained results in terms of precision, recall and F1-score, are presented in Table 10. The overall accuracy of the model, in the presence of this independent dataset for unseen 1583 images, is 85.96%.

Table 10. Performance of the model in external dataset-2 for both classes.

	Precision	Recall	F1-Score	Support (Image)
Not infected	0.9936	0.8574	0.9205	1459
Infected	0.2545	0.8987	0.3966	79
Average of two classes	0.6241	0.8781	0.6586	1538

Once again, the proposed architecture, trained on a different dataset, has been capable of achieving good accuracy.

Contrary to the previous testing, for these datasets, it appears that the performance of the model in identifying COVID-19 improves, which confirms the fundamental difference between datasets. However, the overall accuracy is relatively in acceptable range.

7. Discussion

By analyzing the results, we demonstrate that the proposed model successfully identifies the symptoms of COVID-19, automatically extracting the COVID-19 images from X-rays. The resulting 'accuracy' describes how effectively the values are predicted. The 'precision' determines the reproducibility of the measurement, or how many predictions are correct. The 'recall' shows how many of the correct results are discovered. The 'F1-score' is the harmonic mean of precision and recall.

In Section 5.4, we have compared the most interesting contributions in binary classifications of COVID-19, in which the same dataset of normal X-rays as ours has been used. Table 11 presents another batch of approaches, that have used other identical datasets. The COVID-19 and normal datasets, utilized by the works presented in the Table 11, can be seen in number 2 and 7 in Table 7 respectively.

Table 11. Comparative analysis of binary classification of COVID-19 based on similar datasets.

Network Utilized by Ref.	Accuracy	Sensitivity	Specificity	F1-Score	Precision	AUC
CNN–LSTM [53]	0.9920	0.9930	0.9920	0.9890	0.9850	0.9990
DarkCovidNet [72]	0.9808	0.9513	0.9530	0.9651	0.9803	–
EfficientNet [71]	0.9962	0.9963	0.9963	0.9962	0.9964	0.9949
nCOVnet [57]	0.8810	0.9762	0.7857	0.9762	0.9762	0.8800
Xception [19]	0.9900	0.9930	0.9860	0.9850	0.9830	–
Xception [47]	0.9740	0.9709	0.9729	0.9696	–	–

Table 12 shows some information on the software and hardware that are used in this work.

Table 12. software and hardware information.

OS	OS Architecture	Processor	RAM	System Type
Windows 10	AMD A8-4500M APU	1.90 GHz	6 GB	64-bit

A detailed observation of Tables 8 and 11 shows that the results obtained by the proposed model compete with the state-of-the-art methods. In other words, the proposed model presents a superior set of results in terms of all the validation factors. In spite of the methods based on transfer learning with their complexities, our model delivers a preferable performance with a high level of accuracy accompanied by its simplicity. Those contributions that utilize the same architecture, for example, the works [24,68] in Table 8, appear to produce a quite similar set of results. Although both the papers [19,47] utilize the Xception network, their results appear to be different from each other. The adequate explanation concerning why the two sets of results are different is that the authors of [19] use the Xception method, only as the base model, accompanied by a dropout and two fully-connected layers at the end. The nCOVnet, a VGG-16-based 24-layer network proposed in [57], and the CovXNet, with a large number of convolutional layers introduced by the author of [43] in Table 8, utilize a very deep and complex architecture. Even though we do not replicate the previous methods, our obtained results indicate a high accuracy and a low complexity, compared to all other works in the literature. Obviously, Inception, EfficientNet, ResNet, VGG, and DenseNet involve a computational complexity, considerably greater than our proposed approach. Just for a comparison purpose, Table 13 recaps the number of the trainable parameters of the most common deep architectures used in the approaches compared in Tables 8 and 11 [73]. Table 13 clearly shows the affordability of the proposed idea.

Table 13. Number of trainable parameters of the most common deep models (in millions).

Model	Trainable Parameters
AlexNet	62 M
CapsNet	8 M
DenseNet	25 M
GoogLeNet	4 M
Inception-V3	23.6 M
MobileNet-V2	3.5 M
ResNet18	11.5 M
SqueezeNet	27.5 M
VGG-19	138 M
Xception	22.8 M
Proposed	1.6 M

Therefore, the superior results achieved by our model, along with its simplicity and low computational cost, confirm the efficiency with which the model is able to detect COVID-19 X-rays, with a true positive rate of 99.80%. The accuracy of 99.90%, the AUC of 0.9990, and also the hold out test accuracy of 99% indicate that the model is capable of separating the two classes, indubitably. Moreover, performance of the model, in the presence of different datasets with various characteristics, results in the accuracy of 92.95% and 85.96%. The outcomes of the study indicate that the model is highly capable of classifying the X-rays into COVID-19 and healthy.

8. Conclusions and Hints for Future Research

A fast diagnosis method has a key role in the control of infectious diseases and pandemic situations like the current COVID-19. Some limitations of the PCR test reveal a need for fast alternative methods to be able to serve the front-line experts to make them reach a rapid and accurate diagnosis. Building DNN-based networks, which are capable of identifying COVID-19 symptoms fast and efficiently, and, at the same time, possess uncomplicated architectures, is a major concern to researchers. In this regard, we draw a comparison of the noteworthy approaches devoted to the binary classification of infected images by using Deep Learning techniques with high accuracy (a general framework that we called COVID-in-Depth CoDe). We also propose a variant of a convolutional neural network with optimized parameters that performs very well on a recent dataset. The model presents the average performance accuracy of 99.90% on 5-fold cross validation, and 99.80% for the single recognition of COVID-19. The test accuracy of 99% indicates that the model performs with high precision.

Moreover, we utilize two external datasets to examine the performance of our model, while the obtained results demonstrate that the model achieves 92.95% and 85.96% degrees of accuracy. A hint that could be given here, on the further achievement is pursuing the matter of generalization of the CoDe framework, by providing suitable datasets for training the model that can be large enough and well balanced. In addition, this work can be extended, as a future work, to models capable of recognizing the stages of COVID-19 progression.

Being still in its infancy the topic of this paper, we finally observe that the presented results could be further developed in several directions. In this regard, the (quite recent) contribution in [74] points out that a main promising research direction could be the exploitation of the emerging paradigm of Fog Computing for the distributed implementation and execution of Deep-Learning based analytics engines. Hence, since technological Fog Computing platforms are based on wireless (and possible mobile) technologies [75], a first research direction of potential interest may concern the utilization of massive numbers of transmit/receive antennas at the Fog nodes [76–78] for improving the (possibly, randomly time-varying [79] and a priori unknown [80]) communication capacity of the underlying Fog-based execution platforms, so to shorten the resulting execution time of the supported Deep Learning engines. Motivated by this consideration, we outline a second promising research direction, which can be focused on the utilization of the emerging paradigm of the so-called Conditional Deep Neural Networks (CDNNs) with multiple early-exits to speed up the overall Fog-supported COVID-19 diagnosis process [81].

Author Contributions: The authors contributed equally to this work. All authors have read and agreed to the published version of the manuscript.

Funding: This work has been supported by the projects: "SoFT: Fog of Social IoT" funded by Sapienza University of Rome Bando 2018 and 2019; "End-to-End Learning for 3D Acoustic Scene Analysis (ELeSA)" funded by Sapienza University of Rome Bando Acquisizione di medie e grandi attrezzature scientifiche 2018; and, "DeepFog – Optimized distributed implementation of Deep Learning models over networked multitier Fog platforms for IoT stream applications" funded by Sapienza University of Rome Bando 2020.

Institutional Review Board Statement: Not applicable.

Informed Consent Statement: Not applicable.

Data Availability Statement: Publicly available datasets were analyzed in this study. This data can be found in the web links of Table 7.

Conflicts of Interest: The authors declare no conflict of interest.

References

1. Lai, C.C.; Shih, T.P.; Ko, W.C.; Tang, H.J.; Hsueh, P.R. Severe acute respiratory syndrome coronavirus 2 (SARS-CoV-2) and corona virus disease-2019 (COVID-19): The epidemic and the challenges. *Int. J. Antimicrob. Agents* **2020**, 105924. [CrossRef]
2. Peeling, R.W.; Wedderburn, C.J.; Garcia, P.J.; Boeras, D.; Fongwen, N.; Nkengasong, J.; Sall, A.; Tanuri, A.; Heymann, D.L. Serology testing in the COVID-19 pandemic response. *Lancet Infect. Dis.* **2020**, *20*, 245–249. [CrossRef]
3. Sahar, F.; Iqbal, R.; Maha, H.; Salim, S. Radiological Findings in Patients with COVID-19. *Cureus* **2020**, *12*. [CrossRef]
4. Lomoro, P.; Verde, F.; Zerboni, F.; Simonetti, I.; Borghi, C.; Fachinetti, C.; Natalizi, A.; Martegani, A. COVID-19 pneumonia manifestations at the admission on chest ultrasound, radiographs, and CT: Single-center study and comprehensive radiologic literature review. *Eur. J. Radiol. Open* **2020**, *7*, 100231. [CrossRef]
5. Wong, H.Y.F.; Lam, H.Y.S.; Fong, A.H.T.; Leung, S.T.; Chin, T.W.Y.; Lo, C.S.Y.; Lee, E.Y.P.; Macy Lui, M.; Lee, J.C.Y.; Chiu, K.W.; et al. Frequency and distribution of chest radiographic findings in COVID-19 positive patients. *Radiology* **2020**, *296*, 201160. [CrossRef]
6. Goodfellow, I.; Bengio, Y.; Courville, A. *Deep Learning*; Adaptive Ccomputation and Machine Learning Series; MIT Press: Cambridge, UK, 2016; p. 800.
7. Ai, T.; Yang, Z.; Hou, H.; Zhan, C.; Chen, C.; Lv, W.; Tao, Q.; Sun, Z.; Xia, L. Correlation of chest CT and RT-PCR testing in coronavirus disease 2019 (COVID-19) in China: A report of 1014 cases. *Radiography* **2020**, *296*, 200642. [CrossRef]
8. Song, F.; Shi, N.; Shan, F.; Zhang, Z.; Shen, J.; Lu, H.; Ling, Y.; Jiang, Y.; Shi, Y. Emerging 2019 novel coronavirus (2019-nCoV) pneumonia. *Radiology* **2020**, *295*, 210–217. [CrossRef] [PubMed]
9. Fang, Y.; Zhang, H.; Xie, J.; Lin, M.; Ying, L.; Pang, P.; Ji, W. Sensitivity of chest CT for COVID-19: comparison to RT-PCR. *Radiology* **2020**, *200432*. [CrossRef]
10. Shuja, J.; Alanazi, E.; Alasmary, W.; Alashaikh, A. COVID-19 open source data sets: A comprehensive survey. *Appl. Intell.* **2020**, 1–30. [CrossRef]
11. Goodfellow, I.; Bengio, Y.; Courville, A. *Deep Learning*; MIT Press: Cambridge, MA, USA, 2016.

12. Tartaglione, E.; Barbano, C.A.; Berzovini, C.; Calandri, M.; Grangetto, M. Unveiling COVID-19 from Chest X-ray with deep learning: A hurdles race with small data. *Int. J. Environ. Res. Public Health* **2020**, *17*, 6933. [CrossRef] [PubMed]
13. Shi, F.; Wang, J.; Shi, J.; Wu, Z.; Wang, Q.; Tang, Z.; He, K.; Shi, Y.; Shen, D. Review of artificial intelligence techniques in imaging data acquisition, segmentation and diagnosis for COVID-19. *IEEE Rev. Biomed. Eng.* **2020**. [CrossRef] [PubMed]
14. Albahri, O.; Zaidan, A.; Albahri, A.; Zaidan, B.; Abdulkareem, K.H.; Al-Qaysi, Z.; Alamoodi, A.; Aleesa, A.; Chyad, M.; Alesa, R.; et al. Systematic review of artificial intelligence techniques in the detection and classification of COVID-19 medical images in terms of evaluation and benchmarking: Taxonomy analysis, challenges, future solutions and methodological aspects. *J. Infect. Public Health* **2020**, *13*, 1381–1396. [CrossRef] [PubMed]
15. Karim, M.; Döhmen, T.; Rebholz-Schuhmann, D.; Decker, S.; Cochez, M.; Beyan, O. DeepCOVIDExplainer: Explainable covid-19 predictions based on chest x-ray images. *arXiv* **2020**, arXiv:2004.04582.
16. Victor, U.; Dong, X.; Li, X.; Obiomon, P.; Qian, L. Effective COVID-19 screening using chest radiography images via deep learning. *Training* **2020**, *7*, 152.
17. Selvan, R.; Dam, E.; Detlefsen, N.S.; Rischel, S.; Sheng, K.; Nielsen, M.; Pai, A. Lung Segmentation from Chest X-rays using Variational Data Imputation. In *ICML Workshop on the Art of Learning with Missing Values (ICML Artemiss)*; 2020. Available online: https://arxiv.org/abs/2005.10052 (accessed on 20 November 2020).
18. Alafif, T. Machine and Deep Learning Towards COVID-19 Diagnosis and Treatment: Survey, Challenges, and Future Directions. *engrXiv* **2020**. [CrossRef]
19. Khan, A.I.; Shah, J.L.; Bhat, M.M. CoroNet: A deep neural network for detection and diagnosis of COVID-19 from chest x-ray images. *Comput. Methods Programs Biomed.* **2020**, 105581. [CrossRef]
20. Butt, C.; Gill, J.; Chun, D.; Babu, B.A. Deep learning system to screen coronavirus disease 2019 pneumonia. *Appl. Intell.* **2020**, 1. [CrossRef]
21. Toğaçar, M.; Ergen, B.; Cömert, Z. COVID-19 detection using deep learning models to exploit Social Mimic Optimization and structured chest X-ray images using fuzzy color and stacking approaches. *Comput. Biol. Med.* **2020**, 103805. [CrossRef]
22. Ucar, F.; Korkmaz, D. COVIDiagnosis-Net: Deep Bayes-SqueezeNet based Diagnostic of the Coronavirus Disease 2019 (COVID-19) from X-Ray Images. *Med. Hypotheses* **2020**, 109761. [CrossRef]
23. Elasnaoui, K.; Chawki, Y. Using X-ray images and deep learning for automated detection of coronavirus disease. *J. Biomol. Struct. Dyn.* **2020**, *38*, 1–22. [CrossRef]
24. Vaid, S.; Kalantar, R.; Bhandari, M. Deep learning COVID-19 detection bias: accuracy through artificial intelligence. *Int. Orthop.* **2020**, *44*, 1539–1542. [CrossRef] [PubMed]
25. Krizhevsky, A.; Sutskever, I.; Hinton, G.E. ImageNet classification with deep convolutional neural networks. *Commun. ACM* **2017**, *60*, 84–90. [CrossRef]
26. He, K.; Zhang, X.; Ren, S.; Sun, J. Deep residual learning for image recognition. In Proceedings of the IEEE Conference on Computer Vision and Pattern Recognition, Las Vegas, NV, USA, 27–30 June 2016; pp. 770–778. [CrossRef]
27. Simonyan, K.; Zisserman, A. Very deep convolutional networks for large-scale image recognition. In Proceedings of the International Conference on Learning Representations, San Diego, CA, USA, 7–9 May 2015.
28. Szegedy, C.; Liu, W.; Jia, Y.; Sermanet, P.; Reed, S.; Anguelov, D.; Erhan, D.; Vanhoucke, V.; Rabinovich, A. Going deeper with convolutions. In Proceedings of the IEEE Conference on Computer Vision and Pattern Recognition, Boston, MA, USA, 7–12 June 2015; pp. 1–9. [CrossRef]
29. Tan, M.; Le, Q.V. EfficientNet: Rethinking model scaling for convolutional neural networks. In Proceedings of the 36th International Conference on Machine Learning, Long Beach, CA, USA, 9–15 June 2019; pp. 6105–6114.
30. Huang, G.; Liu, Z.; Van Der Maaten, L.; Weinberger, K.Q. Densely connected convolutional networks. In Proceedings of the IEEE Conference on Computer Vision and Pattern Recognition, Honolulu, HI, USA, 21–26 July 2017; pp. 4700–4708. [CrossRef]
31. Zhang, L.; Schaeffer, H. Forward stability of ResNet and its variants. *J. Math. Imaging Vis.* **2020**, *62*, 328–351. [CrossRef]
32. Véstias, M.P. A survey of convolutional neural networks on edge with reconfigurable computing. *Algorithms* **2019**, *12*, 154. [CrossRef]
33. Qiu, J.; Wang, J.; Yao, S.; Guo, K.; Li, B.; Zhou, E.; Yu, J.; Tang, T.; Xu, N.; Song, S.; et al. Going Deeper with Embedded FPGA Platform for Convolutional Neural Network. In Proceedings of the 2016 ACM/SIGDA International Symposium on Field-Programmable Gate (FPGA '16), Monterey, CA, USA, 21–23 February 2016. [CrossRef]
34. Dieterle, F.J. Multianalyte Quantifications by Means of Integration of Artificial Neural Networks, Genetic Algorithms and Chemometrics for Time-Resolved Analytical Data. Ph.D. Thesis, Eberhard-Karls-Universität Tübingen, Tübingen, Germany, 2003.
35. Turkoglu, M. COVIDetectioNet: COVID-19 diagnosis system based on X-ray images using features selected from pre-learned deep features ensemble. *Appl. Intell.* **2020**, 1–14. [CrossRef]
36. Narin, A.; Kaya, C.; Pamuk, Z. Automatic detection of coronavirus disease (covid-19) using x-ray images and deep convolutional neural networks. *arXiv* **2020**, arXiv:2003.10849.
37. Farooq, M.; Hafeez, A. Covid-ResNet: A deep learning framework for screening of covid19 from radiographs. *arXiv* **2020**, arXiv:2003.14395.
38. Wu, X.; Hui, H.; Niu, M.; Li, L.; Wang, L.; He, B.; Yang, X.; Li, L.; Li, H.; Tian, J.; et al. Deep learning-based multi-view fusion model for screening 2019 novel coronavirus pneumonia: A multicentre study. *Eur. J. Radiol.* **2020**, 109041. [CrossRef]

39. Hall, L.O.; Paul, R.; Goldgof, D.B.; Goldgof, G.M. Finding covid-19 from chest x-rays using deep learning on a small dataset. *arXiv* **2020**, arXiv:2004.02060.
40. Sethy, P.K.; Behera, S.K. Detection of coronavirus disease (COVID-19) based on deep features. *Preprints* **2020**, [CrossRef]
41. Loey, M.; Manogaran, G.; Khalifa, N.E.M. A deep transfer learning model with classical data augmentation and CGAN to detect COVID-19 from chest CT radiography digital images. *Neural Comput. Appl.* **2020**, 1–13. [CrossRef] [PubMed]
42. Khoshbakhtian, F.; Ashraf, A.B.; Khan, S.S. COVIDomaly: A Deep Convolutional Autoencoder Approach for Detecting Early Cases of COVID-19. *arXiv* **2020**, arXiv:2010.02814.
43. Mahmud, T.; Rahman, M.A.; Fattah, S.A. CovXNet: A multi-dilation convolutional neural network for automatic COVID-19 and other pneumonia detection from chest X-ray images with transferable multi-receptive feature optimization. *Comput. Biol. Med.* **2020**, *122*, 103869. [CrossRef]
44. Toraman, S.; Alakus, T.B.; Turkoglu, I. Convolutional CapsNet: A novel artificial neural network approach to detect COVID-19 disease from X-ray images using capsule networks. *Chaos Solitons Fractals* **2020**, *140*, 110122. [CrossRef]
45. Loey, M.; Smarandache, F.; M Khalifa, N.E. Within the Lack of Chest COVID-19 X-ray Dataset: A Novel Detection Model Based on GAN and Deep Transfer Learning. *Symmetry* **2020**, *12*, 651. [CrossRef]
46. Luz, E.; Silva, P.L.; Silva, R.; Moreira, G. Towards an efficient deep learning model for covid-19 patterns detection in x-ray images. *arXiv* **2020**, arXiv:2004.05717.
47. Das, N.N.; Kumar, N.; Kaur, M.; Kumar, V.; Singh, D. Automated deep transfer learning-based approach for detection of COVID-19 infection in chest X-rays. *IRBM* **2020**. [CrossRef]
48. Maguolo, G.; Nanni, L. A critic evaluation of methods for covid-19 automatic detection from x-ray images. *arXiv* **2020**, arXiv:2004.12823.
49. Maghdid, H.S.; Asaad, A.T.; Ghafoor, K.Z.; Sadiq, A.S.; Khan, M.K. Diagnosing COVID-19 pneumonia from X-ray and CT images using deep learning and transfer learning algorithms. *arXiv* **2020**, arXiv:2004.00038.
50. Hemdan, E.E.D.; Shouman, M.A.; Karar, M.E. COVIDX-Net: A framework of deep learning classifiers to diagnose covid-19 in x-ray images. *arXiv* **2020**, arXiv:2003.11055.
51. Hammoudi, K.; Benhabiles, H.; Melkemi, M.; Dornaika, F.; Arganda-Carreras, I.; Collard, D.; Scherpereel, A. Deep Learning on Chest X-ray Images to Detect and Evaluate Pneumonia Cases at the Era of COVID-19. *arXiv* **2020**, arXiv:2004.03399.
52. Waheed, A.; Goyal, M.; Gupta, D.; Khanna, A.; Al-Turjman, F.; Pinheiro, P.R. CovidGAN: Data augmentation using auxiliary classifier gan for improved covid-19 detection. *IEEE Access* **2020**, *8*, 91916–91923. [CrossRef]
53. Islam, M.Z.; Islam, M.M.; Asraf, A. A combined deep CNN-LSTM network for the detection of novel coronavirus (COVID-19) using X-ray images. *Informatics Med. Unlocked* **2020**, *20*, 100412. [CrossRef] [PubMed]
54. Zulkifley, M.A.; Abdani, S.R.; Zulkifley, N.H. COVID-19 Screening Using a Lightweight Convolutional Neural Network with Generative Adversarial Network Data Augmentation. *Symmetry* **2020**, *12*, 1530. [CrossRef]
55. Minaee, S.; Kafieh, R.; Sonka, M.; Yazdani, S.; Jamalipour Soufi, G. Deep-COVID: Predicting COVID-19 from chest X-ray images using deep transfer learning. *Med. Image Anal.* **2020**, *65*, 101794. [CrossRef]
56. Sekeroglu, B.; Ozsahin, I. Detection of COVID-19 from Chest X-Ray Images Using Convolutional Neural Networks. *Slas Technol. Transl. Life Sci. Innov.* **2020**, 1–13. [CrossRef]
57. Panwar, H.; Gupta, P.; Siddiqui, M.K.; Morales-Menendez, R.; Singh, V. Application of Deep Learning for Fast Detection of COVID-19 in X-Rays using nCOVnet. *Chaos Solitons Fractals* **2020**, 109944. [CrossRef]
58. Clark, K.; Vendt, B.; Smith, K.; Freymann, J.; Kirby, J.; Koppel, P.; Moore, S.; Phillips, S.; Maffitt, D.; Pringle, M.; et al. The Cancer Imaging Archive (TCIA): maintaining and operating a public information repository. *J. Digit. Imaging* **2013**, *26*, 1045–1057. [CrossRef]
59. Desai, S.; Baghal, A.; Wongsurawat, T.; Al-Shukri, S.; Gates, K.; Farmer, P.; Rutherford, M.; Blake, G.; Nolan, T.; Powell, T.; et al. Data from Chest Imaging with Clinical and Genomic Correlates Representing a Rural COVID-19 Positive Population [Data set]. *Cancer Imaging Arch.* **2020**. [CrossRef]
60. Kermany, D.; Zhang, K.; Goldbaum, M. Labeled optical coherence tomography (OCT) and Chest X-Ray images for classification. *Mendeley Data* **2018**, *2*. [CrossRef]
61. Perez, L.; Wang, J. The effectiveness of data augmentation in image classification using deep learning. *arXiv* **2017**, arXiv:1712.04621.
62. Ioffe, S.; Szegedy, C. Batch normalization: Accelerating deep network training by reducing internal covariate shift. In Proceedings of the 32nd International Conference on Machine Learning, Lille, France, 6–11 July 2015.
63. Gal, Y.; Ghahramani, Z. A theoretically grounded application of dropout in recurrent neural networks. In Proceedings of the Advances in Neural information processing systems, Barcelona, Spain, 5–10 December 2016; pp. 1019–1027.
64. Pedregosa, F.; Varoquaux, G.; Gramfort, A.; Michel, V.; Thirion, B.; Grisel, O.; Blondel, M.; Prettenhofer, P.; Weiss, R.; Dubourg, V.; et al. Scikit-learn: Machine learning in Python. *J. Mach. Learn. Res.* **2011**, *12*, 2825–2830.
65. Kermany, D.S.; Goldbaum, M.; Cai, W.; Valentim, C.C.; Liang, H.; Baxter, S.L.; McKeown, A.; Yang, G.; Wu, X.; Yan, F.; et al. Identifying medical diagnoses and treatable diseases by image based deep learning. *Cell* **2018**, *172*, 1122–1131. [CrossRef] [PubMed]

66. Dansana, D.; Kumar, R.; Bhattacharjee, A.; Hemanth, D.J.; Gupta, D.; Khanna, A.; Castillo, O. Early diagnosis of COVID-19-affected patients based on X-ray and computed tomography images using deep learning algorithm. *Soft Comput.* **2020**, *28*, 1–9. [CrossRef]
67. Narayanan, B.N.; Hardie, R.C.; Krishnaraja, V.; Karam, C.; Davuluru, V.S.P. Transfer-to-Transfer Learning Approach for Computer Aided Detection of COVID-19 in Chest Radiographs. *AI* **2020**, *1*, 539–557. [CrossRef]
68. Apostolopoulos, I.D.; Mpesiana, T.A. Covid-19: automatic detection from x-ray images utilizing transfer learning with convolutional neural networks. *Phys. Eng. Sci. Med.* **2020**, *43*, 635–640. [CrossRef]
69. Chest X-ray Images (Pneumonia). 2020. Available online: https://www.kaggle.com/paultimothymooney/chest-xray-pneumonia (accessed on 20 November 2020).
70. Cohen, J.P. COVID-19 Image Data Collection. 2020. Available online: https://github.com/ieee8023/COVID-chestxray-dataset (accessed on 20 November 2020).
71. Marques, G.; Agarwal, D.; de la Torre Díez, I. Automated medical diagnosis of COVID-19 through EfficientNet convolutional neural network. *Appl. Soft Comput.* **2020**, *96*, 106691. [CrossRef]
72. Ozturk, T.; Talo, M.; Yildirim, E.A.; Baloglu, U.B.; Yildirim, O.; Acharya, U.R. Automated detection of COVID-19 cases using deep neural networks with X-ray images. *Comput. Biol. Med.* **2020**, *121*, 103792. [CrossRef]
73. Khan, A.; Sohail, A.; Zahoora, U.; Qureshi, A.Q. A survey of the recent architectures of deep convolutional neural networks. *Artif. Intell. Rev.* **2020**, *53*, 5455–5516. [CrossRef]
74. Baccarelli, E.; Scardapane, S.; Scarpiniti, M.; Momenzadeh, A.; Uncini, A. Optimized training and scalable implementation of Conditional Deep Neural Networks with early exits for Fog-supported IoT applications. *Inf. Sci.* **2020**, *521*, 107–143. [CrossRef]
75. Shojafar, M.; Pooranian, Z.; Vinueza Naranjo, P.G.; Baccarelli, E. FLAPS: bandwidth and delay-efficient distributed data searching in Fog-supported P2P content delivery networks. *J. Supercomput.* **2017**, *73*, 5239–5260. [CrossRef]
76. Baccarelli, E.; Biagi, M. Power-allocation policy and optimized design of multiple-antenna systems with imperfect channel estimation. *IEEE Trans. Veh. Technol.* **2004**, *53*, 136–145. [CrossRef]
77. Baccarelli, E.; Biagi, M.; Pelizzoni, C. On the information throughput and optimized power allocation for MIMO wireless systems with imperfect channel estimation. *IEEE Trans. Signal Process.* **2005**, *53*, 2335–2347. [CrossRef]
78. Baccarelli, E.; Biagi, M. Performance and optimized design of space-time codes for MIMO wireless systems with imperfect channel estimates. *IEEE Trans. Signal Process.* **2004**, *52*, 2911–2923. [CrossRef]
79. Baccarelli, E.; Cusani, R.; Galli, S. A novel adaptive receiver with enhanced channel tracking capability for TDMA-based mobile radio communications. *IEEE J. Sel. Areas Commun.* **1998**, *16*, 1630–1639. [CrossRef]
80. Baccarelli, E.; Cusani, R. Recursive Kalman-type optimal estimation and detection of hidden Markov chains. *Signal Process.* **1996**, *51*, 55–64. [CrossRef]
81. Scardapane, S.; Scarpiniti, M.; Baccarelli, E.; Uncini, A. Why should we add early exits to neural networks? *Cogn. Comput.* **2020**, *12*, 954–966. [CrossRef]

computation

MDPI

Article

Causal Modeling of Twitter Activity during COVID-19

Oguzhan Gencoglu [1,*] and Mathias Gruber [2]

1 Faculty of Medicine and Health Technology, Tampere University, 33720 Tampere, Finland
2 LEO Pharma, 2750 Ballerup, Denmark; nano.mathias@gmail.com
* Correspondence: oguzhan.gencoglu@tuni.fi

Received: 26 August 2020; Accepted: 25 September 2020; Published: 29 September 2020

Abstract: Understanding the characteristics of public attention and sentiment is an essential prerequisite for appropriate crisis management during adverse health events. This is even more crucial during a pandemic such as COVID-19, as primary responsibility of risk management is not centralized to a single institution, but distributed across society. While numerous studies utilize Twitter data in descriptive or predictive context during COVID-19 pandemic, causal modeling of public attention has not been investigated. In this study, we propose a causal inference approach to discover and quantify causal relationships between pandemic characteristics (e.g., number of infections and deaths) and Twitter activity as well as public sentiment. Our results show that the proposed method can successfully capture the epidemiological domain knowledge and identify variables that affect public attention and sentiment. We believe our work contributes to the field of infodemiology by distinguishing events that correlate with public attention from events that cause public attention.

Keywords: Twitter; machine learning; causal inference; COVID-19; sentiment analysis; social media

1. Introduction

On 11 March 2020, Coronavirus disease 2019 (COVID-19) was declared a pandemic by the World Health Organization [1] and more than 30 million people have been infected by it as of 19 September 2020 [2]. During such crises, capturing the dissemination of information, monitoring public opinion, observing compliance to measures, preventing disinformation, and relaying timely information is crucial for risk communication and decision-making about public health [3]. Previous national and global adverse health events show that social media surveillance can be utilized successfully for systematic monitoring of public perception in real-time due to its instantaneous global coverage [4–9].

Due to its large number of users, Twitter has been the primary social media platform for acquiring, sharing, and spreading information during global adverse events, including the COVID-19 pandemic [10]. Especially during the early stages of the COVID-19 pandemic, millions of posts have been tweeted in a span of couple of weeks by users, that is, citizens, politicians, corporations, and governmental institutions [11–14]. Consequently, numerous studies proposed and utilized Twitter as a data source for extracting insights on public health as well as insights on public attention during the COVID-19 pandemic. Focus of these studies include content analysis [15], topic modeling [16], sentiment analysis [17], nowcasting or forecasting of the disease [18], early detection of the outbreak [19], quantifying and detecting misinformation, disinformation, or conspiracies [20], and measuring public attitude towards relevant health concepts (e.g., social distancing or working from home) [21].

Despite such abundance of studies on manual or automatic analysis of social media data during COVID-19, *causal* modeling of relationships between characteristics of the pandemic and social media activity has not been investigated at all, as of September 2020. While descriptive statistical analysis (e.g., correlation, cluster, or exploratory analysis) is beneficial for pattern and hypothesis discovery, and standard machine learning methods are effective in predictive modeling of those patterns, causal inference of relevant phenomena will not be possible without causal computational modeling. Causal modeling in the context of social media and pandemic can enable the optimization of onset of risk communication interventions to increase dissemination of accurate information. Similarly, it can be utilized to prevent acute propagation of negative sentiment with timely interventions. Consequently, such causal modeling can help risk communication policies to shift from alerting people to reassuring them. Furthermore, causal modeling enables simulation of what-if scenarios to enhance disaster preparedness. Therefore, as public decision-making can benefit from adequate assessment of public attention and correct understanding of underlying causes affecting it, we hereby propose causal modeling of Twitter activity.

We hypothesize that daily Twitter activity and sentiment during the COVID-19 pandemic has a causal relationship with the characteristics of the pandemic as well as with certain country statistics. We propose a structural causal modeling approach for discovering causal relationships and quantifying likelihood of events under various conditions (i.e., causal queries). To validate our approach, we collect close to 1 million tweets with location information spanning 57 days and identify several attributes of COVID-19 pandemic that might affect Twitter activity. We first employ a structure learning method to automatically construct a graphical causal structure in a data-driven manner. Then, we utilize Bayesian Networks (BNs) to learn conditional probability distributions of daily Twitter activity (number of daily tweets) and average public sentiment with respect to several pandemic characteristics such as total number of deaths and number of new infections. Our results show that the proposed structure discovery method can successfully capture the epidemiological domain knowledge. Furthermore, causal inference of daily Twitter activity with cross-validation across 12 countries show that our approach provides accurate predictions of Twitter activity with interpretable and intuitive results. We have released the full source code of our study (https://github.com/ogencoglu/causal_twitter_modeling_covid19). We believe our study contributes to the field of infodemiology by proposing causal modeling of public attention during the crisis of COVID-19 pandemic.

2. Going Beyond Correlations

Use of observational data from social media was proven to be beneficial in systematic monitoring of public opinion during adverse health events [4–9]. Such utilization of large, publicly available data becomes even more relevant during a global pandemic such as COVID-19, as neither enough time nor a practical way to run variety of randomized control trials for quantifying public opinion exist. Furthermore, as disease containment measures (e.g., lockdowns, quarantines, and curfews), associated financial issues (e.g., due to inability to work), and changes in social dynamics may impact mental health negatively [22–24], opinion surveillance methods that do not carry the risk of further stressing of the participants are pertinent.

Themes of previous studies that focus on exploration of, description of, correlation of, or predictive modeling with Twitter data during COVID-19 pandemic include sentiment analysis [17,25–28], public attitude/interest measurement [21,29–31], content analysis [15,32–36], topic modeling [16,26,27,37–40], analysis of misinformation, disinformation, or conspiracies [20,41–46], outbreak detection or disease nowcasting/forecasting [18,19], and more [47–52]. Similarly, data from other social media channels (e.g., Weibo, Reddit, Facebook) or search engine statistics are utilized for parallel analyses related to COVID-19 pandemic as well [53–69]. While these studies reveal important information and patterns, they do not attempt to uncover or model causal relationships between the attributes of COVID-19 pandemic and social media activity. As *correlation does not imply causation* (e.g., spurious correlations), the ability to identify truly causal relationships between pandemic characteristics and public behaviour (online or not) remains crucial for devising public policies that are more impactful. Without causal understanding, our efforts and decisions on risk communication, public health engagement, health intervention timing, and adjustment of resources for fighting disinformation, fearmongering, and alarmism will stay subpar.

The task of forging causal models comes with numerous challenges in various domains because, typically, domain knowledge and significant amount of time from the experts is required. For substantially complex phenomena such as a pandemic due to a novel virus, diagnosing causal attributions becomes even harder. Therefore, learning causal relationships automatically from observational data has been studied in machine learning. One of the primary challenges for this pursuit is that numerous latent variables that we can not observe exist in real world problems. In fact, numerous other latent variables that we are not even aware of may exist as well. As latent variables can induce statistical correlations between observed variables that do not have a causal relationship, *confounding factors* arise. While this phenomenon may not exhibit a considerable problem in standard probabilistic models, causal modeling suffers from it immensely.

Several machine learning methods are proposed for learning causal structures from observational data and some allow combination of statistical information (learned from the data) and domain expertise [70,71]. Bayesian networks are frequently utilized frameworks for learning models once the causal structure is fixed. As probabilistic graphical models, BNs flexibly unify graphical models, structural equations, and counterfactual logic [71–74]. A causal BN consists of a directed acyclic graph (DAG) in which nodes correspond to random variables and edges correspond to direct causal influence of one node on another [71]. This compact representation of high-dimensional probability spaces (e.g., joint probability distributions) provides intuitive and explainable models for us. In addition, BNs allow not only straightforward observational computations (e.g., calculation of marginal probabilities) but also interventional ones (e.g., *do-calculus*), enabling simulations of various what-if scenarios.

3. Methods

3.1. Data

We primarily utilized two data sources for our study, that is, daily number of officially reported COVID-19 infections and deaths from "COVID-19 Data Repository" by the Center for Systems Science and Engineering at Johns Hopkins University [2] and daily count of COVID-19 related tweets from Twitter [75]. A 57 day period between 22 January–18 March 2020 is chosen for this study to represent the early stages of the pandemic when disease characteristics are less known and public panic is elevated. We collected 954,902 tweets that have location information from Twitter by searching for *#covid19* and *#coronavirus* hashtags. Similar to other studies [18,20,46], geolocation of the tweets is inferred either by using user geo-tagging or geo-coding the information available in users' profiles. Timeline of daily log-distribution of collected tweet counts among 177 countries can be examined from Figure 1. The trend shows an increasing prevalence of high daily number of tweets as the pandemic spreads across the globe with time.

We select the following 12 countries for our causal modeling analysis: Italy, Spain, Germany, France, Switzerland, United Kingdom, Netherlands, Norway, Austria, Belgium, Sweden, and Denmark. These are the countries with substantial number of reported COVID-19 cases (listed in descending order) in Europe as of 18 March 2020, yet still exhibiting a high diversity in terms of the timeline of the pandemic. For instance, while Italy located further in the pandemic timeline due to being hit first in Europe, United Kingdom could be considered in the very initial stages of it for the analysis period of our study. Figure 2 depicts the cumulative number of tweet counts alongside with that of reported infections and deaths for the selected countries. Evident correlations between these variables can be noticed. A sharp increase in Twitter activity is observed after 28–29 February, which corresponds to the period of each country having at least one confirmed COVID-19 case.

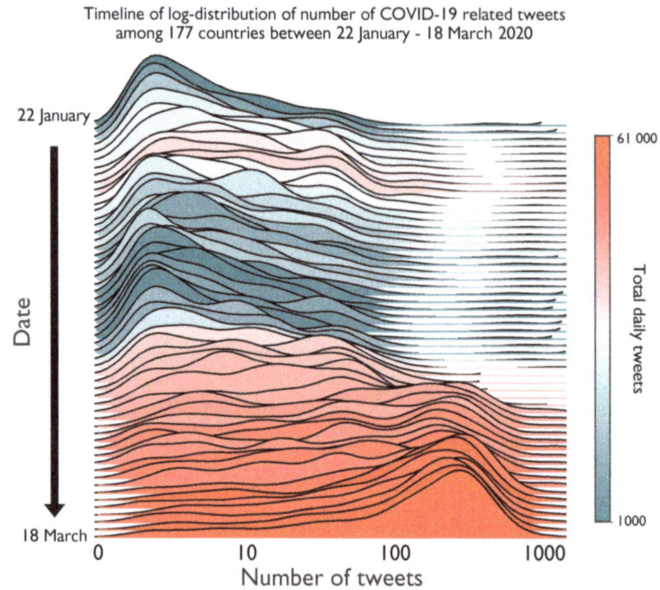

Figure 1. Evolution of COVID-19 related Twitter activity between 22 January–18 March 2020.

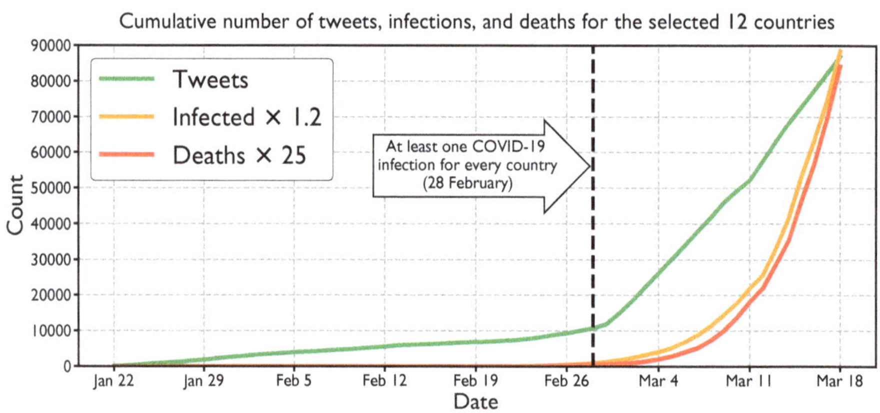

Figure 2. Cumulative counts of Twitter activity and COVID-19 statistics for the selected countries during the study period.

3.2. Feature Selection

In order to characterize the pandemic straightforwardly, we calculate the following six features (attributes) from the official COVID-19 incident statistics for each day for 12 selected countries: (1) *total number of infections up to that day* (normalized by the country's population), (2) *number of new infections* (normalized by the country's population), (3) *percentage increase in infections* (with respect to previous day), and the same three statistics for *deaths* (4-5-6).

Recent epidemiological studies on COVID-19 reveal the following: people over the age 65 are the primary risk group both for infection and mortality [76–79] and human-to-human transmission of the virus is largely occurring among family members or among people who co-reside [77,80,81]. In order to be able to test whether our approach can capture this scientific domain knowledge or not, we collect the following two features for each country: (7) *percentage of population over the age of 65* [82] and (8) *percentage of single-person households* [83]. Finally, as we know that popularity of Twitter in a country and announcement of national lockdown (e.g., closing of schools, banning of gatherings) unequivocally affect the Twitter activity in that country, we add (9) *percentage of population using Twitter* [84] and (10) *is_lockdown_announced?* (3 day period is encoded as `Yes` if government restriction is announced [85], `No` otherwise) features as well. We represent Twitter activity by simply counting the (11) *number of daily tweets* (normalized by the country's population). We also calculate the (12) *average daily sentiment* (in range $[-1, 1]$) of English tweets (corresponding to over 80% of all tweets) by utilizing a pre-trained sentiment classifier (DistilBERT [86]). We treat each day as an observation and represent each day with these 12 attributes ($n = 12$) for structure learning, resulting in a feature matrix of dimensions 684×12. 684 observations come from 12 countries times 57 days.

For the purpose of increasing interpretability, we discretize the daily numerical features by mapping them to 2 categorical levels, namely `High` or `Low`. Features related to the pandemic (infections and deaths) and Twitter activity employ a cut-off value of 75th percentile and remaining numerical features employ a cut-off value of 50th percentile (corresponding to median). Such categorization, for instance, turns the numerical value of "population-normalized increase in deaths of 1.7325×10^{-7}" into a relatively calculated category of `High` for a given day. Sentiment scores are mapped to `Positive` ($\geq$0) or `Negative` (<0) as well.

3.3. Structure Learning and Causal Inference

In structure learning we would like to learn a directed acyclic graph, G, that describes the conditional dependencies between variables in a given data matrix. A typical formulation of this problem is a *structural equation model* (more generally a *generalized linear model*) in which a weighted adjacency matrix, $W \in \mathbb{R}^{n \times n}$, defines the graph. This is essentially a parametric model that enables operations on the continuous space of $n \times n$ matrices instead of discrete space of DAGs. Such formulation enables a score-based learning of DAGs, that is,

$$\begin{aligned} \min_{W \in \mathbb{R}^{n \times n}} \; & L(W) \\ \text{subject to} \;\; & G(W) \in \text{DAGs}, \end{aligned} \tag{1}$$

where $G(W)$ is the n-node graph induced by the weighted adjacency matrix, W, and L is the score/loss function to be minimized. Even though the loss function is continuous, solving Equation (1) is still a non-convex, combinatorial optimization problem as the acyclicity constraint is discrete and difficult to enforce. Note that acyclicity is a strict requirement for causal graphs. In order to tackle this problem efficiently, we utilize the recently proposed NOTEARS (corresponding to *Non-combinatorial Optimization via Trace Exponential and Augmented lagRangian for Structure learning*) algorithm for structure learning [87].

NOTEARS algorithm discovers a directed acyclic graph from the observational data by re-formulating the structure learning problem as a purely continuous optimization. This approach differs significantly from existing work in the field which predominantly operates on discrete space of graphs. Re-formulation is achieved by introducing a continuous measure of "DAG-ness", $h(W)$, which quantifies the severity of violations from acyclicity as W changes. Consequently, the problem formulation becomes

$$\begin{aligned} \min_{W \in \mathbb{R}^{n \times n}} \; & L(W) \\ \text{subject to} \;\; & h(W) = 0, \end{aligned} \tag{2}$$

which enables utilization of standard numerical solving methods and scales cubically, $\mathcal{O}(n^3)$, with the number of variables instead of exponentially as in other structure learning methods. We have chosen the score to be the least squared loss (can be any smooth loss function) with l_1-regularization term to discover a sparse DAG and use a gradient-based minimizer to solve Equation (2). In our context, we discover such an adjacency matrix that the graph it defines encodes the dependencies between our features in a close-to-optimal manner (finding the global optimum is NP-hard [88,89]) and is a DAG. Efficiency of this approach enables structure learning in a scalable manner.

As NOTEARS algorithm allows incorporation of expert knowledge, we also put certain constraints on the structure in our experiment. These constraints correspond to prohibited causal attributions based on simple logical assumptions, for example, Twitter activity on a given day can not have a causal effect on number of deaths from COVID-19 on that day. Full list of these constraints can be found in Table A1 in the Appendix A. Once the structure is learned (both by data and logical constraints), we treat it as a causal model and learn the parameters of a Bayesian network on it with the training data in order to capture the conditional dependencies between variables. During inference on test data, probabilities of each possible state of a node with respect to the given input data is computed from the conditional probability distributions.

Our approach allows straightforward querying of the model with varying observations. For instance for a given day, the probability of Twitter activity being `High`, when total number of infections are `Low` and new deaths are `High`, that is,

$$\Pr(\text{Twitter Activity} = \texttt{H} \mid \text{Total Infections} = \texttt{H}, \text{New Deaths} = \texttt{L}), \tag{3}$$

can be computed by propagating the impact of these queries through the nodes of interest. By utilizing this property of our approach, we compute marginal probabilities for gaining further insights on likelihoods of various events.

Essentially, we expect two observations from our experiment. First, we expect the structure learning algorithm to discover the causal relations verified by domain/expert knowledge (e.g., % of single-person households and % of 65+ people affecting infections) and common sense/elementary algebra (e.g., new deaths affecting percentage change in deaths). Second, we expect the calculated likelihoods from the Bayesian network are in parallel with domain knowledge as well, for example, high % of people over 65 increasing the marginal likelihood of deaths instead of decreasing it or high % of single households (better social isolation) decreasing the marginal likelihood of infections instead of increasing it. Realization of these expectations will show that the proposed method can indeed capture causal relationships and will increase our confidence in discovered relationships between the pandemic attributes and Twitter activity as well as confidence in corresponding likelihoods.

3.4. Evaluation

We validate our approach first by inspecting whether the expected causal relationships (e.g., domain knowledge on COVID-19) are captured or not. Then, we infer the Twitter activity of each day from the learned Bayesian Network. Essentially, this corresponds to a binary classification task, that is, predicting the Twitter activity as `High` or `Low` from the rest of the variables. We utilize a Leave-One-Country-Out (LOCO) cross-validation scheme in which each fold consists of training set from 11 countries (627 samples) and test set (57 samples) from the remaining country. We do not perform standard k-fold cross-validation as we would like to measure the generalization performance across countries and prevent overly optimistic results. Therefore, we ensure that the observations from the same country fall in the same set (either training or test) for every fold. We evaluate the performance of our approach by calculating the average Area Under the Receiver Operating Characteristic curve (AUROC) of the cross-validation runs. For quantifying the causal effect of characteristics of pandemic and relevant country statistics on Twitter activity, we report likelihoods from the model by querying various conditions.

4. Results

The jointly (with statistical learning from data and user-defined logical constraints) discovered causal model by the structure learning algorithm can be examined from Figure 3. Different families of attributes are colored differently for ease of inspection—blue for COVID-19 pandemic related variables, yellow for country-specific statistics, green for government interventions, and red for representing variables related to public attention and sentiment in Twitter. Daily Twitter activity is affected by 4 variables, namely Twitter usage statistics of that country, new infections on that day, new deaths on that day, and whether national lockdown is announced or not. Similarly, 4 variables affecting the average daily sentiment in Twitter are new infections on that day, new deaths on that day, total deaths up to that day, and again lockdown announcements. Total number of infections did not show any causal effect on Twitter activity or on average public sentiment.

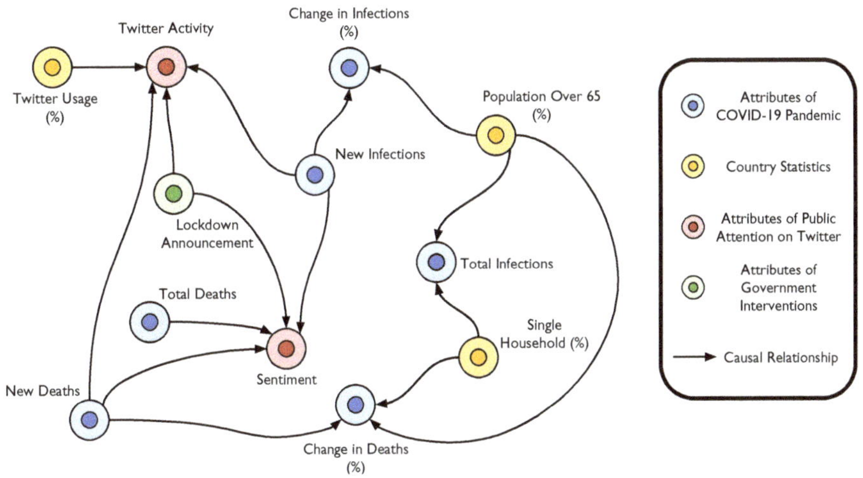

Figure 3. Discovered graph depicting causal relationships between various attributes.

Leave-One-Country-Out cross-validation results in terms of AUROCs can be seen in Table 1. Each row in the table corresponds to a cross-validation fold in which the Twitter activity in that particular country was tried to be predicted. The Bayesian network model achieves an average AUROC score of 0.833 across countries when trying to infer the Twitter activity from the rest of the variables for a given day. Daily Twitter patterns of Germany, Italy, and Sweden show very high predictability with AUROC scores above 0.97. United Kingdom shows the worst predictability with an AUROC of 0.68.

Calculation of marginal probabilities for several queries are presented in Table 2. Public attention and sentiment-related target variables and states are set to `High` Twitter Activity and `Negative` Sentiment.

Table 1. Area Under the Receiver Operating Characteristic curve (AUROC) result for each fold of Leave-One-Country-Out cross-validation.

Cross Validation Test Country	AUROC
Austria	0.798
Belgium	0.728
Denmark	0.831
France	0.776
Germany	0.992
Italy	0.976
Netherlands	0.746
Norway	0.907
Spain	0.766
Sweden	0.998
Switzerland	0.789
United Kingdom	0.684
Average	*0.833*

Table 2. Examples of queries and computed marginal probabilities for Twitter activity and average sentiment.

Query	Variable and State	Pr()
Single-person household (%) = H 65+ (%) = L	Total Infections = H	0.178
Single-person household (%) = L 65+ (%) = H	Total Infections = H	0.241
New Infections = H New Deaths = H	Twitter Activity = H	0.496
New Infections = L New Deaths = L	Twitter Activity = H	0.184
New Infections = H New Deaths = H Twitter Usage = H Lockdown Announcement = Yes	Twitter Activity = H	0.800
New Infections = L New Deaths = L Twitter Usage = L Lockdown Announcement = No	Twitter Activity = H	0.120
New Deaths = H	Sentiment = Neg	0.624
New Deaths = L	Sentiment = Neg	0.277
Total Deaths = H	Sentiment = Neg	0.344
Total Deaths = L	Sentiment = Neg	0.290
Lockdown Announcement = Yes	Sentiment = Neg	0.501
Lockdown Announcement = No	Sentiment = Neg	0.286

5. Discussion

By analyzing observational data, we attempt to discover causal associations between national COVID-19 patterns and Twitter activity as well as public sentiment during the early stages of the pandemic. Some of our findings are expected associations such as popularity of Twitter in a country (Twitter usage) affecting Twitter activity. Other expected causal relationships were new deaths affecting change in deaths and new infections affecting change in infections, due to trivial mathematical definitions. These were captured successfully as well. It is important to note that no causal relationship between infection statistics and death statistics was discovered which might seem against intuition. This is because in this study we treat each day as an observation in our modeling and do not create time-lagged version of variables. While some of our results imply expected associations, we also observe more interesting implications that are in alignment with recent scientific literature on COVID-19. For instance, percentage of single-person households affects the total number of COVID-19 infections. Similarly, the percentage of 65+ population affects the percentage change in deaths (essentially corresponding to rate of deaths). When the queries regarding domain knowledge are examined, we see that low percentage of single-person households (less social isolation) and high percentage of 65+ population increases the probability of total infections being high when compared to the opposite settings. This is in line with recent scientific literature on COVID-19 transmission characteristics [76–81].

By inferring Twitter activity, we show the generalization ability of causal inference across 12 countries with reasonable accuracy. Factors affecting Twitter activity and sentiment are discussion-worthy as well. By observing correlations, Wong et al. hints that there may be a link between announcement of new infections and Twitter activity [17]. Our results in Figure 3 and Table 2 suggest the same with a causal point of view. Similarly, our finding of negative impact of declaration of government measures on public sentiment is also in parallel with recent research. By analyzing Chinese social media, Li et al. show that official declaration of COVID-19 (epidemic at that time) correlates with increased negative emotions such as anxiety, depression, and indignation [56]. When new infections, new deaths, total deaths are high and an announcement of lockdown is made, Twitter activity on that day becomes more than 6 times more likely than when the situation is opposite (probabilities of 0.8 vs. 0.12). High number of new deaths for a given day causes the sentiment to be much more negative than low number of new deaths (probabilities of 0.624 vs. 0.277). Similarly, an announcement of lockdown is causally associated with an increase in negative sentiment in Twitter (probabilities of 0.501 vs. 0.286).

As it is important to observe the countries that are ahead in terms of pandemic timeline and learn the behaviour of the pandemic, it is equally important to understand also the public attention and sentiment characteristics from those countries. Wise et al. show that risk perception of people and their frequency of engagement in protective behaviour change during the early stages of the pandemic [90]. Inference of such patterns in a causal manner from social media can aid us in the pursuit of timely decisions and suitable policy-making, and consequently, high public engagement. After all, primary responsibility of risk management during a global pandemic is not centralized to a single institution, but distributed across society. For example, Zhong et al. shows that people's adherence to COVID-19 control measures is affected by their knowledge and attitudes towards it [91]. In that regard, computational methods such as causal inference and causal reasoning can help us disentangle correlations and causation between the observed variables of the adverse phenomenon.

In real-world scenarios, it is virtually impossible to correctly identify all the causal associations due to presence of numerous confounding factors. As in with all methods in machine learning, a trade-off between false positive associations and false negative ones exists in our approach as well. While we rely on official COVID-19 statistics, testing and reporting methodologies as well as policies can change during the course of the pandemic. Furthermore, in the context of this study, ground truth causal associations do not exist even for a few variables, preventing the direct measurement of performance of causal discovery methods. We would like to emphasize that we acknowledge these and other relevant limitations of our study. Our study has further limitations regarding the simplifications on our problem formulation and data. For instance, we do not attempt to model temporal causal relationships in this study, for example, high deaths numbers having an impact on the public sentiment possibly for several following days. We have not taken into account remarks by famous politicians, public figures, or celebrities which may indeed impact social media discussions. We have not incorporated "retweets" or "likes" into our models either. We would also like emphasize that with this study we wanted to introduce an uncomplicated example of causal modeling perspective to social media analysis during COVID-19.

Future work includes investigating the effect of dynamics of the pandemic on the spreading mechanisms of information, including relevant health topics in Twitter and other social media. As social media can be exploited for deliberately creating panic and confusion [92], causal inference on patterns of misinformation and disinformation propagation in Twitter will be studied as well. Finally, country-specific models with more granular statistics of the country and time-delayed variables will be investigated for a longer analysis period.

6. Conclusions

Distinguishing epidemiological events that correlate with public attention from epidemiological events that cause public attention is crucial for constructing impactful public health policies. Similarly, monitoring fluctuations of public opinion becomes actionable only if causal relationships are identified. We hope our study serves as a first example of causal inference on social media data for increasing our understanding of factors affecting public attention and sentiment during COVID-19 pandemic.

Author Contributions: Conceptualization, O.G.; methodology, O.G.; software, O.G.; validation, O.G.; formal analysis, O.G.; investigation, O.G.; resources, O.G. and M.G.; data curation, O.G. and M.G.; writing–original draft preparation, O.G.; writing–review and editing, O.G.; visualization, O.G.; supervision, O.G.; project administration, O.G.; funding acquisition, O.G. All authors have read and agreed to the published version of the manuscript.

Funding: This research received no external funding.

Conflicts of Interest: The authors declare no conflict of interest.

Abbreviations

The following abbreviations are used in this manuscript:

AUROC	Area Under the Receiver Operating Characteristic curve
COVID-19	Coronavirus Disease 2019
BN	Bayesian Network
DAG	Directed Acyclic Graph
LOCO	Leave-One-Country-Out
NOTEARS	Non-combinatorial Optimization via Trace Exponential and Augmented lagRangian for Structure learning

Appendix A

Prohibited causal associations are listed in Table A1 below. For example, *Twitter activity* can not cause any other variable for a given day. Similarly, *Twitter usage percentage* or *lockdown announcement* can not have a causal relationship with *new deaths* for a given day.

Table A1. Prohibited causal associations (constraints) for structure learning.

From	To
Any node	Population Over 65 (%) Twitter Usage (%) Single Household (%)
Twitter Activity Sentiment	Any node
Twitter Usage (%) Lockdown Announcement	Total Infections New Infections Change in Infections (%) Total Deaths New Deaths Change in Deaths (%)
Population Over 65 (%) Single Household (%)	Twitter Activity Sentiment
Twitter Usage (%)	Sentiment

References

1. Cucinotta, D.; Vanelli, M. WHO Declares COVID-19 a Pandemic. *Acta Bio-Medica Atenei Parm.* **2020**, *91*, 157–160. [CrossRef]
2. Dong, E.; Du, H.; Gardner, L. An Interactive Web-based Dashboard to Track COVID-19 in Real Time. *Lancet Infect. Dis.* **2020**. [CrossRef]
3. Van Bavel, J.J.; Baicker, K.; Boggio, P.S.; Capraro, V.; Cichocka, A.; Cikara, M.; Crockett, M.J.; Crum, A.J.; Douglas, K.M.; Druckman, J.N.; et al. Using Social and Behavioural Science to Support COVID-19 Pandemic Response. *Nat. Hum. Behav.* **2020**, 1–12. [CrossRef]
4. Signorini, A.; Segre, A.M.; Polgreen, P.M. The Use of Twitter to Track Levels of Disease Activity and Public Concern in the US During the Influenza A H1N1 Pandemic. *PLoS ONE* **2011**, *6*. [CrossRef]

5. Ji, X.; Chun, S.A.; Geller, J. Monitoring Public Health Concerns Using Twitter Sentiment Classifications. In Proceedings of the IEEE International Conference on Healthcare Informatics, Philadelphia, PA, USA, 9–11 September 2013; pp. 335–344. [CrossRef]
6. Ji, X.; Chun, S.A.; Wei, Z.; Geller, J. Twitter Sentiment Classification for Measuring Public Health Concerns. *Soc. Netw. Anal. Min.* **2015**, *5*, 13. [CrossRef]
7. Weeg, C.; Schwartz, H.A.; Hill, S.; Merchant, R.M.; Arango, C.; Ungar, L. Using Twitter to Measure Public Discussion of Diseases: A Case Study. *JMIR Public Health Surveill.* **2015**, *1*, e6. [CrossRef]
8. Mollema, L.; Harmsen, I.A.; Broekhuizen, E.; Clijnk, R.; De Melker, H.; Paulussen, T.; Kok, G.; Ruiter, R.; Das, E. Disease Detection or Public Opinion Reflection? Content Analysis of Tweets, Other Social Media, and Online Newspapers During the Measles Outbreak in the Netherlands in 2013. *J. Med. Internet Res. (JMIR)* **2015**, *17*, e128. [CrossRef]
9. Jordan, S.E.; Hovet, S.E.; Fung, I.C.H.; Liang, H.; Fu, K.W.; Tse, Z.T.H. Using Twitter for Public Health Surveillance from Monitoring and Prediction to Public Response. *Data* **2019**, *4*, 6. [CrossRef]
10. Rosenberg, H.; Syed, S.; Rezaie, S. The Twitter Pandemic: the Critical Role of Twitter in the Dissemination of Medical Information and Misinformation During the COVID-19 Pandemic. *Can. J. Emerg. Med.* **2020**, 1–7. [CrossRef]
11. Chen, E.; Lerman, K.; Ferrara, E. Covid-19: The First Public Coronavirus Twitter Dataset. *arXiv* **2020**, arXiv:2003.07372.
12. Gao, Z.; Yada, S.; Wakamiya, S.; Aramaki, E. NAIST COVID: Multilingual COVID-19 Twitter and Weibo Dataset. *arXiv* **2020**, arXiv:2004.08145.
13. Lamsal, R. Corona Virus (COVID-19) Tweets Dataset. *IEEEDataPort* **2020**. [CrossRef]
14. Aguilar-Gallegos, N.; Romero-García, L.E.; Martínez-González, E.G.; García-Sánchez, E.I.; Aguilar-Ávila, J. Dataset on Dynamics of Coronavirus on Twitter. *Data Brief* **2020**, *30*, 105684. [CrossRef] [PubMed]
15. Thelwall, M.; Thelwall, S. Retweeting for COVID-19: Consensus Building, Information Sharing, Dissent, and Lockdown Life. *arXiv* **2020**, arXiv:2004.02793.
16. Sha, H.; Hasan, M.A.; Mohler, G.; Brantingham, P.J. Dynamic Topic Modeling of the COVID-19 Twitter Narrative Among US Governors and Cabinet Executives. *arXiv* **2020**, arXiv:2004.11692.
17. Wong, C.M.L.; Jensen, O. The Paradox of Trust: Perceived Risk and Public Compliance During the COVID-19 Pandemic in Singapore. *J. Risk Res.* **2020**, 1–10. [CrossRef]
18. Turiel, J.; Aste, T. Wisdom of the Crowds in Forecasting COVID-19 Spreading Severity. *arXiv* **2020**, arXiv:2004.04125.
19. Gharavi, E.; Nazemi, N.; Dadgostari, F. Early Outbreak Detection for Proactive Crisis Management Using Twitter Data: COVID-19 a Case Study in the US. *arXiv* **2020**, arXiv:2005.00475.
20. Chary, M.; Overbeek, D.; Papadimoulis, A.; Sheroff, A.; Burns, M. Geospatial Correlation Between COVID-19 Health Misinformation on Social Media and Poisoning with Household Cleaners. *medRxiv* **2020**. [CrossRef]
21. Kayes, A.; Islam, M.S.; Watters, P.A.; Ng, A.; Kayesh, H. Automated Measurement of Attitudes Towards Social Distancing Using Social Media: A COVID-19 Case Study. *Preprints* **2020**. [CrossRef]
22. Wang, C.; Pan, R.; Wan, X.; Tan, Y.; Xu, L.; Ho, C.S.; Ho, R.C. Immediate Psychological Responses and Associated Factors During the Initial Stage of the 2019 Coronavirus Disease (COVID-19) Epidemic Among the General Population in China. *Int. J. Environ. Res. Public Health* **2020**, *17*, 1729. [CrossRef] [PubMed]
23. Cullen, W.; Gulati, G.; Kelly, B. Mental Health in the COVID-19 Pandemic. *QJM An Int. J. Med.* **2020**, *113*, 311–312. [CrossRef] [PubMed]
24. Brooks, S.K.; Webster, R.K.; Smith, L.E.; Woodland, L.; Wessely, S.; Greenberg, N.; Rubin, G.J. The Psychological Impact of Quarantine and How to Reduce It: Rapid Review of the Evidence. *Lancet* **2020**, *395*, 912–920. [CrossRef] [PubMed]
25. Dubey, A.D.; Tripathi, S. Analysing the Sentiments towards Work-From-Home Experience during COVID-19 Pandemic. *J. Innov. Manag.* **2020**, *8*. [CrossRef]
26. Duong, V.; Pham, P.; Yang, T.; Wang, Y.; Luo, J. The Ivory Tower Lost: How College Students Respond Differently than the General Public to the COVID-19 Pandemic. *arXiv* **2020**, arXiv:2004.09968.
27. Medford, R.J.; Saleh, S.N.; Sumarsono, A.; Perl, T.M.; Lehmann, C.U. An "Infodemic": Leveraging High-Volume Twitter Data to Understand Public Sentiment for the COVID-19 Outbreak. *medRxiv* **2020**. [CrossRef]

28. Samuel, J.; Ali, G.M.N.; Rahman, M.M.; Esawi, E.; Samuel, Y. COVID-19 Public Sentiment Insights and Machine Learning for Tweets Classification. *Preprints* **2020**. [CrossRef]
29. Batooli, Z.; Sayyah, M. Measuring Social Media Attention of Scientific Research on Novel Coronavirus Disease 2019 (COVID-19): An Investigation on Article-level Metrics Data of Dimensions. *Prepr. Res. Sq.* **2020**. [CrossRef]
30. Kwon, J.; Grady, C.; Feliciano, J.T.; Fodeh, S.J. Defining Facets of Social Distancing during the COVID-19 Pandemic: Twitter Analysis. *medRxiv* **2020**. [CrossRef]
31. Cinelli, M.; Quattrociocchi, W.; Galeazzi, A.; Valensise, C.M.; Brugnoli, E.; Schmidt, A.L.; Zola, P.; Zollo, F.; Scala, A. The COVID-19 Social Media Infodemic. *arXiv* **2020**, arXiv:2003.05004.
32. Park, H.W.; Park, S.; Chong, M. Conversations and Medical News Frames on Twitter: Infodemiological Study on COVID-19 in South Korea. *J. Med. Internet Res. (JMIR)* **2020**, *22*, e18897. [CrossRef] [PubMed]
33. Thelwall, M.; Thelwall, S. Covid-19 tweeting in English: Gender differences. *arXiv* **2020**, arXiv:2003.11090.
34. Alshaabi, T.; Minot, J.; Arnold, M.; Adams, J.L.; Dewhurst, D.R.; Reagan, A.J.; Muhamad, R.; Danforth, C.M.; Dodds, P.S. How the World's Collective Attention is Being Paid to a Pandemic: COVID-19 Related 1-gram Time Series for 24 Languages on Twitter. *arXiv* **2020**, arXiv:2003.12614.
35. Lopez, C.E.; Vasu, M.; Gallemore, C. Understanding the Perception of COVID-19 Policies by Mining a Multilanguage Twitter Dataset. *arXiv* **2020**, arXiv:2003.10359.
36. Dewhurst, D.R.; Alshaabi, T.; Arnold, M.V.; Minot, J.R.; Danforth, C.M.; Dodds, P.S. Divergent Modes of Online Collective Attention to the COVID-19 Pandemic are Associated with Future Caseload Variance. *arXiv* **2020**, arXiv:2004.03516.
37. Abd-Alrazaq, A.; Alhuwail, D.; Househ, M.; Hamdi, M.; Shah, Z. Top Concerns of Tweeters During the COVID-19 Pandemic: Infoveillance Study. *J. Med. Internet Res. (JMIR)* **2020**, *22*, e19016. [CrossRef]
38. Wicke, P.; Bolognesi, M.M. Framing COVID-19: How We Conceptualize and Discuss the Pandemic on Twitter. *arXiv* **2020**, arXiv:2004.06986.
39. Jarynowski, A.; Wójta-Kempa, M.; Belik, V. Trends in Perception of COVID-19 in Polish Internet. *medRxiv* **2020**. [CrossRef]
40. Ordun, C.; Purushotham, S.; Raff, E. Exploratory Analysis of Covid-19 Tweets Using Topic Modeling, UMAP, and DiGraphs. *arXiv* **2020**, arXiv:2005.03082.
41. Yang, K.C.; Torres-Lugo, C.; Menczer, F. Prevalence of Low-Credibility Information on Twitter During the COVID-19 Outbreak. *arXiv* **2020**, arXiv:2004.14484.
42. Ahmed, W.; Vidal-Alaball, J.; Downing, J.; Seguí, F.L. COVID-19 and the 5G Conspiracy Theory: Social Network Analysis of Twitter Data. *J. Med. Internet Res. (JMIR)* **2020**, *22*, e19458. [CrossRef] [PubMed]
43. Ferrara, E. #COVID-19 on Twitter: Bots, Conspiracies, and Social Media Activism. *arXiv* **2020**, arXiv:2004.09531.
44. Bridgman, A.; Merkley, E.; Loewen, P.J.; Owen, T.; Ruths, D.; Teichmann, L.; Zhilin, O. The Causes and Consequences of COVID-19 Misperceptions: Understanding the Role of News and Social Media. *OSF Prepr.* **2020**. [CrossRef]
45. Ahmed, W.; Vidal-Alaball, J.; Downing, J.; Seguí, F.L. Dangerous Messages or Satire? Analysing the Conspiracy Theory Linking 5G to COVID-19 through Social Network Analysis. *J. Med. Internet Res. (JMIR)* **2020**. [CrossRef]
46. Gallotti, R.; Valle, F.; Castaldo, N.; Sacco, P.; De Domenico, M. Assessing the Risks of "Infodemics" in Response to COVID-19 Epidemics. *medRxiv* **2020**. [CrossRef]
47. Golder, S.; Klein, A.; Magge, A.; O'Connor, K.; Cai, H.; Weissenbacher, D. Extending A Chronological and Geographical Analysis of Personal Reports of COVID-19 on Twitter to England, UK. *medRxiv* **2020**. [CrossRef]
48. Sarker, A.; Lakamana, S.; Hogg-Bremer, W.; Xie, A.; Al-Garadi, M.A.; Yang, Y.C. Self-reported COVID-19 Symptoms on Twitter: An Analysis and a Research Resource. *J. Am. Med. Informat. Assoc.* **2020**. [CrossRef]
49. Li, I.; Li, Y.; Li, T.; Alvarez-Napagao, S.; Garcia, D. What Are We Depressed about When We Talk about COVID19: Mental Health Analysis on Tweets Using Natural Language Processing. *arXiv* **2020**, arXiv:2004.10899.
50. Xu, P.; Dredze, M.; Broniatowski, D.A. The Twitter Social Mobility Index: Measuring Social Distancing Practices from Geolocated Tweets. *arXiv* **2020**, arXiv:2004.02397.

51. Lyu, H.; Chen, L.; Wang, Y.; Luo, J. Sense and Sensibility: Characterizing Social Media Users Regarding the Use of Controversial Terms for COVID-19. *IEEE Trans. Big Data* **2020**. [CrossRef]
52. Schild, L.; Ling, C.; Blackburn, J.; Stringhini, G.; Zhang, Y.; Zannettou, S. "Go Eat A Bat, Chang!": An Early Look on the Emergence of Sinophobic Behavior on Web Communities in the Face of COVID-19. *arXiv* **2020**, arXiv:2004.04046.
53. Rovetta, A.; Bhagavathula, A.S. COVID-19-Related Web Search Behaviors and Infodemic Attitudes in Italy: Infodemiological Study. *JMIR Public Health Surveill.* **2020**, *6*, e19374. [CrossRef] [PubMed]
54. Shahsavari, S.; Holur, P.; Tangherlini, T.R.; Roychowdhury, V. Conspiracy in the Time of Corona: Automatic detection of Covid-19 Conspiracy Theories in Social Media and the News. *arXiv* **2020**, arXiv:2004.13783.
55. Li, J.; Xu, Q.; Cuomo, R.; Purushothaman, V.; Mackey, T. Data Mining and Content Analysis of the Chinese Social Media Platform Weibo during the Early COVID-19 Outbreak: Retrospective Observational Infoveillance Study. *JMIR Public Health Surveill.* **2020**, *6*, e18700. [CrossRef] [PubMed]
56. Li, S.; Wang, Y.; Xue, J.; Zhao, N.; Zhu, T. The Impact of COVID-19 Epidemic Declaration on Psychological Consequences: A Study on Active Weibo Users. *Int. J. Environ. Res. Public Health* **2020**, *17*, 2032. [CrossRef] [PubMed]
57. Velásquez, N.; Leahy, R.; Restrepo, N.J.; Lupu, Y.; Sear, R.; Gabriel, N.; Jha, O.; Johnson, N. Hate Multiverse Spreads Malicious COVID-19 Content Online Beyond Individual Platform Control. *arXiv* **2020**, arXiv:2004.00673.
58. Zhao, Y.; Xu, H. Chinese Public Attention to COVID-19 Epidemic: Based on Social Media. *medRxiv* **2020**. [CrossRef]
59. Li, L.; Zhang, Q.; Wang, X.; Zhang, J.; Wang, T.; Gao, T.L.; Duan, W.; Tsoi, K.K.f.; Wang, F.Y. Characterizing the Propagation of Situational Information in Social Media during COVID-19 Epidemic: A Case Study on Weibo. *IEEE Trans. Comput. Soc. Syst.* **2020**, *7*, 556–562. [CrossRef]
60. Lampos, V.; Moura, S.; Yom-Tov, E.; Cox, I.J.; McKendry, R.; Edelstein, M. Tracking COVID-19 Using Online Search. *arXiv* **2020**, arXiv:2003.08086.
61. Boberg, S.; Quandt, T.; Schatto-Eckrodt, T.; Frischlich, L. Pandemic Populism: Facebook Pages of Alternative News Media and the Corona Crisis–A Computational Content Analysis. *arXiv* **2020**, arXiv:2004.02566.
62. Jelodar, H.; Wang, Y.; Orji, R.; Huang, H. Deep Sentiment Classification and Topic Discovery on Novel Coronavirus or COVID-19 Online Discussions: NLP Using LSTM Recurrent Neural Network Approach. *arXiv* **2020**, arXiv:2004.11695.
63. Liu, D.; Clemente, L.; Poirier, C.; Ding, X.; Chinazzi, M.; Davis, J.T.; Vespignani, A.; Santillana, M. A Machine Learning Methodology for Real-time Forecasting of the 2019-2020 COVID-19 Outbreak Using Internet Searches, News Alerts, and Estimates from Mechanistic Models. *arXiv* **2020**, arXiv:2004.04019.
64. Hou, Z.; Du, F.; Jiang, H.; Zhou, X.; Lin, L. Assessment of Public Attention, Risk Perception, Emotional and Behavioural Responses to the COVID-19 Outbreak: Social Media Surveillance in China. *medRxiv Prepr.* **2020**. [CrossRef]
65. Stokes, D.C.; Andy, A.; Guntuku, S.C.; Ungar, L.H.; Merchant, R.M. Public Priorities and Concerns Regarding COVID-19 in an Online Discussion Forum: Longitudinal Topic Modeling. *J. Gen. Intern. Med.* **2020**. [CrossRef] [PubMed]
66. Shen, C.; Chen, A.; Luo, C.; Liao, W.; Zhang, J.; Feng, B. Reports of Own and Others' Symptoms and Diagnosis on Social Media Predict COVID-19 Case Counts in Mainland China. *arXiv* **2020**, arXiv:2004.06169.
67. Chen, Q.; Min, C.; Zhang, W.; Wang, G.; Ma, X.; Evans, R. Unpacking the Black Box: How to Promote Citizen Engagement Through Government Social Media During the COVID-19 Crisis. *Comput. Hum. Behav.* **2020**, 106380. [CrossRef]
68. Lucas, B.; Elliot, B.; Landman, T. Online Information Search During COVID-19. *arXiv* **2020**, arXiv:2004.07183.
69. Pekoz, E.A.; Smith, A.; Tucker, A.; Zheng, Z. COVID-19 Symptom Web Search Surges Precede Local Hospitalization Surges. *SSRN Prepr.* **2020**. [CrossRef]
70. Ellis, B.; Wong, W.H. Learning Causal Bayesian Network Structures from Experimental Data. *J. Am. Stat. Assoc.* **2008**, *103*, 778–789. [CrossRef]
71. Koller, D.; Friedman, N. *Probabilistic Graphical Models: Principles and Techniques*; MIT Press: Cambridge, MA, USA, 2009.
72. Rubin, D.B. Causal Inference Using Potential Outcomes: Design, Modeling, Decisions. *J. Am. Stat. Assoc.* **2005**, *100*, 322–331. [CrossRef]

73. Pearl, J. An Introduction to Causal Inference. *Int. J. Biostat.* **2010**, *6*. [CrossRef]
74. Pearl, J. *Causality*; Cambridge University Press: Cambridge, UK, 2009.
75. Twitter. Available online: https://twitter.com/ (accessed on 12 May 2020).
76. Dowd, J.B.; Andriano, L.; Brazel, D.M.; Rotondi, V.; Block, P.; Ding, X.; Liu, Y.; Mills, M.C. Demographic Science Aids in Understanding the Spread and Fatality Rates of COVID-19. *Proc. Natl. Acad. Sci. USA* **2020**, *117*, 9696–9698. [CrossRef] [PubMed]
77. Guo, Y.R.; Cao, Q.D.; Hong, Z.S.; Tan, Y.Y.; Chen, S.D.; Jin, H.J.; Tan, K.S.; Wang, D.Y.; Yan, Y. The Origin, Transmission and Clinical Therapies on Coronavirus Disease 2019 (COVID-19) Outbreak-An Update on the Status. *Mil. Med. Res.* **2020**, *7*, 1–10. [CrossRef] [PubMed]
78. Yang, X.; Yu, Y.; Xu, J.; Shu, H.; Liu, H.; Wu, Y.; Zhang, L.; Yu, Z.; Fang, M.; Yu, T.; et al. Clinical Course and Outcomes of Critically Ill Patients with SARS-CoV-2 Pneumonia in Wuhan, China: A Single-centered, Retrospective, Observational Study. *Lancet Respir. Med.* **2020**, *8*, 475–481. [CrossRef]
79. Wang, W.; Tang, J.; Wei, F. Updated Understanding of the Outbreak of 2019 Novel Coronavirus (2019-nCoV) in Wuhan, China. *J. Med. Virol.* **2020**, *92*, 441–447. [CrossRef] [PubMed]
80. WHO. *Report of the WHO-China Joint Mission on Coronavirus Disease 2019 (COVID-19)*; World Health Organization: Geneva, Switzerland, 2020.
81. Li, C.; Ji, F.; Wang, L.; Hao, J.; Dai, M.; Liu, Y.; Pan, X.; Fu, J.; Li, L.; Yang, G.; et al. Asymptomatic and Human-to-Human Transmission of SARS-CoV-2 in a 2-Family Cluster, Xuzhou, China. *Emerg. Infect. Dis.* **2020**, *26*, 1626–1628. [CrossRef] [PubMed]
82. World Bank Open Data—Population Ages 65 and Above. Available online: https://data.worldbank.org/ (accessed on 12 May 2020).
83. Distribution of Households by Household Type from 2003 Onwards—EU-SILC Survey. Available online: https://appsso.eurostat.ec.europa.eu/nui/show.do?dataset=ilc_lvph02&lang=en (accessed on 12 May 2020).
84. Social Media Stats-February 2020. Available online: https://gs.statcounter.com/ (accessed on 12 May 2020).
85. National Responses to the COVID-19 Pandemic—Lockdown Data. Available online: https://en.wikipedia.org/wiki/National_responses_to_the_COVID-19_pandemic (accessed on 12 May 2020).
86. Sanh, V.; Debut, L.; Chaumond, J.; Wolf, T. DistilBERT, A Distilled Version of BERT: Smaller, Faster, Cheaper and Lighter. *arXiv* **2019**, arXiv:1910.01108.
87. Zheng, X.; Aragam, B.; Ravikumar, P.; Xing, E.P. DAGs with NO TEARS: Continuous Optimization for Structure Learning. In Proceedings of the 32nd International Conference on Neural Information Processing Systems, Montreal, QC, Canada, 3–8 December 2018; pp. 9492–9503. [CrossRef]
88. Chickering, D.M. Learning Bayesian Networks is NP-complete. In *Learning from Data*; Springer: Berlin/Heidelberg, Germany, 1996; pp. 121–130. [CrossRef]
89. Chickering, D.M.; Heckerman, D.; Meek, C. Large-sample Learning of Bayesian Networks is NP-hard. *J. Mach. Learn. Res.* **2004**, *5*, 1287–1330.
90. Wise, T.; Zbozinek, T.D.; Michelini, G.; Hagan, C.C.; Mobbs, D. Changes in Risk Perception and Self-reported Protective Behaviour During the First Week of the COVID-19 Pandemic in the United States. *R. Soc. Open Sci.* **2020**, *7*, 200742. [CrossRef]
91. Zhong, B.L.; Luo, W.; Li, H.M.; Zhang, Q.Q.; Liu, X.G.; Li, W.T.; Li, Y. Knowledge, Attitudes, and Practices Towards COVID-19 Among Chinese Residents During the Rapid Rise Period of the COVID-19 Outbreak: A Quick Online Cross-sectional Survey. *Int. J. Biol. Sci.* **2020**, *16*, 1745. [CrossRef]
92. Merchant, R.M.; Lurie, N. Social Media and Emergency Preparedness in Response to Novel Coronavirus. *J. Am. Med. Assoc. (JAMA)* **2020**, *323*. [CrossRef] [PubMed]

computation

MDPI

Article

Particulate Matter and COVID-19 Disease Diffusion in Emilia-Romagna (Italy). Already a Cold Case?

Giovanni Delnevo, Silvia Mirri and Marco Roccetti *

Department of Computer Science and Engineering, University of Bologna, 40127 Bologna, Italy; giovanni.delnevo2@unibo.it (G.D.); silvia.mirri@unibo.it (S.M.)
* Correspondence: marco.roccetti@unibo.it

Received: 11 May 2020; Accepted: 19 June 2020; Published: 23 June 2020

Abstract: As we prepare to emerge from an extensive and unprecedented lockdown period, due to the COVID-19 virus infection that hit the Northern regions of Italy with the Europe's highest death toll, it becomes clear that what has gone wrong rests upon a combination of demographic, healthcare, political, business, organizational, and climatic factors that are out of our scientific scope. Nonetheless, looking at this problem from a patient's perspective, it is indisputable that risk factors, considered as associated with the development of the virus disease, include older age, history of smoking, hypertension and heart disease. While several studies have already shown that many of these diseases can also be favored by a protracted exposure to air pollution, there has been recently an insurgence of negative commentary against authors who have correlated the fatal consequences of COVID-19 (also) to the exposition of specific air pollutants. Well aware that understanding the real connection between the spread of this fatal virus and air pollutants would require many other investigations at a level appropriate to the scale of this phenomenon (e.g., biological, chemical, and physical), we propose the results of a study, where a series of the measures of the daily values of $PM_{2.5}$, PM_{10}, and NO_2 were considered over time, while the Granger causality statistical hypothesis test was used for determining the presence of a possible correlation with the series of the new daily COVID19 infections, in the period February–April 2020, in Emilia-Romagna. Results taken both before and after the governmental lockdown decisions show a clear correlation, although strictly seen from a Granger causality perspective. Moving beyond the relevance of our results towards the real extent of such a correlation, our scientific efforts aim at reinvigorating the debate on a relevant case, that should not remain unsolved or no longer investigated.

Keywords: COVID-19; air pollution; Emilia-Romagna; Granger-causality; time series; correlation

1. Introduction

Although COVID-19 has originated in Wuhan, China in late 2019, several provinces of northern Italy have soon become among the hardest-hit regions in Europe. This virus outbreak spread with a particular intensity to the Italian regions of Lombardy, Veneto, Emilia-Romagna, and Piedmont, in the period from late February to late April, with a severe toll in terms of human deaths. As a simple evidence of this disaster, it suffices to remind that the Italian Institute of Statistics (ISTAT) has recently computed for Italy an average increase of 49.4% in the number of all the fatalities occurred during the month of March 2020, as compared with the number of deaths of March 2019 [1]. Not to mention that, in the same month of March 2020, the official death toll, for some given provinces, like Bergamo and Brescia (in Lombardy), stands at more than five times the value recorded one year before, same period [1].

While it is true that Italy had the bad luck of being the first European country to be devastated by the outbreak, what has gone wrong has motivations in a combination of demographic, political, organizational, industrial, climatic factors, and low intensive care unity (ICU) capacity as well, that need further specific investigations.

Nonetheless, while we are aware that what went wrong will be a subject of studies for years, we are concerned, here, with the fact that COVID-19 manifests as a severe respiratory disease, mostly pneumonia. This motivates why many researchers have focused their attention on the potential relationship between the exposure to particulate pollution and the rapid contagion brought by this virus. With this in view, recently, many international scientific studies were developed to investigate the relationship between particulates of various types and the COVID-19 incidence.

Exemplar is the work by Jiang, Wu and Guan that addresses two relevant issues, with reference to the association between particulate and COVID-19 [2]. They start from the very general consideration that air pollutants raise concerns over their association with infectious diseases, being often the cause of local epidemics [3,4]. This is typical with influenza, since the airborne air pollutants perform as condensation nuclei for the virus to attach, as also confirmed by several other studies [5–9]. Owing to this consideration, Jiang et al. proceed with the following reasoning: since COVID-19 is known to cause human-to-human transmission by infectious secretions [10], these secretions could be transferred in many different ways, including ambient air pollutants. Not only, Jiang, Wu and Guan also observe that is not by chance that $PM_{2.5}$ is the air pollutant constantly associated with an increased COVID-19 incidence in all the Chinese cities of their study, namely: Wuhan, XiaoGan, and HuangGang. Besides the fact that particulate could provide condensation nuclei for viral attachment, Jiang, Wu and Guan add a second biomedical argument which is as follows. It has been discovered that the receptor for COVID binding is the angiotensin-converting enzyme 2, that concentrates on the type II alveolar cells [11]. Since, type II alveolar cells are located in the alveoli, which are only reachable to particles with diameters less than 5 micro meters, it becomes evident that very small airborne pollutants, such as $PM_{2.5}$, have the potential to penetrate, unfiltered, the respiratory tract, down to the alveolar region [12–15].

Similarly, interesting results were found also by Pansini and Fornacca who investigated the incidence of COVID-19 mortality rate in highly polluted areas. They focused their attention on selected areas from different countries (including, among others, China, Italy, and US), and considered also CO and NO2, in addition to particulates. In particular, they collected data about air quality from two kinds of sources: ground monitoring stations and satellite. According to the analysis they performed, they found significant positive correlations between COVID-19 infections and air quality variables. Yet, while in China the strongest correlation was found with the (satellite-derived) CO values, in Italy and in the US the highest correlation values, with the incidence of COVID-19, were those of NO2, derived respectively from satellite (Italy) and ground measurements (US). One of their final observation is that the COVID-19 mortality ratio is higher, regardless of the higher number of infections, in all those areas with poor air quality, that is, where values of CO, NO2, and PM are constantly higher than the acceptable limits [16].

Nevertheless, besides this set of international studies developed in this field (the interested reader can refer also to [17,18]), we scrutinized, with special interest, just those recently conducted by members of the Italian scientific community, for two main reasons. First, the impact of particulate pollution was already being severely felt as a huge health problem in Northern Italy, well before the advent of COVID19 and, second, those studies have been put at the centre of a heated debate in Italy, and considered not convincing under different perspectives.

To be precise, (almost) all those papers at the centre of this controversy have followed two concurrent lines of reasoning, that are typical when one wishes to infer causal relations from data. On one side, they have tried to acquire (through experimentation) the knowledge of the biological/chemical/physical mechanisms at the basis of the possible correlation between the particulate and the virus spread. On the other side, they have tried to confirm the existence of a true causal relation between the two aforementioned phenomena, using some kind of statistical hypothesis testing.

The works conducted by Setti et al., for example, provided a quite convincing contribution to this discussion, by both revealing that traces were found of the COVID-19 RNA in PM_{10} samples in Bergamo [19], and also testing the hypothesis of such a correlation between the daily surplus of that particulate and the consequent contagion between humans by exploiting the statistical model of the coefficient of determination [20]. Daily infections were recorded in the period from 24 February to 13 March, while a surplus of PM_{10} values was considered, on a daily basis, in the period from 9 February to 29 February.

Conticini, Frediani, and Caro, instead, without any statistical testing activity in support of their hypothesis, argued about the fact that poor air quality can lead to a state of permanent body inflammation and chronic respiratory difficulties, along with a hyper-activation of the immune system; being these all circumstances that makes human lungs prone to be attacked by the virus. This is their hypothesis explaining the high mortality rate, recorded in Emilia-Romagna and Lombardy, owing to the virus outbreak [21].

Finally, Becchetti et al. analysed both the PM_{10} and $PM_{2.5}$ values, although recorded on an annual basis, and correlated them to COVID-19 infections and mortality, using a cross-sectional regression statistical method. Theirs is a vast study, where scrutinized are also other factors, including temperature, population density, income, number of lung ventilators, and public transport usage. Nonetheless, the conclusion is that air pollution can be considered as a strong predictor for both virus contagions and mortality [22]. In that paper, again, cited as mechanisms at the basis of the correlation between the particulate pollution and the contagion are, respectively, the hypotheses that: (i) humans living in highly polluted areas have a reduced respiratory capacity to react to the virus, and (ii) the particulate may act as a carrier for the virus.

Unfortunately, all these papers have been severely criticized, mostly based on the considerations that they did not contain any robust evidence of the aforementioned correlation, and that all those discoveries boil down to vague clues, completely preliminary, not yet subject to peer-review by experts in the field [23].

Far from taking a final position, we hold the firm view that all the authors we have cited before share, at least, the merit of having tried to inquire into a vexed problem, that should not go unsolved, or no longer investigated, until a final solution is found.

Hence, our contribution, here, is to provide a further investigation on the possibility that a causal correlation exists between the two cited phenomena (i.e., pollution and spread of the infections). We, as investigators, have to admit that we do not possess any prior knowledge of the researched correlation at a level appropriate to the scale of this phenomenon, e.g., biological, chemical, and physical, and we want to limit our study to an examination of the plausibility of the existence of that correlation at a statistical level. In particular, we are interested in verifying if that correlation comes either confirmed (or rejected) using an alternative statistical model, namely the Granger-causality hypothesis testing model.

Specifically, the Granger causality test is a statistical hypothesis test where a time series X is said to Granger cause Y, if it can be shown, through a series of statistical tests on lagged values of X, that those X values provide statistically significant information about future values of Y. To this aim, it is worth mentioning that we analysed the daily values of the following air pollutants: $PM_{2.5}$, PM_{10}, and NO_2, treated as time series occurring in a given temporal period that has preceded the series of the COVID-19 infections, in all the provinces of the Emilia-Romagna region.

Finally, it is also worth mentioning that we know very well that many believe that some results of the Granger-causality tests can often have a low epistemic utility. Especially, in specific situations when the theoretical background behind the cause–effect correlation is insufficient, or the validation experiments on the field have not been yet conducted. Though, we will argue that the results of our tests, obtained both before and after the Italian government lockdown decisions taken on 8–10 March, posit the correlational structure between pollution and infections well beyond the limit of a weak Humean interpretation of causality [24], with possible implications of practical relevance.

Nevertheless, our study does not have to be treated as the final proof a true causality nexus between the two phenomena, but as an additional strong clue on a case that does not deserve to be already archived.

The remainder of the paper is structured as follows. In the next Section, we describe the methodology behind our approach. Section 3, instead, presents and critically discusses the results we yielded. Finally, Section 4 concludes the paper, with some final considerations.

2. Methods

We now present some preliminary information relevant to our study and a description of the data we have used, along with some reflections on the statistical methodology we have employed.

2.1. Preliminary Information

As already anticipated, in this study we are interested in reasoning around the plausibility of a correlation between air pollution and the spread of COVID-19 infections in the Emilia-Romagna region, by subjecting such hypothesis to a statistical hypothesis testing from a *Granger-causality* perspective.

Prior to beginning, it is important to make clear that we have taken into considerations all the provinces of the Emilia-Romagna region, in the period of interest, namely: Bologna, Ferrara, Forlì-Cesena, Modena, Parma, Piacenza, Reggio nell'Emilia, Rimini, and Ravenna. It is worth mentioning that this Italian region is populated by almost 4,500,000 citizens and has been one of the more seriously affected by this virus, with a total number of infections of 26,719, and as many as 3827 fatalities, as of 9 May 2020.

Of paramount importance to view this process from the right temporal perspective, there are also to consider the events of the chronology according to which restrictions were imposed to human activities in those provinces (with the aim of slowing down the infective diffusion). In particular:

- on 8 March 2020: a full lockdown was imposed for Modena, Parma, Piacenza, Reggio nell'Emilia and Rimini [25].
- on 10 March 2020: a full lockdown was imposed for the remaining provinces, Bologna, Ferrara, Forlì-Cesena and Ravenna [26].

2.2. Data Description

The data on which we performed testing activity were essentially of two types: i) the time series relative to the new daily COVID-19 infections, and ii) the air pollution In Emilia-Romagna, under the form of the measurements of the following pollutants: $PM_{2.5}$, PM_{10}, and $N0_2$, taken on a daily basis at all the aforementioned provinces (Bologna, Ferrara, Forlì-Cesena, Modena, Parma, Piacenza, Reggio nell'Emilia, Rimini and Ravenna).

The amount of daily infections was collected using the GitHub repository of the Italian Civil Protection, for the entire period starting on 24 February and closing on 17 April 2020 [27].

The daily values of the pollutants mentioned before, instead, were collected using the website of the Regional Environmental Protection Agencies (ARPA) of the Emilia-Romagna region, for all the nine provinces we have cited before [28]. Since there were multiple monitoring stations distributed over each province, an average of the values returned by each station was computed, on a daily, provincial basis.

More important is what follows. We have all learnt that this COVID-19 infection can be subjected to an incubation period, whose duration can range from a few days to almost 14, before an infected human begin to manifest some given symptoms. More precisely, authors of [29] maintain that the median incubation period can be estimated to be 5.1 days (with a confidence interval of 95%, it takes from 4.5 to 5.8 days), and that the 97.5% of those who develop symptoms will do so within 11.5 days (with a confidence interval of 95%, it takes from 8.2 to 15.6 days). These estimates imply, at the end, that the 99% of the infected population will develop symptoms within 14 days. Further, other authors also emphasize that a spare delay of 3.6 days can be experienced from the moment in time the result of a virological test is performed, and the time when it is recorded in the correspondent database [30].

These are the reasons why we designed the two different time series:

- the one with the average daily pollution values (say X), and the one with the number of the new daily infections (say Y).
- where X was anticipated in time with respect to Y of 14 days. We decided not to use an offset of sixteen days (as resulting from the sum of 12.5 with 3.6) between (a) and (b), simply because this minimum time difference lag was absorbed by the specific statistical methodology we have employed (i.e., the Granger causality), where we have varied the so-called lag length parameter in a range from 3 to 8 days (as better explained in the Section 2.4 below) [31].

Following this reasoning, the period when me measured the particulate (specifically, $PM_{2.5}$, PM_{10}, and NO_2) started on 10 February and closed on 3 April 2020. As already told, instead, the period for measuring the infections was: 24 February–17 April 2020.

Hence, at the end, it should be clear that an offset has been put that temporally separates these two time-series, due to the consideration that all what can happen on a given day, say x, may have its effect in terms of manifestations of the infection after a period in time which can be as long as x + 14 days.

For the sake of conciseness, we have moved the three Figures, with all the twenty-seven graphs showing how our time-series ($PM_{2.5}$, PM_{10}, and NO_2 vs. infections) evolve in time to the Appendix A at the end of this paper.

2.3. *Methodology*

As already mentioned, we have employed a *Granger causality* testing model to study if a causal correlation may exist between particulate matter and the spread of new COVID-19 infections in Emilia-Romagna [16].

This is a statistical hypothesis testing model typically used to determine if there is a causal relationship between two time-series. In particular, a time series X is said to Granger-causes a time series Y if the prediction of the nth value of Y, using both the past values of X and Y, provides more information rather than the prediction based only on past values of Y [32].

This model typically rests upon two axioms. The former is that past and present may cause the future, but future cannot cause the past. The latter is that the cause contains a unique information about its effects. Usually, the null hypothesis of such a test is set to the fact that the time series X does **not** Granger cause the time series Y, while, consequently, the unique alternative hypothesis is that the time series X Granger causes the time series Y.

In our study, the alternative hypothesis was that the pollutants' time series Granger causes the time-series of the infections. Hence, our aim has been that to verify if we could reject the opposite null hypothesis (i.e., pollution does not Granger cause infections), based on the available data.

To this aim, we set the level of significance at 5%, hence preparing to reject the null hypothesis, only in the case that the corresponding p-values came less than 0.05. Further, as the test assumes that both the time-series under investigation should be stationary, we check and found this condition satisfied using the well-known augmented Dickey–Fuller method [33].

Not only. Since we have designed two time-series where the former (X = pollution) temporally precedes the latter (Y = infection), we did not need to check if the infection Granger causes the pollution, given that the time precedence of Y by X comes naturally.

Nonetheless, it is important to repeat, here again, a concept we have already anticipated in the Introduction. Neither the Granger causality method, nor any other statistical test can provide a final and convincing evidence that two phenomena are correlated, from an epistemological viewpoint, if one has neither a clear knowledge of the motivation that causes that relationship, nor has developed sufficient experiments at a scale that should be appropriate to the observed phenomena.

With this regard, the Granger causality approach suffers from an additional problem. In fact, if both X and Y are driven by a common third process, say W, one might still accept the alternative hypothesis of Granger causality (X Granger causes Y), even though it is evident that both X and Y have a common cause (i.e., W), that determines their mutual correlation [34].

Moving this argument at the center of our specific case, one could even argue that the human activities (playing the role of W, here) have been the common basis for the correlation between pollution and infections (as portrayed in Figure 1a) and, hence, a true causality relation between pollution and infection could not be demonstrated, even when our alternative hypothesis is accepted. Nonetheless, in the next section, we will show results, taken both before and after the lockdown decisions (when almost all human activities were at a minimum), that do seem to confirm the existence of a causal structure similar, instead, to that shown in Figure 1b.

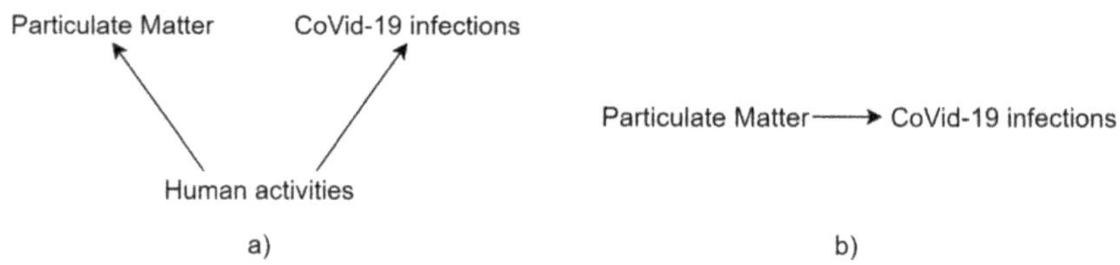

Figure 1. Causality structure: (**a**) mutual interaction, (**b**) causal relation.

2.4. A Computational View

To better understand how a *Granger causality* testing model works from a computational perspective, fundamental is the following explanation.

We start from two time series X and Y (i.e., pollution and infections), whose causal relationship is to be either demonstrated or rejected. In other words, X and Y are the time series under investigation that can be modeled with the following Granger causality equation:

$$Y_t = \sum_{i=1}^{L} \alpha_i \, Y_{t-i} + \sum_{i=1}^{L} \beta_i \, X_{t-i} + \varepsilon_t. \tag{1}$$

Specifically, Y_t and X_t are the single elements of the two series Y and X, and, in our case, they correspond to the values that Y and X can take on, on a daily basis. In essence, with the formula above we can compute current values of Y, based on previous values of both X and Y. How far back one can go with previous values of X and Y, to perform the computation of the current value of Y, is given by the value of L, the so-called lag. To complete the formula, ε_t is a white-noise-random vector.

This said, now comes the turn of explaining how to use this formula for performing a Granger causality hypothesis testing. To this aim, crucial is the role of the β coefficients. In fact, we can say that X Granger-causes Y only if the β coefficients are not zero, since only in this case past values of X (and Y) become useful to compute current values of Y. On the contrary, β coefficients equal to zero make a null contribution to the final sum. It is now easy to understand that modelling a causal relationship with the Granger formula amounts to perform a statistical hypothesis test, where the null hypothesis is that all the β coefficients are zero:

$$H_0 : \beta_1 = \beta_2 = \ldots = \beta_L = 0. \tag{2}$$

The alternative hypothesis being, instead, that at least one of the β coefficients is different from zero.

From a computational perspective, at this point, in a case like that of our study, assigned all the actual values for Y and Y, a vector autoregressive procedure (VAR) is to be run to derive the β coefficients. Upon computation of those β coefficients, a F test procedure must be performed to check if those computed values fit with the all zero distribution of the null hypothesis. This statistical test will return p-values. The higher the returned p-values, the more plausible is the null hypothesis. The lower the p-values, the more plausible is the alternative hypothesis: that is, X Granger causes Y.

Said about the general Granger computational process, now comes the motivation why we have chosen this procedure for our study, rather than other more traditional statistical approaches, like, for the example, the one adopted in [5].

To better understand, consider the following example: Suppose we want to evaluate if a relation exists between the number of viral infections happened in a specific day (e.g., 18 February) and the amount of pollution in the air. To do that, traditional approaches would compute values, based on measurements taken on just two days: the day of the infections vs. the day assumed to be the one when the pollution occurred that was considered at the basis of those infections, say for example February 14th, exactly like in Figure 2a.

With the approach based on the Granger formula, instead, we can take into simultaneous consideration multiple days, each with its amount of measured pollution. This is by virtue of the lag factor (i.e., the L value in the Granger formula above) that allows one to go back as many days as one wants in the computation. For example: three days, like in Figure 2b (or from 3 to 8, like in the case of our study, see Section 2.2).

This is a prominent computational aspect that should not go neglected, since the information on when a given infection precisely occurs comes with a large amount of uncertainty. Still more remarkably, since COVID-19 is manifesting with variable temporal dynamics, we should adopt flexible computational methods to study it. From this point of view, as the series shown in Figure 3 comparatively demonstrate, methods like Granger should be preferred, since they hold the promise to analyze simultaneous contributions to the cause of a unique effect.

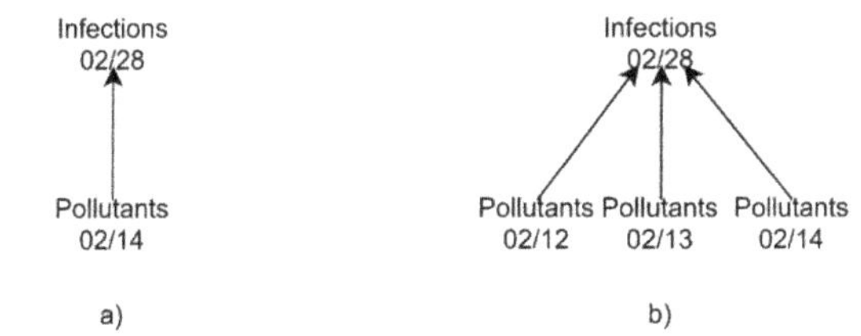

Figure 2. The role of the lag factor in the *Granger* formula: (**a**) without lag, (**b**) with lag.

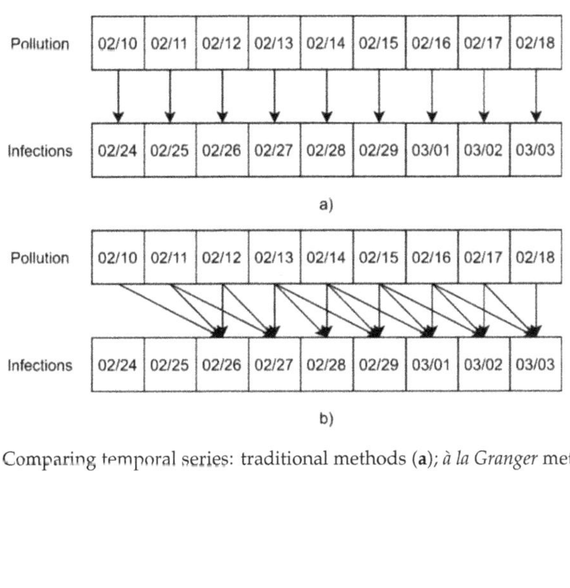

Figure 3. Comparing temporal series: traditional methods (**a**); *à la Granger* methods (**b**).

3. Results

We present the results returned by our Granger causality testing model, differentiating between those illustrating the situation before the lockdown measures were adopted that contained the infection surge, and those showing the ex-post situation.

3.1. Before the Lockdown

The following Figure 4 reports the results of our Granger causality testing campaign, conducted for all the nine aforementioned provinces of the Emilia-Romagna (Bologna-BO, Ferrara-FE, Forlì-Cesena-FC, Modena-MO, Parma-PR, Piacenza-PC, Ravenna-RA, Reggio nell'Emilia-RE, and Rimini-RM).

As already anticipated, we tried to verify if the series X, comprised of all the average daily values of a given pollutant (e.g., $PM_{2.5}$), measured in terms of micrograms per cubic meter, starting on day x_1 and closing on day x_2, Granger causes the series Y of the new daily infections, measured in terms of infected human beings, starting on day y_1 and closing on day y_2, where obviously: $y_1 = x_1 + 14$ and $y_2 = x_2 + 14$, for all the days between x_1 and x_2.

For each of the possible combination pollutant ($PM_{2.5}$, PM_{10}, and NO_2)/infections, our Figure 4 shows in the correspondent cell the *p*-value obtained through a pairwise series computational comparison, using Granger. All this yields a total amount of 189 pairwise series comparisons. In particular, to read well the results: if a cell in Figure 4 reports a *p*-value less than 0.005, we have a confirmation of the causality relation between pollutant and infections (finally, note that if a cell in the Figure reports the value of 0, this means that a *p*-value less than 10^{-4} was computed).

For an easier comprehension of the Figure, one should also notice that the time scale values reported at the left of Figure 4 are the closing days of the two series (respectively, for pollutants and infections), namely the values termed: x_2 and y_2.

Precisely, x_2 ranges in the Figure from 1 to 7 March or from 3 to 9 March, depending on the specific province under consideration with its correspondent lockdown date (8–10 March), while y_2 may range from 15 to 21 March or from 17 to 23 March, due to the 14 days-long temporal shift with which we distanced the two series (pollution precedes infections).

To note, finally, is the fact of prominent importance that all the pairwise series comparisons whose results are reported in Figure 4 were conducted during a period when the lockdown measures were still inactive, since the specific series supposed as the cause of this relation (that is X, the pollutants) starts on 10 February and closes on 7 or 9 March, depending on the province.

All this said, what is clear from an analysis of Figure 4 is that we have got a total amount of 175 (out of 189) statistical confirmations (almost 93%) that *X* Granger causes *Y*; that is. that the pollutants under consideration have some effect on the number of new infections, from a Granger-causality perspective. In particular, this correlation is slightly more evident with $PM_{2.5}$ (yielding 94%), rather than with PM_{10} and NO_2 (92%). Further, to be specified is the fact that there are 189 different pairwise temporal series comparisons, and each was performed with the Granger method explained in Section 2.4.

Nonetheless, before one can come to some final conclusion, we have to remember, here again, the reflection we have anticipated in the previous Section, and that we can repeat, under the alternative form of a question: What about if the human activities carried out in the period from 10 February to 7 or 9 March, were the only common cause for both pollution and infections, exactly like in the causality scheme portrayed in the example a) of Figure 1?

If so, the value of the analysis we have conducted so far would be almost controversial. To respond to this doubt, we ask the reader to refer to the next Subsection.

End Period		PM10->Infections					End Period		PM10 ->Infections			
PM	Infections	MO	PR	PC	RE	RM	PM	Infections	RA	BO	FE	FC
01-mar	15-mar	0.0000	0.0014	0.0862	0.4079	0.1665	03-mar	17-mar	0.0000	0.0000	0.0065	0.0000
02-mar	16-mar	0.0000	0.0000	0.0000	0.0000	0.0703	04-mar	18-mar	0.0000	0.0000	0.0003	0.0000
03-mar	17-mar	0.0042	0.0003	0.0000	0.0004	0.0001	05-mar	19-mar	0.0000	0.0002	0.0006	0.0000
04-mar	18-mar	0.0000	0.0183	0.0000	0.0000	0.0000	06-mar	20-mar	0.0000	0.2441	0.0009	0.0000
05-mar	19-mar	0.0031	0.0040	0.0000	0.0000	0.0000	07-mar	21-mar	0.0000	0.0253	0.0085	0.0001
06-mar	20-mar	0.0000	0.0000	0.0000	0.0041	0.0000	08-mar	22-mar	0.0000	0.0002	0.0020	0.0317
07-mar	21-mar	0.0000	0.0000	0.0000	0.0022	0.0000	09-mar	23-mar	0.0000	0.0001	0.0000	0.0425

End Period		PM2.5->Infections					End Period		PM2.5->Infections			
PM	Infections	MO	PR	PC	RE	RM	PM	Infections	RA	BO	FE	FC
01-mar	15-mar	0.0000	0.0046	0.0487	0.3044	0.0079	03-mar	17-mar	0.0000	0.0000	0.0956	0.0006
02-mar	16-mar	0.0000	0.0000	0.0000	0.0141	0.2789	04-mar	18-mar	0.0000	0.0000	0.0035	0.0001
03-mar	17-mar	0.0000	0.0003	0.0000	0.0194	0.0057	05-mar	19-mar	0.0000	0.0000	0.0028	0.0000
04-mar	18-mar	0.0000	0.0216	0.0000	0.0055	0.0008	06-mar	20-mar	0.0000	0.0009	0.0026	0.0000
05-mar	19-mar	0.0000	0.0153	0.0000	0.0000	0.0000	07-mar	21-mar	0.0000	0.0004	0.0068	0.0001
06-mar	20-mar	0.0000	0.0000	0.0000	0.0001	0.0000	08-mar	22-mar	0.0000	0.0000	0.0504	0.0282
07-mar	21-mar	0.0000	0.0000	0.0000	0.0001	0.0000	09-mar	23-mar	0.0000	0.0000	0.0066	0.0186

End Period		NO2->Infections					End Period		NO2->Infections			
PM	Infections	MO	PR	PC	RE	RM	PM	Infections	RA	BO	FE	FC
01-mar	15-mar	0.0000	0.0000	0.0000	0.0000	0.0000	03-mar	17-mar	0.0016	0.0000	0.0068	0.0000
02-mar	16-mar	0.0000	0.0000	0.0000	0.0000	0.0000	04-mar	18-mar	0.0008	0.0000	0.0017	0.0000
03-mar	17-mar	0.0031	0.0000	0.0000	0.0000	0.0000	05-mar	19-mar	0.0000	0.0000	0.0115	0.0000
04-mar	18-mar	0.0004	0.0000	0.0000	0.0001	0.0000	06-mar	20-mar	0.0000	0.0001	0.0602	0.0000
05-mar	19-mar	0.0000	0.0000	0.0000	0.0369	0.0000	07-mar	21-mar	0.0000	0.0000	0.1248	0.0000
06-mar	20-mar	0.0000	0.0000	0.0000	0.0000	0.0000	08-mar	22-mar	0.1249	0.0000	0.0852	0.0000
07-mar	21-mar	0.0000	0.0000	0.0000	0.0000	0.0000	09-mar	23-mar	0.0803	0.0000	0.0162	0.0000

Figure 4. Particulate matter and COVID-19 infections (before lockdown): Granger-causality and *p*-values.

3.2. After the Lockdown

As already told, the causal modeling method proposed by Granger was designed to handle pairs of variables, and consequently it may suffer from a typical limitation when a third variable is engaged in the relation, as explained in a previous Section. In our specific case, this third variable could be identified with all the variety of human activities that could be the common cause for both the air pollution and the spread of infections in Emilia-Romagna.

Nonetheless, an important factor has come to the scene through which we will try to argue that the relation identified in the previous Subsection still holds. This factor amounts to the lockdown decisions taken either on either 8 or 10 March, depending on the specific province under investigation.

As a result of these decisions, human activities had fallen down to a minimum starting again on either 8 or 10 March, depending on the province under consideration. This has a precise meaning with an impact onto the rationale behind our analysis, which is as follows: All what happens after those dates can no longer be ascribed to the activity carried out by humans (if not minimally).

Nonetheless, looking at this from an opposite perspective, one should also argue that this new factor (i.e., the lockdown) can also have a confusing effect on the researched phenomena, since the absence of humans in the scene could open the way to new unexpected implications, and hence to a variety of different possible interpretations.

To avoid this possible pitfall, we have redesigned our experiments with a specific care to select for our analysis only those provinces whose general characteristics could be considered to be more easily observable, with less external interferences. Two design principles drove us for this new set of experiments. The first was that to exclude from our analysis all those provinces with a too high number of infected individuals per population, with respect to the average value of the region under investigation. This way, Piacenza, Reggio nell'Emilia, Parma and Rimini were excluded, yielding the highest percentages of infected individuals per population, namely: 1.509%, 0.906%, 0.723% and 0.606% (as recorded on 9 May 2020). For an analogous reason, we excluded the largest province in the region, precisely Bologna, since it is suffering a very high number of infected individuals, which are currently as many as 4751. For an opposite motivation, we cut off from the second part of our study also the province of Ferrara, which for a long time, fortunately, had hit the lowest rate of infected individuals per population (even though it has recently recorded higher values, thus reaching currently the percentage of 0.281%).

Finally, excluded went also the province of Forlì-Cesena, in this case due to the fact that we measured a marked decrease in the amount of the values of the particulate measured during the new period of investigation.

To this aim, it is interesting to notice that the difference between the amount of particulate matter taken both before and after the lockdown, computed as an average of the daily measurements of the two two-weeks long periods that preceded and followed the lockdown date, ranged in an interval from +8.65 micrograms per cubic meter (Parma) to –3.33 micrograms per cubic meter (Rimini) for the $PM_{2.5}$ pollutant, and from +6.30 micrograms per cubic meter (Parma) to –10.47 micrograms per cubic meter (Rimini) for the PM_{10} pollutant. (At this point, it is also interesting to remind to the reader that the acceptable daily limit considered for PM_{10} pollutant is set to be 50 micrograms per cubic meter).

All this considered, both the province of Modena and Ravenna were rather stable under this perspective, with incremental values amounting to: +7.86 ($PM2_{2.5}$) and +6.11 ($PM_{2.5}$) micrograms per cubic meter for Modena, and +2.09 (PM_{10}) and −3.37 (PM_{10}) micrograms per cubic meter for Ravenna.

In essence, our post-lockdown analysis was confined to just the two provinces of Modena and Ravenna, because they both satisfy all the following requirements:

- a rate of infected individuals ranging from moderate to mild (Modena, 0.538% or 3792; Ravenna, 0.281% or 995);
- a quantity of infected individuals not hitting the highest values in absolute, like instead Reggio nell'Emilia (4835) and Bologna (4751), for example;
- a relative stability in the in/decrease of the particulate matter after the restrictions imposed by the lockdown.

Summing up, our choice towards these two provinces have been orientated by the fact that they looked like to us as the only provinces on which the changes induced by the lockdown had a minimal external impact, even though the human activities were prohibited. In some sense, they were those provinces less affected by interferences whose causal factors remain unobservable and unknowable.

All this said, Figure 5 reports the results of our Granger causality analysis conducted for the provinces of both Modena (MO) and Ravenna (RA).

For a full comprehension of the Figure, one should notice that all has remained unchanged here, with respect to Figure 4, as to how the experiments were developed, with just these three natural considerations:

- Each observed series closes in a period ranging, respectively, from 8 March (Modena) and 10 March (Ravenna) for the pollutants' series, and from 22 March (Modena) and from 24 March (Ravenna) for the infections, up to 1 April (Modena) and to 13 April (Ravenna) for the pollutants' series, and up to 15 April (Modena) and to 17 April (Ravenna) for the infections;
- The beginning day for both series (pollution and infections) remains the same as in the comments provided for Figure 4.
- The analysis, this time, was conducted just for the particulate matter of type: $PM_{2.5}$ and $PM_{10,}$ not being available at that time stable measurements for NO_2.

In essence, our scientific target, here, was to verify if the pairwise series correlation observed before was still confirmed, even if we have been adding some more 25 days at each series, with all the 25 days happened after that the lockdown took place.

To this aim, an analysis of the p - values of Figure 5 shows that we have got a total amount of 97 (out of 100) statistical confirmations (yielding a 97% value) that X Granger causes Y; that is, that some given pollutants have some effect on the number of infections, from a Granger-causality perspective.

To be precise, interesting is the fact that a similar analysis conducted for all the other provinces (Bologna, Ferrara, Forlì-Cesena, Parma, Piacenza, Reggio nell'Emilia and Rimini) provides a more controversial result, with a lower number of statistical confirmations (approximately around 50%), probably depending on all those interferences, happened as a consequence of the lockdown, which we mentioned before as the motivation of our decision for the exclusion.

Nonetheless, at the end of this study, we can maintain that strong statistical clues emerge in favor of a causal correlation between pollution and infections, at least in Emilia-Romagna. This should be confirmed by those readers who are taking into serious consideration the fact that we have conducted a careful study, based on an analysis of time series, considered both before and after the lockdown, and aimed at screening off all the typical limitations that can afflict the Granger causality hypothesis testing method.

End Period		MO		End Period		RA	
PM	Infections	PM10->Infections	PM2.5->Infections	PM	Infections	PM10->Infections	PM2.5->Infections
08-mar	22-mar	0.0000	0.0000	10-mar	24-mar	0.0000	0.0000
09-mar	23-mar	0.0000	0.0000	11-mar	25-mar	0.0000	0.0000
10-mar	24-mar	0.0000	0.0000	12-mar	26-mar	0.0001	0.0000
11-mar	25-mar	0.0000	0.0001	13-mar	27-mar	0.0144	0.0039
12-mar	26-mar	0.0000	0.0001	14-mar	28-mar	0.0077	0.0006
13-mar	27-mar	0.0000	0.0000	15-mar	29-mar	0.0089	0.0178
14-mar	28-mar	0.0267	0.0500	16-mar	30-mar	0.0073	0.0113
15-mar	29-mar	0.0305	0.0668	17-mar	31-mar	0.0126	0.0120
16-mar	30-mar	0.0272	0.0603	18-mar	01-apr	0.0134	0.0213
17-mar	31-mar	0.0438	0.0472	19-mar	02-apr	0.0126	0.0351
18-mar	01-apr	0.0133	0.0158	20-mar	03-apr	0.0139	0.0382
19-mar	02-apr	0.0165	0.0180	21-mar	04-apr	0.0283	0.0330
20-mar	03-apr	0.0113	0.0116	22-mar	05-apr	0.0282	0.0323
21-mar	04-apr	0.0112	0.0109	23-mar	06-apr	0.0236	0.0273
22-mar	05-apr	0.0254	0.0242	24-mar	07-apr	0.0182	0.0205
23-mar	06-apr	0.0238	0.0197	25-mar	08-apr	0.0142	0.0157
24-mar	07-apr	0.0320	0.0230	26-mar	09-apr	0.0127	0.0154
25-mar	08-apr	0.0358	0.0265	27-mar	10-apr	0.0152	0.0244
26-mar	09-apr	0.0387	0.0285	28-mar	11-apr	0.0073	0.0132
27-mar	10-apr	0.0332	0.0238	29-mar	12-apr	0.0063	0.0143
28-mar	11-apr	0.0402	0.0247	30-mar	13-apr	0.0043	0.0256
29-mar	12-apr	0.0436	0.0243	31-mar	14-apr	0.0027	0.0150
30-mar	13-apr	0.0334	0.0203	01-apr	15-apr	0.0002	0.0085
31-mar	14-apr	0.0260	0.0328	02-apr	16-apr	0.0000	0.0137
01-apr	15-apr	0.0177	0.0431	03-apr	17-apr	0.0000	0.0110

Figure 5. Particulate matter and COVID-19 infections (after lockdown): Granger-causality and *p*-values.

4. Conclusions

We have conducted a statistical analysis that confirms, under a Granger causality perspective, that a causal correlation may exist between the two researched phenomena of: pollution and COVID-19 infections, in Emilia-Romagna, Italy. Here, we survey, at the end of the paper, the possible limitations of our study (as well as, its potentials).

As to this issue of possible fallacies and limitations of our investigations, we feel necessary to discuss, at least, on the three following points: (i) the robustness of the scientific methodology we adopted, (ii) the choice of the Emilia-Romagna region as the primary subject of our study, and finally (iii) the scientific validity of the data we used.

As far as the Granger causality method is concerned, we have already admitted that neither Granger, nor any other statistical testing procedure, can provide a final evidence that the two phenomena we have studied (i.e., pollution vs. COVID-19 infections) are definitely correlated in nature. In fact, to achieve an ultimate knowledge of this correlation, statistical evidences, like those demonstrated in this paper, should be always accompanied by additional experiments at a scale that is appropriate to the observed phenomena; that is, in this case, at a biomedical, chemical or even physical level. Apart from this issue, our study has demonstrated that using Granger may be a valid solution, over alternative computational methodologies, to infer statistical evidences from sets of data subjected to high levels of temporal uncertainty [35]. In this case, in particular, fundamental has been the idea of testing the correlation hypothesis with data taken both before and after the lockdown.

To move on to the second issue, we understand very well that the choice to limit our study to the Italian region of Emilia-Romagna can be a source of controversy, and a limitation, as well. However, none should forget that the COVID-19 pandemic spread to Italy very early in 2020, and that the virus hit this nation with a number of active cases (i.e., infections), and deaths, that were unmatched in Europe, at least at that time of the year. It is also another truth that the region hit hardest in Italy was Lombardy (with almost the 48% of all the fatalities in Italy). Nonetheless, we all know very well that Lombardy is still Italy's COVID-19 hotspot, probably due to a combination of factors, including wrong medical, governmental, and industrial policies which are controversial, yet not negligible [36].

The Emilia-Romagna region was severely devasted, too (with almost the 12% of all the fatalities in Italy). However, regardless the size of the investigated sample, the relative absence of dispute on external factors, like overwhelmed hospitals and controversial decision making, at both a political and an industrial level, made this specific region a subject of study where the phenomena of interest (that is, pollution and infections) could emerge without an annoying level of external interferences.

Finally, it is the turn of the data. First, we want to emphasize that all the used data and statistics were publicly available, at the time of our investigation, on Italian governmental sites, precisely [1,27,28]. It is also worth noticing that all our experiments are reproducible using the data available in the public repositories we have mentioned. Nevertheless, it is also a fact that COVID-19 infections are by now assumed to be more widespread than initially expected, thus making many of the studies conducted so far (including ours) a poor proxy for understanding the extension of this infection, with all the relative implications [37]. Anyway, as already mentioned at the beginning of this paper, if we move beyond the relevance of our results, towards the real extent of the correlation we have statistically demonstrated, our ultimate aim is that of reinvigorating the debate on a scientific case (pollution vs. COVID-19), that should not go unsolved or remain uninvestigated.

As a very final note, it is worth mentioning that, for the sake of completeness, we have provided a graphical summarization of all the data we have used on our study in the Appendix A that follows this paper. .

Author Contributions: All authors contributed equally to the manuscript. All authors have read and agreed to the published version of the manuscript.

Funding: This research received no external funding.

Conflicts of Interest: The authors declare no conflict of interest.

Appendix A

We provide a summarization of all the data we used for our experiments. These graphs simultaneously show the curves for both the pollutants (black) and the COVID-19 infections (blue). On the *x axis* of all graphs reported are the timelines for the pollutants' series (in black) and for the infections (in blue). Figures A1–A3 are, respectively, concerned with: $PM_{2.5}$, PM_{10} and NO_2.

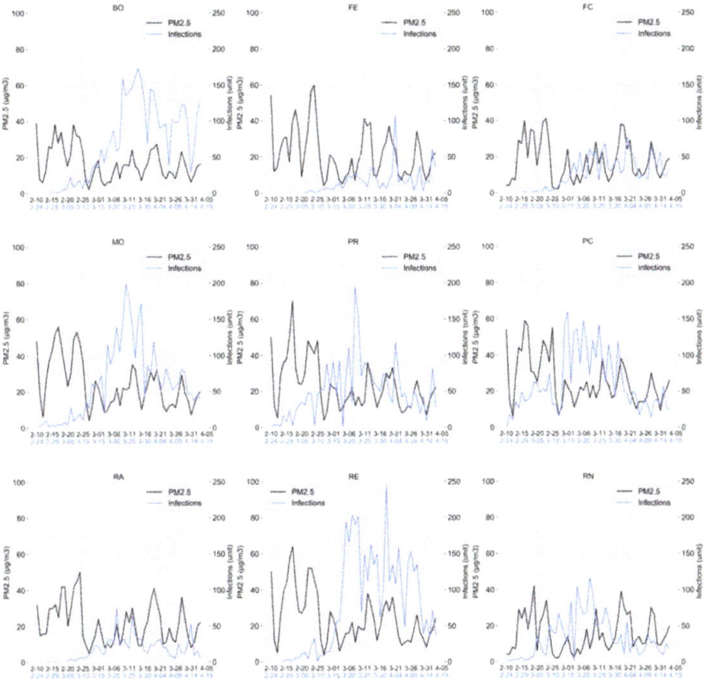

Figure A1. Particulate matter ($PM_{2.5}$) and COVID-19 infections (all the examined periods).

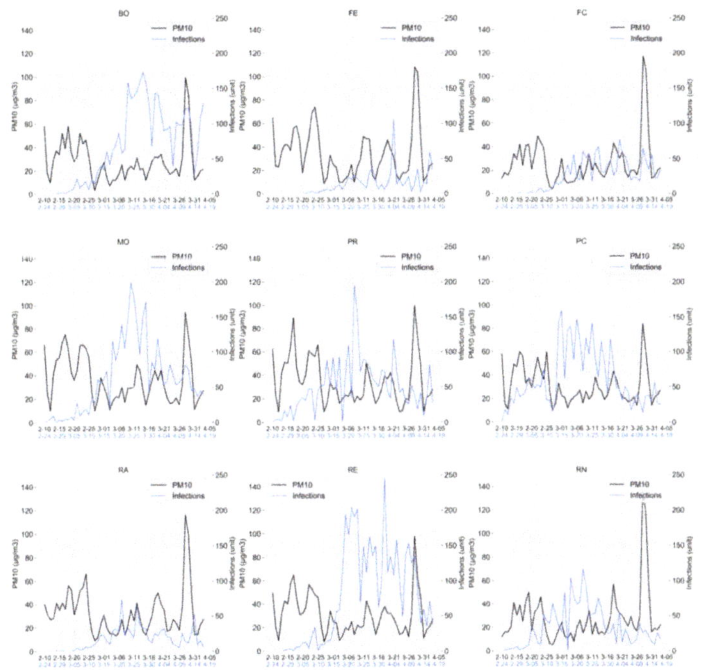

Figure A2. Particulate matter (PM_{10}) and COVID-19 infections (all the examined periods).

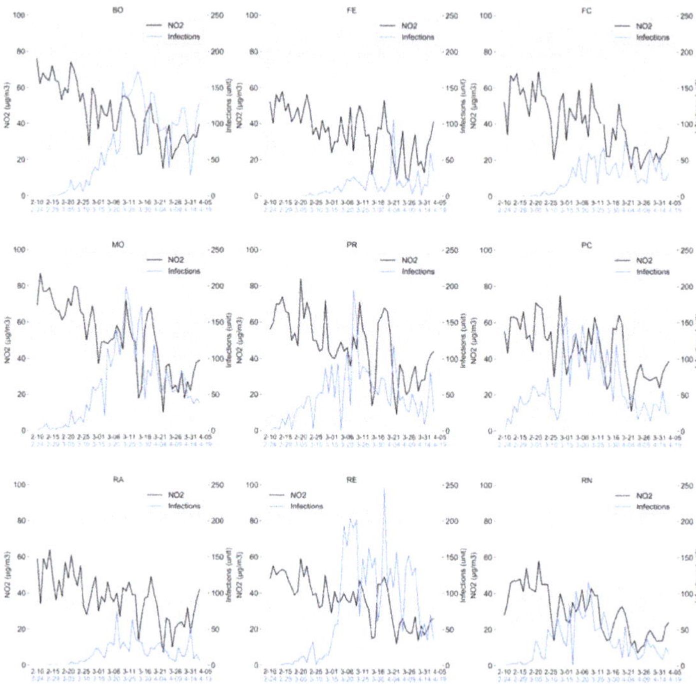

Figure A3. Particulate matter (NO_2) and COVID-19 infections (all the examined periods).

References

1. Istituto Nazionale di Statistica. Impatto Dell'epidemia COVID-19 Sulla Mortallità Totale Della Popolazione Residente Primo Trimestre 2020. Available online: https://www.istat.it/en/files//2020/05/Rapporto_Istat_ISS.pdf (accessed on 6 May 2020).
2. Jiang, Y.; Wu, X.J.; Guan, Y.J. Effect of ambient air pollutants and meteorological variables on COVID-19 incidence. *Infect. Control Hosp. Epidemiol.* **2020**, 1–11. [CrossRef] [PubMed]
3. You, S.; Tong, Y.W.; Neoh, K.G.; Dai, Y.; Wang, C.H. On the association between outdoor PM2.5 concentration and the seasonality of tuberculosis for Beijing and Hong Kong. *Environ. Pollut.* **2016**, *218*, 1170–1179. [CrossRef] [PubMed]
4. Horne, B.D.; Joy, E.A.; Hofmann, M.G.; Gesteland, P.H.; Cannon, J.B.; Lefler, J.S.; Blagev, D.P.; Korgenski, E.K.; Torosyan, N.; Hansen, G.I.; et al. Short-term elevation of fine particulate matter air pollution and acute lower respiratory infection. *Am. J. Respir. Crit. Care Med.* **2018**, *198*, 759–766. [CrossRef] [PubMed]
5. Su, W.; Wu, X.; Geng, X.; Zhao, X.; Liu, Q.; Liu, T. The short-term effects of air pollutants on influenza-like illness in Jinan, China. *BMC Public Health* **2019**, *19*, 1319. [CrossRef]
6. Lee, G.I.; Saravia, J.; You, D.; Shrestha, B.; Jaligama, S.; Hebert, V.Y.; Dugas, T.R.; Cormier, S.A. Exposure to combustion generated environmentally persistent free radicals enhances severity of influenza virus infection. *Part. Fibre Toxicol.* **2014**, *11*, 57. [CrossRef]
7. Carbone, M.; Green, J.B.; Bucci, E.M.; Lednicky, J.A. Coronaviruses: Facts, Myths, and Hypotheses. *J. Thorac. Oncol.* **2020**, *15*, 675–678. [CrossRef] [PubMed]
8. Liang, Y.; Fang, L.; Pan, H.; Zhang, K.; Kan, H.; Brook, J.R.; Sun, Q. PM 2.5 in Beijing–temporal pattern and its association with influenza. *Environ. Health* **2014**, *13*, 102. [CrossRef]
9. Feng, C.; Li, J.; Sun, W.; Zhang, Y.; Wang, Q. Impact of ambient fine particulate matter (PM 2.5) exposure on the risk of influenza-like-illness: A time-series analysis in Beijing, China. *Environ. Health* **2016**, *15*. [CrossRef]
10. Li, Q.; Guan, X.; Wu, P.; Wang, X.; Zhou, L.; Tong, Y.; Ren, R.; Leung, K.S.M.; Lau, E.H.Y.; Wong, J.Y.; et al. Early transmission dynamics in Wuhan, China, of novel coronavirus–infected pneumonia. *New Engl. J. Med.* **2020**, *382*, 1199–1207. [CrossRef]

11. Zhou, P.; Yang, X.L.; Wang, X.G.; Hu, B.; Zhang, L.; Zhang, W.; Si, H.R.; Zhu, Y.; Li, B.; Huang, C.L.; et al. A pneumonia outbreak associated with a new coronavirus of probable bat origin. *Nature* **2020**, *579*, 270–273. [CrossRef] [PubMed]
12. Brankston, G.; Gitterman, L.; Hirji, Z.; Lemieux, C.; Gardam, M. Transmission of influenza A in human beings. *Lancet Infect. Dis.* **2007**, *7*, 257–265. [CrossRef]
13. Tellier, R. Aerosol transmission of influenza A virus: A review of new studies. *J. R. Soc. Interface* **2009**, *6*, S783–S790. [CrossRef] [PubMed]
14. Hinds, W.C. Properties, behavior, and measurement of airborne particles. In *Aerosol Technology*, 2nd ed.; John Wiley and Sons Press: New York, NY, USA, 1999; pp. 182–204.
15. Nunez Soza, L.; Jordanova, P.; Nicolis, L.; Strelec, L.; Stehlik, M. Small sample robust approach to outliers and correlation of Atmospheric Pollution and Health Effects in Santiago de Chile. *Chemom. Intell. Lab. Syst.* **2019**, *185*, 73–84. [CrossRef]
16. Pansini, R.; Fornacca, D. Initial evidence of higher morbidity and mortality due to SARS-CoV-2 in regions with lower air quality. *MedRxiv Preprint* **2020**. [CrossRef]
17. Wu, X.; Nethery, R.C.; Sabath, B.M.; Braun, D.; Dominici, F. Exposure to air pollution and COVID-19 mortality in the United States. 2020. Available online: https://www.medrxiv.org/content/10.1101/2020.04.05.20054502v2 (accessed on 25 April 2020).
18. Ogen, Y. Assessing nitrogen dioxide (NO_2) levels as a contributing factor to the coronavirus (COVID-19) fatality rate. *Sci. Total Environ.* **2020**, *726*, 138605. [CrossRef] [PubMed]
19. Setti, L.; Passarini, F.; De Gennaro, G.; Baribieri, P.; Perrone, M.G.; Borelli, M.; Palmisani, J.; Di Gilio, A.; Torboli, V.; Pallavicini, A.; et al. SARS-Cov-2 RNA Found on Particulate Matter of Bergamo in Northern Italy: First Preliminary Evidence. Available online: https://www.medrxiv.org/content/10.1101/2020.04.15.20065995v2 (accessed on 2 May 2020).
20. Setti, L.; Passarini, F.; De Gennaro, G.; Barbieri, P.; Perrone, M.G.; Piazzalunga, A.; Borelli, M.; Palmisani, J.; Di Giglio, A.; Piscitelli, P.; et al. The Potential role of Particulate Matter in the Spreading of COVID-19 in Northern Italy: First Evidence-based Research Hypotheses. 2020. Available online: https://www.medrxiv.org/content/10.1101/2020.04.11.20061713v1 (accessed on 25 April 2020).
21. Conticini, E.; Frediani, B.; Caro, D. Can atmospheric pollution be considered a co-factor in extremely high level of SARS-CoV-2 lethality in Northern Italy? *Environ. Pollut.* **2020**, *261*, 114465. [CrossRef]
22. Becchetti, L.; Conzo, G.; Conzo, P.; Salustri, F. Understanding the Heterogeneity of Adverse COVID-19 Outcomes: The Role of Poor Quality of Air and Lockdown Decisions. Available online: https://papers.ssrn.com/sol3/papers.cfm?abstract_id=3572548 (accessed on 25 April 2020).
23. Caserini, S.; Perrino, C.; Forastiere, F.; Poli, G.; Vicenzi, E.; Carra, L. Pollution and COVID. Two Vague Clues don't Make an Evidence. Available online: http://www.scienceonthenet.eu/articles/pollution-and-COVID-two-vague-clues-dont-make-evidence/stefano-caserini-cinzia-perrino (accessed on 29 April 2020).
24. Maziarz, M. A review of the Granger-causality fallacy. *J. Philos. Econ.* **2015**, *8*, 86–105.
25. Gazzetta Ufficiale della Repubblica Italiana. Decreto del Presidente del Consiglio dei Ministri 8 Marzo 2020. Available online: https://www.gazzettaufficiale.it/eli/id/2020/03/08/20A01522/sg (accessed on 20 April 2020).
26. Governo Italiano Presidenza del Consiglio dei Ministri. Available online: http://www.governo.it/it/articolo/firmato-il-dpcm-9-marzo-2020/14276 (accessed on 20 April 2020).
27. COVID-19 Italia—Monitoraggio Situazione. Available online: https://github.com/pcm-dpc/COVID-19 (accessed on 18 April 2020).
28. Arpae Emilia-Romagna. Available online: https://arpae.it/mappa_qa.asp?idlivello=1682&tema=stazioni (accessed on 18 April 2020).
29. Lauer, S.A.; Grantz, K.H.; Bi, Q.; Jones, F.K.; Zheng, Q.; Meredith, H.R.; Azman, A.S.; Reich, N.G.; Lessler, J. The incubation period of coronavirus disease 2019 (COVID-19) from publicly reported confirmed cases: Estimation and application. *Ann. Intern. Med.* **2020**, *172*, 577–582. [CrossRef]
30. Cereda, D.; Tirani, M.; Rovida, F.; Demicheli, V.; Ajelli, M.; Poletti, P.; Trentini, F.; Guzzetta, G.; Marziano, V.; Barone, A.; et al. The early Phase of the COVID-19 Outbreak in Lombardy, Italy. Available online: https://arxiv.org/abs/2003.09320 (accessed on 25 April 2020).
31. Granger, C. Investigating Causal Relations by Econometric Models and Cross-spectral Methods. *Econometrica* **1969**, *37*, 424–438. [CrossRef]
32. Granger, C.W. Testing for causality: A personal viewpoint. *J. Econ. Dyn. Control* **1980**, *2*, 329–352. [CrossRef]

33. Dickey, D.; Fuller, W. Distribution of the estimators for autoregressive time series with a unit root. *J. Am. Stat. Assoc.* **1979**, *74*, 427–431. [CrossRef]
34. Arntzenius, F. The common cause principle. In *1992 Biennial Meeting of the Philosophy of Science Association*; The University of Chicago Press: Chicago, IL, USA, 1992; Volume 2, pp. 227–237.
35. Delnevo, G.; Roccetti, M.; Mirri, S. Modeling Patients' Online Medical Conversations: A Granger Causality Approach. In Proceedings of the 2018 IEEE/ACM International Conference on Connected Health: Applications, Systems and Engineering Technologies, Washington, DC, USA, 26–28 September 2018; IEEE: Piscataway, NJ, USA, 2018; pp. 40–44. [CrossRef]
36. Privitera, G. First in, Last Out: Why Lombardy is Still Italy's Coronavirus Hotspot. Politico. 2020. Available online: https://www.politico.eu/article/first-in-last-out-why-lombardy-is-still-italys-coronavirus-COVID19-hotspot-italy/ (accessed on 13 June 2020).
37. Sood, N.; Simon, P.; Ebner, P. Seroprevalence of SARS-CoV-2–Specific Antibodies Among Adults in Los Angeles County, California, on April 10–11, 2020. *J. Am. Med Assoc.* **2020**, *323*, 2425–2427. [CrossRef] [PubMed]

computation

Article

Is a COVID-19 Second Wave Possible in Emilia-Romagna (Italy)? Forecasting a Future Outbreak with Particulate Pollution and Machine Learning

Silvia Mirri, Giovanni Delnevo and Marco Roccetti *

Department of Computer Science and Engineering, University of Bologna, 40127 Bologna, Italy; silvia.mirri@unibo.it (S.M.); giovanni.delnevo2@unibo.it (G.D.)
* Correspondence: marco.roccetti@unibo.it

Received: 24 July 2020; Accepted: 20 August 2020; Published: 24 August 2020

Abstract: The Nobel laureate Niels Bohr once said that: "Predictions are very difficult, especially if they are about the future". Nonetheless, models that can forecast future COVID-19 outbreaks are receiving special attention by policymakers and health authorities, with the aim of putting in place control measures before the infections begin to increase. Nonetheless, two main problems emerge. First, there is no a general agreement on which kind of data should be registered for judging on the resurgence of the virus (e.g., infections, deaths, percentage of hospitalizations, reports from clinicians, signals from social media). Not only this, but all these data also suffer from common defects, linked to their reporting delays and to the uncertainties in the collection process. Second, the complex nature of COVID-19 outbreaks makes it difficult to understand if traditional epidemiological models, such as susceptible, infectious, or recovered (SIR), are more effective for a timely prediction of an outbreak than alternative computational models. Well aware of the complexity of this forecasting problem, we propose here an innovative metric for predicting COVID-19 diffusion based on the hypothesis that a relation exists between the spread of the virus and the presence in the air of particulate pollutants, such as $PM_{2.5}$, PM_{10}, and NO_2. Drawing on the recent assumption of 239 experts who claimed that this virus can be airborne, and further considering that particulate matter may favor this airborne route, we developed a machine learning (ML) model that has been instructed with: (i) all the COVID-19 infections that occurred in the Italian region of Emilia-Romagna, one of the most polluted areas in Europe, in the period of February–July 2020, (ii) the daily values of all the particulates taken in the same period and in the same region, and finally (iii) the chronology according to which restrictions were imposed by the Italian Government to human activities. Our ML model was then subjected to a classic ten-fold cross-validation procedure that returned a promising 90% accuracy value. Finally, the model was used to predict a possible resurgence of the virus in all the nine provinces of Emilia-Romagna, in the period of September–December 2020. To make those predictions, input to our ML model were the daily measurements of the aforementioned pollutants registered in the periods of September–December 2017/2018/2019, along with the hypothesis that the mild containment measures taken in Italy in the so-called Phase 3 are obeyed. At the time we write this article, we cannot have a confirmation of the precision of our predictions. Nevertheless, we are projecting a scenario based on an original hypothesis that makes our COVID-19 prediction model unique in the world. Its accuracy will be soon judged by history—and this, too, is science at the service of society.

Keywords: COVID-19; predictions; second wave; machine learning models; air pollution; Emilia- Romagna; Italy

1. Introduction

"The virus remains the public enemy number one", World Health Organization (WHO), Director General, Tedros Adhanom Ghebreyesus maintained at a recent press conference, and he also added that: "If basics are not followed, the only way the pandemic is going to go, it is going to get worse and worse and worse" [1]. These threatening words are justified in light of the current pandemic numbers. As of 17 July 2020, global COVID-19 cases exceed 13.5 million, and 584,940 people have died of it in almost seven months, with the current biggest rises in the United States, Brazil, India, and South Africa [2].

When COVID-19 first struck several provinces of Northern Italy in early 2020 (especially in Lombardy and in Emilia-Romagna), the conditions there made it a perfect storm. The virus outbreak spread with an unusual violence (in the period from late February to April 2020), with a catastrophic toll in terms of human deaths. Still now, several months after that last virus surge, and a severe subsequent lockdown period, the consequences are profound. Italy counts a total number of 252,235 registered infections, and as many as 35,231 human deaths (as of 13 August 2020) [3]. Not only that, but some recent financial studies also estimate that Italy's Gross Domestic Product (GDP) could drop significantly in 2020 due to the impact of the pandemic, with some industrial sectors severely hit, including textile, train and air transport, hotels, restaurants, entertainment, and automotive [4].

The proportion of this disaster is key to understanding why policymakers, health officials, and media in general have an increasing interest in making use of computational models that can forecast possible resurgences of the virus, in order to put in place containment measures [5]. Unfortunately, there are several problems here, primarily linked to collecting data, and then using them to feed an adequate forecasting model.

Along this line of reasoning, we propose a clear direction. We do believe that a relationship exists between particulate matter (of various types) and COVID-19 incidence, and that this favors the spread of the contagion. We have devoted a previous study to verifying the presence of such a possible correlation between the series of the new daily COVID-19 infections in the period February–April 2020 in Emilia-Romagna (Italy) and the correspondent series of the daily values of the $PM_{2.5}$, PM_{10}, and NO_2 pollutants [6]. A specific statistical hypothesis testing method was then employed (i.e., the Granger causality statistical methodology [7,8]), which returned a positive response to our question based on a complex set of experiments that extended before, during, and after the lockdown periods decided by the Italian Government on 8–10 March 2020.

Obviously, it is not our intention to run through, again, all the technical and epistemic issues behind this hypothesis here. The interested reader can refer to [6]; nonetheless, some of the basics that lie behind our decision to use this hypothesis to select the data used to make predictions on future COVID-19 outbreaks need to be discussed.

First, it is out of discussion that poor air quality easily brings one to a state of permanent inflammation and chronic respiratory difficulties, along with a hyper-activation of the immune system. All these circumstances make human lungs prone to be attacked by respiratory viral infections [9]. Owing to this condition, it has been demonstrated that humans living in highly polluted areas have a reduced respiratory capacity to react to virus attacks [10]. In addition to these general considerations, which are confirmed by an impressive wealth of recent literature [11–13], more interesting is the biological phenomenon at the center of the following controversy: Can particulate matter be a carrier for COVID-19?

To respond to this question, it would be enough to remind a recent claim of 239 experts who maintained that this virus can be airborne [14], united with the information of the presence of the COVID-19 RNA, found in the particulate matter of Bergamo (Italy) [15]. All these seem to confirm that this virus can create clusters with particulate matter, and that it can be carried by this type of microscopic pollutants.

To close this issue: Although we are aware that there is an ongoing scientific controversy, concerning the link between that first experimental finding (i.e., [15]) and the degree of severity with which a COVID-19 outbreak can spread [16], we believe that both our previous study and those detailed in [17,18] provide a support to the hypothesis that the presence of COVID-19 on outdoor air samples can represent a potential early indicator of the diffusion of the virus in a given area.

Hence, based on the hypothesis that this virus can be airborne and assuming that particulate matter may favor this airborne route, we developed a machine learning model (ML, for short), with special attention to the Italian region of Emilia-Romagna (Italy). Our model was instructed with:

- All the COVID-19 infections that occurred in Emilia-Romagna, one of the most polluted areas in Europe, in the period of February–July 2020;
- The daily values of all the aforementioned particulates taken in the same period and in the same region; and finally,
- The chronology according to which restrictions were imposed by the Italian Government to human activities in the same period under observation.

Our ML model was then subjected to a classic testing procedure that has returned a promising accuracy value of approximately 90%. Finally, the model was used to predict a possible second wave of the virus for all the nine provinces of Emilia-Romagna, in the period of September–December 2020.

To get the predictions, input to our ML model were the daily measurements of the aforementioned pollutants registered in the periods of September–December 2017/2018/2019, along with the hypothesis that the mild containment measures taken by Italy in the so-called *Phase* 3 are obeyed [19].

Having covered the reasoning behind our choice, we shall return now to possible alternatives.

Inspired by a wealth of recent literature, new techniques have been proposed to aggregate data that could predict the pandemic's next moves. For example, drawing on the use of new information technologies, including search engines, news reports, crowdsourced infoveillance, Twitter feeds, travel data, tele-traffic measurements, and many others again, the authors of [20] exploited a Bayesian model that calculates, in near-real time, the probability of an exponential growth or subsequent decay of the virus spread, based on data collected in the USA, between January and June 2020. Interesting in this kind of study is the fact that data from Twitter and Google searches emerge as the earliest uptrend signals to anticipate a virus surge (with a median earliness of 2–3 weeks), while UpToDate (an evidence-based clinical decision support system, developed by the health division of Wolters Kluwer [21]) was capable of providing early signals of uptrend for deaths (earliness of 4.5 weeks). Additionally, Google searches, united with the elaboration of some form of mobility data from citizens, provided early downtrend signals to anticipate a virus decay (median earliness of 2 weeks).

This type of proposal appears as an advancement to the state-of-the-art, especially if one considers that, as far as data are concerned, the problem is that virus case counts, hospitalized patients, number of deaths, reports from clinicians, etc. all suffer from reporting delays (as well as from uncertainties in the data collection process).

While we refrain from expressing non-positive comments on this research, we have to admit that it is certainly true that combining many streams of real time information may lead decision makers to be responsive to sudden changes; nevertheless, crucial remains the issue of how reliable and precise those streams of observations are when it comes to describing a pandemic spread, especially if no working hypothesis lies behind. Told in simpler words, the strength of the approach here is also its weakness: What the authors of [20] are doing is observing, without making any assumptions. This could be just a little bit extreme, since we all know that it is not the first time in the history of the science of data that one realizes, just at the end of the process, that too many data can be a bad thing. Making useful predictions requires something more than data, in fact—for example, some strong conceptual insights, as discussed at length in [22].

Let us talk now about forecasting models in more detail. Once, it was the SIR (susceptible, infectious, or recovered) model that dominated this scenery. A survey on this model is out of the scope of this paper, and the interested reader can refer to [23]. The actual value and importance of this traditional model is, obviously, out of the question in the epidemiological field; nonetheless, new proposals are emerging for modeling the COVID-19 pandemic that share similar goals, such as making predictions on the disease spread, yet adopting different computational methodologies.

Among these new proposals, the lion's share is played by machine learning models. The majority of the ML models used in practice are supervised. Learning, with supervision, involves learning a function that maps an input to an output based on examples of input–output pairs [24]. Providing a very simple example: If we had a set of data, regarding children with age in the range of 0–10 years, along with their correspondent weight, we could implement a very simple supervised ML model that predicts the weight of a child, based on their exact age.

Returning to the use of ML models for predicting a COVID-19 emergence, exemplary is the case of the work done in [25]. There, the authors provide a comparative analysis of various ML models to predict COVID-19 outbreaks. After a study of different ML models, based on the collection of data on infectious cases for 30 days from five different countries (Italy, China, Iran, Germany, and the USA), their most prominent finding is that the multilayered perceptron (MLP) model delivers the most accurate results, in terms of predicting an outbreak, without the assumptions that epidemiological models typically require.

Nevertheless, a criticism that we pose to articles of similar tenor is that all these studies can be assimilated to a process that starts from a bunch of example data and learns to point to the most likely output; where the meaning of likely is usually vague or fuzzy [26–30] or stochastic at best [31]. While we agree on the fundamental role played by data in these models, our belief is that, at least, a conceptual hypothesis should exist that drives one in their choice and selection.

Returning to our approach, we would like to conclude this section certain that the reader has a clear vision of our position, before they proceed with the article. To this aim, we have already stated in our previous work that neither Granger nor any other statistical testing procedure can provide final evidence that the two phenomena, between which we conjecture a relationship (i.e., pollution and COVID-19 infection spread), are correlated in nature. Additionally, the same holds for the predictions of our ML model. Nonetheless, with our predictions, we are projecting a scenario based on an original assumption that makes our COVID-19 ML model unique, as it selects the data to be used based on a well-defined and unambiguous hypothesis. Whatever will happen in September–December 2020, we will have learnt an important truth about the validity of our hypothesis—and this, too, is science at the service of society.

The remainder of the paper is structured as follows. In the next section, we describe the methodology behind our approach. Section 3 presents and critically discusses the results we yielded. Finally, Section 4 concludes the paper, with some final considerations.

2. Methods

We now present, first, some preliminary information relevant to the present study, second, a description of the dataset we used, and third, the reasoning we used to precisely decide what kind of predictions we are looking for. Finally, we provide some reflections on how we have selected the ML model that could fit squarely into our COVID-19 scenario of interest.

2.1. Preliminary Assumptions

As already anticipated, we have based this study on the precise idea that the correlation between air pollution (specifically, the $PM_{2.5}$, PM_{10}, and NO_2 pollutants) and the spread of COVID-19 infections in the Emilia-Romagna region is a very plausible hypothesis. Using that hypothesis, we consequently selected the data of interest.

We do not return, again, to this main assumption; it suffices here to remind that the presence of COVID-19 on air samples can represent an early indicator of the diffusion of the virus in a given geographical area [18]. To further summarize this concept, one should consider that, whatever the origin of this virus is, there are clear indications that COVID-19 transmission occurs from infected people, either through virus-laden droplets or aerosol transmissions. In this second case of airborne transportation, pollutants may help the diffusion, playing the role of additional carriers. All this is graphically summarized, with relative simplifications, in the following Figure 1, where it is crystal clear that the arrows in the figure should not be intended as a means to represent a direct causation, but they amount to a simple indication of a conceptual path; that is, infection propensity is favored by the transmission of droplets and aerosols, with air pollutants as further carriers.

Figure 1. Infection propensity.

All this anticipated, it is also important to give some details about the Italian region we took into consideration in our studies: Emilia-Romagna. The region is situated in the northeast section of the country and is divided into nine provinces: Bologna, Ferrara, Forlì-Cesena, Modena, Parma, Piacenza, Reggio nell'Emilia, Rimini, and Ravenna. It is populated by almost 4,500.000 citizens and was one of the more seriously hit by this virus in Italy, with a total number of infections of 30,342, and as many as 4298 fatalities, as of 13 August 2020. Its death toll linked to the virus is second, in Italy, only to Lombardy, where, on the same date, more than 97,000 persons were registered as infected, and the fatalities had almost reached the number of 17,000.

Relevant for considering this process from the right temporal perspective is also the chronology according to which restrictions were first imposed to human activities in those provinces, and then released after a substantial decay of the virus incidence. In particular, we can count four subsequent phases:

- Phase 0: Prior to 8 March 2020, no specific restriction was imposed, which was valid for all the nine provinces of Emilia-Romagna, except for some local control measures (for example, for schools and universities);
- Phase 1: A full lockdown was first imposed to the provinces of Modena, Parma, Piacenza, Reggio nell'Emilia, and Rimini, as of 8 March 2020 [32], and then extended to the remaining provinces of Bologna, Ferrara, Forlì-Cesena, and Ravenna on 10 March 2020 [33];
- Phase 2: On 4 May 2020, the lockdown was partially released, though with several commercial and industrial activities still suspended, as well as the obligation for people to stay in quarantine if found or suspected ill, wear cloth face covering in public settings, wash hands frequently, etc., and where a social distancing of at least 1 meter and a half was difficult to maintain [34];
- Phase 3: On 14 June 2020, the lockdown was almost completely removed, with almost all activities resuming, provided that the personal protection measures mentioned above were obeyed [19].

2.2. Dataset Description

Based on the hypothesis that a relation exists between pollutants and infections, at least in Emilia-Romagna, the data we used to instruct our ML model were essentially of two types:

- The measurements of the particulate pollutants: $PM_{2.5}$, PM_{10}, and $N0_2$; taken on a daily basis, for all the aforementioned provinces (Bologna, Ferrara, Forlì-Cesena, Modena, Parma, Piacenza, Reggio nell'Emilia, Rimini, and Ravenna).
- The number of the daily COVID-19 infections, again for all the provinces mentioned above.

The amount of daily infections was collected using the GitHub repository of the Italian Civil Protection, for the entire period, starting on 24 February and closing on 7 July 2020 [35].

The daily values of the pollutants, by contrast, were collected using the website of the Regional Environmental Protection Agencies (ARPA) of the Emilia-Romagna region for all the nine provinces [36]. With various ARPA monitoring stations distributed over each province, an average of the values returned by each station was computed on a daily/provincial basis. The period along which those data were collected was from 10 February up to 30 June 2020.

The two periods have the same temporal length, but there is a discrepancy in the starting/closing dates. This is due to the fact that a typical COVID-19 infection can be subjected to an incubation period, whose duration can range from a few days to almost 14 before a human begins to manifest some symptoms. Many papers provide evidence of this fact, concluding, at the end, that 99% of the infected population develop symptoms within 14 days [37].

Following this reasoning, the period during which we measured the particulates started on 10 February and closed on 30 June 2020, while the infections were registered in the period of 24 February–7 July 2020. Simply told, if we want to instruct an ML model with a function that maps input (particulates) into output (infections), based on examples of the input–output pairs we have collected, an offset has to be introduced that temporally separates these two time-series. This stems from the simple consideration that all that can happen on a given day, say x, in terms of augmented spread of the virus due to pollution, may have its effects in terms of manifestations of the infections up to day $x + 14$.

This is not still enough, though: In fact, the function that our ML model has to learn is a little bit more complex than usual, as we need to also take into consideration the specific period during which a given event (for example, an infection) has occurred. It makes a great difference, in fact, whether we consider events occurring during either Phase 0, or Phase 1, or Phase 2, or finally Phase 3. In conclusion, in addition to the data that are part of the relationship between particulates and infections, input to the ML model should also be the various phases through which the management of the spread of the infections has passed.

To conclude and summarize this complex situation, the following three figures show the entire dataset we have used, in a graphical form. All these graphs simultaneously show the curves for both the pollutants (black, measured in micrograms per cubic meter) and COVID-19 infections (gray, measured in units). On the x axis of all graphs reported are the timelines for the pollutants' series (in black) and for the infections (in gray). Important to note is the temporal offset explained above. Figures 2–4 are, respectively, relative to $PM_{2.5}$, PM_{10}, and NO_2. Moreover, in each graph, with the colors orange, yellow, light blue, and green, the passage through the different four phases we have mentioned is demarcated.

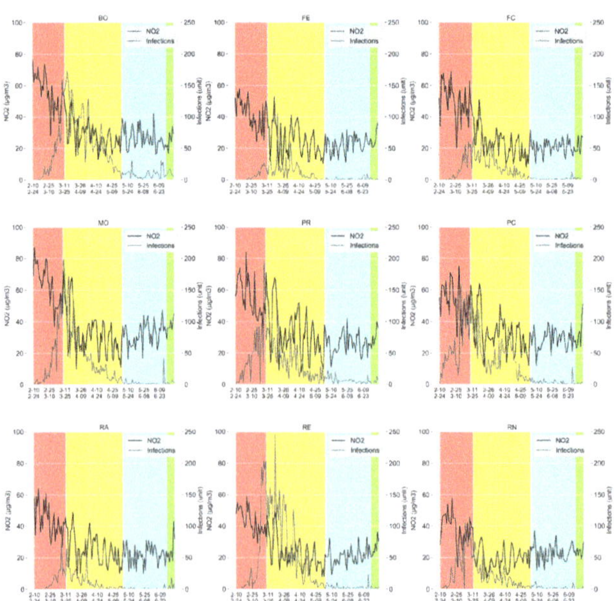

Figure 2. The dataset (I): infections, pollutant (NO_2), and phases.

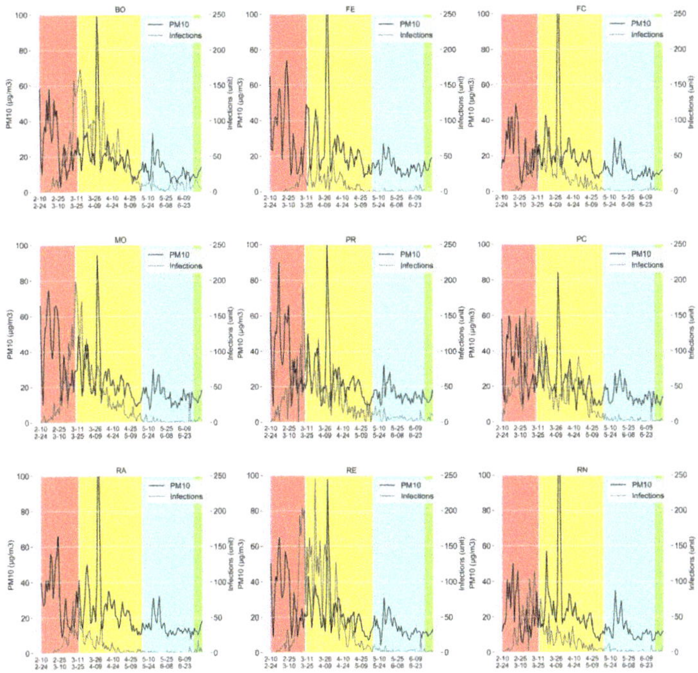

Figure 3. The dataset (II): infections, pollutant (PM_{10}), and phases.

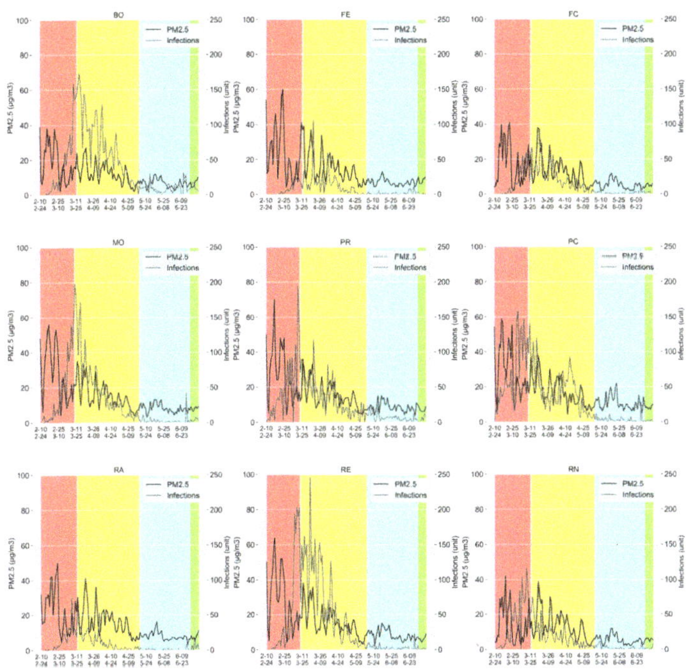

Figure 4. The dataset (III): infections, pollutant ($PM_{2.5}$), and phases.

2.3. *What Kind of Predictions Are We Looking for?*

In essence, the point is: If we have an ML model that has learnt from data the possible relationship between the presence of particulate in the air and the incidence of the virus, what kind of predictions should we ask of our model?

Let us say that the situation could become complex. In fact, while it is true that we are interested in knowing whether a second wave of COVID-19 could hit the provinces of the Emilia-Romagna region, we cannot ignore that trying to extract, from an ML model, a prediction on the exact number of new infected people, per each province, on a daily basis, is something more like a puzzle, rather than a scientific investigation.

To simplify this problem, we resorted to a more effective procedure which was as follows. The idea was to count the number of daily infections registered per each province, in all the nine provinces of Emilia-Romagna, during the four days that preceded the lockdown decision taken by the Italian Government on 8–10 March 2020 (the specific day depends on the specific province).

Once those infections counts were obtained, we computed an average value of those daily numbers on a per-province basis for those 4 days. We then got nine numbers that were finally aggregated on a regional basis, under the form of a further average count, thus yielding the average number of infections per-province on a regional basis in Emilia-Romagna. The result was 17 (from now on, the so-called threshold). Told differently, the daily number of infected people, in Emilia-Romagna, averaged over those four days, amounts to 17 times 9 = 153.

Now, please follow the reasoning. If the Italian Government, using its own decisional models, opted for a lockdown decision, as soon as the average regional number of daily infections on a per-province basis in Emilia-Romagna had surpassed the threshold of 17, then we could use that number as a key to design the predictions scheme of our ML model. Not to forget also the fact that Emilia-Romagna was, at that time, the region with the largest number of infections after Lombardy. Hence, the number of infections that occurred in this region has had an important role in that lockdown decision.

To conclude this reasoning, our intention is to replace the initial idea to predict if, in a given future day, Emilia-Romagna is under the risk of a second wave of a COVID-19 resurgence with the more concrete and effective prediction of whether the number of infected people will surpass that threshold of 17, on that day, on a per-province, regional basis. More precisely, we ask our ML model to compute the probability that, in a given future day, each province in Emilia-Romagna will count a number of infections larger than 17—and, then, we look at the regional picture with all its nine provinces, and the probability that the number of infections for each exceeds 17. The higher this probability is, the higher the risk of a second regional wave will be, especially if various provinces simultaneously surpass that threshold on a certain given day.

For the sake of completeness, in Figure 5, we provide a graph with the cumulative quantities of infected people, per day, for all the nine provinces of interest, plus the cumulative values of the regional and the national averages, registered during the four days prior to 8–10 March 2020.

In Figure 5, one can read: Bologna, bo; Ferrara, fe; Forlì-Cesena, fc; Modena, mo; Parma, pr; Piacenza, pc; Ravenna, ra; Reggio nell'Emilia, re; Rimini, rn; Emilia-Romagna, er; and Italy, ita.

Important to note is the fact that, in Figure 5, our regional infection average, being cumulative over those four days, amounts to 17 times 4 = 68 (as read at the rightmost end of the figure).

By contrast, if one takes into consideration the national average, they can notice that the following value of 8 times 4 = 32 can be computed (as read at the rightmost end of the figure). This smaller quantity at a national level is due to the fortunate fact that many regions in Southern Italy were not severely affected by the virus, thus providing a smaller contribution to the national average.

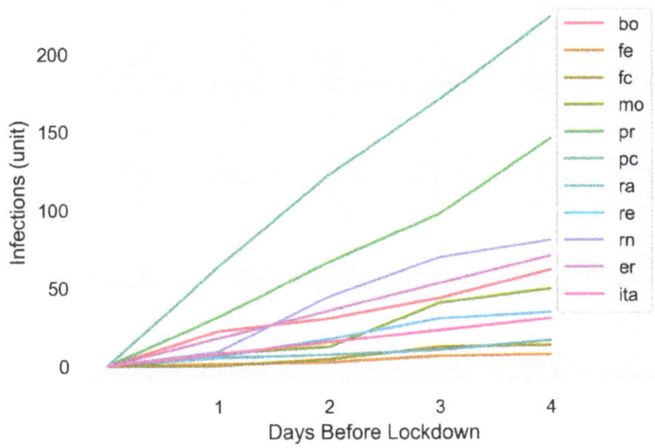

Figure 5. 4–7 March 2020—cumulative number of infections: Modena, Parma, Piacenza, Reggio nell'Emilia, and Rimini; 7–10 March 2020—cumulative number of infections: Bologna, Ferrara, Forlì-Cesena, and Ravenna; cumulative regional and national infections averages.

Interesting to remind is also the average number of infections per day in Lombardy, computed in a similar way (i.e., the average per-province number of infections, on a regional basis), which was as high as 38. This latter number is important. It is well known that in that periodm Lombardy was really the hardest-hit Italian area, thus becoming a sort of hotspot for COVID-19 diffusion in Italy. This is why we decided not to choose this number (i.e., 38) in our scheme. It would have been somewhat misleading, especially in consideration of the fact that we want predictions that are valid for the Emilia-Romagna region.

At this point, it is important to mention that, using the value of 17, we have essentially split our initial dataset into two separate portions:

(i) The former, with all those days with a number y of daily infections, equal or smaller than 17; and
(ii) The latter, with those days registering a number of new infected people larger than 17.

Not only this, but also, to properly manage the hypothesis of a relationship between pollutants and infection spread, crucial is also the concept of *lag*. In particular, with *lag*, we account for the following fact: On a given day, say z, we may have registered a certain number of infections, say y. Those y infections could have manifested themselves after exactly fourteen days, since the original contagion happened exactly on the day: $z - 14$. Nonetheless, we also know that there is a degree of uncertainty, affecting the exact number of days that should be taken into account for this count.

To take this fact into account, with a *lag* equal to 4, for example, we reason as if all the y infections, which occurred on day z, originated from the contribution of pollutants that were in the air during a longer time interval of length 4 (starting from day: $z - 14$), in this specific case, our interval would go from day $z - 14$ up to day $z - (14 - 4 + 1)$, that is, day: $z - 11$.

This is an important fact, giving rise to an important implication: With the concept of *lag*, which can range from 1 to 8 in our model, we try to mitigate the uncertainty concerning the exact day when people get infected (as also discussed in [37]).

To conclude, considering the nine provinces of Emilia-Romagna, each one observed for a period of 135 days, the number of examples we counted where the number of daily infections was equal to or smaller than 17 amounted to 789. By contrast, the number of examples where the number of daily infections was larger than 17 amounted to 426.

Finally, it is important also to note that 80% of all these data were employed for instructing our ML models, while the remaining 20% were retained to test the performance those models could reach upon completion of the learning phase.

2.4. Model Selection

We have reached the following point: We have collected a special set of data, based on the hypothesis of a relation between pollutants and infections. Those data represent each single day, with its infections and quantities of measured pollution in the air. Then, we have divided them into two separate sets of examples, specifically: (i) the first one comprising those days with a number of infections equal to or smaller than 17, and (ii) the second one with those days with a number of infected people larger than 17. We have also introduced the concept of *lag* and have exactly computed how large the two aforementioned sets of examples are (789 vs. 426).

What we still need to decide now is the ML model to adopt that should be instructed with all those data.

To choose our ML model, we proceeded as follows. Without any initial preference, we tried to instruct a wide range of possible ML models, suitable to learn the function pollutants/infections. We started with the following ML models:

- K nearest neighbor (KNN) [38];
- Classification and regression tree (CART) [39];
- Support vector machine (SVM) [40];
- Multilayer perceptron (MLP) [41];
- Ada boosting with decision tree (AB) [42];
- Gradient boosting (GB) [43];
- Random forest (RF) [44];
- Extra tree (ET) [45].

The procedure with which we selected our ML model went through two separate and subsequent phases, aimed at measuring their performance in terms of accuracy of the predictions they made, more precisely, a ten-fold cross-validation and a testing phase.

First, we allowed all the eight models mentioned above to learn the function we described before, and then we subjected each one to a classic ten-fold cross-validation procedure, yielding an F1 score. Before we proceed, we shall briefly remind what a ten-fold cross-validation procedure and a F1-score are.

Simply put, cross-validation is a procedure that evaluates predictive models by partitioning the original dataset into two portions. With the training portion, the model is trained, while with the validation portion, the model is evaluated. In a ten-fold cross-validation, the original dataset is randomly partitioned into 10 subsamples of equal size. Of the 10 portions, a single portion is kept separate to validate the model, while the model is trained with the remaining nine portions of data. We use the term cross as this validation procedure is reiterated 10 times, with each of the 10 portions used exactly once to validate the model. The ten obtained results coming from the validation portions can then be averaged to produce a final evaluation.

As regards the F1 score, in a classic classification problem (comprising true and false positives, and true and false negatives), it is intended to be the harmonic mean of the precision and recall values, where such a score reaches its best at 1. In turn, precision is the number of true positives divided by the number of true positives plus the number of false positives, while recall is the number of true positives divided by the number of true positives plus the number of false negatives (i.e., all the samples that should have been identified as positive).

All this anticipated, in Table 1, we show the results we have obtained with 80% of our data, and a ten-fold cross-validation conducted with all the eight ML models mentioned before. All the results are in terms of the F1 score, which was measured on average, plus its standard deviation.

Important to note is the fact that we allowed the models to learn our function both with each single pollutant (i.e., $PM_{2.5}$, PM_{10}, and NO_2) in isolation, and then with all the pollutants considered together; not only this, but we also varied the *lag*, as already anticipated, from 1 to 8.

In essence, each cell in Table 1 tells us how accurate, on average, the prediction was that a given model has made that the threshold of 17 infections was either surpassed or not, for a given day, with a certain amount of pollutants in the air.

If one accurately analyzes Table 1, they can find that almost all the ML models have comparable performances, except for CART and AB (highlighted with the red color). This convinced us to proceed with the next step of testing, at the end of which only one model was to be selected to be used to make COVID-19 predictions for the period of September–December 2020, excluding the CART and the AB candidates.

Table 1. Ten-fold cross-validation for all the eight ML models: F1 score (average and standard deviation).

Pollution	Lag	KNN	CART	SVC	MLP	AB	GB	RF	ET
All	1	0.81 ± 0.03	0.75 ± 0.05	0.81 ± 0.04	0.84 ± 0.03	0.78 ± 0.05	0.81 ± 0.04	0.81 ± 0.03	0.78 ± 0.04
	2	0.81 ± 0.03	0.81 ± 0.04	0.83 ± 0.03	0.84 ± 0.04	0.79 ± 0.05	0.83 ± 0.03	0.84 ± 0.05	0.83 ± 0.04
	3	0.83 ± 0.05	0.78 ± 0.05	0.83 ± 0.04	0.84 ± 0.04	0.81 ± 0.04	0.83 ± 0.02	0.85 ± 0.03	0.85 ± 0.03
	4	0.82 ± 0.04	0.78 ± 0.05	0.84 ± 0.04	0.84 ± 0.05	0.81 ± 0.03	0.84 ± 0.03	0.84 ± 0.03	0.85 ± 0.03
	5	0.85 ± 0.03	0.82 ± 0.05	0.85 ± 0.03	0.86 ± 0.04	0.82 ± 0.05	0.86 ± 0.04	0.87 ± 0.04	0.86 ± 0.03
	6	0.86 ± 0.03	0.82 ± 0.05	0.87 ± 0.03	0.86 ± 0.03	0.82 ± 0.04	0.86 ± 0.03	0.85 ± 0.04	0.88 ± 0.03
	7	0.86 ± 0.03	0.83 ± 0.04	0.87 ± 0.03	0.87 ± 0.02	0.84 ± 0.03	0.85 ± 0.03	0.86 ± 0.03	0.87 ± 0.04
	8	0.87 ± 0.02	0.84 ± 0.04	0.89 ± 0.03	0.89 ± 0.02	0.82 ± 0.03	0.86 ± 0.03	0.86 ± 0.04	0.90 ± 0.03
PM2.5	1	0.79 ± 0.03	0.77 ± 0.04	0.80 ± 0.04	0.81 ± 0.03	0.78 ± 0.06	0.81 ± 0.03	0.76 ± 0.03	0.77 ± 0.04
	2	0.80 ± 0.04	0.78 ± 0.05	0.82 ± 0.04	0.82 ± 0.04	0.78 ± 0.04	0.82 ± 0.04	0.81 ± 0.03	0.79 ± 0.03
	3	0.81 ± 0.03	0.79 ± 0.05	0.82 ± 0.03	0.82 ± 0.04	0.81 ± 0.03	0.84 ± 0.03	0.83 ± 0.03	0.82 ± 0.03
	4	0.80 ± 0.03	0.81 ± 0.04	0.85 ± 0.02	0.83 ± 0.03	0.81 ± 0.03	0.85 ± 0.04	0.85 ± 0.03	0.83 ± 0.03
	5	0.84 ± 0.03	0.82 ± 0.04	0.86 ± 0.03	0.85 ± 0.04	0.81 ± 0.04	0.86 ± 0.04	0.87 ± 0.03	0.85 ± 0.02
	6	0.85 ± 0.04	0.84 ± 0.04	0.87 ± 0.03	0.87 ± 0.03	0.82 ± 0.04	0.87 ± 0.04	0.87 ± 0.03	0.87 ± 0.03
	7	0.85 ± 0.04	0.85 ± 0.03	0.87 ± 0.03	0.86 ± 0.02	0.82 ± 0.05	0.86 ± 0.04	0.88 ± 0.03	0.88 ± 0.03
	8	0.88 ± 0.03	0.84 ± 0.04	0.88 ± 0.03	0.87 ± 0.03	0.83 ± 0.05	0.86 ± 0.03	0.87 ± 0.04	0.89 ± 0.04
PM10	1	0.79 ± 0.05	0.77 ± 0.03	0.80 ± 0.04	0.81 ± 0.04	0.78 ± 0.05	0.81 ± 0.04	0.78 ± 0.05	0.79 ± 0.04
	2	0.81 ± 0.04	0.79 ± 0.05	0.81 ± 0.04	0.82 ± 0.04	0.78 ± 0.04	0.82 ± 0.03	0.82 ± 0.05	0.82 ± 0.04
	3	0.80 ± 0.03	0.77 ± 0.03	0.82 ± 0.03	0.83 ± 0.04	0.80 ± 0.03	0.83 ± 0.03	0.83 ± 0.04	0.83 ± 0.03
	4	0.83 ± 0.03	0.78 ± 0.03	0.84 ± 0.03	0.84 ± 0.04	0.80 ± 0.02	0.85 ± 0.04	0.84 ± 0.02	0.84 ± 0.03
	5	0.84 ± 0.04	0.81 ± 0.06	0.85 ± 0.03	0.86 ± 0.03	0.80 ± 0.05	0.87 ± 0.04	0.86 ± 0.03	0.85 ± 0.03
	6	0.85 ± 0.03	0.82 ± 0.04	0.86 ± 0.03	0.87 ± 0.03	0.82 ± 0.04	0.86 ± 0.04	0.86 ± 0.04	0.85 ± 0.04
	7	0.87 ± 0.03	0.85 ± 0.04	0.88 ± 0.03	0.87 ± 0.03	0.83 ± 0.05	0.87 ± 0.04	0.87 ± 0.03	0.88 ± 0.03
	8	0.87 ± 0.02	0.85 ± 0.03	0.88 ± 0.03	0.88 ± 0.03	0.82 ± 0.04	0.88 ± 0.04	0.87 ± 0.04	0.89 ± 0.03
NO2	1	0.80 ± 0.04	0.78 ± 0.03	0.81 ± 0.03	0.81 ± 0.04	0.78 ± 0.04	0.80 ± 0.04	0.77 ± 0.03	0.78 ± 0.03
	2	0.79 ± 0.03	0.76 ± 0.04	0.81 ± 0.02	0.82 ± 0.02	0.79 ± 0.05	0.81 ± 0.03	0.80 ± 0.04	0.79 ± 0.04
	3	0.82 ± 0.03	0.77 ± 0.03	0.82 ± 0.03	0.83 ± 0.03	0.80 ± 0.03	0.82 ± 0.04	0.83 ± 0.02	0.83 ± 0.02
	4	0.85 ± 0.02	0.80 ± 0.02	0.83 ± 0.03	0.84 ± 0.04	0.80 ± 0.04	0.83 ± 0.03	0.86 ± 0.03	0.85 ± 0.02
	5	0.86 ± 0.02	0.81 ± 0.03	0.84 ± 0.03	0.85 ± 0.04	0.82 ± 0.04	0.83 ± 0.03	0.87 ± 0.03	0.85 ± 0.02
	6	0.86 ± 0.03	0.83 ± 0.04	0.85 ± 0.03	0.86 ± 0.04	0.80 ± 0.04	0.84 ± 0.04	0.87 ± 0.03	0.86 ± 0.03
	7	0.85 ± 0.03	0.82 ± 0.05	0.85 ± 0.02	0.85 ± 0.03	0.81 ± 0.04	0.84 ± 0.04	0.87 ± 0.04	0.87 ± 0.03
	8	0.86 ± 0.03	0.81 ± 0.03	0.86 ± 0.02	0.86 ± 0.03	0.81 ± 0.04	0.85 ± 0.04	0.87 ± 0.03	0.88 ± 0.02

In this second step were, hence, included only the six models that exhibited good comparable performances during the ten-fold cross-validation, namely: KNN, SVC, MLP, GB, RF, and ET.

All these six models were subjected to this final testing step, conducted with 20% of the dataset we had retained for this specific aim. Results from this final testing phase are reported in Table 2.

As expected, all the six models under consideration yielded a reasonably good performance; nonetheless, the one with the best F1 score was GB, gradient boosting, as Table 2 reveals. The simple reason we expected quite good performances from almost all those six models is that they had already done well during the phase of the ten-fold cross-validation. Nonetheless, gradient boosting (GB) manifested as the best model in this specific circumstance (with its F1 score equal to 0.893). In other words, GB is the model that better learned the function pollution/infections on which our hypothesis is based. Consequently, it is the best candidate for making accurate predictions for the future.

Table 2. Testing phase: gradient boosting (GB) selected by virtue of its F1 score.

Algorithm	Testing			
	Class	**Precision**	**Recall**	**F1 Score**
KNN	<=17	0.90	0.85	0.845
	>17	0.75	0.82	
SVC	<=17	0.95	0.87	0.890
	>17	0.80	0.92	
MLP	<=17	0.93	0.90	0.890
	>17	0.82	0.87	
GB	<=17	0.92	0,91	0.893
	>17	0.84	0.86	
RF	<=17	0.93	0.87	0.878
	>17	0.79	0.88	
ET	<=17	0.91	0.91	0.881
	>17	0.83	0.84	

In addition to the F1 score, where the GB surpasses all the other five candidates, with regard to the other metrics of precision and recall, it is interesting to note that it achieves a better accuracy (91–92%) in predicting whether a given future day will be classified in the class of those days with a number of infections equal or smaller than 17, and a slightly lower accuracy (84–86%) in predicting whether a given day will be classified in the class of those days with a number of infections larger than 17. This slight difference is probably due to the fact that it has been instructed with a larger number of examples of the former class.

To conclude this section, a simple explanation on how the GB model computationally works is in order now.

In very simple words, the gradient boosting method tries to find an approximation to the function $\hat{F}$ that we are letting our model learn (i.e., the relationship of pollutants vs. infections). To do that, a value is computed based on a weighted sum of M functions h_i, which are, in some sense, the estimators of the number y of infected people we expect to have for each given day, given that we have registered a certain value x of some pollutant. All this is based on the following formula (where a_i is the additional parameter to be learnt, and ε is a given predefined constant value).

$$\hat{F}(x) = \sum_{i=1}^{M} \alpha_i h_i(x) + \varepsilon. \quad (1)$$

Technically speaking, we are minimizing loss function L, given a training set composed of couples of known values of x (pollutant) and y (infections), where the final target is to make the estimation as close as possible to the real value of y. All this is based on the following minimization procedure:

$$\hat{F}(x) = \ min_F \ E_{x,y}[L(y, F(x))]. \quad (2)$$

With this clarified, in the next section, we present the predictions that our GB model has made, regarding the plausibility of a second wave of COVID-19 infections in Emilia-Romagna.

3. Results: Predictions

We come now to the final step. Upon completion of the activities that led us to instruct our GB model using data from the period of February–July 2020, selected based on the assumption of a relationship between pollutants and infections, we now need to ask our model to make the predictions, on a daily and provincial basis, for the Emilia-Romagna region, for all the future days from 21 September to 31 December 2020.

The motivation behind the choice of this precise prediction period is obvious. We are all worried about the possibility that a second wave of COVID-19 will coincide with the end of the summer period (21 September), when many human activities will resumed in Italy, including schools and universities, for example. As for the closing period for our predictions, we deem it natural not to extend the scope too much, thus reaching the end of the current year 2020.

Nonetheless, one element is still missing, which is relevant to our prediction activity. Our model has learnt the function that maps pollutant values into the number of infected people. Nevertheless, if we want it to try to predict what can happen on, say, day z, (with, for example, z = 21 September 2020, in the province of Bologna), we need to give our model as an input the value of the pollutants circulating on that day z in Bologna.

Obviously, at the time we write our article, we do not have the precise value of those pollutants for that future day z. What we can do to mitigate that factor is to try to estimate those values, based on the amounts of pollutants circulating in the air in Bologna, as measured on the same day z some given years ago, for example, 21 September 2019.

Put simply, we have exploited the amount of the pollution registered in some previous years in Emilia-Romagna to have an estimate of those pollutants that need then to be given as an input to the ML model. We have done this following two alternative strategies. In the first case, we used all the values of the pollutants registered in the period 21 September–31 December 2019. In the second case, we used all the values of the pollutants measured in the period 21 September–31 December, yet averaged on three different previous years, namely: 2017–2019.

We present the obtained results in the following two subsections, in isolation.

Before we proceed, it is very important to remind that *all* the predictions presented in the two following subsections were made under the hypothesis that all the control/containment measures of the so-called Phase 3 are strictly obeyed. If those measures are not obeyed (or even partially disregarded), our model would return predictions very different from those shown in Sections 3.1 and 3.2.

3.1. Predictions: 2019->2020

We report in Table 3 the predictions that our GB model has made based on all the assumptions described in the previous sections, including that Phase 3 is obeyed.

Important are the following instructions to better read those results. Along the columns, we have all the nine provinces (Bologna, Ferrara, Forlì-Cesena, Modena, Parma, Piacenza, Reggio nell'Emilia, Rimini, and Ravenna), while on the rows, the prediction is given for each single day. A mixture of the values of the pollutants $PM_{2.5}$, PM_{10}, and NO_2 was considered as input to the model, in this case measured in the period 21 September–31 December 2019.

Each cell in the table shows the value of the probability that the number of infections, on that day per that province, exceeds the quantity of 17 (with the maximum probability value set equal to 1). The higher that probability, the higher the risk that we will have a number of infected people surpassing 17, on that day in that province, thus raising the relative concerns.

In a red color, we highlighted those days, on a per-province basis, where that threshold is surpassed. A quick comment to this table is that concerns arise, especially for two provinces (Parma and Piacenza), in the two periods of mid-October/mid-November 2020, as well as at the end of November–end of December 2020. Again, we insist on the fact that these predictions were obtained based on the assumption that the personal protection measures of Phase 3 are respected.

We have deliberately moved a more detailed discussion of these results to the final section of this paper.

Here, we just input evidence that the following provinces seem to have the following total number of crucial days during the observed period, whose length is 135 days:

- Bologna (0);
- Ferrara (1);
- Forlì-Cesena (2);
- Modena (1);
- Parma (16);
- Piacenza (23);
- Ravenna (0);
- Reggio Emilia (1);
- Rimini (1).

Moreover, to allow the reader to a have a simpler and more comprehensive view of the results presented above, we have also reported them under an alternative format. In particular, in Figure 6, we present the same results as those of Table 3, but portrayed as a heatmap. Simply put, high probability values turn lean toward red, while low probability values are depicted in white. Different orange color gradations represent intermediate situations. Predictions are grouped on a weekly basis, per each province in the region.

Table 3. Predictions (2019->2020): probability values.

Day	Bo	Fe	Fc	Mo	Pr	Pc	Ra	Re	Rn
9/21	0.00	0.00	0.00	0.02	0.01	0.35	0.00	0.01	0.00
9/22	0.04	0.00	0.04	0.11	0.11	0.38	0.03	0.03	0.00
9/23	0.1	0.08	0.07	0.18	0.3	0.15	0.08	0.22	0.07
9/24	0.14	0.17	0.17	0.27	0.11	0.31	0.14	0.12	0.11
9/25	0.01	0.01	0.01	0.19	0.24	0.09	0.00	0.07	0.01
9/26	0.00	0.01	0.01	0.01	0.11	0.12	0.00	0.03	0.00
9/27	0.00	0.01	0.00	0.1	0.06	0.07	0.03	0.02	0.01
9/28	0.00	0.02	0.01	0.05	0.2	0.09	0.02	0.06	0.01
9/29	0.14	0.00	0.04	0.23	0.04	0.02	0.16	0.04	0.02
9/30	0.03	0.01	0.01	0.04	0.01	0.02	0.06	0.01	0.02
10/1	0.01	0.01	0.01	0.02	0.01	0.04	0.01	0.02	0.00
10/2	0.00	0.01	0.05	0.23	0.53	0.07	0.00	0.12	0.00
10/3	0.00	0.02	0.03	0.13	0.54	0.46	0.02	0.25	0.01
10/4	0.04	0.04	0.01	0.17	0.16	0.19	0.00	0.05	0.09
10/5	0.01	0.03	0.00	0.03	0.02	0.15	0.02	0.00	0.00
10/6	0.01	0.02	0.00	0.02	0.16	0.19	0.00	0.03	0.00
10/7	0.01	0.00	0.1	0.25	0.59	0.73	0.00	0.03	0.00
10/8	0.01	0.00	0.00	0.12	0.09	0.39	0.02	0.19	0.02
10/9	0.00	0.00	0.01	0.02	0.21	0.15	0.03	0.00	0.01
10/10	0.00	0.00	0.00	0.01	0.01	0.07	0.00	0.00	0.00
10/11	0.07	0.01	0.00	0.01	0.01	0.04	0.00	0.00	0.00
10/12	0.03	0.18	0.02	0.21	0.04	0.42	0.02	0.01	0.00
10/13	0.01	0.01	0.01	0.06	0.08	0.37	0.01	0.04	0.00
10/14	0.00	0.00	0.01	0.01	0.03	0.02	0.00	0.00	0.00

Table 3. *Cont.*

Day	Bo	Fe	Fc	Mo	Pr	Pc	Ra	Re	Rn
10/15	0.01	0.01	0.01	0.05	0.16	0.02	0.00	0.01	0.00
10/16	0.08	0.04	0.4	0.35	0.29	0.1	0.1	0.15	0.00
10/17	0.11	0.37	0.09	0.39	0.47	0.45	0.25	0.18	0.08
10/18	0.06	0.01	0.05	0.04	0.02	0.02	0.02	0.04	0.06
10/19	0.01	0.04	0.01	0.16	0.45	0.14	0.2	0.09	0.02
10/20	0.04	0.4	0.04	0.2	0.74	0.81	0.06	0.38	0.07
10/21	0.07	0.03	0.06	0.07	0.4	0.64	0.09	0.07	0.09
10/22	0.01	0.02	0.01	0.08	0.45	0.51	0.02	0.1	0.01
10/23	0.03	0.07	0.01	0.14	0.44	0.46	0.02	0.04	0.02
10/24	0.04	0.06	0.09	0.17	0.56	0.49	0.02	0.03	0.02
10/25	0.03	0.02	0.02	0.22	0.14	0.35	0.01	0.04	0.01
10/26	0.03	0.04	0.09	0.17	0.26	0.64	0.04	0.1	0.00
10/27	0.01	0.01	0.00	0.01	0.06	0.07	0.01	0.00	0.00
10/28	0.01	0.02	0.04	0.19	0.53	0.13	0.01	0.04	0.01
10/29	0.03	0.59	0.68	0.33	0.79	0.74	0.28	0.35	0.18
10/30	0.03	0.07	0.04	0.04	0.19	0.09	0.02	0.03	0.05
10/31	0.02	0.01	0.02	0.02	0.42	0.44	0.02	0.11	0.03
11/1	0.12	0.34	0.12	0.09	0.22	0.13	0.45	0.03	0.05
11/2	0.06	0.33	0.05	0.4	0.82	0.71	0.1	0.3	0.07
11/3	0.03	0.24	0.02	0.17	0.38	0.78	0.06	0.03	0.03
11/4	0.14	0.06	0.08	0.25	0.42	0.55	0.09	0.03	0.03
11/5	0.09	0.17	0.14	0.13	0.62	0.4	0.09	0.16	0.15
11/6	0.04	0.06	0.11	0.05	0.12	0.15	0.05	0.02	0.07
11/7	0.05	0.08	0.25	0.08	0.72	0.51	0.13	0.14	0.13
11/8	0.03	0.06	0.08	0.08	0.45	0.32	0.09	0.02	0.04
11/9	0.13	0.03	0.04	0.13	0.19	0.08	0.35	0.06	0.21
11/10	0.03	0.00	0.03	0.06	0.03	0.01	0.03	0.01	0.1
11/11	0.01	0.00	0.00	0.00	0.05	0.07	0.01	0.06	0.02
11/12	0.04	0.00	0.07	0.02	0.18	0.09	0.03	0.05	0.13
11/13	0.04	0.05	0.03	0.01	0.02	0.03	0.01	0.02	0.05
11/14	0.01	0.01	0.01	0.01	0.01	0.03	0.01	0.01	0.00
11/15	0.00	0.00	0.00	0.01	0.01	0.08	0.00	0.01	0.00
11/16	0.00	0.00	0.01	0.23	0.22	0.22	0.01	0.22	0.00
11/17	0.06	0.03	0.02	0.1	0.05	0.07	0.02	0.01	0.00
11/18	0.32	0.1	0.22	0.06	0.09	0.03	0.39	0.05	0.06
11/19	0.14	0.00	0.11	0.08	0.06	0.05	0.02	0.52	0.01
11/20	0.00	0.00	0.01	0.16	0.08	0.12	0.01	0.01	0.00
11/21	0.09	0.03	0.17	0.01	0.02	0.07	0.12	0.01	0.00
11/22	0.15	0.06	0.11	0.12	0.04	0.05	0.11	0.02	0.06
11/23	0.01	0.01	0.01	0.02	0.03	0.01	0.00	0.02	0.00
11/24	0.03	0.03	0.02	0.09	0.1	0.57	0.01	0.04	0.01
11/25	0.02	0.00	0.55	0.03	0.01	0.54	0.08	0.01	0.16
11/26	0.03	0.02	0.09	0.01	0.02	0.04	0.01	0.00	0.03
11/27	0.29	0.13	0.15	0.11	0.12	0.1	0.03	0.11	0.00
11/28	0.08	0.05	0.03	0.11	0.1	0.00	0.09	0.01	0.01
11/29	0.01	0.00	0.03	0.07	0.24	0.05	0.01	0.07	0.01
11/30	0.02	0.00	0.37	0.02	0.02	0.16	0.01	0.13	0.04

Table 3. *Cont.*

Day	Bo	Fe	Fc	Mo	Pr	Pc	Ra	Re	Rn
12/1	0.03	0.05	0.03	0.04	0.13	0.16	0.05	0.01	0.03
12/2	0.05	0.02	0.02	0.05	0.47	0.04	0.01	0.01	0.01
12/3	0.01	0.12	0.08	0.21	0.15	0.07	0.04	0.07	0.16
12/4	0.01	0.00	0.37	0.03	0.33	0.02	0.02	0.24	0.01
12/5	0.01	0.01	0.02	0.04	0.02	0.01	0.01	0.00	0.01
12/6	0.06	0.03	0.01	0.21	0.08	0.01	0.16	0.01	0.01
12/7	0.01	0.12	0.48	0.17	0.39	0.28	0.1	0.05	0.29
12/8	0.05	0.05	0.2	0.2	0.41	0.19	0.05	0.25	0.04
12/9	0.07	0.06	0.03	0.02	0.07	0.06	0.05	0.04	0.00
12/10	0.02	0.01	0.23	0.11	0.21	0.05	0.02	0.02	0.16
12/11	0.13	0.06	0.05	0.29	0.38	0.61	0.05	0.1	0.01
12/12	0.07	0.18	0.05	0.11	0.12	0.2	0.07	0.03	0.02
12/13	0.23	0.14	0.13	0.13	0.03	0.26	0.21	0.1	0.05
12/14	0.14	0.03	0.29	0.3	0.47	0.57	0.1	0.08	0.03
12/15	0.25	0.09	0.25	0.16	0.69	0.84	0.3	0.34	0.47
12/16	0.12	0.18	0.08	0.14	0.52	0.78	0.18	0.07	0.4
12/17	0.09	0.1	0.11	0.08	0.46	0.63	0.1	0.11	0.08
12/18	0.06	0.05	0.04	0.18	0.21	0.8	0.06	0.04	0.02
12/19	0.18	0.16	0.15	0.12	0.44	0.87	0.19	0.19	0.18
12/20	0.17	0.3	0.07	0.28	0.39	0.92	0.28	0.04	0.26
12/21	0.04	0.13	0.17	0.31	0.61	0.69	0.15	0.27	0.28
12/22	0.14	0.34	0.14	0.15	0.61	0.77	0.45	0.42	0.05
12/23	0.21	0.23	0.23	0.13	0.46	0.49	0.21	0.39	0.14
12/24	0.32	0.12	0.08	0.26	0.48	0.46	0.2	0.08	0.06
12/25	0.09	0.13	0.08	0.16	0.4	0.33	0.15	0.16	0.02
12/26	0.1	0.04	0.09	0.18	0.55	0.57	0.11	0.11	0.09
12/27	0.06	0.04	0.01	0.04	0.33	0.48	0.03	0.02	0.01
12/28	0.07	0.02	0.12	0.12	0.27	0.21	0.03	0.02	0.07
12/29	0.38	0.14	0.36	0.55	0.71	0.19	0.15	0.26	0.53
12/30	0.4	0.32	0.27	0.44	0.23	0.29	0.04	0.14	0.29
12/31	0.02	0.01	0.01	0.01	0.04	0.4	0.1	0.01	0.00

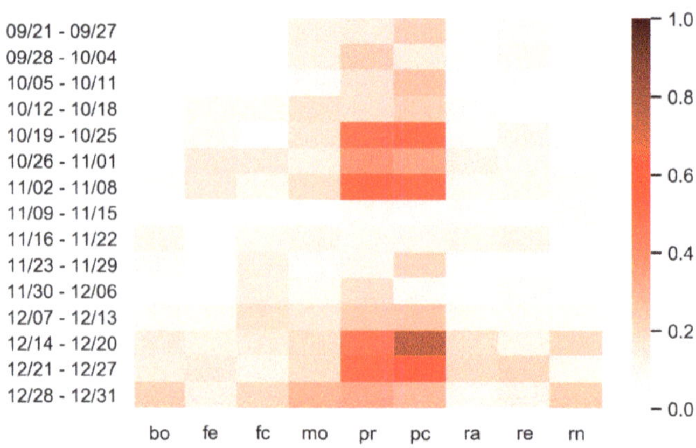

Figure 6. Predictions (2019->2020): *heatmap*.

3.2. Predictions: 2017–2019->2020

In this section, we report the predictions our GB model has made in the case that the pollutants, at the basis of the relationship that we assume them to have, are considered as a mixture obtained with an average of the three previous years, more specifically, 2017, 2018, and 2019, and finally provided as input to the model. As before, the predictions are given in Table 4 on both a daily and a per-province basis.

All the considerations we have already anticipated in the previous subsections are all still valid, including the one on the personal protection measures.

It is worth noting here that extending to the three previous years (2017–2019) has not brought to more positive predictions. This can be due to several factors, including the fact that 2019 may have been a quite favorable year, in terms of registered pollution. In any case, the general trend of our predictions comes confirmed. Two provinces, in particular, Parma and Piacenza, seem to run larger risks in terms of the number of infections that exceed the threshold of 17. This is clearly visible both in the probability values reported in Table 4 and in the heatmap of Figure 7. Again, we have input evidence that during the observed period, as long as 135 days, the numbers of crucial days, on a per-province basis, are as follows:

Bologna (1);
Ferrara (0);
Forlì-Cesena (0);
Modena (0);
Parma (29);
Piacenza (43);
Ravenna (0);
Reggio Emilia (1);
Rimini (0).

Table 4. Predictions (2017-2018-2019->2020): probability values.

Day	Bo	Fe	Fc	Mo	Pr	Pc	Ra	Re	Rn
9/21	0.00	0.01	0.00	0.02	0.02	0.33	0.00	0.12	0.00
9/22	0.00	0.03	0.00	0.12	0.05	0.12	0.02	0.04	0.00
9/23	0.11	0.02	0.07	0.08	0.09	0.08	0.03	0.07	0.05
9/24	0.04	0.01	0.15	0.1	0.09	0.08	0.01	0.02	0.1
9/25	0.00	0.02	0.00	0.06	0.06	0.06	0.01	0.02	0.00
9/26	0.04	0.01	0.00	0.11	0.08	0.05	0.01	0.01	0.00
9/27	0.01	0.01	0.01	0.05	0.05	0.03	0.03	0.04	0.05
9/28	0.00	0.03	0.00	0.12	0.04	0.09	0.03	0.06	0.00
9/29	0.01	0.07	0.00	0.15	0.06	0.03	0.02	0.04	0.00
9/30	0.02	0.01	0.01	0.03	0.01	0.13	0.00	0.01	0.00
10/1	0.00	0.01	0.00	0.26	0.05	0.07	0.04	0.04	0.00
10/2	0.01	0.09	0.1	0.08	0.09	0.14	0.02	0.04	0.00
10/3	0.07	0.02	0.06	0.1	0.15	0.19	0.01	0.07	0.01
10/4	0.14	0.03	0.11	0.22	0.24	0.15	0.02	0.03	0.12
10/5	0.01	0.04	0.01	0.11	0.05	0.19	0.05	0.01	0.01
10/6	0.00	0.06	0.00	0.06	0.08	0.05	0.00	0.02	0.00

Table 4. *Cont.*

Day	Bo	Fe	Fc	Mo	Pr	Pc	Ra	Re	Rn
10/7	0.00	0.00	0.02	0.12	0.4	0.12	0.01	0.09	0.00
10/8	0.07	0.04	0.13	0.06	0.24	0.23	0.04	0.06	0.02
10/9	0.1	0.07	0.11	0.09	0.22	0.3	0.08	0.12	0.05
10/10	0.03	0.07	0.01	0.22	0.19	0.19	0.03	0.06	0.01
10/11	0.01	0.04	0.01	0.13	0.32	0.18	0.05	0.07	0.01
10/12	0.01	0.13	0.15	0.36	0.28	0.21	0.04	0.09	0.02
10/13	0.00	0.01	0.03	0.04	0.26	0.04	0.05	0.07	0.00
10/14	0.03	0.04	0.08	0.06	0.21	0.52	0.12	0.08	0.00
10/15	0.04	0.11	0.04	0.16	0.44	0.47	0.05	0.06	0.06
10/16	0.07	0.08	0.23	0.37	0.51	0.34	0.05	0.12	0.04
10/17	0.16	0.09	0.12	0.19	0.41	0.28	0.09	0.23	0.04
10/18	0.02	0.21	0.04	0.06	0.14	0.43	0.15	0.05	0.07
10/19	0.03	0.47	0.05	0.22	0.67	0.35	0.11	0.6	0.23
10/20	0.11	0.11	0.13	0.16	0.56	0.66	0.1	0.24	0.15
10/21	0.06	0.16	0.05	0.05	0.28	0.86	0.07	0.07	0.03
10/22	0.05	0.09	0.13	0.1	0.38	0.82	0.07	0.07	0.03
10/23	0.05	0.12	0.06	0.19	0.62	0.81	0.03	0.18	0.02
10/24	0.12	0.21	0.21	0.2	0.53	0.9	0.22	0.25	0.03
10/25	0.34	0.27	0.11	0.18	0.36	0.92	0.1	0.19	0.07
10/26	0.18	0.11	0.04	0.18	0.49	0.81	0.08	0.19	0.05
10/27	0.16	0.1	0.4	0.17	0.62	0.86	0.23	0.16	0.11
10/28	0.27	0.18	0.26	0.15	0.59	0.85	0.28	0.22	0.19
10/29	0.17	0.3	0.05	0.19	0.84	0.82	0.05	0.26	0.04
10/30	0.02	0.08	0.07	0.18	0.77	0.6	0.06	0.11	0.02
10/31	0.04	0.27	0.25	0.13	0.4	0.64	0.14	0.12	0.01
11/1	0.04	0.14	0.11	0.23	0.68	0.91	0.09	0.12	0.02
11/2	0.09	0.06	0.15	0.18	0.43	0.64	0.15	0.1	0.04
11/3	0.07	0.47	0.14	0.2	0.33	0.7	0.25	0.21	0.04
11/4	0.23	0.16	0.21	0.38	0.64	0.84	0.2	0.33	0.05
11/5	0.17	0.17	0.06	0.15	0.53	0.48	0.25	0.13	0.02
11/6	0.03	0.02	0.04	0.15	0.62	0.35	0.04	0.03	0.01
11/7	0.05	0.05	0.03	0.02	0.3	0.43	0.03	0.03	0.03
11/8	0.03	0.05	0.06	0.11	0.42	0.27	0.01	0.08	0.02
11/9	0.02	0.07	0.03	0.28	0.68	0.23	0.02	0.05	0.03
11/10	0.07	0.06	0.03	0.07	0.41	0.18	0.1	0.12	0.07
11/11	0.04	0.01	0.17	0.15	0.49	0.22	0.04	0.14	0.01
11/12	0.06	0.09	0.05	0.28	0.67	0.07	0.04	0.18	0.05
11/13	0.02	0.02	0.02	0.01	0.14	0.1	0.04	0.01	0.02
11/14	0.03	0.01	0.03	0.02	0.13	0.15	0.01	0.02	0.05
11/15	0.01	0.02	0.01	0.08	0.47	0.1	0.02	0.05	0.05
11/16	0.08	0.00	0.01	0.09	0.41	0.03	0.06	0.03	0.00
11/17	0.01	0.01	0.02	0.03	0.09	0.05	0.02	0.03	0.01
11/18	0.01	0.09	0.03	0.16	0.22	0.03	0.02	0.05	0.01
11/19	0.01	0.08	0.02	0.32	0.52	0.22	0.13	0.18	0.01
11/20	0.03	0.04	0.16	0.04	0.17	0.05	0.07	0.04	0.04
11/21	0.05	0.02	0.08	0.21	0.44	0.18	0.05	0.01	0.09
11/22	0.03	0.01	0.03	0.09	0.13	0.04	0.03	0.02	0.04

Table 4. *Cont.*

Day	Bo	Fe	Fc	Mo	Pr	Pc	Ra	Re	Rn
11/23	0.04	0.04	0.02	0.08	0.26	0.1	0.02	0.05	0.01
11/24	0.01	0.04	0.03	0.16	0.4	0.13	0.02	0.07	0.00
11/25	0.07	0.06	0.03	0.16	0.4	0.26	0.09	0.05	0.05
11/26	0.14	0.02	0.08	0.22	0.52	0.21	0.13	0.07	0.08
11/27	0.05	0.06	0.04	0.15	0.15	0.19	0.06	0.04	0.05
11/28	0.22	0.28	0.03	0.24	0.46	0.29	0.25	0.08	0.07
11/29	0.12	0.04	0.14	0.23	0.31	0.51	0.15	0.22	0.11
11/30	0.48	0.15	0.15	0.13	0.19	0.38	0.23	0.11	0.15
12/1	0.03	0.07	0.09	0.11	0.39	0.16	0.12	0.05	0.12
12/2	0.05	0.09	0.03	0.09	0.52	0.63	0.07	0.1	0.02
12/3	0.04	0.04	0.03	0.04	0.32	0.67	0.03	0.05	0.04
12/4	0.04	0.08	0.02	0.09	0.38	0.32	0.07	0.04	0.02
12/5	0.14	0.08	0.06	0.27	0.76	0.66	0.13	0.04	0.01
12/6	0.14	0.08	0.22	0.1	0.38	0.5	0.12	0.06	0.07
12/7	0.01	0.36	0.35	0.25	0.66	0.45	0.25	0.17	0.3
12/8	0.03	0.32	0.19	0.38	0.38	0.48	0.2	0.08	0.03
12/9	0.15	0.22	0.03	0.13	0.35	0.55	0.2	0.08	0.01
12/10	0.12	0.12	0.17	0.1	0.45	0.56	0.09	0.05	0.34
12/11	0.24	0.08	0.11	0.09	0.36	0.73	0.08	0.08	0.06
12/12	0.1	0.06	0.04	0.18	0.22	0.77	0.09	0.05	0.03
12/13	0.23	0.19	0.25	0.19	0.53	0.9	0.2	0.22	0.09
12/14	0.09	0.12	0.18	0.22	0.39	0.79	0.22	0.19	0.12
12/15	0.15	0.27	0.18	0.24	0.72	0.84	0.23	0.26	0.07
12/16	0.07	0.38	0.1	0.29	0.73	0.85	0.29	0.13	0.07
12/17	0.05	0.18	0.21	0.12	0.36	0.82	0.16	0.09	0.12
12/18	0.12	0.18	0.08	0.26	0.53	0.86	0.18	0.18	0.05
12/19	0.16	0.36	0.42	0.38	0.55	0.92	0.24	0.33	0.06
12/20	0.13	0.05	0.18	0.13	0.42	0.71	0.24	0.15	0.07
12/21	0.18	0.28	0.33	0.2	0.51	0.71	0.27	0.34	0.17
12/22	0.06	0.48	0.15	0.23	0.37	0.44	0.4	0.17	0.00
12/23	0.59	0.08	0.41	0.13	0.57	0.5	0.25	0.19	0.04
12/24	0.12	0.14	0.14	0.08	0.43	0.66	0.1	0.08	0.15
12/25	0.07	0.11	0.08	0.11	0.24	0.48	0.14	0.12	0.06
12/26	0.09	0.23	0.1	0.18	0.52	0.8	0.1	0.08	0.16
12/27	0.1	0.1	0.11	0.18	0.49	0.82	0.11	0.16	0.2
12/28	0.05	0.15	0.03	0.21	0.49	0.63	0.1	0.12	0.02
12/29	0.09	0.22	0.13	0.19	0.45	0.86	0.18	0.15	0.04
12/30	0.27	0.22	0.19	0.13	0.27	0.7	0.15	0.12	0.1
12/31	0.12	0.24	0.05	0.12	0.55	0.5	0.12	0.04	0.06

3.3. Predictions: What Happens If Personal Protection Measures Are Not Respected?

In this final section, we report on the predictions our GB model has made in the case that the personal protection measures indicated by the Italian Government are not respected.

In some sense, this is a special kind of sensitivity analysis where we have varied the unique model parameter that can be significantly touched (i.e., the personal protection measures).

In this specific case, we present the predictions, returned by our model, only under the form of heatmaps, where again, exactly like before, a lot of red color in the map corresponds to a very likely occurrence of a resurgence of the virus, on that week, in that province.

Once again, we have provided two separate sets of predictions and two correspondent heatmaps. The first one is that where we have used only the pollution measured in the year 2019 (Figure 8), while the second one averages the pollutants over three different years, 2017–2019 (Figure 9).

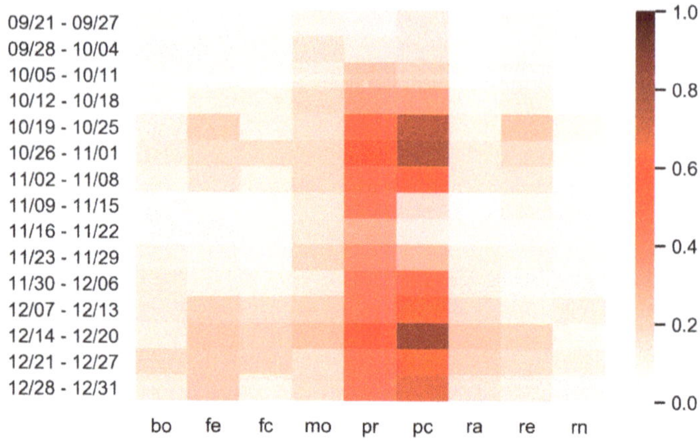

Figure 7. Predictions (2017-2018-2019->2020): heatmap.

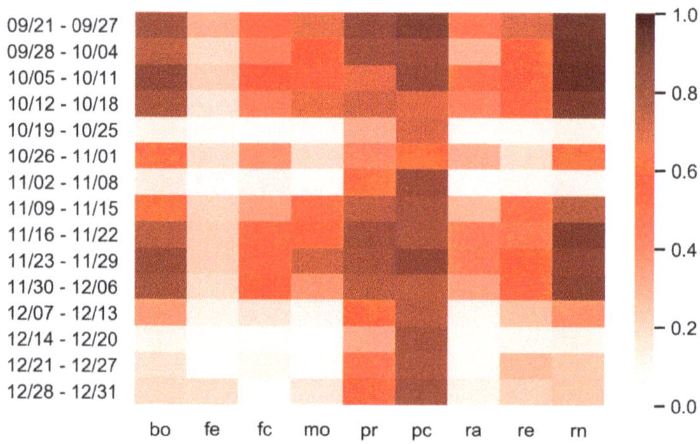

Figure 8. Predictions: no personal protection measure (2019->2020).

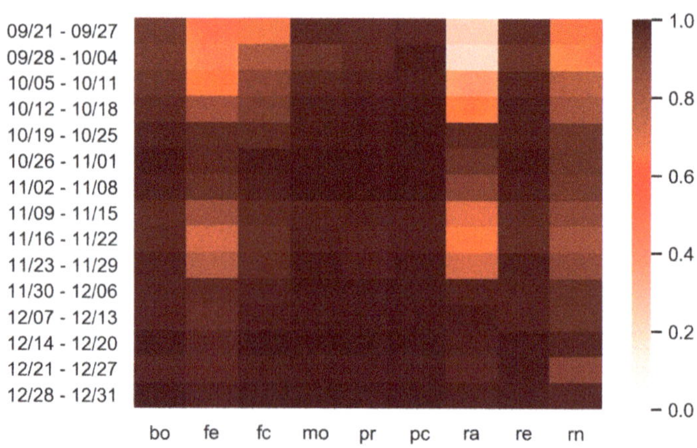

Figure 9. Predictions: no personal protection measure (2017-2018-2019->2020).

We believe that no other comment is needed here, as the plausibility of a resurgence of the virus is highly evident under the given circumstances. We could only add the consideration that one could recognize that the current rising infective trend of August 2020 could be just the trigger of a new virus explosion that those heatmaps clearly display.

4. Discussion and Conclusions

We have developed a scientific study that aims at making predictions on a possible resurgence of a COVID-19 incidence in the Italian region of Emilia-Romagna (which was one of the most hardly hit during the first phase of contagion in the period of February–April 2020).

We have based our study on a precise, given hypothesis, the most important being that of a correlation existing between the presence of circulating pollutants in the air, such as $PM_{2.5}$, PM_{10}, and NO_2, and the number of infected people. Believing in the existence of this relationship, data were collected, on a daily basis, for a period as long as mid-February 2020–end of July 2020.

These data amounted both to the measurements of the values of the aforementioned pollutants, as well as to the registered number of infections. This was carried out for all the nine provinces comprising the Italian region of Emilia-Romagna. Not only that, but data which were useful to instruct our predictive models were also represented by the restrictions that were imposed to the region of Emilia-Romagna by the Italian Government, during four different and subsequent phases, which happened during that period.

Upon completion of the data collection activity, we moved on to the selection of a computational model. Among many possible alternatives, we resorted to machine learning (ML) models, suitable for learning the function we believe can be at the basis of our hypothesis. After having conducted a numerical comparative study among several ML models, based on the available data, we found that the gradient boosting (GB) model was the one that fit squarely to the situation under observation, reaching an accuracy of almost 90% in a preliminary testing phase.

With that model, we then moved to the predictions, considering as possible estimates of the pollution than could happen in the future period of 21 September–31 December 2019 the values of the pollutants measured in previous years, namely, 2017–2019 (for the same temporal period of interest).

Relevant is also the consideration that the predictions were made by inputting to our model the situation demarcated by the measures decided on Phase 3 by the Italian Government.

At the end of all this long process, we have got our predictions provided under the form of a probability value. In essence, our model predicts the probability of surpassing a threshold of infected people in a given province, and on a certain day. Based on those probability values, we finally depicted heatmaps that could better give a general picture of the possible COVID-19 resurgence in the region of interest.

To summarize these results, the risk of a very strong second wave of COVID-19 in Emilia-Romagna seems moderate, even if those predictions also express the concern that at least two single provinces (namely, Parma and Piacenza) could be subjected to a more complex situation.

To conclude the set of our predictions, we also conducted a special kind of sensitivity analysis where our model was run, yet with a variation on the parameter concerning the use of personal protection measures. In such a case (i.e., the personal protection measures are not adopted by people), the situation becomes very different, and a risk of a resurgence of the virus becomes very plausible, for almost all the nine provinces in Emilia-Romagna.

Before we can conclude this paper, we feel the duty to finally discuss possible fallacies and limitations of our investigation regarding, at least, the three following points: (i) the scientific methodology we adopted, (ii) the choice of the data, the adopted model, and the decisional threshold of 17 infections, and finally (iii) the extensibility of the model to different COVID-19 situations.

As far as the scientific method is concerned, we have already discussed, at length, in a previous paper on the hypothesis of the existence of a correlation between pollutants and infections in our region [6]. We do not have any intention to retrace here the entire scientific path that led us to believe to this hypothesis. It suffices here to remind that we have already subjected that assumption to a statistical testing procedure (i.e., a Granger causality testing), whose results were essentially confirmative. Moreover, we know very well that, while this scientific issue is still at the center of a controversy [46], it is also true that various papers (and numerous researchers) have claimed that this virus can be airborne, and that particulate matter may further favor an airborne route, as various already cited papers have confirmed.

To move on to the second point, we would like to discuss, first, the issue of the employed data. We understand the reasons valued researchers have decided to resort to multiple sources of data to be used as early indicators of a second wave of the virus (see [20], for example). Nonetheless, we, as experienced data scientists, believe in the actual validity of data only when they are accompanied by a well-defined hypothesis. This always brings to a positive result. If experimental data provide confirmative results, in fact, one gets a kind of confirmation that can also be extended, by some measure, to the theory in general; otherwise, the hypothesis needs to be rejected, or at least revised. On the other side, with a lot of data, yet without a working hypothesis, one could also get good/bad results in some circumstances, but they would ignore what the real motivations are behind that success/failure. This justifies our approach as to the choice of our data.

As for the computational model, it should be clear that with our work, we do not want to refuse to acknowledge the importance of more traditional predictive methodologies, such as SIR, for example. They are well-founded epidemiologic models whose validity is out of discussion [47]. Nonetheless, the incidence of a quite unknown virus, like COVID-19, has put all of us into the difficult position of dealing with new alternatives. From this point of view, we are confident that ML models can provide great help, provided that they are used by experts, who are perfectly aware of all the implications they carry [48].

The issue regarding the threshold value of 17 infections may be the source of much controversy. Nevertheless, first, we would like to work with a parameter that was both simple to calculate and also a clear direct indicator of how many people got infected on a daily per-province basis. Following this reasoning, one could suggest working with a separate prediction model for each province, based on the average value of those infections that occurred only in that province. Nonetheless, In Emilia-Romagna, our experience was that we had provinces (such as Ferrara, for example) with a constantly low value of that average, even during the hardest part of the COVID-19 outbreak, while the situation in the region was generally very bad, hence our decision to use a unique value, computed as an average over all the total number of infected people in the region, yet to be applied to each province, as an early indicator. To strengthen this argument, one should consider as generally alarming, and needing to be taken into serious consideration, a situation where many provinces in a region simultaneously reach, or surpass, a given predefined value of infections. Less worrying, by contrast, would be that situation where just a very few of them surpass even a high value of infected people. In this case, more plausible is the occurrence of a local isolated outbreak, whose management is usually easier. In simpler words, this latter situation would not raise any serious concerns about the plausibility of a second wave running over the majority of the region, and over all its provinces.

Finally, allow us to address the extensibility of our model to different COVID-19 scenarios. In some sense, we recognize that this could be one of the main weaknesses of our approach. Just to cite one, for example, the hypothesis that links infections with pollution can be applied just to those geographical areas that mostly suffer from this unpleasant condition. Nevertheless, it is also true that pieces of evidence have begun to emerge that this virus hits harder in those geographical areas where the general climatic and environmental conditions are somewhat complex.

This said, our model could be generalized, provided that our GB algorithm is instructed, validated, and finally tested with the values of the pollutants and the number of infections coming from the region of interest. Not only this: As one of the inputs of the model is also the type of personal protection measures which are adopted (or even enforced) in that given region, this parameter is also needed to allow the model to make the predictions.

Again, one could criticize our study on the basis of the fact that there are multiple possible factors that have led to the devastations brought by the virus in many areas in the world. Nonetheless, we respond to this criticism with the consideration that many traditional studies have been already conducted that have proven to be a very poor proxy for understanding the extension of this contagion. Our investigation, by contrast, is projecting a new scenario based on an original hypothesis that makes our prediction model unique in the world. At the time we write this article, we cannot have a confirmation of the precision of our predictions, but they will be soon confirmed/rejected by history—and this, too, is science at the service of society.

We want to conclude with a final, but important, consideration. All the experiments we have conducted are reproducible using the data available in the public repositories we have mentioned.

Author Contributions: Conceptualization, M.R. and G.D.; methodology, M.R. and S.M.; software, G.D.; validation, M.R., S.M. and G.D.; formal analysis, M.R.; investigation, G.D.; resources, S.M.; data curation, G.D.; writing—original draft preparation, M.R.; writing—review and editing, M.R.; visualization, G.D.; supervision, S.M.; project administration, S.M.; funding acquisition, M.R. All authors have read and agreed to the published version of the manuscript.

Funding: This research received no external funding.

Conflicts of Interest: The authors declare no conflict of interest.

References

1. Jamieson, A. Coronavirus Crisis May Get Worse and Worse and Worse, Warns WHO. Available online: https://www.independent.co.uk/news/uk/home-news/coronavirus-cases-deaths-who-infection-rate-global-latest-a9616366.html?fbclid=IwAR1rTs52bD1jZBjNEYNt63OuN_DweUkCHlB5oQAAExD2JAR-TXpc5pL2-QA (accessed on 18 July 2020).
2. World Health Organization (WHO). Coronavirus Disease (COVID-2019) Situation Reports. Available online: https://www.who.int/emergencies/diseases/novel-coronavirus-2019/situation-reports (accessed on 18 July 2020).
3. Dipartimento Della Protezione Civile, Italia. Aggiornamento Casi COVID-19. Available online: http://opendatadpc.maps.arcgis.com/apps/opsdashboard/index.html#/b0c68bce2cce478eaac82fe38d4138b1 (accessed on 18 July 2020).
4. Mehta, D. Moody's: Italy's GDP to Contract 9.3% in 2020. FXStreet. 2020. Available online: https://www.fxstreet.com/news/moodys-italys-gdp-to-contract-93-in-2020-202004300541 (accessed on 18 July 2020).
5. Carey, B. Can an Algorithm Predict the Pandemic's Next Moves? The New York Times. 2020. Available online: https://www.nytimes.com/2020/07/02/health/santillana-coronavirus-model-forecast.html?smid=fb-share&fbclid=IwAR15B7tGHRL8oyL1NHgjXyGojTSYbHpoO0ww8hG85B2bN7NVMxJVK2da5wU (accessed on 18 July 2020).
6. Delnevo, G.; Mirri, S.; Roccetti, M. Particulate Matter and COVID-19 disease diffusion in Emilia-Romagna (Italy). Already a cold case? *Computation* **2020**, *8*, 59. [CrossRef]
7. Granger, C. Investigating causal relations by econometric models and cross-spectral methods. *Econometrica* **1969**, *37*, 424–438. [CrossRef]
8. Maziarz, M. A review of the Granger-causality fallacy. *J. Philos. Econ.* **2015**, *8*, 86–105.
9. Conticini, E.; Frediani, B.; Caro, D. Can atmospheric pollution be considered a co-factor in extremely high level of SARS-CoV-2 lethality in Northern Italy? *Environ. Pollut.* **2020**, *261*, 114465. [CrossRef] [PubMed]
10. Becchetti, L.; Conzo, G.; Conzo, P.; Salustri, F. Understanding the Heterogeneity of Adverse COVID-19 Outcomes: The Role of Poor Quality of Air and Lockdown Decisions. Available online: https://papers.ssrn.com/sol3/papers.cfm?abstract_id=3572548 (accessed on 18 July 2020).
11. Li, Q.; Guan, X.; Wu, P.; Wang, X.; Zhou, L.; Tong, Y.; Ren, R.; Leung, K.S.M.; Lau, E.H.Y.; Wong, J.Y.; et al. Early transmission dynamics in Wuhan, China, of novel coronavirus–infected pneumonia. *N. Engl. J. Med.* **2020**, *382*, 1199–1207. [CrossRef] [PubMed]
12. Zhou, P.; Yang, X.L.; Wang, X.G.; Hu, B.; Zhang, L.; Zhang, W.; Si, H.R.; Zhu, Y.; Li, B.; Huang, C.L.; et al. A pneumonia outbreak associated with a new coronavirus of probable bat origin. *Nature* **2020**, *579*, 270–273. [CrossRef]

13. Nunez Soza, L.; Jordanova, P.; Nicolis, L.; Strelec, L.; Stehlik, M. Small sample robust approach to outliers and correlation of Atmospheric Pollution and Health Effects in Santiago de Chile. *Chemom. Intell. Lab. Syst.* **2019**, *185*, 73–84. [CrossRef]
14. Morawska, L.; Milton, D.K. It is time to address airborne transmission of COVID-19. *Clin. Infect. Dis.* **2020**. [CrossRef]
15. Setti, L.; Passarini, F.; De Gennaro, G.; Barbieri, P.; Perrone, M.G.; Borelli, M.; Palmisani, J.; Di Gilio, A.; Priscitelli, P.; Miani, A. SARS-Cov-2RNA found on particulate matter of Bergamo in Northern Italy: T First evidence. *Environ. Res.* **2020**, *188*, 109754. [CrossRef]
16. Ferrarotti, M.J.; Cavalli, A. *On the Hypothesis of Particulate Matter as Carrier of SARS-CoV-2. Numerical Findings with a Novel Package for Smoluchowski Aggregation Via Direct Monte Carlo*; Preprint, n 1; Department of Chemistry, University of Bologna: Bologna, Italy, 2020.
17. Setti, L.; Passarini, F.; De Gennaro, G.; Barbieri, P.; Perrone, M.G.; Borelli, M.; Palmisani, J.; Di Gilio, A.; Priscitelli, P.; Miani, A. Airborne Transmission Route of COVID-19: Why 2 Meters/6 Feet of Inter-Personal Distance Could Not Be Enough. *Int. J. Environ. Res. Public Health* **2020**, *17*, 2932. [CrossRef]
18. Setti, L.; Passarini, F.; De Gennaro, G.; Barbieri, P.; Perrone, M.G.; Borelli, M.; Palmisani, J.; Di Gilio, A.; Priscitelli, P.; Miani, A. Searching for SARS-COV-2 on Particulate Matter: A Possible Early Indicator of COVID-19 Epidemic Recurrence. *Int. J. Environ. Res. Public Health* **2020**, *17*, 2986. [CrossRef] [PubMed]
19. Gazzetta Ufficiale della Repubblica Italiana. Decreto del Presidente del Consiglio dei Ministri 11 Giugno 2020. Available online: https://www.gazzettaufficiale.it/eli/id/2020/06/11/20A03194/sg (accessed on 19 July 2020).
20. Kogan, N.E.; Clemente, L.; Liautaud, P.; Kaashoek, J.; Link, N.B.; Nguyen, A.T.; Lu, F.S.; Huybers, P.; Resch, B.; Havas, C.; et al. An early warning approach to monitor COVID-19 activity with multiple digital traces in near real-time. *arXiv* **2020**, arXiv:2007.00756.
21. Fox, G.N.; Moawad, N.S. UpToDate: A comprehensive clinical database. *J. Family Pract.* **2003**, *52*, 706–710.
22. Buchanan, M. The limits of machine prediction. *Nat. Phys.* **2019**, *15*, 304. [CrossRef]
23. Kermack, W.O.; McKendrick, A.G. Contributions to the mathematical theory of epidemics—I. *Bull. Math. Boil.* **1991**, *53*, 33–55.
24. Russell, S.J.; Norvig, P. Artificial Intelligence: A Modern Approach. *Prentice Hall* **2010**. Available online: https://www.google.com.hk/url?sa=t&rct=j&q=&esrc=s&source=web&cd=&ved=2ahUKEwi2m73KubPrAhVPPnAKHeURBS8QFjACegQIBhAB&url=https%3A%2F%2Ffaculty.psau.edu.sa%2Ffiledownload%2Fdoc-7-pdf-a154ffbcec538a4161a406abf62f5b76-original.pdf&usg=AOvVaw0i7pLrlBs9LMW296xeV6b0 (accessed on 18 July 2020).
25. Ardabili, S.F.; Mosavi, A.; Ghamisi, P.; Ferdinand, F.; Varkonyi-Koczy, A.R.; Reuter, U.; Rabczuck, T.; Atkinson, P.M. COVID-19 outbreak prediction with machine learning. *medRxiv Prepr.* **2020**. Available online: https://www.medrxiv.org/content/10.1101/2020.04.17.20070094v1.full.pdf (accessed on 18 July 2020).
26. Tuli, S.; Tuli, S.; Tuli, R.; Gill, S.S. Predicting the Growth and Trend of COVID-19 Pandemic using Machine Learning and Cloud Computing. *Internet Things* **2020**, *2020*, 100222. [CrossRef]
27. Liu, Y.; Wang, Z.; Ren, J.; Tian, Y.; Zhou, M.; Zhou, T.; Ye, K.; Zhao, Y.; Qiu, Y.; Li, J. A COVID-19 Risk Assessment Decision Support System for General Practitioners: Design and Development Study. *J. Med. Internet Res.* **2020**, *22*, e19786. [CrossRef]
28. Nguyen, T.T. Artificial Intelligence in the Battle Against Coronavirus (COVID-19): A Survey and Future Research Directions. *Preprint* **2020**. Available online: https://figshare.com/articles/Artificial_Intelligence_in_the_Battle_against_Coronavirus_COVID-19_A_Survey_and_Future_Research_Directions/12127020 (accessed on 18 July 2020).
29. Alimadadi, A.; Aryal, S.; Manandhar, I.; Munroe, P.B.; Joe, B.; Cheng, X. Artificial intelligence and machine learning to fight COVID-19. *Physiol. Genomics* **2020**, *52*, 200–202. [CrossRef]
30. Naudé, W. Artificial Intelligence against COVID-19: An Early Review. Available online: https://www.iza.org/publications/dp/13110/artificial-intelligence-against-covid-19-an-early-review (accessed on 18 July 2020).
31. Kucharski, A.J.; Russell, T.W.; Diamond, C.; Liu, Y.; Edmunds, J.; Funk, S.; Eggo, R.M. Early dynamics of transmission and control of COVID-19: A mathematical modelling study. *Lancet Infect. Dis.* **2020**, *20*, 553–558. [CrossRef]
32. Gazzetta Ufficiale della Repubblica Italiana. Decreto del Presidente del Consiglio dei Ministri 8 Marzo 2020. Available online: https://www.gazzettaufficiale.it/eli/id/2020/03/08/20A01522/sg (accessed on 18 July 2020).

33. Governo Italiano Presidenza del Consiglio dei Ministri. Available online: http://www.governo.it/it/articolo/firmato-il-dpcm-9-marzo-2020/14276 (accessed on 18 July 2020).
34. Gazzetta Ufficiale della Repubblica Italiana. Decreto del Presidente del Consiglio dei Ministri 26 Aprile 2020. Available online: https://www.gazzettaufficiale.it/eli/id/2020/04/27/20A02352/sg (accessed on 18 July 2020).
35. COVID-19 Italia—Monitoraggio Situazione. Available online: https://github.com/pcm-dpc/COVID-19 (accessed on 18 July 2020).
36. Arpae Emilia-Romagna. Available online: https://arpae.it/mappa_qa.asp?idlivello=1682&tema=stazioni (accessed on 18 July 2020).
37. Lauer, S.A.; Grantz, K.H.; Bi, Q.; Jones, F.K.; Zheng, Q.; Meredith, H.R.; Azman, A.S.; Reich, N.G.; Lessler, J. The incubation period of coronavirus disease 2019 (COVID-19) from publicly reported confirmed cases: Estimation and application. *Ann. Intern. Med.* **2020**, *172*, 577–582. [CrossRef]
38. K Keller, J.M.; Gray, M.R.; Givens, J.A. A fuzzy k-nearest neighbor algorithm. *IEEE Trans. Syst. Man Cybern.* **1985**, *4*, 580–585. [CrossRef]
39. Lawrence, R.L.; Wright, A. Rule-based classification systems using classification and regression tree (CART) analysis. *Photogramm. Eng. Remote Sens.* **2001**, *67*, 1137–1142.
40. Cortes, C.; Vapnik, V. Support-vector networks. *Mach. Learn.* **1995**, *20*, 273–297. [CrossRef]
41. Hastie, T.; Tibshirani, R.; Friedman, J.; Franklin, J. The elements of statistical learning: Data mining, inference and prediction. *Math. Intell.* **2005**, *27*, 83–85.
42. Dietterich, T.G. Ensemble methods in machine learning. In Proceedings of the First International Workshop on Multiple Classifier Systems, Cagliari, Italy, 21–23 June 2000; Springer: Berlin, Germany, 2000; pp. 1–15.
43. Friedman, J.H. Stochastic gradient boosting. *Comput. Stat. Data Anal.* **2002**, *38*, 367–378. [CrossRef]
44. Barandiaran, I. The random subspace method for constructing decision forests. *IEEE Trans. Pattern Anal. Mach. Intell.* **1998**, *20*, 1–22.
45. Geurts, P.; Ernst, D.; Wehenkel, L. Extremely randomized trees. *Mach. Learn.* **2006**, *63*, 3–42. [CrossRef]
46. Cohen, A.J.; Brauer, M.; Burnett, R.; Ross Anderson, H.; Frostad, J.; Estep, K.; Balakrishnan, K.; Brunekreef, B.; Dandona, L.; Dandona, R.; et al. Estimates and 25-year trends of the global burden of disease attributable to ambient air pollution: an analysis of data from the Global Burden of Diseases. *Lancet* **2017**, *389*, 215. [CrossRef]
47. Carbone, M.; Green, J.B.; Bucci, E.M.; Lednicky, J.A. Coronaviruses: Facts, Myths, and Hypotheses. *J. Thorac. Oncol.* **2020**, *15*, 675–678. [CrossRef]
48. Delnevo, G.; Roccetti, M.; Mirri, S. Modeling Patients' Online Medical Conversations: A Granger Causality Approach. In Proceedings of the 2018 IEEE/ACM International Conference on Connected Health: Applications, Systems and Engineering Technologies, Washington, DC, USA, 26–28 September 2018; IEEE: Piscataway, NJ, USA, 2018; pp. 40–44.

computation

MDPI

Article

Coarse-Grained Modeling of the SARS-CoV-2 Spike Glycoprotein by Physics-Informed Machine Learning

David Liang [1,*], Ziji Zhang [2], Miriam Rafailovich [3], Marcia Simon [4], Yuefan Deng [2] and Peng Zhang [2,*]

[1] Department of Chemistry, University of Chicago, Chicago, IL 60637, USA
[2] Departments of Applied Mathematics and Statistics, Stony Brook University, Stony Brook, NY 11790, USA
[3] Materials Science and Chemical Engineering, Stony Brook University, Stony Brook, NY 11790, USA
[4] Oral Biology and Pathology, Stony Brook University, Stony Brook, NY 11790, USA
* Correspondence: dliang7234@uchicago.edu (D.L.); peng.zhang@stonybrook.edu (P.Z.); Tel.: +1-(631)-377-8211 (D.L.)

Abstract: Coarse-grained (CG) modeling has defined a well-established approach to accessing greater space and time scales inaccessible to the computationally expensive all-atomic (AA) molecular dynamics (MD) simulations. Popular methods of CG follow a bottom-up architecture to match properties of fine-grained or experimental data whose development is a daunting challenge for requiring the derivation of a new set of parameters in potential calculation. We proposed a novel physics-informed machine learning (PIML) framework for a CG model and applied it, as a verification, for modeling the SARS-CoV-2 spike glycoprotein. The PIML in the proposed framework employs a force-matching scheme with which we determined the force-field parameters. Our PIML framework defines its trainable parameters as the CG force-field parameters and predicts the instantaneous forces on each CG bead, learning the force field parameters to best match the predicted forces with the reference forces. Using the learned interaction parameters, CGMD validation simulations reach the microsecond time scale with stability, at a simulation speed 40,000 times faster than the conventional AAMD. Compared with the traditional iterative approach, our framework matches the AA reference structure with better accuracy. The improved efficiency enhances the timeliness of research and development in producing long-term simulations of SARS-CoV-2 and opens avenues to help illuminate protein mechanisms and predict its environmental changes.

Keywords: coarse-grained modeling; SARS-CoV-2; molecular dynamics; machine learning

Citation: Liang, D.; Zhang, Z.; Rafailovich, M.; Simon, M.; Deng, Y.; Zhang, P. Coarse-Grained Modeling of the SARS-CoV-2 Spike Glycoprotein by Physics-Informed Machine Learning. *Computation* **2023**, *11*, 24. https://doi.org/10.3390/computation11020024

Academic Editors: Simone Brogi and Vincenzo Calderone

Received: 1 January 2023
Revised: 26 January 2023
Accepted: 30 January 2023
Published: 2 February 2023

1. Introduction

All-atomic molecular dynamics (AAMD) simulations have defined a foundational basis for molecular modeling, providing both atomic- and femtosecond-level resolutions into the dynamic evolution of systems. However, its computational cost often limits its practical and large-scale applications beyond microsecond simulations of millions of atoms. Multiscale coarse-grained (CG) modeling defines a well-established approach within literature for simulating complex, high-definition systems using simplified, lower-resolution representations, often by aggregating groups of atoms into a single CGMD "bead," thus increasing computational efficiency [1–4]. Popular methods of CG strive to match structure properties or energy distributions of fine-grained or experimental data which center around describing a new force field, consisting of the system's parameters and potential calculations, to reproduce the properties of all-atomic (AA) reference simulations. In practice, the CG approaches do not aim to fully reproduce the distributions of the reference data, instead focusing on optimizing, and thus sacrificing complexity in favor of accessing more relevant simulation spatial and temporal scales. The optimization of an accurate and consistent CG model remains an active and significant challenge in the field [5–7].

Recent advances in machine learning (ML) have proven their strength to accelerate both in vitro and in silico biological studies [8–11]. In this work, we develop a novel physics-informed machine learning (PIML) framework for parameterization and optimization of CG force fields, resulting in the development of physics-informed CG models from fine-grained molecular dynamics (MD) to enable simulation across greater spatial and temporal scales that are inaccessible to conventional AAMD simulations. As an example, we focus on the SARS-CoV-2 spike glycoprotein in practical application. The outbreak of SARS-CoV-2 in 2019 and its continued persistence have led to millions of deaths globally [12], prompting investigations of its molecular structure and mechanisms of infection. The outer surface of the virion is covered by numerous unique spike proteins, largely responsible for the binding of the virus to the host cell receptor angiotensin-converting enzyme 2, thus mediating cell entry [13]. Hence, understanding this protein is crucial to investigating the infectivity of the virus and taking steps toward better therapeutics and vaccines. In this study, the protein serves as a prime example of both a timely and significant application for our proposed methodology. While studies are currently underway in uncovering specific mechanisms of action of the SARS-CoV-2 virion or possible therapeutics [14,15], many practical and large-scale applications of AAMD simulations are challenged by the computational expense when dealing with this S-protein of over twenty thousand atoms [16].

Efforts have been made to develop CG models with the corresponding force fields to simulate the S-protein; for instance, a hetero-elastic network model [17,18] was used to optimize bonded energy calculations, while relative-entropy minimization was applied to learn nonbonded interactions and an empirical approach was taken to refine the CG model [18]. Another study [2] utilized the iterative Boltzmann inversion method (IBIM) to reproduce the reference atomic fluctuations. We propose a novel ML-based parameterization approach that goes beyond the existing approaches by defining physics-informed force field parameters and learning the CG free energy functions that account for the entire network of bonded and nonbonded interactions. We unify the optimization task for the CG force field for more efficient parameter determination. The model, trained by a force-matching scheme, corroborates the CG forces and associated effective potential with the AAMD simulation data. This approach, offering an easily generalizable means of parametrization to different proteins and applications, differentiates from other schemes that rely on empirical or user-defined parameters.

While there exist ML-based force fields in other studies, most notably CGNet [19], and its variants CGSchNet [20], as well as TorchMD [21], they are different from our approach. While they were developed for application on smaller proteins such as alanine dipeptide or chignolin, this study aims to tackle a more challenging application with a significantly larger protein, and hence we rely on ML to derive and parameterize a force field.

The interactions of the bottom-up CG model, in our approach, use a combination of iterative and PIML strategies. The AAMD simulations, producing the ground truth, are conducted on powerful supercomputers to help obtain massive data to derive the associated CG model. Our main contributions are:

- An innovative application of supervised ML is proposed to derive a physics-informed CG model.
- The supervised ML is combined with molecular dynamics towards greater efficiency, achieving a speed-up of CGMD simulations of 40,000 over the conventional AAMD simulations while retaining structural accuracy.
- The greater efficiency enhances the timeliness of the research in producing long-term simulations and blazes a path for new applications and further investigation, i.e., protein binding and prediction of environmental changes.

The remainder of this paper is organized as follows. Section 2 describes the physics-informed CG model and its implementation. Section 3 reports the experimental results of the CGMD simulations and corroborates them with the AAMD simulations. Section 4 provides discussions and future direction in multiscale modeling of biomolecular systems.

2. Materials and Methods

2.1. Coarse-Grained Structure

The full SARS-CoV-2 S-protein model was obtained from the protein data bank 6VXX and was run through NAMD software [22,23] on the AA system consisting of 22,815 atoms (a total of 45,153 atoms including the hydrogens). The coarse-grain structure follows the established aggressive Shape-Based Coarse Graining (SBCG) approach [6], which reduced the model to 60 representing particles, maintaining the homotrimeric structure with 20 atoms per chain (Figure 1 and Table 1). Atoms were assigned to beads based on the overall topology of the macromolecule. This involved the use of a topology-preserving neural network, where each CG bead corresponds to a node in the network and the coordinates of the atoms are inputted to adapt the neural network [24]. The hyperparameters used in the SBCG GUI are as follows: initial eps = 0.3, final eps = 0.05, initial lambda = 5.0, and final lambda = 0.01, with bonds formed from the all-atom structure. Beads are uncharged, but the CG model is fitted to reproduce the electrostatics present in the AAMD simulations.

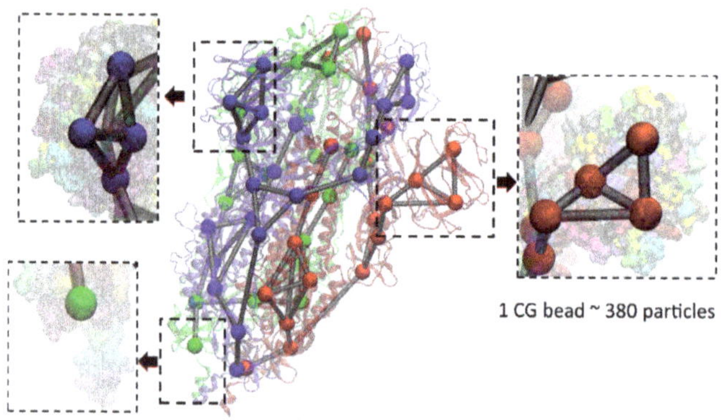

Figure 1. Structural visualization for AA vs. CG model. Red, blue, and green denote 3 chains.

Table 1. Statistics for AA vs. CG model.

	AA Model	CG Model
Atoms	22,815 (45,153 w/hydrogens)	60
Bonds	23,385	81
Angles	31,887	159
Dihedrals	37,872	231

2.2. Coarse-Grained Force Field

Our physics-informed CG modeling follows a multiscale approach, characterized roughly by the transfer of high resolution AA data to the CG scale through the parameterization of a CG model [25]. The approach is shown in Figure 2. In the first box of "Data Collection," spatial and temporal mapping schemes are employed to map the AAMD simulations to the reduced-resolution CG structure, representing the ground truth. In the second box of "Parameter Optimization," a new CG force field is parameterized to conform to this ground truth. This is carried out by first employing IBIM on the bonded parameters [26,27], which iteratively scales parameters and simulates trials to match the reference radial distribution function (RDF). Visual Molecular Dynamics (VMD) software is used to initialize the non-bonded terms [28] based on approximated values and solvent-accessible-surface area (SASA) calculations [29]. In the last box of "Validation Analysis," the learned parameters are implemented in a CGMD to corroborate the proposed method with the baseline reference. The simulations are evaluated in terms of simulation accuracy and computation speed. Specific details are provided in the following sections.

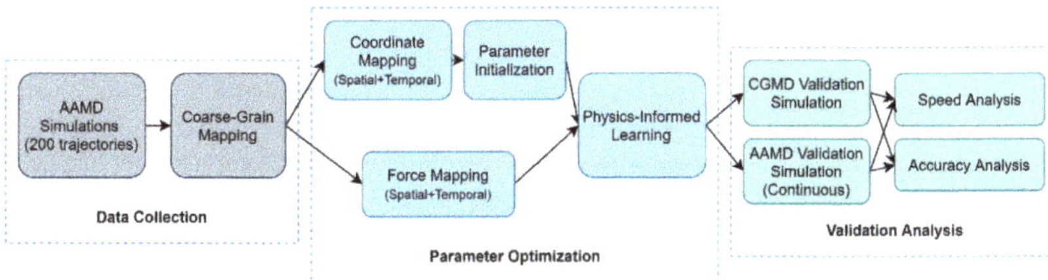

Figure 2. Illustration of the proposed CG modeling pipeline.

We converted our reference data to the CG scale to use the AAMD validation simulation data for training. This reference trajectory obtained in Appendix A.1. (Appendix A) was processed by mapping the extracted coordinate and force data to the CG scale both spatially and temporally. Spatial mapping was conducted by computing the center of mass and the sum of forces for each atom group, constituting a bead, according to:

$$X_{I,CG} = \frac{\sum_i w_i x_{i,AA}}{\sum_i w_i}, \tag{1}$$

$$F_{I,CG} = \sum_i f_{i,AA}, \tag{2}$$

where $X_{I,CG}$ and $F_{I,CG}$ represent the calculated position and force of bead I, $x_{i,AA}$ and $f_{i,AA}$ represent the position and force of atom i within the atom group constituting bead I, and w_i represents the mass of atom i as a weighting factor. In addition to the spatial mapping, temporal averaging is performed to account for the greater temporal scales used in CGMD simulations. We averaged both coordinates and forces across the temporal dimension every 100 frames.

We initialized the parameters with traditional CG force field parameterization methods with bonded and nonbonded potentials. The bonded potentials are based on fixed lists of 2-, 3-, and 4-body interactions (bonds, angles, and dihedrals) modeled as spring harmonics with parameters as spring harmonic constants. The nonbonded potential is modeled with a Lennard-Jones (LJ) potential accounting for the weak dipole attraction between distant atoms and the hard-core repulsion between close atoms. The IBIM method is employed to initialize the new CG model force-field parameters, specifically the bonded parameters. Diverging from the original implementation, we incorporated the refinement of dihedral parameters in addition to the bonds and angles. From the ground truth, we extracted distribution functions $P(x)$ of variable x representing the bond lengths, bond angles, or torsion angles. The potential function $U(x)$ is constructed using the Boltzmann relation:

$$U(x) = -k_B T ln P(x), \tag{3}$$

where k_B is a parameter and T represents the temperature. Furthermore, the bonded parameters can be modeled as harmonics:

$$U(x) = \frac{1}{2}k(x - x_0)^2, \tag{4}$$

where x_0 represents the respective equilibrium measurement and k represents the harmonic constant. Thus, the Boltzmann inversion relationship between distribution functions and harmonic constants can be illustrated as follows:

$$\langle x^2 \rangle - \langle x^2 \rangle = \frac{k_B T}{2k}, \tag{5}$$

where the equilibrium measurement x_0 is equal to the average position $\langle x \rangle$. For a network of these bonded interactions, these bonds, angles, and dihedrals are not independent, and thus when parameters for each of them are derived individually using this Boltzmann inversion relationship, the stiffness of the structure may be overestimated. Hence, there is necessity in further optimization to better match the reference distributions.

The parameters for the non-bonded LJ potential are initialized and approximated by VMD and are based on the SASA calculations of the beads. Further detail into this procedure and its calculations are given in (A3) in Appendix A.

With the LJ potential, U_{LJ}, between pairs of beads (denoted by i and j subscripts) is defined as shown in Equation (6):

$$U_{LJ} = \epsilon_{ij}\left[\left(\frac{R_{min_{ij}}}{r_{ij}}\right)^{12} - 2\left(\frac{R_{min_{ij}}}{r_{ij}}\right)^{6}\right] \quad (6)$$

The relations between the ϵ_{ij} and $R_{min_{ij}}$ pair parameters with their respective trainable parameters for individual beads are defined below in Equations (7) and (8), respectively.

$$\epsilon_{ij} = \sqrt{\epsilon_i * \epsilon_j}, \quad (7)$$

$$R_{min_{ij}} = \frac{R_{min_i}}{2} + \frac{R_{min_j}}{2}. \quad (8)$$

2.3. Physics-Informed ML Model

A force-matching approach helps preserve thermodynamic consistency by minimizing the error between the instantaneous ground-truth forces and predicted forces [19,20,29,30]. Our PIML model defines its trainable parameters as the CG force field parameters. The CG coordinates serve as the input to the model, and the model further predicts the total potential energy of the system. All physically relevant invariances are thus preserved. Leveraging an automatic differentiation function, we take the negative gradient of this energy with respect to the input coordinates, and thus effectively obtain the instantaneous predicted forces. The task is thus to learn the parameters to minimize the error between these predicted forces and ground-truth forces in the loss function.

The model architecture, shown in Figure 3, is detailed further below. The model contains an initial featurization layer that converts the input coordinates to the pairwise distances, bond lengths, bond angles, and torsion angles, as displayed in Figure 3. The model uses two physics-informed layers, containing the trainable parameters, for the prediction of energy: one is the Harmonic layer comprised of bond, angle, and dihedral terms as bonded potentials; and the other is the Lennard-Jones layer.

Within the Harmonic layer, the trainable parameters include the harmonic constants, whereas, in the LJ layer, the trainable parameters are the bead strength ϵ_i and the minimum radius, R_{min_i}, for each unique bead i. There exist 471 and 40 trainable parameters that comprise the bonded and non-bonded interactions, respectively, in the physics-informed model. For the dihedral potentials, the periodic representation accounts for the periodicity of dihedrals, where the phase shift angle was adjusted to fit the equilibrium value as the potential minima. The resulting energy governing the CG force field can be calculated as:

$$U_{CG} = \sum_{bonds} k_b(r - r_0)^2 + \sum_{angles} k_a(\theta - \theta_0)^2 + \sum_{dihedrals} k_d(1 + cos(n\psi - \phi)) + \sum_{i<j}^{atoms} \epsilon_{ij}\left[\left(\frac{R_{min_{ij}}}{r_{ij}}\right)^{12} - 2\left(\frac{R_{min_{ij}}}{r_{ij}}\right)^{6}\right], \quad (9)$$

where k_a, k_b, and k_d are spring factors, r is bond distance, θ is bond angle, ψ is torsion angle, and ϕ is defined as the torsion phase shift angle, which acts as an equilibrium angle in the periodical representation.

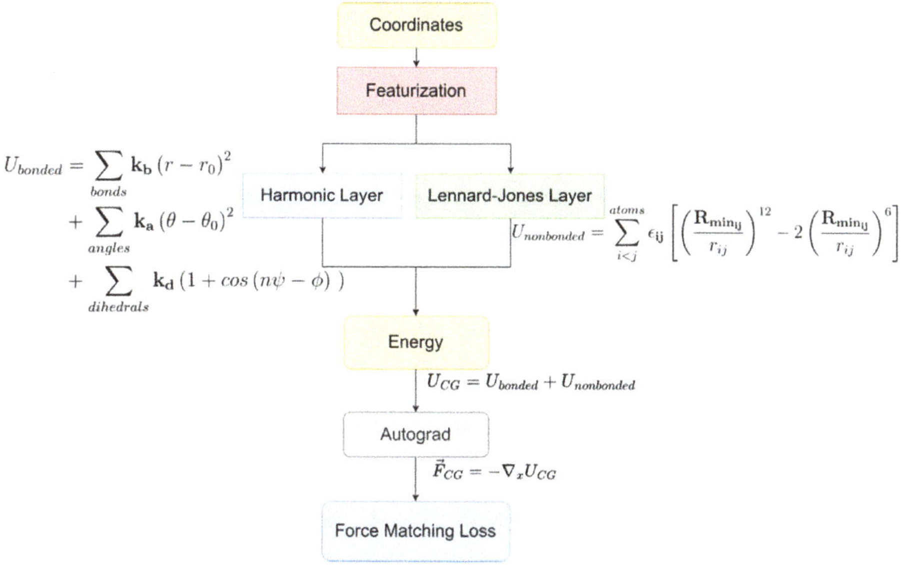

Figure 3. The proposed physics-informed model architecture.

The force $\vec{F}_{CG}$ can be calculated by the gradient of the potential

$$\vec{F}_{CG} = -\nabla_x U_{CG}. \tag{10}$$

with the loss function defined as

$$Loss = \left\langle \left(\vec{F}_{CG} + \nabla U_{CG} \right)^2 \right\rangle, \tag{11}$$

where F_{CG} represents the predicted instantaneous force, and U_{CG} represents the CG potential. This loss as a mean-squared error function between the predicted and the mapped ground-truth forces provides a means of minimizing their difference.

2.4. Validation and Verification

A simulation for CGMD validation is carried out using the learned parameters, together with a separate AAMD simulation, to measure the performance of our approach across the metrics of accuracy and speed. With regards to accuracy analysis, the RDFs are applied in providing insight into the distance distribution of particles around certain particles. The torsional analysis is applied in the form of free energy surface plots and the free energy was plotted along two dihedral quadruples, providing insight into the conformational states. From the plots, validation simulations are compared with the ground-truth training data using the dihedral pairs belonging to the S-protein receptor-binding domain (RBD) and S2 domain. Additionally, root-mean-square-deviation (RMSD) and root-mean-square-fluctuation (RMSF) are analyzed to monitor the structural stability of the compared models throughout their respective trajectories.

In addition to the simulated accuracy, we examined the speeds to measure our model's efficiency. The CGMD simulation was run for one microsecond and its simulation speed was carefully compared with the continuous AAMD validation simulation.

We extended the study to a solvated application beyond the solvent-free simulation environment. Using the same learned forces, we explicitly solvate the CG S-protein into a 18 nm × 18 nm × 18 nm MARTINI water box [31]. In this hybrid system, each MARTINI water molecule is represented by a single bead (of mass 72 amu). To evaluate the accuracy of this solvated experiment, we ran an AAMD simulation of the S-protein solvated in a water box of TIP3 water molecules at the same 310K temperature in canonical (NVT) ensembles.

3. Results

With the learned parameters, the accuracy and speed of the CGMD simulations vs. the AAMD validation simulations are reported. Using the 97,905 coordinates and force frames, the parameter initialization for bonds, angles, and dihedrals, respectively, proceeded with 3 IBIM iterations. For each iteration, the trial simulations were conducted with 10 femtosecond timesteps, minimized for 500 picoseconds, and simulated for 4 ns. There exist 511 total learnable parameters that are learned with the physics-informed model configured with the Adam optimizer with a learning rate of 0.001 and a batch size of 256 for 10 epochs.

3.1. Accuracy Analysis

The CGMD and the AAMD simulations start from the same structure; the visualized protein structures of the starting and ending conformations after microsecond-level simulation in Figure 4 show their good alignment.

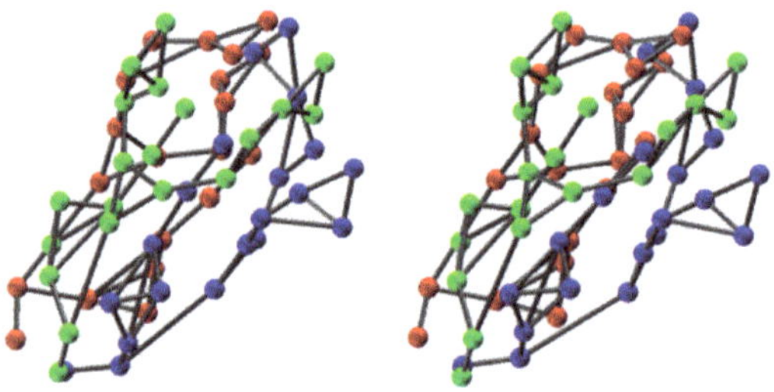

Figure 4. CG structure visualization: AAMD validation simulation final frame state (**left**). CGMD simulation final frame state (**right**).

The RDF measures the distribution of distances between the pairs of particles of two specified atom groups. For instance, Figure 5 defines these two groups to be some given "Atom #" and all "Atoms," respectively. Comparing the RDFs of our CGMD simulations with the ground-truth data, we measure the deviation between the mapped ground truth and the proposed CGMD simulations. As illustrated in Figures 5 and 6, the proposed PIML approach reproduces the structure in reference plots with reasonable accuracy, as it can capture the peaks in RDF.

To quantitatively measure the accuracy of the RDF plots, we incorporate Spearman's correlation coefficient [32] to measure the correlation between the CGMD and reference AAMD RDF plots. In Figures 5 and 6, the Spearman's correlation coefficients for each RDF plot are 0.7472, 0.6560, 0. 6278, 0.5690, and 0.9692 for atoms 7, 11, 14, and 19 and all atom pairs, respectively. This incorporation of a quantitative metric of Spearman's correlation coefficient confirms this reasonable correlation in Figure 5, and strong overall correlation in Figure 6.

Four representations are chosen in Figure 5: "Atom 7" and "Atom 14" plots present regions on the N-terminal domain (NTD), whereas the "Atom 11" plot references a bead located on the receptor-binding domain of the S-protein. "Atom 19" represents the base of the S2 subunit, closer to the stalk of the S-protein.

The free energy profiles are plotted as a function of dihedral angles. The plots are used to analyze and compare the torsion angles as a representation of the protein conformational states. Two separate pairs of torsional angles are displayed for such analysis: one is located on the receptor-binding domain, and the other is in the S2 subunit. Figure 7 shows that the proposed CGMD simulations match precisely the ground-truth training data. The proposed physics-informed CG model captures the positions and peaks in the respective pairs with comparable accuracy to the AA model.

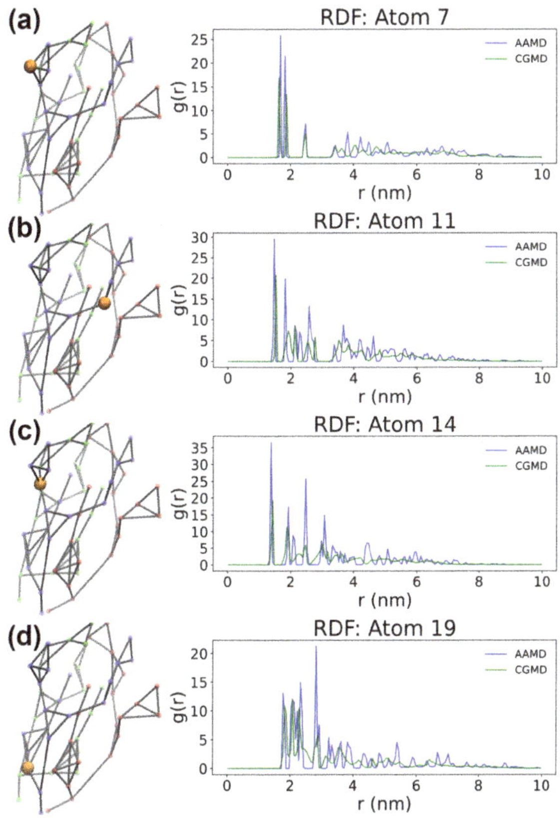

Figure 5. RDF plot from single reference atom comparison of the CGMD simulations vs. the AAMD validation simulations. In the colored beads visualizations: blue—chain A; red—chain B; green—chain C; orange—selected atoms. Spearman's correlation coefficients: (**a**) 0.7472; (**b**) 0.6560; (**c**) 0.6278; (**d**) 0.5690.

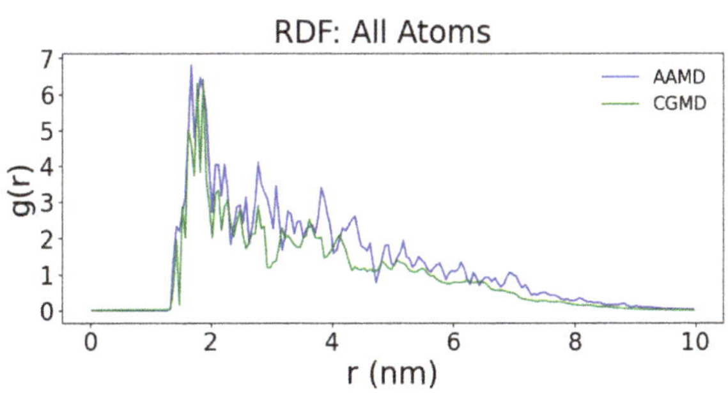

Figure 6. RDF plot of all atoms for comparison of CGMD vs. AAMD. Spearman's correlation coefficient: 0.9692.

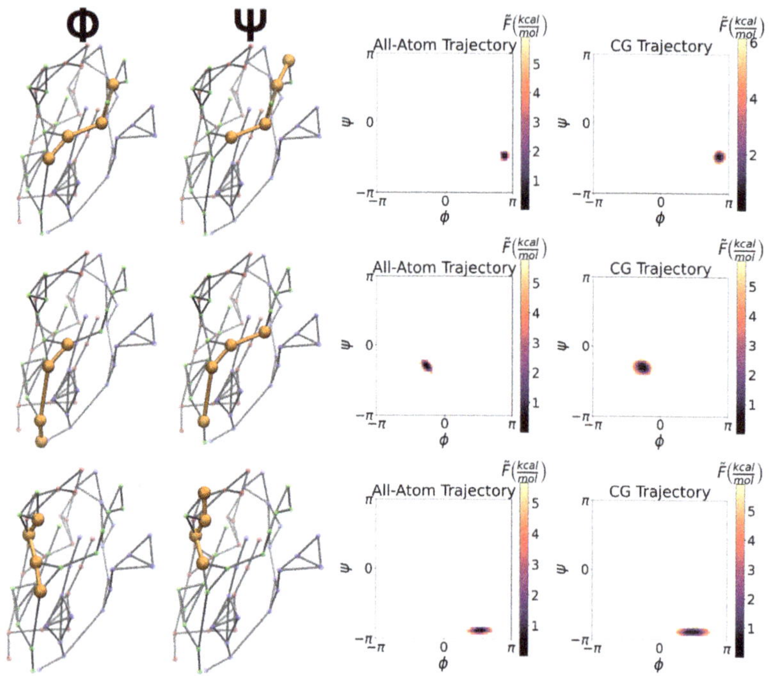

Figure 7. Free energy profiles of RBD pair (**top**), NTD pair (**middle**), and S2 subunit pair (**bottom**). Blue—chain A; red—chain B; green—chain C; orange—selected dihedral quadruplet.

Further analysis, showing the stability for the entirety of the microsecond, suggests the proposed physics-informed CG approach is feasible for long-term modeling of the SARS-CoV-2 S-protein. The evolution of the proposed physics-informed CG model trajectory was analyzed by calculating the RMSD values using the starting structure as a reference frame. The RMSD reveals the overall stability and conformational change of the whole protein. Protein coordinates are recorded every 10 picoseconds and the RMSD was calculated on the aligned trajectory. Figure 8 presents the RMSD of the proposed CGMD simulations alongside the AAMD validation simulations. The CGMD RMSD remains consistent throughout the full microsecond of simulation, indicating long-term structural stability.

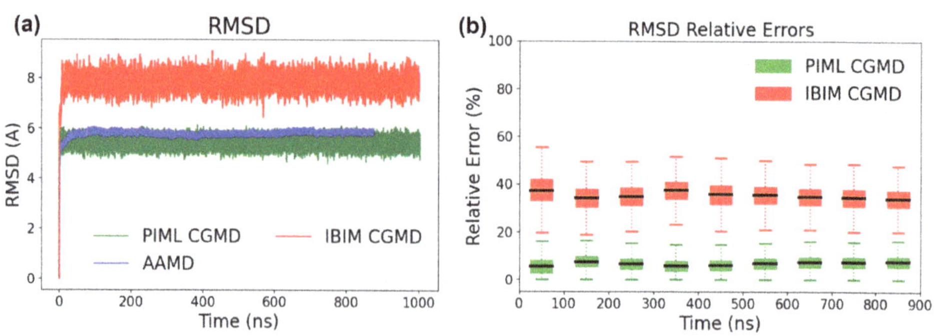

Figure 8. RMSD comparison between the proposed CGMD simulation and the AAMD validation simulation. RMSD in angstroms (**a**). RMSD relative errors (**b**).

The RMSD relative errors are included as well. Each error bar is normalized and extracted for statistics within a time period of 100 ns. All three simulations plotted below start with the same structure, and the relative error represents the relative error of our PIML and IBIM methods, respectively, with respect to the AAMD structures throughout their CGMD simulations. The calculation for such relative error is defined below:

$$e_t = \frac{|RMSD_{CG}(t) - RMSD_{AA}(t)|}{RMSD_{AA}(t)} \tag{12}$$

where $RMSD_{CG}(t)$ and $RMSD_{AA}(t)$ represent the RMSD of the CGMD and AAMD simulations, respectively, at time t.

The presented CGMD simulations appear to have greater fluctuations in comparison with the AAMD validation simulations, which indicates the CGMD is likely exploring a greater distribution of conformations. This is expected from the CG procedure, specifically how the averaging procedure smooths effective potentials, and thus how it facilitates enhanced sampling of the underlying phase space [33]. Our CGMD appears to have reached structures with RMSD values consistently closer to the RMSD values of the validation AAMD simulation compared with the IBIM approach. The animated trajectories of the AAMD validation simulation and the CGMD simulation are provided in the Supplementary Materials.

3.2. Speed Analysis

Both the AAMD validation simulations and the presented CGMD simulations are conducted on a local cluster, where each computing node consists of two Intel Xeon E5-2690v3 CPUs. By using the parallel NAMD package on 1 node with 24 CPU cores, the AAMD validation simulations with 1 femtosecond as the time step size produced 0.243 nanoseconds/day while the CGMD simulations with 10 femtoseconds as the time step size produced 9532.6 nanoseconds/day. This CGMD timestep was determined experimentally as the optimal speed that would maintain stable simulation. Specifically, we experimented with an array of timestep sizes ranging from 4 fs to 100 fs, and we settled on 10 fs for stability and speed. The experimental outcomes indicate that the presented CGMD validation simulations have a speed nearly 40,000 times faster than that of the AAMD validation simulations. Detailed measurements are presented in Table 2.

Table 2. Validation simulation comparisons using 24 CPU cores.

Simulations	Time Step Size	Total Steps	Simulated Time	Simulating Time	Simulation Speed
AAMD	1 fs	100,000	0.1 ns	35,557 s	0.243 ns/day
CGMD	10 fs	500,000,000	5 μs	45,318 s	9532.6 ns/day

3.3. Solvation Application

We assimilated our CG S-protein model with the MARTINI solvent using two separate cutoff configurations for nonbonded interactions. In this new configuration, the nonbonded interactions within the S-protein group are configured with a cutoff of 4.5 nm and smooth switching starting at 2.0 nm. Nonbonded interactions within the MARTINI solvent and between the solvent and the S-protein are configured with a cutoff of 1.2 nm and a smooth switching starting at 0.9 nm. For the RDF results as illustrated in Figure 9, our solvated model reproduces the structure in AAMD validation simulations collected in [14], resembling the significant peaks and retaining the overall structure of the protein.

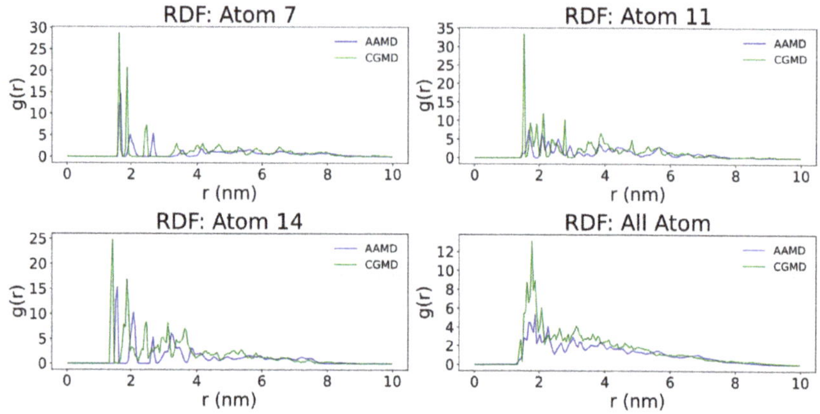

Figure 9. RDF plot from single reference atom comparison of the solvated CGMD simulations vs. the solvated AAMD validation simulations. "Atom 7" and "Atom 14" reflect on the NTD, and "Atom 11" reflects the receptor-binding domain.

4. Discussion

We presented an artificial intelligence-enabled model for multiscale CGMD simulations. The PIML approach to the model parameterization includes two phases: (1) using AAMD simulations to generate the ground truth for learning parameters and (2) using the learned parameters to run long-term CGMD simulations. The physics-informed bottom-up CGMD model simulations are compared with the ground truth AAMD simulations, the gold standard in accuracy, indicating a resemblance of the conformation. The proposed CG model is significantly faster than the AAMD simulation model. With the aggressive CG approach, the proposed model achieves nearly 40,000× the speed of the AAMD simulations.

The work underscores the following contributions toward more efficient multiscale modeling:

- The approach demonstrates the superiority of the supervised ML in deriving a CG model.
- In combining ML with molecular dynamics, our approach immensely accelerates simulations compared with the conventional AA models while maintaining stability and structural accuracy.
- The gained efficiency can elucidate protein mechanisms and render a great impact on future simulation studies by relieving the ongoing concerns about timeliness.

The application of our model into a solvated environment was presented and no term in our CG model was calibrated to reproduce the solvated reference. While rough structural accuracy was preserved, most clearly seen with the RDF plots, a limitation was noticed in our solvated simulation; the protein is observed to contract more than the reference. It is likely that calibration to the cutoffs, as well as the switching value, could yield better accuracy, and we intend to explore this as a future work. The proposed method underscores an important step forward in extending these large systems to actual applications in cases that it was not explicitly parametrized to reproduce, and in future works, we intend to adapt this proposed approach to binding of the S-protein with the ACE-2 receptor.

Supplementary Materials: The following supporting information can be downloaded at: https://www.mdpi.com/article/computation11020024/s1.

Author Contributions: Conceptualization, methodology, software, validation, formal analysis, investigation, data curation, and visualization, D.L., Z.Z. and P.Z.; writing—original draft preparation, D.L.; writing—review and editing, D.L., Z.Z., P.Z., Y.D., M.R. and M.S.; supervision, project administration, and funding acquisition, P.Z., Y.D., M.R. and M.S. All authors have read and agreed to the published version of the manuscript.

Funding: The project is sponsored by Stony Brook University's OVPR and IEDM COVID-19 Seed Grant, PIs: P.Z., Y.D., M.R. and M.S.

Data Availability Statement: Data generation was conducted through NAMD while machine learning and data analysis were conducted by our own code that is available upon request.

Acknowledgments: The project is supported by the SUNY-IBM Consortium Award, IPDyna: Intelligent Platelet Dynamics, FP00004096 (PI: Y.D., Co-I: P.Z.). All simulations were conducted on the AiMOS supercomputer at Rensselaer Polytechnic Institute through an IBM Faculty Award FP0002468 (PI: Y.D.) and the Seawulf cluster at Stony Brook University.

Conflicts of Interest: The authors declare no conflict of interest.

Appendix A

Appendix A.1. All-Atomic Simulations

To obtain the reference data, we first conducted AAMD simulations on the AiMOS supercomputer, a heterogeneous system architecture that includes IBM POWER9 CPUs connected to NVIDIA TESLA V100 GPUs, and the local computing cluster Seawulf at Stony Brook University. We utilized the CHARMM-36 force field [34] in describing the system in a vacuum canonical ensemble at 310K. Using NAMD software, conjugate gradient and line search energy minimization (10 picoseconds) was run prior to 400 picoseconds of simulation (1 fs timestep). From the stable simulation range, we randomly generated 200 different initial positions and orientations to branch off into separate, unique simulations. This was carried out to include replicas to address the chaotic component of MD simulations. From these simulations, frames containing coordinate and force data were collected every fs. A total of 9.7905 ns of the simulation data were accumulated, which upon mapping yielded 97,905 frames of coordinates and forces. From here on, this data constitutes our ground-truth data that represents the reference data the CG model aims to match.

Appendix A.2. Dihedral Potential Term

For the dihedral potentials, they can be represented in two ways: quadratic representation of Equation (A1) and periodic representation of Equation (A2). The quadratic form represents the dihedral potential in the same manner as bonded potentials, where the trainable constants are analogous to spring constants. The periodic representation accounts for the periodicity of dihedrals, where the phase shift angle was adjusted to fit the equilibrium value as the potential minima.

$$U_{CG} = \sum_{bonds} k_b(r - r_0)^2 + \sum_{angles} k_a(\theta - \theta_0)^2 + \sum_{dihedrals} k_d(\psi - \phi)^2 + \sum_{i<j}^{atoms} \epsilon_{ij} \left[\left(\frac{R_{min_{ij}}}{r_{ij}} \right)^{12} - 2\left(\frac{R_{min_{ij}}}{r_{ij}} \right)^{6} \right], \tag{A1}$$

$$U_{CG} = \sum_{bonds} k_b(r - r_0)^2 + \sum_{angles} k_a(\theta - \theta_0)^2 + \sum_{dihedrals} k_d(1 + cos(n\psi - \phi)) + \sum_{i<j}^{atoms} \epsilon_{ij} \left[\left(\frac{R_{min_{ij}}}{r_{ij}} \right)^{12} - 2\left(\frac{R_{min_{ij}}}{r_{ij}} \right)^{6} \right], \tag{A2}$$

The torsion angle distribution can be plotted to depict the unimodality in Figure A1 and thus confirm the choice of n = 1 as the multiplicity for the periodic representations. The distributions in Figure A1 present more common conformations with the yellow color, where the means are the respective equilibrium states. While quadratic, or n = 0 representation also fits this unimodality, we understand that long-term secondary structural changes are unlikely to be modeled properly with this quadratic dihedral form. Thus, we favor the use of the periodic form, which lends itself to more flexibility in case of additional states.

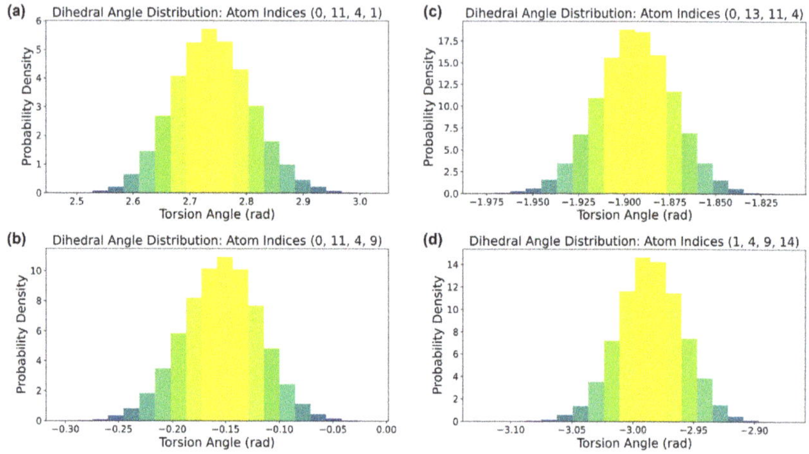

Figure A1. Randomly selected examples of torsion angle distributions for 4 dihedrals of atom indices of (**a**): (0, 11, 4, 1); (**b**): (0, 11, 4, 9); (**c**): (0, 13, 11, 14); and (**d**): (1, 4, 9, 14).

Appendix A.3. Parameter Initialization

The traditional IBIM used to initialize the bonded parameters process follows the following procedure: reference distributions extracted from ground-truth AAMD simulations. Initial bonded parameter "guesses" are obtained through the relation between references and bonded parameters in Equation (5). A trial simulation is run by configuring a short CG simulation with the aforementioned parameter guesses under the environment setup specified in Section 3. Distributions are extracted and compared with the reference AAMD distributions, and we then scale the bonded parameters accordingly to better match the distributions. This procedure of trial simulations and scaling parameters is iterated until the distributions match within reasonable tolerance. Our procedure involved 3 iterations until the parameters (denoting stiffness) extracted from its distributions are roughly within a 25% average deviation from that of reference [6]. Figure A2 illustrates the IBIM refinement of the parameters to initialize our parameters and match the reference distributions to reasonable accuracy after three iterations.

The nonbonded LJ parameter initialization based on SASA calculations is described as follows [29]. In this procedure, each bead i was assigned an LJ strength ϵ_i based on:

$$\epsilon_i = \epsilon_{max} \left(\frac{SASA_i^{hphob}}{SASA_i^{tot}} \right)^2, \tag{A3}$$

where $SASA_i^{hphob}$ and $SASA_i^{tot}$ represent the hydrophobic and total solvent-accessible surface areas of domain i, respectively, and ϵ_{max} is the user-controlled maximum energy for the LJ potential well depth. The reasoning behind using the SASA to determine ϵ_i is to allow hydrophobic beads to aggregate and hydrophilic beads to dissolve in the solvent, which is implicitly present in the CGMD simulations. The user-controlled ϵ_{max} was selected to be 20 kcal/mol based on approximations from findings in previous studies [29]. It is noted that while the user-defined constants are often tested for closest agreement with AAMD simulations in other studies [5], they will be later refined in the methodology as parameters by the ML model. The LJ potential radius r_i (with the minimum of R_{min_i}) is given by the radius of gyration of the group of atoms constituting bead i, which is increased by a user-defined addition, e.g., an increment of 1 Å was selected in this work which accounts for the fact that each atom has a radius typically of 1–2 Å.

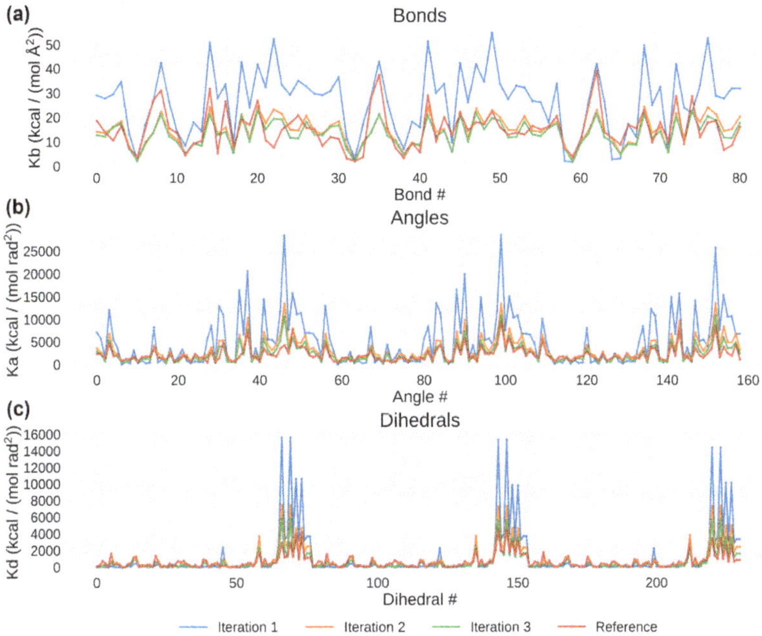

Figure A2. Illustration of IBI method's initialization of parameters for 3 iterations. Bonds (**a**); angles (**b**); dihedrals (**c**).

Appendix A.4. Parameter Learning

Figure A3 displays the loss plot over the training process. Both training and validation losses approached convergence after 4 epochs. The optimization of each individual bonded and non-bonded parameter over the 10 epochs is visualized in Figure A4. Hyperparameters of the network were determined experimentally to reach lower and faster convergence of the training loss. In our PIML, there exists two different groups of hyperparameters: the layers and trainable parameter count that were determined by the physics knowledge and the protein structure; and the learning rate, optimizer, learning rate decay scheduler, and batch size which were tuned experimentally. The range of learning rates we experimented with was 0.0005 to 0.003, and we settled on 0.001. The range of batch sizes experimented with was 16 to 512, and we settled on 256. The learning rate decay scheduler was experimented with along the full 10 epochs; the rate of decay ranged from 0.1 to 0.3, and we settled on 0.3.

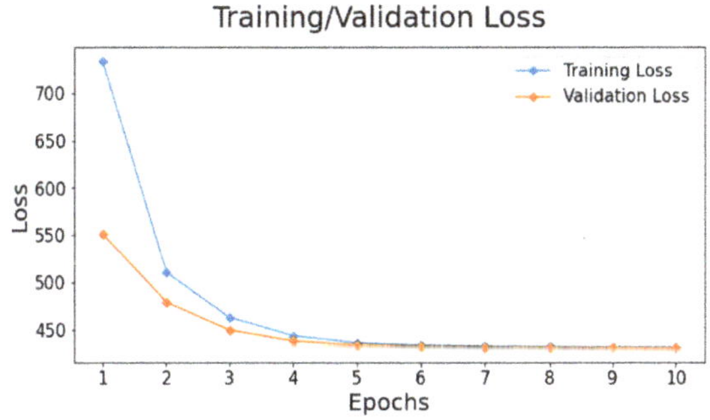

Figure A3. Training and validation loss vs. epochs.

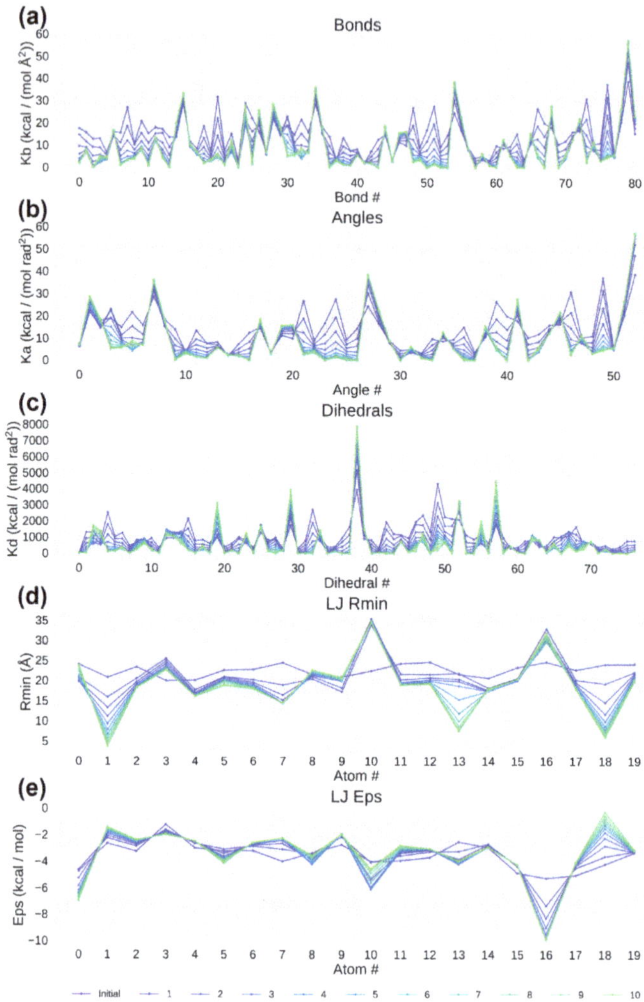

Figure A4. Physics-informed parameter learning for over 10 epochs. Bonds (**a**); angles (**b**); dihedrals (**c**); LJ Rmin (**d**); LJ Epsilon (**e**).

Appendix A.5. ML Refinement on LJ Terms

We delved further into some specific changes reflected in the RDF measurements because of our ML design. The ML procedure indicates significant refinements in the model parameterization, particularly on the non-bonded LJ potential terms. Within this refinement is the very noticeable decrease in both the epsilon and the associated well-depth terms. Upon further investigation, it is shown that the model's calculated energies begin as positive (repulsion) and gradually become negative (attraction) by the end of the training, demonstrating the proper optimization to match the distances of the ground-truth data. In comparison with the IBIM trial results, specifically on the atom pair between atom numbers 17 and 46, the learned distances are more consistent with the ground-truth result, as shown in Figure A5. Furthermore, the incorporation of a quantitative metric of Spearman's correlation coefficient, which is 0.5831 for the PIML CGMD and −0.0376 for the IBIM CGMD, confirms this advantage.

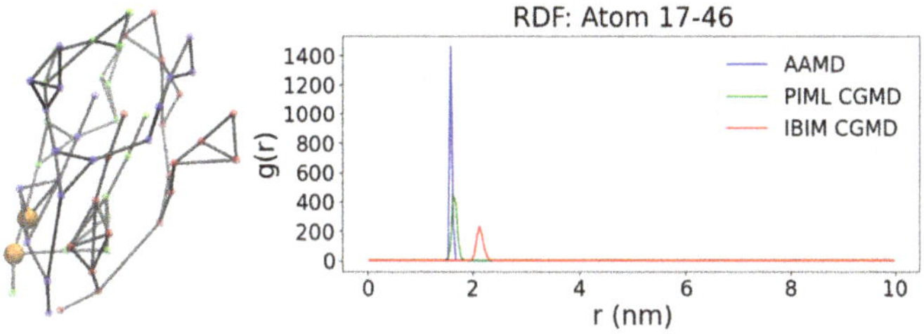

Figure A5. RDF plot between the atoms 17 and 46. Blue—chain A; red—chain B; green—chain C; orange—selected nonbonded atom pair. Spearman's correlation coefficients: (PIML CGMD) 0.5831; (IBIM CGMD) −0.0376.

Appendix A.6. RMSFs for Bonded Interactions

The comparison RMSFs of our CGMD simulation and the ground-truth AAMD simulations is shown in Figure A6. The results indicate that the PIML yielded a relatively accurate fit to the AAMD fluctuations. The difference between the ground-truth and continuous validation data in this case mainly stems from the temporal averaging in the ground-truth data, which may have dampened some fluctuations.

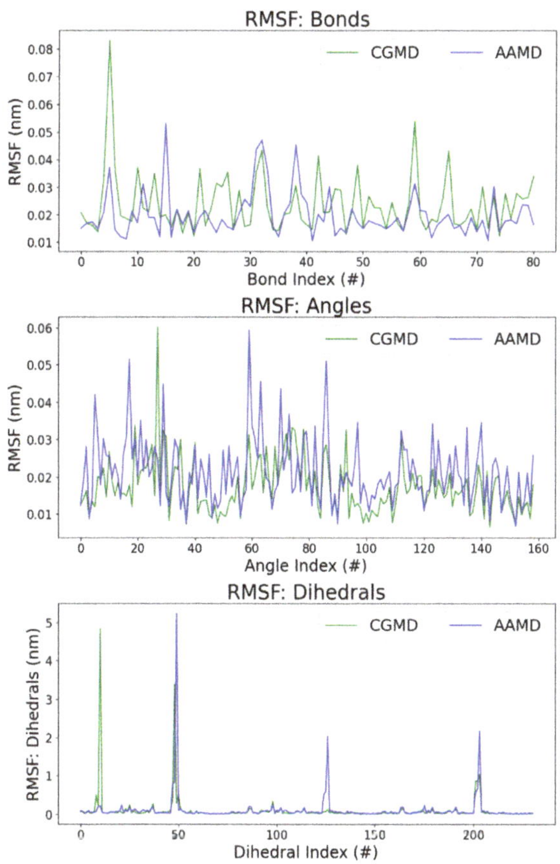

Figure A6. RMSF comparison between the proposed CGMD simulation and the AAMD validation simulations.

References

1. Moore, T.C.; Iacovella, C.R.; McCabe, C. Derivation of coarse-grained potentials via multistate iterative Boltzmann inversion. *J. Chem. Phys.* **2014**, *140*, 224104. [CrossRef] [PubMed]
2. Leong, T.; Voleti, C.; Peng, Z. Coarse-Grained Modeling of Coronavirus Spike Proteins and ACE2 Receptors. *Front. Phys.* **2021**, *9*, 680983. [CrossRef]
3. Yu, A.; Pak, A.J.; He, P.; Monje-Galvan, V.; Casalino, L.; Gaieb, Z.; Dommer, A.C.; Amaro, R.E.; Voth, G.A. A multiscale coarse-grained model of the SARS-CoV-2 virion. *Biophys. J.* **2021**, *120*, 1097–1104. [CrossRef] [PubMed]
4. Izvekov, S.; Voth, G.A. Multiscale coarse graining of liquid-state systems. *J. Chem. Phys.* **2005**, *123*, 134105. [CrossRef] [PubMed]
5. Izvekov, S.; Voth, G.A. A Multiscale Coarse-Graining Method for Biomolecular Systems. *J. Phys. Chem. B* **2005**, *109*, 2469–2473. [CrossRef] [PubMed]
6. Voth, G.A. *Coarse-Graining of Condensed Phase and Biomolecular Systems*; CRC Press: Boca Raton, FL, USA, 2009.
7. Liang, D.; Zhang, Z.; Rafailovich, M.; Simon, M.; Deng, Y.; Zhang, P. Beyond the Scales: A physics-informed machine learning approach for more efficient modeling of SARS-CoV-2 spike glycoprotein. *Res. Sq.* **2021**. [CrossRef]
8. Zhang, Z.; Zhang, P.; Wang, P.; Sheriff, J.; Bluestein, D.; Deng, Y. Rapid analysis of streaming platelet images by semi-unsupervised learning. *Comput. Med. Imaging Graph.* **2021**, *89*, 101895. [CrossRef]
9. Zhang, Z.; Zhang, P.; Han, C.; Cong, G.; Yang, C.-C.; Deng, Y. AI Meets HPC: Learning the Cell Motion in Biofluids. In Proceedings of the Supercomputing Conference 2020 (SC20), Atlanta, GA, USA, 16–19 November 2020. Research Posters Track. [CrossRef]
10. Zhang, Z.; Zhang, P.; Han, C.; Cong, G.; Yang, C.-C.; Deng, Y. Online Machine Learning for Accelerating Molecular Dynamics Modeling of Cells. *Front. Mol. Biosci.* **2022**, *8*, 812248. [CrossRef] [PubMed]
11. Sheriff, J.; Wang, P.; Zhang, P.; Zhang, Z.; Deng, Y.; Bluestein, D. In Vitro Measurements of Shear-Mediated Platelet Adhesion Kinematics as Analyzed through Machine Learning. *Ann. Biomed. Eng.* **2021**, *49*, 3452–3464. [CrossRef]
12. Dong, E.; Du, H.; Gardner, L. An interactive web-based dashboard to track COVID-19 in real time. *Lancet Infect. Dis.* **2020**, *20*, 533–534. [CrossRef] [PubMed]
13. Letko, M.; Marzi, A.; Munster, V. Functional assessment of cell entry and receptor usage for SARS-CoV-2 and other lineage B betacoronaviruses. *Nat. Microbiol.* **2020**, *5*, 562–569. [CrossRef]
14. Niu, Z.; Hasegawa, K.; Deng, Y.; Zhang, Z.; Rafailovich, M.; Simon, M.; Zhang, P. Modeling of the thermal properties of SARS-CoV-2 S-protein. *Front. Mol. Biosci.* **2022**, *9*, 953064. [CrossRef]
15. Song, M.; Zhang, P.; Han, C.; Zhang, Z.; Deng, Y. Long-time simulation of temperature varying conformations of COVID-19 spike glycoprotein on IBM supercomputers. In Proceedings of the Supercomputing Conference 2020 (SC20), Atlanta, GA, USA, 16–19 November 2020. Research Posters Track.
16. Liang, D.; Song, M.; Niu, Z.; Zhang, P.; Rafailovich, M.; Deng, Y. Supervised machine learning approach to molecular dynamics forecast of SARS-CoV-2 spike glycoproteins at varying temperatures. *MRS Adv.* **2021**, *6*, 362–367. [CrossRef]
17. Lyman, E.; Pfaendtner, J.; Voth, G.A. Systematic Multiscale Parameterization of Heterogeneous Elastic Network Models of Proteins. *Biophys. J.* **2008**, *95*, 4183–4192. [CrossRef]
18. Pak, A.J.; Yu, A.; Ke, Z.; Briggs, J.A.G.; Voth, G.A. Cooperative multivalent receptor binding promotes exposure of the SARS-CoV-2 fusion machinery core. *Nat. Commun.* **2022**, *13*, 1002. [CrossRef]
19. Wang, J.; Olsson, S.; Wehmeyer, C.; Pérez, A.; Charron, N.E.; de Fabritiis, G.; Noé, F.; Clementi, C. Machine Learning of Coarse-Grained Molecular Dynamics Force Fields. *ACS Cent. Sci.* **2019**, *5*, 755–767. [CrossRef]
20. Husic, B.E.; Charron, N.E.; Lemm, D.; Wang, J.; Pérez, A.; Majewski, M.; Krämer, A.; Chen, Y.; Olsson, S.; de Fabritiis, G.; et al. Coarse graining molecular dynamics with graph neural networks. *J. Chem. Phys.* **2020**, *153*, 194101. [CrossRef]
21. Doerr, S.; Majewski, M.; Pérez, A.; Krämer, A.; Clementi, C.; Noe, F.; Giorgino, T.; De Fabritiis, G. TorchMD: A Deep Learning Framework for Molecular Simulations. *J. Chem. Theory Comput.* **2021**, *17*, 2355–2363. [CrossRef]
22. Walls, A.C.; Park, Y.-J.; Tortorici, M.A.; Wall, A.; McGuire, A.T.; Veesler, D. Structure, Function, and Antigenicity of the SARS-CoV-2 Spike Glycoprotein. *Cell* **2020**, *181*, 281–292. [CrossRef]
23. Phillips, J.C.; Braun, R.; Wang, W.; Gumbart, J.; Tajkhorshid, E.; Villa, E.; Chipot, C.; Skeel, R.D.; Kalé, L.; Schulten, K. Scalable molecular dynamics with NAMD. *J. Comput. Chem.* **2005**, *26*, 1781–1802. [CrossRef]
24. Arkhipov, A.; Freddolino, P.L.; Schulten, K. Stability and Dynamics of Virus Capsids Described by Coarse-Grained Modeling. *Structure* **2006**, *14*, 1767–1777. [CrossRef]
25. Ayton, G.S.; Noid, W.G.; Voth, G.A. Multiscale modeling of biomolecular systems: In serial and in parallel. *Curr. Opin. Struct. Biol.* **2007**, *17*, 192–198. [CrossRef]
26. Reith, D.; Pütz, M.; Müller-Plathe, F. Deriving effective mesoscale potentials from atomistic simulations: Mesoscale Potentials from Atomistic Simulations. *J. Comput. Chem.* **2003**, *24*, 1624–1636. [CrossRef]
27. Agrawal, V.; Arya, G.; Oswald, J. Simultaneous Iterative Boltzmann Inversion for Coarse-Graining of Polyurea. *Macromolecules* **2014**, *47*, 3378–3389. [CrossRef]
28. Humphrey, W.; Dalke, A.; Schulten, K. VMD: Visual molecular dynamics. *J. Mol. Graph.* **1996**, *14*, 33–38. [CrossRef]
29. Arkhipov, A.; Yin, Y.; Schulten, K. Four-Scale Description of Membrane Sculpting by BAR Domains. *Biophys. J.* **2008**, *95*, 2806–2821. [CrossRef]
30. Noid, W.G.; Chu, J.-W.; Ayton, G.S.; Krishna, V.; Izvekov, S.; Voth, G.A.; Das, A.; Andersen, H.C. The multiscale coarse-graining method. I. A rigorous bridge between atomistic and coarse-grained models. *J. Chem. Phys.* **2008**, *128*, 244114. [CrossRef]

31. Marrink, S.J.; Risselada, H.J.; Yefimov, S.; Tieleman, D.P.; de Vries, A.H. The MARTINI Force Field: Coarse Grained Model for Biomolecular Simulations. *J. Phys. Chem. B* **2007**, *111*, 7812–7824. [CrossRef]
32. Spearman Rank Correlation Coefficient. In *The Concise Encyclopedia of Statistics*; Springer: New York, NY, USA, 2008; pp. 502–505. [CrossRef]
33. Zhou, J.; Thorpe, I.F.; Izvekov, S.; Voth, G.A. Coarse-Grained Peptide Modeling Using a Systematic Multiscale Approach. *Biophys. J.* **2007**, *92*, 4289–4303. [CrossRef]
34. Huang, J.; MacKerell, A.D. CHARMM36 all-atom additive protein force field: Validation based on comparison to NMR data. *J. Comput. Chem.* **2013**, *34*, 2135–2145. [CrossRef]

computation

MDPI

Article

Bioinformatics, Computational Informatics, and Modeling Approaches to the Design of mRNA COVID-19 Vaccine Candidates

Olugbenga Oluseun Oluwagbemi [1,2,3,*], Elijah K. Oladipo [4,5], Olatunji M. Kolawole [6], Julius K. Oloke [7], Temitope I. Adelusi [8], Boluwatife A. Irewolede [5], Emmanuel O. Dairo [5,9], Ayodele E. Ayeni [5,10], Kehinde T. Kolapo [5], Olawumi E. Akindiya [5,11], Jerry A. Oluwasegun [5], Bamigboye F. Oluwadara [5] and Segun Fatumo [12,13,*]

1 Department of Computer Science and Information Technology, Faculty of Natural and Applied Sciences, Sol Plaatje University, Kimberley 8301, South Africa
2 Department of Mathematical Sciences, Stellenbosch University, Stellenbosch 7602, South Africa
3 National Institute of Theoretical and Computational Sciences (NiThECs), Stellenbosch 7602, South Africa
4 Laboratory of Molecular Biology, Immunology and Bioinformatics, Department of Microbiology, Adeleke University, Ede 232104, Nigeria; koladipo2k3@yahoo.co.uk
5 Genomics Unit, Helix Biogen Institute, Ogbomoso 210214, Nigeria; boluwatifeboluene@gmail.com (B.A.I.); edairo7538@stu.ui.edu.ng (E.O.D.); ayenieugene@gmail.com (A.E.A.); kolapokehinde95@gmail.com (K.T.K.); akindiya.liz@gmail.com (O.E.A.); jerryoluwasegun3@gmail.com (J.A.O.); favourbamigboye1@gmail.com (B.F.O.)
6 Department of Microbiology, University of Ilorin, Ilorin 234031, Nigeria; tomak7475@gmail.com
7 Department of Natural Science, Precious Cornerstone University, Ibadan 200223, Nigeria; jkoloke@yahoo.co.uk
8 Computational Biology/Drug Discovery Laboratory, Biochemistry Department, Ladoke Akintola University of Technology, (LAUTECH), Ogbomoso 210214, Nigeria; tiadelusi@lautech.edu.ng
9 Department of Virology, College of Medicine, University of Ibadan, Ibadan 200132, Nigeria
10 Department of Medical Microbiology and Parasitology, University of Ibadan, Ibadan 200132, Nigeria
11 Microbiology Programme, Department of Biological Science, Olusegun Agagu University of Science and Technology, Okitipupa 350113, Nigeria
12 The African Computational Genomics (TACG) Research Group, MRC/UVRI and LSHTM, Entebbe 7545, Uganda
13 Department of Non-Communicable Disease Epidemiology (NCDE), London School of Hygiene and Tropical Medicine, London WC1E 7HT, UK
* Correspondence: olugbenga.oluwagbemi@fulbrightmail.org (O.O.O.); segun.fatumo@lshtm.ac.uk (S.F.); Tel.: +27-663926506 (O.O.O.)

Citation: Oluwagbemi, O.O.; Oladipo, E.K.; Kolawole, O.M.; Oloke, J.K.; Adelusi, T.I.; Irewolede, B.A.; Dairo, E.O.; Ayeni, A.E.; Kolapo, K.T.; Akindiya, O.E.; et al. Bioinformatics, Computational Informatics, and Modeling Approaches to the Design of mRNA COVID-19 Vaccine Candidates. *Computation* **2022**, *10*, 117. https://doi.org/10.3390/computation10070117

Academic Editors: Simone Brogi and Vincenzo Calderone

Received: 4 April 2022
Accepted: 27 June 2022
Published: 8 July 2022
Corrected: 2 December 2022

Abstract: This article is devoted to applying bioinformatics and immunoinformatics approaches for the development of a multi-epitope mRNA vaccine against the spike glycoproteins of circulating SARS-CoV-2 variants in selected African countries. The study's relevance is dictated by the fact that severe acute respiratory syndrome coronavirus 2 (SARS-CoV-2) began its global threat at the end of 2019 and since then has had a devastating impact on the whole world. Measures to reduce threats from the pandemic include social restrictions, restrictions on international travel, and vaccine development. In most cases, vaccine development depends on the spike glycoprotein, which serves as a medium for its entry into host cells. Although several variants of SARS-CoV-2 have emerged from mutations crossing continental boundaries, about 6000 delta variants have been reported along the coast of more than 20 countries in Africa, with South Africa accounting for the highest percentage. This also applies to the omicron variant of the SARS-CoV-2 virus in South Africa. The authors suggest that bioinformatics and immunoinformatics approaches be used to develop a multi-epitope mRNA vaccine against the spike glycoproteins of circulating SARS-CoV-2 variants in selected African countries. Various immunoinformatics tools have been used to predict T- and B-lymphocyte epitopes. The epitopes were further subjected to multiple evaluations to select epitopes that could elicit a sustained immunological response. The candidate vaccine consisted of seven epitopes, a highly immunogenic adjuvant, an MHC I-targeting domain (MITD), a signal peptide, and linkers. The molecular weight (MW) was predicted to be 223.1 kDa, well above the acceptable threshold of

110 kDa on an excellent vaccine candidate. In addition, the results showed that the candidate vaccine was antigenic, non-allergenic, non-toxic, thermostable, and hydrophilic. The vaccine candidate has good population coverage, with the highest range in East Africa (80.44%) followed by South Africa (77.23%). West Africa and North Africa have 76.65% and 76.13%, respectively, while Central Africa (75.64%) has minimal coverage. Among seven epitopes, no mutations were observed in 100 randomly selected SARS-CoV-2 spike glycoproteins in the study area. Evaluation of the secondary structure of the vaccine constructs revealed a stabilized structure showing 36.44% alpha-helices, 20.45% drawn filaments, and 33.38% random helices. Molecular docking of the TLR4 vaccine showed that the simulated vaccine has a high binding affinity for TLR-4, reflecting its ability to stimulate the innate and adaptive immune response.

Keywords: bioinformatics; COVID-19; SARS-CoV-2; immunoinformatic; mRNA; vaccine; modeling; computational

1. Introduction

Severe acute respiratory syndrome (SARS) and Middle East respiratory syndrome (MERS), of the viral coronavirus family, have ravaged the world in the last two decades [1]. The World Health Organization (WHO, henceforth) declared COVID-19 a global pandemic. This declaration was made open in the year 2020. As of 20 January 2022, 336,790,193 cases of COVID-19 and 5,560,718 deaths [2] were confirmed. The RNA of the virus SARS-CoV-2, of the family coronaviridae, possesses a spike (S) glycoprotein, which extends over the surface of the virus to initiate the insemination of coronavirus into the host cells [3,4]. On this glycoprotein, there are 14 residue-binding receptors, which communicate with the angiotensin-converting-enzyme 2 (ACE2) receptor [5]. Coronavirus spike glycoprotein has acceptable antigenicity and immunogenicity [3,6].

As of 5th of August 2021, over 6000 deaths were recorded within a week, with 19% increases in the confirmed cases of SARS-CoV at almost 300,000 [7,8]. About 6000 delta variants have been recognized in more than 20 countries in Africa, with South Africa having the greatest percentage [9].

Vaccine administration holds a great promise to successfully combat the menace of the COVID-19 global pandemic [10–12]. The adoption of messenger RNA (mRNA) in vaccine development is characterized with great flexibility. Messenger RNA encodes and expresses all types of proteins, and by rule it enables the production of vaccines for combating diverse diseases and protein replacement remedy [13]. The scientific significance of mRNA vaccine development cannot be overemphasized. The production of the Moderna and Pfizer/BioNTech COVID-19 vaccines followed this large-scale vaccine production pattern [14].

Messenger RNA vaccines provide a novel method of building immunity against pathogens [15]. One of the distinct roles of mRNA vaccines in the fight against SARS-CoV-2 is the provision of the blueprint of genes for the spike protein of COVID-19 [16]. Unlike peptide-based vaccines, mRNA vaccines do not have restraints of the MHC haplotype; however, as an advantage, mRNA vaccines have a self-adjuvanting property, which is lacking in protein-based vaccines. The mRNA also binds to pattern-recognition receptors [13]. The fundamental principle in the operations of the mRNA vaccines is to provide transcription, which assists in encoding wanted antigens. This is closely followed by the synthesis of RNA. The sequence that encodes the immunogens is present, and the technique can effortlessly be implemented for mRNA production [17].

Messenger-RNA-based vaccines have been found to have better biosafety characteristics in comparison to DNA-based vaccines because the translation of antigens and immunogens occurs in the cytoplasm instead of the nucleus. Therefore, it is almost impossible for mRNA to fuse into the host genome as opposed to DNA-based vaccines [10]. In addition, mRNA is safe for use as a vector in comparison with DNA because it conveys a small line of sequence for translation (a transient molecule) and does not communicate with the genes present in the host [10]. The methods of administering mRNA vaccines vary. Moreover, the effectiveness of the vaccine is sometimes influenced by its route of administration [18]. Furthermore, mRNA vaccines are effective and safe [18]. The most common method of administering the mRNA vaccine is by injection [19]. Available mRNA vaccines such as Moderna and Pfizer, 1–2 days after administering, have related side effects such as [16], pain, redness, fatigue, fever, myalgias, and arthralgias.

Immunoinformatics is an aspect of bioinformatics that is involved with the computational analysis of biological and immunological data; it also involves the designing of vaccine candidates by predicting the best usable antigens, adjuvant, carriers, and epitopes for a vaccine. Immunoinformatics approaches have reduced the needed time and cost for vaccine development [1].

The aim of this study is to apply an integrated knowledge of bioinformatics, computational informatics, and modeling approaches towards the design of mRNA COVID-19 vaccine candidates. Specifically, this study is aimed at designing a multi-epitope mRNA vaccine based on the genome sequences of circulating SARS-CoV-2 variants in Africa. The human leucocyte antigen (HLA) allele's supertypes were also analyzed to ensure a wide population coverage for the designed vaccine. This is the scientific novelty of this research paper.

2. Materials and Methods

2.1. Study Design

The systematic workflow diagram for the mRNA vaccine design is shown in Figure 1. The design has 13 different sections, which are as follows: (1) retrieval of the whole genome sequences of SARS-CoV-2; (2) prediction and evaluation of CTL epitopes (See Table 1); (3) prediction and evaluation of HTL epitopes (See Table 1); (4) prediction and evaluation of LBL epitopes (See Table 1); (5) multiple sequence alignment (MSA) (See Figure 2); (6) docking between T-lymphocyte epitopes and MHC alleles (See Table 2); (7) population coverage prediction (See Table 3); (8) construction of mRNA vaccine (See Figure 3); (9) prediction of the toxicity, allergenicity, antigenicity, and physicochemical properties (See Table 4); (10) structure modeling, assessment, and validation; (11) conformational B-cell epitopes prediction; (12) molecular docking of vaccine with TLR receptor; (13) molecular dynamic simulation; (13) computational or *in silico* simulation.

2.2. Retrieval of SARS-CoV-2 Nucleotide Sequence

The data used for this research were retrieved from the Global Initiative for Sharing All Influenza Data (GISAID) database [20]. The data retrieved were targeted towards five African countries, namely Angola, Botswana, Mozambique, Lesotho, and Namibia. These data are the relevant genomic sequence data needed for the experiment and analysis. The data retrieved for these five African countries were based on criteria such as complete genome, low coverage exclusion, high coverage level, host, and date of submission. These criteria were considered for retrieving our sequences, and a total of 189 SARS-CoV-2 whole genome sequences deposited on the GISAID database between 1 December 2020 and 5 March 2021 were retrieved. Based on the criteria, three out of the five African countries in the study area—Angola (54), Botswana (26) and Mozambique (109)—had records for SARS-CoV-2 submitted to the GISAID database. There were no records found for the other two African countries (Lesotho and Namibia). The retrieved nucleotide sequences were annotated with the Wuhan reference sequence (accession number NC_045512.2) downloaded from NCBI database to establish the existing surface glycoprotein in the downloaded sequences.

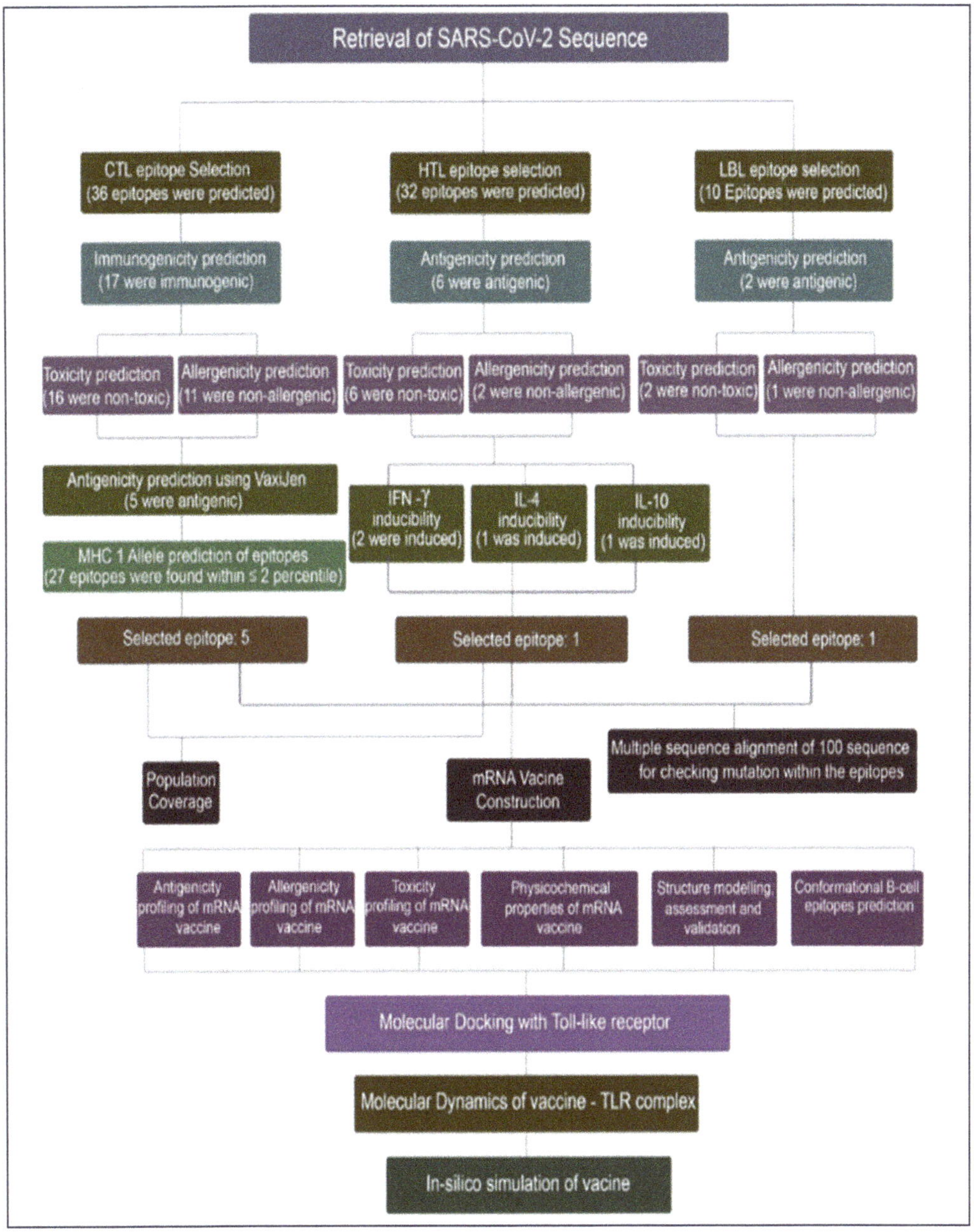

Figure 1. Systematic workflow diagram of the vaccine design for mRNA vaccine.

Table 1. Epitopes selected for vaccine construction.

Recognizing Cell	Epitope Sequence
Cytotoxic T lymphocyte	WTAGAAAYY HRHLRFLTL YQPYRVVVL YPQILLLVL SPRRARSVA
Helper T lymphocyte	ISFHVLTKLRLKCKL
B lymphocyte	WVFITTKTTKVGWKVSSEF

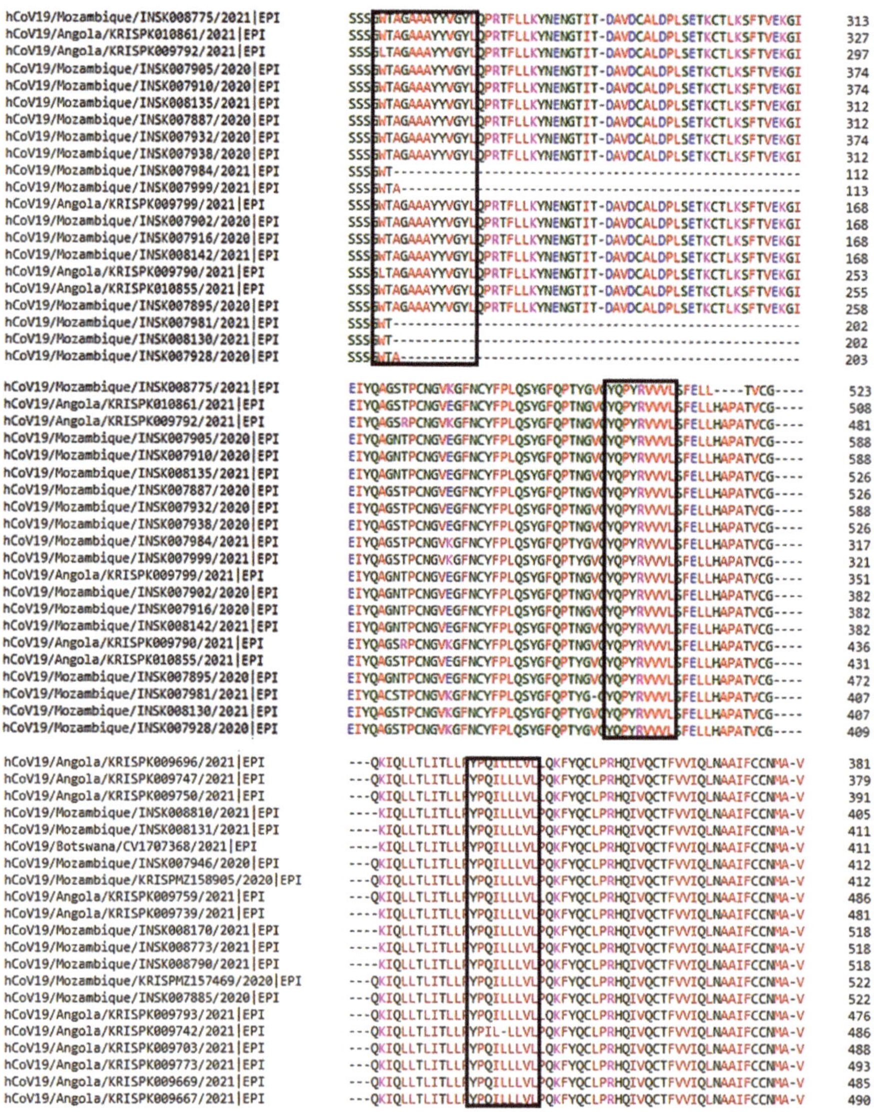

Figure 2. Multiple sequence alignment of spike glycoprotein sequences. The epitope sequences selected for vaccine design have been identified by boxes.

Table 2. Selected T-lymphocyte epitopes and their associated MHC alleles.

T-Lymphocyte Type	CTL Epitopes	MHC Binding ALLELES
CTL	WTAGAAAYY	HLA-A*29:02, HLA-A*30:02, HLA-B*15:01, HLA-B*46:01, HLA-B*58:01, HLA-B*53:01, HLA-B*35:01, HLA-C*07:01, HLA-C*03:03
	HRHLRFLTL	HLA-B*48:01, HLA-C*06:02, HLA-C*07:01
	YQPYRVVVL	HLA-A*02:06, HLA-A*32:01, HLA-B*48:01, HLA-B*46:01, HLA-C*06:02, HLA-C*07:01, HLA-C*03:03
	YPQILLLVL	HLA-B*51:01, HLA-B*53:01, HLA-B*35:01
	SPRRARSVA	HLA-B*51:01
HTL	ISFHVLTKLRLKCKL	HLA-DRB1*11:01

Table 3. IEDB server predicted results.

Population/Region	MHC Class Combined		
	Coverage Area	Average Hit	PC90
Central Africa	75.64%	2.21	0.41
East Africa	80.44%	2.33	0.51
North Africa	76.13%	2.29	0.42
South Africa	77.23%	2.23	0.44
West Africa	76.65%	2.22	0.43
Average	77.22	2.26	0.44
Standard deviation	1.7	0.05	0.04

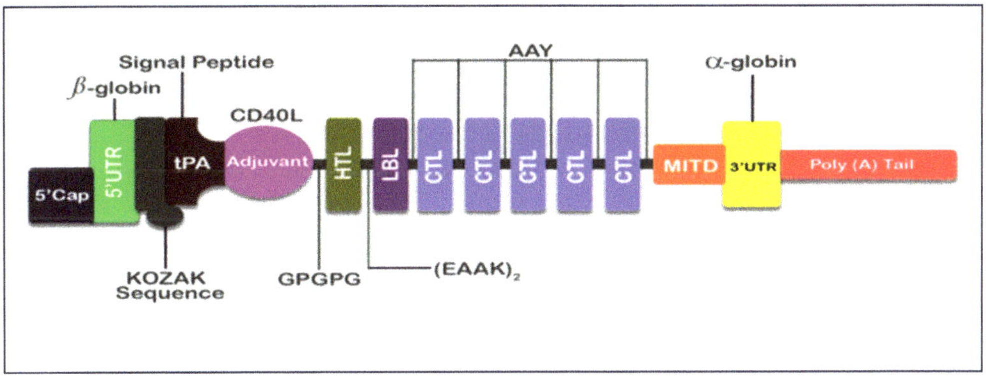

Figure 3. Scheme of the formulation of the mRNA vaccine against COVID-19.

2.3. Prediction and Evaluation of Cytotoxic T-Lymphocytes (CTL) Epitopes

Innate and adaptive defenses are some of the mechanisms utilized by host cells to neutralize viral replication [21]. One of such important adaptive defenses is the CD8+ cytotoxic T lymphocyte (CTL)'s response, which controls infection by a few mechanisms, along with the secretion of anti-viral cytokines and Fas/FasL-mediated lysis [21]. Many studies have featured CTL-mediated cytotoxicity, but the rate of fatality of virus-infected cells by CTL response in vivo is yet to be understood [21]. NetCTL v1.2 server was used to predict CTL epitopes [22]. A 9-mer CTL epitope was generated by the server for all the twelve major histocompatibility (MHC) class I supertypes available on its database, which includes A1, A2, A3, A24, A26, B7, B8, B27, B39, B44, B58, and B62, respectively, with the threshold set at 0.75. Epitopes predicted in the CTL supertypes for each country were inspected to determine overlapping epitopes. Epitopes having frequency overlap between 60 and above for each country were subjected for further analysis. Furthermore, the prediction of the CTL epitopes' immunogenicity was performed by using the IEDB analysis resources. This tool showed negative and positive values. The epitope with the positive value was selected for further studies. Furthermore, Toxinpred [23] and AllerTOP 2.0 servers [24] were utilized to determine the toxicity and allergenicity of the immunogenic epitopes, respectively. Those that met the criteria were subjected to antigenicity assessment through the VaxiJen server [25]. Epitopes with value ≥ 0.5 were classified as antigenic. The antigenic epitopes were subsequently predicted for their MHC class I allelic partners by adopting the IEDB consensus algorithm, with human selected as the host.

Table 4. Allergenic, antigenic, physicochemical assessments, and toxicity of the vaccine construct.

Features	Result	Assessment
Number of amino acids	1995	Suitable
Molecular weight	223.1 kDa	Average
Theoretical pI	8.69	Slightly basic
Total number of negatively charged residues (Asp + Glu)	196	-
Total number of positively charged residues (Arg + Lys)	223	-
Total number of atoms	312178	-
Chemical formula	$C_{9908}H_{15532}N_{2774}O_{2906}S_{97}$	-
Instability index (II)	48.78	Unstable
Aliphatic index	82.13	Thermostable
Grand average of hydropathicity (GRAVY)	−0.296	Hydrophilic
Antigenicity	0.5059 (VaxiJen) 0.7334 (ANTIGENPro)	Antigenic Antigenic
Allergenicity	Probable non-allergen (AllerTOP 2.0 and AllergenFP)	Non-allergen
Toxicity	Non-toxin (ToxinPred)	Non-toxic

2.4. Prediction and Evaluation of Helper T-Lymphocytes (HTL) Epitopes

The prediction of HTL epitopes is part of immunoinformatics approaches for vaccine development [1]. HTL aid the activity of other immune cells by binding to specific HTL epitopes using MHC class II molecules. Prospective HTL epitopes were predicted using IEDB MHC-II consensus algorithm [26]. Mouse was selected as the host species, and the epitopes were filtered using a percentile rank less than 0.25. The VaxiJen server was then used to compute the epitopes' antigenicity [25]. Epitopes that were non-toxic and epitopes that were non-allergenic were selected by the ToxinPred and AllerTOP 2.0 servers [24]. After utilizing the IFNepitope [26], IL4pred [26], and IL10pred [27] servers, the remaining epitopes were assessed for interferon- (IFN-) inducibility and interleukin-4 (IL-4) and interleukin-10 (IL-10) inducibility. Three antigenic and cytokine-producing epitopes were shortlisted for the construction of the vaccine.

2.5. Projection and Evaluation of Linear B-Cells Lymphocytes (LBL) Epitopes

B cells form a core component of the adaptive immune system. One of their characteristics is the ability to identify and grant lasting protection against pathogens. They perform these functions by expressing proteins and producing antibodies on B cells' surfaces. They identify antigens and bind to a section of an antigen in a very selective fashion. The knowledge of the identification procedure is adapted in vaccine design to develop more effective and long-lasting vaccines against pathogens [28]. IBCE-EL server [29] was utilized to predict the LBL epitopes in this study. Only epitopes that were positive were selected for the prediction by the server for probable LBL and occurrence of 25 times or more across the three countries. An in-house-developed R-script program was used to perform further analysis on the selected epitopes. The antigenicity of possible LBL epitopes was tested by applying the VaxiJen server [25]. The evaluation of the toxicity and the antigenic epitopes' allergenicity was performed by using ToxinPred server and the AllerTOP 2.0 server.

2.6. Multiple Sequence Alignment of SARS-CoV-2 Nucleotide Sequence

Unlike DNA viruses, SARS-CoV-2 (an RNA virus) has great propensity of undergoing repetitive mutation [30]. To authenticate the probability of the selected region of our epitopes not having undergone mutation, 100 randomly selected spike glycoprotein sequences from the study area were fed into Clustal Omega software for analysis [31].

2.7. Designing of mRNA Vaccine

The methods adopted in the research conducted by Ahammad and colleagues [32] were applied to construct an mRNA against SARS-CoV-2 virus. The epitopes used for the vaccine construct were selected based on criteria such as the antigenicity, non-toxicity, non-allergenicity, and cytokine-inducing properties (HTL only). Linkers were used to link CTL, HTL, and LBL epitopes. These epitopes were selected to construct the mRNA vaccine. HTL and LBL epitopes were linked by using (EAAK)2. (EAAK)2 was used to space intra-LBL epitopes. AAY linker was used to link LBL and CTL epitopes. AAY linkers were also used to combine intra-CTL epitopes. Adjuvants play important roles in the design of effective vaccines for increased immunogenicity [33]. In this study, CD40 ligand (CD40L), a co-stimulatory molecule, which functions as an agonist to the human immune receptor by interacting with the antigen presenting cells [33], was utilized. CD40L sequence (UniProt ID: P29965) was retrieved from the UniProt database and putatively linked together with the HTL epitopes using the GPGPG linkers.

Furthermore, MHC I-targeting domain (MITD) was used to direct CTL epitopes to MHC I compartment of the endoplasmic reticulum and tissue plasminogen activator (tPA) [32]. The sequences of tPA (UniProt ID: P00750) and MITD (UniProt ID: Q8WV92) were retrieved from the Uniprot database. It is evident that instability has been a major problem encountered in the production of many mRNA-based therapeutics. Therefore, it is pertinent to include elements that naturally find expressions in eukaryotic mRNA, which is very important for mRNA stabilization [32]. We integrated the sequences of 5′ cap and poly (A) tail of our vaccine design with the sequence of 5′ and 3′ untranslated regions (UTRs) flanking its protein encoding ORF for mRNA molecules' stability, accessibility of ribosomes, and interactions with the translation machinery [34]. It has been established that the length of poly (A) tail is significant in the translation efficiency of mRNA [32]. The length of the poly (A) tail utilized in our study was from 120–150 base long. This range has been considered as the optimal length according to existing studies. For instance, according to Ahammad and colleagues [32], Poly (A) tail functions effectively alongside 5′ m7G cap sequences within this range.

2.8. Prediction of Class I and Class II Epitopes' Population Coverage

One of the important features of an epitope is its ability to closely bind to an HLA molecule. Human population coverage is another significant feature that is very crucial in selecting epitopes for vaccine design [35]. Population coverage is the expected percentage of individuals in a population having the likelihood of inducing an immune response to not less than one T-cell epitope in a set [36].

We utilized the IEDB population coverage tool [37] to determine the population of the screened epitopes and their MHC alleles. This tool was used to compute the distribution of persons predicted to respond to epitopes that have been selected with known HLA background [37]. The genotypic frequencies of HLA were also computed. The T-cell epitopes were queried based on certain variables such as area, ethnicity, and country [37].

The total global population was selected, and this was followed by the selection of the population associated with the subcontinents.

2.9. Vaccine Construct's Predicted Antigenicity, Toxicity, Physicochemical Properties, and Allergenicity

One of the significant features considered essential in vaccine development is the propensity of designated vaccine candidates to possess antigenic property [33] and the capability of inducing immune response leading to the formation of B and T lymphocytes after administration. VaxiJen 2.0 and ANTIGENpro servers [38] were used for predicting the antigenicity of the final vaccine construct [25].

ANTIGENpro server is dependent on sequence, alignment free, and independent of pathogens in its predictive operations of protein antigenicity [38]. AllerTOP 2.0 [24] and AllergenFP 1.0 servers [39] were used to check the allergenicity of the final vaccine construct to determine if a vaccine construct is allergen or non-allergen.

AllerTOP's prediction was based on ACC transformation and E-descriptors [24], while AllergenFP's prediction was based on the classification of allergens and non-allergens datasets into five E-descriptors and then using auto-cross covariance (ACC) to transform its strings into uniform vectors [39]. ToxinPred server [23] was used to predict the vaccine construct's toxicity. The operation of this server is dependent on the support vector machine (SVM) model. This helps in toxicity and non-toxicity classification [23]. An online web server, ExPASy ProtParam [40], was used to examine the physicochemical properties of the vaccine construct.

2.10. Prediction of the Secondary Structure of the Vaccine Construct

Protein secondary structure helps to determine the protein folding orientation [41]. SOPMA online server [42] was applied to assess the vaccine construct's secondary structure [42]. The vaccine construct's secondary structure prediction by SOPMA yielded an accuracy of 69.5%. Three-state structure (B-sheet, coil, and a-helix,) description was produced [43].

2.11. 3D Structural Modeling, Assessment, and Validation

The vaccine construct's 3D structure was evaluated using the Phyre2 server [44]. Phyre2 built 3D models by utilizing and applying remote homology detection techniques that have advanced capabilities. Phyre2 also assists with the prediction of binding sites for ligands and the analysis of amino acids variants' effects [44].

Despite making use of advanced template-based methods for modeling the 3D structure of an unknown protein, inaccuracies may still exist within the model structure [45]. This is expected especially when the template proteins do not share enough corresponding homology with target proteins and thus may cause a deviation from the overall target structure due to differences in their sequences [45].

GalaxyRefine web server [46] was used to refine the vaccine construct's 3D structure built by the Phyre2 server. The operations of the GalaxyRefine web server are as follows: rebuilding of side chains, repacking side chains, and the relaxation of structure by the molecular dynamics' simulation. It has been proven that the web server approach has produced the best performance according to the assessment of CASP10 [46].

ProSA-web server [47] was used to validate the refined 3D model of the vaccine construct. This process helps to check the constructed 3D models of the vaccine structure for any potential errors. Some of the other applications of this web server include identification of errors in experimentally determined structures, protein engineering, and computation of the total score for specific input structure [47]. A structure is said to contain errors if its scores fall outside the score range of a native protein.

2.12. Prediction of Conformational B-Cell Epitope

ElliPro server was used for predicting [48] the conformational B-cell epitopes of the final vaccine 3D model structure. As a web-based tool, ElliPro can assist with predicting epitope antibodies inherent in a sequence's protein antigens. It performs implementation of existing methods for protein structure as an ellipsoid. It computes protein residues' protrusion indexes that lie outside ellipsoid [48].

2.13. Molecular Docking of Vaccine with TLR Receptor

Molecular docking is a significant bioinformatics technique widely adopted for the prediction of the binding affinity and orientation between a receptor and its ligand [32]. Our study examined the possible binding affinity between the vaccine's construct and its receptor. ClusPro 2.0 server was used to conduct molecular docking [49]. There was molecular docking between the refined 3D model of the final vaccine construct and an immune toll-like receptor (TLR 4, PDB ID: 4G8A) obtained from Protein Data Bank [50].

2.14. Molecular Dynamics Simulations

The molecular dynamics was simulated by utilizing the iMODs server [51] to assess the physical movements and stability of the vaccine's TLR4 docked complex. The iMOD server performs evaluation of protein stability by applying the normal mode analysis (NMA) towards the computation of the interior coordinates. The eigenvalue, elastic network model, covariance matrix, main-chain deformability plot, and B-factor values were used to depict the stability of the protein [51].

2.15. Immune Response Simulation

An assessment of the immunogenicity of all the predicted conjugate vaccine peptides and the attributes of the immune response was conducted by utilizing the C-ImmSim online server [52]. Associated immune interactions and epitopes are predicted by the server after utilizing a machine-learning-based technique. Three anatomical compartments are automatically simulated. These include: (i) bone, in a situation where there is stimulation of the hematopoietic stem cells, accompanied by the production of the myeloid cells, (ii) thymus, and (iii) lymphatic organ. The administering of three injections having the designed peptide vaccine was simulated at intervals of four weeks, i.e., day 0, day 28, and day 56. This prime–booster–booster method was adopted at an interval of 4 weeks to accomplish a lasting protective immune response.

Default parameter settings indicate the positioning at 1, 84, and 168. This implies that each time step has an interval of 8 h. Time step 1 depicts the administration of injection at time zero. There were administrations of three injections at intervals of four weeks. However, there were administrations of eight injections at four-week intervals to cause stimulation of repetitive reactions to the antigen. This scenario subjects the T-cell memory to continuous examination. The plot analysis provided a platform to graphically interpret the Simpson index [53].

3. Results

3.1. Prediction and Evaluation of CTL Epitopes

To obtain the best and choicest epitopes suitable for our vaccine construction from among the large number of CTL epitopes predicted across the three countries, an overlapping procedure was employed to avoid repetition of predicted epitopes.

There were 36 unique CTL epitopes of 12 MHC class I supertypes, with frequencies above 60 times predicted by the NetCTL v1.2 server [22]. Seventeen of the epitopes predicted, when subjected to evaluation, were positive for immunogenicity. IEDB class I immunogenicity tool was used for this evaluation. ToxinPred [23] and AllerTOP 2.0 servers [24] predicted that of the 17 immunogenic epitopes, 16 were non-toxic and 11 were non-allergenic. The 11 epitopes passed the three stages of prediction (i.e., immunogenicity, toxicity and allergenicity), and they were further assessed for antigenicity by utilizing the VaxiJen server [25]. The results predicted revealed that 5 out of the 11 epitopes were successful in scaling above the antigenicity threshold of 0.5, which were then selected for the vaccine construction (Table 1).

3.2. Prediction and Evaluation of HTL Epitopes

Predicted results from the IEDB MHC-II allele tool revealed that 32 distinctive epitopes with frequency of occurrence above 25 were predicted, and these spanned the three countries. Six of these epitopes assumed the VaxiJen threshold (≥0.5) for antigenicity. Results from the ToxinPred server revealed that the six antigenic epitopes were non-toxic [23]. AllerTOP 2.0 server predicted two epitopes to be non-allergenic [24].

Following a careful examination of the interferon-γ (IFN-γ), interleukin-4 (IL-4), and interleukin-10 (IL-10) inducibility using the IFNepitopes, IL4pred, and ILpred servers, respectively, there was only one epitope that fulfilled all the criteria (see Table 1).

3.3. Assessment and Prediction of LBL Epitopes

A rigorous assessment of the translated nucleotide sequences for the possible existence of B-cells epitopes was conducted using the BCpred server [54]. Epitopes that had percentile ranks higher than 0.9 were selected. Further evaluation was conducted for the possible existence of linear B cells by utilizing the IBCE-EL server. A total of 10 probable LBL epitopes were predicted. Two of the ten epitopes predicted for possible presence of LBL epitopes were proven to be antigenic (they met the criteria of being antigenic (≥0.5)) on the VaxiJen server. The two epitopes were predicted to be non-toxic by ToxinPred while one epitope was predicted to be non-allergic by the AllerTOP 2.0 server (see Table 1).

3.4. Multiple Sequence Alignment of SARS-CoV-2 Sequences

Multiple sequence alignment (MSA) of 100 spike glycoproteins of coronavirus was conducted by using Clustal Omega. Interestingly, no mutation was observed within the seven epitopes selected amongst the 100 SARS-CoV-2 spike glycoproteins (see Figure 2).

3.5. Designing of mRNA Vaccine

Analysis was performed on all seven selected epitopes (WTAGAAAYY, HRHLRFLTL, YQPYRVVVL, YPQILLLVL, SPRRARSVA, ISFHVLTKLRLKCKL, and WVFITTKTTKVG-WKVSSEF) to examine their interactions and their subsequent potentials for possible development of an mRNA vaccine construct. The 5′ Cap, 5′ UTR, Kozak sequence, and tPA (signal peptide) were merged with the adjuvant (CD40 ligand) and then linked to the HTL epitope by the assistance of the GPGPG linkers from the beginning of the N-terminal of the mRNA vaccine.

Epitopes were bonded together depending on their degree of interactional compatibility in a sequential manner with EAAKEAAK (HTL to LBL) and AAY (LBL to CTL and intra CTL) linkers, respectively. The AAY linkers were used to connect the C-terminal end to the CTL epitopes (see Figure 3).

3.6. Population Coverage

It is very important during epitope-based vaccine design to select epitopes that have diverse HLA binding specificities and to ensure a broad population coverage. This is particularly paramount because an epitope can effectively evoke an immunological response in individuals for cases that only find expression for a particular MHC molecule that forms a complex with it [35].

Analysis on population coverage was performed with MHC class I and MHC class II epitopes and with associated HLA alleles within five African geographic regions found in the IEDB database. The result obtained revealed that all epitopes combined class I and II to have an average coverage of 77.22% (Table 3). Maximum coverage (80.44%) was found in East Africa, followed by South Africa (77.23%). West Africa and North Africa have a coverage of 76.65% and 76.13%, respectively, while Central Africa has the minimum coverage (75.64%), (See Table 3 and Figure 4).

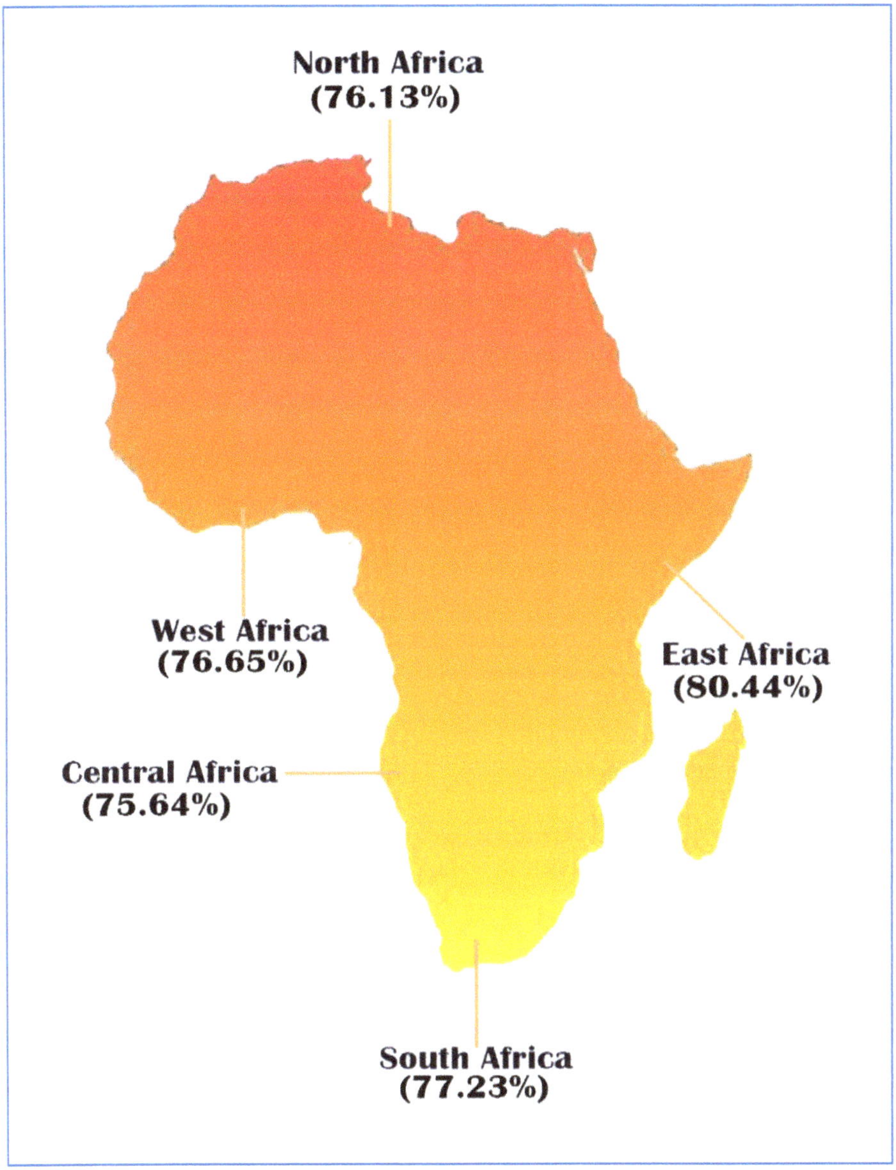

Figure 4. Population coverage of utilized T-lymphocyte epitopes. The highest region and the lowest region of coverage are, respectively, East Africa (80.44%) and Central Africa (75.64%).

3.7. Prediction of Allergenicity, Antigenicity, Physicochemical Properties, and Toxicity of the Vaccine Construct

The results of the allergenicity, antigenicity, physicochemical properties, and toxicity analyses are depicted in Table 4. The mRNA vaccine candidate's antigenicity was examined using the VaxiJen and AntigenPro servers. Both servers revealed that the vaccine construct has high antigenicity scores of 0.5059 and 0.7334, respectively. These results indicate the vaccine's ability to induce a robust immune response. Afterward, the vaccine's allergenicity was evaluated by the Allertop server and the AllergenFP 1.0 server to determine its nature.

The results revealed that the vaccine construct was non-allergenic and devoid of toxicity. Subsequently, the ExPaSy Protparam server [40] was applied to determine the vaccine's physicochemical behavior. The results from the assessment of the vaccine construct's physicochemical properties revealed that it consists of 1995 amino acids having a molecular weight of 223.1 kDalton. Results showed that the vaccine construct has a theoretical isoelectric point (pI) of 8.69. This indicates that it is slightly basic in nature.

The PH is defined by the theoretical pI in cases where the total charge of the peptide is zero and computed based on the pK of all amino acids' resident in the peptides [54]. In addition, an index of 48.78 depicted an unstable vaccine construct. An instability score of 40 or less is a steady and a stable score [40]. Furthermore, the vaccine was characterized with a globular structure with 82.13 as its aliphatic index score. The vaccine construct has an estimated existence of 30 h in vitro for analysis in mammalian reticulocytes, more than >20 h in vivo in yeast, and >10 h in vivo for Escherichia coli, which connotes the stability of the vaccine's in vitro and vivo phases. Furthermore, the coefficient of extinction was computed and was 273,135 $M^{-1}cm^{-1}$, with absorption values (0.1%) (g/1) 1.224 consisting of all cysteine pairs under aqueous conditions at 280 nm. The computed grand average of hydropathicity (GRAVY) score was −0.296. This shows that the vaccine construct is hydrophilic. It can foster interaction with water and blood and easily identify targets. It is clear from physicochemical analysis results that the vaccine construct's contents meet the necessary criteria for vaccine formulation (See Table 4).

3.8. Secondary Structure Prediction

The vaccine's secondary structure was analyzed using the SOPMA online server [42]. The results obtained revealed a stabilized structure for the vaccine construct with 36.44% alpha-helix, 20.45% extended strands, and 33.38% random coils. This result also showed that the vaccine construct's secondary structure is of good flexibility, stability, and globular conformation (see Figure 5A,B).

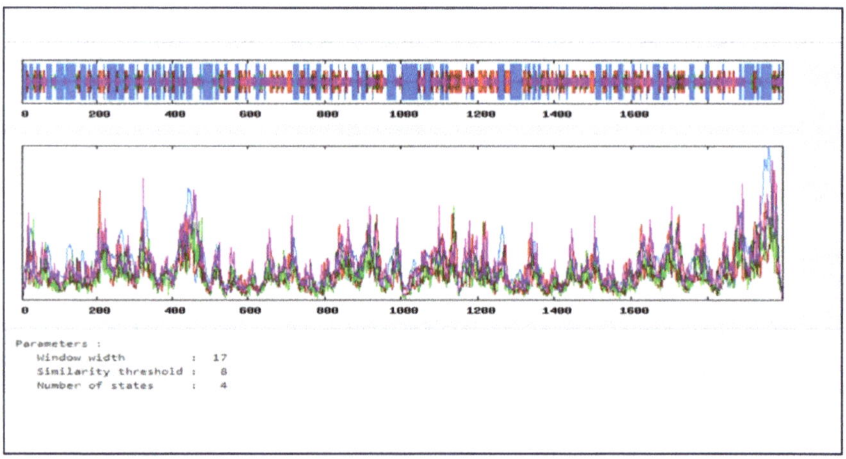

(A)

Figure 5. *Cont.*

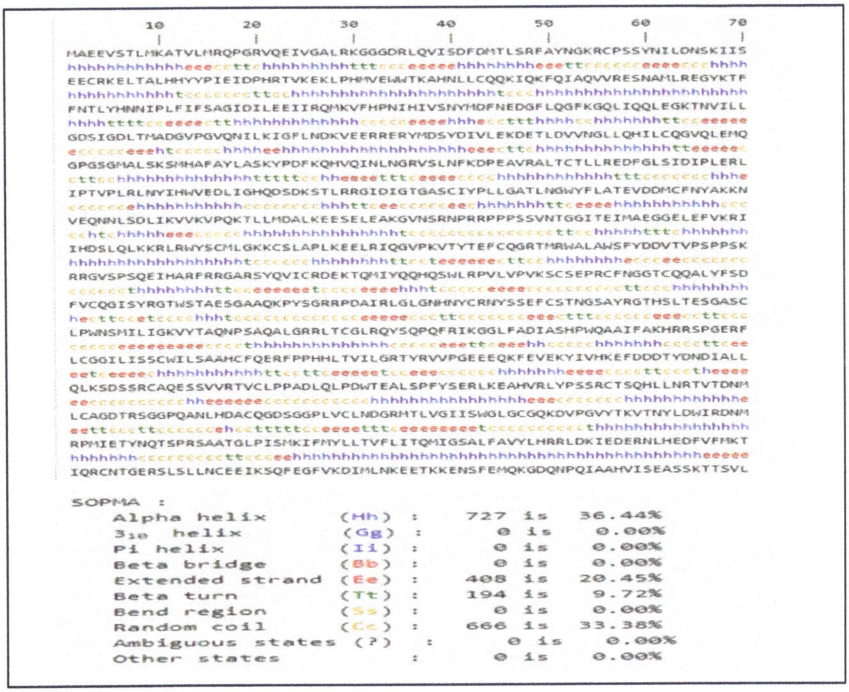

(**B**)

Figure 5. (**A**) Vaccine construct's secondary structure. (**B**) Designed vaccine's secondary structural analysis, revealing the fluctuations of its structural atoms, within a minimal range, depicting the stability of its structure.

3.9. Three-Dimensional Structural Modeling, Refining, and Validation

We utilized phyre2 [44] to complete the buildup of the 3D model structure of the final vaccine construct. The modelled structure based on template c6b92A was predicted by phyre2 to be the best template. This was based on several constructs of the 3D structural model of the vaccine (Figure 6A). In total, 13% (250) of the amino acids residues in the construct were modelled with 100% confidence in a single high scoring template. The GalaxyRefine server [46] was utilized to accomplish the refinement of the 3D structural model of the vaccine construct. The server predicted five (5) refined models. Of these models, model 2 (see Figure 6B) was selected as our final vaccine model because of its quality scores (see Table 5). A yellow highlight depicts predicted B-cell epitopes, which reflects a good surface accessibility. The measurement of similarities between two protein structures is depicted by the global distance test high accuracy (GDT-HA) score [46]. A value of 0.9717 depicts the GDT-HA score. This is a high value, which indicates that there is a high level of similarity between the two models. The root-mean-square deviation (RMSD) score computes the distance between atoms. A low RMSD value depicts a better level of stability. Acceptable RMSD score ranges from 0 to 1.2 [46]. The RMSD score of this model is 0.364. This depicts a good level of protein structure stability. Our vaccine model's MolProbity score is 1.733 (a value lower than that of the initial model). This shows that critical errors in the 3D model have been reduced. The clash score depicts all unfavorable numbers of overlapping atoms. The model's clash score was reduced from 48 to 10.6 (an evidence of increased stability to high level). The surface areas of energetically favored regions are depicted by the Ramachandran plot score. An acceptable Ramachandran plot score is that which is greater than 85% [46].

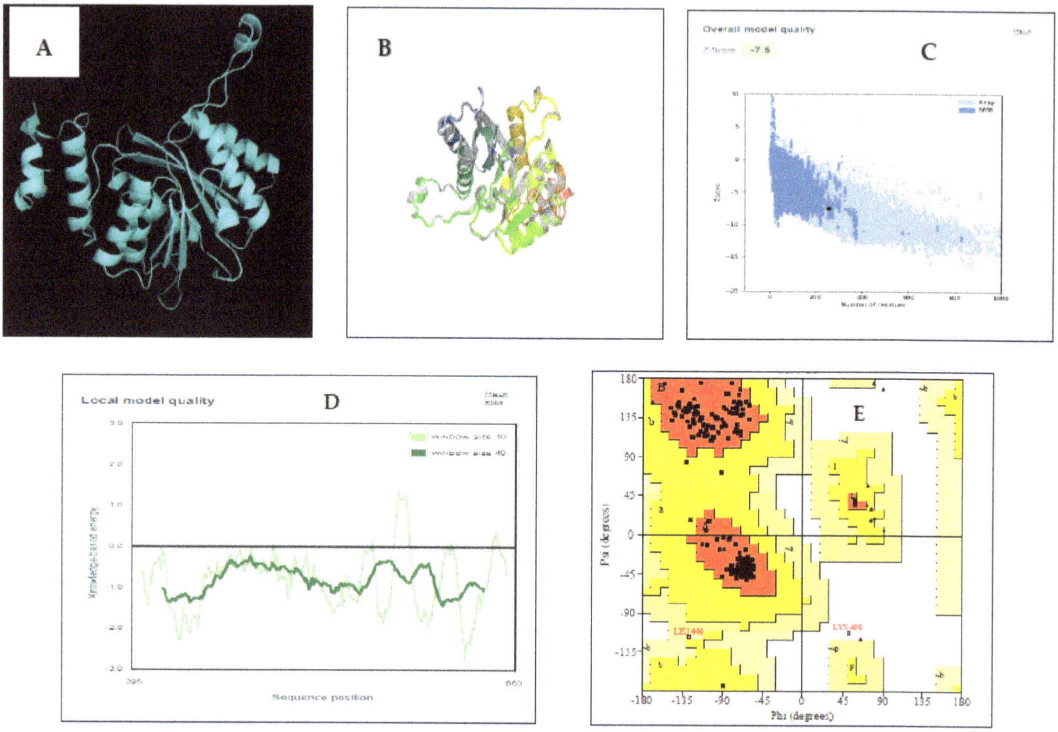

Figure 6. (**A**)Vaccine construct's 3D crystal structure, (**B**) refined vaccine 3D structure model, (**C**) assessment of the Ramachandran plot for the multi-epitope vaccine construct, and (**D**) a ProSA-web validation of the vaccine 3D structure. The Z-score of the refined model is −7.5, which lies inside the score range (**E**) residue's score plot by ProSA-web to verify the local model quality.

Table 5. Predictions of the models' quality scores by GalaxyRefine.

Model	GDT-HA	RMSD	MolProbity	Clash Score	Poor Rotamers	Rama Favored
Initial	1.0000	0.000	2.457	48.0	0.0	95.7
Model 1	0.9707	0.353	1.527	10.2	0.4	98.4
Model 2	0.9717	0.364	1.733	10.6	1.8	98.8
Model 3	0.9658	0.371	1.570	11.3	0.4	98.8
Model 4	0.9707	0.365	1.594	12.1	0.0	98.4
Model 5	0.9697	0.359	1.545	10.6	0.4	98.4

There was an improvement in the Ramachandran plot score from 95.7% to 98.8% after refinement. To validate the overall refined prototype vaccine quality, ProSA-web was utilized. A Z-score of −7.5 was predicted by ProSA (see Figure 6C), which depicts a good quality model.

The local quality of the model is also checked by ProSA. Figure 6D shows the plotted graph of the residue scores. Negative values depict that there is an absence of erroneous parts in the structure of the model. The results of the ProSA-web server presented a Ramachandran plot analysis score of 97.8%, which is like that obtained by GalaxyRefine, which can be found in Figure 6E.

3.10. Conformational B-Cell Epitopes Prediction

The conformational B-cell epitope of the 3D-refined model was predicted using the Ellipro server [48]. The server predicted eight new conformational B-cell epitopes, which consisted of 110 residues having scores between 0.589 and 0.856. Figure 7 shows the detailed information of the eight epitopes and the 3D model.

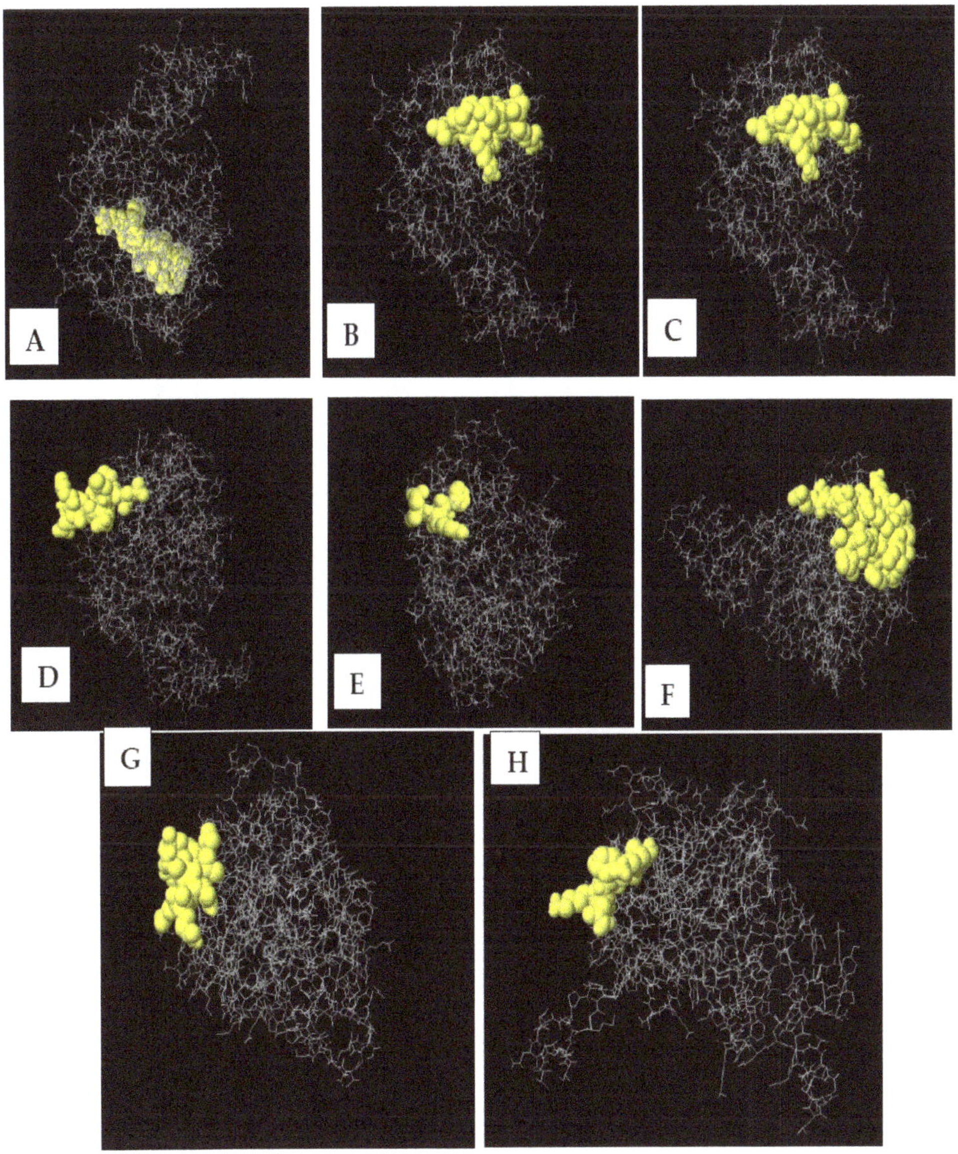

Figure 7. The 3D model of the 8 predicted conformational B-cell epitopes. The yellow regions are the conformational B-cell epitopes, while the grey regions are the residue remnant. (**A**) 18 residues with 0.856 score. (**B**) 32 residues with 0.825 score. (**C**) 12 residues with 0.76 score. (**D**) 11 residues with 0.686 score. (**E**) 4 residues with 0.681 score. (**F**) 19 residues with 0.624 score. (**G**) 8 residues with 0.613 score. (**H**) 6 residues with 0.589 score.

3.11. Molecular Docking of Vaccine with TLR Receptor

The evaluation of interactions between a ligand molecule and a receptor molecule was conducted through molecular docking. This was carried out to verify the binding affinity of the docked complex. TLR4 was used as the immune receptor for this study to conduct the molecular docking. Toll-like receptor 4 is a very significant human protein that helps with immune response and the recognition of pathogens.

Molecular docking was performed by utilizing the ClusPro 2.0 server [49]. Molecular docking was conducted between the final refined 3D vaccine and the TLR4 (PDB ID: 4G8A) immune receptor. Ten different models were produced from the docking process. These models were characterized by low-energy docked structures [49]. A selection was made of the lowest-energy docked model. This result indicates that the vaccine model has good binding affinity and fully occupies the receptor (see Figure 8).

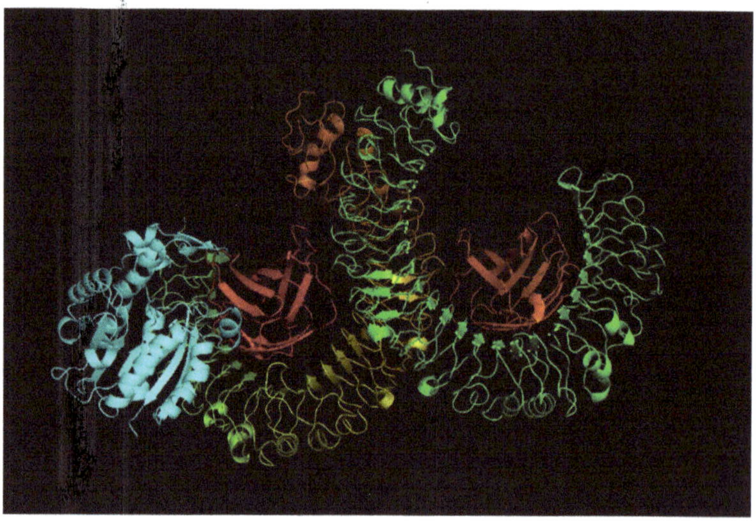

Figure 8. The docked complex of the vaccine model and the TLR4 immune receptor. The vaccine protein is shown in blue, while the rest of the residues are the TLR4 receptor.

3.12. Molecular Dynamics Simulations

The i-Mode server was used to perform analysis of the molecular interaction of the vaccine target with the target TLR-4 receptor. The evaluation of the vaccine–TLR-4 docked complex was carried out by using NMA. The i-Mode suite was used for simulating access to the internal coordinates of the complexity of the system. An examination of the trajectory of the system was performed to determine the deformability. Figure 9 depicts the vaccine–TLR4 complex's molecular dynamics simulation results showing the spin prediction of the ligand–receptor interaction and other results. The results of the complex trajectory revealed a minimal deformation in the coordinates within the range 0 to 1°A. This depicts that the vaccine has a steady binding with minimal deformation (see Figure 9b). Traces of some atomic fluctuations were observed by NMR in the system trajectory of the TLR-4 and the vaccine. Figure 9c shows the computed eigen score of 1.843800×10^{-05}. Furthermore, the covariance matrix analysis revealed the vaccine–TLR-4's atomic pairs. The analysis depicted the correlated segments in red color, non-correlated segments in blue color, and the uncorrelated segments in white color. Figure 9d shows the integration of the TLR-4 protein residues with the construct of the vaccine and the changes in the TLR-4 binding groove. The model's elastic network revealed pairs of atomic coordinates through distance-based spring analysis. Each dot in the network plot represents a spring and is colored based on the complex's stiffness in relation with corresponding atomic pairs.

Grey-colored spring models depict the level of compactness and stability of the binding complex system (see Figure 9e). These important results reveal the stable binding and complex rigidity of the vaccine coupled with some atomic fluctuations, characterized with a low deformation index.

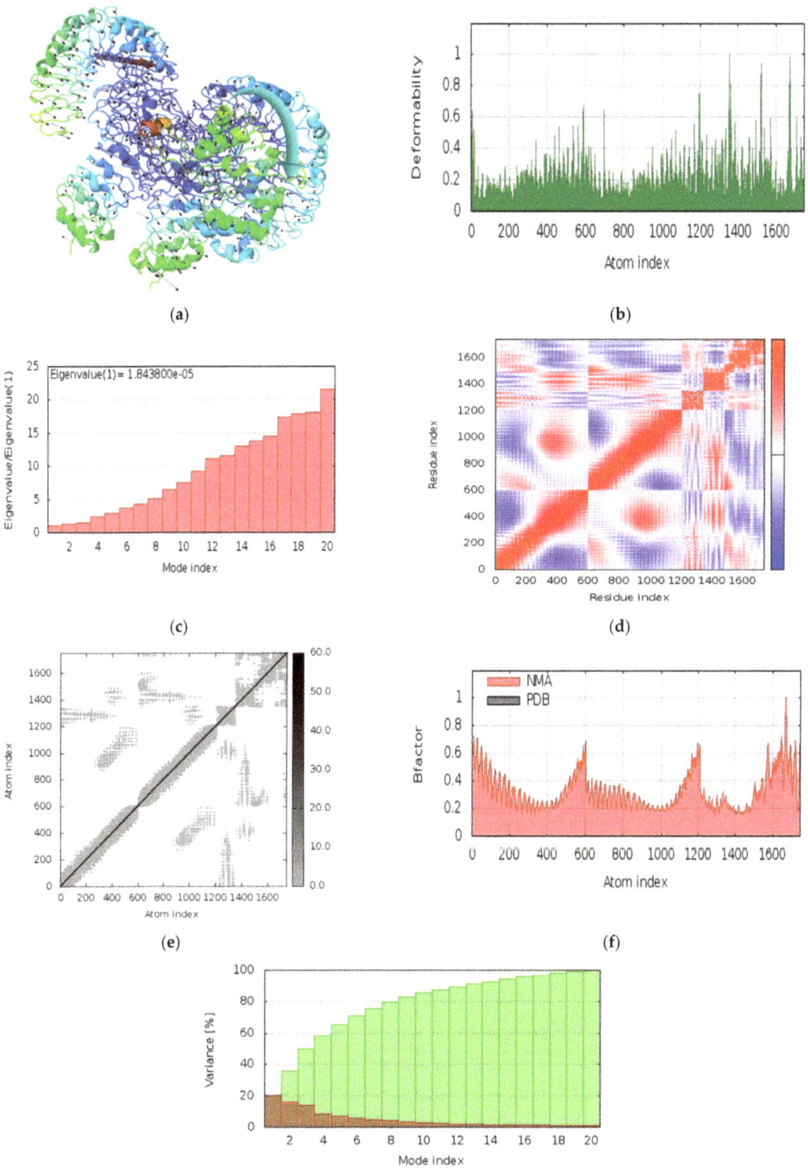

Figure 9. Vaccine–TLR4 complex molecular dynamics simulation, showing (**a**) spin prediction of the ligand–receptor interaction; (**b**) deformability, (**c**) eigenvalue; (**d**) covariance matrix depicting the coupling between pairs of residues (red), uncorrelated (white), or anti-correlated (blue) motions; (**e**) elastic network analysis defining which pairs of atoms are connected by springs; (**f**) B-factor; and (**g**) variance.

3.13. Immune Response Simulation

We assessed the vaccine construct's immune response elicitation by adopting an *in silico* immune simulation technique for 100 steps of simulation. This method is used for the analysis of the capability of the vaccine construct's immune response elicitation and antigens, amongst others. B cells, T cells, and memory cells that generate immune responses that fight viral infections were assessed by exploring the vaccine candidate. Results obtained from our *in silico* experiments revealed the potency of our designed vaccine candidate. Results revealed that primary and secondary immune responses were elicited through very important players such as the T-cell populations (helper T cells and cytotoxic T cells) and sustainable memory cells (see Figure 11). Figure 11 depicts the induced immune cells as illustrated by the mRNA vaccine.

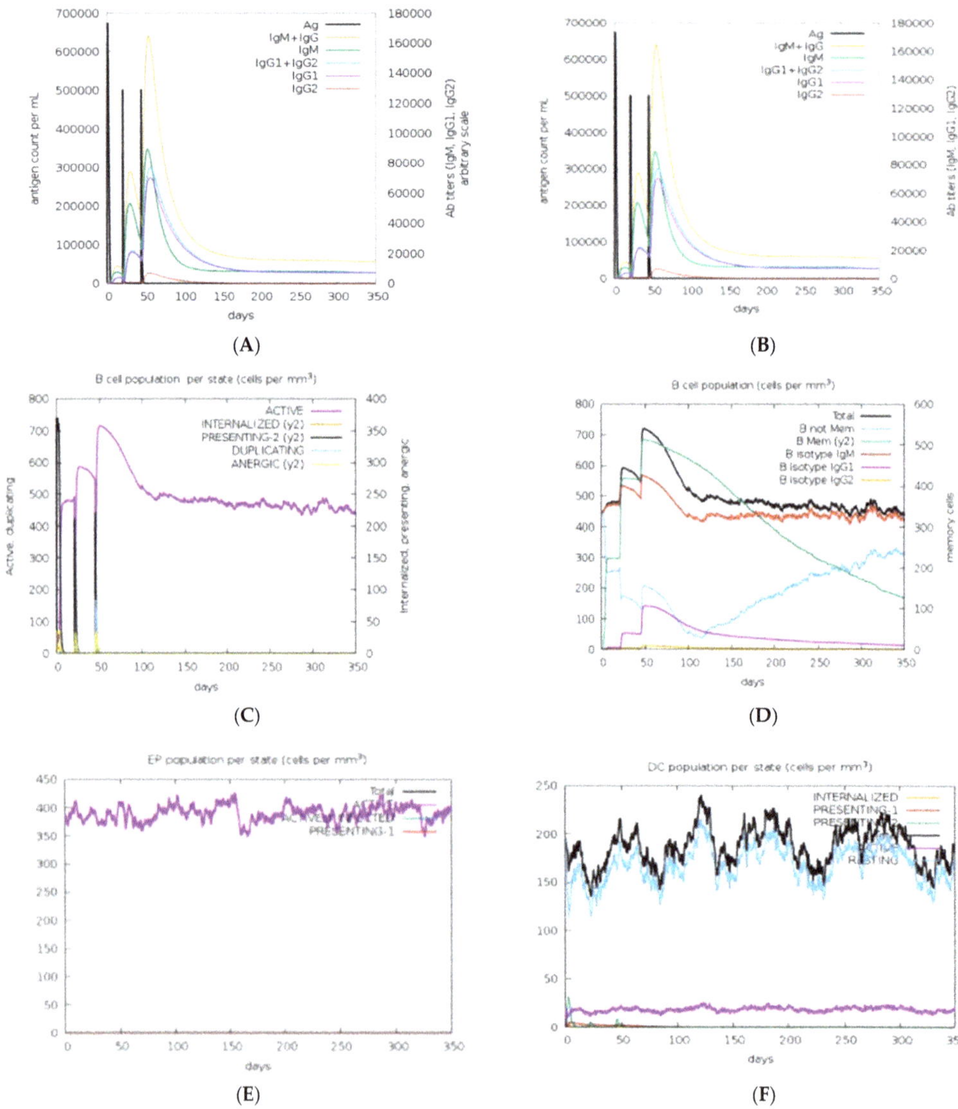

Figure 10. *Cont.*

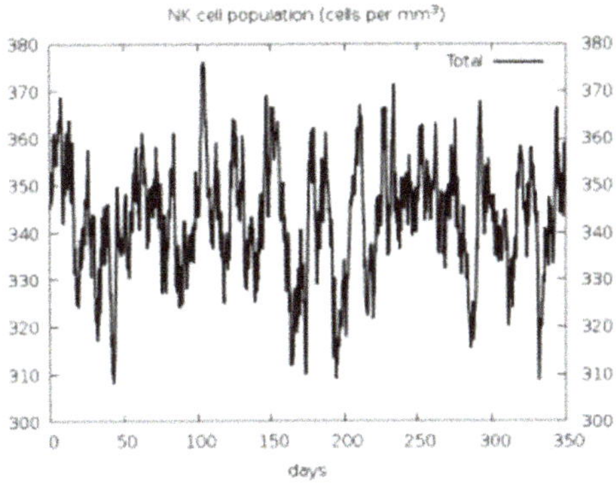

(**G**)

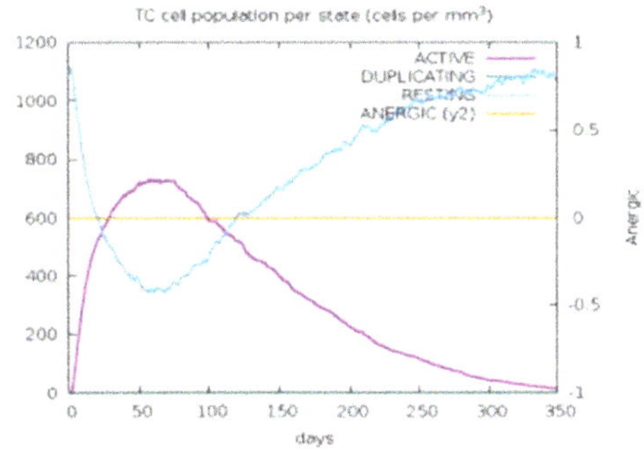

(**H**)

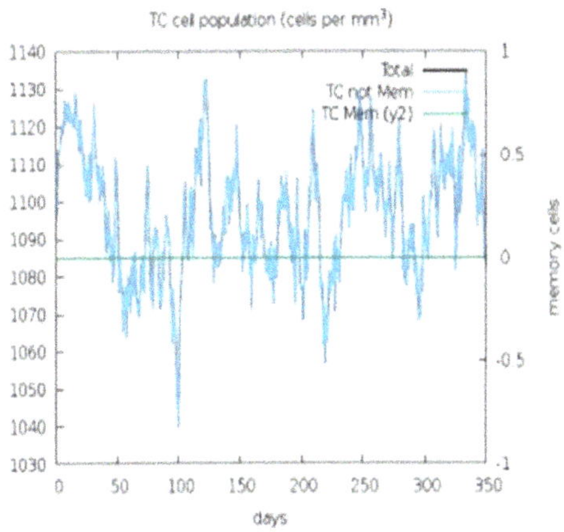

(**I**)

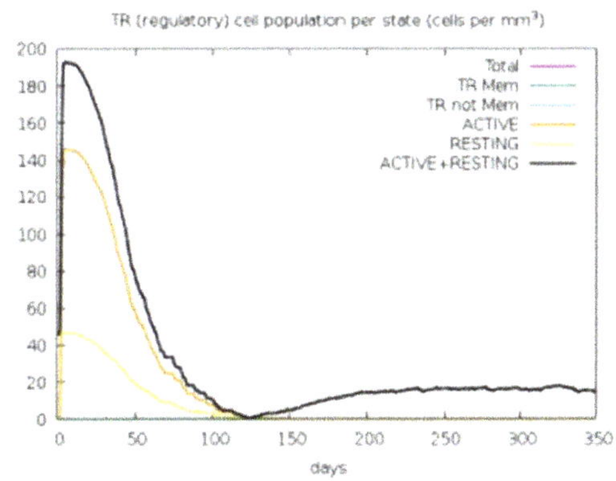

(**J**)

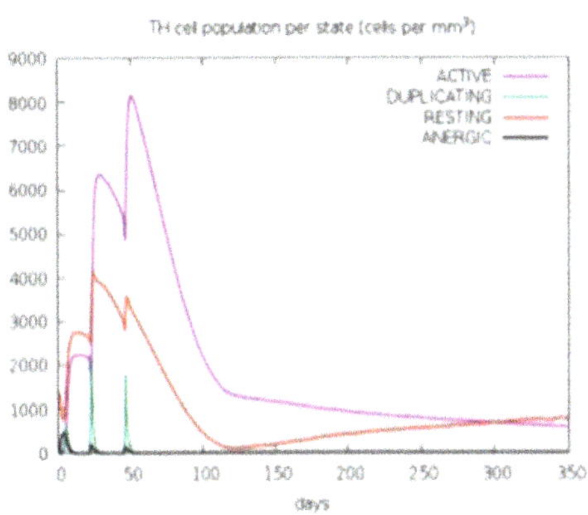

(**K**)

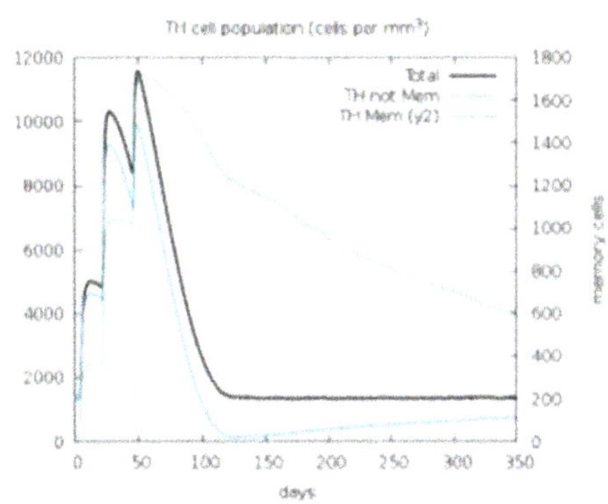

(**L**)

Figure 11. *Cont.*

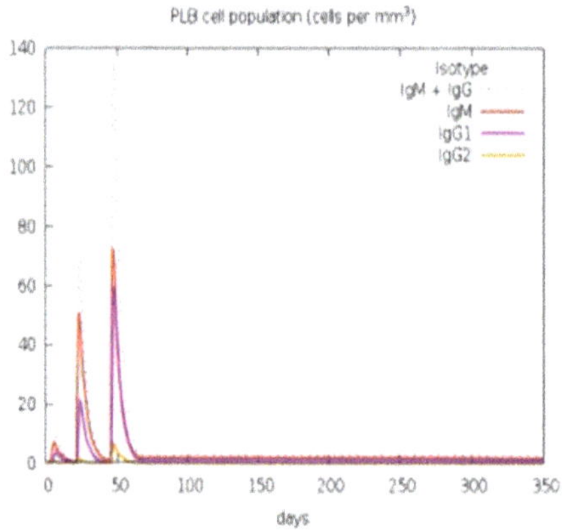

(**M**)

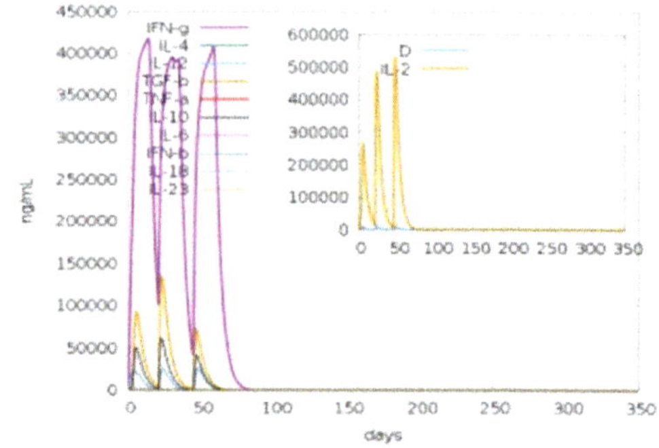

(**N**)

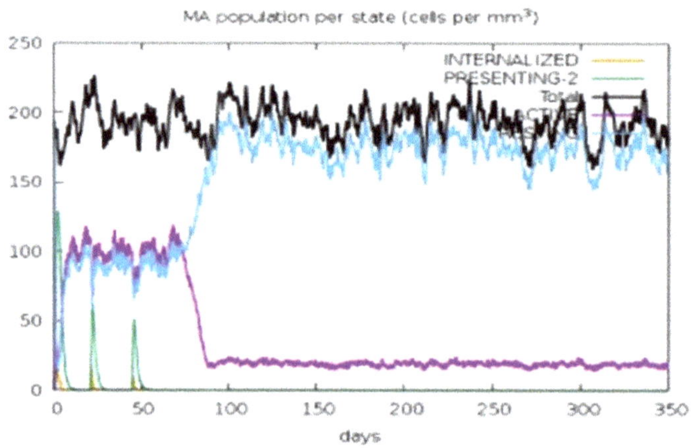

(**O**)

Figure 11. The induced immune cells by the mRNA vaccine. (**A**) Antigen and immunoglobulins of control. (**B**) Antigen and immunoglobulins of vaccine construct with antibodies subdivided per isotype. (**C**) B-lymphocytes population per entity-state (showing counts for active presenting on class II, internalized the Ag. Duplicating and anergic). (**D**) B lymphocytes showing total count and memory cells. (**E**) Epithelia cells population per state (showing total count broken down to active, virus-infected, and presenting on class I MHC molecule. (**F**) Dendritic cells population per state (showing the total number broken down to active, resting, internalized, and presenting the Ag). (**G**) Natural killer cells population showing total count. (**H**) CD8 T-cytotoxic lymphocytes count per entity-state. (**I**) CD8 T-cytotoxic lymphocytes count showing total and memory. (**J**) CD4 T-regulatory lymphocytes count with both total, memory, and per entity-state count plotted. (**K**) CD4 T-helper lymphocytes count subdivided per entity-state. (**L**) CD4 T-helper lymphocytes count with plot showing total and memory counts. (**M**) Plasma B-lymphocytes count subdivided per isotype (IgM, IgG1, and IgG2). (**N**) Cytokines plot showing the concentration of cytokines and interleukins. (**O**) Macrophages population per state (showing total count, internalized, presenting on MHC class II, active, and resting macrophages).

High measures of IgM, IgG1, and IgG3 antibodies were discovered after administering the vaccine, and prolonged immune responses against the virus were evident through high measures of IgG and IgM immunoglobulins. Furthermore, simulation statistics revealed that an amino acid sequence revealed that cytotoxic T-cells' elevation, after 13 days of administering vaccine, attained a maximum of 1155 cells per mm^3. This value gradually decreased to 1120 cells per mm^3 after 33 days. An increase in helper T cells to 5400–6000 cells per mm^3 was observed after 5–6 days. An elongated concentration subsisted up to 35 days. Increased immune T cells evoked a high number of memory cells. Adaptive immunity was strengthened against the virus infections because of high levels of HTLs and CTLs in both active and passive states as a response to the vaccine.

Furthermore, there was an increase in the population of the B cells. Similarly, the concentrations of IgM and IgG isotypes increased to around 460–480 cells per mm^3 and was sustained over a long period. There were also increased levels of cytokines, interleukins, and natural killer cells by the vaccine in *in silico* immunization experimentation (Figure 11). All these results depict the potency and efficacy of the designed vaccine to combat the virus. A mechanism of action for the designed vaccine was proposed. The binding of mRNA vaccine to MHCs and TLR receptors activates key players against the virus (Figure 11). After vaccine administration, there was proliferation of HTLs, CTLs, and other regulatory immune cells to destroy the virus.

4. Discussion

The outbreak of the SARS-CoV-2 virus is a major global pandemic [55]. Prior the evolution of vaccines in past decades, vaccines have brought about the complete extermination or near eradication of some infectious diseases [34]. These diseases include measles, rubella, smallpox, mumps, polio diphtheria, pertussis, and tetanus [56]. Vaccination has been the most successful and effective public health strategy adopted for the eradication of various infectious diseases. Presently, vaccination has become an effective tool for preventing diseases and drastically reducing the negative impacts of different dreaded diseases.

However, the increased transmission of COVID-19 (SARS-CoV-2) has resulted in millions of deaths worldwide and caused wreckage on the economies of many nations of the world [57]. These therefore call for the development of effective and safe prophylactics or therapeutics that could be administered to either mitigate the effects of the menace caused by the deadly virus or protect against its ever evolving and mutating new variants. Several methods have been devised to develop an effective medical therapy, such as vaccines to prevent virus transmission; however, many of the methods have been quite laborious and time-consuming and may ultimately slow down efforts in the development of an effective vaccine, thus contributing less towards mollifying the recent spread of the disease [57,58].

One of the most popular approaches adopted in the past is the conventional method of vaccine design such as live attenuated and inactivated viral vaccines, which utilizes the traditional vaccine development pathway, based on cultivation and inactivation of pathogenic organisms. Although this approach has successfully provided an enduring protection against infectious diseases, mRNA vaccines nonetheless possess great promise for the future [59], as they have been proven to have many merits over conventional vaccine platforms [60]. Besides safety and potency, one of the important benefits of mRNA is the flexibility of its design. Its antigen-coding sequence (open reading frame, ORF) can be easily modified at specific locations and/or codon-optimized to bring about improvements in translation or help channel antigens to the right compartment [60]. Our study was centered on designing an mRNA vaccine against COVID-19 using an array of bioinformatics and immunoinformatics tools to predict epitopes inducing the immune system.

The first step in the design of a novel prophylactic and immunotherapeutic vaccine involved predicting the T-cell and B-cell epitopes. Identifying epitopes is a very important process for the development of effective antibodies that can help neutralize bioactive proteins. Identifying the correct epitopes helps to select high affinity antibodies for immunotherapy and immunodiagnostics [61]. T-cell epitopes are very important for the purpose of adaptive immune simulation, and they interact with MHC molecules [33]. Therefore, when selecting an mRNA vaccine, it is expedient to ensure that a target is immunogenic and can elicit a protective immune response [62].

In this study, highly immunogenic epitopes for B and T cells, humoral prime molecules, and immunity as mRNA vaccine candidates were determined to combat COVID-19 disease. Checking through the various parameters, five CTL epitopes, one HTL epitope, and one LBL epitope were extracted and connected by using the EAAKEAAK and AAY linkers (see Table 1). EAAKEAAK and AAY linkers were connected between the selected epitopes to enable a rational design of mRNA vaccine. The GPGPG linker was also embedded between the adjuvant and the epitope sequences to produce bioactivity improvement for the vaccine [33].

During vaccination, quite a reasonable amount of antibody and T-cell responses are produced, and these required administration of multiple doses of the purified antigens to elicit sufficient antibody response [63]. To address these challenges, significant efforts have been put in by researchers to identify components defined as adjuvants capable of increasing the immunogenic response of antigens in vaccines. Adjuvants are very important in increasing the potency and efficacy of a vaccine [64]. Incorporating them into vaccine design has many advantages, which include provision of stronger immune responses [63]. In this study, we included the co-stimulatory molecule CD40L as our adjuvant. Its involvement in this study was considered due to its inherent efficiency to stimulate the professional antigen-presenting cells (pAPCs), which could invariably lead to the induction of immune response molecules. Although several studies have revealed that mRNA possesses a self-adjuvanting property when administered naked, including an adjuvant will nevertheless contribute more to its efficacity. CD40L is a cell-surface interaction molecule whose expression is pronounced in a CD4+ T-cell subset [65].

Shortly after activating the T cells, its expression is induced. This depicts an early activation marker of T lymphocytes. After a careful and detailed study of its pathway, it was revealed that CD40L plays multiple roles in ensuring a healthy immune system. These include enhancing antigen-specific T-cell response by activating the dendritic cells and the induction of interleukin 12 (IL-12) production by the cells [65].

This response could be sustained for the duration of time the antigen's presence is felt within the system and the time it takes to interact with CD40+ target cells [65].

The previous mRNA approaches in the design of vaccines have produced remarkable results in the past decades. Although there have been shortcomings in their production, notably in their stability and delivery [54,66] in the production of an RNA vaccine, stability and translation of mRNA is crucial [67]. The fact that the half-lives of mRNA molecules are relatively short and tend to be easily degraded in the body calls the need for improvement on the mode of mRNA vaccine production before its administration for proper stability and the efficient promotion of mRNA therapy [68]. 5′ and 3′ UTRs Five prime and three prime UTRs were incorporated into the vaccine ORF to ensure the sufficient production of antigens and effective vaccination of host. The 5′ untranslated region, or 5′ caps, carry out efficient protein production, while the 3′ untranslated region determines mRNA stability and increased protein translation [67].

The frequent transmission of SARS-CoV-2 across the globe has created a platform for making its RNA sequence subsequently undergo mutations, which invariably lead to the translation of different viral proteins (Zikun et al., 2021). Although, these types of mutations can have influence on the epitope-based vaccines, because a change in a single amino acid can alter the results predicted from the epitope analyses (Zikun et al., 2021). However, the proposed final vaccine candidate can tackle the mutations. A multiple sequence alignment was performed for 100 randomly selected SARS-CoV-2 spike glycoproteins from the study area. The results obtained showed no occurrence of mutation in the selected epitope area (Figure 2), thereby indicating the effectiveness of our vaccine construct.

In the context of genetically heterogeneous human populations, HLA polymorphism and its consequent population coverage has been the major concern in epitope-based vaccine design. To address the situation, a careful consideration of the population coverage of the T-lymphocytes epitopes is needed because individuals will likely react to different sets of peptides from a given pathogen [69]. The coverage of the CTL and HTL epitopes was assessed to predict the vaccine construct's effectiveness within the study areas. The epitopes showed a good population coverage (77.22% in average), and a high degree of coverage was predicted for all the regions under study (Table 3), thus indicating the possibility that the vaccine construct can promote an immunological reaction within the population in the study areas. These high values are needed to reduce the complexity of including different epitopes in the vaccine development.

The physicochemical properties of the vaccine construct prove its ability to be a good potential candidate for a vaccine. The molecular weight (MW) was 223.1 kDa, higher above the acceptable threshold value of 110 kDa for a good vaccine candidate (Chukwudozie et al., 2021), thus signifying the efficacy of the vaccine construct. The estimated theoretical pI was 8.69, suggesting the vaccine to be slightly basic. The score of the instability index (48.78) is slightly above the standard threshold. This suggests that the protein would be unstable upon expression, therefore validating the problem of instability, which is majorly encountered in the production of mRNA vaccines. The aliphatic index shows that the vaccine construct is thermostable. The GRAVY score obtained (−0.296) proposes that the vaccine construct is hydrophilic, representing its ability to be highly soluble upon expression as seen in Table 4.

In the development of a vaccine, having the knowledge of the secondary and tertiary structure of the target proteins is crucial to gaining a better understanding of the constructed vaccine candidate. The analyses of our vaccine's secondary structure revealed that the protein contained mainly 33.38% coils. Secondary structures have been shown to be recognized by a few innate immune receptors, and this recognition most times tends to inhibit protein translation. To avoid being recognized by these immune receptors, incorporation of modified nucleosides, such as 5-methylcytidine (5 mC) and pseudouridine (ψ), optimized codons, and a cap-1 structure are important, as it may in turn improve the efficiency of the protein. The 3D structure improved well after refinement. The Ramachandran plot indicates that 97.8% of the residues lie within the favored regions, and 1.3% are allowed regions with less (0.4%) residues in the outlier region. This has provided more evidence that the model's quality is acceptable.

The analysis of antibody–antigen interactions is a very important modeling and docking concept required in vaccine design [49].

The application of protein docking is essential to determine the structure of the antibody–antigen complexes. This interaction is very crucial in understanding the basic function of cells and larger biological systems in all living organisms [70]. To determine the ability of the vaccine construct to bind with TLR on immune cells, TLR-4 was docked with the vaccine considering its importance for easy recognition of pathogens and stimulation of immune response. The results revealed a constructed vaccine with high binding affinity towards the TLR-4. This interface of vaccine with TLR-4 significantly indicates the probability of the vaccine to have the potential of stimulating innate and adaptive immune response. Subsequently, in an advent to explore the stability and dynamics performance of the TLR-4–vaccine docked complex; a molecular dynamic simulation was conducted. The RMSD plot depicts a steady binding of the complex.

The *in silico* immune response simulation depicted a consistency in the immune response. There was an increase in the generated immune responses before its repeated exposure to the antigen. There was evident development of memory B cells and T cells, with memory in B cells lasting several months. This indicates humoral immunity and is essential for complimenting the immune response. Moreover, helper T cells were particularly stimulated, therefore establishing the capacity of the vaccine construct to protect against SARS-CoV-2.

The negative impacts of COVID-19 cannot be overemphasized. Such impacts have affected the economy of many countries [71]; stock markets have been affected by COVID-19 [72,73]; COVID-19 outbreaks have affected the mental health of many categories of people [74]; it has had a negative impact on public health and on addressing non-COVID-19 diseases [75,76]; the COVID-19 pandemic has negatively impacted on health workers by causing anxiety and high levels of stress [77,78]; COVID-19 has had mental and psychological impacts on different individuals [79–81].

Prior the emergence of the SARs-CoV-2 virus, several bioinformatics, computational informatics, and modeling approaches have been applied towards proffering solutions to existing infectious and non-infectious diseases such as HIV, Ebola, malaria, and hereditary diseases, amongst others [82–91]. In this current work, bioinformatics, computational, and modeling approaches are being harnessed towards developing a potent and effective vaccine candidate against SARS-CoV-2 virus.

Data collection within the COVID-19 pandemic has been from different (diverse) sources. Real-time dashboard COVID-19 data have indicated or revealed the impact of COVID-19 on human health and human lives [92–95]. Some of the sources are in real time, for instance web-dashboards [92–95], while others are diverse. COVID-19 data are of different forms, namely genomic sequences of different variants of the SARS-CoV-2 virus [92–95], chest X-rays of COVID-19 patients [96], kidney replacement therapy data for COVID-19 patients [97], lungs data of patients [98], blood data of COVID-19 patients [99–104], medical data of patients that were infected and recovered [105], medical data of patients that were infected and died [106], patients' demographic data, patients' bio-data, COVID-19 health facility data for different regions of the world, and COVID-19 phylogenetic data, amongst others. Genomic data collected for the SARS-CoV-2 virus were stored in bioinformatics databases such as NCBI, EBL, and GISAID [20,107,108].

To examine the binding stability, conformation, and interaction modes of the vaccine–TLR4 docked complex, the molecular dynamics were simulated using the online WEBGRO macromolecular simulations platform [109]. WebGro is an entirely automated online tool for simulating macromolecules (proteins) alone or in complexes with ligands (small molecules) using molecular dynamics modeling [109]. For comprehensive solvated molecular dynamics simulations, WebGro utilizes the GROMACS simulation program [110]. The energy of the complex formed was first minimized using the steepest descent integrator at every 5000 steps for molecular dynamics simulation. Afterwards, enough of an amount of Na+ and Cl- counter ions were added to maintain a salt concentration of 0.15 M in the complex system. The NVT/ NPT equilibration was performed at 300 K and 1 bar pressure. In addition, leapfrog was selected as the MD integrator for a simulation time of 100 ns and 1000 frames per MD simulation [110]. To better understand the formation of the complex, a trajectory analysis of the root-mean-square deviation (RMSD), root-mean-square fluctuation (RMSF), radius of gyration (Rg), solvent accessible surface area (SASA), and hydrogen bonds (HBs) was performed [110].

One of the main functions of the root-mean-square deviation (RMSD) is to depict the average distance between the backbone atoms of the starting structure and the simulated structures when superimposed [111]. As a useful parameter, the RMSD can be utilized to study the equilibration of MD trajectories as well as checking the stability of complex systems during simulation. This could be achieved by plotting the RMSD of the protein backbone atoms against time to see how the structural shape of the protein changes over time [111]. The RMSD plot was significantly dynamic, with fluctuations occurring after 5 ns. The stable conformation was attained at a time range between 75 and 100 nanoseconds, with no significant changes in the results (see Figure 12A).

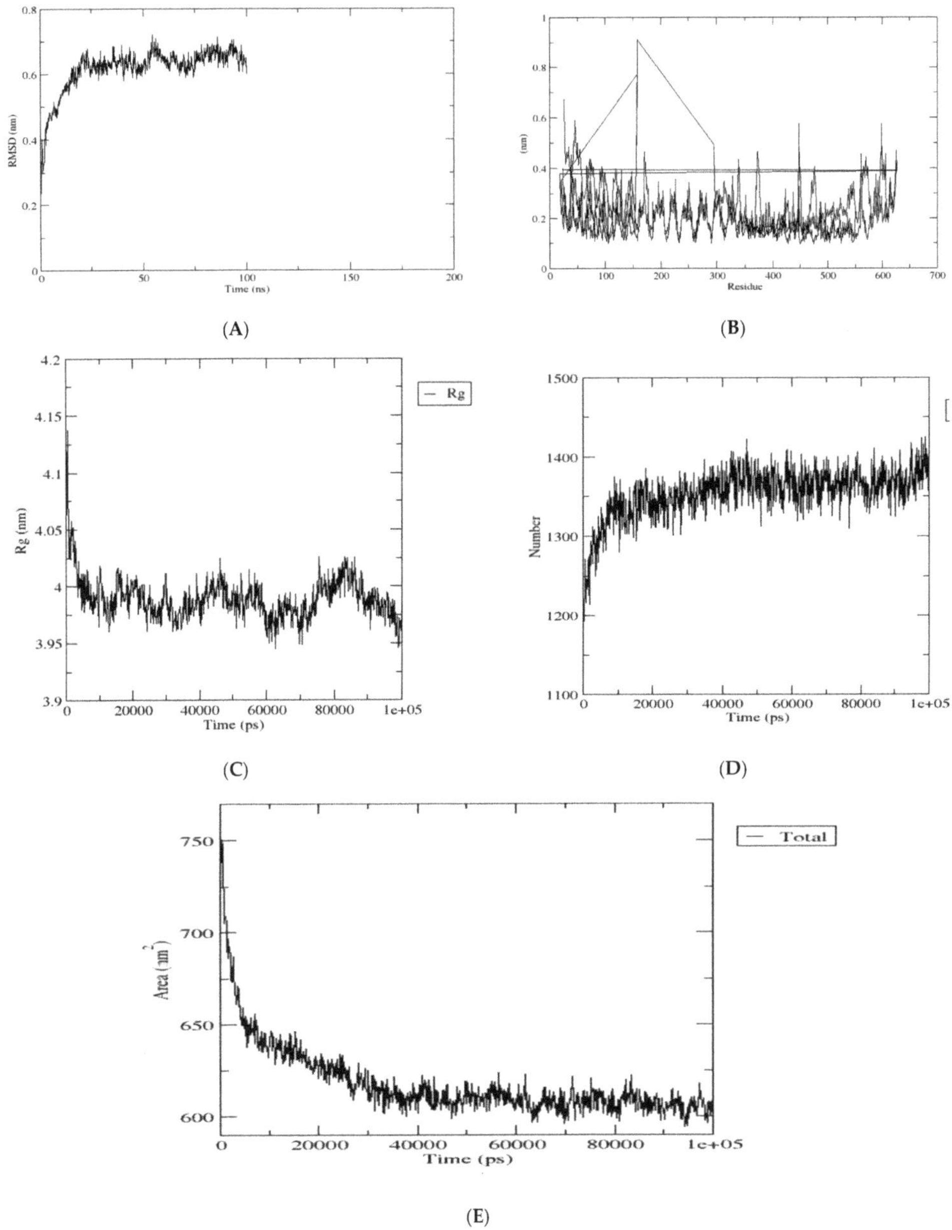

Figure 12. (**A**) RMSD study plot for 100 ns MD simulation of vaccine–TLR. (**B**) RMSF study plots for 100 ns MD simulation. (**C**) Radius of gyration study plot for 100 ns MD simulation vaccine–TLR. (**D**) Solvent accessible surface area study plot for 100 ns MD simulation of vaccine–TLR. (**E**) Solvent accessible surface area study plot for 100 ns MD simulation of vaccine–TLR.

Furthermore, the RMSF values of the protein atoms were calculated and plotted against the residues. When evaluating the stability and flexibility of a complex system, another important parameter to consider is the root-mean-square fluctuation (RMSF).

This parameter is useful because it can be used to study how well the behavior of amino acid residues in a target protein changes as it binds to a ligand [111]. Throughout the simulation, the amino acid showed very little fluctuation (Figure 12B). Additionally, the complex systems' gyration radius (Rg) was investigated. This is the distance between the rotational axis and the mass center [112]. It is essential and important to know and understand how structural variation affects the compactness of the protein after binding with the ligands when examining the stability and flexibility of the complex structure during simulation [113], and this can be accomplished by analyzing the complex structure's radius of gyration (Rg). Higher Rg values indicate that the protein is less compact and flexible, whereas low values indicate that the protein packing has not changed much (see Figure 12C), thus exemplifying the high degree of compactness and stiffness. To investigate changes in structural compactness, the Rg values of protein backbone atoms were plotted versus time. The backbone Rg values gradually declined until they reached 10 ns. There were no significant variations in the time between 11 and 100 ns, and a nearly constant value of about 4.0 nm was maintained, indicating that the protein packing did not vary considerably.

Similarly, we examined the formation of hydrogen bonds in the complex structure by plotting the number of hydrogen atoms against time. This is necessary for a better understanding of the protein's structural integrity, catalytic region, and protein–ligand interaction in the complex structure [113]. Within the complex structure of the protein, there is a significant change in the hydrogen bond interaction (see Figure 12D).

We also calculated the interaction area between the solvent and the protein complexes to implement the solvent accessible surface area (SASA) of the complex structure. To assess changes in surface area, the protein's values were plotted against a function of time. SASA is a significant parameter for determining the extent of receptor exposure to surrounding solvent molecules during simulation [111]. SASA with a higher value indicates more hydrophilicity [113]. The SASA complex trajectory values gradually decreased till 400 ns. Throughout the simulation period, minute changes were noticed, except for a few time intervals (see Figure 12E). The average SASA value was 610 nm^2, with values ranging from 625–600 nm^2.

5. Conclusions

In conclusion, the process involved in the production of an effective traditional vaccine often takes several months or years of trial before it can be accomplished. Moreover, these vaccines are quite expensive. The integration of the bioinformatics approach into the development of vaccines has helped overcome many of these challenges by focusing mainly on the selection of appropriate antigens or antigenic structures, carriers, and adjuvants used in the design. In the face of the current pandemic, which has ravaged the world, the development of vaccines is an urgent need. This is especially true for African countries, which lack critical infrastructure for vaccine development to combat the circulating variants within the region. Our results show that the vaccine candidate consisted of seven epitopes, namely a highly immunogenic adjuvant, an MHC I-targeting domain (MITD), a signal peptide, and linkers. The vaccine candidates' molecular weight (MW) was predicted to be 223.1 kDa, which is greater than the acceptable threshold of 110 kDa on an excellent vaccine candidate. The summary of the results obtained from the experiments revealed that the vaccine candidate was antigenic, non-allergenic, non-toxic, thermostable, and hydrophilic. The vaccine candidate has good population coverage, with the highest range in East Africa (80.44%) followed by South Africa (77.23%). West Africa and North Africa have 76.65% and 76.13%, respectively, while Central Africa (75.64%) has minimal coverage. Evaluation of the secondary structure of the vaccine construct revealed a stabilized structure showing 36.44% alpha-helices, 20.45% drawn filaments, and 33.38% random helices. Molecular docking of the TLR4 vaccine showed that the simulated vaccine has a high binding affinity for TLR-4, reflecting its ability to stimulate the innate and adaptive immune response. Bioinformatics, computational, and immunoinformatic approaches for a multi-epitope mRNA vaccine design against the circulating variants of SARS-CoV-2 within the African population have shown that this vaccine candidate can be a useful therapeutic in fighting the deadly virus. This is because the designed construct has been shown to meet the requisite threshold for each of the physicochemical properties that make a candidate vaccine effective. According to our findings, the designed construct is antigenic, non-toxic, non-allergenic, slightly basic, thermostable with wide population coverage, and capable of tackling any mutation. Further work can be carried out after the results and performances from this computational research have been subjected to in vitro and in vivo validations.

Author Contributions: Conceived and conceptualized: O.O.O.; study design: O.O.O., E.K.O. and S.F.; methodology: O.O.O., E.K.O., B.A.I., E.O.D., A.E.A., K.T.K., O.E.A., J.A.O. and B.F.O.; curation of data: O.O.O., E.K.O., B.A.I., E.O.D., O.M.K., J.K.O., T.I.A., A.E.A., K.T.K., O.E.A., J.A.O. and B.F.O.; data analysis: O.O.O., E.K.O., B.A.I., E.O.D., O.M.K., J.K.O., T.I.A., A.E.A., K.T.K., O.E.A., J.A.O. and B.F.O.; resources: O.O.O.; writing—original draft preparation: O.O.O.; writing—critical review, revised version, and re-editing: O.O.O. and S.F.; project administration: O.O.O. and E.K.O.; funding acquisition: O.O.O.; principal investigator: O.O.O. The funders had no role in study design, data collection and analysis, decision to publish, or preparation of the manuscript. All authors have read and agreed to the published version of the manuscript.

Funding: This study was funded by the German Deutscher Akademischer Austauschdienst (DAAD) climapAfrica grant (with grant/scholarship personal reference number ST32/91769426) and by the Oppenheimer Memorial Trust (OMT) personal research grant (with grant award/scholarship reference number OMT Ref. 21563/01) awarded to O.O.O. This work was partly supported by the Wellcome Trust grant number 220740/Z/20/Z to Segun Fatumo.

Institutional Review Board Statement: Not applicable.

Informed Consent Statement: Not applicable.

Data Availability Statement: Experimental data sequences for this research are freely available but can be found in the GISAID database (https://www.gisaid.org/); Access date: 30 March 2021. GISAID data can be shared only with GISAID registered database users and not with the non-registered. All relevant data for this article are within the manuscript and its Supporting Information files. This project information can be found at: https://github.com/oluwagbemi/Bioinformatics-and-Computational-Approaches-to-the-Design-of-mRNA-COVID-19-vaccine-candidates-. This project's information can also be found on the research project page with weblink: https://olugbengaoluwagbemi.weebly.com/research-projects.html.

Acknowledgments: The research of this article was supported by German Deutscher Akademischer Austauschdienst (DAAD), Germany, personal grant to O.O.O. The research was also supported by the Oppenheimer Memorial Trust (OMT), South Africa personal grant to O.O.O. We thank Helix Biogen Institute for some technical support. This work was partly supported by the Wellcome Trust grant number 220740/Z/20/Z to Segun Fatumo.

Conflicts of Interest: The authors declare no conflict of interest.

References

1. Safavi, A.; Kefayat, A.; Mahdevar, E.; Abiri, A.; Ghahremani, F. Exploring the out of sight antigens of SARS-CoV-2 to design a candidate multi-epitope vaccine by utilizing immunoinformatics approaches. *Vaccine* **2020**, *38*, 7612–7628. [CrossRef] [PubMed]
2. World Health Organization. WHO Coronavirus Disease (COVID-19) Dashboard. 2022. Available online: https://covid19.who.int (accessed on 20 January 2022).
3. Oladipo, E.K.; Ajayi, A.F.; Ariyo, O.E.; Onile, S.O.; Jimah, E.M.; Ezediuno, L.O.; Adebayo, O.I.; Adebayo, E.T.; Odeyemi, A.N.; Oyeleke, M.O.; et al. Exploration of surface glycoprotein to design multi-epitope vaccine for the prevention of COVID-19. *Inform. Med. Unlocked* **2020**, *21*, 100438. [CrossRef] [PubMed]
4. Oluwagbemi, O.O.; Oladipo, E.K.; Dairo, E.O.; Ayeni, A.E.; Irewolede, B.A.; Jimah, E.M.; Oyewole, M.P.; Olawale, B.M.; Adegoke, H.M.; Ogunleye, A.J. Computational construction of a glycoprotein multi-epitope subunit vaccine candidate for old and new South-African SARS-CoV-2 virus strains. *Inform. Med. Unlocked* **2022**, *28*, 100845. [CrossRef] [PubMed]
5. Walls, A.C.; Park, Y.-J.; Tortorici, M.A.; Wall, A.; McGuire, A.T.; Veesler, D. Structure, Function, and Antigenicity of the SARS-CoV-2 Spike Glycoprotein. *Cell* **2020**, *181*, 281–292.e6. [CrossRef]
6. Kumar, A.; Sharma, B.P. *In silico* study to predict and characterize of SARS CoV 2 Surface glycoprotein. *Vaccine Res.* **2020**, *7*, 10–16. [CrossRef]

7. WHO Africa. Eight in 10 African Countries to Miss Crucial COVID-19 Vaccination Goal. World Health Organization. 2021. Available online: https://www.afro.who.int/news/eight-10-african-countries-miss-crucial-covid-19-vaccination-goal (accessed on 3 September 2021).
8. United Nations. Record Weekly COVID-19 Deaths in Africa. African Renewal (2021). Available online: https://www.un.org/africarenewal/news/record-weekly-covid-19-deaths-Africa (accessed on 6 August 2021).
9. Faria, J. Number of COVID-19 Delta Variant Cases in Africa 2021, by Country. Statista. Available online: https://www.statista.com/statistics/1249798/number-of-sars-cov-2-delta-variant-cases-in-africa-by-country/ (accessed on 28 March 2021).
10. Park, J.W.; Lagniton, P.N.; Liu, Y.; Xu, R.-H. mRNA vaccines for COVID-19: What, why and how. *Int. J. Biol. Sci.* **2021**, *17*, 1446–1460. [CrossRef]
11. Schmidt, S.T.; Foged, C.; Korsholm, K.S.; Rades, T.; Christensen, D. Liposome-Based Adjuvants for Subunit Vaccines: Formulation Strategies for Subunit Antigens and Immunostimulators. *Pharmaceutics* **2016**, *8*, 7. [CrossRef] [PubMed]
12. Blakney, A.; Ip, S.; Geall, A. An Update on Self-Amplifying mRNA Vaccine Development. *Vaccines* **2021**, *9*, 97. [CrossRef]
13. Schlake, T.; Thess, A.; Fotin-Mleczek, M.; Kallen, K. Developing mRNA-vaccine technologies. *RNA Biol.* **2012**, *9*, 1319–1330. [CrossRef]
14. Ho, W.; Gao, M.; Li, F.; Li, Z.; Zhang, X.Q.; Xu, X. Next-generation vaccines nanoparticle-mediated DNA and mRNA delivery. *Adv. Healthc. Mater.* **2021**, *10*, e2001812. [CrossRef]
15. Xu, S.; Yang, K.; Li, R.; Zhang, L. mRNA vaccine era-mechanisms, drug platform and clinical prospection. *Int. J. Mol. Sci.* **2020**, *21*, 6582. [CrossRef] [PubMed]
16. Anand, P.; Stahel, V.P. The safety of Covid-19 mRNA vaccines: A review. *Patient Saf. Surg.* **2021**, *15*, 20. [CrossRef] [PubMed]
17. Jackson, N.A.C.; Kester, K.E.; Casimiro, D.; Gurunathan, S.; DeRosa, F. The promise of mRNA vaccines: A biotech and industrial perspective. *NPJ Vaccines* **2020**, *5*, 11. [CrossRef] [PubMed]
18. Zeng, C.; Zhang, C.; Walker, P.G.; Dong, Y. Formulation and Delivery Technologies for mRNA Vaccines. *Curr. Top. Microbiol. Immunol.* **2020**, *Epub ahead of print*. [CrossRef]
19. Kowalzik, F.; Schreiner, D.; Jensen, C.; Teschner, D.; Gehring, S.; Zepp, F. mRNA-Based Vaccines. *Vaccines* **2021**, *9*, 390. [CrossRef] [PubMed]
20. The Global Initiative for Sharing All Influenza Data (GISAID) Database. Available online: https://www.gisaid.org/ (accessed on 30 March 2021).
21. Maarouf, M.; Rai, K.R.; Goraya, M.U.; Chen, J.-L. Immune Ecosystem of Virus-Infected Host Tissues. *Int. J. Mol. Sci.* **2018**, *19*, 1379. [CrossRef]
22. Larsen, M.V.; Lundegaard, C.; Lamberth, K.; Buus, S.; Lund, O.; Nielsen, M. Large-scale validation of methods for cytotoxic T-lymphocyte epitope prediction. *BMC Bioinform.* **2007**, *8*, 424. [CrossRef]
23. Gupta, S.; Kapoor, P.; Chaudhary, K.; Gautam, A.; Kumar, R.; Raghava, G.P.S. *In silico* approach for predicting toxicity of peptides and proteins. *PLoS ONE* **2013**, *8*, e73957. [CrossRef]
24. Dimitrov, I.; Bangov, I.; Flower, D.R.; Doytchinova, I. AllerTOP v.2—A server for *in silico* prediction of allergens. *J. Mol. Model.* **2014**, *20*, 2278. [CrossRef]
25. Doytchinova, I.A.; Flower, D.R. VaxiJen: A server for prediction of protective antigens, tumour antigens and subunit vaccines. *BMC Bioinform.* **2007**, *8*, 4. [CrossRef]
26. Wang, P.; Sidney, J.; Kim, Y.; Sette, A.; Lund, O.; Nielsen, M.; Peters, B. Peptide binding predictions for HLA DR, DP and DQ molecules. *BMC Bioinform.* **2010**, *11*, 568. [CrossRef]
27. Nagpal, G.; Usmani, S.S.; Dhanda, S.; Kaur, H.; Singh, S.; Sharma, M.; Raghava, G.P.S. Computer-aided designing of immunosuppressive peptides based on IL-10 inducing potential. *Sci. Rep.* **2017**, *7*, srep42851. [CrossRef] [PubMed]
28. Jespersen, M.C.; Peters, B.; Nielsen, M.; Marcatili, P. BepiPred-2.0: Improving sequence-based B-cell epitope prediction using conformational epitopes. *Nucleic Acids Res.* **2017**, *45*, W24–W29. [CrossRef] [PubMed]
29. Manavalan, B.; Govindaraj, R.G.; Shin, T.H.; Kim, M.O.; Lee, G. iBCE-EL: A New Ensemble Learning Framework for Improved Linear B-Cell Epitope Prediction. *Front. Immunol.* **2018**, *9*, 1695. [CrossRef] [PubMed]
30. Abdelmageed, M.I.; Abdelmoneim, A.H.; Mustafa, M.I.; Elfadol, N.M.; Murshed, N.S.; Shantier, S.W.; Makhawi, A.M. Design of a Multiepitope-Based Peptide Vaccine against the E Protein of Human COVID-19: An Immunoinformatics Approach. *BioMed Res. Int.* **2020**, *2020*, 2683286. [CrossRef]
31. Sievers, F.; Higgins, D.G. Clustal omega. *Curr. Protoc. Bioinform.* **2014**, *48*, 1–16. [CrossRef]
32. Ahammad, I.; Lira, S.S. Designing a novel mRNA vaccine against SARS-CoV-2: An immunoinformatics approach. *Int. J. Biol. Macromol.* **2020**, *162*, 820–837. [CrossRef]
33. Bibi, S.; Ullah, I.; Zhu, B.; Adnan, M.; Liaqat, R.; Kong, W.-B.; Niu, S. *In silico* analysis of epitope-based vaccine candidate against tuberculosis using reverse vaccinology. *Sci. Rep.* **2021**, *11*, 1249. [CrossRef]
34. Maruggi, G.; Zhang, C.; Li, J.; Ulmer, J.B.; Yu, D. mRNA as a Transformative Technology for Vaccine Development to Control Infectious Diseases. *Mol. Ther.* **2019**, *27*, 757–772. [CrossRef]
35. Zahroh, H.; Ma'Rup, A.; Tambunan, U.S.F.; Parikesit, A.A. Immunoinformatics Approach in Designing Epitope-based Vaccine against Meningitis-inducing Bacteria (*Streptococcus pneumoniae*, *Neisseria meningitidis*, and *Haemophilus influenzae* Type b). *Drug Target Insights* **2016**, *10*, DTI-S38458. [CrossRef]

36. Kedzierska, K.; Thomas, P.G. Count on us: T cells in SARS-CoV-2 infection and vaccination. *Cell Rep. Med.* **2022**, *3*, 100562. [CrossRef]
37. Bui, H.-H.; Sidney, J.; Dinh, K.; Southwood, S.; Newman, M.J.; Sette, A. Predicting population coverage of T-cell epitope-based diagnostics and vaccines. *BMC Bioinform.* **2006**, *7*, 153. [CrossRef]
38. Magnan, C.N.; Zeller, M.; Kayala, M.A.; Vigil, A.; Randall, A.; Felgner, P.L.; Baldi, P. High-throughput prediction of protein antigenicity using protein microarray data. *Bioinformatics* **2010**, *26*, 2936–2943. [CrossRef] [PubMed]
39. Dimitrov, I.; Naneva, L.; Doytchinova, I.; Bangov, I. AllergenFP: Allergenicity prediction by descriptor fingerprints. *Bioinformatics* **2013**, *30*, 846–851. [CrossRef] [PubMed]
40. Gasteiger, E.; Hoogland, C.; Gattiker, A.; Duvaud, S.; Wilkins, M.R.; Appel, R.D.; Bairoch, A. Protein Identification and Analysis Tools on the ExPASy Server. In *The Proteomics Protocols Handbook*; John, W.M., Ed.; Humana Press: Totowa, NJ, USA, 2005; pp. 571–607. Available online: http://www.expasy.org/tools/protparam.html; (accessed on 2 July 2021).
41. Rehman, A.; Ahmad, S.; Shahid, F.; Albutti, A.; Alwashmi, A.; Aljasir, M.; Alhumeed, N.; Qasim, M.; Ashfaq, U.; Qamar, M.T.U. Integrated Core Proteomics, Subtractive Proteomics, and Immunoinformatics Investigation to Unveil a Potential Multi-Epitope Vaccine against Schistosomiasis. *Vaccines* **2021**, *9*, 658. [CrossRef] [PubMed]
42. Geourjon, C.; Deléage, G. SOPMA: Significant improvement in protein secondary structure prediction by consensus prediction from multiple alignments. *C2abios* **1995**, *11*, 681–684. Available online: https://npsa-prabi.ibcp.fr/cgi-bin/npsa_automat.pl?page=/NPSA/npsa_sopma.html (accessed on 5 August 2021). [CrossRef] [PubMed]
43. Deléage, G. ALIGNSEC: Viewing protein secondary structure predictions within large multiple sequence alignments. *Bioinformatics* **2017**, *33*, 3991–3992. [CrossRef]
44. Kelley, L.A.; Mezulis, S.; Yates, C.M.; Wass, M.N.; Sternberg, M.J.E. The Phyre2 web portal for protein modeling, prediction and analysis. *Nat. Protoc.* **2015**, *10*, 845–858. [CrossRef]
45. Lee, G.R.; Heo, L.; Seok, C. Effective protein model structure refinement by loop modeling and overall relaxation. *Proteins Struct. Funct. Bioinform.* **2015**, *84*, 293–301. [CrossRef]
46. Heo, L.; Park, H.; Seok, C. GalaxyRefine: Protein structure refinement driven by side-chain repacking. *Nucleic Acids Res.* **2013**, *41*, W384–W388. [CrossRef]
47. Wiederstein, M.; Sippl, M.J. ProSA-web: Interactive web service for the recognition of errors in three-dimensional structures of proteins. *Nucleic Acids Res.* **2007**, *35*, W407–W410. [CrossRef]
48. Ponomarenko, J.V.; Bui, H.-H.; Li, W.; Fusseder, N.; Bourne, P.E.; Sette, A.; Peters, B. ElliPro: A new structure-based tool for the prediction of antibody epitopes. *BMC Bioinform.* **2008**, *9*, 514. [CrossRef] [PubMed]
49. Kozakov, D.; Hall, D.R.; Xia, B.; Porter, K.A.; Padhorny, D.; Yueh, C.; Beglov, D.; Vajda, S. The ClusPro web server for protein–protein docking. *Nat. Protoc.* **2017**, *12*, 255–278. [CrossRef] [PubMed]
50. Berman, H.M.; Westbrook, J.; Feng, Z.; Gilliland, G.; Bhat, T.N.; Weissig, H.; Shindyalov, I.N.; Bourne, P.E. The Protein Data Bank. *Nucleic Acids Res.* **2020**, *28*, 235–242. [CrossRef] [PubMed]
51. López-Blanco, J.R.; Aliaga, J.I.; Quintana-Ortí, E.S.; Chacón, P. iMODS: Internal coordinates normal mode analysis server. *Nucleic Acids Res.* **2014**, *42*, W271–W276. [CrossRef] [PubMed]
52. Rapin, N.; Lund, O.; Bernaschi, M.; Castiglione, F. Computational Immunology Meets Bioinformatics: The Use of Prediction Tools for Molecular Binding in the Simulation of the Immune System. *PLoS ONE* **2010**, *5*, e9862. [CrossRef] [PubMed]
53. Chukwudozie, O.S.; Gray, C.M.; Fagbayi, T.A.; Chukwuanukwu, R.C.; Oyebanji, V.O.; Bankole, T.T.; Adewole, R.A.; Daniel, E.M. Immuno-informatics design of a multimeric epitope peptide-based vaccine targeting SARS-CoV-2 spike glycoprotein. *PLoS ONE* **2021**, *16*, e0248061. [CrossRef]
54. Chen, J.; Liu, H.; Yang, J.; Chou, K. Prediction of linear B-cell epitopes using amino acid pair antigenicity scale. *Amino Acids* **2007**, *33*, 423–428. [CrossRef]
55. Rauch, S.; Jasny, E.; Schmidt, K.E.; Petsch, B. New Vaccine Technologies to Combat Outbreak Situations. *Front. Immunol.* **2018**, *9*, 1963. [CrossRef]
56. Roush, S.W.; Murphy, T.V.; The Vaccine-Preventable Disease Table Working Group. Historical Comparisons of Morbidity and Mortality for Vaccine-Preventable Diseases in the United States. *J. Am. Med Assoc.* **2007**, *298*, 2155–2163. [CrossRef]
57. Yang, Z.; Bogdan, P.; Nazarian, S. An *in silico* deep learning approach to multi-epitope vaccine design: A SARS-CoV-2 case study. *Sci. Rep.* **2021**, *11*, 3238. [CrossRef]
58. Kalita, P.; Padhi, A.K.; Zhang, K.Y.; Tripathi, T. Design of a peptide-based subunit vaccine against novel coronavirus SARS-CoV-2. *Microb. Pathog.* **2020**, *145*, 104236. [CrossRef] [PubMed]
59. Pardi, N.; Hogan, M.J.; Weissman, D. Recent advances in mRNA vaccine technology. *Curr. Opin. Immunol.* **2020**, *65*, 14–20. [CrossRef] [PubMed]
60. Alameh, M.-G.; Weissman, D.; Pardi, N. *Messenger RNA-Based Vaccines Against Infectious Diseases*; Springer: Berlin/Heidelberg, Germany, 2020; pp. 1–35. [CrossRef]
61. Kozlova, E.E.G.; Cerf, L.; Schneider, F.S.; Viart, B.T.; Nguyen, C.; Steiner, B.T.; Lima, S.D.A.; Molina, F.; Guerra-Duarte, C.; Felicori, L.; et al. Computational B-cell epitope identification and production of neutralizing murine antibodies against Atroxlysin-I. *Sci. Rep.* **2018**, *8*, 14904. [CrossRef] [PubMed]
62. Bettini, E.; Locci, M. SARS-CoV-2 mRNA Vaccines: Immunological Mechanism and Beyond. *Vaccines* **2021**, *9*, 147. [CrossRef]

63. Pellegrino, P.; Clementi, E.; Radice, S. On vaccine's adjuvants and autoimmunity: Current evidence and future perspectives. *Autoimmun. Rev.* **2015**, *14*, 880–888. [CrossRef]
64. Bastola, R.; Noh, G.; Keum, T.; Bashyal, S.; Seo, J.-E.; Choi, J.; Oh, Y.; Cho, Y.; Lee, S. Vaccine adjuvants: Smart components to boost the immune system. *Arch. Pharmacal Res.* **2017**, *40*, 1238–1248. [CrossRef]
65. Daoussis, D.; Andonopoulos, A.P.; Liossis, S.-N.C. Targeting CD40L: A Promising Therapeutic Approach. *Clin. Vaccine Immunol.* **2004**, *11*, 635–641. [CrossRef]
66. Tsui, N.B.; Ng, E.K.; Lo, Y.D. Stability of Endogenous and Added RNA in Blood Specimens, Serum, and Plasma. *Clin. Chem.* **2002**, *48*, 1647–1653. [CrossRef]
67. Zhang, C.; Maruggi, G.; Shan, H.; Li, J. Advances in mRNA Vaccines for Infectious Diseases. *Front. Immunol.* **2019**, *10*, 594. [CrossRef]
68. Adibzadeh, S.; Fardaei, M.; Takhshid, M.A.; Miri, M.R.; Dehbidi, G.R.; Farhadi, A.; Ranjbaran, R.; Alavi, P.; Nikouyan, N.; Seyyedi, N.; et al. Enhancing Stability of Destabilized Green Fluorescent Protein Using Chimeric mRNA Containing Human Beta-Globin 5′ and 3′ Untranslated Regions. *Avicenna J. Med Biotechnol.* **2019**, *11*, 112–117.
69. Oyarzun, P.; Kobe, B. Computer-aided design of T-cell epitope-based vaccines: Addressing population coverage. *Int. J. Immunogenet.* **2015**, *42*, 313–321. [CrossRef] [PubMed]
70. Desta, I.T.; Porter, K.A.; Xia, B.; Kozakov, D.; Vajda, S. Performance and Its Limits in Rigid Body Protein-Protein Docking. *Structure* **2020**, *28*, 1071–1081.e3. [CrossRef] [PubMed]
71. Ferrel, M.N.; Ryan, J.J. The Impact of COVID-19 on Medical Education. *Cureus* **2020**, *12*, e7492. [CrossRef]
72. Liu, H.; Manzoor, A.; Wang, C.; Zhang, L.; Manzoor, Z. The COVID-19 Outbreak and Affected Countries Stock Markets Response. *Int. J. Environ. Res. Public Health* **2020**, *17*, 2800. [CrossRef] [PubMed]
73. Khan, K.; Zhao, H.; Zhang, H.; Yang, H.; Shah, M.H.; Jahanger, A. The Impact of COVID-19 Pandemic on Stock Markets: An Empirical Analysis of World Major Stock Indices. *J. Asian Financ. Econ. Bus.* **2020**, *7*, 463–474. [CrossRef]
74. Caparros-Gonzalez, R.A.; Ganho-Ávila, A.; De La Torre-Luque, A. The COVID-19 Pandemic Can Impact Perinatal Mental Health and the Health of the Offspring. *Behav. Sci.* **2020**, *10*, 162. [CrossRef]
75. Bell, D.; Hansen, K.S.; Kiragga, A.N.; Kambugu, A.; Kissa, J.; Mbonye, A.K. Predicting the Impact of COVID-19 and the Potential Impact of the Public Health Response on Disease Burden in Uganda. *Am. J. Trop. Med. Hyg.* **2020**, *103*, 1191–1197. [CrossRef]
76. Coccia, M. The impact of first and second wave of the COVID-19 pandemic in society: Comparative analysis to support control measures to cope with negative effects of future infectious diseases. *Environ. Res.* **2021**, *197*, 111099. [CrossRef]
77. Malesza, M.; Kaczmarek, M.C. Predictors of anxiety during the COVID-19 pandemic in Poland. *Pers. Individ. Differ.* **2020**, *170*, 110419. [CrossRef]
78. Vasudevan, M.; Mehrolia, S.; Alagarsamy, S. Battle fatigue of Covid 19 warriors—Heal the healers. *J. Affect. Disord.* **2021**, *294*, 477–478. [CrossRef]
79. Koçak, O.; Koçak, Ö.; Younis, M. The Psychological Consequences of COVID-19 Fear and the Moderator Effects of Individuals' Underlying Illness and Witnessing Infected Friends and Family. *Int. J. Environ. Res. Public Health* **2021**, *18*, 1836. [CrossRef] [PubMed]
80. Usher, K.; Durkin, J.; Bhullar, N. The COVID-19 pandemic and mental health impacts. *Int. J. Ment. Health Nurs.* **2020**, *29*, 315–318. [CrossRef] [PubMed]
81. Jungmann, S.M.; Witthöft, M. Health anxiety, cyberchondria, and coping in the current COVID-19 pandemic: Which factors are related to coronavirus anxiety? *J. Anxiety Disord.* **2020**, *73*, 102239. [CrossRef]
82. Oluwagbemi, O. Development of a prototype hybrid-grid-based computing framework for accessing bioinformatics databases and resources. *Sci. Res. Essays* **2012**, *7*, 730–739. [CrossRef]
83. Oluwagbemi, O.; Adeoye, E.T.; Fatumo, S. Building a Computer-Based Expert System for Malaria Environmental Diagnosis: An Alternative Malaria Control Strategy. *Egypt. Comput. Sci. J.* **2009**, *33*, 55–69.
84. Madeira, F.; Park, Y.M.; Lee, J.; Buso, N.; Gur, T.; Madhusoodanan, N.; Basutkar, P.; Tivey, A.R.N.; Potter, S.C.; Finn, R.D.; et al. The EMBL-EBI search and sequence analysis tools APIs in 2019. *Nucleic Acids Res.* **2019**, *47*, W636–W641. [CrossRef] [PubMed]
85. Oluwagbemi, O.; Oluwagbemi, F.; Abimbola, O. Ebinformatics: Ebola fuzzy informatics systems on the diagnosis, prediction and recommendation of appropriate treatments for Ebola virus disease (EVD). *Inform. Med. Unlocked* **2016**, *2*, 12–37. [CrossRef]
86. Oluwagbemi, O.O.; Fornadel, C.M.; Adebiyi, E.F.; Norris, D.E.; Rasgon, J. AnoSpEx: A Stochastic, Spatially -Explicit Computational Model for Studying Anopheles Metapopulation Dynamics. *PLoS ONE* **2013**, *8*, e68040. [CrossRef]
87. Oluwagbemi, O.; Oluwagbemi, F.; Ughamadu, C. Android Mobile Informatics Application for some Hereditary Diseases and Disorders (AMAHD): A complementary framework for medical practitioners and patients. *Inform. Med. Unlocked* **2016**, *2*, 38–69. [CrossRef]
88. Oluwagbemi, O.; Oluwagbemi, F.; Fagbore, O. Malavefes: A computational fuzzy voice-enabled anti-malarial drug informatics software for correct dosage prescription of anti-malaria drugs. *J. King Saud Univ.—Comput. Inf. Sci.* **2017**, *30*, 185–197. [CrossRef]
89. Oluwagbemi, O.; Awe, O. A comparative computational genomics of Ebola Virus Disease strains: In-silico Insight for Ebola control. *Inform. Med. Unlocked* **2018**, *12*, 106–119. [CrossRef]
90. Oluwagbemi, O.; Jatto, A. Implementation of a TCM-based computational health informatics diagnostic tool for Sub-Saharan African students. *Inform. Med. Unlocked* **2019**, *14*, 43–58. [CrossRef]

91. Oluwagbemi, O.O.; Oluwagbemi, F.E.; Jatto, A.; Hui, C. MAVSCOT: A fuzzy logic-based HIV diagnostic system with indigenous multi-lingual interfaces for rural Africa. *PLoS ONE* **2020**, *15*, e0241864. [CrossRef]
92. Center for Systems Science and Engineering (CSSE) at Johns Hopkins University (JHU). Available online: https://coronavirus.jhu.edu/map.html (accessed on 15 May 2022).
93. Japan COVID-19 Coronavirus Tracker. Available online: https://covid19japan.com/ (accessed on 3 April 2022).
94. China COVID-19 Dashboard. Available online: https://www.zoho.com/covid/china/ (accessed on 3 April 2022).
95. European COVID-19 Data Portal. Available online: https://www.covid19dataportal.org/ (accessed on 15 May 2022).
96. Cushnan, D.; Bennett, O.; Berka, R.; Bertolli, O.; Chopra, A.; Dorgham, S.; Favaro, A.; Ganepola, T.; Halling-Brown, M.; Imreh, G.; et al. An overview of the National COVID-19 Chest Imaging Database: Data quality and cohort analysis. *GigaScience* **2021**, *10*, giab076. [CrossRef] [PubMed]
97. Noordzij, M.; Duivenvoorden, R.; Pena, M.; De Vries, H.; Kieneker, L.M.; Franssen, C.F.M.; Hemmelder, M.H.; Hilbrands, L.B.; Jager, K.J.; Gansevoort, R.T.; et al. ERACODA: The European database collecting clinical information of patients on kidney replacement therapy with COVID-19. *Nephrol. Dial. Transplant.* **2020**, *35*, 2023–2025. [CrossRef] [PubMed]
98. Mittal, S.; Venugopal, V.K.; Agarwal, V.K.; Malhotra, M.; Chatha, J.S.; Kapur, S.; Gupta, A.; Batra, V.; Majumdar, P.; Malhotra, A.; et al. A Novel Abnormality Annotation Database for COVID-19 Affected Frontal Lung X-rays. *MedRxiv* **2021**. [CrossRef]
99. Latz, C.A.; DeCarlo, C.; Boitano, L.; Png, C.Y.M.; Patell, R.; Conrad, M.F.; Eagleton, M.; Dua, A. Blood type and outcomes in patients with COVID-19. *Ann. Hematol.* **2020**, *99*, 2113–2118. [CrossRef]
100. Wu, Y.; Feng, Z.; Li, P.; Yu, Q. Relationship between ABO blood group distribution and clinical characteristics in patients with COVID-19. *Clin. Chim. Acta* **2020**, *509*, 220–223. [CrossRef]
101. Xie, G.; Ding, F.; Han, L.; Yin, D.; Lu, H.; Zhan, M. The role of peripheral blood eosinophil counts in COVID-19 patients. *Allergy* **2021**, *76*, 471–482. [CrossRef]
102. Zhao, X.; Wang, K.; Zuo, P.; Liu, Y.; Zhang, M.; Xie, S.; Zhang, H.; Chen, X.; Liu, C. Early decrease in blood platelet count is associated with poor prognosis in COVID-19 patients—indications for predictive, preventive, and personalized medical approach. *EPMA J.* **2020**, *11*, 139–145. [CrossRef]
103. Brinati, D.; Campagner, A.; Ferrari, D.; Locatelli, M.; Banfi, G.; Cabitza, F. Detection of COVID-19 Infection from Routine Blood Exams with Machine Learning: A Feasibility Study. *J. Med. Syst.* **2020**, *44*, 135. [CrossRef] [PubMed]
104. Joob, B.; Wiwanitkit, V. Blood viscosity of COVID-19 patient: A preliminary report. *Am. J. Blood Res.* **2021**, *11*, 93–95. [PubMed]
105. Lan, L.; Xu, D.; Ye, G.; Xia, C.; Wang, S.; Li, Y.; Xu, H. Positive RT-PCR Test Results in Patients Recovered From COVID-19. *J. Am. Med Assoc.* **2020**, *323*, 1502. [CrossRef]
106. Elezkurtaj, S.; Greuel, S.; Ihlow, J.; Michaelis, E.G.; Bischoff, P.; Kunze, C.A.; Sinn, B.V.; Gerhold, M.; Hauptmann, K.; Ingold-Heppner, B.; et al. Causes of death and comorbidities in hospitalized patients with COVID-19. *Sci. Rep.* **2021**, *11*, 4263. [CrossRef]
107. National Center for Biotechnology Information (NCBI). Available online: https://www.ncbi.nlm.nih.gov/ (accessed on 21 May 2022).
108. EMBL's European Bioinformatics Institute (EMBL-EBI). Available online: https://www.ebi.ac.uk/ (accessed on 21 May 2022).
109. WebGRO. Available online: https://simlab.uams.edu/index.php. (accessed on 21 May 2022).
110. Abraham, M.J.; Murtola, T.; Schulz, R.; Páll, S.; Smith, J.C.; Hess, B.; Lindahl, E. GROMACS: High performance molecular simulations through multi-level parallelism from laptops to supercomputers. *SoftwareX* **2015**, *1–2*, 19–25. [CrossRef]
111. Kalimuthu, A.K.; Panneerselvam, T.; Pavadai, P.; Pandian, S.R.K.; Sundar, K.; Murugesan, S.; Ammunje, D.N.; Kumar, S.; Arunachalam, S.; Kunjiappan, S. Pharmacoinformatics-based investigation of bioactive compounds of Rasam (South Indian recipe) against human cancer. *Sci. Rep.* **2021**, *11*, 21488. [CrossRef]
112. Vishvakarma, V.K.; Singh, M.B.; Jain, P.; Kumari, K.; Singh, P. Hunting the main protease of SARS-CoV-2 by plitidepsin: Molecular docking and temperature-dependent molecular dynamics simulations. *Amino Acids* **2021**, *54*, 205–213. [CrossRef]
113. Gorai, S.; Junghare, V.; Kundu, K.; Gharui, S.; Kumar, M.; Patro, B.S.; Nayak, S.K.; Hazra, S.; Mula, S. Synthesis of Dihydrobenzofuro [3, 2-b] chromenes as Potential 3CLpro Inhibitors of SARS-CoV-2: A Molecular Docking and Molecular Dynamics Study. *ChemMedChem* **2022**, *17*, e202100782. [CrossRef]

computation

Article

Effect of Key Phytochemicals from *Andrographis paniculata*, *Tinospora cordifolia*, and *Ocimum sanctum* on PLpro-ISG15 De-Conjugation Machinery—A Computational Approach

Prachi Singh, Shruthi S. Bhat, Ardra Punnapuzha, Amrutha Bhagavatula, Babu U. Venkanna, Rafiq Mohamed * and Raghavendra P. Rao *

R&D, Himalaya Wellness Company, Makali, Bangalore 562162, India; prachi.singh@himalayawellness.com (P.S.); shruthi.bhat@himalayawellness.com (S.S.B.); histopathology.lab@himalayawellness.com (A.P.); amrutha.bh@himalayawellness.com (A.B.); dr.babu@himalayawellness.com (B.U.V.)
* Correspondence: dr.rafiq@himalayawellness.com (R.M.); raghavendra.pr@himalayawellness.com (R.P.R.)

Abstract: ISGylation is an important process through which interferon-stimulated genes (ISGs) elicit an antiviral response in the host cells. Several viruses, including the SARS-CoV-2, suppress the host immune response by reversing the ISGylation through a process known as de-ISGylation. The PLpro of SARS-CoV-2 interacts with the host ISG15 and brings about de-ISGylation. Hence, inhibiting the de-ISGylation to restore the activity of ISGs can be an attractive strategy to augment the host immune response against SARS-CoV-2. In the present study, we evaluated several phytochemicals from well-known immunomodulatory herbs, viz. *Andrographis paniculata* (AG), *Tinospora cordifolia* (GU), and *Ocimum sanctum* (TU) for their effect on deISGylation that was mediated by the PLpro of SARS-CoV2. For this purpose, we considered the complex 6XA9, which represents the interaction between SARS-CoV-2 PLpro and ISG15 proteins. The phytochemicals from these herbs were first evaluated for their ability to bind to the interface region between PLpro and ISG15. Molecular docking studies indicated that 14-deoxy-15-isopropylidene-11,12-didehydroandrographolide (AG1), Isocolumbin (GU1), and Orientin (TU1) from AG, GU, and TU, respectively possess better binding energy. The molecular dynamic parameters and MMPBSA calculations indicated that AG1, GU1, and TU1 could favorably bind to the interface and engaged key residues between (PLpro-ISG15)-complex. Protein–protein MMPBSA calculations indicated that GU1 and TU1 could disrupt the interactions between ISG15 and PLpro. Our studies provide a novel molecular basis for the immunomodulatory action of these phytochemicals and open up new strategies to evaluate drug molecules for their effect on de-ISGylation to overcome the virus-mediated immune suppression.

Keywords: innate immunity; interferon-stimulated genes (ISGs); ISGylation; phytochemicals; PLpro; immunomodulation

Citation: Singh, P.; Bhat, S.S.; Punnapuzha, A.; Bhagavatula, A.; Venkanna, B.U.; Mohamed, R.; Rao, R.P. Effect of Key Phytochemicals from *Andrographis paniculata*, *Tinospora cordifolia*, and *Ocimum sanctum* on PLpro-ISG15 De-Conjugation Machinery—A Computational Approach. *Computation* **2022**, *10*, 109. https://doi.org/10.3390/computation10070109

Academic Editors: Simone Brogi and Vincenzo Calderone

Received: 9 May 2022
Accepted: 13 June 2022
Published: 30 June 2022

Publisher's Note: MDPI stays neutral with regard to jurisdictional claims in published maps and institutional affiliations.

1. Introduction

The host innate immune system acts as the first line of defense during viral infections. Following the initial viral infection, a type-I interferon response is activated in the host, and this subsequently leads to the upregulation of several interferon-stimulated genes (ISGs). ISGs act as effectors of the interferon-mediated antiviral host immune response, which leads to an appropriate antiviral host immune response.

Among the ISGs, ISG15 is a well-studied ISG. It is shown to be robustly induced by type-I interferons [1] following various viral infections. ISG15 is a ubiquitin-like protein [2]. ISG15 is conjugated to target proteins upon its induction through a reaction called ISGylation. The ISGylation of target proteins is a key mechanism through which ISG15 mediates its effect. It plays critical roles in various phases of the host innate immune response against viruses [3]. Several important immune regulatory transcription factors and receptors have been established as substrates for ISGylation by ISG15. ISG15 is shown to limit the viral replication in many viruses [4]. Hence, ISG15 is an important component in the host immune response pathway against viruses.

Viruses have developed mechanisms to defend against the ISG-mediated immune response. Many viruses, such as herpes simplex virus, norovirus, chikungunya virus, and HIV evade the host immune system by counteracting the ISG15-mediated pathways [1]. Studies indicate that many viruses express proteins that possess ISG15 de-conjugating activity [5]. De-ISGylation refers to the process of the deconjugation of ISG from target proteins. It is considered an important means through which viruses evade the interferon-mediated innate immune response. Even in the case of SARS-CoV-2, the PLpro protein is shown to exhibit de-ISGylation activity, potentially leading to a diminished early-host immune response [6]. Suppressing the early-phase host immune system by SARS-CoV-2 is proposed as a key mechanism that can further lead to exaggerated viral replication, viral dominance over the host, and cytokine response at the later stages of infection [7–9]. Hence, the therapeutic strategies aimed at enhancing the early-stage host immune response against the SARS-CoV-2 or other viruses should consider targeting the de-ISGylation mechanism. Although vaccination is one of the effective strategies to elicit a specific (adaptive), immune response against the viruses, additional interventions to enhance the early-phase host innate immune response will be an added advantage.

Many medicinal herbs and their extracts have been used to treat and manage viral infections. Various phytochemicals that are present in *Andrographis paniculata* (AG), *Tinospora cordifolia* (GU), and *Ocimum sanctum* (TU) have been shown to possess antiviral and immune-potentiating activity [10–12]. For example, Andrographolide can modulate the innate and adaptive immune responses by regulating macrophage phenotypic polarization [11]. Furthermore, the phytochemicals from *Tinospora cordifolia* show association with immune pathways and act as immunomodulators [12,13]. Previous studies have identified phytochemicals from AG, GU, and TU as potent inhibitors of SARS-CoV-2 [14,15]. Treatment with these herbs in the initial phases of viral infection is shown to have beneficial effects. Most of these studies describe the immune modulatory mechanism of these herbs at the cellular level, involving immune cells and their respective pathways. However, the nature of the molecular targets that are engaged by these phytochemicals is unclear from the literature. A molecular-level understanding of the targets that are engaged by the bioactive compounds will add value for evaluating the immunomodulatory herbs as broad spectrum antiviral immunomodulators. Since de-ISGylation is an important mechanism that is employed by viruses to suppress the host immunity, we hypothesized that phytochemicals from these herbs might act on this arm to bring about their immune-potentiating activity.

In order to evaluate this hypothesis, in the current study, we considered a protein complex 6XA9, which represents the interaction between the SARS-CoV2 PLpro and C-terminal of ISG15 [16]. We refer to this structure as (PLpro-ISG15)-complex throughout the manuscript. We considered key interacting residues between PLpro and ISG15 in this complex for evaluating the potential phytochemicals for their ability to bind to this region and affect the protein–protein interaction. A systematic approach was followed in which the phytochemicals from AG, GU, and TU herbs were first screened by molecular docking. Following this, the top-ranked phytochemicals from these herbs, as per the docking scores, were further evaluated by MD simulation for 300 nanoseconds. In addition, we analysed MM/PBSA (molecular mechanics/Poisson–Boltzmann Surface Area) for protein-ligand and protein–protein interactions. GRL0617, which is known to inhibit de-ISGylation, was also included in our studies for a comparison of the parameters. Our results indicated that 14-deoxy-15-isopropylidene-11,12-didehydroandrographolide (AG1), Isocolumbin (GU1), Orientin (TU1), and GRL0617 could favorably bind to the interface region between PLpro and ISG15. Our results also indicated that among these ligands, TU1 and GU1 could potentially disrupt the PLpro and ISG15 interactions.

2. Results and Discussion

2.1. Screening of Phytochemicals of Andrographis paniculata (AG), Tinospora cordifolia (GU), and Ocimum sanctum (TU) against SARS-CoV-2 PLpro ISG15 Site at UIM (PDB:6XA9)

The key phytochemical constituents of AG, GU, and TU were selected based on the literature search, viz. PubMed (https://www.ncbi.nlm.nih.gov/pmc/) (accessed on 9 July 2021), Google Scholar (https://scholar.google.com/) (accessed on 9 July 2021), and DOAJ (https://doaj.org/) (accessed on 9 July 2021) and are given in Supplementary Table S1. A total of 90 phytochemicals from AG, GU, and TU were docked against the target protein of SARS-CoV-2 (PDB) ID: 6XA9) using AutoDock Vina [17]. The AutoDock Vina results represent the docking scores as the Gibbs free energy of binding (ΔG (kcal/mol), which approximates the sum of all interactions between ligand and receptor minus the desolvation energies. The binding energies of the top five phytochemicals from AG, GU, and TU in order of increasing docking scores are given in Table 1. The lesser the value, the better the binding affinity.

Docking against SARS-CoV-2 PLpro ISG15 Interacting Site

The dual role of PLpro in viral peptide cleavage and immune regulation has made it an important target for inhibiting SARS-CoV-2 infectivity. PLpro inhibits the host innate immune response by reversing ISG15 modifications from the proteins. The PLpro of the coronavirus family has a ubiquitin-interacting motif (UIM) that can recognize and hydrolyze ubiquitin (Ub) and the ubiquitin-like protein ISG15 (interferon-induced gene 15). However, SARS-CoV-2 PLpro preferentially catalyzes de-ISGylation over de-ubiquitylation [18,19]. SARS-CoV-2 PLpro UIM accommodates both ubiquitin and ISG15 binding sites. In the current work, 6XA9 crystal structure (SARS-CoV-2 PLpro in complex with ISG15 C-terminal domain) was taken for the study. This complex represents the interaction between PLpro and ISG15. In this complex, Ser170, Tyr171, Phe216, Gln195, Thr225, Lys232, Asn151, and Asn156 from PLpro and Trp123, and Pro130/Glu132 from ISG15 are key interacting amino acid residues that play critical roles in de-ISGylation. In addition, Met208, Glu167, and Arg166 are also important residues in the S2 palm domain of the PLpro that interact with the substrate [16]. In Supplementary Figure S1, we have mapped the respective amino acid of the interface in a 3D structure format. The rationale of the study was to look for the phytochemicals that bind to the key residues representing the interface between PLpro and ISG15. The effective binding of the phytochemicals to the interface region or interfering with the PLpro and ISG15 interaction would potentially inhibit the de-ISGylation activity.

GRL0617 binds to the ISG15 interacting site of PLpro [20]. Hence, we included this molecule as a positive control for docking studies. Our docking results indicate that GRL0617 shows ΔG of −8.5 kcal/mol. The top three phytochemicals from AG, 14-deoxy-15-isopropylidene-11,12-didehydroandrographolide (−9.4 kcal/mol); Andrographolactone (−9.2 kcal/mol); and Neoandrographolide (−9 kcal/mol) showed the highest binding affinity amongst all the screened phytochemicals and were better than the control drug molecule. The top three phytochemicals from AG are diterpenes which possess immunomodulatory and antiviral activity [21,22]. The top-ranked phytochemicals from GU and TU were Isocolumbin (−9.9) and Orientin (−9.4 kcal/mol), respectively.

The docking of phytochemicals shows that they form hydrogen bonds and other non-covalent and electrostatic interactions with major amino acid residues of PLpro at the ISG15 interacting site. The two-dimensional binding interactions of the top phytochemicals from AG, GU, and TU with the PLpro ISG15 binding site are shown in Figure 1.

Table 1. Docking scores of various phytochemicals from *A. paniculata* (AG), *T. cordifolia* (GU), *O. sanctum* (TU) against SARS-CoV-2 (PLpro-ISG15) complex interacting site (PDB ID:6XA9).

S.No.	Herb	Phytochemical	PubChem CID	Canonical SMILES	ΔG (kcal/mol)
					Plpro ISG15 Interaction Site at UIM (PDB: 6XA9)
1		14-deoxy-15-isopropylidene-11,12-didehydroandrographolide	637300	CC(=C1C=C(C(=O)O1)C=CC2C(=C)CCC3C2(CCC(C3(C)CO)O)C)C	−9.4
2		Andrographolactone	44206466	CC1=CC2=C(CCC1)C(=C(C(=C2)C)CCC3=CCOC3=O)C	−9.2
3	Andrographis paniculata	Neoandrographolide	9848024	CC1(CCCC2(C1CCC(=C)C2CCC3=CCOC3=O)C)COC4C(C(C(C(O4)CO)O)O)O	−9
4		14-Deoxy-11,12-didehydroandrographolide	5708351	CC12CCC(C(C1CCC(=C)C2C=CC3=CCOC3=O)(C)CO)O	−8.8
5		Andrographolide	5318517	CC12CCC(C(C1CCC(=C)C2CC=C3C(COC3=O)O)(C)CO)O	−8.7
6		Isocolumbin	24721165	CC12CCC3C(=O)OC(CC3(C1C4C=CC2(C(=O)O4)O)C)C5=COC=C5	−9.9
7		Berberin	2353	COC1=C(C2=C[N+]3=C(C=C2C=C1)C4=CC5=C(C=C4CC3)OCO5)OC	−9.4
8	Tinospora cordifolia	Ecdysterone	12304165	CC12CCC3C(=CC(=O)C4C3(CC(C(C4)O)O)C)C1(CCC2C(C)(C(CCC(C)(C)O)O)O)O	−9
9		Magnoflorine	73337	C[N+]1(CCC2=CC(=C(C3=C2C1CC4=C3C(=C(C=C4)OC)O)O)OC)C	−9
10		Beta-Sitosterol	222284	CCC(CCC(C)C1CCC2C1(CCC3C2CC=C4C3(CCC(C4)O)C)C)C(C)C	−8.4
11		Orientin	5281675	C1=CC(=C(C=C1C2=CC(=O)C3=C(O2)C(=C(C=C3O)O)C4C(C(C(C(O4)CO)O)O)O)O)O	−9.4
12		Isoorientin	114776	C1=CC(=C(C=C1C2=CC(=O)C3=C(O2)C=C(C(=C3O)C4C(C(C(C(O4)CO)O)O)O)O)O)O	−9.2
13	Ocimum sanctum	Vitexin	5280441	C1=CC(=CC=C1C2=CC(=O)C3=C(O2)C(=C(C=C3O)O)C4C(C(C(C(O4)CO)O)O)O)O	−9.1
14		Isovitexin	162350	C1=CC(=CC=C1C2=CC(=O)C3=C(O2)C=C(C(=C3O)C4C(C(C(C(O4)CO)O)O)O)O)O	−9
15		Molludistin	44258315	COC1=C(C2=C(C(=C1)O)C(=O)C=C(O2)C3=CC=C(C=C3)O)C4C(C(C(CO4)O)O)O	−8.8
16	Positive control	GRL0617	24941262	CC1=C(C=C(C=C1)N)C(=O)NC(C)C2=CC=CC3=CC=CC=C32	−8.5

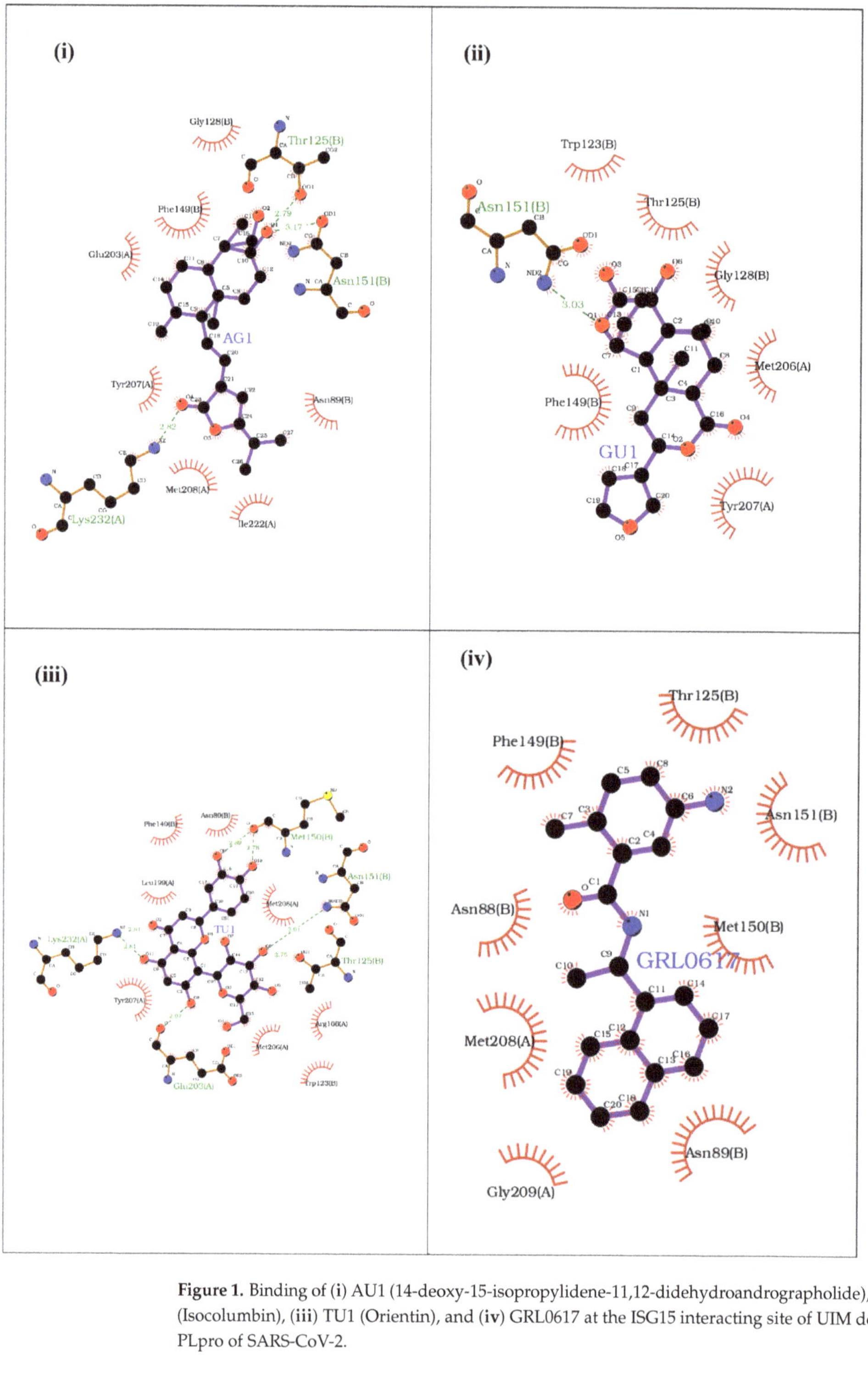

Figure 1. Binding of (**i**) AU1 (14-deoxy-15-isopropylidene-11,12-didehydroandrographolide), (**ii**) GU1 (Isocolumbin), (**iii**) TU1 (Orientin), and (**iv**) GRL0617 at the ISG15 interacting site of UIM domain of PLpro of SARS-CoV-2.

To analyze the molecular interactions (H-bond, van der Waals bonds, pi–pi interactions, salt bridges) between various phytochemicals against PLpro ISG15 binding site, LIGPLOT software (Ligplot version 4.15.0-142 generic, E.M.B.L., Hinxton, Cambridgeshire, UK) was used [23]. The 14-deoxy-15-isopropylidene-11,12-didehydroandrographolide phytochemical from AG interacts with the UIM through hydrogen bond interactions with Lys232 (PLpro), Thr125 (ISG15), and Asn151 (ISG15), whereas Met208, an important residue in the S2 palm domain, lies in the binding pocket. (Figure 1i). Isocolumbin from GU interacts with the UIM through hydrogen bond interactions with Asn151 (ISG15) and other important amino acid residues surrounding the binding interface site (Figure 1ii). Orientin from TU forms hydrogen bonds with Lys232 (PLpro), Glu203 (PLpro), Met150 (ISG15), and Asn151 (ISG15), along with other non-covalent interactions that further stabilize the binding (Figure 1iii). The positive control GRL0617 perfectly fits at the ISG15 binding site of UIM, being surrounded by the key residues, such as Met208 (PLpro) and Gly209 (PLpro) that stabilizes the binding of the drug molecule (Figure 1iv). Our results suggest that the top ranked phytochemicals from AG, GU and TU interact with the critical residues of the ISG15 binding site of UIM and, therefore, can destabilize the PLpro and ISG15 interaction, thereby potentially inhibiting the de-ISGylation activity of the PLpro.

2.2. *Physiochemical Properties Analysis of the Phytoactives Affirms Their Drug-Likeness*

The ADME property of a drug is important in determining its safety and efficacy. The molecular weight, hydrogen bond acceptors and donors, lipophilicity, etc., are important parameters for the generation of an effective and successful drug. The Lipinski rule of five can be applied for filtering out the best potential drug. According to the Lipinski rule, a molecule with a molecular weight of <500 Dalton, a maximum of five hydrogen bond donors, and 10 hydrogen bond acceptors with a log p value of <5 have a better drug-likeness than others that fail these parameters. The drug-likeness and ADMET properties of top-ranked phytochemicals AG1, GU1, TU1, and GRL0617 were calculated using web tools SWISS ADME and PreADMET, and the results are represented in Supplementary Table S2i,ii. The top-ranked phytochemical from AG or GU or GRL0617 showed a drug-likeness property, making them potential lead molecules against the PLpro of SARS-CoV-2. Orientin from TU has the molecular weight and LogP values within the threshold range but violates the rule in terms of hydrogen-bonding properties. However, in vivo studies indicate that Orientin is quickly distributed to the kidney, liver, and lung [24].

2.3. *Molecular Dynamics (MD) Simulation Study*

The effectiveness of the screened phytochemicals was further analyzed by performing all-atom MD simulations for 300 ns using GROMACS. MD simulation studies provide insights on the dynamic state of the ligands at the interaction site of the target protein in the presence of an ionic aqueous environment. In addition, they provide an elaborate understanding not only of the molecular dynamics of ligand-protein complexes, but also evaluate the crucial interactions during the time scale of few nanoseconds.

2.3.1. Stability and Fluctuations of the Protein: RMSD and RMSF Analysis of the Protein Complex

To understand the predicted binding modes of the candidate phytochemicals that were selected from docking studies, we illustrated the detailed interactions of both the (PLpro-ISG15)-complex and (PLpro-ISG15)-complex with the phytochemicals, viz. AG1, GU1, TU1 or GRL0617 over the 300 ns MD simulation (Figure 2). Figure 2i–iv(A) represents the 3D and 2D images of the phytochemicals with (PLpro-ISG15)-complex at 0 ns and 300 ns. As clear from the Figures, the phytochemicals could bind to the interface at 0 ns and 300 ns. Once the proper binding of the phytochemicals within the interface cavity was confirmed, the other parameters such as RMSD, RMSF, SASA, Rg, hydrogen bond, PCA analysis, and MM-PBSA were evaluated over the entire 300 ns simulation time. PCA analysis data can be found in supplementary information (Figure S2).

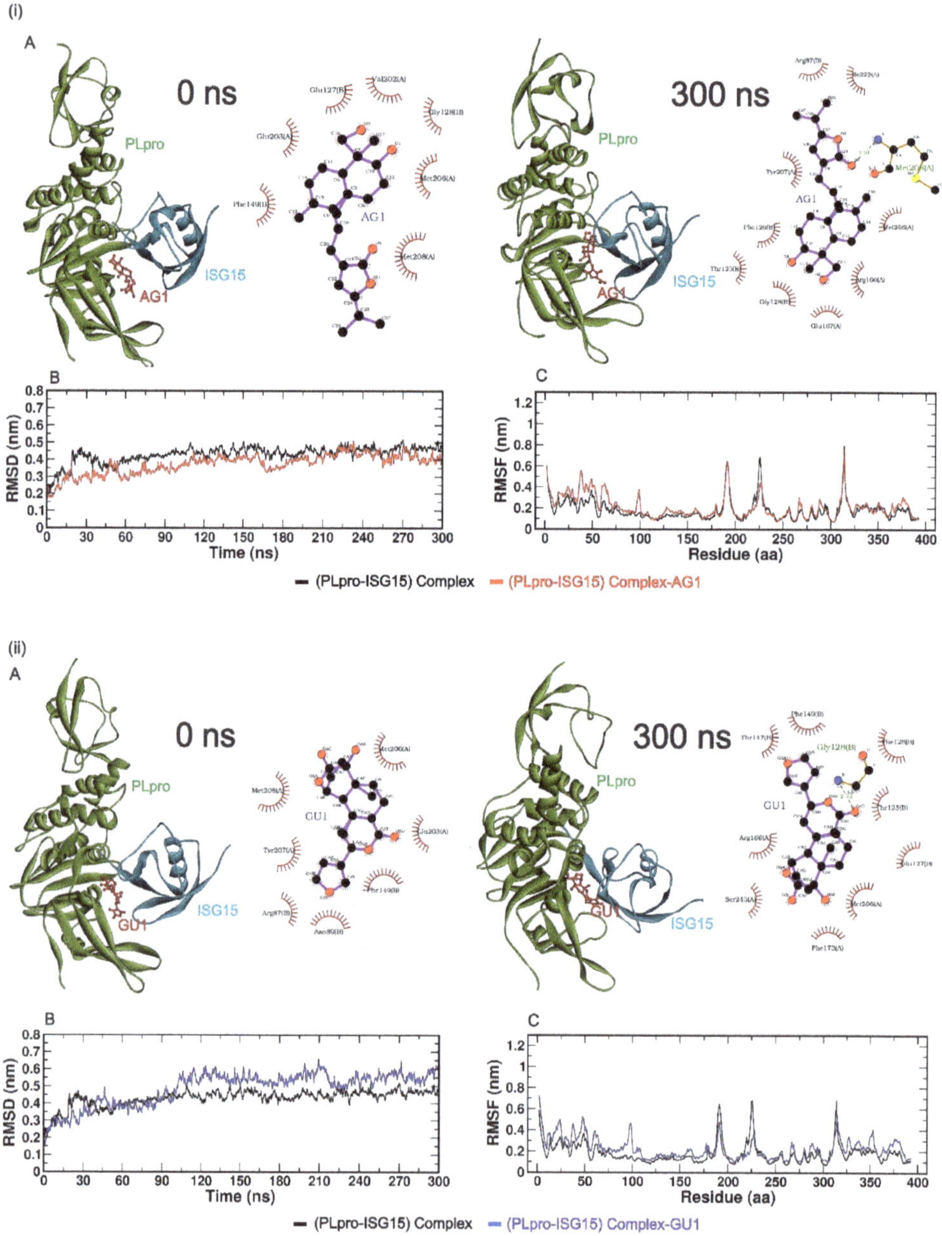

Figure 2. *Cont.*

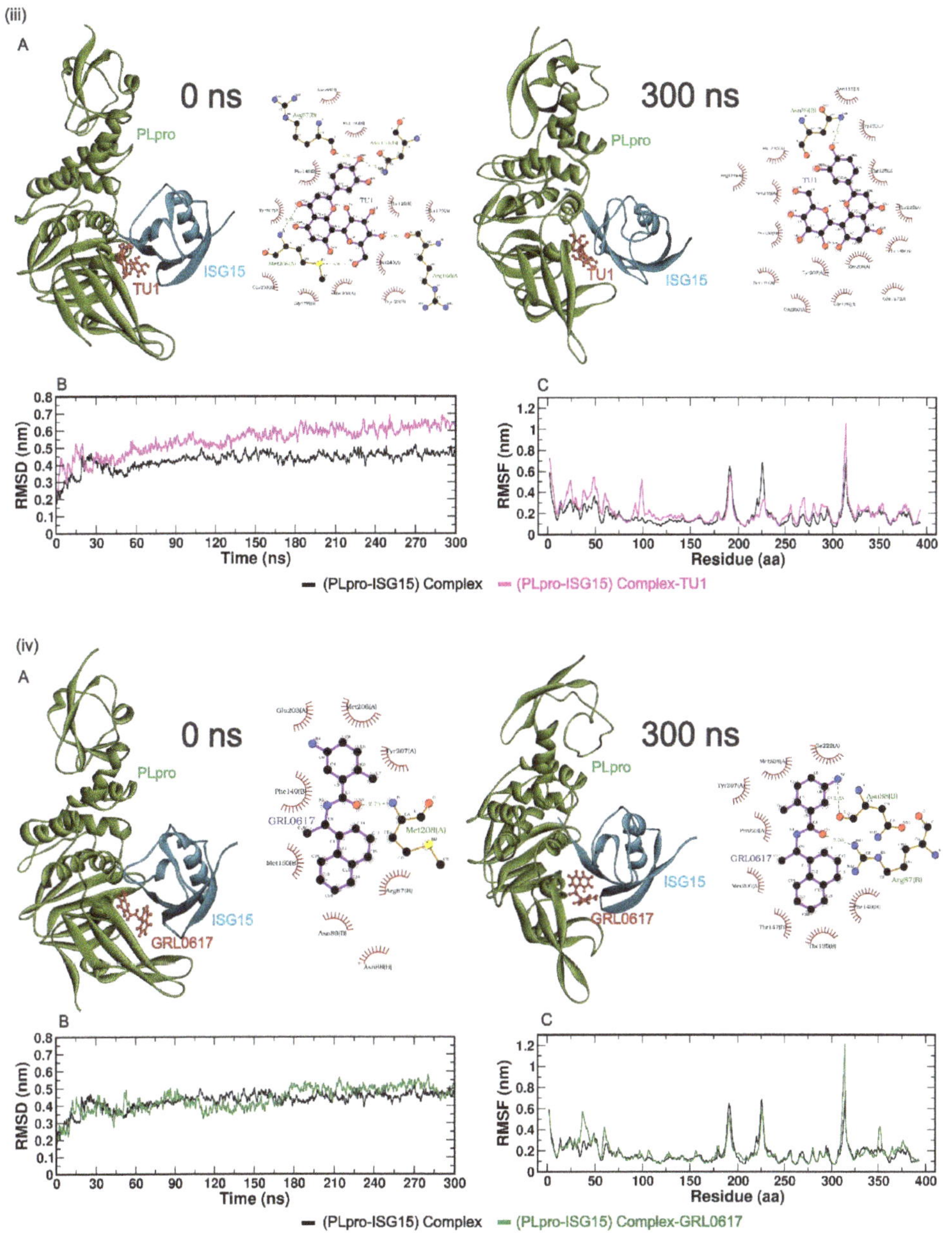

Figure 2. (**i–iv** (**A**))—The best poses of the phytochemicals (AG1 (red), GU1 (blue), TU1 (magenta), and GRL0617 (green)) binding to the interface of (PLpro-ISG15)-complex at 0 ns (*left*) and 300 ns (*right*) during molecular dynamics (MD) simulation. Each inset shows the detailed interactions of each drug candidate docked to the ISG15 interacting site of the PLpro, indicating the amino acids involved in the interaction and the type of interaction (hydrogen bonds, hydrophilic interactions, salt bridges, Π-stacking, etc.). (**i–iv** (**B**)) show RMSD, and (**i–iv** (**C**)) show RMSF values over 300 ns simulation for different complexes with phytochemicals.

In order to evaluate the stability profiles of the (PLpro-ISG15) complex with phytochemicals, RMSD was calculated for C_α backbone atoms of protein over the entire period of 300 ns simulation. RMSD values provide information on the extent of deviation of a given protein–ligand complex, compared to a reference structure over the simulation time. A lower deviation of the given protein–ligand complex from respective reference structures indicates a suitable accommodation of the ligand within the binding pocket.

The time evolution of RMSD for different protein–ligand complexes relative to the initial structure of (PLpro-ISG15)-complex is presented in (Figure 2i–iv(B)). As seen in Figure 2, the RMSD values for the (PLpro-ISG15)-complex increased until 50 ns, after which the values stabilized and leveled off. The same trend was observed for (PLpro-ISG15)-AG1 and (PLpro-ISG15)-GRL0617. The average RMSD values between 50 and 300 ns for (PLpro-ISG15)-complex, (PLpro-ISG15)-AG1, and (PLpro-ISG15)-GRL0617 were 0.44 nm, 0.39 nm, and 0.45 nm, respectively. In the case of (PLpro-ISG15)-GU1 complex, the system equilibrated between 50 and 100 ns (RMSD = 0.40 nm) but later in the trajectory, the RMSD values showed fluctuation. However, this structure maintained an average RMSD of 0.55 nm between 100 and 300 ns. In the case of (PLpro-ISG15)-TU1 complex, the ligand-bound complex deviated from the (Plpro-ISG15)-complex structure, and its average RMSD value was 0.57 nm (between 50 and 300 ns). The higher RMSD values and fluctuations in the RMSD trajectory indicate that the binding of GU1 and TU1 might affect the protein–protein interactions between PLpro and ISG15.

Next, in order to investigate the local fluctuations at the residue level before and after binding with the phytochemicals, the root mean square fluctuation (RMSF) of $C\alpha$ atom for the entire 300 ns was predicted (Figure 2i–iv(C)). The residues from 1 to 314 correspond to PLpro, and 315 to 400 correspond to ISG15. As indicated in (Figure 2i–iv(C)), there were no significant fluctuations between residues 160–175 and 200–210. This region corresponds to the PLpro and ISG15 interface region. On the other hand, fluctuations were observed in the loop regions for all the complexes. In the case of (PLpro-ISG15)-complex and the (PLpro-ISG15)-complexes with AG1, TU1, and GRL0617, there were fluctuations in the region between 180 and 200 residues. Similarly, (PLpro-ISG15)-complex and (PLpro-ISG15)-complex with GRL0617 showed some fluctuations in the region corresponding to 220–230 residues. However, these fluctuations were not significantly higher than those that were observed for loop regions. The fluctuations were observed between the 310 and 320 residues, between the chains of PLpro and ISG15.

2.3.2. Compactness of Protein Complex: Radius of Gyration (Rg) and Solvent Accessible Surface Area (SASA)

Rg is a parameter that scores for the compactness of protein. We evaluated the compactness of (PLpro-ISG15)-complex upon binding with the phytochemicals over the course of MD simulation Figure 3i(A–D). The average Rg value of (PLpro-ISG15)-complex, (PLpro-ISG15)-complex with AG1, (PLpro-ISG15)-complex with GU1, (PLpro-ISG15)-complex with TU1, and (PLpro-ISG15)-complex with GRL0617 were found to be 2.35 nm, 2.32 nm, 2.33 nm, 2.31 nm, and 2.31 nm, respectively. As shown in Figure 3i(A–D), the Rg values of both (PLpro-ISG15)-complex and (Plpro-ISG15)-complex with phytochemicals did not significantly change over 300 ns simulation, suggesting that all the systems were compact.

Flexibility and compactness are correlated with each other. The SASA is a useful parameter for understanding the conformational dynamics of a protein in a solvent environment. In the current study, we evaluated the SASA of the selected docked complexes over 300 ns MD simulation. Figure 3i(A–D) shows the time-dependent SASA plot. The average SASA values of (Plpro-ISG15)-complex, (Plpro-ISG15)-complex with AG1, Plpro-ISG15 complex with GU1, (Plpro-ISG15)-complex with TU1, and (Plpro-ISG15)-complex with GRL0617 were found to be 179.77 nm^2, 187.73 nm^2, 190.56 nm^2, 188.58 nm^2 and 188.92 nm^2, respectively. All the complexes with phytochemicals showed slightly higher values of SASA compared to (PLpro-ISG15)-complex.

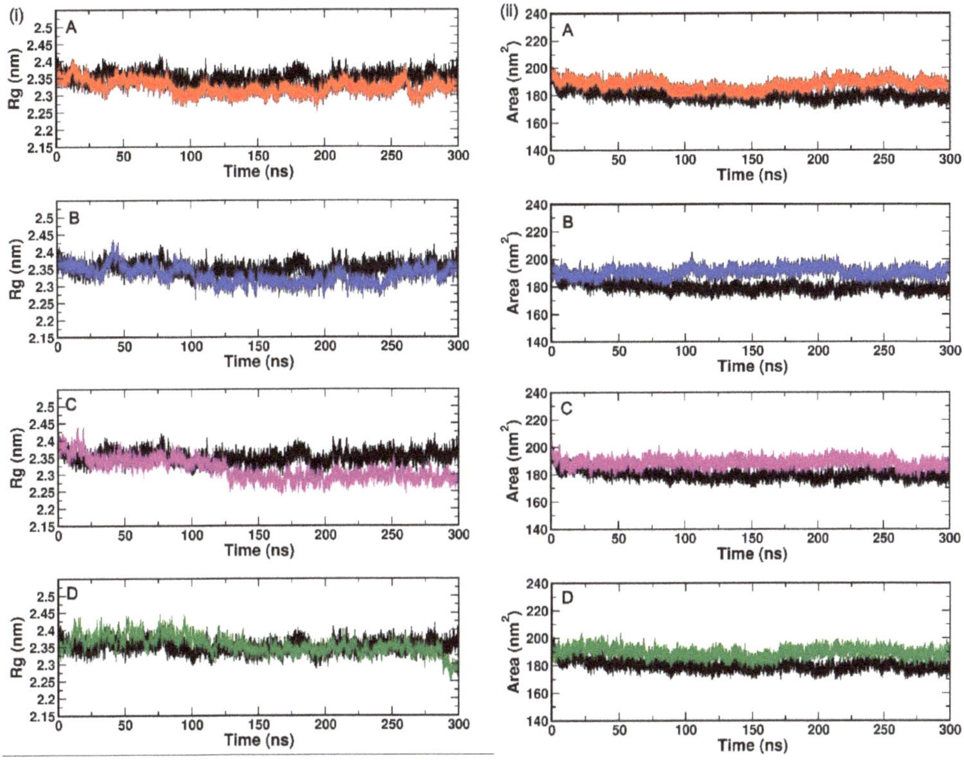

Figure 3. (**i**(**A–D**))—Rg values of backbone atoms over 300 ns simulation of different (Plpro-ISG15)-complex with phytochemicals (AG1 (red), GU1 (blue), TU1 (magenta), and GRL0617 (green)) (*left*). (**ii**(**A–D**))—SASA values of backbone atoms over 300 ns simulation for different (Plpro-ISG15)-complexes with phytochemicals (AG1 (red), GU1 (blue), TU1 (magenta), and GRL0617 (green)) (*right*). The black line represents (Plpro-ISG15)-complex without phytochemicals.

2.3.3. Interactions between the Protein–Ligand Complex: Hydrogen Bond (H-Bond)

Hydrogen bonding plays a critical role in stabilizing the protein–ligand interactions. In our study, the number of H-bonds were calculated over a simulation time of 300 ns for all the complexes. According to Figure 4, (PLpro-ISG15)-complex-AG1, (PLpro-ISG15)-complex-GU1, (PLpro-ISG15)-complex-TU1, and (PLpro-ISG15)-complex-GRL0617 all had an average of four to eight H-bonds each. In addition, the H-bond analysis indicated that the protein–ligand complexes remained stable during simulation.

2.3.4. Binding Affinities of Phytochemicals to the Interface of PLpro-ISG15 Complex: MMPBSA Based Calculations

A detailed understanding of the interactions between (PLpro-ISG15)-complex and the phytochemicals (AG1, GU1, TU1, and GRL0617) is feasible by investigating the thermodynamic parameters that were calculated by the simulations. The MMPBSA method has been widely used to quantify protein–ligand affinities [25]. During 300 ns MD simulation, we extracted an ensemble containing conformations corresponding to the last 50 ns and investigated the details of protein–ligand interactions. The amino acids that were located within the vicinity of 3.5 A distance from the ligand were identified in this average structure. These amino acids represent the critical residues that are involved in binding and interactions, and they were chosen for calculating the free energy of binding. Figure 5 provides more details on these critical residues and their binding energy contributions to each interacting partner of (PLpro-ISG15)-complex.

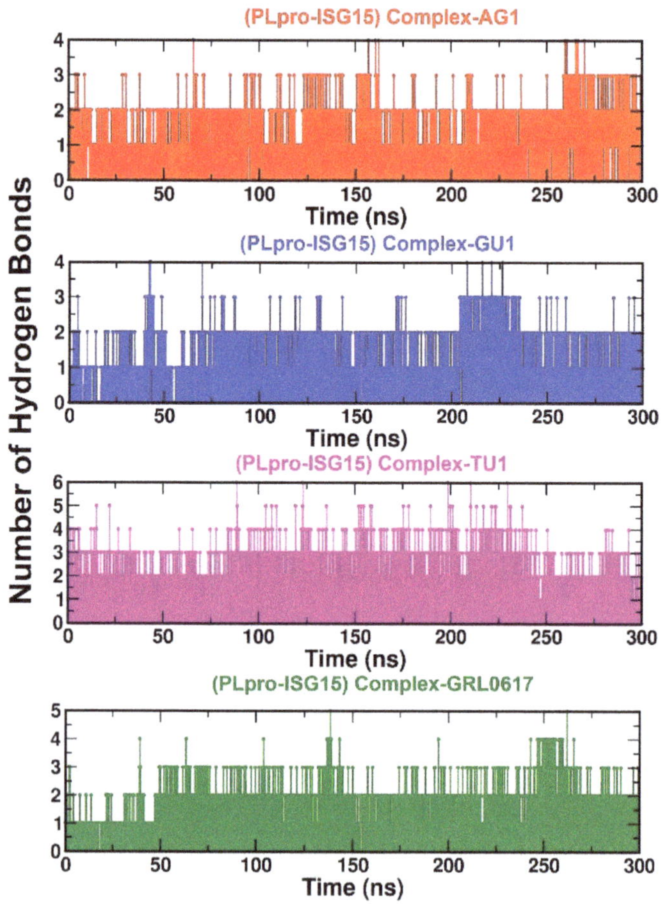

Figure 4. Number of hydrogen bonds formed between the phytochemicals (AG1 (red) or GU1 (blue) or TU1 (magenta) or GRL0617 (green)) and the (PLpro-ISG15) complex during 300 ns MD simulation.

As detailed in this figure, GRL0617, a known inhibitor of de-ISGylation, engages Val-202 Glu-203 Met-206 Tyr-207 Met-208 Gln-221 Ile-222 Pro-223 from PLpro, and Arg-87, Asn-88, Asn-89, Lys-90, Gly-91, Arg-92, Thr-125, Phe-126, Glu-127, Gly-128, Thr-147, Val-148, Phe-149 from ISG-15. Since GRL0617 is shown to compete with ISG15 for the binding site on PLpro, the amino acids that GRL0617 binds may play an important role in inhibiting de-ISGylation. In order to gain more insight into the similar mechanism of action for AG1, GU1, and TU1 in comparison to GRL0617, we compared the number of amino acids shared between different phytochemicals. As indicated in Table 2, all the phytochemicals (AG1, GU1, and TU1) shared four amino acids, namely Glu-203, Met-206, Tyr-207, and Met-208, in common with GRL0617 for binding to PLpro. Similarly, when compared to GRL0617, all the phytochemicals shared Glu-127, Gly-128, Phe-149, Thr-125, and Phe-126 in common with GRL0617 for binding to ISG15. These analyses indicate that AG1, GU1, and TU1 interact with critical residues from the (PLpro-ISG15)-complex, similar to GRL0617.

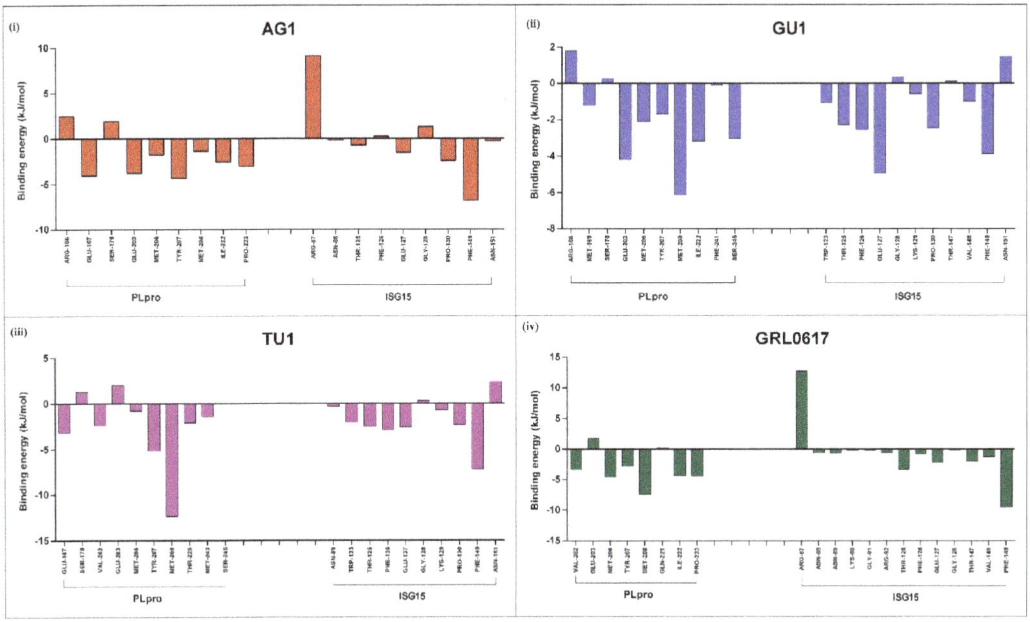

Figure 5. Binding energies (kJ/mol) at the individual amino acid levels were calculated for the average structure of the last 50 ns for each of the complexes. Key interacting residues from PLpro or ISG15 were considered for calculating binding energy for AG1 (**i**), GU1 (**ii**), TU1 (**iii**) and GRL0617 (**iv**) in each case.

Table 2. Common amino acids engaged by different phytochemicals for binding to PLpro and ISG-15 complex.

Interacting Partner	Ligands/Phytochemicals Bound	Common Amino Acids	Number of Common Amino Acids
PLpro	AG1, GRL0617, GU1 TU1	Glu-203, Met-208, Met-206, Tyr-207	4
	AG1, GU1, TU1	Ser-170	1
	AG1, GRL0617, GU1	Ile-222	1
	AG1, GU1	Arg-166	1
	AG1, TU1	Glu-167	1
	AG1, GRL0617	Pro-223	1
	GU1, TU1	Ser-245	1
	GRL0617, TU1	Val-202	1
	GU1	Met-169, Phe-241	2
	TU1	Thr-225, Met-243	2
	GRL0617	Gln-221	1
ISG15	AG1, GRL0617, GU1, TU1	Glu-127, Gly-128, Phe-149, Thr-125, Phe-126	5
	AG1, GU1, TU1	Pro-130, Asn-151	2
	AG1, GRL0617	Arg-87, Asn-88	2
	GU1, TU1	Lys-129, Trp-123	2
	GRL0617, GU1	Thr-147, Val-148	2
	GRL0617, TU1	Asn-89	1
	GRL0617	Arg-92, Gly-91, Lys-90	3

Next, to gain insight into the total binding free energy (kJ/mol) for protein–ligand interactions, we analyzed the total binding energy (kJ/mol) of the phytochemicals binding to the key interacting residues from PLpro or ISG15 or (PLpro-ISG15)-complex. As indicated in Figure 6, the binding energy for AG1, GU1, and TU1 were −17.8 kJ/mol, −36.14 kJ/mol, and −41.9 kJ/mol, respectively. GRL0617 showed a value of −33.50 kJ/mol. Overall, GU1 and TU1 had binding affinities that were comparable to or higher than the positive control GRL0617.

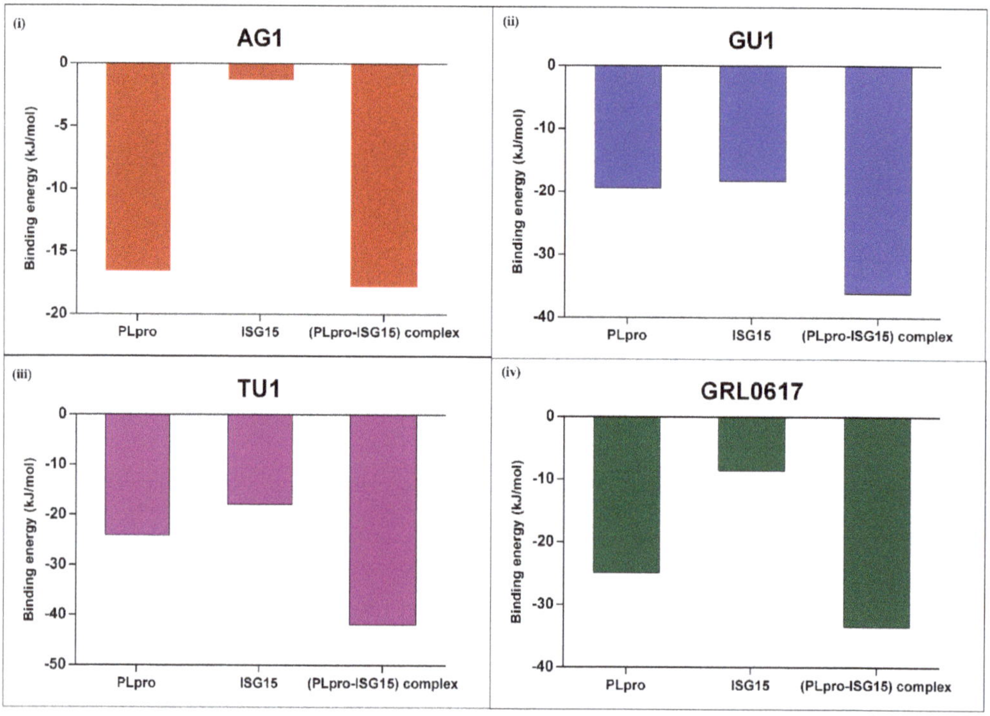

Figure 6. Binding energies (kJ/mol) of the complexes were calculated by MMPBSA method. Key interacting residues from PLpro or ISG15 or (PLpro-ISG15) complexes were considered for calculating binding energy for AG1 (**i**), GU1 (**ii**), TU1 (**iii**) and GRL0617 (**iv**).

2.3.5. Effect of Phytochemicals on Protein–Protein Interactions between (PLpro-ISG15)-Complex

As mentioned earlier, PLpro possesses de-ISGylation activity through which it inhibits the immunomodulatory activity of ISG15. For the effective de-ISGylation activity of PLpro, the interaction between PLpro and ISG15 should be maintained stably. Our observations with RMSD values hint that at least GU1 and TU1 may result in disruption of this protein–protein interaction. Work that was carried out by other researchers [26] used the total MMPBSA values to score for protein–protein interactions, and an increase in the MMPBSA for protein–protein complexes was used as an indicator of the destabilization of protein–protein interactions. We employed a similar approach and calculated the MMPBSA for the average structure for the final 50 ns during our simulation periods. Figure 7 provides the details on this. The binding energy for the interaction between PLpro and ISG15 was 437.58 kJ/mol. At the same time, binding the ligands GU1 and TU1 to this complex increased the free energy. This indicates that the binding of GU1 and TU1 to the (PLpro-ISG15)-complex decreased the binding affinity between PLpro and ISG15 proteins. The binding of GRL0617 to (PLpro-ISG15)-complex had a similar trend, but the effect may be less. AG1, on the other hand, did not increase the free energy of binding, instead it slightly decreased it.

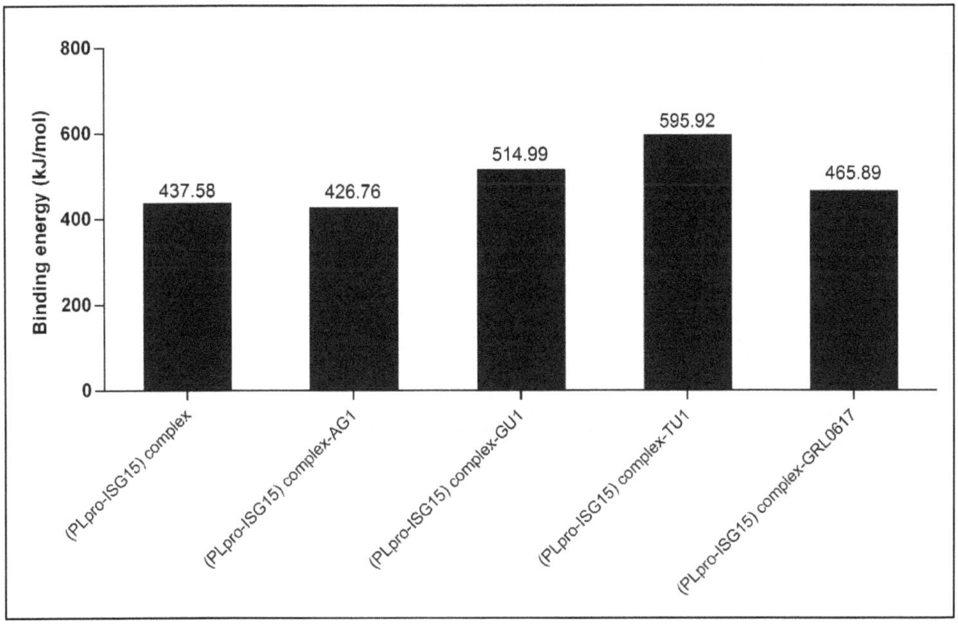

Figure 7. Protein–protein MMPBSA value (kJ/mol) for PLpro-ISG15 complex, PLpro-ISG15 complex with AG1, PLpro-ISG15 complex with GU1, PLpro-ISG15 complex with TU1 and PLpro-ISG15 complex with GRL0617.

In summary, our studies indicate that 14-deoxy-15-isopropylidene-11,12-didehydroandrographolide from AG, Isocolumbin from GU, and Orientin from TU all accommodated favorably in the interface region of the complex between PLpro and ISG15 proteins. The amino acids that were engaged by these phytochemicals were similar to those that were engaged by GRL0617, indicating that these phytochemicals, similar to GRL0617, interfere with the interaction between PLpro and ISG15 complex. The critical residues of the ISG15 interacting site of the PLpro were well engaged in the interaction with all these ligands, as indicated by docking studies. The efficient binding of the ligands to this interface may potentially interfere with the function of PLpro, i.e., de-ISGylation. In the case of Isocolumbin from GU and Orientin from TU, the protein–protein interaction between PLpro and ISG15 was affected.

3. Materials and Methods

3.1. Target Enzyme Preparation

The PDB structure of a complex between SARS-CoV-2 PL^{pro} and ISG15 (Protein Data Bank (PDB) ID: 6XA9) [16] was obtained from PDB (https://www.rcsb.org/) (accessed on 9 July 2021) and saved as a PDB file (.pdb). AutoDock Tools 1.5.6 was used to prepare the protein targets [17]. Chain A representing PLpro and Chain B representing ISG15 of the 6XA9 complex were retained for the study. Where any water molecules or ligands were removed, polar hydrogen atoms and Kollman charges were added to the protein and saved as a PDBQT file (.pdbqt).

3.2. Ligand Preparation

The 3D structure of the phytochemical compounds (total 90 phytochemicals) was obtained from PubChem (https://pubchem.ncbi.nlm.nih.gov/) (accesses on 9 July 2021) in an SDF file (.sdf) and were converted to a PDB file (.pdb) using Open Babel GUI tool [27]. The ligands were prepared by adding Gasteiger charges, merging non-polar hydrogen, and setting torsion root and then converted to PDBQT (.pdbqt) files using AutoDock Tools 1.5.6 and saved.

3.3. Docking and Visualization

AutoDock Vina 1.1.2 software was used for the molecular docking experiments [17]. Phytochemical compounds were used as ligands. The protein target was prepared as mentioned above. The grid spacing was set to 0.5 Å. The number of grid points along x, y and z dimensions were set as $22 \times 40 \times 34$, respectively, and centered at (x = −30.71, y = −0.002, z = −42.148). The AutoDock Vina output file gives docking scores corresponding to Gibbs free energy of binding (ΔG) (kcal/mol) for each conformation of the ligand. It represents the efficiency of ligand binding to the designated protein–protein interaction interface. Further, the output file of optimal ligand conformations and their 2D interaction with interface residues were visualized using LIGPLOT software [23]. The 3D structure representing the key residues that were involved in PLpro and ISG15 interaction was visualized using a Discovery Studio visualizer version v20.1.0.19295 (BIOVIA, San Diego, CA, USA) [28].

3.4. Drug Likeness Study and ADME Screening

A SwissADME web tool was used to predict the ADME parameters and drug-like nature of GRL0617 and top screened phytochemicals from AG, GU, and TU [29]. A PreADMET web tool was used to predict the ADMET and toxicity parameters of the chemical compound [30]. The ADMET (Absorption, Distribution, Metabolism, Excretion, and Toxicity) properties were determined according to the Lipinski rule, which includes the molecular weight, H-bond acceptors, and H-bond donors, and lipophilicity.

3.5. Molecular Dynamic Simulations and Free Energy Calculation (MM-PBSA)

A molecular dynamics simulation of the selected top protein–ligand complexes was run using Gromacs-2019.4 [31]. For the force field coordinates, the ligand topology was downloaded from the PRODRG server [32]. Using the steepest descent algorithm, 1500 steps were used to prepare the system with the vacuum minimized. Using a water simple point charge (SPC) water model, complex structures were solvated into cubic periodic boxes of 0.5 nm. A salt concentration of 0.15 M was subsequently maintained by adding appropriate numbers of Na^+ and Cl^- counterions. The system preparation was referred from the previously published paper [33]. Each resultant structure from the NPT equilibration phase was subjected to a final production run in an NPT ensemble for 300 ns of simulation time. The six systems that were considered for simulations include:

(i) (PLpro-ISG15)-complex
(ii) (PLpro-ISG15)-complex with AG1
(iii) (PLpro-ISG15)-complex with GU1
(iv) (PLpro-ISG15)-complex withTU1 and
(v) (PLpro-ISG15)-complex with GRL0617

A trajectory analysis was performed using the GROMACS simulation package for proteins RMSD, RMSF, RG, SASA, H-Bond, and PCA [31]. The molecular mechanics Poisson–Boltzmann surface area (MM-PBSA) approach was employed to understand the free energy of binding (ΔG binding) between the phytochemicals and the target protein complex over the simulation time. A GROMACS utility g_mmpbsa was employed to estimate the binding free energy. To obtain an accurate result, we computed ΔG for the last 50 ns with dt 1000 frames [34].

4. Conclusions

The recent outbreak of SARS-CoV-2 and its subsequent mutations have posed serious problems for disease management. Many therapeutic intervention methods employ strategies to directly inhibit different classes of viral targets, such as proteases and nucleases. However, the host immune system is of equal importance for an effective antiviral response, which in many cases is evaded by viruses using different strategies. ISG15 is established as one of the key mediators of the immunomodulating effects of interferons, and ISGylation is an important host cellular defense against viruses. This is further evident by the observations that several viruses evolved enzyme activities to execute the de-ISGylation activity [35]. Therefore, molecules that interfere with the de-ISGylation activity can offer a very good strategy to boost the host immune system against viruses. They can also be effective in overcoming the virus-mediated host immune suppression. The herbs that we considered in our current study (*A. paniculata, T. cordifolia, O. sanctum*) have been used for decades as immune modulators, and several phytochemicals from these herbs act at various levels as immunomodulators. In the current study, we provide a molecular basis for a specific mechanism through which the phytochemicals from these herbs can augment the host antiviral immune response. De-ISGylation activity is not restricted only to coronaviruses, and it appears to be a general strategy that is employed by many viruses to overcome the host immune system. While highlighting the mechanism behind the action of these immunomodulatory phytochemicals, our studies also provide a computational strategy to screen for molecules interfering with ISGylation.

Supplementary Materials: The following supporting information can be downloaded at: https://www.mdpi.com/article/10.3390/computation10070109/s1, Figure S1: Key interacting residues between PLpro and ISG15 ; Table S1: Docking scores of various phytochemicals. Table S2 (i), (ii): Drug-likeness and ADMET properties of top-ranked phytochemicals. Figure S2: PCA analysis.

Author Contributions: Conceptualization, P.S. and R.P.R.; Data curation, A.P.; Formal analysis, S.S.B., R.M. and R.P.R.; Investigation, P.S.; Methodology, A.P.; Resources, A.B. and B.U.V.; Supervision, R.P.R.; Validation, S.S.B. and R.M.; Writing—original draft, P.S.; Writing—review & editing, R.P.R. All authors have read and agreed to the published version of the manuscript.

Funding: The research work was supported by Himalaya Wellness Company, Bangalore, India.

Institutional Review Board Statement: Not applicable.

Informed Consent Statement: Not applicable.

Acknowledgments: The authors sincerely thank Sciomics LLP for offering infrastructure support for conducting the simulation studies.

Conflicts of Interest: The authors declare no conflict of interest.

Abbreviations

ADME	Absorption, distribution, metabolism, elimination
AG	Andrographis paniculata
AG1	14-deoxy-15-isopropylidene-11,12-didehydroandrographolide
GU	Tinospora cordifolia
GU1	Isocolumbin
ISGs	Interferon-stimulated genes
MM/PBSA	Molecular mechanics/Poisson–Boltzmann surface area
MD	Molecular dynamics
PDB	Protein data bank
SPC	Simple point charge
TU	Ocimum sanctum
TU1	Orientin
UIM	Ubiquitin-interacting motif

References

1. Jeon, Y.J.; Yoo, H.M.; Chung, C.H. ISG15 and immune diseases. *Biochim. Biophys. Acta* **2010**, *1802*, 485–496. [CrossRef] [PubMed]
2. Zhang, D.; Zhang, D.E. Interferon-Stimulated Gene 15 and the Protein ISGylation System. *J. Interf. Cytokine Res.* **2011**, *31*, 119. [CrossRef] [PubMed]
3. Freitas, B.T.; Scholte, F.E.M.; Bergeron, É.; Pegan, S.D. How ISG15 combats viral infection. *Virus Res.* **2020**, *286*, 198036. [CrossRef] [PubMed]

4. Zhang, M.; Li, J.; Yan, H.; Huang, J.; Wang, F.; Liu, T.; Zeng, L.; Zhou, F. ISGylation in Innate Antiviral Immunity and Pathogen Defense Responses: A Review. *Front. Cell Dev. Biol.* **2021**, *9*, 3196. [CrossRef] [PubMed]
5. Perng, Y.C.; Lenschow, D.J. ISG15 in antiviral immunity and beyond. *Nat. Rev. Microbiol.* **2018**, *16*, 423–439. [CrossRef] [PubMed]
6. Li, D.; Luan, J.; Zhang, L. Molecular docking of potential SARS-CoV-2 papain-like protease inhibitors. *Biochem. Biophys. Res. Commun.* **2021**, *538*, 72–79. [CrossRef]
7. Fung, S.Y.; Yuen, K.S.; Ye, Z.W.; Chan, C.P.; Jin, D.Y. A tug-of-war between severe acute respiratory syndrome coronavirus 2 and host antiviral defence: Lessons from other pathogenic viruses. *Emerg. Microbes Infect.* **2020**, *9*, 558–570. [CrossRef] [PubMed]
8. Shemesh, M.; Aktepe, T.E.; Deerain, J.M.; McAuley, J.L.; Audsley, M.D.; David, C.T.; Purcell, D.F.J.; Urin, V.; Hartmann, R.; Moseley, G.W.; et al. SARS-CoV-2 suppresses IFNβ production mediated by NSP1, 5, 6, 15, ORF6 and ORF7b but does not suppress the effects of added interferon. *PLoS Pathog.* **2021**, *17*, e1009800. [CrossRef]
9. Kim, Y.M.; Shin, E.C. Type I and III interferon responses in SARS-CoV-2 infection. *Exp. Mol. Med.* **2021**, *53*, 750–760. [CrossRef]
10. Gupta, S.; Mishra, K.P.; Ganju, L. Broad-spectrum antiviral properties of andrographolide. *Arch. Virol.* **2017**, *162*, 611–623. [CrossRef]
11. Wang, W.; Wang, J.; Dong, S.F.; Liu, C.H.; Italiani, P.; Sun, S.H.; Xu, J.; Boraschi, D.; Ma, S.P.; Qu, D. Immunomodulatory activity of andrographolide on macrophage activation and specific antibody response. *Acta Pharmacol. Sin.* **2010**, *31*, 191–201. [CrossRef] [PubMed]
12. Niraj, S.; Varsha, S. A review on scope of immuno-modulatory drugs in Ayurveda for prevention and treatment of COVID-19. *Plant. Sci. Today* **2020**, *7*, 417–423. [CrossRef]
13. Borse, S.; Joshi, M.; Saggam, A.; Bhat, V.; Walia, S.; Marathe, A.; Sagar, S.; Chavan-Gautam, P.; Girme, A.; Hingorani, L.; et al. Ayurveda botanicals in COVID-19 management: An in silico multi-target approach. *PLoS ONE* **2021**, *16*, e0248479. [CrossRef] [PubMed]
14. Swain, S.S.; Panda, S.K.; Luyten, W. Phytochemicals against SARS-CoV as potential drug leads. *Biomed. J.* **2021**, *44*, 74–85. [CrossRef] [PubMed]
15. Krupanidhi, S.; Peele, K.A.; Venkateswarulu, T.C.; Ayyagari, V.S.; Bobby, M.N.; Babu, D.J.; Narayana, A.V.; Aishwarya, G. Screening of phytochemical compounds of Tinospora cordifolia for their inhibitory activity on SARS-CoV-2: An in silico study. *J. Biomol. Struct. Dyn.* **2021**, *39*, 5799–5803. [CrossRef]
16. Klemm, T.; Ebert, G.; Calleja, D.J.; Allison, C.C.; Richardson, L.W.; Bernardini, J.P.; Lu, B.G.; Kuchel, N.W.; Grohmann, C.; Shibata, Y.; et al. Mechanism and inhibition of the papain-like protease, PLpro, of SARS-CoV-2. *EMBO J.* **2020**, *39*, e106275. [CrossRef]
17. Anil, K.T.J.W. Autodock vina: Improving the speed and accuracy of docking. *J. Comput. Chem.* **2019**, *31*, 455–461. [CrossRef]
18. Shin, D.; Mukherjee, R.; Grewe, D.; Bojkova, D.; Baek, K.; Bhattacharya, A.; Schulz, L.; Widera, M.; Mehdipour, A.R.; Tascher, G.; et al. Papain-like protease regulates SARS-CoV-2 viral spread and innate immunity. *Nature* **2020**, *587*, 657–662. [CrossRef]
19. Freitas, B.T.; Durie, I.A.; Murray, J.; Longo, J.E.; Miller, H.C.; Crich, D.; Hogan, R.J.; Tripp, R.A.; Pegan, S.D. Characterization and Noncovalent Inhibition of the Deubiquitinase and deISGylase Activity of SARS-CoV-2 Papain-Like Protease. *ACS Infect. Dis.* **2020**, *6*, 2099–2109. [CrossRef]
20. Fu, Z.; Huang, B.; Tang, J.; Liu, S.; Liu, M.; Ye, Y.; Liu, Z.; Xiong, Y.; Zhu, W.; Cao, D.; et al. The complex structure of GRL0617 and SARS-CoV-2 PLpro reveals a hot spot for antiviral drug discovery. *Nat. Commun.* **2021**, *12*, 488. [CrossRef]
21. Islam, M.T.; Bardaweel, S.K.; Mubarak, M.S.; Koch, W.; Gaweł-Beben, K.; Antosiewicz, B.; Sharifi-Rad, J. Immunomodulatory Effects of Diterpenes and Their Derivatives Through NLRP3 Inflammasome Pathway: A Review. *Front. Immunol.* **2020**, *11*, 2234. [CrossRef] [PubMed]
22. Wardana, A.P.; Aminah, N.S.; Rosyda, M.; Abdjan, M.I.; Kristanti, A.N.; Tun, K.N.W.; Choudhary, M.I.; Takaya, Y. Potential of diterpene compounds as antivirals, a review. *Heliyon* **2021**, *7*, e07777. [CrossRef] [PubMed]
23. Wallace, A.C.; Laskowski, R.A.; Thornton, J.M. LIGPLOT: A program to generate schematic diagrams of protein-ligand interactions. *Protein. Eng. Des. Sel.* **1995**, *8*, 127–134. [CrossRef] [PubMed]
24. Li, D.; Wang, Q.; Yuan, Z.F.; Zhang, L.; Xu, L.; Cui, Y.; Duan, K. Pharmacokinetics and tissue distribution study of orientin in rat by liquid chromatography. *J. Pharm. Biomed. Anal.* **2008**, *47*, 429–434. [CrossRef]
25. Wang, C.; Nguyen, P.H.; Pham, K.; Huynh, D.; Le, T.B.N.; Wang, H.; Ren, P.; Luo, R. Calculating protein-ligand binding affinities with MMPBSA: Method and error analysis. *J. Comput. Chem.* **2016**, *37*, 2436–2446. [CrossRef]
26. Martin, W.R.; Lightstone, F.C.; Cheng, F. In Silico Insights into Protein–Protein Interaction Disruptive Mutations in the PCSK9-LDLR Complex. *Int. J. Mol. Sci.* **2020**, *21*, 1550. [CrossRef]
27. O'Boyle, N.M.; Banck, M.; James, C.A.; Morley, C.; Vandermeersch, T.; Hutchison, G.R. Open Babel: An Open chemical toolbox. *J. Cheminform.* **2011**, *3*, 33. [CrossRef]
28. Biovia: Discovery Studio Modeling Environment-Google Scholar. Available online: https://scholar.google.com/scholar?cluster=17675170202455151209&hl=en&as_sdt=2005&sciodt=0,5 (accessed on 8 June 2022).
29. Daina, A.; Michielin, O.; Zoete, V. SwissADME: A free web tool to evaluate pharmacokinetics, drug-likeness and medicinal chemistry friendliness of small molecules. *Sci. Rep.* **2017**, *7*, 42717. [CrossRef]
30. PreADME in EuroQSAR 2004–PreADMET | Prediction of ADME/Tox. 2004. Available online: https://preadmet.webservice.bmdrc.org/2004/09/27/200409-preadme-in-euroqsar-2004/ (accessed on 7 June 2022).

31. Abraham, M.J.; Murtola, T.; Schulz, R.; Páll, S.; Smith, J.C.; Hess, B.; Lindah, E. GROMACS: High performance molecular simulations through multi-level parallelism from laptops to supercomputers. *SoftwareX* **2015**, *1–2*, 19–25. [CrossRef]
32. Schüttelkopf, A.W.; Van Aalten, D.M.F. PRODRG: A tool for high-throughput crystallography of protein-ligand complexes. *Acta Crystallogr. D Biol. Crystallogr.* **2004**, *60*, 1355–1363. [CrossRef]
33. Gangadharappa, B.S.; Sharath, R.; Revanasiddappa, P.D.; Chandramohan, V.; Balasubramaniam, M.; Vardhineni, T.P. Structural insights of metallo-beta-lactamase revealed an effective way of inhibition of enzyme by natural inhibitors. *J. Biomol. Struct. Dyn.* **2020**, *38*, 3757–3771. [CrossRef] [PubMed]
34. Kumari, R.; Kumar, R.; Lynn, A. g_mmpbsa–a GROMACS tool for high-throughput MM-PBSA calculations. *J. Chem. Inf. Model.* **2014**, *54*, 1951–1962. [CrossRef] [PubMed]
35. Villarroya-Beltri, C.; Guerra, S.; Sánchez-Madrid, F. ISGylation-a key to lock the cell gates for preventing the spread of threats. *J. Cell Sci.* **2017**, *130*, 2961–2969. [CrossRef] [PubMed]

computation

MDPI

Article

Virtual Combinatorial Library Screening of Quinadoline B Derivatives against SARS-CoV-2 RNA-Dependent RNA Polymerase

Simone Brogi [1,*], Mark Tristan Quimque [2,3,4], Kin Israel Notarte [5], Jeremiah Gabriel Africa [2], Jenina Beatriz Hernandez [2], Sophia Morgan Tan [6], Vincenzo Calderone [1] and Allan Patrick Macabeo [2,*]

1 Department of Pharmacy, University of Pisa, Via Bonanno 6, 56126 Pisa, Italy; vincenzo.calderone@unipi.it
2 Laboratory for Organic Reactivity, Discovery and Synthesis (LORDS), Research Center for the Natural and Applied Sciences, University of Santo Tomas, España Blvd., Manila 1015, Philippines; mjtquimque@gmail.com (M.T.Q.); jeremiahgabriel.africa.sci@ust.edu.ph (J.G.A.); jeninabeatriz.hernandez.sci@ust.edu.ph (J.B.H.)
3 The Graduate School, University of Santo Tomas, España Blvd., Manila 1015, Philippines
4 Chemistry Department, College of Science and Mathematics, Mindanao State University—Iligan Institute of Technology, Tibanga, Iligan City 9200, Philippines
5 Faculty of Medicine and Surgery, University of Santo Tomas, Espana Blvd., Manila 1015, Philippines; kinotarte@gmail.com
6 Department of Biological Sciences, College of Science, University of Santo Tomas, España Blvd., Manila 1015, Philippines; sophiamorgan.tan.sci@ust.edu.ph
* Correspondence: simone.brogi@unipi.it (S.B.); agmacabeo@ust.edu.ph (A.P.M.)

Citation: Brogi, S.; Quimque, M.T.; Notarte, K.I.; Africa, J.G.; Hernandez, J.B.; Tan, S.M.; Calderone, V.; Macabeo, A.P. Virtual Combinatorial Library Screening of Quinadoline B Derivatives against SARS-CoV-2 RNA-Dependent RNA Polymerase. *Computation* **2022**, *10*, 7. https://doi.org/10.3390/computation10010007

Academic Editor: Rainer Breitling

Received: 13 December 2021
Accepted: 10 January 2022
Published: 12 January 2022

Abstract: The unprecedented global health threat of SARS-CoV-2 has sparked a continued interest in discovering novel anti-COVID-19 agents. To this end, we present here a computer-based protocol for identifying potential compounds targeting RNA-dependent RNA polymerase (RdRp). Starting from our previous study wherein, using a virtual screening campaign, we identified a fumiquinazolinone alkaloid quinadoline B (Q3), an antiviral fungal metabolite with significant activity against SARS-CoV-2 RdRp, we applied in silico combinatorial methodologies for generating and screening a library of anti-SARS-CoV-2 candidates with strong in silico affinity for RdRp. For this study, the quinadoline pharmacophore was subjected to structural iteration, obtaining a Q3-focused library of over 900,000 unique structures. This chemical library was explored to identify binders of RdRp with greater affinity with respect to the starting compound Q3. Coupling this approach with the evaluation of physchem profile, we found 26 compounds with significant affinities for the RdRp binding site. Moreover, top-ranked compounds were submitted to molecular dynamics to evaluate the stability of the systems during a selected time, and to deeply investigate the binding mode of the most promising derivatives. Among the generated structures, five compounds, obtained by inserting nucleotide-like scaffolds (**1**, **2**, and **5**), heterocyclic thiazolyl benzamide moiety (compound **3**), and a peptide residue (compound **4**), exhibited enhanced binding affinity for SARS-CoV-2 RdRp, deserving further investigation as possible antiviral agents. Remarkably, the presented in silico procedure provides a useful computational procedure for hit-to-lead optimization, having implications in anti-SARS-CoV-2 drug discovery and in general in the drug optimization process.

Keywords: quinadoline B; SARS-CoV-2; RNA-dependent RNA polymerase inhibitors; virtual screening; combinatorial screening; molecular dynamics

1. Introduction

The continued rise in COVID-19 cases worldwide despite the availability of vaccines sustains the demand to discover treatment and prophylactic regimens, particularly through natural products' repurposing and design [1–3]. Computational strategies play a crucial role in accelerating the discovery of effective anti-SARS-CoV-2 agents [4–8], as in silico experiments are vital in the screening of biologically active compounds, offering a rapid, low-cost, and effective adjunct to in vitro and in vivo experiments. Such methods can facilitate the iteration of known potential compounds to further enhance their biological and pharmacokinetic activities, capable of constructing virtually all possible permutational derivatives from a single parent compound [9].

In COVID-19 drug discovery, several possible drug targets, comprising structural and non-structural proteins, have been exploited in searching novel chemical entities as anti-SARS-CoV-2 agents [10–13]. Among these targets is the RNA-dependent RNA polymerase (RdRp), which is a multi-domain SARS-CoV-2 protein playing a crucial role in the viral life cycle. In particular, RdRp is involved in the replication and transcription of the viral genome [14,15]. Structurally, RdRp is deemed a conserved protein within coronaviruses and carries an accessible region as its active site. Thus, RdRp represents an attractive drug target to inhibit viral replication [14,16]. In our framework, we combined several computational approaches for optimizing a previously described compound targeting SARS-CoV-2 RdRp.

In our recent work, we performed a series of computer-based approaches, employing RdRp as one of the target proteins against fungal secondary metabolites with profound antiviral activity against various known pathogenic viruses. Our work allowed the identification of quinadoline B (Q3, Figure 1), an anti-influenza (H1N1) metabolite isolated from the mangrove-derived fungus *Cladosporium* sp. The fumiquinazoline alkaloid was shown to exhibit a high binding affinity to RdRp, with dynamic stability and favorable pharmacokinetic properties [17]. These results inspired us to further investigate the identified scaffold employing computational drug design methodologies, including structure-based methods such as molecular docking and molecular dynamics, in order to enhance the activity of quinadoline B against SARS-CoV-2 RdRp. Thus, in this study, we structurally redesigned quinadoline B to generate a focused library of derivatives with potentially enhanced antagonism to RdRp through combinatorial in silico techniques.

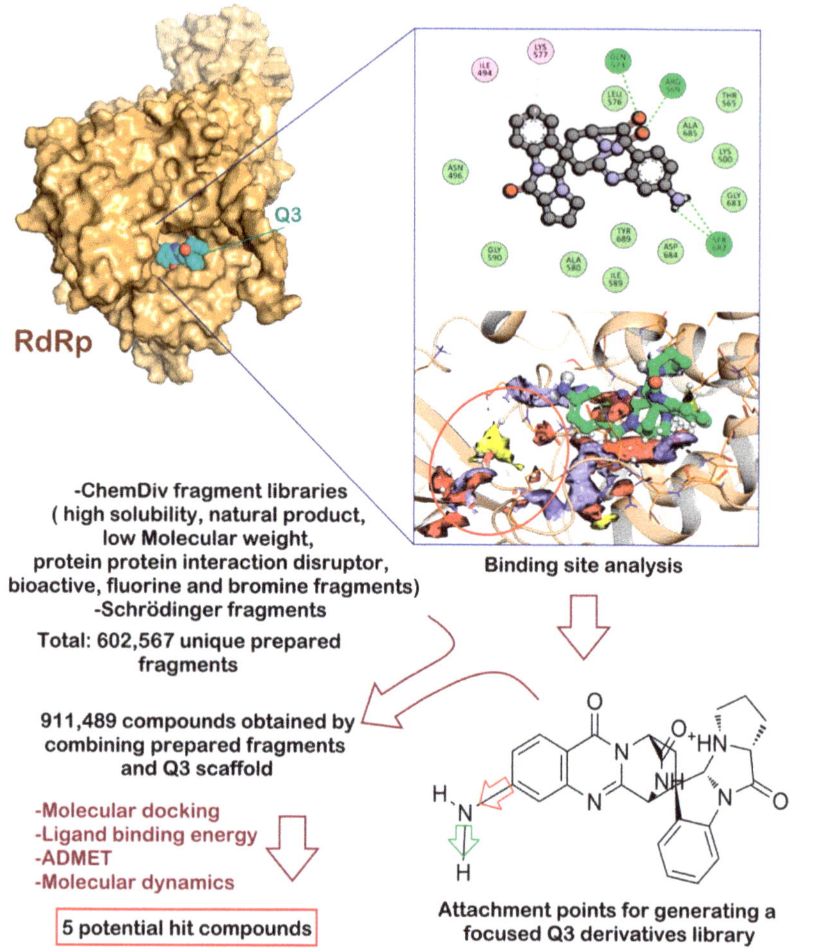

Figure 1. Schematic representation of the computational protocol adopted in this study for finding Q3 derivatives with improved in silico affinity for SARS-CoV-2 RdRp.

2. Materials and Methods

2.1. Computational Details

2.1.1. Ligand and Protein Preparation

Quinadoline B (Q3) was treated by LigPrep (LigPrep release 2018, Schrödinger, LLC, New York, NY, USA, 2018) for identifying the most probable ionization state at cellular pH value (7.4 ± 0.5), and minimized using MacroModel (MacroModel release 2018, Schrödinger, LLC, New York, NY, USA, 2018) implemented in Maestro software (Maestro release 2018, Schrödinger, LLC, New York, NY, USA, 2018), employing OPLS3 as a force field [18]. To simulate the solvent effects, the GB/SA model was employed, selecting "no cutoff" for non-bonded interactions. The PRCG technique (5000 maximum iterations and threshold for gradient convergence = 0.001) was employed to minimize the potential energy.

The structure of the RdRp enzyme of SARS-CoV-2 enzyme was downloaded from the Protein Data Bank (PDB ID 6M71 [19]; crystal structure of RdRp in complex with cofactors) and imported into Maestro suite 2018 and prepared using the protein preparation wizard protocol to acquire an appropriate starting structure for further in silico studies [20,21]. Using this protocol, we performed different computational steps to (1) add hydrogens;

(2) optimize the orientation of hydroxyl groups, Asn, and Gln, as well as the protonation state of His; and (3) perform a constrained minimization refinement using the *impref* utility. At first, the protein was pre-processed by adding all hydrogen atoms to the structure, assigning bond orders, creating disulfide bonds, and filling missing side chains and loops. To optimize the hydrogen bond network, His tautomers and ionization states were predicted; 180° rotations of the terminal angle of Asn, Gln, and His residues were assigned; and hydrogen atoms of the hydroxyl and thiol groups were sampled. Finally, a restrained minimization was performed using the Impact Refinement (*impref*) module, employing an OPLS3 force field to optimize the geometry and minimize the energy of the protein. The minimization was terminated when the energy converged, or the RMSD reached a maximum cutoff of 0.30 Å.

2.1.2. Binding Site Analysis

A comprehensive analysis of the binding site of SARS-CoV-2 RdRp was performed using the protein prepared as reported in Section 2.1.1 and the software SiteMap (SiteMap, release 2018, Schrödinger, LLC, New York, NY, USA, 2018).

2.1.3. Molecular Docking and Ligand-Energy Evaluation

Glide software (Glide release 2018, Schrödinger, LLC, New York, NY, USA, 2018) employing the XP-scoring function was used to perform all docking studies conducted in this work [22]. The energy grid for docking was prepared using the default value of the protein atom-scaling factor (1.0 Å), with a cubic box centered on the previously identified binding site. The docked poses considered for the post-docking minimization step were 1000, evaluating the Glide XP docking score.

To improve the quality of the screening, we also evaluated the ligand binding energies from the complexes derived by the docking calculation. For this purpose, Prime/MM-GBSA method available in Prime software (Prime release 2018, Schrödinger, LLC, New York, NY, USA, 2018) was used. This technique computes the variation between the free and complex state of both the ligand and enzyme after energy minimization [23,24].

2.1.4. Q3-Focused Library Generation

The library was generated as previously reported [25], using several series of fragments obtained from ChemDiv (https://store.chemdiv.com/ accessed on 20 March 2021) in SDF file format. These fragments were treated by LigPrep, in order to convert the 2D structure into the 3D one, and added to Q3 in a side chain hopping approach, considering the selected attachment points that comprise bonds, belonging to the Q3 core structure, replaced in the build process. This strategy allowed to obtain a Q3-focused library that consists of 991,489 compounds. This resulting library was employed in further computational experiments.

2.1.5. Evaluation of Drug-like Profile

The drug-like profile was evaluated using SwissADME [26], OSIRIS property explorer, and our in-house cardiotoxicity tool (3D-chERGi) [27]. PAINS assessment was executed employing SwissADME web-server [26], as previously reported [17,28].

2.1.6. Molecular Dynamics Simulation Details

Desmond 5.6 academic version, provided by D. E. Shaw Research ("DESRES"), was used to perform MD simulation experiments via Maestro graphical interface (Desmond Molecular Dynamics System, version 5.6, D. E. Shaw Research, New York, NY, USA, 2018. Maestro-Desmond Interoperability Tools, Schrödinger, New York, NY, USA, 2018). MD was performed using the compute unified device architecture (CUDA) API [29] on two NVIDIA GPUs. The complexes derived from docking studies (Figure 2) were imported in Maestro and, using the Desmond system builder, were solvated into an orthorhombic box filled with water, simulated by the TIP3P model [25,30]. An OPLS force field [18] was used for MD calculations. OPLS-aa (all atom) includes every atom explicitly with specific functional groups and types of molecules, including several biomacromolecules. A distinctive feature of the OPLS parameters is that they were optimized to fit the experimental properties of liquids, such as density and heat of vaporization, in addition to fitting gas-phase torsional profiles. Moreover, it is also largely used by us for performing MD simulations of protein/ligand complexes [25,31,32]. Na^+ and Cl^- ions were added to provide a final salt concentration of 0.15 M to simulate the physiological concentration of monovalent ions. Constant temperature (300 K) and pressure (1.01325 bar) were employed with the NPT (constant number of particles, pressure, and temperature) as an ensemble class. RESPA integrator [33] was used to integrate the equations of motion, with an inner time step of 2.0 fs for bonded and non-bonded interactions within the short-range cutoff. Nose–Hoover thermostats [34] were used to maintain the constant simulation temperature, and the Martyna–Tobias–Klein method [35] was applied to control the pressure. Long-range electrostatic interactions were calculated by particle-mesh Ewald method (PME) [36]. The cutoff for van der Waals and short-range electrostatic interactions was set at 9.0 Å. The equilibration of the system was performed using the default protocol provided in Desmond, which consists of a series of restrained minimization and MD simulations applied to slowly relax the system. Consequently, one individual trajectory for each complex of 100 ns was calculated. The trajectory files were analyzed by MD analysis tools implemented in the software package. The same application was used to generate all plots concerning MD simulation presented in this study. Accordingly, the *RMSD* was calculated using the following equation:

$$RMSD_x = \sqrt{\frac{1}{N}\sum_{i=1}^{N}\left(r'_i(t_x) - r_i\left(t_{ref}\right)\right)^2}$$

where the $RMSD_x$ refers to the calculation for a frame x; N is the number of atoms in the atom selection; t_{ref} is the reference time (typically, the first frame is used as the reference and it is regarded as time $t = 0$); and r' is the position of the selected atoms in frame x, after superimposing on the reference frame, where frame x is recorded at time t_x. The procedure is repeated for every frame in the simulation trajectory. Regarding the *RMSF*, the following equation was used for the calculation:

$$RMSF_i = \sqrt{\frac{1}{T}\sum_{t=1}^{T} < \left(r'_i(t) - r_i\left(t_{ref}\right)\right)^2 >}$$

where $RMSF_i$ refers to a generic residue i, T is the trajectory time over which the *RMSF* is calculated, t_{ref} is the reference time, r_i is the position of residue i, r' is the position of atoms in residue i after superposition on the reference, and the angle brackets indicate that the average of the square distance is taken over the selection of atoms in the residue.

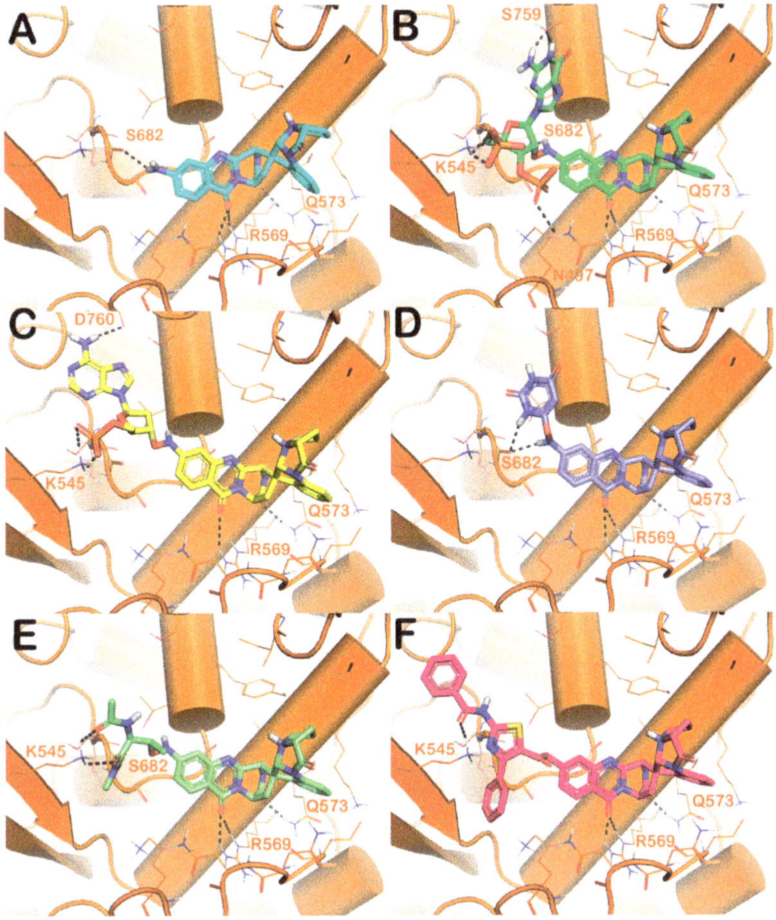

Figure 2. Putative binding mode of Q3 (cyan sticks, panel (**A**)) and Q3 derivatives **1–5** (colored sticks, panel (**B–F**), respectively) within the SARS-CoV-2 binding site (PDB ID 6M71, orange cartoon). Interacting amino acids are represented by lines, while the H-bonds are indicated by grey-dotted lines. Pictures were generated by PyMOL (The PyMOL Molecular Graphics System, v1.8; Schrödinger, LLC, New York, NY, USA, 2015).

3. Results and Discussion

SARS-CoV-2 and its predecessor SARS-CoV have significant similarities in their gene sequence, including the spike (S) glycoprotein, RdRp, and the two cysteine proteases: PL^{pro} and $3CL^{pro}$ [37]. Among these viral target proteins, RdRp plays a crucial role in viral replication, and is thus considered an exceptional molecular target for developing anti-SARS-CoV-2 drugs. Accordingly, different fungal derivatives, in particular quinaxoline alkaloids identified from the mangrove-derived fungus *Cladosporium* sp., were identified as possible SARS-CoV-2 RdRp inhibitors [17]. Among them, the ligand quinadoline B (Q3) showed the most interesting inhibitory profile in silico against RdRp. Q3 was found to tightly bind to the active site of RdRp by a series of polar and non-polar interactions. Three H-bonds were observed between the following: (a) the amino group and S682 and (b) carbonyl oxygens of the quinazolinone core and Q573 and R569. The indoline moiety was also involved in π-alkyl interactions with I494 and K577. Several van der Waals interactions against N496, G590, A580, I589, Y689, D684, G683, K500, A685, T565, and L576 were also noted [17]. The identified binding mode accounted for a binding energy of −9.5 kcal/mol, as found by AutoDock software, highlighting Q3 as one of the most promising derivatives of the series (Figure 1). To further explore the potential of quinadoline B as a drug prototype, in silico combinatorial techniques were employed to generate novel derivatives and enhance the previously reported antagonistic potential to RdRp. To this purpose, we used Schrödinger Drug-discovery Suite. As the first step, we retrieved the previously described binding mode of Q3 within the RdRp binding site by using Glide software (Figure S1). After establishing that the docking protocol was able to correctly locate the quinadoline B scaffold, we deeply investigated the RdRp binding site. The SiteMap analysis revealed the existence of a druggable sub-pocket that can be targeted by modifying Q3 derivatives (Figure 1). In particular, examining the orientation of the compound, we hypothesized that, by introducing appropriate moiety to Q3, possibly linked to the NH_2, it could be possible to reach the mentioned sub-pocket at the RdRp binding site. To accomplish this task, we used an in silico structure-based combinatorial library design approach, successfully employed by us, to generate focused libraries targeting specific binding site regions [25]. In the first step, we downloaded several sets of chemical fragments from ChemDiv, including high solubility fragments, natural product fragments, low molecular weight fragments, protein–protein interaction disruptor fragments, bioactive fragments, fluorine and bromine fragments, and other synthetic fragments. These fragments were properly prepared (see Section 2) and added to an existing library available from Schrödinger environment, obtaining 602,567 unique fragments to use in the side chain hopping approach. We selected two possible attachment points on the Q3 derivative exploiting NH_2 group (Figure 1). By combining the generated fragments and Q3 at the defined attachment points, we generated a focused library containing 991,489 Q3 derivatives.

The Q3-focused chemical library was employed in a virtual screening protocol based on molecular docking experiments and ligand-binding energy evaluation to identify Q3 derivatives that were able to bind RdRp with greater affinity compared with the starting compound Q3. For this purpose, compounds were docked into the binding site of SARS-CoV-2 RdRp [17] using Glide (Glide release 2018, Schrödinger, LLC, New York, NY, USA, 2018), employing XP as the scoring function and Prime software (Prime release 2018, Schrödinger, LLC, New York, NY, USA, 2018). The output of this step is reported in Table 1. Only Q3 derivatives showing a GlideScore value lower than −6.22 kcal/mol were considered. The threshold was chosen based on the value obtained by performing a docking calculation of Q3 into RdRp. The selected chemical entities were further examined by visual inspection to select molecules displaying a proper binding mode. By employing the above-mentioned computational protocol, we obtained 26 compounds showing improved affinities for the RdRp binding site with respect to the starting compound Q3 (structures are reported in Table S1).

Table 1. Final hits and their computational parameters derived from in silico studies.

Cpd	GlideScore (kcal/mol)	ΔG_{bind} (kcal/mol)	Main Contacts	$LogP_{o/w}$ [a]	Solubility [b]	GI abs. [c]	PAINS [d]	Tumorigenic [e]	pK*i* *h*ERG [f]
1	−8.71	−51.1	H−bonds R569, Q573, S682, N497, S759 salt bridges K545	−3.72	High	Low	No	No	5.03
2	−8.47	−52.3	H−bonds R569, Q573, K545, D760	−1.82	High	Low	No	No	5.24
3	−8.12	−43.9	H−bonds R569, Q573, S682	−0.23	Moderate	Low	No	No	5.06
4	−7.51	−44.8	H−bonds R569, Q573, S682, K545	−0.27	Moderate	Low	No	No	5.35
5	−7.46	−46.3	H−bonds R569, Q573, K545 cation−π K500, R555	3.07	Poor	Low	No	No	5.11

Table 1. *Cont.*

Cpd	GlideScore (kcal/mol)	ΔG_{bind} (kcal/mol)	Main Contacts	$LogP_{o/w}$ [a]	Solubility [b]	GI abs. [c]	PAINS [d]	Tumorigenic [e]	pK*i* *h*ERG [f]
6	−7.42	−41.5	H−bonds R569, Q573, S682 double cation−π K500	2.75	Moderate	High	No	No	5.32
7	−7.38	−40.6	H−bonds R569, Q573, S682 double cation−π K500	1.67	Poor	Low	No	No	5.17
8	−7.14	−41.2	H−bonds R569, Q573, S682, D684 cation−π K500	2.45	Moderate	High	No	No	5.51
9	−7.08	−39.1	H−bonds R569, Q573, S682	0.80	Moderate	Low	No	No	5.84
10	−7.03	−43.7	H−bonds R569, Q573, A685, A688 cation−π K545	1.32	Moderate	Low	No	No	5.63
11	−6.97	−38.8	H−bonds R569, Q573, S682 cation−π K500 π−π Y689	2.50	Poor	Low	No	No	4.92
12	−6.88	−40.2	H−bonds R569, Q573, K545, R555 halogen bonds R624	3.02	Poor	Low	No	No	5.68
13	−6.84	−42.3	H−bonds R569, Q573, S682 cation−π K500	2.61	Poor	Low	No	No	5.26
14	−6.81	−39.4	H−bonds R569, Q573, S682 cation−π K500	1.10	Moderate	High	No	No	5.15
15	−6.77	−42.9	H−bonds R569, Q573, D684 cation−π K545	1.05	Moderate	Low	No	No	4.93
16	−6.71	−41.0	H−bonds R569, Q573, S682, K545	1.01	Moderate	High	No	No	6.21
17	−6.59	−37.1	H−bonds R569, Q573, S682 π−π Y689	2.55	Poor	Low	No	No	5.24
18	−6.51	−47.2	H−bonds R569, Q573, S501	1.96	Poor	Low	No	No	5.67
19	−6.44	−41.3	H−bonds R569, Q573, S682 salt bridges D760	0.24	Moderate	High	No	alert: anil_di_alk_A	5.79
20	−6.39	−33.8	H−bonds R569, Q573, S682	1.84	Poor	Low	No	No	
21	−6.37	−34.9	H−bonds R569, Q573, S682, K545	3.18	Poor	Low	No	No	5.47
22	−6.36	−34.3	H−bonds R569, Q573, S682 halogen bonds K545	0.81	Moderate	Low	No	No	5.18
23	−6.34	−39.7	H−bonds R569, Q573 halogen bonds N497	1.53	Poor	Low	No	No	5.60
24	−6.30	−35.4	H−bonds R569, Q573, S682	0.14	Moderate	Low	No	No	5.51
25	−6.29	−40.2	H−bonds R569, Q573, R553, R555 salt bridges R553, R555	1.37	Moderate	Low	No	No	4.89
26	−6.24	−39.5	H−bonds R569, Q573, S682 cation−π K500	3.34	Poor	Low	No	No	5.54
Q3	−6.22	−32.3	H−bonds R569, Q573, S682	−0.12	Moderate	High	No	No	5.77

[a] Consesus LogP (lipophilicity)—average of five predictions using different algorithms (recommended value < 5); [b] water solubility assessed using three different methods; [c] gastrointestinal (GI) absorption; [d] PAINS (pan-assay interference compounds) predict the possibility of a given compound to behave as PAINS and, consequently, to interfere with biological assay; [e] tumorigenic—the evaluation was performed employing OSIRIS property explorer [38]; [f] predicted activity on seven PLS factors derived from our in-house 3D-QSAR model for predicting *h*ERG K^+ channel affinity (3D-chERGi) (pKi (M); pKi > 6, Ki < 1 μM) [27].

The analysis of docking output demonstrated an improvement in the number of contacts (polar and/or hydrophobic contacts) within the selected binding site for all selected compounds along with a greater binding affinity with respect to the starting molecule. The docking results for the five top-ranked compounds are illustrated in Figure 2 in comparison with Q3.

Briefly, starting from compound **1**, obtained by inserting a guanosine-like moiety on Q3 scaffold, we detected the same contacts found for Q3 (H-bonds R569, Q573, and S682) (Figure 2A and Table 1). Additionally, the novel substituent can target the hypothesized region of the RdRp binding site, producing strong interactions with N497, S759, and K545, by polar contacts (Figure 2B and Table 1). This molecular arrangement conferred a strong improvement in binding affinity with respect to the Q3 derivative, showing a GlideScore of −8.71 kcal/mol and a ΔG_{bind} of −51.1 kcal/mol (Q3, GlideScore −6.22 kcal/mol and ΔG_{bind} of −32.3 kcal/mol). Interestingly, compound **2** is also modified with a nucleotide moiety. In this case, Q3 was modified by inserting an adenine-like moiety (Figure 2C and Table 1). The docking output revealed that compound **2** similarly interacted within the RdRp binding site compared with compound **1**, except for the lack of H-bonds with N497 and S759 replaced with an H-bond with D760. This strong targeting observation accounted for a significant improvement in the computational score of compound **2** (GlideScore −8.47 kcal/mol and ΔG_{bind} −52.3 kcal/mol). Compound **3** lacks the previously described contacts, maintaining only the contacts found for Q3 with the addition of an additional H-bond with S682, strongly stabilizing the binding mode (Figure 2D and Table 1), as highlighted by in silico scores (GlideScore −8.12 kcal/mol and ΔG_{bind} −43.9 kcal/mol) compared with that found for Q3. For compound **4**, the insertion of a peptidic tail allowed to target the residue K545, in addition to the previously described contacts (H-bonds R569, Q573, and S682) (Figure 2E and Table 1). Moreover, in this case, the inserted substituent is well-tolerated by the RdRp binding site, as indicated by the satisfactory computational scores found for compound **4** (GlideScore −7.51 kcal/mol and ΔG_{bind} −44.8 kcal/mol). Inserting a bulky region with a stronger aromatic nature, as in compound **5**, allowed improvement of hydrophobic contacts within the RdRp binding site. In fact, compound **5** is able to form two cation-π interactions with residues K500 and R555, in addition to the maintained contacts (Figure 2E and Table 1). Compound **5** showed a GlideScore −7.46 kcal/mol and ΔG_{bind} −46.3 kcal/mol.

To validate the docking output, we conducted MD simulation on the top-five ranked compounds (**1–5**), investigating the evolution of biological systems for 100 ns. In this regard, the resulting trajectories for all complexes were completely examined through different standard simulation parameters including root mean square deviation (RMSD) analysis for all backbone atoms and ligands, and the root mean square fluctuation (RMSF) of individual amino acid residue. The selected complexes showed a general stability from the early stages of the simulation, as indicated by the results found by calculating the RMSD for each complex. In fact, we did not observe any major expansion and/or contraction, after the binding of these compounds during the entire simulation period (Figure 3A–E regarding the simulation of compounds **1–5**, respectively). This stability was also substantiated by observing the RMSF calculated for the selected complexes. RMSF indicates the difference between the atomic Cα coordinates of the protein from its average position during the MD simulation. This calculation is mainly helpful to characterize the flexibility of individual residues in the protein backbone. The considered systems did not show significant fluctuation phenomena, with the exclusion of a restricted number of residues at the N- and C-terminal regions of RdRp (Figure S2). In contrast, the conformational alterations of critical residues in the RdRp binding cleft (lowest RMSF values for all complexes) confirmed the capacity of compounds to form stable interactions within the protein.

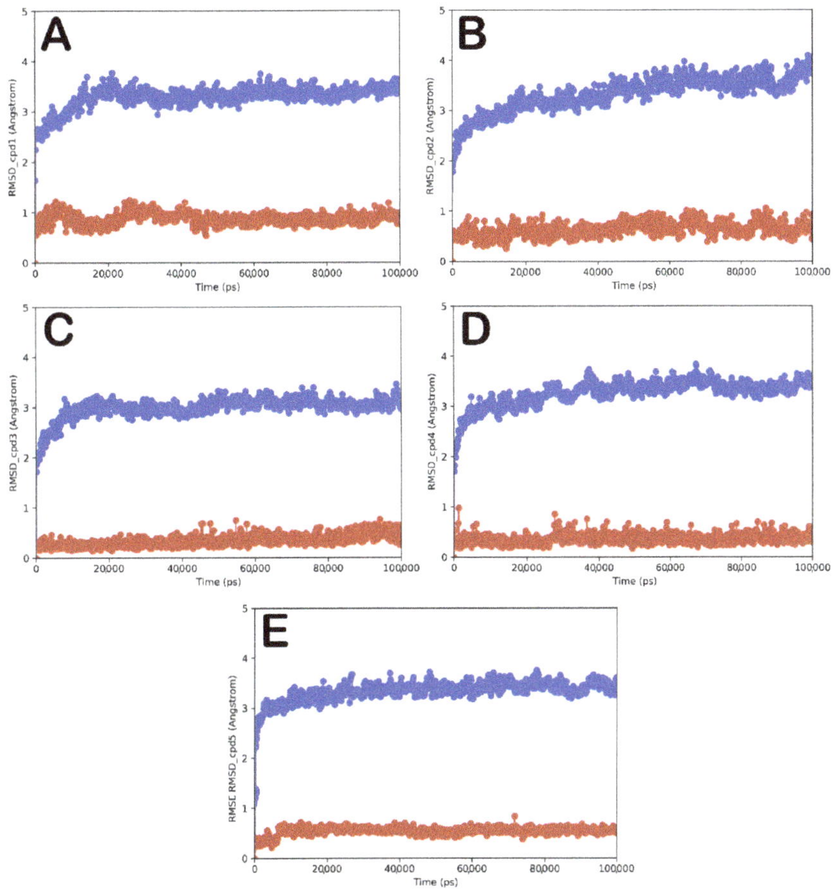

Figure 3. RMSD calculation for each complex and for each ligand. RdRp/compounds **1–5**, panel (**A–E**) (blue line), respectively. Compounds **1–5**, panel (**A–E**) (red line), respectively.

In order to better understand the behavior of compounds **1–5** in the SARS-CoV-2 RdRp binding site, we performed a detailed analysis of the MD simulation investigating the contacts established by compounds in the active site. The output of the analysis performed on the complex RdRp/compound **1** is reported in Figure 4. Compound **1** maintained the contacts found by docking calculation, interacting with R569 and Q573 during the MD simulation, while we observed a decrease in targeting S682. The interactions found by residues N497, S759, and K545 were evident through the time of simulation, as well as the salt bridges. In addition, interactions with A558, T556, R555, and N496 became apparent, while sporadic contacts were observed with residues S681, A685, and D760 considering the 100 ns of the simulation. Analysing the trajectory of compound **2**, we observed that the main contacts established with residues R569, Q573, K545, and D760 were maintained and N496, N497, K500, D623, and S759 were formed, although with no great potency. The output for compound **2** is illustrated in Figure 5. Compound **3** is able to strongly interact with S759 and D760, while less apparent contacts were detected with N496, N497, and D684 in addition to the contacts with the residues R569, Q573, and S682 (Figure 6). The results of this analysis for compounds **4** and **5** are found in the Supplementary Material file (Figures S3 and S4). Compound **4** maintained the contacts through H-bonds with R569, Q573, S682, and K545, while it formed additional contacts with N497, K500, G683, and D684 (Figure S3). Finally, compound **5** was still able to target R569, Q573, K500, and K545, while the interaction with R555 became sporadic. In contrast, compound **5** strongly targeted N496 and N497 (Figure S3).

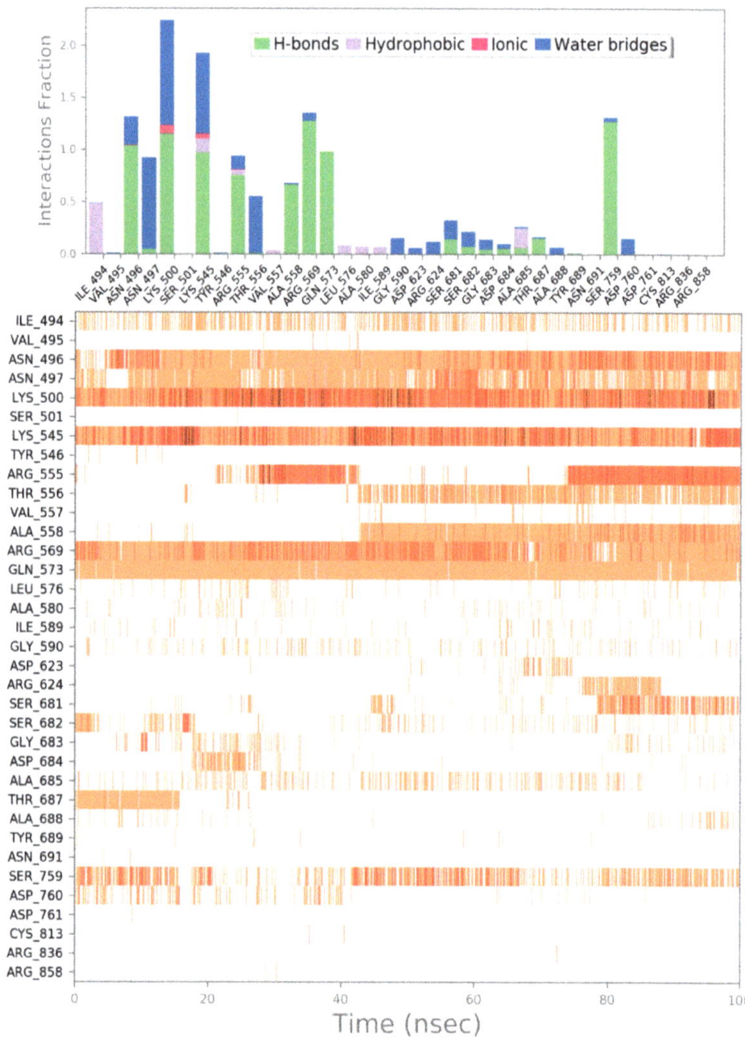

Figure 4. Compound **1** monitored during the simulation. The contacts can be grouped by type and summarized, as shown in the plots. Grouping protein–ligand interactions into four types: H-bonds (green), hydrophobic (grey), ionic (magenta), and water bridges (blue). The second graph of the picture displays a timeline representation of the contacts. Some residues make more than one specific contact with the ligand, which is represented by a darker shade of orange.

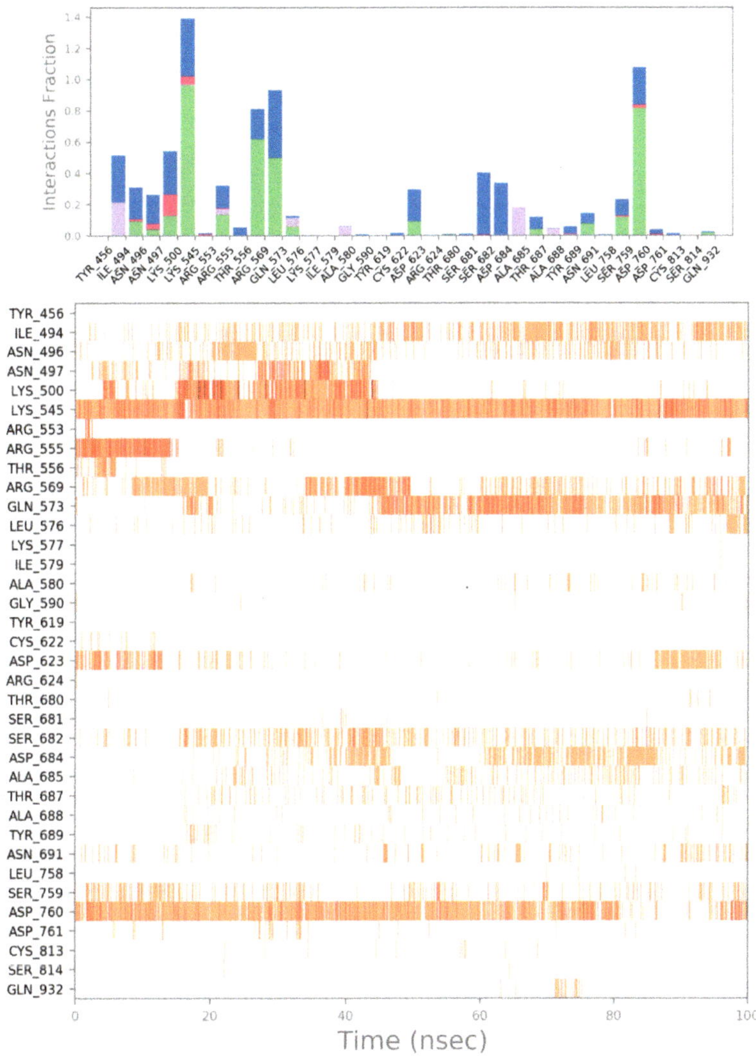

Figure 5. Compound **2** monitored during the simulation. The contacts can be grouped by type and summarized, as shown in the plots. Grouping protein–ligand interactions into four types: H-bonds (green), hydrophobic (grey), ionic (magenta), and water bridges (blue). The second graph of the picture displays a timeline representation of the contacts. Some residues make more than one specific contact with the ligand, which is represented by a darker shade of orange.

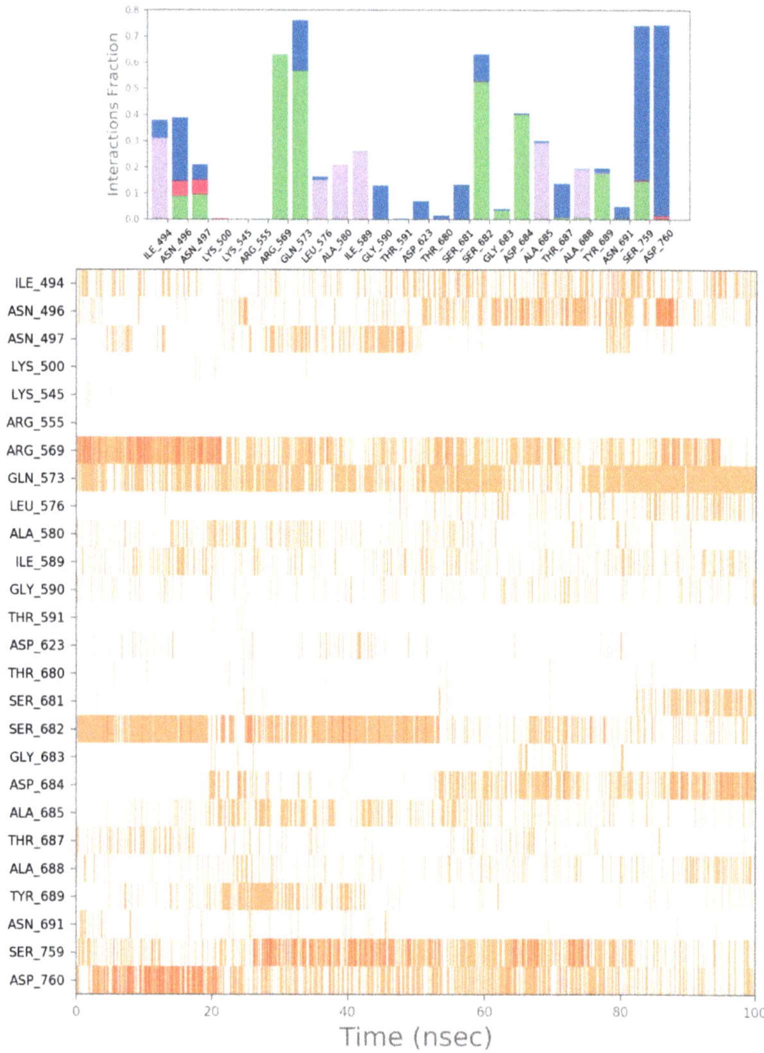

Figure 6. Compound **3** monitored during the simulation. The contacts can be grouped by type and summarized, as shown in the plots. Grouping protein–ligand interactions into four types: H-bonds (green), hydrophobic (grey), ionic (magenta), and water bridges (blue). The second graph of the picture displays a timeline representation of the contacts. Some residues make more than one specific contact with the ligand, which is represented by a darker shade of orange.

Overall, the MD simulation outcomes undoubtedly validated the advantageous interactions of five top-ranked compounds screened compounds, showing satisfactory thermodynamic stability in the RdRp binding site, suggesting that they can act as possible SARS-CoV-2 RdRp inhibitors. Furthermore, despite the fact that the addition of bulky moiety results in compounds with a high molecular weight, they showed an acceptable ADMET profile with logP and solubility in acceptable ranges, although the gastrointestinal (GI) absorption was found to be low. They were also found to be non-tumorigenic and devoid of cardiotoxicity, as assessed by our in-house tool, 3D-chERGi [27]; finally, the selected compounds did not have substructural features that allow to behave as pan-assay interference compounds (PAINS) (Table 1). PAINS compounds are chemical compounds that tend to display activity against numerous targets by nonspecific interactions or by altering the results of the biological tests. Compounds containing such moieties, which are often present in PAINS compounds, could be false positive hits and in general should be removed from the designed series [39]. Accordingly, our computational investigation provided five compounds as potential RdRp inhibitors and, more importantly, suggested guidelines for optimizing compounds considering the binding site of interest, showing improved binding affinity with respect to quinadoline B. In fact, such a structure-based methodology can be easily applied to other ligand–protein complexes for optimizing existing hit compounds.

4. Conclusions

In summary, we presented a computer-aided investigation for identifying possible SARS-CoV-2 RdRp inhibitors based on the quinadoline B scaffold, previously identified as possible RdRp ligands [17]. In particular, we used Q3 derivatives to explore the RdRp binding site by inserting several chemical fragments, obtained from the ChemDiv database, obtaining a Q3-focused library of over 900,000 unique structures. This library was used in a virtual screening protocol, employing the crystal structure of SARS-CoV-2 RdRp, for identifying Q3 derivatives with improved binding affinity with respect to quinadoline B. Moreover, the top-ranked compounds were subjected to MD simulations, in order to evaluate the stability of the systems during a selected time, and to deeply investigate the binding mode of the most promising derivatives. Finally, the in silico searching protocol allowed the identification of five compounds with improved affinity for SARS-CoV-2 RdRp, ushering interests for further investigation as possible antiviral agents. Notably, the developed computational protocol has implications in anti-SARS-CoV-2 drug discovery and in general in the drug optimization process, providing a convenient computational procedure for hit-to-lead optimization.

Supplementary Materials: The following supporting information can be downloaded at https://www.mdpi.com/article/10.3390/computation10010007/s1, Figure S1: Superposition between the docked pose of Q3 obtained by AutoDock and by Glide into RdRp binding site; Figure S2: RMSF calculation for each complex, selected by docking studies, after 100 ns of MD simulation; Figure S3: Compound **4** monitored during the simulation. The contacts can be grouped by type and summarized, as shown in the plots. Grouping protein–ligand interactions into four types: H-bonds, hydrophobic, ionic, and water bridges; Figure S4: Compound **5** monitored during the simulation. The contacts can be grouped by type and summarized, as shown in the plots. Grouping protein–ligand interactions into four types: H-bonds, hydrophobic, ionic, and water bridges; Table S1: Structure of selected compounds reported as SMILES string.

Author Contributions: Conceptualization, S.B., and A.P.M.; methodology, S.B., M.T.Q., K.I.N., V.C., and A.P.M.; software, S.B., M.T.Q., J.G.A., J.B.H., and S.M.T.; validation, S.B., M.T.Q., K.I.N., and A.P.M.; formal analysis, S.B., M.T.Q., K.I.N., V.C., and A.P.M.; investigation, S.B., M.T.Q., K.I.N., J.G.A., J.B.H., S.M.T., V.C., and A.P.M.; writing—original draft preparation, S.B.; writing—review and editing, S.B., M.T.Q., K.I.N., V.C., and A.P.M.; supervision, S.B., and A.P.M. All authors have read and agreed to the published version of the manuscript.

Funding: This research received no external funding.

Institutional Review Board Statement: Not applicable.

Informed Consent Statement: Not applicable.

Data Availability Statement: Not applicable.

Conflicts of Interest: The authors declare no conflict of interest.

References

1. Zhou, Y.; Wang, F.; Tang, J.; Nussinov, R.; Cheng, F. Artificial intelligence in COVID-19 drug repurposing. *Lancet Digit. Health* **2020**, *2*, e667–e676. [CrossRef]
2. Wang, Z.; Yang, L. Turning the Tide: Natural Products and Natural-Product-Inspired Chemicals as Potential Counters to SARS-CoV-2 Infection. *Front. Pharmacol.* **2020**, *11*, 1013. [CrossRef] [PubMed]
3. Abdallah, H.M.; El-Halawany, A.M.; Sirwi, A.; El-Araby, A.M.; Mohamed, G.A.; Ibrahim, S.R.M.; Koshak, A.E.; Asfour, H.Z.; Awan, Z.A.; Elfaky, M.A. Repurposing of Some Natural Product Isolates as SARS-COV-2 Main Protease Inhibitors via In Vitro Cell Free and Cell-Based Antiviral Assessments and Molecular Modeling Approaches. *Pharmaceuticals* **2021**, *14*, 213. [CrossRef] [PubMed]
4. Pitsillou, E.; Liang, J.; Karagiannis, C.; Ververis, K.; Darmawan, K.K.; Ng, K.; Hung, A.; Karagiannis, T.C. Interaction of small molecules with the SARS-CoV-2 main protease in silico and in vitro validation of potential lead compounds using an enzyme-linked immunosorbent assay. *Comput. Biol. Chem.* **2020**, *89*, 107408. [CrossRef] [PubMed]
5. Meng, X.Y.; Zhang, H.X.; Mezei, M.; Cui, M. Molecular docking: A powerful approach for structure-based drug discovery. *Curr. Comput. Aided Drug Des.* **2011**, *7*, 146–157. [CrossRef]
6. Bharadwaj, S.; El-Kafrawy, S.A.; Alandijany, T.A.; Bajrai, L.H.; Shah, A.A.; Dubey, A.; Sahoo, A.K.; Yadava, U.; Kamal, M.A.; Azhar, E.I.; et al. Structure-Based Identification of Natural Products as SARS-CoV-2 M(pro) Antagonist from Echinacea angustifolia Using Computational Approaches. *Viruses* **2021**, *13*, 305. [CrossRef]
7. Hajbabaie, R.; Harper, M.T.; Rahman, T. Establishing an Analogue Based In Silico Pipeline in the Pursuit of Novel Inhibitory Scaffolds against the SARS Coronavirus 2 Papain-Like Protease. *Molecules* **2021**, *26*, 1134. [CrossRef]
8. Fakhar, Z.; Khan, S.; AlOmar, S.Y.; Alkhuriji, A.; Ahmad, A. ABBV-744 as a potential inhibitor of SARS-CoV-2 main protease enzyme against COVID-19. *Sci. Rep.* **2021**, *11*, 234. [CrossRef]
9. Brogi, S.; Ramalho, T.C.; Kuca, K.; Medina-Franco, J.L.; Valko, M. Editorial: In silico Methods for Drug Design and Discovery. *Front. Chem.* **2020**, *8*, 612. [CrossRef]
10. Yadav, R.; Chaudhary, J.K.; Jain, N.; Chaudhary, P.K.; Khanra, S.; Dhamija, P.; Sharma, A.; Kumar, A.; Handu, S. Role of Structural and Non-Structural Proteins and Therapeutic Targets of SARS-CoV-2 for COVID-19. *Cells* **2021**, *10*, 821. [CrossRef]
11. Quimque, M.T.; Notarte, K.I.; Adviento, X.A.; Cabunoc, M.H.; de Leon, V.N.; Delos Reyes, F.S.L.; Lugtu, E.J.; Manzano, J.A.; Monton, S.N.; Munoz, J.E.; et al. Polyphenolic Natural Products Active In Silico against SARS-CoV-2 Spike Receptor Binding Domains and Non-Structural Proteins—A Review. *Comb. Chem. High Throughput Screen.* **2022**, *25*. [CrossRef]
12. de Leon, V.N.O.; Manzano, J.A.H.; Pilapil, D.Y.H.t.; Fernandez, R.A.T.; Ching, J.; Quimque, M.T.J.; Agbay, J.C.M.; Notarte, K.I.R.; Macabeo, A.P.G. Anti-HIV reverse transcriptase plant polyphenolic natural products with in silico inhibitory properties on seven non-structural proteins vital in SARS-CoV-2 pathogenesis. *J. Genet. Eng. Biotechnol.* **2021**, *19*, 104. [CrossRef]
13. Fernandez, R.A.; Quimque, M.T.; Notarte, K.I.; Manzano, J.A.; Pilapil, D.Y.t.; de Leon, V.N.; San Jose, J.J.; Villalobos, O.; Muralidharan, N.H.; Gromiha, M.M.; et al. Myxobacterial depsipeptide chondramides interrupt SARS-CoV-2 entry by targeting its broad, cell tropic spike protein. *J. Biomol. Struct. Dyn.* **2021**, 1–12. [CrossRef]
14. Aftab, S.O.; Ghouri, M.Z.; Masood, M.U.; Haider, Z.; Khan, Z.; Ahmad, A.; Munawar, N. Analysis of SARS-CoV-2 RNA-dependent RNA polymerase as a potential therapeutic drug target using a computational approach. *J. Transl. Med.* **2020**, *18*, 275. [CrossRef]
15. Ahmad, J.; Ikram, S.; Ahmad, F.; Rehman, I.U.; Mushtaq, M. SARS-CoV-2 RNA Dependent RNA polymerase (RdRp)–A drug repurposing study. *Heliyon* **2020**, *6*, e04502. [CrossRef] [PubMed]
16. Mondal, S.K.; Mukhoty, S.; Kundu, H.; Ghosh, S.; Sen, M.K.; Das, S.; Brogi, S. In silico analysis of RNA-dependent RNA polymerase of the SARS-CoV-2 and therapeutic potential of existing antiviral drugs. *Comput. Biol. Med.* **2021**, *135*, 104591. [CrossRef] [PubMed]
17. Quimque, M.T.J.; Notarte, K.I.R.; Fernandez, R.A.T.; Mendoza, M.A.O.; Liman, R.A.D.; Lim, J.A.K.; Pilapil, L.A.E.; Ong, J.K.H.; Pastrana, A.M.; Khan, A.; et al. Virtual screening-driven drug discovery of SARS-CoV2 enzyme inhibitors targeting viral attachment, replication, post-translational modification and host immunity evasion infection mechanisms. *J. Biomol. Struct. Dyn.* **2021**, *39*, 4316–4333. [CrossRef] [PubMed]
18. Jorgensen, W.L.; Maxwell, D.S.; Tirado-Rives, J. Development and Testing of the OPLS All-Atom Force Field on Conformational Energetics and Properties of Organic Liquids. *J. Am. Chem. Soc.* **1996**, *118*, 11225–11236. [CrossRef]
19. Gao, Y.; Yan, L.; Huang, Y.; Liu, F.; Zhao, Y.; Cao, L.; Wang, T.; Sun, Q.; Ming, Z.; Zhang, L.; et al. Structure of the RNA-dependent RNA polymerase from COVID-19 virus. *Science* **2020**, *368*, 779–782. [CrossRef] [PubMed]
20. Di Capua, A.; Sticozzi, C.; Brogi, S.; Brindisi, M.; Cappelli, A.; Sautebin, L.; Rossi, A.; Pace, S.; Ghelardini, C.; Di Cesare Mannelli, L.; et al. Synthesis and biological evaluation of fluorinated 1,5-diarylpyrrole-3-alkoxyethyl ether derivatives as selective COX-2 inhibitors endowed with anti-inflammatory activity. *Eur. J. Med. Chem.* **2016**, *109*, 99–106. [CrossRef] [PubMed]
21. Brindisi, M.; Senger, J.; Cavella, C.; Grillo, A.; Chemi, G.; Gemma, S.; Cucinella, D.M.; Lamponi, S.; Sarno, F.; Iside, C.; et al. Novel spiroindoline HDAC inhibitors: Synthesis, molecular modelling and biological studies. *Eur. J. Med. Chem.* **2018**, *157*, 127–138. [CrossRef]
22. Testai, L.; Piragine, E.; Piano, I.; Flori, L.; Da Pozzo, E.; Miragliotta, V.; Pirone, A.; Citi, V.; Di Cesare Mannelli, L.; Brogi, S.; et al. The Citrus Flavonoid Naringenin Protects the Myocardium from Ageing-Dependent Dysfunction: Potential Role of SIRT1. *Oxid. Med. Cell. Longev.* **2020**, *2020*, 4650207. [CrossRef]

23. Brogi, S.; Brindisi, M.; Butini, S.; Kshirsagar, G.U.; Maramai, S.; Chemi, G.; Gemma, S.; Campiani, G.; Novellino, E.; Fiorenzani, P.; et al. (S)-2-Amino-3-(5-methyl-3-hydroxyisoxazol-4-yl)propanoic Acid (AMPA) and Kainate Receptor Ligands: Further Exploration of Bioisosteric Replacements and Structural and Biological Investigation. *J. Med. Chem.* **2018**, *61*, 2124–2130. [CrossRef]
24. Frydenvang, K.; Pickering, D.S.; Kshirsagar, G.U.; Chemi, G.; Gemma, S.; Sprogoe, D.; Kaern, A.M.; Brogi, S.; Campiani, G.; Butini, S.; et al. Ionotropic Glutamate Receptor GluA2 in Complex with Bicyclic Pyrimidinedione-Based Compounds: When Small Compound Modifications Have Distinct Effects on Binding Interactions. *ACS Chem. Neurosci.* **2020**, *11*, 1791–1800. [CrossRef] [PubMed]
25. Sirous, H.; Chemi, G.; Gemma, S.; Butini, S.; Debyser, Z.; Christ, F.; Saghaie, L.; Brogi, S.; Fassihi, A.; Campiani, G.; et al. Identification of Novel 3-Hydroxy-pyran-4-One Derivatives as Potent HIV-1 Integrase Inhibitors Using in silico Structure-Based Combinatorial Library Design Approach. *Front. Chem.* **2019**, *7*, 574. [CrossRef]
26. Daina, A.; Michielin, O.; Zoete, V. SwissADME: A free web tool to evaluate pharmacokinetics, drug-likeness and medicinal chemistry friendliness of small molecules. *Sci. Rep.* **2017**, *7*, 42717. [CrossRef] [PubMed]
27. Chemi, G.; Gemma, S.; Campiani, G.; Brogi, S.; Butini, S.; Brindisi, M. Computational Tool for Fast in silico Evaluation of hERG K(+) Channel Affinity. *Front. Chem.* **2017**, *5*, 7. [CrossRef] [PubMed]
28. Zaccagnini, L.; Brogi, S.; Brindisi, M.; Gemma, S.; Chemi, G.; Legname, G.; Campiani, G.; Butini, S. Identification of novel fluorescent probes preventing PrP(Sc) replication in prion diseases. *Eur. J. Med. Chem.* **2017**, *127*, 859–873. [CrossRef]
29. Nickolls, J.; Buck, I.; Garland, M.; Skadron, K. Scalable parallel programming with CUDA. *Queue* **2008**, *6*, 40. [CrossRef]
30. Jorgensen, W.L.; Chandrasekhar, J.; Madura, J.D.; Impey, R.W.; Klein, M.L. Comparison of simple potential functions for simulating liquid water. *J. Chem. Phys.* **1983**, *79*, 926–935. [CrossRef]
31. Brindisi, M.; Ulivieri, C.; Alfano, G.; Gemma, S.; de Asis Balaguer, F.; Khan, T.; Grillo, A.; Chemi, G.; Menchon, G.; Prota, A.E.; et al. Structure-activity relationships, biological evaluation and structural studies of novel pyrrolonaphthoxazepines as antitumor agents. *Eur. J. Med. Chem.* **2019**, *162*, 290–320. [CrossRef] [PubMed]
32. Brogi, S.; Butini, S.; Maramai, S.; Colombo, R.; Verga, L.; Lanni, C.; De Lorenzi, E.; Lamponi, S.; Andreassi, M.; Bartolini, M.; et al. Disease-modifying anti-Alzheimer's drugs: Inhibitors of human cholinesterases interfering with beta-amyloid aggregation. *CNS Neurosci. Ther.* **2014**, *20*, 624–632. [CrossRef] [PubMed]
33. Humphreys, D.D.; Friesner, R.A.; Berne, B.J. A Multiple-Time-Step Molecular Dynamics Algorithm for Macromolecules. *J. Phys. Chem.* **1994**, *98*, 6885–6892. [CrossRef]
34. Hoover, W.G. Canonical dynamics: Equilibrium phase-space distributions. *Phys. Rev. A* **1985**, *31*, 1695–1697. [CrossRef] [PubMed]
35. Martyna, G.J.; Tobias, D.J.; Klein, M.L. Constant pressure molecular dynamics algorithms. *J. Chem. Phys.* **1994**, *101*, 4177–4189. [CrossRef]
36. Essmann, U.; Perera, L.; Berkowitz, M.L.; Darden, T.; Lee, H.; Pedersen, L.G. A smooth particle mesh Ewald method. *J. Chem. Phys.* **1995**, *103*, 8577–8593. [CrossRef]
37. Cevik, M.; Bamford, C.G.G.; Ho, A. COVID-19 pandemic-a focused review for clinicians. *Clin. MicroBiol. Infect.* **2020**, *26*, 842–847. [CrossRef]
38. Organic Chemistry Portal. 2021. Available online: http://www.organic-chemistry.org/prog/peo/ (accessed on 25 July 2021).
39. Baell, J.B.; Holloway, G.A. New substructure filters for removal of pan assay interference compounds (PAINS) from screening libraries and for their exclusion in bioassays. *J. Med. Chem.* **2010**, *53*, 2719–2740. [CrossRef] [PubMed]

computation

MDPI

Article

In Silico Analysis of Peptide-Based Derivatives Containing Bifunctional Warheads Engaging Prime and Non-Prime Subsites to Covalent Binding SARS-CoV-2 Main Protease (M^{pro})

Simone Brogi [1,*], Sara Rossi [2], Roberta Ibba [2], Stefania Butini [2], Vincenzo Calderone [1], Giuseppe Campiani [2] and Sandra Gemma [2,*]

1 Department of Pharmacy, University of Pisa, Via Bonanno 6, 56126 Pisa, Italy; vincenzo.calderone@unipi.it
2 Department of Biotechnology, Chemistry and Pharmacy, DoE Department of Excellence 2018–2022, University of Siena, Via Aldo Moro 2, 53100 Siena, Italy; rossi115@student.unisi.it (S.R.); roberta.ibba@unisi.it (R.I.); butini3@unisi.it (S.B.); campiani@unisi.it (G.C.)
* Correspondence: simone.brogi@unipi.it (S.B.); gemma@unisi.it (S.G.)

Abstract: Despite the progress of therapeutic approaches for treating COVID-19 infection, the interest in developing effective antiviral agents is still high, due to the possibility of the insurgence of viable SARS-CoV-2-resistant strains. Accordingly, in this article, we describe a computational protocol for identifying possible SARS-CoV-2 M^{pro} covalent inhibitors. Combining several in silico techniques, we evaluated the potential of the peptide-based scaffold with different warheads as a significant alternative to nitriles and aldehyde electrophilic groups. We rationally designed four potential inhibitors containing difluorstatone and a Michael acceptor as warheads. In silico analysis, based on molecular docking, covalent docking, molecular dynamics simulation, and FEP, indicated that the conceived compounds could act as covalent inhibitors of M^{pro} and that the investigated warheads can be used for designing covalent inhibitors against serine or cysteine proteases such as SARS-CoV-2 M^{pro}. Our work enriches the knowledge on SARS-CoV-2 M^{pro}, providing a novel potential strategy for its inhibition, paving the way for the development of effective antivirals.

Keywords: SARS-CoV-2; main protease (M^{pro}); computer-aided drug design; molecular docking; molecular dynamics

Citation: Brogi, S.; Rossi, S.; Ibba, R.; Butini, S.; Calderone, V.; Campiani, G.; Gemma, S. In Silico Analysis of Peptide-Based Derivatives Containing Bifunctional Warheads Engaging Prime and Non-Prime Subsites to Covalent Binding SARS-CoV-2 Main Protease (M^{pro}). *Computation* **2022**, *10*, 69. https://doi.org/10.3390/computation10050069

Academic Editor: Brendan Howlin

Received: 19 March 2022
Accepted: 27 April 2022
Published: 1 May 2022

Publisher's Note: MDPI stays neutral with regard to jurisdictional claims in published maps and institutional affiliations.

1. Introduction

Severe acute respiratory syndrome coronavirus-2, widely known as SARS-CoV-2, is the etiological agent of COVID-19 that caused several epidemic outbreaks since its first appearance in 2019 in the city of Wuhan, China [1]. Since then, rapid vaccination campaigns have been implemented at the global level to protect the population from the most severe symptoms. Moreover, very recently novel antivirals such as molnupiravir and paxlovid (PF-07321332 + ritonavir) have been added to the COVID-19 therapeutic armamentarium to treat the infection in patients with high risk of severe symptoms [2–5]. However, the research of effective antivirals remains a priority, both for the current and future pandemics. SARS-CoV-2 belongs to the *Coronaviridae* subfamily which is composed of alpha-, beta-, gamma-, and delta-CoVs [6]. The SARS-CoV-2 genome comprises approximately 30,000 nucleotides that feature genes for the production of nonstructural proteins (enzymes required for viral transcription and replication) and structural proteins. The life cycle of the virus begins when the spike glycoprotein (S) binds the host receptor, which, in the case of SARS-CoV-2, is the ACE2 enzyme. This interaction determines the fusion of the cell membrane with the viral one and allows the entry of the virus inside the cell. Once inside the host cell, the virus disassembles to release the nucleocapsid and viral RNA into the cytoplasm. Afterward, translation of the ORF1a/b takes place to form the large replicase polyprotein 1a (pp1a) and pp2ab, and the replication of genomic RNA occurs. The pp1ab polyprotein is processed by two viral proteases, 3-chymotrypsin-like protease (3CL^{pro} or M^{pro}) and papain-like protease enzyme (PL^{pro}), to release nonstructural proteins such as RNA-dependent RNA polymerase and helicase, which are involved in viral transcription and replication and the structural proteins that will form the new viral particles. The virion is assembled in the endoplasmic reticulum and Golgi and is finally released into the extracellular compartment by exocytosis.

The M^{pro} enzyme as a target for developing new antiviral drugs: The M^{pro} enzyme is one of the best characterized and validated as a drug target among those known for coronaviruses [7]. Together with PL^{pro}, it is essential for the maturation of the polyprotein, which is translated from viral RNA and cleaved by the proteases. The M^{pro} enzyme operates no less than 11 hydrolytic breaks on the polyprotein 1ab in correspondence with specific recognition sequences. Most of the cleavage sites hold Leu-Gln ↓ (Ser, Ala, Gly) as the recognition sequence. Hence, M^{pro} inhibition blocks viral replication [8].

Among the three antivirals currently approved for the treatment of SARS-CoV-2 infection, PF-07321332 (**1**, Figure 1) is an M^{pro} inhibitor [9,10]. Its structure is characterized by a peptidomimetic scaffold bearing a nitrile moiety as an electrophilic warhead.

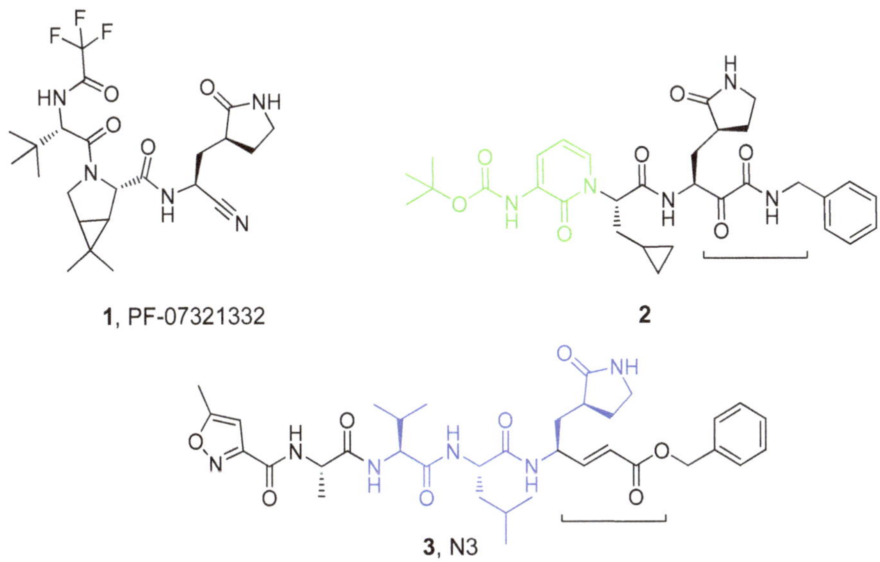

Figure 1. Chemical structures of reported SARS-CoV-2 M^{pro} inhibitors (**1**–**3**) showing different electrophilic warheads.

In general, the design of serine and cysteine protease inhibitors involves the insertion of electrophilic groups (warheads) that are reversibly or irreversibly attacked by nucleophilic serine or cysteine catalytic residues to form covalent adducts. The selectivity of the inhibitors is guaranteed by warhead flanking moieties able to specifically interact with the subsites of the enzyme mimicking the endogenous peptidic substrates. For these reasons, serine/cysteine protease inhibitors are usually characterized by a peptidic or peptidomimetic structure, albeit nonpeptidic M^{pro} inhibitors have also been reported [11]. Compounds **2** and **3** in Figure 1 are examples of different mono- or bifunctional warheads [8,12].

The electrophilic warhead plays a critical role in the development of M^{pro} inhibitors since it has to be characterized by sufficient reactivity to react with active-site residues, but stable enough not to engage in unwanted and aspecific reactions with other nucleophiles;

it should be readily accommodated inside the active site and be able to appropriately orientate the flanking moieties toward the enzyme subsites. In order to better understand the potential binding mode of electrophilic warheads and the role of the affinity of flanking substituents, it is important to implement appropriate computational protocols able to fully elucidate the parameters involved in covalent and noncovalent interactions. Here, we report an in silico protocol aimed at investigating the potential binding mode and reactivity of two different bifunctional warheads (compounds **4–7**, Figure 2). We chose to investigate in silico bifunctional electrophilic moieties that can be functionalized at both sides in order to engage with residues at both the prime and nonprime subsites of the enzyme or to be exploited to modulate drug-like properties. In particular, the difluorostatone-based warhead has been demonstrated by us and others to be able to engage in reversible covalent interactions with different serine proteases [13–15].

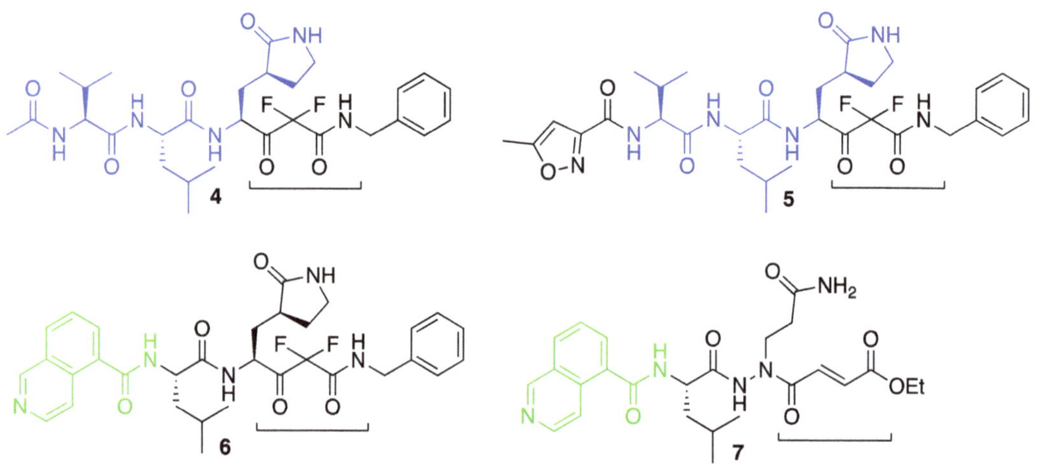

Figure 2. Chemical structures of potential inhibitors of SARS-CoV-2 M^{pro} (**4–7**) reported in this study.

On the other hand, Michael-based acceptor electrophilic moieties have been previously reported as alternatives to nitriles and aldehyde warheads. Starting from inhibitor **3**, the structural models **4–6** used for our computational investigation were designed by keeping constant residues P1–P3 and replacing the Michael acceptor group with a difluorostatone moiety. In particular, we wanted to verify if, in our computational protocol, compound **6** could be potentially able to form H-bond interactions inside the S2 subsite, similarly to what is described for reference inhibitor **2**. Here, we report a preliminary in silico investigation aimed at assessing the potential binding mode of the difluorostatone/aza-Michael moieties.

Thus, we conducted an extensive computational investigation based on molecular docking, molecular dynamics, and covalent docking approaches for determining the potential of the conceived compounds in inhibiting SARS-CoV-2 M^{pro}.

2. Materials and Methods

2.1. Computational Details

2.1.1. Protein and Ligand Preparation

Peptide-based derivatives were built in Maestro Molecular Modeling Suite (Maestro release 2020-3, Schrödinger, LLC, New York, NY, USA, 2020) using the available drawing tools as described [13,16]. Energy minimization was performed using the MacroModel application with the OPLS-2005 force field [17]. The resulting compounds were treated by LigPrep software (LigPrep release 2020, Schrödinger, LLC, New York, NY, USA, 2020) to provide the most probable ionization state at physiological pH (7.4 ± 0.2). To simulate solvent effects, the GBSA model was used with "no cutoff" for nonbonded interactions. The

PRCG method (5000 maximum iterations and threshold for gradient convergence = 0.001) was used to minimize the potential energy. The experimental structure of the SARS-CoV-2 M^{pro} enzyme was downloaded from the Protein Data Bank (PDB ID: 6Y2G [12]; crystal structure of M^{pro} in complex with α-ketoamide-based covalent inhibitor) and imported into Maestro Suite 2020. The first step was to break the covalent bond between C145 and the α-ketoamide derivative to restore the native arrangement of the enzyme. Next, to refine the structure, we applied the Protein Preparation Wizard protocol available in Maestro for performing various computational steps to (1) add hydrogens; (2) optimize the orientation of hydroxyl groups of residues, Asn and Gln, and the protonation state of His; and (3) perform a constrained minimization refinement using the *impref* utility. At first, the protein was preprocessed by adding all hydrogen atoms to the structure, assigning bond orders, creating disulfide bonds, and filling missing sidechains and loops. To optimize the hydrogen bond network, His tautomers and ionization states were predicted, 180° rotations of the terminal angle of Asn, Gln, and His residues were assigned, and hydrogen atoms of the hydroxyl and thiol groups of residues were sampled. Finally, a restrained minimization was performed using the Impact Refinement (*impref*) module, employing the OPLS3 force field to optimize the geometry and minimize the energy of the protein. The minimization was terminated when the energy converged or the root-mean-square deviation (RMSD) reached a maximum cutoff of 0.30 Å.

2.1.2. Molecular Docking

Glide software (Glide release 2020, Schrödinger, LLC, New York, NY, USA, 2020) employing the SP scoring function was used to perform all docking studies conducted in this work [18]. The energy grid for docking was prepared using the default value of the protein atom-scaling factor (1.0 Å), with a cubic box centered on the crystallized ligand. The docked poses considered for the post-docking minimization step were 1000.

To improve the quality of the investigation, we also evaluated the ligand-binding energies from the complexes derived by the docking calculation. For this purpose, the Prime/MM-GBSA method, available in Prime software (Prime release 2020, Schrödinger, LLC, New York, NY, USA, 2020), was used. This technique computes the variation between the free and the complex states of both the ligand and enzyme after energy minimization [19,20].

2.1.3. Molecular Dynamics

Desmond 5.6 academic version, provided by D. E. Shaw Research ("DESRES"), was utilized to perform MD simulation experiments via the Maestro graphical interface (Desmond Molecular Dynamics System, version 5.6, D. E. Shaw Research, New York, NY, USA, 2018. Maestro-Desmond Interoperability Tools, Schrödinger, New York, NY, USA, 2018). MD was performed using the Compute Unified Device Architecture (CUDA) API [21] on two NVIDIA GPUs. The Desmond system builder available via Maestro was employed for solvating the complexes derived from the docking studies into an orthorhombic box filled with water, simulated by the TIP3P model [22,23]. The OPLS force field [17] was used for MD calculations as reported [23–25]. To simulate the physiological concentration of monovalent ions, we added Na^+ and Cl^- ions to obtain a final salt concentration of 0.15 M. Constant temperature (300 K) and pressure (1.01325 bar) were employed with the NPT (constant number of particles, pressure, and temperature) as the ensemble class. The RESPA integrator [26] was used to integrate the equations of motion, with an inner time step of 2.0 fs for bonded and nonbonded interactions within the short-range cutoff. Nose–Hoover thermostats [27] were used to keep the constant simulation temperature, and the Martyna–Tobias–Klein method [28] was applied to control the pressure. Long-range electrostatic interactions were calculated by the particle-mesh Ewald method (PME) [29]. The cutoff for van der Waals and short-range electrostatic interactions was set at 9.0 Å. The equilibration of the system was performed using the default protocol provided in Desmond, which consists of a series of restrained minimization and MD simulations applied to slowly relax the system. Consequently, one individual trajectory for each complex of 100 ns was calculated. The trajectory files were analyzed by MD analysis tools available in Maestro. The same applications were used for generating all plots regarding MD simulations presented in this article. Therefore, the RMSD was calculated using the following equation:

$$RMSD_x = \sqrt{\frac{1}{N}\sum_{i=1}^{N}(r_i'(t_x)-r_i\,(t_{ref}))^2}$$

where $RMSD_x$ refers to the calculation for a frame x; N is the number of atoms in the atom selection; t_{ref} is the reference time (typically the first frame is used as the reference, and it is regarded as time $t = 0$); and r' is the position of the selected atoms in frame x, after superimposing on the reference frame, where frame x is recorded at time t_x. The procedure is repeated for every frame in the simulation trajectory. Regarding the root-mean-square fluctuation (*RMSF*), the following equation was used for the calculation:

$$RMSF_i = \sqrt{\frac{1}{T}\sum_{t=1}^{T}<(r_i'(\mathrm{t})-r_i\,(t_{ref}))^2>}$$

where $RMSF_i$ refers to a generic residue i, T is the trajectory time over which the *RMSF* is calculated, t_{ref} is the reference time, r_i is the position of residue I, r' is the position of atoms in residue i after superposition on the reference, and the angle brackets indicate that the average of the square distance is taken over the selection of atoms in the residue.

Free-energy perturbation (FEP) was performed using the FEP module available in the Desmond package using the complexes obtained by docking calculations, employing the default setting of the FEP protocol. The simulation was split into 12 λ-windows, with replica exchange attempted every 1.2 ps.

2.1.4. Covalent Docking

Covalent docking studies were executed in Maestro Suite 2020 applying the Covalent Docking protocol (CovDock) [30] as previously reported by us [14,31]. The algorithm utilizes both the Glide docking algorithm and Prime structure refinement. The CovDock application considers custom reactions enclosed in a list of possible covalent reactions (implemented in the software) using the SMARTS pattern. In this way, it is possible to automatically recognize the reactive residue and the portion of the ligand that are involved in the reaction. If the desired reaction is not present in that list, it is possible to write the reaction that involves the correct atoms. In this study, since the desired reaction considering the difluorostatone derivatives was not present in the list of reactions provided by CovDock, the reaction of the SMARTS pattern was customized [CC(C)=O] to obtain a reliable reaction for the compounds. Instead, the reaction involving a Michael acceptor is present in the reaction list. To start the calculation, the reactive residue of the receptor was selected (C145) and matched to the one defined in the custom chemistry file to specify the reaction type. The grid center was positioned at the centroid of the selected docked ligand, and the size of the grid box was automatically determined. No constraints were used, and the pose prediction option was selected for obtaining more accurate output results. Following the docking procedure, the obtained poses were filtered using default parameters, and the scoring option MM/GBSA was selected.

2.1.5. Physicochemical Properties Evaluation

QikProp (QikProp release 2020, Schrödinger, LLC, New York, NY, USA, 2020) was used for assessing logP and logS, while the possible pan-assay interference compounds (PAINS) issue was evaluated employing the online server FAFDrugs4 (https://fafdrugs4.rpbs.univ-paris-diderot.fr/ accessed on 25 February, 2022).

3. Results and Discussion

3.1. Molecular Docking Studies

In order to assess the tendency of our conceived compounds, reported in Figure 2, to bind the SARS-CoV-2 M^{pro} enzyme, we conducted a series of in silico experiments mainly based on molecular docking and molecular dynamics (Table 1). The protein (PDB ID: 6Y2G) and the ligands prepared as reported in the Materials and Methods section were docked into the well-established M^{pro} binding site using Glide software, employing the SP scoring function. Furthermore, we also calculated a relative binding affinity (ΔG_{bind}) using the MM/GBSA method. The output of this calculation is reported in Table 1, while the docking results are illustrated in Figure 3.

Table 1. Computational analysis regarding compounds **4–7** as potential M^{pro} inhibitors, along with reference compounds **2** and **3** (N3).

Compound	Docking Score (kcal/mol)	ΔG_{bind} (kcal/mol)	QPlogP [B]	QPlogS [C]	PAINS [D]
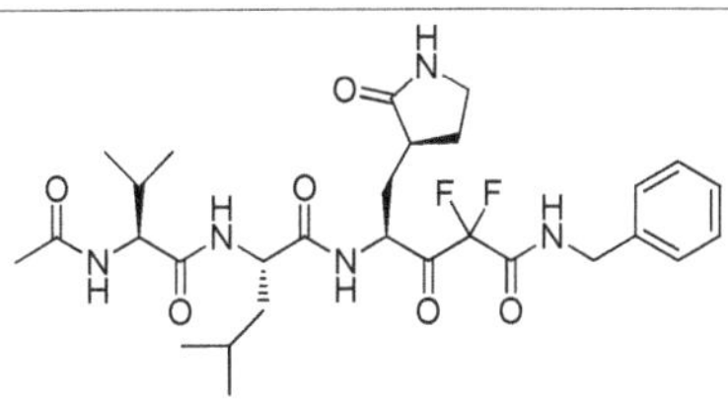 **4**	−10.779	−123.15	1.82	−3.68	No
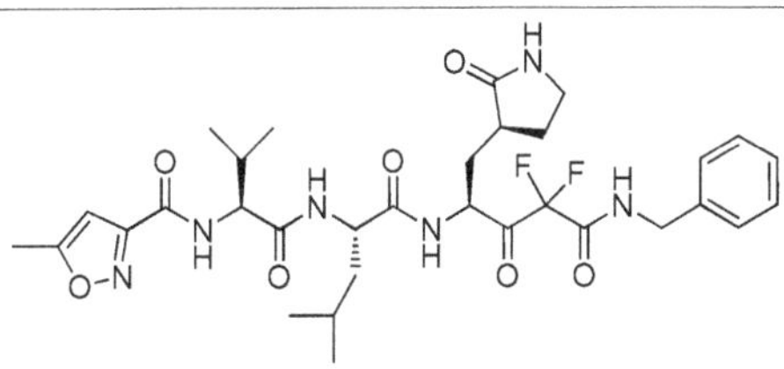 **5**	−10.027	−109.41	2.32	−4.68	No
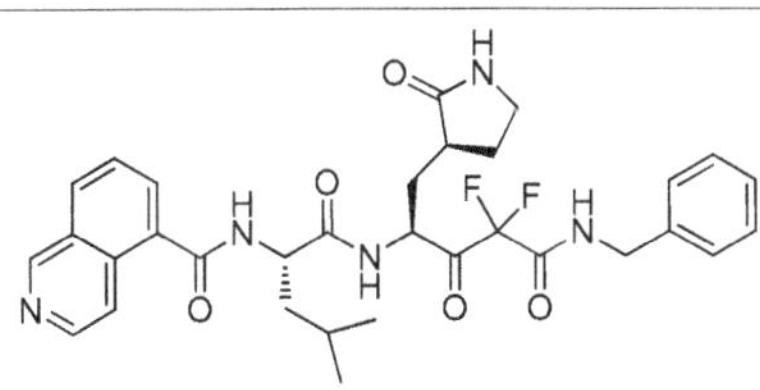 **6**	−11.269	−114.26	3.11	−4.70	No
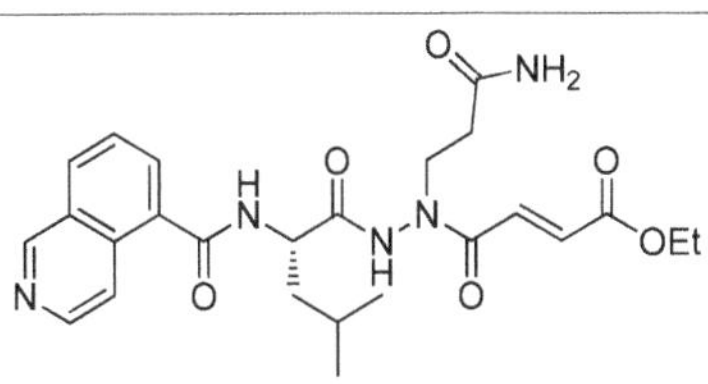 **7**	−9.540	−114.04	1.72	−3.54	No

Table 1. *Cont.*

Compound	Docking Score (kcal/mol)	ΔG_{bind} (kcal/mol)	QPlogP [B]	QPlogS [C]	PAINS [D]
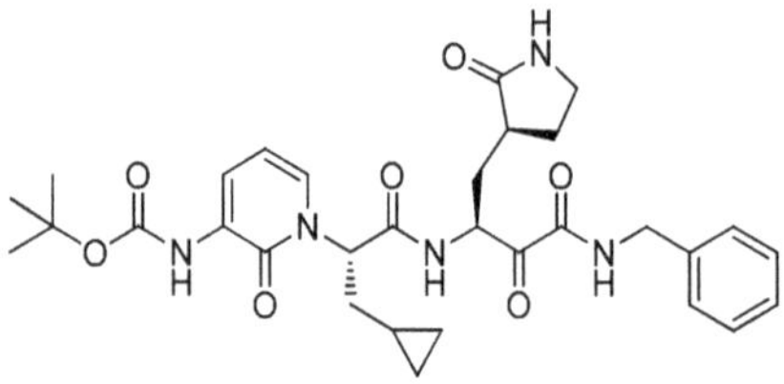 **2**	−9.976	−110.55	3.23	−6.04	No
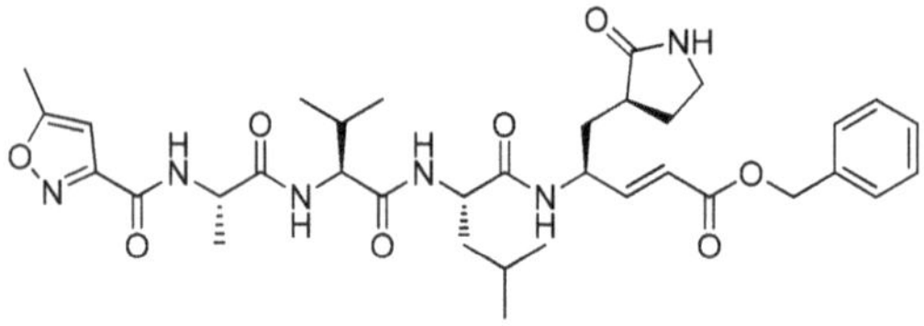 **3**, N3	−10.138	−108.36	3.18	−7.25	No

[A] QPlogP predicted octanol/water partition coefficient (range or recommended value for 95% of known drugs −2–6.5); [B] QPlogS predicted aqueous solubility in mol/dm^3 (range or recommended value for 95% of known drugs: −6.5–0.5); PAINS assessment was performed by FAFDrugs4 online server (accessed on 25 February 2022).

Based on docking results, we observed a significant binding affinity of the developed compounds for the selected target. Considering the retrieved binding mode, we observed that compound **4** (Table 1 and Figure 3A) spanned and interacted with all regions S1–S4 of the M^{pro} binding site (S1–S4). In fact, the difluorostatone moiety established a polar contact with the backbone of G143 (S1′ region), and the pyrrolidinone moiety strongly targeted residues belonging to the S1 region, establishing a series of H-bonds with the backbone of F140 and the sidechains of E166 and H163. The central region of the molecule targeted the backbone of H164 and E166. The P1-moiety of the peptide-based derivative **4** formed H-bonds with the backbone of E166 and with the sidechain of Q192 (S4 region).

The detailed binding mode of compound **4**, represented by a ligand interaction diagram, is visible in Figure 4A. This binding mode accounted for a docking score of −10.779 kcal/mol and a ΔG_{bind} of −123.15 kcal/mol.

The introduction of the oxazole moiety to replace the methyl group of compound **4** led to the peptide-based derivative **5**. As observed for its parent molecule, it can establish a strong H-bond network within the active site of the enzyme (Figures 3B and 4B). Compound **5** could establish interactions with the backbone of G143 and C145 by its difluorostatone portion, and the pyrrolidinone moiety could strongly target F140, E166, and H163, establishing the same contacts described for compound **4**. The oxazole moiety did not form polar contacts, while it established hydrophobic interactions within the S4 region of the enzyme. This binding mode accounted for a docking score comparable to that found for derivative **4** (GlideScore compound **5**: −10.027 kcal/mol; ΔG_{bind} −109.41 kcal/mol). A further modification of the tail of compound **5** by introducing a quinoline group aimed at maximizing the number of contacts within the S4 region of the binding site led to the design of the peptidomimetic **6**. The results of the modeling study on derivative **6** are depicted in Figures 3C and 5A. Gratifyingly, our hypothesis was confirmed by the molecular docking calculation. In effect, in addition to the contacts previously described for compound **5**, compound **6** could target the backbone of T190, increasing the hydrophobic contacts within the S4 region. This binding mode, with an improved number of contacts, accounts for a docking score of −11.269 kcal/mol and a ΔG_{bind} of −114.26 kcal/mol.

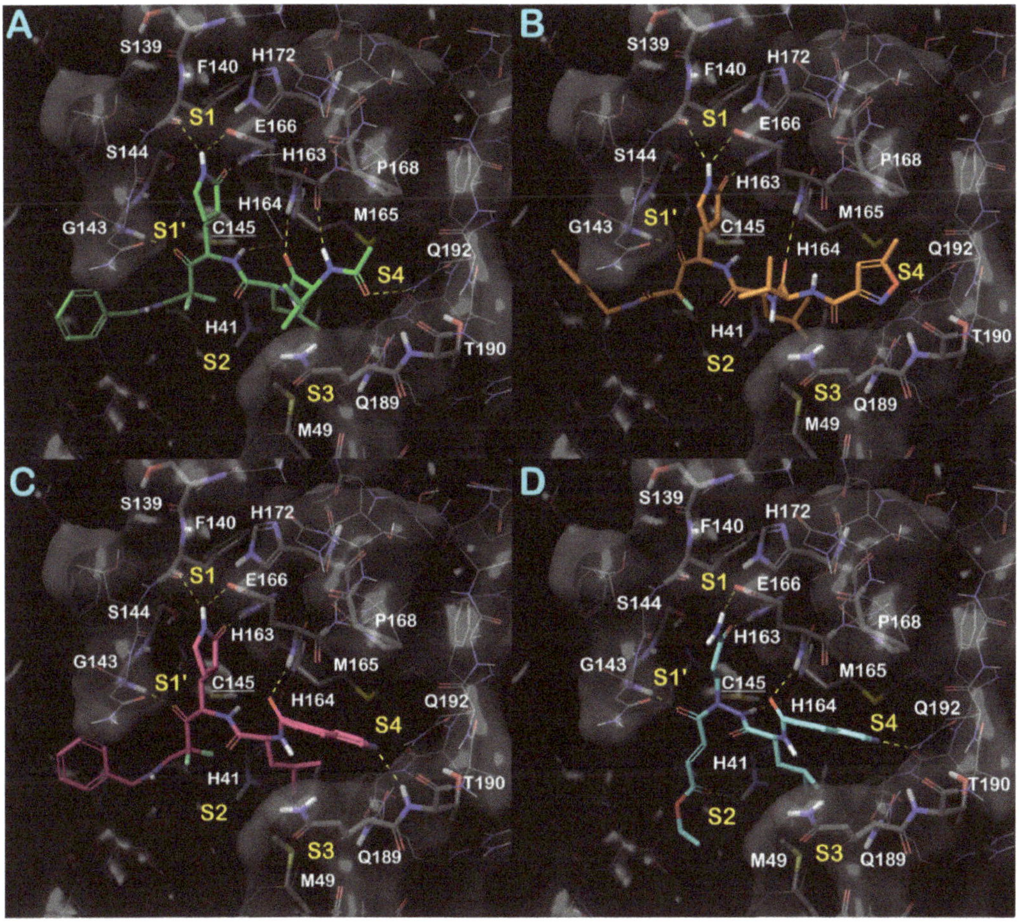

Figure 3. Docked pose of compounds **4–7** (panels **A–D**, respectively) into M^{pro}-SARS-CoV-2 (PDB ID: 6Y2G). Key interacting residues from different regions are represented by sticks and labeled. H-bonds are represented as yellow dotted lines. Pictures were generated by Maestro (Maestro, Schrödinger LLC, release 2020-3).

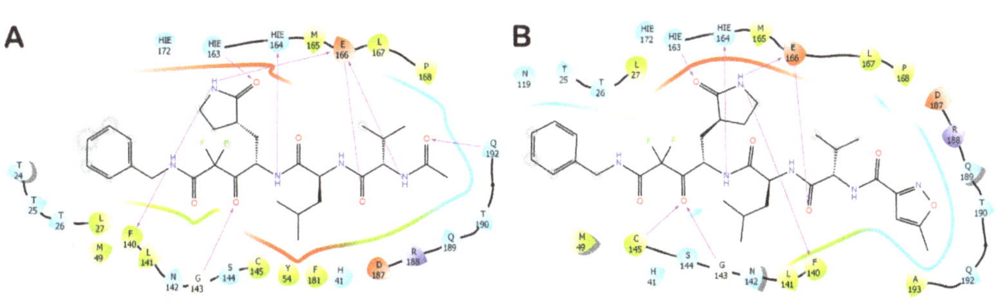

Figure 4. Detailed binding modes of compound **4** (panel **A**) and compound **5** (panel **B**). Pictures were generated by ligand interaction diagram available in Maestro (Maestro, Schrödinger LLC, release 2020-3).

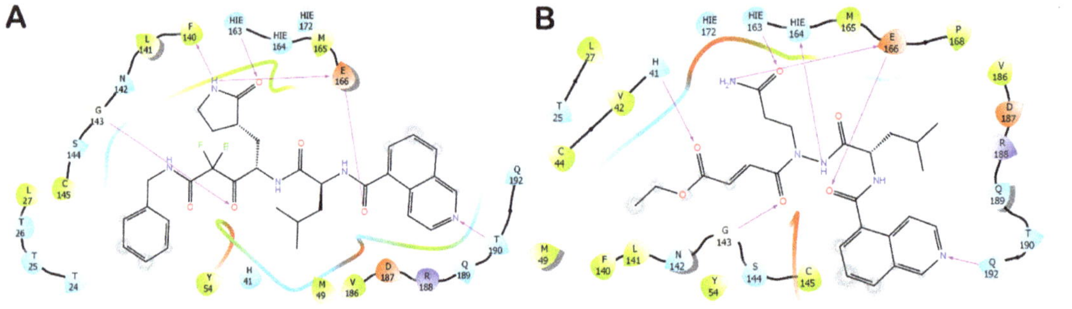

Figure 5. Detailed binding mode of compound **6** (panel **A**) and compound **7** (panel **B**). Pictures were generated by ligand interaction diagram available in Maestro (Maestro, Schrödinger LLC, release 2020-3).

Finally, we attempted to modify the head of the molecule by introducing a Michael acceptor and replacing the pyrrolidinone group with glutamine to evaluate a different warhead in the quinoline-based derivative. Interestingly, the resulting compound, derivative **7**, showed a comparable binding mode with respect to the previously discussed molecules. As reported in Figures 3D and 5B, compound **7** could establish the same above-described interactions at the S1 and S1′ regions of the enzyme, targeting G143, H163, H164, and E166. Additionally, we observed an H-bond with the sidechain of H141. Notably, the quinoline moiety at the S4 site established an H-bond with the sidechain of Q192. Although there was a slight decrease in the docking score (−9.540 kcal/mol), the binding affinity (ΔG_{bind} of −114.04 kcal/mol) is in line with the values estimated for the discussed derivatives **4**–**6**. Accordingly, the docking studies confirmed the potential of the selected peptide-based derivatives to target SARS-CoV-2 M^{pro}.

Because of the mechanism of the enzyme, it is crucial to evaluate the distance between the reactive residues of the enzyme and the possible atoms of the compound susceptible to the attack for covalent bonding. In particular, the M^{pro} C145 residue represents the pivotal residue to form a covalent adduct. Therefore, we measured the distance between the sulfur atom of C145 and the carbon atom of the compound susceptible to the nucleophilic attack. As reported in Figure S1, the measured distances are for all compounds under 3 Å (compound **4**: −2.87 Å; compound **5**: −2.69 Å; compound **6**: −2.84 Å; compound **7**: −2.93 Å). Remarkably, the findings agree with the possibility that these compounds can form a covalent adduct within the active site of the enzyme, precluding its function.

To compare the mentioned results, we performed further docking calculations, using the same computational protocol, employing two reference compounds, **2** and **3** (N3) (Table 1). According to the crystal structures of the reference compounds, the docking protocol was able to correctly accommodate these ligands within the M^{pro} binding site (Figure S2). Furthermore, these calculations also provided computational scores (Table 1) that can be compared to those obtained for compounds **4**–**7** (Table 1). The analysis of docking scores and ΔG_{bind} indicated that our compounds can bind the M^{pro} binding site with affinities comparable to those observed for the reference compounds **2** and **3**, establishing similar contacts, targeting crucial residues for enzyme activity. Moreover, as reported in Figure S3, as expected, also for reference compounds, the distance between the reactive residues of the enzyme (C145) and the possible atom of the compounds susceptible to the attack for a covalent bonding was found to be compatible with the formation of a covalent adduct within the active site of the enzyme, in line with the experimental activity of reference compounds.

3.2. Molecular Dynamics Simulations

After docking studies, we validated the retrieved binding modes by conducting MD simulations in the explicit solvent. We employed M^{pro}/ligands docking-derived complexes to investigate the evolution of biological systems for 100 ns. The resultant MD trajectories for all complexes were deeply examined through several standard simulation parameters, including RMSD analysis for all backbone atoms and ligands and RMSF of individual amino acid residue. The selected complexes displayed reasonable stability from the early stages of the simulation, as indicated by observing the RMSD. Considering the entire simulation time, we did not observe any major expansion and/or contraction, caused by the binding of the investigated compounds (Figure 6, regarding the simulation of compounds **4–7**). This stability was also corroborated by examining the RMSF determined for the selected complexes. RMSF denotes the variation between the atomic Cα coordinates of the enzyme from its average position during the MD simulation. This computation is profitable to characterize the flexibility of individual residues of the protein backbone. The systems under study did not show considerable fluctuation events, with the exclusion of an extremely limited number of residues at the N- and C-terminal regions of M^{pro} (Figure S4). Likewise, the conformational adaptations of critical residues in the active site (lowest RMSF values for all complexes) confirmed the ability of compounds to form stable interactions within the protein.

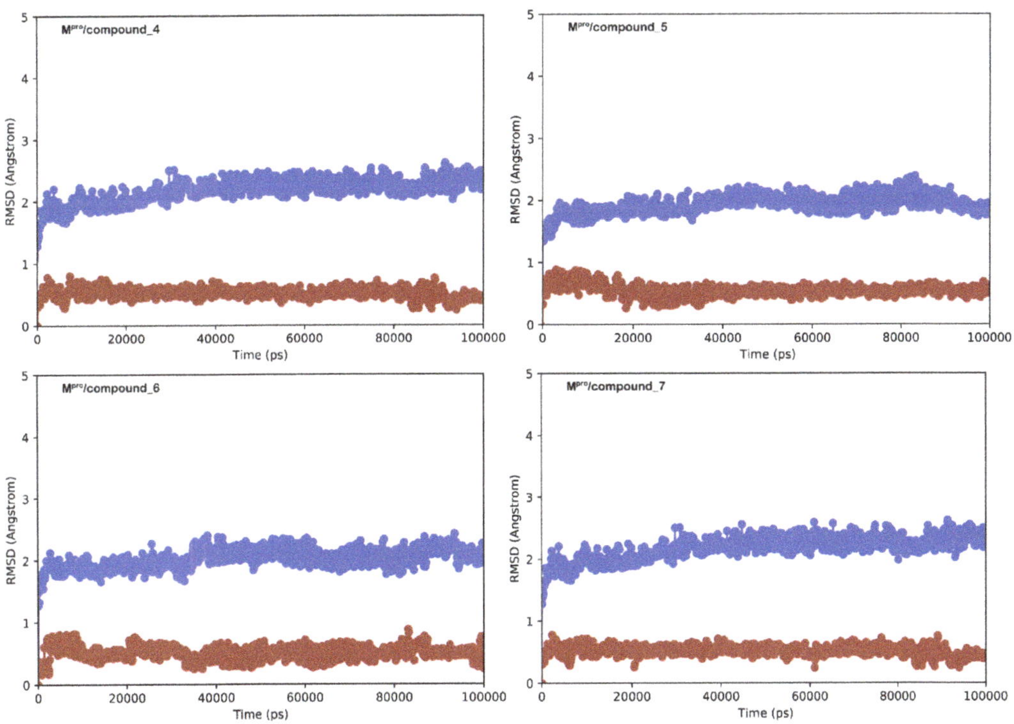

Figure 6. RMSD calculation for each complex investigated in this study: protein (blue line) and ligand (red line). Pictures were generated by Maestro (Maestro, Schrödinger LLC, release 2020-3).

To better comprehend the behavior of derivatives **4–7** into the SARS-CoV-2 M^{pro} binding site, we performed a comprehensive analysis of MD simulations, exploring the established contacts. In general, compound **4** (Figures 7A and S5A) maintained the contacts found by docking calculation. Interestingly, we observed a stronger network of interactions at the S1′ region since contacts with S144 and the backbone of C145 were detectable. Furthermore, the tail of compound **4**, in addition to the H-bond with Q192, established additional H-bonds with Q189 and T190 (S3 region), sometimes water-mediated. The analysis conducted on the trajectory of MD simulation for compound **5** is illustrated in Figures 7B and S5B. Here again, the crucial contacts established by compound **5** within the M^{pro} binding site were maintained. Notably, the strong network of contacts at the S1 and S1′ regions was conserved during the simulation with the addition of contacts with H41. The tail becomes able to effectively target Q189, T190, and Q192 at the S3 and S4 regions, resulting in a more tightly binding within the active site of the enzyme.

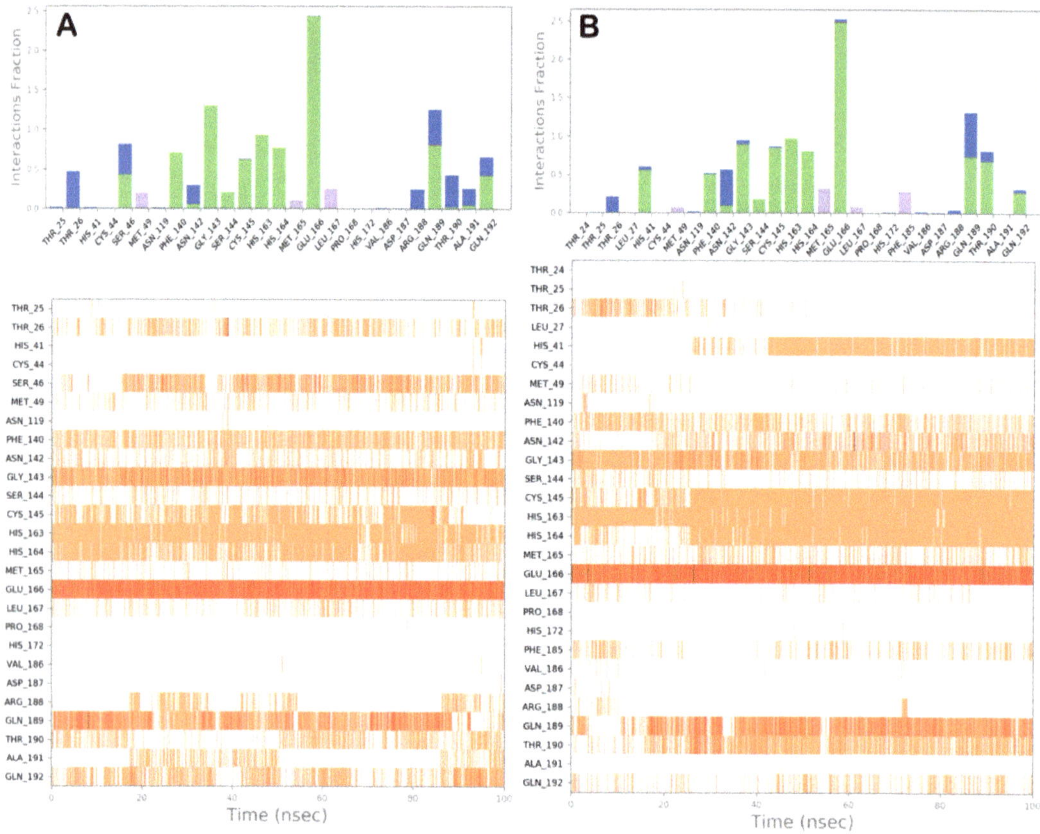

Figure 7. Compounds **4** (panel **A**) and **5** (panel **B**) monitored during the simulation. The contacts can be grouped by type and summarized, as shown in the plots. Grouping protein–ligand interactions into four types: H-bonds (green), hydrophobic (gray), ionic (magenta), and water bridges (blue). In the second graph of the picture is reported a timeline representation of the contacts. Some residues make more than one specific contact with the ligand, which is represented by a darker shade of orange. Pictures were generated by the simulation interaction diagram available in Desmond via Maestro (Maestro, Schrödinger LLC, release 2020-3).

Additionally, the MD analysis of compound **6** (Figures 8A and S5C) and compound **7** (Figures 8B and S5D) revealed a similar trend. Regarding compound **6**, interactions with H41 at the S2 site and Q189 and Q192 at site S3 in addition to the contacts found by molecular docking studies were observed. Compound **7** showed a comparable behavior since it can strengthen the interaction within the active site, establishing a strong network of contacts at S3 with Q189 and T190 and increasing the contacts with Q192. Overall, the MD simulation analysis indicated a high stability of the binding mode found by molecular docking for each selected complex. In addition to the existing contacts, we found a reasonable number of novel contacts that can further stabilize the binding modes.

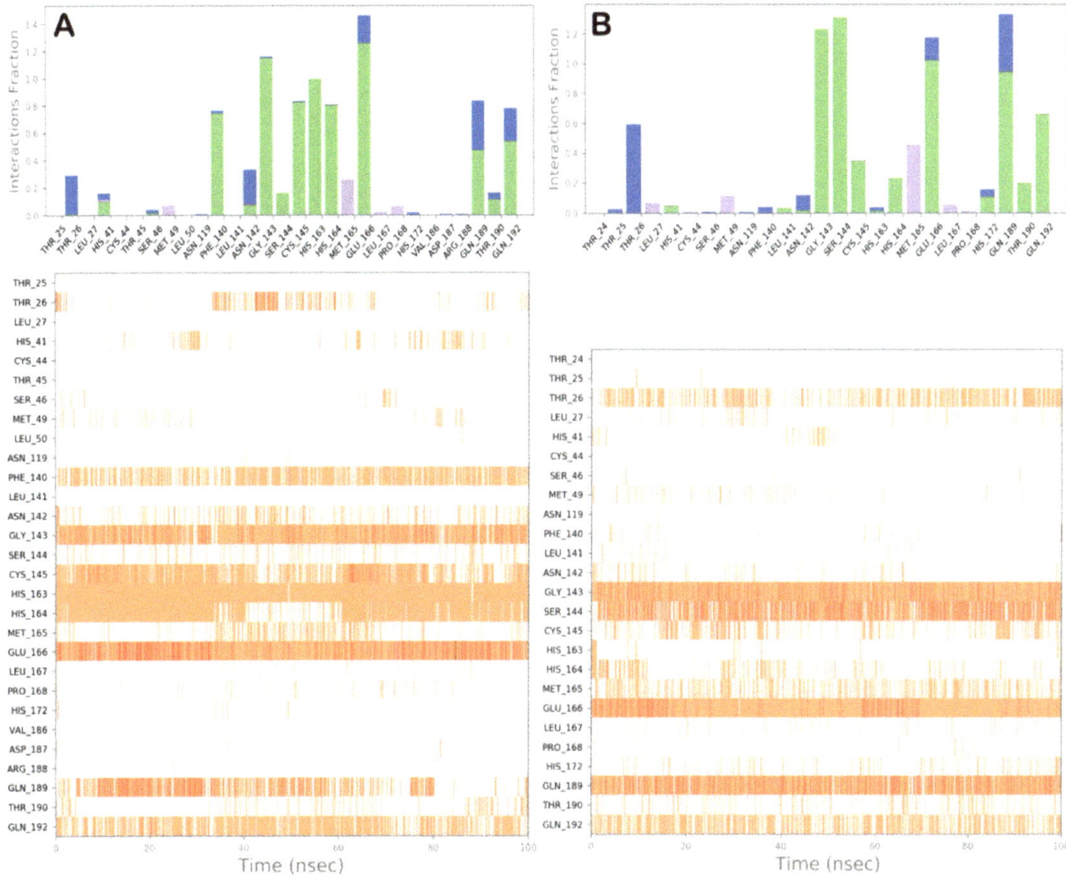

Figure 8. Compounds **6** (panel **A**) and **7** (panel **B**) monitored during the simulation. The contacts can be grouped by type and summarized, as shown in the plots. Grouping protein–ligand interactions into four types: H-bonds (green), hydrophobic (gray), ionic (magenta), and water bridges (blue). In the second graph of the picture is reported a timeline representation of the contacts. Some residues make more than one specific contact with the ligand, which is represented by a darker shade of orange. Pictures were generated by the simulation interaction diagram available in Desmond via Maestro (Maestro, Schrödinger LLC, release 2020-3).

We then monitored the distance between the sulfur atom of C145 and the electrophilic carbon atom of the ligand, susceptible to nucleophilic attack for each complex. As reported in Figure S6, the distance between the selected atoms remained mainly constant with very small variations, as expected due to the high stability of the complexes. Accordingly, the measures indicated that the considered electrophilic carbon atom remained susceptible to a possible nucleophilic attack from C145 during the simulation time.

In addition, to further validate our computational protocol, we performed MD simulations also for the ligand/enzyme complexes of the reference compounds previously described (Figure S2). As observed for compounds **4**–**7**, the investigated systems were reasonably stable with small fluctuations (Figure S7), and the contacts found by docking studies were maintained during the MD simulations (Figure S8). As expected, the distances between the reactive residues of the enzyme (C145) and the possible atom of the compounds susceptible to the attack for a covalent bonding were also found to be stable during the simulations (Figure S9), indicating the reliability of the computational approach.

Finally, to further corroborate the obtained results, we performed additional calculations, using the FEP technique to compute the differences in protein–ligand-binding free energies from MD simulations. The output of this calculation in terms of $\Delta\Delta G_{bind}$ is reported in Table 2, with compound **2** (crystallized ligand in the structure 6Y2G, used in this study) employed as the reference compound. As indicated by the results, compounds **4–6** showed an improved binding affinity with respect to the reference molecule (compound **2**), while compound **7** showed a slight decrease in binding affinity consistent with lower computational scores, found by other methods, with respect to the best performing compounds. Notably, FEP calculation confirms the potency of compound **3** in inhibiting M^{pro} with a slight improvement with respect to the value found for the 6Y2G ligand (compound **2**) [8,12].

Table 2. Computational scores (covalent docking score and ΔG_{bind} derived from docking studies, and $\Delta\Delta G_{bind}$ derived by FEP calculation) obtained for compounds **4–7** compared to the reference compounds **2** and **3**.

Compound	Covalent Docking Score (kcal/mol)	Covalent Docking ΔG_{bind} (kcal/mol)	FEP/MD $\Delta\Delta G_{bind}$ (kcal/mol)
4	−10.834	−128.29	−0.18 ± 0.11
5	−10.232	−119.17	−0.45 ± 0.21
6	−11.681	−116.49	−0.73 ± 0.32
7	−9.828	−115.96	0.12 ± 0.09
2	−10.174	−113.87	–
3,N3	−10.043	−114.74	−0.13 ± 0.12

3.3. Covalent Docking Approach

To gain further insight into the formation of the tetrahedral intermediate and predict the binding mode of peptide-based derivatives, different molecular models of the selected complexes were generated using a covalent docking protocol, namely CovDock, available in Maestro. Once the correct reaction is written and the software recognizes all the residues involved, CovDock initially combines the Glide docking algorithm and Prime structure refinement to determine whether the ligand can be accommodated into the selected binding site (standard docking). In this way, as a constraint, the ligand should sit in a position close enough to the nucleophilic group of the reactive residue. The reactive residue, cysteine, is mutated with an alanine residue to generate an initial association in which the ligand is noncovalently bound to the target protein. Subsequently, the receptor is restored, and the reaction occurs. Once the covalent bond is formed, the complex is minimized. Now, the obtained poses are clustered and ranked after a complete minimization. The output of this calculation is illustrated in Figure 9. The docked poses of compounds **4–7** into the catalytic site of M^{pro} were chosen as the starting point for the covalent docking procedure. As displayed in Figure 9, the tetrahedral intermediates can be stabilized by the formation of novel H-bonds with specific amino acid residues, thus resulting in an overall fine-tuning of the binding conformation of compounds **4–7** within the binding cleft and in the generation of more stable complexes. Accordingly, based on this computational approach, the conceived compounds can act as covalent inhibitors of SARS-CoV-2 M^{pro}.

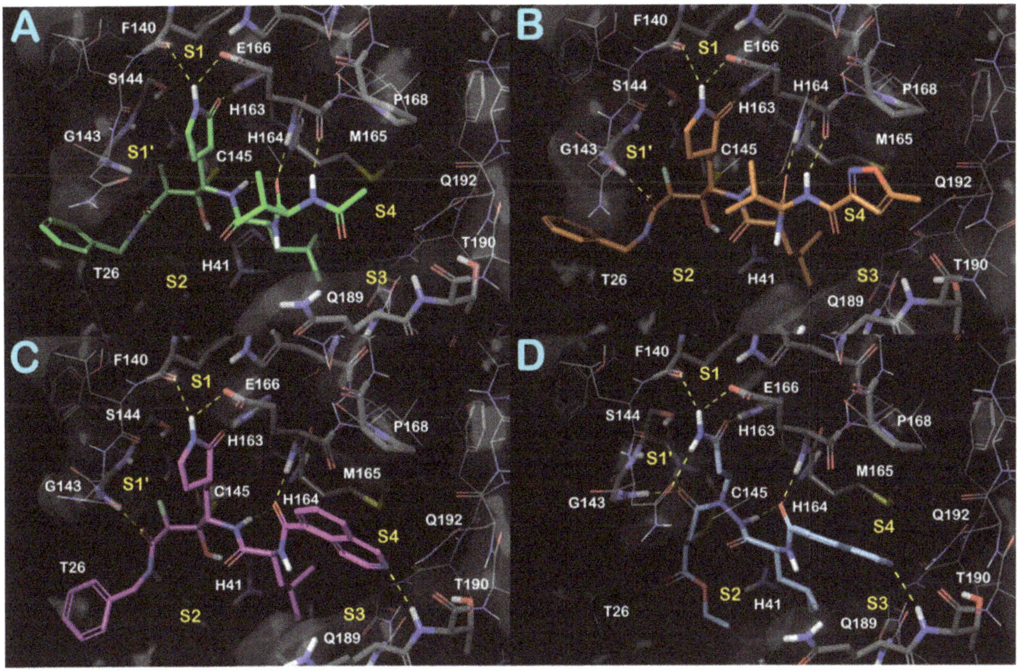

Figure 9. Output regarding the covalent docking investigation considering compounds **4–7** (panels **A–D**, respectively). Key interacting residues from different regions are represented by sticks and labeled. H-bonds are represented as yellow dotted lines. Pictures were generated by Maestro (Maestro, Schrödinger LLC, release 2020-3).

In this case, we conducted the same computational study on reference compounds. The adopted covalent docking protocol is effectively able to correctly accommodate the reference compounds within the M^{pro} binding site with high accuracy, reproducing the crystal structure conformation of reference compounds when they are covalently bound to the binding site (Figure S10). Remarkably, the geometry obtained for the reference ligands covalently bound to M^{pro} is very close to that observed in the crystal structures [8,12]. Gratifyingly, the computational docking scores reported in Table 2 further confirmed the susceptibility of compounds **4–7** to react within the M^{pro} binding site, forming a covalent adduct with C145 due to the comparable scores found for the reference compounds.

4. Conclusions

In summary, we have described a computational protocol aimed at designing novel SARS-CoV-2 M^{pro} covalent inhibitors. The work was focused on the evaluation of bifunctional warheads engaging prime and nonprime subsites of the active site of the enzyme. To this end, we designed, considering the binding site of M^{pro}, peptide-based inhibitors based on the difluorostatone scaffold that has been demonstrated to be effective in inhibiting other proteases [13–15]. In addition, a peptide-based inhibitor containing a Michael acceptor has been designed. All these compounds were computationally investigated using several in silico techniques such as molecular docking, covalent docking, MD simulation, and FEP, for evaluating their potential as covalent inhibitors against SARS-CoV-2 M^{pro}. Computational hints indicated that the proposed compounds can be effective in inhibiting the enzyme, deserving further experimental studies to confirm these findings to expand the armamentarium for fighting this virus. Moreover, our work provides a rational computer-driven approach for developing covalent inhibitors of the M^{pro} enzyme. This approach could also be extended to the inhibition of other drug targets.

Supplementary Materials: The following supporting information can be downloaded at: https://www.mdpi.com/article/10.3390/computation10050069/s1, Figure S1: Measured distances between the sulfur of the catalytic residue C145 and the electrophilic carbon of compounds **4** (panel A), **5** (panel B), **6** (panel C), and **7** (panel D) that can be susceptible of nucleophilic attack; Figure S2: Docked pose of compound **2** and compound **3** (N3) (panels A,B, respectively) into M^{pro}-SARS-CoV-2 (PDB ID: 6Y2G); Figure S3: Measured distances between the sulfur of the catalytic residue C145 and the electrophilic carbon of compound **2** (panel A) and compound **3** (N3) (panel B) that can be susceptible to nucleophilic attack; Figure S4: RMSF calculation for each complex, selected by docking studies, after 100 ns of MD simulation; Figure S5: Dynamic ligand interaction diagram regarding compounds **4** (panel A), **5** (panel B), **6** (panel C), and **7** (panel D), calculated through 100 ns of MD simulation; Figure S6: Monitored distances between the sulfur of the catalytic residue C145 and the electrophilic carbon of compounds **4** (panel A), **5** (panel B), **6** (panel C), and **7** (panel D) that can be susceptible to nucleophilic attack; Figure S7: RMSD calculation for each complex investigated in this study: protein (blue line) and ligand (red line); RMSF calculation for each complex, selected by docking studies, after 100 ns of MD simulation (panel A, compound **2**; panel B, compound **3** (N3)); Figure S8: Compound **2** (panel A) and compound **3** (N3) (panel B) monitored during the simulation; Figure S9: Monitored distances between the sulfur of the catalytic residue C145 and the electrophilic carbon of compound **2** (panel A) and compound **3** (N3) (panel B) that can be susceptible to nucleophilic attack; Figure S10: Output regarding the covalent docking investigation considering compound **2** (panel A) and compound **3** (N3) (panel B).

Author Contributions: Conceptualization, S.B. (Simone Brogi) and S.G.; methodology, S.B. (Simone Brogi), R.I., S.B. (Stefania Butini) and S.G.; software, S.B. (Simone Brogi); validation, S.B. (Simone Brogi), S.B. (Stefania Butini), V.C., G.C. and S.G.; investigation, S.B. (Simone Brogi), S.R., S.B. (Stefania Butini) and S.G.; data curation, S.B. (Simone Brogi), S.R., R.I., S.B. (Stefania Butini), V.C., G.C. and S.G.; writing—original draft preparation, S.B. (Simone Brogi) and S.G.; writing—review and editing, S.B. (Simone Brogi), S.R., R.I., S.B. (Stefania Butini), V.C., G.C. and S.G.; supervision, S.B. (Simone Brogi) and S.G.; funding acquisition, S.B. (Simone Brogi) and S.G. All authors have read and agreed to the published version of the manuscript.

Funding: This research was funded by "Fondo di Beneficenza di Intesa Sanpaolo" (Grant Number B/2020/0113 to S.B. (Simone Brogi) and S.G.).

Institutional Review Board Statement: Not applicable.

Informed Consent Statement: Not applicable.

Data Availability Statement: Not applicable.

Conflicts of Interest: The authors declare no conflict of interest.

References

1. Pollard, C.A.; Morran, M.P.; Nestor-Kalinoski, A.L. The COVID-19 pandemic: A global health crisis. *Physiol. Genomics* **2020**, *52*, 549–557. [CrossRef] [PubMed]
2. Mahase, E. COVID-19: UK becomes first country to authorise antiviral molnupiravir. *BMJ* **2021**, *375*, n2697. [CrossRef] [PubMed]
3. Mahase, E. COVID-19: Pfizer's paxlovid is 89% effective in patients at risk of serious illness, company reports. *BMJ* **2021**, *375*, n2713. [CrossRef] [PubMed]
4. Painter, W.P.; Holman, W.; Bush, J.A.; Almazedi, F.; Malik, H.; Eraut, N.; Morin, M.J.; Szewczyk, L.J.; Painter, G.R. Human Safety, Tolerability, and Pharmacokinetics of Molnupiravir, a Novel Broad-Spectrum Oral Antiviral Agent with Activity Against SARS-CoV-2. *Antimicrob. Agents Chemother.* **2021**, *65*, e02428-20. [CrossRef]
5. Fischer, W.; Eron, J.J.; Holman, W.; Cohen, M.S.; Fang, L.; Szewczyk, L.J.; Sheahan, T.P.; Baric, R.; Mollan, K.R.; Wolfe, C.R.; et al. Molnupiravir, an Oral Antiviral Treatment for COVID-19. *medRxiv* **2021**. [CrossRef]
6. Cui, J.; Li, F.; Shi, Z.L. Origin and evolution of pathogenic coronaviruses. *Nat. Rev. Microbiol.* **2019**, *17*, 181–192. [CrossRef]
7. Banerjee, R.; Perera, L.; Tillekeratne, L.M.V. Potential SARS-CoV-2 main protease inhibitors. *Drug Discov. Today* **2021**, *26*, 804–816. [CrossRef]
8. Jin, Z.; Du, X.; Xu, Y.; Deng, Y.; Liu, M.; Zhao, Y.; Zhang, B.; Li, X.; Zhang, L.; Peng, C.; et al. Structure of M(pro) from SARS-CoV-2 and discovery of its inhibitors. *Nature* **2020**, *582*, 289–293. [CrossRef]

9. Owen, D.R.; Allerton, C.M.N.; Anderson, A.S.; Aschenbrenner, L.; Avery, M.; Berritt, S.; Boras, B.; Cardin, R.D.; Carlo, A.; Coffman, K.J.; et al. An oral SARS-CoV-2 M(pro) inhibitor clinical candidate for the treatment of COVID-19. *Science* **2021**, *374*, 1586–1593. [CrossRef]
10. Zhao, Y.; Fang, C.; Zhang, Q.; Zhang, R.; Zhao, X.; Duan, Y.; Wang, H.; Zhu, Y.; Feng, L.; Zhao, J.; et al. Crystal structure of SARS-CoV-2 main protease in complex with protease inhibitor PF-07321332. *Protein Cell* **2021**. [CrossRef]
11. Narayanan, A.; Narwal, M.; Majowicz, S.A.; Varricchio, C.; Toner, S.A.; Ballatore, C.; Brancale, A.; Murakami, K.S.; Jose, J. Identification of SARS-CoV-2 inhibitors targeting Mpro and PLpro using in-cell-protease assay. *Commun. Biol.* **2022**, *5*, 169. [CrossRef] [PubMed]
12. Zhang, L.; Lin, D.; Sun, X.; Curth, U.; Drosten, C.; Sauerhering, L.; Becker, S.; Rox, K.; Hilgenfeld, R. Crystal structure of SARS-CoV-2 main protease provides a basis for design of improved alpha-ketoamide inhibitors. *Science* **2020**, *368*, 409–412. [CrossRef] [PubMed]
13. Brogi, S.; Giovani, S.; Brindisi, M.; Gemma, S.; Novellino, E.; Campiani, G.; Blackman, M.J.; Butini, S. In silico study of subtilisin-like protease 1 (SUB1) from different Plasmodium species in complex with peptidyl-difluorostatones and characterization of potent pan-SUB1 inhibitors. *J. Mol. Graph. Model.* **2016**, *64*, 121–130. [CrossRef] [PubMed]
14. Giovani, S.; Penzo, M.; Brogi, S.; Brindisi, M.; Gemma, S.; Novellino, E.; Savini, L.; Blackman, M.J.; Campiani, G.; Butini, S. Rational design of the first difluorostatone-based PfSUB1 inhibitors. *Bioorg. Med. Chem. Lett.* **2014**, *24*, 3582–3586. [CrossRef]
15. Giovani, S.; Penzo, M.; Butini, S.; Brindisi, M.; Gemma, S.; Novellino, E.; Campiani, G.; Blackman, M.J.; Brogi, S. Plasmodium falciparum subtilisin-like protease 1: Discovery of potent difluorostatone-based inhibitors. *RSC Adv.* **2015**, *5*, 22431–22448. [CrossRef]
16. Brogi, S.; Maramai, S.; Brindisi, M.; Chemi, G.; Porcari, V.; Corallo, C.; Gennari, L.; Novellino, E.; Ramunno, A.; Butini, S.; et al. Activation of the Wnt Pathway by Small Peptides: Rational Design, Synthesis and Biological Evaluation. *ChemMedChem* **2017**, *12*, 2074–2085. [CrossRef]
17. Jorgensen, W.L.; Maxwell, D.S.; Tirado-Rives, J. Development and Testing of the OPLS All-Atom Force Field on Conformational Energetics and Properties of Organic Liquids. *J. Am. Chem. Soc.* **1996**, *118*, 11225–11236. [CrossRef]
18. Testai, L.; Piragine, E.; Piano, I.; Flori, L.; Da Pozzo, E.; Miragliotta, V.; Pirone, A.; Citi, V.; Di Cesare Mannelli, L.; Brogi, S.; et al. The Citrus Flavonoid Naringenin Protects the Myocardium from Ageing-Dependent Dysfunction: Potential Role of SIRT1. *Oxid. Med. Cell. Longev.* **2020**, *2020*, 4650207. [CrossRef]
19. Brogi, S.; Brindisi, M.; Butini, S.; Kshirsagar, G.U.; Maramai, S.; Chemi, G.; Gemma, S.; Campiani, G.; Novellino, E.; Fiorenzani, P.; et al. (S)-2-Amino-3-(5-methyl-3-hydroxyisoxazol-4-yl)propanoic Acid (AMPA) and Kainate Receptor Ligands: Further Exploration of Bioisosteric Replacements and Structural and Biological Investigation. *J. Med. Chem.* **2018**, *61*, 2124–2130. [CrossRef]
20. Frydenvang, K.; Pickering, D.S.; Kshirsagar, G.U.; Chemi, G.; Gemma, S.; Sprogoe, D.; Kaern, A.M.; Brogi, S.; Campiani, G.; Butini, S.; et al. Ionotropic Glutamate Receptor GluA2 in Complex with Bicyclic Pyrimidinedione-Based Compounds: When Small Compound Modifications Have Distinct Effects on Binding Interactions. *ACS Chem. Neurosci.* **2020**, *11*, 1791–1800. [CrossRef]
21. Nickolls, J.; Buck, I.; Garland, M.; Skadron, K. Scalable parallel programming with CUDA. *Queue* **2008**, *6*, 40. [CrossRef]
22. Jorgensen, W.L.; Chandrasekhar, J.; Madura, J.D.; Impey, R.W.; Klein, M.L. Comparison of simple potential functions for simulating liquid water. *J. Chem. Phys.* **1983**, *79*, 926–935. [CrossRef]
23. Sirous, H.; Chemi, G.; Gemma, S.; Butini, S.; Debyser, Z.; Christ, F.; Saghaie, L.; Brogi, S.; Fassihi, A.; Campiani, G.; et al. Identification of Novel 3-Hydroxy-pyran-4-One Derivatives as Potent HIV-1 Integrase Inhibitors Using in silico Structure-Based Combinatorial Library Design Approach. *Front Chem.* **2019**, *7*, 574. [CrossRef] [PubMed]
24. Brindisi, M.; Ulivieri, C.; Alfano, G.; Gemma, S.; de Asis Balaguer, F.; Khan, T.; Grillo, A.; Chemi, G.; Menchon, G.; Prota, A.E.; et al. Structure-activity relationships, biological evaluation and structural studies of novel pyrrolonaphthoxazepines as antitumor agents. *Eur. J. Med. Chem.* **2019**, *162*, 290–320. [CrossRef] [PubMed]
25. Brogi, S.; Sirous, H.; Calderone, V.; Chemi, G. Amyloid beta fibril disruption by oleuropein aglycone: Long-time molecular dynamics simulation to gain insight into the mechanism of action of this polyphenol from extra virgin olive oil. *Food Funct.* **2020**, *11*, 8122–8132. [CrossRef]
26. Humphreys, D.D.; Friesner, R.A.; Berne, B.J. A Multiple-Time-Step Molecular Dynamics Algorithm for Macromolecules. *J. Phys. Chem.* **1994**, *98*, 6885–6892. [CrossRef]
27. Hoover, W.G. Canonical dynamics: Equilibrium phase-space distributions. *Phys. Rev. A* **1985**, *31*, 1695–1697. [CrossRef]
28. Martyna, G.J.; Tobias, D.J.; Klein, M.L. Constant pressure molecular dynamics algorithms. *J. Chem. Phys.* **1994**, *101*, 4177–4189. [CrossRef]
29. Essmann, U.; Perera, L.; Berkowitz, M.L.; Darden, T.; Lee, H.; Pedersen, L.G. A smooth particle mesh Ewald method. *J. Chem. Phys.* **1995**, *103*, 8577–8593. [CrossRef]
30. Zhu, K.; Borrelli, K.W.; Greenwood, J.R.; Day, T.; Abel, R.; Farid, R.S.; Harder, E. Docking covalent inhibitors: A parameter free approach to pose prediction and scoring. *J. Chem. Inf. Model* **2014**, *54*, 1932–1940. [CrossRef]
31. Brogi, S.; Fiorillo, A.; Chemi, G.; Butini, S.; Lalle, M.; Ilari, A.; Gemma, S.; Campiani, G. Structural characterization of Giardia duodenalis thioredoxin reductase (gTrxR) and computational analysis of its interaction with NBDHEX. *Eur. J. Med. Chem.* **2017**, *135*, 479–490. [CrossRef] [PubMed]

computation

MDPI

Article

A Computational Study to Identify Potential Inhibitors of SARS-CoV-2 Main Protease (Mpro) from Eucalyptus Active Compounds

Ibrahim Ahmad Muhammad [1], Kanikar Muangchoo [2,*], Auwal Muhammad [3], Ya'u Sabo Ajingi [4], Ibrahim Yahaya Muhammad [3], Ibrahim Dauda Umar [3] and Abubakar Bakoji Muhammad [5,6]

1 Department of Biochemistry, Faculty of Science, Kano University of Science and Technology (KUST), Wudil, Kano 713281, Nigeria; ibrahiem4real@gmail.com
2 Faculty of Science and Technology, Rajamangala University of Technology Phranakhon (RMUTP), Bang Sue, Bangkok 10300, Thailand
3 Department of Physics, Faculty of Science, Kano University of Science and Technology (KUST), Wudil, Kano 713281, Nigeria; auwal@kustwudil.edu.ng (A.M.); ibrahimyahayamuhammad@gmail.com (I.Y.M.); dauda6776@gmail.com (I.D.U.)
4 Department of Biology, Faculty of Science, Kano University of Science and Technology (KUST), Wudil, Kano 713281, Nigeria; yausaboajingi@gmail.com
5 Faculty of Natural Sciences II, Institute of Mathematics, Martin Luther University Halle-Wittenberg, 06099 Halle, Germany; abubakar.muhammad@mathematik.uni-halle.de
6 Department of Mathematics, Faculty of Science, Gombe State University, Gombe 760214, Nigeria
* Correspondence: kanikar.m@rmutp.ac.th

Received: 10 August 2020; Accepted: 6 September 2020; Published: 9 September 2020

Abstract: Severe acute respiratory syndrome coronavirus 2 (SARS-CoV-2) was found to be a severe threat to global public health in late 2019. Nevertheless, no approved medicines have been found to inhibit the virus effectively. Anti-malarial and antiviral medicines have been reported to target the SARS-CoV-2 virus. This paper chose eight natural eucalyptus compounds to study their binding interactions with the SARS-CoV-2 main protease (Mpro) to assess their potential for becoming herbal drugs for the new SARS-CoV-2 infection virus. In-silico methods such as molecular docking, molecular dynamics (MD) simulations, and Molecular Mechanics Poisson Boltzmann Surface Area (MM/PBSA) analysis were used to examine interactions at the atomistic level. The results of molecular docking indicate that Mpro has good binding energy for all compounds studied. Three docked compounds, α-gurjunene, aromadendrene, and allo-aromadendrene, with highest binding energies of −7.34 kcal/mol (−30.75 kJ/mol), −7.23 kcal/mol (−30.25 kJ/mol), and −7.17 kcal/mol (−29.99 kJ/mol) respectively, were simulated with GROningen MAchine for Chemical Simulations (GROMACS) to measure the molecular interactions between Mpro and inhibitors in detail. Our MD simulation results show that α-gurjunene has the strongest binding energy of −20.37 kcal/mol (−85.21 kJ/mol), followed by aromadendrene with −18.99 kcal/mol (−79.45 kJ/mol), and finally allo-aromadendrene with −17.91 kcal/mol (−74.95 kJ/mol). The findings indicate that eucalyptus may be used to inhibit the Mpro enzyme as a drug candidate. This is the first computational analysis that gives an insight into the potential role of structural flexibility during interactions with eucalyptus compounds. It also sheds light on the structural design of new herbal medicinal products against Mpro.

Keywords: binding energy; eucalyptus compounds; molecular docking; molecular dynamics; SARS-CoV-2

1. Introduction

Unspecified pneumonia was reported in the Wuhan region of the Hubei Province, China, towards the end of 2019. Medically, it was quite comparable to viral pneumonia. After the screening of clinical samples, the disease control unit specialist reported that it was pneumonia associated with the severe acute respiratory syndrome (SARS). Eventually, the World Health Organization (WHO) officially labeled it COVID-19 and it has quickly spread from its original area to nearly all of China, and over 200 nations and regions worldwide today. The International Committee on Virus Taxonomy called the new coronavirus "severe acute respiratory syndrome coronavirus 2" (SARS-CoV-2) [1,2]. The SARS-CoV-2 infection leads to difficulty breathing, fever, chronic respiratory failure, and dry cough, which might also result in death [3]. A total of 14,263,202 SARS-CoV-2 cases were recorded as of 20 July 2020, with 220,026 new confirmed cases and 602,244 deaths worldwide [4]. Nigeria alone recorded 36,663 confirmed cases and 789 fatalities. Cases increased exponentially between April and June, with the highest number of cases reported on 15 June (904 confirmed cases) and 45 deaths reported on 18 June 2020 [5].

SARS-CoV-2 is part of the Coronaviridae family that consists of the main positive-sense single-strand RNA viruses. These viruses are categorized into α, β, γ, and δ genera. SARS-CoV-2, SARS-CoV, and Middle East respiratory syndrome coronavirus (MERS-CoV) all belong to β-coronaviruses. A study of the genome sequences of these viruses showed that SARS-CoV-2 encompasses a higher nucleotide homology of 89.1% with SARS-CoV compared with MERS-CoV [1,3,6]. In spite of the scientific community's immediate and unprecedented research efforts worldwide, no successful antiviral or vaccine is currently available for SARS-CoV-2. Nevertheless, substantial steps have been taken to produce vaccines and treatment drugs undergoing early clinical studies. Antiviral medicines such as remdesivir developed for Ebola, and anti-HIV drugs such as lopinavir and ritonavir, and popular anti-malaria medicine hydroxychloroquine are currently in mega clinical testing for COVID-19 treatments [7]. Some studies have revealed that chloroquine phosphate inactivates SARS-CoV-2 [8–10], and others revealed that SARS-CoV-2 in-vitro is inhibited by hydroxychloroquine sulfate [8,10]. In addition, computational scientists used the in-silico strategy to identify potential drug targets by investigating complex atomistic interactions [11–15]. One of the most described drug targets for SARS-CoV-2 is the main protease (Mpro, also called 3CLpro), an enzyme necessary for the viral replication. The Mpro works in at least 11 digestive sites in the gigantic polypeptide 1ab (replicase 1ab, approximately 790 kDa) [16].

Based on recent economic impacts on the financial markets, vaccine manufacturing funding appears to be a considerable investment in the coming days [17]. Natural drug treatment may help to avert the spread of the virus in this setting. Nature offers a vast library of chemical compounds that have yet to be researched and established as medicines for the therapy of many viral infections [18]. The eucalyptus tree is one of several plant species used in Nigeria to prevent certain diseases. Customarily, the leaves are boiled, and the fumes are breathed to increase the respiratory tract's effectiveness. East Africa's Mozambique report suggested that eucalyptus could assist in combating malaria, flu, and even fever, reducing the transmission of the disease outbreak. However, national health specialists in the country are often warning of eucalyptus vapor inhalation [19]. Since prehistoric times eucalyptus has been used for many reasons because it has anti-cancer, anti-inflammatory, antiseptic, antioxidant, and antibacterial properties. Therefore, common colds, flu, sinus congestion, and respiratory ailments are cured with eucalyptus [20]. Bahare Salehi et al. (2019) also reported that eucalyptus had gained a great deal of global interest due to its antimicrobial, anti-inflammatory, and insect repellent properties for therapeutic and furniture purposes. The most significant medical benefits of eucalyptus include improving respiratory health, boosting the immune system, lowering blood pressure, and combating bacterial infection. Traditionally, it is being used to promote mucus secretion in the respiratory system [21].

The head of the Indonesian Ministry of Agriculture has recently admitted that eucalyptus-dependent treatment has been established, stating that the spread of COVID-19 has therefore been decreased. Influenza, β, and γ coronaviruses were screened, and 80–100% of the viruses had been destroyed [22]. However, in less than a month, another analysis revealed that the efficacy of eucalyptus oil in COVID-19 therapy still requires extra study, since COVID-19 (SARS-CoV-2) was not included in earlier studies, but other forms of coronavirus were. Hence, as a result of this limited research data, eucalyptus cannot merely be referred to as the SARS-CoV-2 drug [23]. For this research work, eight eucalyptus compounds were selected for use against the target SARS-CoV-2 Mpro. As an in-silico technique, molecular docking simulation was used to understand the positional binding and interaction mechanisms with the target molecule. To further investigate the nature of the interactions and the energy contribution per amino acid residue, the top three compounds with the best (lowest in terms of kcal/mol) binding energies were subjected to classical molecular dynamics simulations. To the best of our knowledge, this is the first computational report that explores the potential inhibitory effects of eucalyptus compounds against the Mpro protein.

2. Methods

2.1. Ligand Preparations

The information about eight compounds selected from eucalyptus with antiviral activities was reported from the literature [20,21,24,25]. Such compounds were found in the PubChem database [26] and saved in .sdf format, then translated into three-dimensional structures with Avogadro software [27]. ChemSketch was used to create two-dimensional structures of the phytochemicals (Figure 1). PubChem ID and the molecular weight are listed in Table 1.

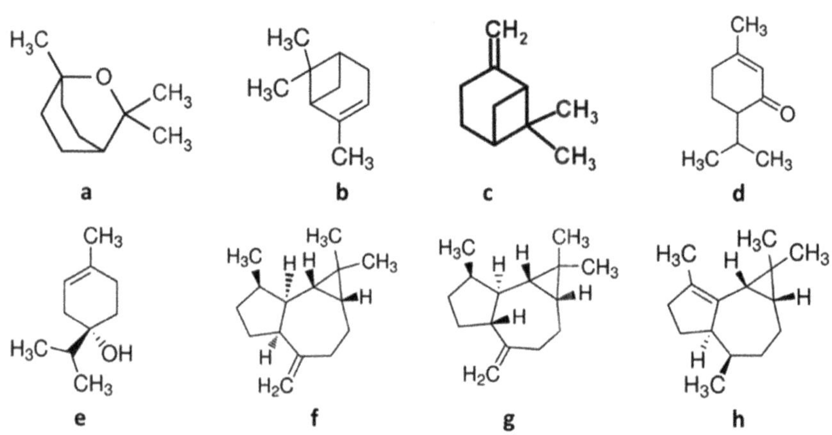

Figure 1. Two molecular dimensional structures of the eucalyptus compounds used in this study: (**a**) 1, 8-cineole (eucalyptus), (**b**) α-pinene, (**c**) β-pinene, (**d**) terpinen-4-ol, (**e**) piperitone, (**f**) allo-aromadendrene, (**g**) aromadendrene, and (**h**) α-gurjunene.

Table 1. PubChem ID and molecular weight of the selected compounds.

Eucalyptol Compounds	PubChem ID	Molecular Weight (g/mol)
1,8-cineole (eucalyptol)	2758	154.25
α-pinene	440968	136.23
β-pinene	440967	136.23
Terpinen-4-ol	11230	154.25
Piperitone	92998	168.23
Allo-aromadendrene	42608158	204.35
Aromadendrene	91354	204.35
α-gurjunene	15560276	204.35

2.2. Protein Model Preparations

The SARS-CoV-2 Mpro 3D structure (Figure 2) with PDB ID: 6LU7 [28] was from the protein database library (https://www.rcsb.org). The structure was imported into visual molecular software, eliminating a ligand (N3) in complex with protein and water molecules. Hydrogen atoms were added prior to docking to correct the ionization and tautomeric states of amino acid residues. There are three different domains in the Mpro structure: residues from domain I (8–100), residues from domain II (101–183), and residues from domain III (200–303). N-terminal amino acids 1 through 7 constitute the N-finger, which plays an essential role in dimerizing and forming the Mpro active site. Domains I and II, known collectively as the N-terminal domain, have an anti-parallel β-sheet structure with 14 β-strands. The substrate-binding site is positioned inside a cleft between domains I and II. A loop from amino acids 184 to 199 connects the N-terminal domain and domain III, also known as the C-terminal domain, and forms a five α-helix anti-parallel cluster [29].

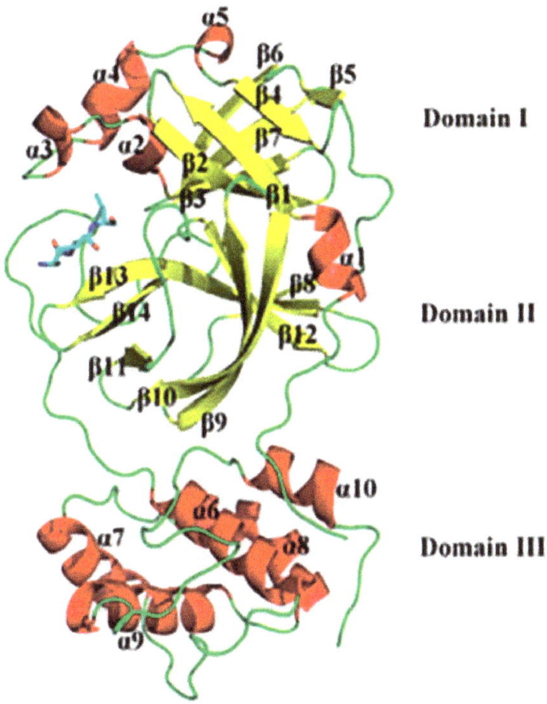

Figure 2. Three-dimensional structure of Mpro protein in complex with N3 inhibitor. It comprises three different domains.

2.3. Molecular Docking

Molecular docking was performed using Autodock4.2 [30] to test the binding affinity of eucalyptus compounds toward the SARS-CoV-2 Mpro protein molecule. The receptor molecule remained rigid, and ligands were versatile to achieve a degree of freedom associated with rotational parameters. Protein and ligand PDB were transformed into .pdbqt after merging the nonpolar hydrogen. The cubical grid box had a size of 126 × 126 × 126 and a spacing of 0.375 Å. A rigid grid box was used for the Autogrid4 parameter. In addition to Autogrid, Autodock4 with Lamarckian genetic algorithms [31] was used to achieve optimal docking conformations. Default docking parameters were used except for the docking run. There was a total of one hundred docking runs per compound. The binding affinity more clearly explains the inhibitor's interaction with the protein molecule. The compound's most desirable binding poses were examined by choosing the lowest free energy of binding (ΔG) and the lowest inhibition constant (Ki). The inhibition constant was calculated theoretically with the help of Autodock4.2. Between a protein and ligand, a stable complex was formed, exhibiting more negative free energy from binding and low Ki indicating high potency of an inhibitor [32]. For further analysis, three compounds with the lowest binding energies were selected for the starting structure to set up molecular dynamics simulations. The interactions between the compounds and the target enzyme were studied using Ligplot+ [33].

2.4. Molecular Dynamics (MD) Simulations and Analysis

The GROMACS 2019.3 package was used to perform MD simulations using selected complexes with the lowest binding energies. The complexes were solvated with TIP3P water molecules, which were constrained by LINCS [34] and SETTLE [35] algorithms. Four Na+ counter-ions were added to neutralize the simulation system's charge, and energy minimization was performed using the steepest descent algorithm and GROMOS54A7 force field [36] with a corresponding equilibration of 1 ns. MD simulation was performed for 100 ns per system. Our previous MD studies described a thorough procedure [37]. Molecular mechanics of Poisson–Boltzmann surface area (MM-PBSA) techniques were used to calculate the free binding energies of the complexes. MM-PBSA enthalpy was calculated using molecular mechanics. The effects of both polar and nonpolar solvent components on free energy were evaluated using the Poisson–Boltzmann equation. For the calculation of energy, the GROMACS built-in tools g_mmpbsa and APBSA [38] were used. The last 30 ns of MD simulations were taken in each complex with 3000 frames in each. The parameters used in g_mmpbsa calculations included a protein dielectric constant of 4, solvent dielectric constant set to 80, vacuum dielectric constant set to 1, temperature of 303 K, and SASA constant and surface tension set to 3.84982 kJ/mol and 0.0226778 kJ/(mol Å^2) respectively.

3. Results

3.1. Molecular Docking Results

The molecular docking technique has become one of the most used methods for determining the drug targets for ligand-based computer-aided drug discovery (LB-CADD). This approach has now been used to analyze vast data from drug repositories and easily register, which can save enormous resources, time, and expense associated with LB-CADD [39]. Until now, successful drug treatment for the SARS-CoV-2 virus has not been approved, and it is urgently necessary to identify possible drug targets. We used in-silico Autodock4 to identify possible binding sites and interaction mechanisms of eight potential natural eucalyptus compounds against the Mpro protein of SARS-CoV-2. The tested compounds will pave the way for the development of drugs against SARS-CoV-2. After the docking simulation, 100 different poses of small molecules (ligands) were produced among which the pose with the strongest binding affinity was considered the best pose.

The findings obtained after the docking analysis are described in Table 2 regarding ligand binding energy (kcal/mol), inhibition constant (K_i), and Mpro amino acid residues interacting with natural compounds. The active site of Mpro protein was found to bind all compounds with a range of amino acid residues engaged in interactions. These interactions have been linked to proof of the in-silico protein-ligand interactions. Figure 3 demonstrates the docked natural compound molecules in complexes with Mpro protein. Three of the natural compounds, α-gurjunene, aromadendrene, and allo-aromadendrene, showed significant binding with binding energies of −7.34 kcal/mol (−30.71 kJ/mol), −7.23 kcal/mol (−30.25 kJ/mol), and −7.17 kcal/mol (−29.99 kJ/mol), respectively.

Table 2. Estimated lowest binding energies of main protease (Mpro) in complex with eucalyptus compounds obtained from molecular docking calculations along with inhibition constants and interaction residues.

Eucalyptol Compounds	Binding Energy (kJ/mol)	Inhibition Constant (Ki) (μM)	Mpro Residues Interacting with Natural Compounds
1,8-cineole (eucalyptol)	−26.90	19.5	His41, Met49, Tyr54, His164, Met165, Asp187, Arg188, and Gln189
α-pinene	−26.23	25.55	His41, Met49, His164, Met165, Asp187, Arg188, and Gln189
β-pinene	−26.57	22.34	His41, Met49, Tyr54, His164, Met165, Asp187, Arg188, and Gln189
Terpinen-4-ol	−23.89	65.5	His41, Met49, Pro52, Tyr54, His164, Arg188, and Gln189
Piperitone	−25.52	33.95	His41, Met49, Tyr54, Cys145, His164, Met165, Glu166, Asp187, and Arg188
Allo-aromadendrene	−29.99	5.54	His41, Met49, Tyr54, Cys145, His164, Met165, Glu166, Asp187, Arg188, and Gln189
Aromadendrene	−30.25	5.06	His41, Met49, Tyr54, Cys145, His164, Met165, Asp187, Arg188, and Gln189
α-gurjunene	−30.71	4.15	His41, Met49, Tyr54, Cys145, His164, Met165, Glu166, Asp187, Arg188, and Gln189

The residue of amino acids that led to the binding of Mpro and natural molecules was achieved through hydrophobic and hydrogen bond interactions. Molecular interactions generally play a significant role in forming and stabilizing docking complexes [40,41]. Hydrophobic interactions with eight amino acid residues (His41, Met49, Tyr54, His164, Met165, Asp187, Arg188, and Gln189) have been observed involving eucalyptus with the binding energy of −6.43 kcal/mol (−26.90 kJ/mol) (Figure 4a). The α-pinene compound made a complex through AutoDock with the binding energy of −6.27 kcal/mol (−26.23 kJ/mol). The hydrophobic interactions were formed by seven amino acid residues His41, Met49, His164, Met165, Asp187, Arg188, and Gln189 (Figure 4b). Nearly the same activity with the β-Pinene compound was observed in contrast to α-pinene with the addition of one amino acid residue, Tyr54 (Figure 4c). The binding energy of the terpinen-4-ol molecule was −5.71 kcal/mol (−23.89 kJ/mol), His164 residue formed a hydrogen bond, and six residues of His41, Met49, Pro52, Tyr54, Arg188, and Gln189 were associated in hydrophobic interactions (Figure 4d). Even though the terpinen-4-ol compound had less binding energy in all, it was found relatively useful to bind with active residues. The docked binding energy of the piperitone substrate was −6.1 kcal/mol (−25.52 kJ/mol). At the active site of Mpro, there was one hydrogen bond formation with His164, while eight amino acids were engaged in the formation of hydrophobic interactions (Figure 4e). The binding energy of allo-aromadendrene was −7.17 kcal/mol (−29.99 kJ/mol) and it interacted with nine amino acid residues (His41, Met49, Tyr54, Cys145, His164, Met165, Glu166, Asp187, Arg188, and Gln189) in the active site of SARS-CoV-2 Mpro. These residues participated in hydrophobic interactions (Figure 4f). The compounds aromadendrene and α-gurjunene with binding energies of −7.23 kcal/mol (−30.25 kJ/mol) and −7.34 kcal/mol (−30.71 kJ/mol) formed van der Waals interactions with seven specific amino acid residues (His41, Met49, Cys145, Met165, Asp187, Arg188, and Gln189). Such molecules had the strongest binding energies in contrast with six other molecules (Figure 4g,h). Earlier studies support our findings by reporting similar amino acid residues to SARS-CoV-2 Mpro protein interactions with other ligands, as confirmed in our study [16,29,42–44].

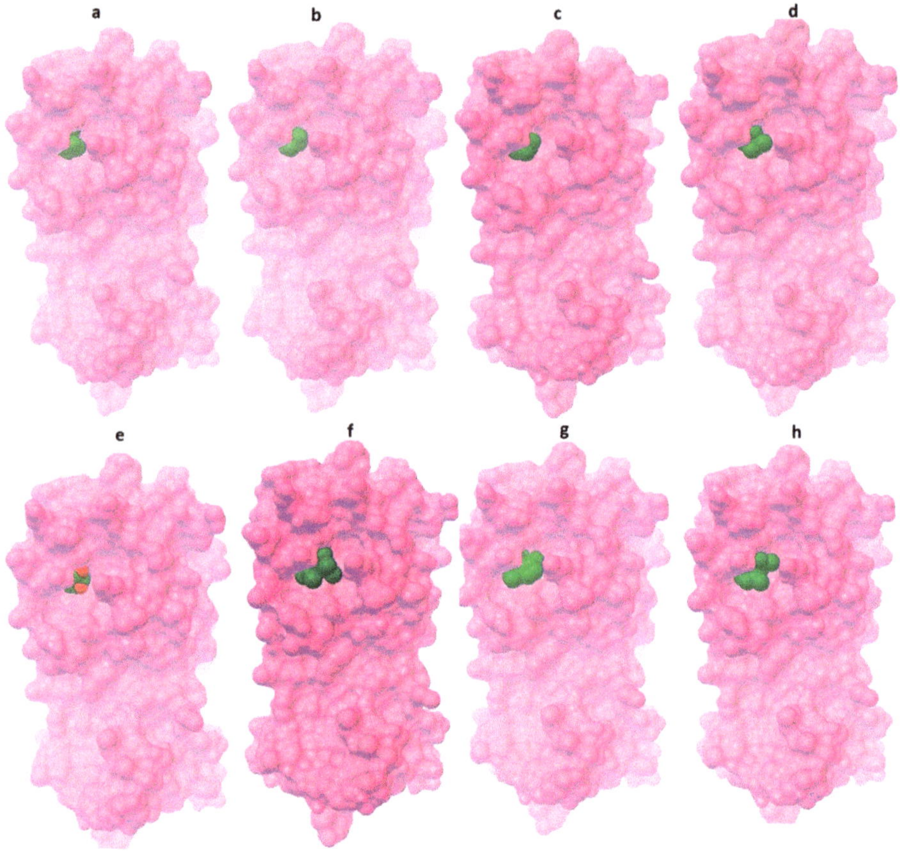

Figure 3. Docking structure of Mpro in complex with eucalyptus compounds: (**a**) 1, 8-cineole (eucalyptus), (**b**) α-pinene, (**c**) β-pinene, (**d**) terpinen-4-ol, (**e**) piperitone, (**f**) allo-aromadendrene, (**g**) aromadendrene, and (**h**) α-gurjunene.

3.2. MD Simulations

Classical MD simulations are one of the computer simulation techniques for atomic resolution dynamic molecular data [45]. The three natural compounds (α-gurjunene, aromadendrene, and allo-aromadendrene) in complex with Mpro protein underwent MD simulations of 100 ns to evaluate the interactions in more detail and the per residue energy contributions of each amino acid. To evaluate the stability of the selected structures, the root mean square deviation (RMSD) of protein and ligand was calculated for 100 ns trajectories. In addition, the root mean square fluctuation (RMSF) values for each residue were analyzed to assess the local flexibility.

Changes were observed, as expected, for both protein conformations and positional ligand binding. Conformational changes as a function of time were monitored by RMSD measurements, in which α-gurjunene induced major conformational changes. Different trends have been identified in the measurement of RMSF, which indicate that domain I, part of domain II, and domain III regions were significantly influenced by protein conformation variations. Aromadendrene and allo-aromadendrene RMSD values were lower than α-gurjunene, suggesting that these ligands may stabilize protein conformations. Further, in the early stages of MD (around 10 ns), the behavior of α-gurjunene shifted and increased suddenly from 0.2 nm to almost 0.8 nm. In about 60 ns, the structure stabilized and had higher RMSD values than aromadendrene and allo-aromadendrene (Figure 5a). The overall average RMSF values were shown to be unstable at residues (8–100), (163–200), and (200–305), respectively (Figure 5b). For α-gurjunene, a lower RMSF value was noticed, followed by aromadendrene and allo-aromadendrene. Lower RMSF values suggest more excellent stability and the natural compound's possibility to inhibit the target molecule [37,46]. Changes in ligand positioning have also been monitored by the ligand RMSD measurement, as shown in Figure 5c. The α-gurjunene molecule (red) was transferred to domain II of the Mpro protein in the earliest stage of MD simulation and lasted for 15.6 ns. The RMSD of the ligand reached its peak at around 4.8 nm, falling immediately to 3.5 nm and stabilized over the remaining time (ns) and bound to domain III for the last 84 ns. The behavior of the α-gurjunene might be due to its unique structure compared to the other two ligands. This structural peculiarity may account for the instability of the ligand at the earlier stage of the simulation. Although in the case of aromadendrene and allo-aromadendrene ligands, they remained intact during the simulation time with the Mpro molecule.

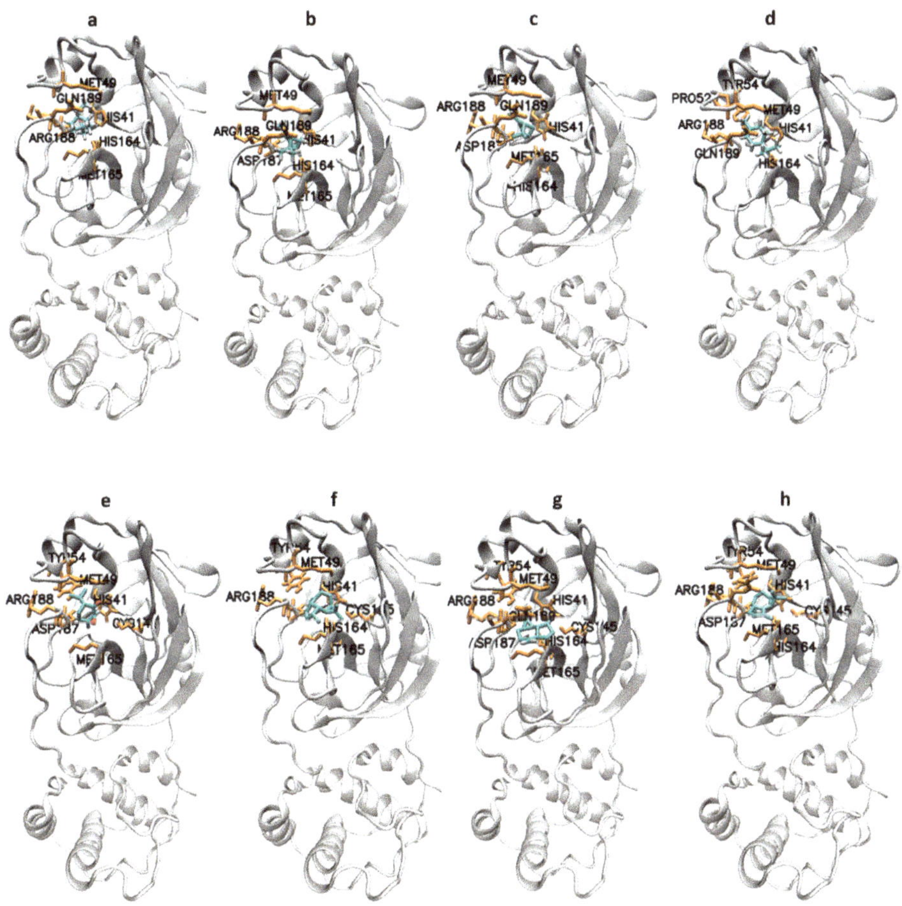

Figure 4. Predicted binding amino acid residues from the docking simulation of Mpro in complex with (**a**) 1, 8-cineole (eucalyptus), (**b**) α-pinene, (**c**) β-pinene, (**d**) terpinen-4-ol, (**e**) piperitone, (**f**) allo-aromadendrene, (**g**) aromadendrene, and (**h**) α-gurjunene.

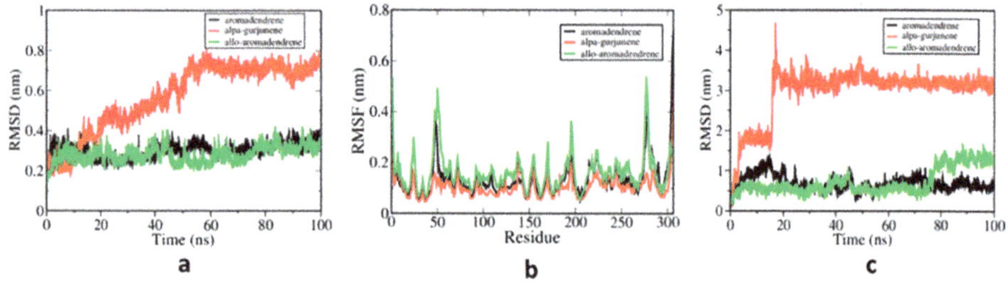

Figure 5. (**a**) Root mean square deviation (RMSD) of backbone carbon (Cα) atomic positions of the Mpro relative to the starting structure measured from the 100 ns trajectories, (**b**) per residue root mean square fluctuations (RMSF) of Cα atomic positions measured over the last 30 ns from protein conformations, and (**c**) RMSD of ligand atomic positions relating to the starting structure measured from the 100 ns trajectories with allo-aromadendrene (green), aromadendrene (black), and α-gurjunene (red).

MM/PBSA calculations were carried out on the last 30 ns of all simulations to examine further inhibitory effects and the interaction networks between Mpro protein and natural molecules (ligands). Table 3 presents a description of the MM/PBSA energy, van der Waals, electrostatic, polar, and apolar solvation energy contributions. Figure 6 shows per residue energy contribution within the binding site given by MM/PBSA energy decomposition. In all, the contributions of van der Waals were stronger, suggesting substantially higher hydrophobic energy contributions with energies of −23.66 ± 2.04 kcal/mol (−99.00 ± 8.53 kJ/mol), −21.23 ± 2.89 kcal/mol (−88.82 ± 12.08 kJ/mol), and −21.66 ± 2.33 kcal/mol (−90.61 ± 9.73 kJ/mol) respectively for α-gurjunene, aromadendrene, and allo-aromadendrene. The strongest total binding energy of −20.37 ± 2.26 kcal/mol (−85.21 ± 9.44 kJ/mol) was found in α-gurjunene, which was higher in van der Waals energy than in aromadendrene and allo-aromadendrene with binding energies of −18.99 ± 3.02 kcal/mol (−79.45 ± 12.62 kJ/mol) and −17.91 ± 2.12 kcal/mol (−74.95 ± 8.88 kJ/mol) respectively. The results for total binding energy were lower than the experimental binding energies of protein in complex with small organic molecules, as reported by [47]. The authors determined the experimental binding energy values ranging from −20.80 kcal/mol (−87.03 kJ/mol) to −37.80 kcal/mol (−158.16 kJ/mol).

Table 3. Binding energy obtained using MM/PBSA technique of each complex along with contributions from van der Waals, electrostatic, polar and apolar solvation energies.

Complex Structures	Van der Waals Energy (±SD) (kJ/mol)	Electrostatic Energy (±SD) (kJ/mol)	Polar Solvation Energy (±SD) (kJ/mol)	Apolar Energy (±SD) (kJ/mol)	Total Binding Energy (±SD) (kJ/mol)
Mpro-allo-aromadendrene	−90.61 (±9.73)	−2.06 (±6.57)	27.31 (±5.34)	−9.59 (±1.23)	−74.95 (±8.88)
Mpro-aromadendrene	−88.82 (±12.08)	−3.42 (±4.77)	21.89 (±4.92)	−9.11 (±1.25)	−79.45 (±12.62)
Mpro-α-gurjunene	−99.00 (±8.53)	−0.33 (±1.91)	25.55 (±6.93)	−11.42 (±1.06)	−85.21 (±9.44)

We also validated the binding energies of the compounds tested in our study with some selected co-crystallized ligands, as presented in Table 4. Compared to the estimated binding energy of co-crystallized ligands, darunavir has a better estimated binding energy value. Based on this finding, ΔG of darunavir was used as a standard in calculating the change in binding energy (ΔΔG) of other ligands. The result showed that the selected eucalyptus compounds could be favorable inhibitors of the Mpro enzyme. This suggests that these small molecules could serve as potential drug candidates for SARS-CoV-2 treatment.

Figure 6. Per residue energy contributions (in kJ/mol) of the inhibitors. Red bars represent amino acids with unfavorable binding energies while light green bars represent amino acids with favorable binding energies; (right) final molecular dynamics (MD) snapshots of the major residues that contributed to the binding of Mpro in complex with (**a**) allo-aromadendrene, (**b**) aromadendrene, and (**c**) α-gurjunene.

Table 4. Estimated binding energies of co-crystalized ligands against selected eucalyptus compounds.

Ligands	Estimated Binding Energy (kJ/mol)	Estimated ΔΔG (kJ/mol)	Reference
N3	−51.07	−44.46	[48]
Indinavir	−72.11	−23.42	[48]
Darunavir	−95.53	-	[48]
Favipiravir	−36.07	−59.46	[49]
Fosfomycin	−60.63	−34.90	[49]
Aspirin	−78.37	−17.16	[49]
Allo-aromadendrene	−74.95	−20.58	This study
Aromadendrene	−79.45	−16.08	This study
α-gurjunene	−85.21	−10.32	This study

In addition, MM/PBSA energy decomposition was used to obtain the individual contributions of amino acids to the binding energy, revealing significant active site interaction residues. Mpro-allo-aromadendrene bound Thr25, Leu27, His41, Met49, Gly143, Ser144, Cys145, Met165, and Pro168 (Figure 5a). Residues such as Glu166 showed higher positive values, implying unfavorable interactions due to the steric obstruction effect caused by repulsive forces. In aromadendrene, the primary amino acid contributors were Thr25, Leu27, His41, Cys44, Met49, Leu50, Cys145, Met165, Glu166, and Gln189 respectively (Figure 5b). The energy per residue was comparable to previous analyses of MD simulations with several SARS-CoV-2 Mpro inhibitors and according to our findings [50–52].

Mpro-α-gurjunene found Lys237, Val204, Phe230, Val233, Tyr237, Tyr239, Leu268, and Leu272 as the most desirable residues (Figure 5c). Further, the residues dominated by the amino acids with hydrophobic and polar uncharged side chains were actively involved. Thus, the strongest values of van der Waals interaction and the nonpolar component of the solvation energy component correspond to the favorable strength of α-gurjunene. Nevertheless, Lys236, Lys269, Glu288, Asp289, and Glu290 displayed unfavorable (positive) interactions, which may be attributable to steric impact and binding opposition. The amino acid residues found in MD simulations were not identical to docking. The principal explanation for this may be the shift of the α-gurjunene in the simulations. This investigation is the first to elucidate the detailed atomistic interactions of eucalyptus phytochemical compounds with the SARS-CoV-2 main protease. For the first time, some new participating amino acids have been reported in another binding site surplus to the active site residues found in Mpro. Eucalyptus has been a traditional medicinal plant for decades, and as a result of this study it could be used as a potential therapeutic drug candidate to suppress the replicative function of the main protease.

4. Conclusions

Due to severe outbreaks and lack of effective drugs, the new coronavirus has become a global concern. Therefore, recovery strategies need to be identified and tested more efficiently. In this regard, in-silico processes are very efficient and helpful. Throughout this research, various computational techniques such as molecular docking, MD simulations, and MM-PBSA calculations have been used to classify novel natural compounds as potential inhibitors for Mpro, the SARS-CoV-2 protein. Eight eucalyptus phytochemical compounds were used here for screening purposes. Based on molecular docking assessments, all eight compounds bound to the binding pocket with strong binding affinities. For further analysis, three molecules (α-gurjunene, aromadendrene, and allo-aromadendrene) with the lowest inhibition constant values were chosen. Eventually, from MD simulation results, we found that all molecules could bind to the target protein with the strongest binding affinities. Such findings have indicated that these compounds could be considered as novel natural molecules for the possible development of appropriate SARS-CoV-2 drug candidates. These results are in line with the recent research that shows that eucalyptus is successful in treating the new coronavirus.

Author Contributions: Conceptualization, K.M.; methodology, A.M.; software, A.M.; validation, I.A.M. and Y.S.A.; formal analysis, A.B.M.; investigation, Y.S.A. and I.Y.M.; resources, K.M.; writing—original draft preparation, I.A.M.; writing—review and editing, A.M. and Y.S.A.; visualization, I.Y.M. and I.D.U.; supervision, K.M.; project administration, K.M.; funding acquisition, K.M. All authors have read and agreed to the published version of the manuscript.

Funding: This research received no external funding.

Acknowledgments: The second author was financially supported by Rajamangala University of Technology Phra Nakhon (RMUTP) research scholarship.

Conflicts of Interest: The authors declare no conflict of interest.

References

1. Macchiagodena, M.; Pagliai, M.; Procacci, P. Identification of potential binders of the main protease 3CLpro of the COVID-19 via structure-based ligand design and molecular modeling. *Chem. Phys. Lett.* **2020**, *750*, 137489. [CrossRef] [PubMed]
2. Wang, L.; Wang, Y.; Ye, D.; Liu, Q. Review of the 2019 novel coronavirus (SARS-CoV-2) based on current evidence. *Int. J. Antimicrob. Agents* **2020**, *55*, 105948. [CrossRef] [PubMed]
3. Jin, Y.-H.; Cai, L.; Cheng, Z.-S.; Cheng, H.; Deng, T.; Fan, Y.-P.; Fang, C.; Huang, D.; Huang, L.-Q.; Huang, Q.; et al. A rapid advice guideline for the diagnosis and treatment of 2019 novel coronavirus (2019-nCoV) infected pneumonia (standard version). *Mil. Med. Res.* **2020**, *7*, 4. [CrossRef] [PubMed]
4. WHO. WHO Coronavirus Disease (COVID-19) Dashboard. Available online: https://covid19.who.int/ (accessed on 20 July 2020).
5. WHO. Global Nigeria WHO (COVID-19) Homepage. Available online: https://covid19.who.int/region/afro/country/ng (accessed on 20 July 2020).
6. He, J.; Hu, L.; Huang, X.; Wang, C.; Zhang, Z.; Wang, Y.; Zhang, D.; Ye, W. Potential of coronavirus 3C-like protease inhibitors for the development of new anti-SARS-CoV-2 drugs: Insights from structures of protease and inhibitors. *Int. J. Antimicrob. Agents* **2020**, *56*, 106055. [CrossRef]
7. Delang, L.; Neyts, J. Medical treatment options for COVID-19. *Eur. Heart J. Acute Cardiovasc. Care* **2020**, *9*, 209–214. [CrossRef]
8. Liu, J.; Cao, R.; Xu, M.; Wang, X.; Zhang, H.; Hu, H.; Li, Y.; Hu, Z.; Zhong, W.; Wang, M. Hydroxychloroquine, a less toxic derivative of chloroquine, is effective in inhibiting SARS-CoV-2 infection in vitro. *Cell Discov.* **2020**, *6*, 16. [CrossRef]
9. Wang, M.; Cao, R.; Zhang, L.; Yang, X.; Liu, J.; Xu, M.; Shi, Z.; Hu, Z.; Zhong, W.; Xiao, G. Remdesivir and chloroquine effectively inhibit the recently emerged novel coronavirus (2019-nCoV) in vitro. *Cell Res.* **2020**, *30*, 269–271. [CrossRef]
10. Yao, X.; Ye, F.; Zhang, M.; Cui, C.; Huang, B.; Niu, P.; Liu, X.; Zhao, L.; Dong, E.; Song, C.; et al. In Vitro Antiviral Activity and Projection of Optimized Dosing Design of Hydroxychloroquine for the Treatment of Severe Acute Respiratory Syndrome Coronavirus 2 (SARS-CoV-2). *Clin. Infect. Dis.* **2020**, *71*, 732–739. [CrossRef]
11. Eleftheriou, P.; Amanatidou, D.; Petrou, A.; Geronikaki, A. In Silico Evaluation of the Effectivity of Approved Protease Inhibitors against the Main Protease of the Novel SARS-CoV-2 Virus. *Molecules* **2020**, *25*, 2529. [CrossRef]
12. Hall, D.C., Jr.; Ji, H.-F. A search for medications to treat COVID-19 via in silico molecular docking models of the SARS-CoV-2 spike glycoprotein and 3CL protease. *Travel. Med. Infect. Dis.* **2020**, *35*, 101646. [CrossRef]
13. Kumar, Y.; Singh, H.; Patel, C.N. In silico prediction of potential inhibitors for the Main protease of SARS-CoV-2 using molecular docking and dynamics simulation based drug-repurposing. *J. Infect. Public Health* **2020**, *13*, 1210–1223. [CrossRef] [PubMed]
14. Prasanth, D.S.N.B.K.; Murahari, M.; Chandramohan, V.; Panda, S.P.; Atmakuri, L.R.; Guntupalli, C. In silico identification of potential inhibitors from Cinnamon against main protease and spike glycoprotein of SARS CoV-2. *J. Biomol. Struct. Dyn.* **2020**, 1–15. [CrossRef]
15. Yu, R.; Chen, L.; Lan, R.; Shen, R.; Li, P. Computational screening of antagonists against the SARS-CoV-2 (COVID-19) coronavirus by molecular docking. *Int. J. Antimicrob. Agents* **2020**, *56*, 106012. [CrossRef] [PubMed]
16. Zhang, L.; Lin, D.; Sun, X.; Curth, U.; Drosten, C.; Sauerhering, L.; Becker, S.; Rox, K.; Hilgenfeld, R. Crystal structure of SARS-CoV-2 main protease provides a basis for design of improved α-ketoamide inhibitors. *Science* **2020**, *368*, 409–412. [CrossRef] [PubMed]
17. Amanat, F.; Krammer, F. SARS-CoV-2 Vaccines: Status Report. *Immunity* **2020**, *52*, 583–589. [CrossRef] [PubMed]
18. Denaro, M.; Smeriglio, A.; Barreca, D.; De Francesco, C.; Occhiuto, C.; Milano, G.; Trombetta, D. Antiviral activity of plants and their isolated bioactive compounds: An update. *Phytother. Res.* **2020**, *34*, 742–768. [CrossRef]
19. Covid-19: Race to eucalyptus in Maputo–Watch. *Club of Mozambique*, 13 April 2020; p. 1.

20. Vecchio, M.G.; Loganes, C.; Minto, C. Beneficial and Healthy Properties of Eucalyptus Plants: A Great Potential Use. *Open Agric. J.* **2016**, *10*, 52–57. [CrossRef]
21. Salehi, B.; Sharifi-Rad, J.; Quispe, C.; Llaique, H.; Villalobos, M.; Smeriglio, A.; Trombetta, D.; Ezzat, S.M.; Salem, M.A.; Zayed, A.; et al. Insights into Eucalyptus genus chemical constituents, biological activities and health-promoting effects. *Trends Food Sci. Technol.* **2019**, *91*, 609–624. [CrossRef]
22. The Jakarta Post. Available online: https://www.thejakartapost.com/news/2020/05/09/agriculture-ministry-claims-to-have-developed-eucalyptus-based-covid-19-treatment.html (accessed on 21 July 2020).
23. Tempo Misleading: COVID-19 Can Be Cured with Eucalyptus Oil. Available online: https://www.poynter.org/?ifcn_misinformation=covid-19-can-be-cured-with-eucalyptus-oil (accessed on 17 July 2020).
24. Adeniyi, B.A.; Ayepola, O.O.; Adu, F.D. The antiviral activity of leaves of Eucalyptus camaldulensis (Dehn) and Eucalyptus torelliana (R. Muell). *Pak. J. Pharm. Sci.* **2015**, *28*, 1773–1776.
25. Usachev, E.V.; Pyankov, O.V.; Usacheva, O.V.; Agranovski, I.E. Antiviral activity of tea tree and eucalyptus oil aerosol and vapour. *J. Aerosol Sci.* **2013**, *59*, 22–30. [CrossRef]
26. Kim, S.; Chen, J.; Cheng, T.; Gindulyte, A.; He, J.; He, S.; Li, Q.; Shoemaker, B.A.; Thiessen, P.A.; Yu, B.; et al. PubChem 2019 update: Improved access to chemical data. *Nucleic Acids Res.* **2019**, *47*, D1102–D1109. [CrossRef] [PubMed]
27. Hanwell, M.D.; Curtis, D.E.; Lonie, D.C.; Vandermeersch, T.; Zurek, E.; Hutchison, G.R. Avogadro: An advanced semantic chemical editor, visualization, and analysis platform. *J. Cheminform.* **2012**, *4*, 17. [CrossRef] [PubMed]
28. Jin, Z.; Du, X.; Xu, Y.; Deng, Y.; Liu, M.; Zhao, Y.; Zhang, B.; Li, X.; Zhang, L.; Peng, C.; et al. Structure of Mpro from SARS-CoV-2 and discovery of its inhibitors. *Nature* **2020**, *582*, 289–293. [CrossRef]
29. Tahir ul Qamar, M.; Alqahtani, S.M.; Alamri, M.A.; Chen, L.-L. Structural basis of SARS-CoV-2 3CLpro and anti-COVID-19 drug discovery from medicinal plants. *J. Pharm. Anal.* **2020**, *10*, 313–319. [CrossRef] [PubMed]
30. Morris, G.M.; Huey, R.; Lindstrom, W.; Sanner, M.F.; Belew, R.K.; Goodsell, D.S.; Olson, A.J. AutoDock4 and AutoDockTools4: Automated docking with selective receptor flexibility. *J. Comput. Chem.* **2009**, *30*, 2785–2791. [CrossRef]
31. Morris, G.M.; Goodsell, D.S.; Halliday, R.S.; Huey, R.; Hart, W.E.; Belew, R.K.; Olson, A.J. Automated docking using a Lamarckian genetic algorithm and an empirical binding free energy function. *J. Comput. Chem.* **1998**, *19*, 1639–1662. [CrossRef]
32. Du, X.; Li, Y.; Xia, Y.-L.; Ai, S.-M.; Liang, J.; Sang, P.; Ji, X.-L.; Liu, S.-Q. Insights into Protein–Ligand Interactions: Mechanisms, Models, and Methods. *Int. J. Mol. Sci.* **2016**, *17*, 144. [CrossRef]
33. Laskowski, R.A.; Swindells, M.B. LigPlot+: Multiple Ligand–Protein Interaction Diagrams for Drug Discovery. *J. Chem. Inf. Model.* **2011**, *51*, 2778–2786. [CrossRef]
34. Hess, B.; Bekker, H.; Berendsen, H.J.C.; Fraaije, J.G.E.M. LINCS: A linear constraint solver for molecular simulations. *J. Comput. Chem.* **1997**, *18*, 1463–1472. [CrossRef]
35. Miyamoto, S.; Kollman, P.A. Settle: An analytical version of the SHAKE and RATTLE algorithm for rigid water models. *J. Comput. Chem.* **1992**, *13*, 952–962. [CrossRef]
36. Schmid, N.; Eichenberger, A.P.; Choutko, A.; Riniker, S.; Winger, M.; Mark, A.E.; van Gunsteren, W.F. Definition and testing of the GROMOS force-field versions 54A7 and 54B7. *Eur. Biophys. J.* **2011**, *40*, 843–856. [CrossRef] [PubMed]
37. Muhammad, A.; Khunrae, P.; Sutthibutpong, T. Effects of oligolignol sizes and binding modes on a GH11 xylanase inhibition revealed by molecular modeling techniques. *J. Mol. Model.* **2020**, *26*, 124. [CrossRef] [PubMed]
38. Kumari, R.; Kumar, R.; Lynn, A. g_mmpbsa—A GROMACS Tool for High-Throughput MM-PBSA Calculations. *J. Chem. Inf. Model.* **2014**, *54*, 1951–1962. [CrossRef] [PubMed]
39. Kumar, A.; Choudhir, G.; Shukla, S.K.; Sharma, M.; Tyagi, P.; Bhushan, A.; Rathore, M. Identification of phytochemical inhibitors against main protease of COVID-19 using molecular modeling approaches. *J. Biomol. Struct. Dyn.* **2020**, 1–11. [CrossRef]
40. Meng, X.-Y.; Zhang, H.-X.; Mezei, M.; Cui, M. Molecular docking: A powerful approach for structure-based drug discovery. *Curr. Comput. Aided Drug Des.* **2011**, *7*, 146–157. [CrossRef]
41. Sethi, A.; Joshi, K.; Sasikala, K.; Alvala, M. Molecular docking in modern drug discovery: Principles and recent applications. In *Drug Discovery and Development-New Advances*; IntechOpen: Rijeka, Croatia, 2019.

42. Das, P.; Majumder, R.; Mandal, M.; Basak, P. In-Silico approach for identification of effective and stable inhibitors for COVID-19 main protease (Mpro) from flavonoid based phytochemical constituents of Calendula officinalis. *J. Biomol. Struct. Dyn.* **2020**, 1–16. [CrossRef]
43. Mpiana, P.T.; Ngbolua, K.-T.-N.; Tshibangu, D.S.T.; Kilembe, J.T.; Gbolo, B.Z.; Mwanangombo, D.T.; Inkoto, C.L.; Lengbiye, E.M.; Mbadiko, C.M.; Matondo, A.; et al. Identification of potential inhibitors of SARS-CoV-2 main protease from Aloe vera compounds: A molecular docking study. *Chem. Phys. Lett.* **2020**, *754*, 137751. [CrossRef]
44. Gurung, A.B.; Ali, M.A.; Lee, J.; Farah, M.A.; Al-Anazi, K.M. Unravelling lead antiviral phytochemicals for the inhibition of SARS-CoV-2 Mpro enzyme through in silico approach. *Life Sci.* **2020**, *255*, 117831. [CrossRef]
45. Gajula, M.N.V.P.P.; Kumar, A.; Ijaq, J. Protocol for Molecular Dynamics Simulations of Proteins. *Bio-Protocol* **2016**, *6*, e2051. [CrossRef]
46. Ning, X.; Zhang, Y.; Yuan, T.; Li, Q.; Tian, J.; Guan, W.; Liu, B.; Zhang, W.; Xu, X.; Zhang, Y. Enhanced Thermostability of Glucose Oxidase through Computer-Aided Molecular Design. *Int. J. Mol. Sci.* **2018**, *19*, 425. [CrossRef]
47. English, A.C.; Groom, C.R.; Hubbard, R.E. Experimental and computational mapping of the binding surface of a crystalline protein. *Protein Eng. Des. Sel.* **2001**, *14*, 47–59. [CrossRef] [PubMed]
48. Sang, P.; Tian, S.-H.; Meng, Z.-H.; Yang, L.-Q. Anti-HIV drug repurposing against SARS-CoV-2. *RSC Adv.* **2020**, *10*, 15775–15783. [CrossRef]
49. Al-Khafaji, K.; Al-Duhaidahawi, D.; Taskin Tok, T. Using integrated computational approaches to identify safe and rapid treatment for SARS-CoV-2. *J. Biomol. Struct. Dyn.* **2020**, 1–9. [CrossRef] [PubMed]
50. Bhardwaj, V.K.; Singh, R.; Sharma, J.; Rajendran, V.; Purohit, R.; Kumar, S. Identification of bioactive molecules from tea plant as SARS-CoV-2 main protease inhibitors. *J. Biomol. Struct. Dyn.* **2020**, 1–10. [CrossRef] [PubMed]
51. Dash, J.J.; Purohit, P.; Muya, J.T.; Meher, B.R. Drug Repurposing of Allophenylnorstatine Containing HIV-Protease Inhibitors Against SARS-CoV-2 Mpro: Insights from Molecular Dynamics Simulations and Binding Free Energy Estimations. *ChemRxiv* **2020**. [CrossRef]
52. Joshi, T.; Sharma, P.; Joshi, T.; Pundir, H.; Mathpal, S.; Chandra, S. Structure-based screening of novel lichen compounds against SARS Coronavirus main protease (Mpro) as potentials inhibitors of COVID-19. *Mol. Divers.* **2020**, 1–13. [CrossRef]

computation

Article

Exploring the SARS-CoV-2 Proteome in the Search of Potential Inhibitors via Structure-Based Pharmacophore Modeling/Docking Approach

Giulia Culletta [1,2], Maria Rita Gulotta [1,3], Ugo Perricone [3], Maria Zappalà [2], Anna Maria Almerico [1] and Marco Tutone [1,*]

1 Dipartimento di Scienze e Tecnologie Biologiche Chimiche e Farmaceutiche (STEBICEF), Università degli Studi di Palermo, Via Archirafi 32, 90123 Palermo, Italy; gculletta@unime.it (G.C.); mrgulotta@fondazionerimed.com (M.R.G.); annamaria.almerico@unipa.it (A.M.A.)
2 Dipartimento di Scienze Chimiche, Biologiche, Farmaceutiche ed Ambientali, Università di Messina, Viale Annunziata, 98168 Messina, Italy; maria.zappala@unime.it
3 Drug Discovery Unit, Fondazione Ri.MED, 90133 Palermo, Italy; uperricone@fondazionerimed.com
* Correspondence: marco.tutone@unipa.it

Received: 27 July 2020; Accepted: 29 August 2020; Published: 8 September 2020

Abstract: To date, SARS-CoV-2 infectious disease, named COVID-19 by the World Health Organization (WHO) in February 2020, has caused millions of infections and hundreds of thousands of deaths. Despite the scientific community efforts, there are currently no approved therapies for treating this coronavirus infection. The process of new drug development is expensive and time-consuming, so that drug repurposing may be the ideal solution to fight the pandemic. In this paper, we selected the proteins encoded by SARS-CoV-2 and using homology modeling we identified the high-quality model of proteins. A structure-based pharmacophore modeling study was performed to identify the pharmacophore features for each target. The pharmacophore models were then used to perform a virtual screening against the DrugBank library (investigational, approved and experimental drugs). Potential inhibitors were identified for each target using XP docking and induced fit docking. MM-GBSA was also performed to better prioritize potential inhibitors. This study will provide new important comprehension of the crucial binding hot spots usable for further studies on COVID-19. Our results can be used to guide supervised virtual screening of large commercially available libraries.

Keywords: COVID-19; SARS-CoV-2; computational chemistry; structure-based; pharmacophore; docking; MM-GBSA

1. Introduction

Coronaviruses (CoVs) are one of the major pathogens that primarily targets the human respiratory system which caused previous outbreaks such as the severe acute respiratory syndrome (SARS)-CoV and the Middle East respiratory syndrome (MERS)-CoV. The novel coronavirus SARS-CoV-2 has become a pandemic threat (COVID-19) to public health. It is a respiratory diseasecausing fever, fatigue, dry cough, muscle aches, shortness of breath and some instances lead to pneumonia [1]. The SARS-CoV-2 genome comprises 29,903 nucleotides, with 10 Open Reading Frames (ORFs). The 3′ terminal regions encode structural viral proteins: whereas the 5′ terminal ORF1ab encodes two viral replicasepolyproteins pp1a and pp1b. The proteolytic cleavage of pp1a and pp1b produces 16 nonstructural proteins (nsp1 to nsp16). Among these, there are nsp3, the papain-like protease (PLpro) and nsp5, the 3-chymotrypsin-like protease (3CLpro, also known as the main proteaseMpro). The viral polyprotein processing is essential for maturation and infectivity of the virus (Figure 1) [2]. Because of the crucial roles, these two proteases are important targets for antiviral drug design. Moreover, the virus encoded for other proteins that could be potential targets of antiviral drugs. The mature proteins of SARS-CoV-2 are: host translation inhibitor nsp1 (nsp1); nonstructural protein 2 (nsp2); papain-like proteinase (PLpro); nonstructural protein 4 (nsp4); 3C-like proteinase (3CLpro), nonstructural protein 6 (nsp6), nonstructural protein 7 (nsp7), nonstructural protein 8 (nsp8), nonstructural protein 9 (nsp9), nonstructural protein 10 (nsp10), RNA-directed RNA polymerase (Pol/RdRp), helicase (Hel), guanine-N7 methyltransferase (ExoN/nsp14), uridylate-specific endoribonuclease (NendoU/nsp15), 2′-O-ribose methyltransferase (nsp16), Spike glycoprotein (S glycoprotein), protein 3a, Envelope small membrane protein (E protein), Membrane protein (M protein), nonstructural protein 6 (nsp6), protein 7a, nonstructural protein 7b (nsp7b), nonstructural protein 8 (nsp8), nucleoprotein (NC), ORF10 protein. These proteins can form hetero-oligomeric complexes such as: nsp7/nsp8 hetero-oligomeric complex; nsp7/nsp8/Pol hetero-oligomeric complex; nsp10/nsp14 hetero-oligomeric complex; nsp10/nsp16 hetero-oligomeric complex; Spike glycoprotein/hACE2 hetero-oligomeric complex.Anti-coronavirus therapies can be split into two main approaches: the first approach is to act on the human immune system or human cells level, and the other approach is to focus on coronavirus itself [3]. In exploring novel therapies for COVID-19, researchers are using computational approaches to aid in the discovery of potential candidates [4]. In particular, in silico drug repurposing, also named drug repositioning, is a strategy used to identify novel uses for existing approved and investigational drugs. This strategy offers numerous advantages over traditional drug development pipelinesthat suffer risks failure in preclinical or early stage clinical trials due to safety and/or toxicological issues. On the contrary, the drug repurposing strategy reduces this risk by using drugs that have demonstrated safety records from previous trials. The real advantage of drug repurposing is that preclinical and early stage clinical trials do not need to be repeated. This determines cost reductions compared to traditional drug development [5–18]. The number of in silico studies on drug repositioning against SARS-CoV2 is growing rapidly in these last months. A major part of these studies is focused on the repurposing of approved and investigational drugs against the 3CLpro or Mpro by using both ligand-based approaches and structure-based approaches. Structure-based approaches are related to different docking analysis [19–29]. In another work, Battisti and coworkers used two different approaches related to docking and pharmacophore combined with molecular dynamics to perform virtual screening of a large database of compounds on 10 different SARS-CoV-2 proteins [30]. To our knowledge, Touret and coworkers performed, to date, the only in vitro screening of an FDA approved chemical library which revealed potential inhibitors of SARS-CoV-2 replication [21]. Nevertheless, the identification of potential inhibitors is still challenging for all the researchers involved in the field. In this study, a computational analysis of the proteins encoded by the SARS-CoV-2 genes was performed. Such an analysis was used as a starting point for a druggability assessment and a computational drug repurposing work-frame. First, high-quality protein structures were built employing homology modeling or exploiting existing experimental structures. Starting from the models, a computational assessment was done to find out a druggable binding pocket for those proteins of which catalytic site is not known in the literature. The best druggable sites found in the previous analysis, together with the catalytic sites reported in the literature, were then used to build structure-based pharmacophore models. In the end, these models were used to screen the DrugBank library (approved and investigational drugs) [31] as a first screening approach.

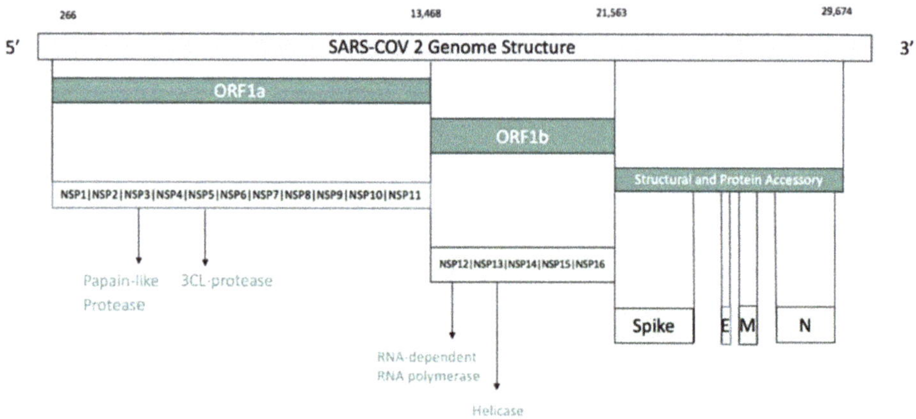

Figure 1. SARS-COV2 genome structure.

2. Materials and Methods

2.1. Library Preparation

A total of 8752 experimental, investigational and approved molecules were downloaded from the DrugBank database (www.drugbank.ca). First, the database molecules were prepared using Schrödinger LigPrep v. 2018-4. The force field adopted was OPLS3e and Epik [32] was selected as an ionization tool at pH 7.0 ± 2.0. Tautomers generation was flagged and the maximum number of conformers generated was set at 32. The database obtained was prepared as a Pharmacophore Screening database, in *.lbd format, through Idbgen (extension present in the LigandScout 4.3 [17] package), which allowed obtaining the best conformation of the ligand (at low energy) between the 200 the application can calculate. The tautomers were considered as separate molecules and those molecules that were duplicated or whose conformation calculation had failed were eliminated.

2.2. Homology Modeling and Protein Preparation

The full SARS-CoV-2 proteome based on the NCBI reference sequence NC_045512, which is identical to GenBank entry MN908947 and annotations from UniProt, was modeled in the SWISS-MODEL [33] workspace (swissmodel.expasy.org/workspace). Only for 10 proteins, it was possible to obtain high-quality models and experimental structures that were considered for further analysis. Investigated proteins are 3C-like protease (3CLpro), papain-like protease (PLpro), guanine-N7 methyltransferase (nsp14), uridylate-specific endoribonuclease (NendoU/nsp15), nsp4, nsp7/nsp8 supercomplex, nsp9, nsp7/nsp8/nsp12 hetero-oligomeric complex, helicase (Hel), 2′-O-ribose methyltransferase (nsp16).For each structure, templates with the highest identity available at the time of this study (25 March 2020) were selected and respective models were generated.

For 3C-like protease (3CLpro) the crystal structure of the COVID-19 main protease (PDB ID: 6LU7) was available. The structure of papain-like protease (PLpro) of SARS virus (PDB ID: 3E9S) was used as a template of the human coronavirus papain-like model (82.86% sequence identity). This one was the best available experimental structure at the time of the study (25 March 2020). On 27 May 2020, the crystal structure of PLpro of SARS-CoV-2 was released (PDB ID: 6WZU). We performed the overlapping of our model and the experimental structure. The RMSD value of 3.99 Åshows that the two structures are identical unless few residues in the C-terminal (See Supplementary Information). For guanine-N7 methyltransferase (nsp14) we used as template the SARS-related coronavirus (PDB ID: 5C8S) that shows 95.07% of sequence identity. For uridylate-specific endoribonuclease (NendoU/nsp15), we used the experimental structure as reported in the Protein Data Bank [34] (PDB ID: 6W01). The crystal structure of nsp4 from mouse hepatitis virus A59 (PDB ID: 3VCB) was used as a template of SARS-CoV-2 nsp4 (61.36% sequence identity). The crystal structure of SARS-CoV super complex of nonstructural proteins (PDB ID: 2AHM) was chosen as a template of nsp7/nsp8 supercomplex (97.86% sequence identity). For nsp9 the template of nsp9 from SARS-coronavirus (PDB ID: 1UW7) was used. It shares a sequence identity of 97.35%. The X-ray structure of SARS coronavirus nsp7/8/12 (PDB ID: 6NUR) was selected as a template of nsp7/nsp8/nsp12 hetero-oligomeric complex (96.70% sequence identity). The crystal structure of SARS-coronavirus helicase (PDB ID: 6JYT) was used as template for SARS-CoV-2 helicase (Hel). It shows a high sequence identity (99.83%). On 29 July 2020, the experimental structure of SARS-CoV-2 helicase (PDB ID: 6ZSL) was released. The overlapping of our model and the experimental structure shows a RMSD value of 4.17 Å. This means a quite identical structure unless some loops (See Supplementary Information). The crystal structure of nsp16/nsp10 SARS coronavirus complex (PDB ID: 2XYQ) was chosen as a template of the model of 2′-O-ribose methyltransferase (nsp16) with 93.45% sequence identity. The models obtained and the PDBs were refined using the protein preparation wizard tool of Maestro Suite Software [35]. This tool allowed the protein structure optimization, including missing loops, side chains and hydrogens, optimization of the protonation state in a pH range 7.0 ± 2.0 and analysis of atomic clashes. For PDBs containing co-crystallized ligands, Epik [32] was used to predict ionization and tautomeric state of ligand, while PROPKA was used to check for the protonation state of ionizable protein groups. Protein was refined using restrained minimization with OPLS3e as force field.

2.3. Pharmacophore Modeling

Pharmacophore model generation was performed using LigandScout 4.3. The structures were imported into LigandScout. 3C-like proteinase, PLpro, nsp14, nsp15, nsp16–nsp10 are protein–ligand complexes, while, nsp4, nsp9, nsp10–nsp14, helicase, nsp7–nsp8 supercomplex, nsp12are targets without ligand-bound. For protein–ligand complexes, a structure-based pharmacophore model was generated [31]. When the model showed more features, to improve the performance of virtual screening, we considered the features for the binding, in other cases the features were omitted until hits were found. The calculate pockets tool has been used to find the binding pockets for the structures without ligand-bound. A grid was calculated over the entire protein structure and grid points were evaluated according to their buriedness and their number of neighboring grid points. Isocontour surfaces were generated. Then, a model was created by selecting the nature and number of six features according to the features showed in the protein–ligand complexes utilizing "Create Apo Site Grids". Next, the pharmacophore model was generated for each one. The obtained pharmacophore models were used as a query to screen the DrugBank library. For apo protein, such an approach allow to evaluate if a putative binding site is suitable for ligand binding.

Pharmacophore screening was preferred to be used prior to docking for two reasons. First, it exploits a rapid screening techniques that is crucial in the first stage of virtual screening cascade. Indeed, this is very common to use it as a first step in virtual screening campaign on large databases [36]. Second, the structure-based pharmacophore uses a static conformation of protein side chains, while the docking funnel here used was set to have a gradually increasing precision with a final step of IFD that allow user to simulate side-chains-induced fit based on the ligand.

2.4. Docking

The hits identified by the virtual screening were submitted to a docking study using Glide [37] in standard precision (SP) with the OPLS3e [38] force field. The crystal structures were optimized using protein preparation wizard in Maestro [35] adding bond orders and hydrogen atoms to the crystal structure using the OPLS3e force field. Next prime [39] was used to fix missing residues or atoms in the protein and to remove co-crystallized water molecules. The protonation state, pH 7.2 ± 0.2 of the protein and the ligand were evaluated using Epik 3.1[32]. The hydrogen bonds were optimized through by reorientation of hydroxyl bonds, thiol groups and amide groups. In the end, the systems were minimized with the value of convergence of the RMSD of 0.3 Å [40,41]. For protein–ligand complexes, the grid boxes were built considering the ligands as a centroid. In contrast, for apoproteins, the amino acid residues, previously identified by LigandScout as crucial, were considered for centering the docking grid. The docking study was performed using the Glide docking tool, in extra precision (XP) using no constraints. Van der Waals radii were set at 0.8 and the partial cutoff was 0.15 and flexible ligand sampling. Bias sampling torsion penalization for amides with nonplanar conformation and Epik state penalties were added to the docking score.

2.5. Induced-Fit Docking and MM-GBSA

The induced-fit protocol (IFD)—developed by Schrödinger [24]—is a method for modeling the conformational changes induced by ligand binding. This protocol models induced-fit docking of one ormore ligands using the following steps as also reported in [42]. The protocol starts with an Initial docking of each ligand using a softened potential (van der Waals radii scaling). Then, a side-chain prediction within a given distance of any ligand pose (5 Å) is performed. Subsequently, a minimization of the same set of residues and the ligandfor each protein/ligand complex pose is performed. After this stage, any receptorstructurein each pose reflect an induced fit to the ligand structureand conformation. Finally, the ligand is rigorously docked, using XP Glide, into the induced-fit receptor structure.

IFD was performed using a standard protocol and OPLS3e force field was chosen [38]. Receptor box was centered on the co-crystallized ligands on the crucial residues identified within the binding site. During the initial docking procedure, the van der Waals scaling factor was set at 0.5 for both receptor and ligand. Prime refinement step was set on side chains of residues within 5Å of the ligand. For each ligand docked, a maximum of 20 poses was retained to be then redocked at XP mode. IFD calculation was followed by prime/MM-GBSA for the estimation of $\Delta G_{binding}$. The MM-GBSA approach employs molecular mechanics, the generalized Born model and the solvent accessibility method to elicit free energies from structural information circumventing the computational complexity of free-energy simulations wherein the net free energy is treated as a sum of a comprehensive set of individual energy components, each with a physical basis [41,43–45]. The conformational entropy change—TΔS—can be computed by normal-mode analysis on docking poses, but many authors have reported that the lack of the evaluation of the entropy is not critical for calculating the MM-GBSA (or MM-PBSA) free energies for similar systems [46–49]. For these reasons, the entropy term–TΔS was not calculated to reduce computational time. In our study, the VSGB solvation model was chosen using OPLS3e force field with a minimized sampling method.

3. Results and Discussion

Recently, SARS-CoV-2 caused the outbreak of coronavirus disease 2019 (COVID-19) threatening global health security. To date, no approved antiviral drugs or vaccines are available against COVID-19 although several clinical trials are underway. In this framework, computational methods offer an immediate and scientifically sound basis to potentially design highly specific inhibitors against important viral proteins and guide the antiviral drug discovery process [50]. In this work, SARS-CoV-2 encoded proteins were analyzed from PDB structures and homology models were generated by using the most similar PDB crystal structures as templates. For the homology models created, starting from the high similarity between SARS-CoV-2 proteins and some available crystal structures from SARS-CoV, ligand coordinates of the available most similar crystals were exploited for the structure-based pharmacophore creation. Below we report the analyzed proteins and the related pharmacophore maps composition.

3C-like proteinase (3CLpro), also termed the main protease, cleaves most of the sites in the polyproteins and the products are nonstructural proteins (NSPS), which assemble into the replicase–transcriptase complex (RTC). The binding site of the main protease consists of a conserved catalytic dyad, i.e., Cys145 and His41 with other crucial residues, which is Phe140, Leu141 Asn142,

Gly143, Ser144, Cys145, Met165, Glu166, Gln189 and Thr190 [27] (Figure 2A). The pharmacophore model was developed on the co-crystallized ligand (N3) that is present in the PDB ID 6LU7; this ligand was covalently bound to Cys145. We modified N3, by breaking the covalent bond and filling in open valence. The final pharmacophore showed 12 features: 2 H-bond acceptors (HBAs) interacting one with Glu166 and the other with Gly143; 4 H-bond donors (HBDs) which interact, respectively, with Phe140, His164, Glu166, Gln189 and Thr190; 4 hydrophobic features interacting with Thr25, Thr26, Met49 and Ala191; and a negative ionizable area with Glu166 (Figure 2B,C).

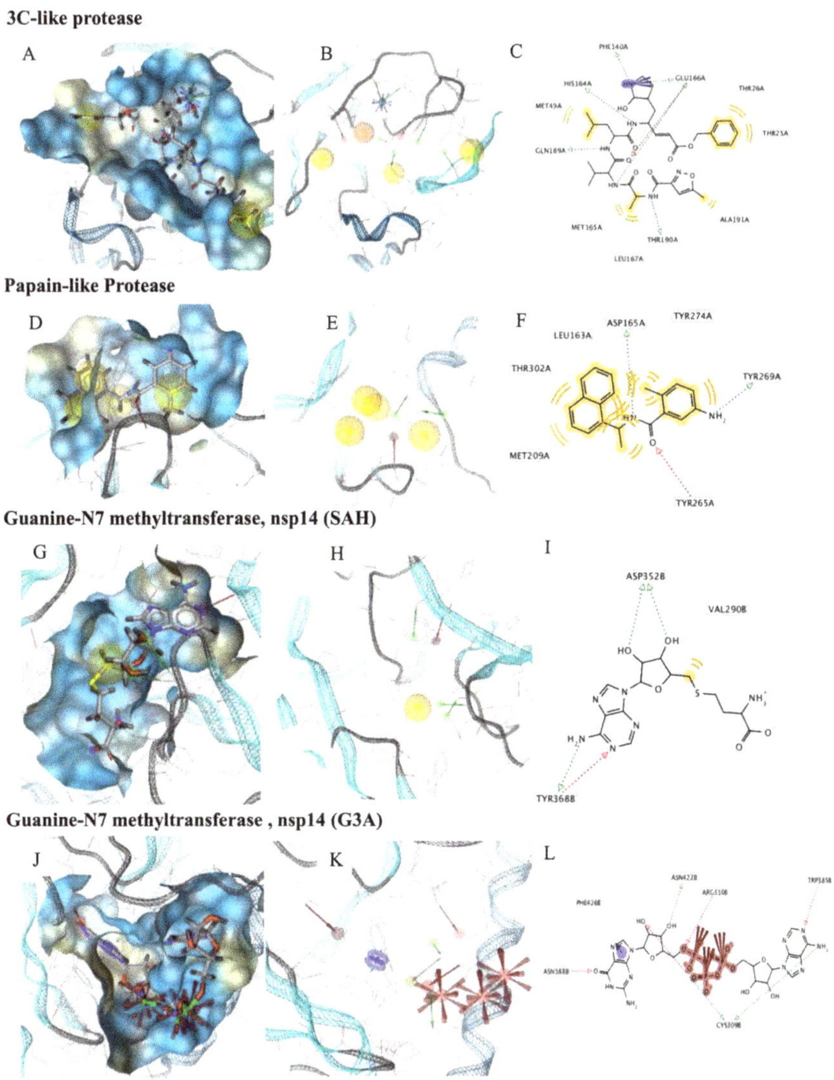

Figure 2. *Cont.*

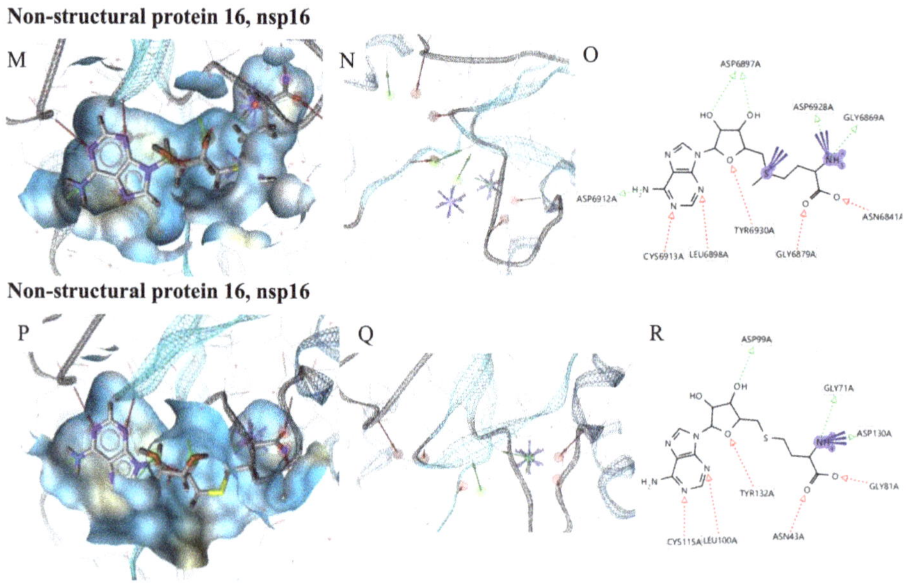

Figure 2. Pharmacophore modeling of ligand–protein complexes. For each structure molecular surface of the active site the co-crystallized ligand (**A**,**D**,**G**,**J**,**M**,**P**), structure-based pharmacophore model (**B**,**E**,**H**,**K**,**N**,**Q**) and ligand interactions (**C**,**F**,**I**,**L**,**O**,**R**) are shown.

Papain-like protease (PLpro) cleaves the nsp1/2, nsp2/3 and nsp3/4 boundaries. It works with 3CLpro to cleave the polyproteins into NSPS [51]. It showed in the active site residues Gly164, Asp165, Arg166, Glu168, Pro248, Pro249, Tyr 265, Gly267, Asn268, Tyr 269, Gln270, Cys271, Gly272, Tyr274 and Thr302 (Figure 2D). The pharmacophore model was developed on the co-crystallized ligand present in the PDB ID 3E9S. The pharmacophore map was composed of 7 features: 1 HBA with Gln270; 2 HBDs, one with Tyr265 and the other with Tyr269; and 4 hydrophobic interactions with Leu163, Met209, Tyr274 and Thr302 (Figure 2E,F).

Guanine-N7 methyltransferase (nsp14) is important for viral replication and transcription. The N-terminal exoribonuclease (ExoN) domain plays a proofreading role in the prevention of lethal mutagenesis and the C-terminal domain functions as a guanine-N7 methyltransferase (N7-MTase) for mRNA capping [52]. The models were developed using as template the PDB ID 5C8S, which shows nsp14 in complex with its activator Nonstructural protein10 (nsp10) and two functional ligands: S-adenosyl-L-homocysteine (SAH) and guanosine-P3-adenosine-5,5′-triphosphate (G3A). One molecule of nsp10 interacts with ExoN of nsp14 to stabilize it and stimulate its activity. SAH and G3A bind the guanine-N7 methyltransferase site. The SAH binding pocket contains residues Trp292, Gly333, Asp352, Phe367 and Tyr368 [29] (Figure 2G). The derived pharmacophore model showed 5 features: 1 HBA with Tyr368, 3 HBDs, two with Asp352 and one with Tyr368 and hydrophobic interaction with Val290 (Figure 2H,I). The binding pocket engaging G3A contains the following residues: Trp292, Arg310, Gly333, Pro335, Lys336, Asn386, Asn388, Tyr420 and Phe426 (Figure 2J). Therefore, the derived pharmacophore model showed 10 features: 4 HBAs which interacted, respectively, with Cys309, Arg310, Trp385, Asn388, 3 HBDs, two with Cys309 and one with Asn422, 3 negative ionizable features at the 3 phosphate groups and an aromatic ring with Phe426 (Figure 2K,L).

Nonstructural protein 16 (nsp16) also termed 2′-O-methyltransferase is activated only by the binding of nsp10. We considered the structure of the nsp16–nsp10 complex from SARS-COV-2 with 1.80 Å of resolution (PDB ID: 6W4H). This complex shows S-adenosylmethionine (SAM) in the binding site. It forms hydrogen bonds with Asp6928, Tyr6930, Asp6897 and Cys6913 (Figure 2P). The derived pharmacophore model on the co-crystallized ligand showed 9 features: 4 HBAs with Gly248 and Thr341, 1 HBD with His250 and 2 negative ionizable areas with Gly248 and Lys290 (Figure 2Q,R).

Moreover, we used the nsp16–nsp10 SARS coronavirus complex (PDB ID: 2XYQ), which shows S-adenosyl-L-homocysteine (SAH) in the binding site. SAH forms hydrogen bonds with Lys46, Asp130, Lys170 e Glu203 (Figure 2S). The derived pharmacophore model showed 9 features: 4 HBAs with Asn43, Leu100, Tyr132, Cys115, 4 HBDs with Gly71 and Asp99, 2 negative ionizable areas with Asp130 (Figure 2T,U).

The other pharmacophore models were developed exploring the apoprotein surfaces as follows: uridylate-specific endoribonuclease (NendoU/nsp15) forms a hexameric endoribonuclease, that preferentially cleaves 3′ of uridines. It is one of the RNA-processing enzymes encoded by the coronavirus [52]. Exploring the apoprotein surface, a potential active site was found, and a pharmacophore model was generated (Figure 3A). It contained the following residues: Thr166, Arg198, Asp267 and Ser273. The pharmacophore model showed 3 features: 2 HBDs and one hydrophobic feature.

Nonstructural protein 4 (nsp4) is localized at the endoplasmic reticulum membrane when expressed alone, but this protein can be recruited into the replication complex in infected cells [52]. After scanning the protein surface, a potential binding pocket was identified containing residues Leu417, Thr460 and Arg464. The derived pharmacophore model showed 6 features: 2 HBAs, 2 HBDs and a hydrophobic feature (Figure 3B).

Nonstructural protein 9 (nsp9), encoded by ORF1a, does not present a designated function, but is most likely involved with viral RNA synthesis. The crystal structure suggests that the protein is dimeric, whereas nsp9 binds RNA and interacts with nsp8 [53]. The potential identified binding site contains the following residues: Gly38, Arg39, Ser59 and Thr64. The derived pharmacophore model showed 6 features: 2 HBAs, 2 HBDs and one hydrophobic feature (Figure 3C).

Helicase (hel) catalyzes the unwinding of duplex oligonucleotides into single strands in an NTP-dependent manner. The structure of SARS-CoV-2 nsp13 adopted a triangular pyramid shape comprising five domains. Among these, there are two "RecA-like" domains, 1A (261–441 a.a.) and 2A (442–596 a.a.) and 1B domain (150–260 a.a.) forming the triangular base, while N-terminal zinc-binding domain (ZBD) (1–99 a.a.) and stalk domain (100–149 a.a.), which connects ZBD and 1B domain, are arranged at the apex of the pyramid [27]. Exploring the apoprotein surface, two putative binding sites were found, pocket A and pocket B. Pocket A contained residues from the stalk domain (Lys139, Lys146), 1B domain (Asn179) and 1A domain (Cys309, Arg339) and domain 1B (Thr228-Thr231) important for helicase activity. Pocket B contained residue from the N-terminal zinc-binding domain, ZBD domain, (Ile20, Arg21, Arg22) and stalk domain (Arg129). The pharmacophore models obtained for each pocket have the same 6 features: 2 HBAs, 2 HBDs and two hydrophobic features.

Nonstructural protein 7 and 8 (nsp7–nsp8) supercomplex are essential cofactors for Nsp12 polymerase [33]. Two putative active sites were found: pocket A and pocket B. Pocket A between chains C, G and H, pocket B between chainG–H of nsp8. The pocket A showed as residues: Glu50 of chain C; Thr124 and Arg190 of chain G; Glu5, Arg57, of chain H. The pocket B of chains G–H of nsp8 showed the residues: Arg57 and Asp64 of chain G; Leu122 and Thr123 of chain H. The pharmacophore model showed 6 features each: 2 HBAs, 2 HBDs and two hydrophobic features (Figure 3F,G).

Nonstructural protein 12 bound to nsp7-8 co-factors (nsp7–nsp8–nsp12) hetero-oligomeric complex is an RNA-dependent RNA polymerase. It is bound to its essential co-factors nsp7 and nsp8 greatly stimulates the replication and transcription activities of the polymerase. The nsp12 contains a polymerase domain (a.a. 398–919) that assumes a structure resembling a cupped "right hand". The polymerase domain consists of a finger domain (a.a. 398–581, 628–687), a palm domain (a.a. 582–627, 688–815) and a thumb domain (a.a. 816–919). CoV nsp12 also contains a nidovirus-unique N-terminal extension (a.a. 1–397) [27]. The putative active sites, pocket A and pocket B were found into conserved motif regions (A–G) possessed of all polymerases [33]. Pocket A contained residues of N-terminal extension Thr246 and Arg249; pocket B contained residues of N-terminal extension Tyr129, His133, Asn138 and motif D (Ala706–Asp711), the pharmacophore model showed 6 features: 2 HBAs, 2 HBDs and two hydrophobic features (Figure 3H,I).

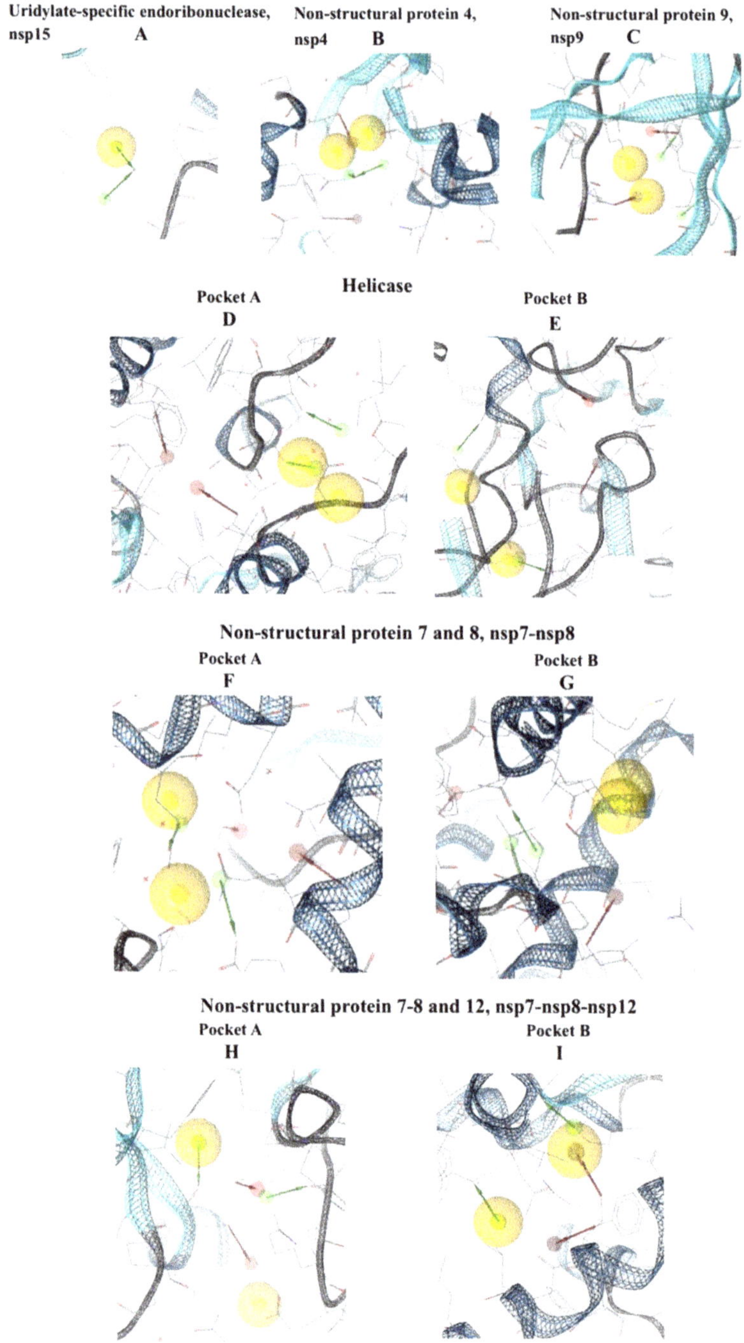

Figure 3. Pharmacophore modeling of the apoproteins. The first step was to find putative active sites, then the pharmacophore models were built in the identified pockets. Four of the structures had two active sites. (**A**) Uridylate-specific endoribonuclease, nsp15; (**B**) nsp4; (**C**) nsp9; (**D**) Helicase pocket A; (**E**) Helicase pocket B; (**F**) nsp7-nsp8 pocket A; (**G**) nsp7-nsp8 pocket B; (**H**) nsp7-8-12 pocket A; (**I**) nsp7-8-12 pocket B.

The identified pharmacophore models were used to perform a virtual screening against the DrugBankdatabase of experimental, investigational and approved drugs considering as a first filter.

The hits found were submitted to docking studies to evaluate the poses and interactions at the putative active site. First, XP docking was performed and subsequently, the highest-ranked hits were submitted to induced fit docking analysis and MM-GBSA calculation to further filter. For just one protein (nsp16), no hits were identified in the DrugBank database. At the end of the computational exploration, we have identified a total of 34 hits for all the explored targets. Among these compounds, 26 are experimental drugs, 5 investigational drugs and 3 approved drugs. The summary results were reported in the Supporting Information. In the main text, we will discuss the molecular recognition analysis for the best binder hits for each target. The rest of the identified hits, docking scores, ΔGbinding, and protein–ligand interactions is reported in a table in Supplementary Information as well as 2D ligand interaction diagrams of the best binders.

The best docked hit molecule for 3CL-protease is the experimental drug DB082309, a phenyl pyrroline derivative (ΔG = −72.56 kcal/mol). This compound is characterized by an H-bond between the carbonyl oxygen with Asn142, but the principal contribution to the binding is given by the ΔGvdW = −52.56 kcal/mol and the ΔGlipo = −23.65 due to the 2 aromatic rings (phenyl and O-difluorophenyl) of the molecules which are located in two hydrophobic pockets (Leu140, Phe141, Leu167, Pro168) and the piperazine moiety interacting with His41 and Met49 (Figure 4A).

The most promising drug candidate for papain-like protease is the experimental drug DB07358 (ΔG= −50.662 kcal/mol), a benzamide derivative. In our study, the experimental drug DB07358 forms three H-bonds with Tyr269, Gln270 and Tyr274. Moreover, the binding is characterized by a strong pi-stacking of the thiazol moiety with the phenyl ring of Tyr269 and phenylamino moiety with the phenyl ring of Tyr274 (ΔGlipo = −19.47 kcal/mol, ΔGvdW = −38.50 kcal/mol) (Figure 4B).

Top-ranked guanine-N7-methyltransferase (nsp14) hit is the experimental drug DB02933 as known as 5′-deoxy-5′-(methylthio)-tubercidin (ΔG = −65.07 kcal/mol). This compound was previously identified as an inhibitor of the h–S-methyl-5′-thioadenosine phosphorylase and bacterial methylthioadenosinenucleosidase. The compound 5′-deoxy-5′-(methylthio)-tubercidin showed 3 H-bond interactions with Asn386, Asn388 and Glu302, but the most contribution to the binding energy is due to pyrrole pyrimidine moiety, which establishes strong pi-stacking interaction with Tyr420 and Phe426 (Figure 4C).

Considering the NendoU/nsp15 protein, the most promising compound is the experimental drug DB01792 as known Adenylyl-(3′-5′)-uridine 3′-monophosphate (ΔG = −63.169 kcal/mol). The compound showed a high number of H-bond interactions with several different residues (Thr166, Ser197, Glu264, Asp272, Tyr278) (Figure 4D).

The experimental drug DB01859 resulted in the hit related to the nsp16. The compound is also known as 4-diphosphocytidyl-2-C-methyl-D-erythritol 2-phosphate (ΔG = −25.204 kcal/mol). It showed 9 H-bond interactions with Gly71, Ala72, Gly81, Ser98, Asp99, Asp130 and Asp133. Residue Cys115 showed 2 H-bonds (Figure 4E).

The top-ranked compound for nsp4 is the experimental drug sinapoyl-coA (DG = −80.73 kcal/mol). The binding of sinapoyl-CoA in the nsp4 pocket is influenced by a high number of H-bonds with several different residues (Leu417, Thr419, Arg464, Thr460) (Figure 4F).

The experimental drug DB02794 resulted in the best binding hit related to the nsp9. Due to the presence in the scaffold of many oxygen atoms, DB02794 establishes many H-bond interactions involving Lys36, Gly38, Arg39, Ser59, Asp60, Glu68. Other H-bond interactions involve some nitrogen of the experimental drug and the residues Gly38, Ser59 and Lys92. The strong net of H-bond interactions is reflected by a ΔGcoul = −84.35 kcal/mol, partially compensated by a loss of binding energy due to the solvation contribution ΔG = +68.45 kcal/mol. It is worthy to note that the next top-ranked hits for nsp9 are 3 approved drugs (ioxilan, Pemetrexed, and isoprenaline), which could be of particular interest due to the status "approved", which would allow to use them in clinical trials (Figure 4G).

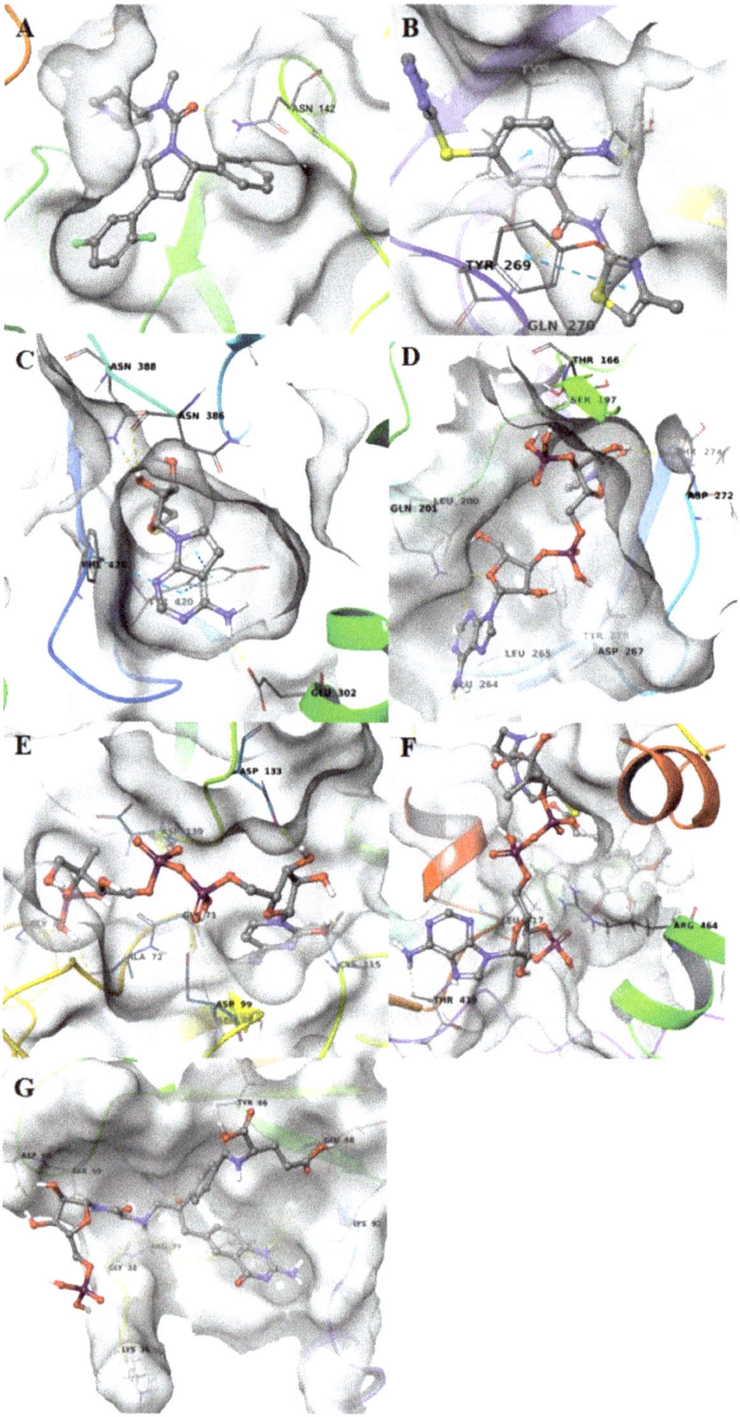

Figure 4. *Cont.*

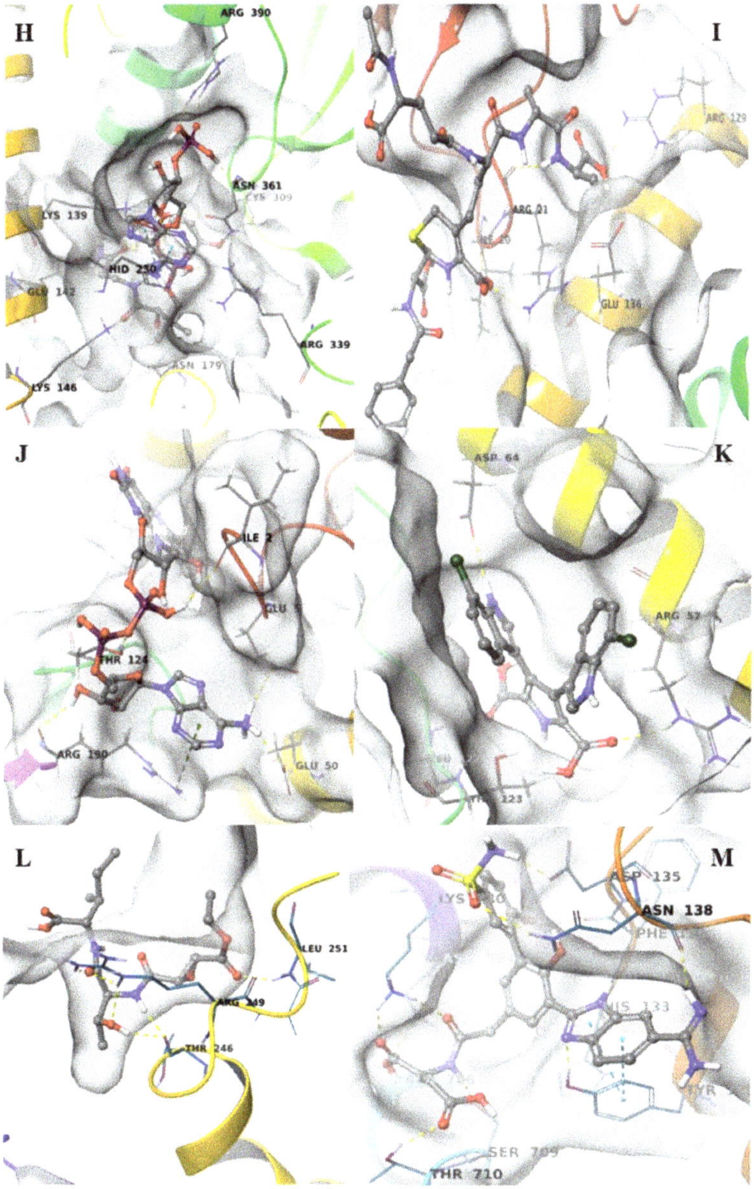

Figure 4. (**A**) DB08239 binding pose in 3C-like protease; (**B**) DB07358 binding pose in papain-like protease; (**C**) DB02933 binding pose in guanine-N7-methyltransferase (nsp14); (**D**) DB01792 binding pose in NendoU/nsp15; (**E**) DB01859 binding pose in nsp16; (**F**) synapoylCoenzyme A binding pose in nsp4; (**G**) DB02794 binding pose in nsp9.(**H**) DB04579 binding pose in helicase, pocket A; (**I**) PCI-27483 binding pose in helicase, pocket B; (**J**) flavin-N7 protonated-adenine dinucleotide binding pose in nsp7-8, pocket A; (**K**) DB06955 binding pocket in nsp7-8, pocket B; (**L**) DB04579 binding pose in nsp7-8–12, pocket A; (**M**) PCI-27483 in nsp7-8–12, pocket B.

For the helicase, the apo binding pocket analysis identified 2 different putative binding sites. The most promising candidate drug-binding pocket A is the experimental drug 4-hydroxybenzoyl-coA ($\Delta G = -91.90$ kcal/mol). The interactions that this compound establishes with the pocket A are characterized by several H-bonds, most of which formed by the three phosphate moieties with Lys139, Arg339, Asn361, Arg390. Other H-bond interactions are among the hydroxyl and carbonyl oxygens and Lys139, Glu142, Lys146, Asp179, His230, Cys309, Arg339, Arg390. Moreover, the purine moiety establishes pi-stacking interactions with the imidazole moiety of His230. Regarding the top-ranked compound in pocket B, this is the experimental drug DB02136, a cephalosporin analog, ($\Delta G = -75.81$ kcal/mol). This compound interacts with the residues Ile20, Arg21, Arg22, Arg129, Glu136 forming H-Bonds with carbonyl and hydroxyl oxygen atoms, but the binding mode is strengthened by an important contribution of $\Delta GvdW = -71.94$ kcal/mol (Figure 4H,I).

Furthermore, for the supercomplex nsp7–nsp8, two different pockets were found. The most promising candidate for pocket A is the experimental drug flavin-N7 protonated-adenine dinucleotide ($\Delta G = 78.86$ kcal/mol). The flavin moiety interacts with the residue Arg57 forming 2 H-bonds. These latter are also formed among the phosphate and Thr190, the ribose moiety and Arg190 and the purine moiety and Ile2, Ile3, Ile4. Moreover, the binding interaction is strengthened by ionic interactions among the NH_3^+ and the glutamic residues 5 and 50. The residue Arg190 interacts with the purine moiety employing pi-stacking interactions. The top-ranked compound for pocket B is the experimental drug DB06955 ($\Delta G = -58.14$ kcal/mol), a pyrrole-indole derivative, interacting with Arg57, Asp64, Leu122 and Thr123 employing H-bond interactions (Figure 4J,K).

Last, but not least, for the hetero-oligomeric complex nsp7–nsp8–nsp12 two different pockets were identified. In pocket A, the most promising compound is the experimental peptide analog DB04579 ($\Delta G = -57.10$ kcal/mol) interacting with the residues Thr246, Arg249, Leu251, Ser255 through H-bond interactions. The most promising compound for the pocket B is the investigational drug PCI-27483, a phenyl benzimidazole derivative to date used for the treatment of the pancreatic adenocarcinoma. The binding mode is characterized by several H-bond interactions involving His133, Phe134, Asp135, Asn138, Ala708, Ser709, Thr710, Lys780 and Asn781. The indole moiety is further involved in pi-stacking interactions with Tyr129 (Figure 4L,M).

4. Conclusions

The recently emerged SARS-CoV-2 caused a major outbreak of COVID-19 and instigated a widespread fear and has threatened global health security because there are no approved therapies for treating. In the attempt to try to speed up the search for new inhibitors of the virus replication, in this study, we performed a computational drug repositioning campaign on the DrugBank database of experimental, investigational and approved drugs. The aim of using such a restricted database had the rationale to identify potential lead compounds to quickly test in vitro and in vivo as they passed toxicity tests. We analyzed the proteome of SARS-CoV-2 and using homology modeling we identified the high-quality models of proteins. A structure-based pharmacophore modeling study was performed to identify pharmacophore features for each target. Successively, the pharmacophore models were used to perform a virtual screening against the DrugBank library. After a docking study, we identified a total of 34 hits for all the explored targets (3CL-protease, papain-like protease, guanine-N7-methyltransferase nsp14, nsp16, NendoU/nsp15, nsp4, nsp9, helicase, nsp7–nsp8 supercomplex and nsp7–nsp8–nsp12 hetero-oligomeric complex). Among these compounds, 26 are experimental drugs, fiveinvestigational drugs and three approved drugs. The final selection of the potential inhibitors was made considering the best binding energy for each compound obtained utilizing MM-GBSA calculation. Molecular recognition analysis showed that these compounds interact with the residues found as crucial for each target. These drugs can be further explored against the successful inhibition of COVID-19. Moreover, a set of hot spot residues and pharmacophore features for each target, which makes substantial contributions to the protein–ligand binding are also identified. This achievement can facilitate us to rationally design novel selective inhibitors targeting SARS-CoV-2, not comprised in the

DrugBank. The results of this study offer a double important hint for anti-COVID19 drug discovery campaigns. On one side, it shows putative repurposing drugs to be adopted as a single therapy or in combination with other therapies. On the other side, our deep studies attempted to map out the main binding hot spots for the most important SARS-CoV-2 proteins, opening an important route to the design of new molecules to test.

Supplementary Materials: The following are available online at http://www.mdpi.com/2079-3197/8/3/77/s1. Table S1: Docking scores, ΔGbinding and protein-ligand interactions of 36 inhibitors; Figure S1: Ligand interaction diagram of the (a) DB08239 binding pose in 3C-like Protease; (b) DB07358 binding pose in Papain-like Protease; (c) DB02933 binding pose in nsp14; Figure S2: Ligand interaction diagram of the (a) DB1792 binding pose in nsp15; (b) DB01859 binding pose in nsp16; Figure S3: Ligand interaction diagram of the (a) Synapoyl Coenzyme A binding pose in nsp4; (b) the DB02794 binding pose in nsp9; Figure S4: Ligand interaction diagram of the (a) DB04579 binding pose in helicase, pocket A; (b) PCI-27483 binding pocket in helicase, pocket B; Figure S5: Ligand interaction diagram of the (a) Flavin-N7 Protonated-Adenine Dinucleotide binding pose in nsp7-8, pocket A; (b) DB06955 binding pocket in nsp7-8, pocket B; Figure S6: Ligand interaction diagram of the (a) DB04579 binding pose in nsp7-8-12, pocket A; (b) PCI-27483 in nsp7-8-12, pocket B.

Author Contributions: Conceptualization, M.T.; methodology, U.P. and M.T.; formal analysis, G.C.; investigation and data curation U.P. and G.C; writing—original draft preparation, M.T. and U.P.; writing—review and editing, M.T., U.P., A.M.A., M.R.G., M.Z. All authors have read and agreed to the published version of the manuscript.

Funding: Regional Assessorship of Productive Activities—Department of Productive Activities, funds: FSC 2014/2020. Project name: Computational Molecular Design e Screening–CheMISt- CUPG77B17000110001, Scientific Research within the "Patto per il sud" of the Sicily Region.

Conflicts of Interest: The authors declare no conflict of interest.

References

1. Rothan, H.A.; Byrareddy, S.N. The epidemiology and pathogenesis of coronavirus disease (COVID-19) outbreak. *J. Autoimmun.* **2020**, *109*, 102433. [CrossRef] [PubMed]
2. Yu, R.; Chen, L.; Lan, R.; Shen, R.; Li, P. Computational screening of antagonists against the SARS-CoV-2 (COVID-19) coronavirus by molecular docking. *Int. J. Antimicrob. Agents* **2020**, *2*, 3–8.
3. Wu, C.; Liu, Y.; Yang, Y.; Zhang, P.; Zhong, W.; Wang, Y.; Wang, Q.; Xu, Y.; Li, M.; Li, X.; et al. Analysis of therapeutic targets for SARS-CoV-2 and discovery of potential drugs by computational methods. *Acta Pharm. Sin. B.* **2020**, *10*, 766–788. [CrossRef] [PubMed]
4. Ciliberto, G.; Cardone, L. Boosting the arsenal against COVID-19 through computational drug repurposing. *Drug Discov. Today* **2020**, *26*, 946–948. [CrossRef]
5. Pushpakom, S.; Iorio, F.; Eyers, P.A.; Escott, K.J.; Hopper, S.; Wells, A.; Doig, A.; Guilliams, T.; Latimer, J.; McNamee, C.; et al. Drug repurposing: Progress, challenges and recommendations. *Nat. Rev. Drug Discov.* **2018**, *18*, 41–58. [CrossRef]
6. Lauria, A.; Tutone, M.; Almerico, A.M. Virtual lock-and-key approach: The in silico revival of Fischer model by means of molecular descriptors. *Eur. J. Med. Chem.* **2011**, *46*, 4274–4280. [CrossRef]
7. Oprea, T.I.; Mestres, J. Drug repurposing: Far beyond new targets for old drugs. *AAPS J.* **2012**, *14*, 759–763. [CrossRef]
8. Vanhaelen, Q.; Mamoshina, P.; Aliper, A.M.; Artemov, A.; Lezhnina, K.; Ozerov, I.; Labat, I.; Zhavoronkov, A. Design of efficient computational workflows for in silico drug repurposing. *Drug Discov. Today* **2017**, *22*, 210–222. [CrossRef]
9. March-Vila, E.; Pinzi, L.; Sturm, N.; Tinivella, A.; Engkvist, O.; Chen, H.; Rastelli, G. On the integration of in silico drug design methods for drug repurposing. *Front. Pharmacol.* **2017**, *8*, 298. [CrossRef]
10. Liu, Z.; Fang, H.; Reagan, K.; Xu, X.; Mendrick, D.L.; Slikker, W.; Tong, W. In silico drug repositioning-what we need to know. *Drug Discov. Today* **2013**, *18*, 110–115. [CrossRef]
11. Oprea, T.I.; Bauman, J.E.; Bologa, C.G.; Buranda, T.; Chigaev, A.; Edwards, B.S.; Jarvik, J.W.; Gresham, H.D.; Haynes, M.K.; Hjelle, B.; et al. Drug repurposing from an academic perspective. *Drug Discov. Today Ther. Strateg.* **2011**, *8*, 61–69. [CrossRef] [PubMed]
12. Lauria, A.; Tutone, M.; Barone, G.; Almerico, A.M. Multivariate analysis in the identification of biological targets for designed molecular structures: The BIOTA protocol. *Eur. J. Med. Chem.* **2014**, *75*, 106–110. [CrossRef] [PubMed]

13. Corsello, S.M.; Bittker, J.A.; Liu, Z.; Gould, J.; McCarren, P.; Hirschman, J.E.; Johnston, S.E.; Vrcic, A.; Wong, B.; Khan, M.; et al. The Drug Repurposing Hub: A next-generation drug library and information resource. *Nat. Med.* **2017**, *23*, 405–408. [CrossRef] [PubMed]
14. Farha, M.A.; Brown, E.D. Drug repurposing for antimicrobial discovery. *Nat. Microbiol.* **2019**, *4*, 565–577. [CrossRef]
15. Sleire, L.; Førde-Tislevoll, H.E.; Netland, I.A.; Leiss, L.; Skeie, B.S.; Enger, P.Ø. Drug repurposing in cancer. *Pharmacol. Res.* **2017**, *124*, 74–91. [CrossRef]
16. Cha, Y.; Erez, T.; Reynolds, I.J.; Kumar, D.; Ross, J.; Koytiger, G.; Kusko, R.; Zeskind, B.; Risso, S.; Kagan, E.; et al. Drug repurposing from the perspective of pharmaceutical companies. *Br. J. Pharmacol.* **2018**, *175*, 168–180. [CrossRef]
17. Tutone, M.; Perricone, U.; Almerico, A.M. Conf-VLKA: A structure-based revisitation of the Virtual Lock-and-key Approach. *J. Mol. Graph. Model.* **2017**, *71*, 50–57. [CrossRef]
18. Tutone, M.; Almerico, A.M. The In Silico Fischer Lock-and-Key Model: The Combined Use of Molecular Descriptors and Docking Poses for the Repurposing of Old Drugs. Targeting Enzymes for Pharmaceutical Development. *Methods Mol. Biol.* **2020**, *2089*, 29–39.
19. Gao, J.; Zhang, L.; Liu, X.; Li, F.; Ma, R.; Zhu, Z.; Zhang, J.; Wu, J.; Shi, Y.; Pan, Y.; et al. Repurposing Low-Molecular-Weight Drugs Against the Main Protease of Severe Acute Respiratory Syndrome Coronavirus 2. *J. Phys. Chem. Lett.* **2020**. [CrossRef]
20. Meyer-Almes, F.J. Repurposing approved drugs as potential inhibitors of 3CL-protease of SARS-CoV-2: Virtual screening and structure based drug design. *Comput. Biol. Chem.* **2020**, *88*. [CrossRef]
21. Touret, F.; Gilles, M.; Barral, K.; Nougairède, A.; van Helden, J.; Decroly, E.; de Lamballerie, X.; Coutard, B. In vitro screening of a FDA approved chemical library reveals potential inhibitors of SARS-CoV-2 replication. *Sci. Rep.* **2020**, *10*, 13093. [CrossRef] [PubMed]
22. Shyr, Z.A.; Gorshkov, K.; Chen, C.Z.; Zheng, W. Drug discovery strategies for SARS- CoV-2. *J. Pharmacol. Exp. Ther.* **2020**, *374*. [CrossRef] [PubMed]
23. Cavasotto, C.; Di Filippo, J. In silico Drug Repurposing for COVID-19: Targeting SARS-CoV-2 Proteins through Docking and Consensus Ranking. *Mol. Inform.* **2020**. [CrossRef] [PubMed]
24. Wang, J. Fast Identification of Possible Drug Treatment of Coronavirus Disease-19 (COVID-19) through Computational Drug Repurposing Study. *J. Chem. Inf. Model.* **2020**, *60*, 3277–3286. [CrossRef]
25. Ferraz, W.R.; Gomes, R.A.; S Novaes, A.L.; Goulart Trossini, G.H. Ligand and structure- based virtual screening applied to the SARS-CoV-2 main protease: An in silico repurposing study. *Future Med. Chem.* **2020**. [CrossRef]
26. Zhou, Y.; Hou, Y.; Shen, J.; Huang, Y.; Martin, W.; Cheng, F. Network-based drug repurposing for novel coronavirus 2019-nCoV/SARS-CoV-2. *Cell. Discov.* **2020**, *6*, 14. [CrossRef]
27. Mirza, M.U.; Froeyen, M. Structural elucidation of SARS-CoV-2 vital proteins: Computational methods reveal potential drug candidates against main protease, Nsp12 polymerase and Nsp13 helicase. *J. Pharm. Anal.* **2020**. [CrossRef]
28. Harrison, C. Coronavirus puts drug repurposing on the fast track. *Nat. Biotechnol.* **2020**, *38*, 379–381. [CrossRef]
29. Gordon, D.E.; Jang, G.M.; Bouhaddou, M.; Xu, J.; Obernier, K.; White, K.M.; O'Meara, M.J.; Rezelj, V.V.; Guo, J.Z.; Swaney, D.L.; et al. A SARS-CoV-2 protein interaction map reveals targets for drug repurposing. *Nature* **2020**, *583*, 459–468. [CrossRef]
30. Battisti, V.; Wieder, O.; Garon, A.; Seidel, T.; Urban, E.; Langer, T. A Computational Approach to Identify Potential Novel Inhibitors against the Coronavirus SARS-CoV-2. *Mol. Inform.* **2020**. [CrossRef]
31. Wishart, D.S.; Knox, C.; Guo, A.C.; Shrivastava, S.; Hassanali, M.; Stothard, P.; Chang, Z.; Woolsey, J. DrugBank: A comprehensive resource for in silico drug discovery and exploration. *Nucleic. Acids Res.* **2006**, *34*, D668–D672. [CrossRef] [PubMed]
32. *Schrödinger Epik*; Schrödinger: München, Germany, 2018.
33. Waterhouse, A.; Bertoni, M.; Bienert, S.; Studer, G.; Tauriello, G.; Gumienny, R.; Heer, F.T.; de Beer, T.A.P.; Rempfer, C.; Bordoli, L.; et al. SWISS-MODEL: Homology modelling of protein structures and complexes. *Nucleic. Acids Res.* **2018**, *46*, W296–W303. [CrossRef] [PubMed]
34. Berman, H.M.; Battistuz, T.; Bhat, T.N. The protein data bank. *Acta Crystallogr. Sect. D. Biol. Crystallogr.* **2002**, *58*, 899–907. [CrossRef] [PubMed]

35. *Schrödinger Protein Preparation Wizard*; Schrödinger: München, Germany, 2018.
36. Schaller, D.; Šribar, D.; Noonan, T.; Lihua Deng, L.; Nguyen, T.N.; Pach, S.; Machalz, D.; Bermudez, M.; Wolber, G. Next generation 3D pharmacophore modeling. *WIREsComput. Mol. Sci.* **2020**, *10*, e1468. [CrossRef]
37. *Schrödinger Glide*; Schrödinger: München, Germany, 2018.
38. Sherman, W.; Day, T.; Jacobson, M.P.; Friesner, R.A.; Farid, R. Novel procedure for modeling ligand/receptor induced fit effects. *J. Med. Chem.* **2006**, *49*, 534–553. [CrossRef]
39. *Schrödinger LLC Prime*; Version 3; Schrödinger: New York, NY, USA, 2018.
40. Perricone, U.; Wieder, M.; Seidel, T.; Langer, T.; Padova, A.; Almerico, A.M.; Tutone, M. A molecular dynamics–shared pharmacophore approach to boost early enrichment virtual screening. A case study on PPAR alpha. *ChemMedChem* **2017**, *12*, 1399–1407. [CrossRef]
41. Almerico, A.M.; Tutone, M.; Lauria, A. Receptor-guided 3D-QSAR approach for the discovery of c-kit tyrosine kinase inhibitors. *J. Mol. Model.* **2012**, *18*, 2885–2895. [CrossRef]
42. Almerico, A.M.; Tutone, M.; Pantano, L.; Lauria, A. Molecular dynamics studies on Mdm2 complexes: An analysis of the inhibitor influence. *Biochem. Biophys. Res. Commun.* **2012**, *424*, 341–347. [CrossRef]
43. Genheden, S.; Ryde, U. The MM/PBSA and MM/GBSA methods to estimate ligand-binding affinities. *Expert Opin. Drug Discov.* **2015**, *10*, 449–461. [CrossRef]
44. Tutone, M.; Pibiri, I.; Lentini, L.; Pace, A.; Almerico, A.M. Deciphering the Nonsense Readthrough Mechanism of Action of Ataluren: An in Silico Compared Study. *ACS Med. Chem. Lett.* **2019**, *10*, 522–527. [CrossRef]
45. Tutone, M.; Virzì, A.; Almerico, A.M. Reverse screening on indicaxanthin from Opuntia ficus-indica as natural chemoactive and chemopreventive agent. *J. Theor. Biol.* **2018**, *455*, 147–160. [CrossRef]
46. Hou, T.; Yu, R. Molecular dynamics and free energy studies on the wild-type and double mutant HIV-1 protease complexed with amprenavir and two amprenavir-related inhibitors: Mechanism for binding and drug resistance. *J. Med. Chem.* **2007**, *50*, 1177–1188. [CrossRef] [PubMed]
47. Massova, I.; Kollman, P.A. Combined molecular mechanical and continuum solvent approach (MM-PBSA/GBSA) to predict ligand binding. *Perspect. Drug Discov.* **2000**, *18*, 113–135. [CrossRef]
48. Wang, J.; Morin, P.; Wang, W.; Kollman, P.A. Use of MM-PBSA in Reproducing the Binding Free Energies to HIV-1 RT of TIBO Derivatives and Predicting the Binding Mode to HIV-1 RT of Efavirenz by Docking and MM-PBSA. *J. Am. Chem. Soc.* **2001**, *123*, 5221–5230. [CrossRef] [PubMed]
49. Khan, S.A.; Zia, K.; Ashraf, S.; Uddin, R.; Ul-Haq, Z. Identification of chymotrypsin-like protease inhibitors of SARS-CoV-2 via integrated computational approach. *J. Biomol. Struct. Dyn.* **2020**, 1–10. [CrossRef] [PubMed]
50. Arya, R.; Das, A.; Prashar, V.; Kumar, M. Potential inhibitors against papain-like protease of novel coronavirus (SARS-CoV-2) from FDA approved drugs. *Chemrxiv. Org.* **2020**. [CrossRef]
51. Gil, C.; Ginex, T.; Maestro, I.; Nozal, V.; Barrado-Gil, L.; Cuesta-Geijo, M.A.; Urquiza, J.; Ramírez, D.; Alonso, C.; Campillo, N.E.; et al. COVID-19: Drug targets and potential treatments. *J. Med. Chem.* **2020**. [CrossRef]
52. Xu, X.; Lou, Z.; Ma, Y.; Chen, X.; Yang, Z.; Tong, X.; Zhao, Q.; Xu, Y.; Deng, H.; Bartlam, M.; et al. Crystal Structure of the C-Terminal Cytoplasmic Domain of Non-Structural Protein 4 from Mouse Hepatitis Virus A59. *PLoS ONE* **2009**, *4*, e6217. [CrossRef]
53. Sutton, G.; Fry, E.; Carter, L.; Sainsbury, S.; Walter, T.; Nettleship, J.; Berrow, N.; Owens, R.; Gilbert, R.; Davidson, A.; et al. The nsp9 Replicase Protein of SARS-Coronavirus, Structure and Functional Insights. *Structure* **2004**, *12*, 341–353. [CrossRef]

computation

Article

Computational View toward the Inhibition of SARS-CoV-2 Spike Glycoprotein and the 3CL Protease

Zhen Qiao [1], Hongtao Zhang [2], Hai-Feng Ji [3,*] and Qian Chen [1]

1 Department of Orthopaedics, The Warren Alpert Medical School of Brown University, Rhode Island Hospital, Providence, RI 02903, USA; zhen_qiao@brown.edu (Z.Q.); qian_chen@brown.edu (Q.C.)
2 Departments of Pathology and Laboratory Medicine, Perelman School of Medicine, University of Pennsylvania, Philadelphia, PA 19301, USA; zhanghon@pennmedicine.upenn.edu
3 Department of Chemistry, Drexel University, Philadelphia, PA 19104, USA
* Correspondence: hj56@drexel.edu; Tel.: +1-215-895-2562; Fax: +1-215-895-1265

Received: 16 April 2020; Accepted: 27 May 2020; Published: 31 May 2020

Abstract: Since the outbreak of the 2019 novel coronavirus disease (COVID-19), the medical research community is vigorously seeking a treatment to control the infection and save the lives of severely infected patients. The main potential candidates for the control of viruses are virally targeted agents. In this short letter, we report our calculations on the inhibitors for the SARS-CoV-2 3CL protease and the spike protein for the potential treatment of COVID-19. The results show that the most potent inhibitors of the SARS-CoV-2 3CL protease include saquinavir, tadalafil, rivaroxaban, sildenafil, dasatinib, etc. Ergotamine, amphotericin b, and vancomycin are most promising to block the interaction of the SARS-CoV-2 S-protein with human ACE-2.

Keywords: COVID-19; coronavirus; protease; spike protein; computational; inhibition

1. Introduction

As of 24 May 2020, over 5 millions people in the world have been confirmed as having the 2019 novel coronavirus disease (COVID-19), an infection with Severe Acute Respiratory Syndrome coronavirus 2 (SARS-CoV-2) (initially called 2019-nCoV before 11 February 2020) which is part of the Coronaviridae family of positive-sense single-stranded RNA viruses that includes SARS-CoV and MERS-CoV (Middle East Respiratory Syndrome coronavirus), both of which also cause severe respiratory infections. The death count in China so far has been over 1700, but the number is expected to go higher with the increasing number of confirmed and non-confirmed cases. The medical research community is vigorously seeking a treatment to control the infection and save the lives of severely infected patients.

Just a few weeks after the COVID-19 outbreak, the complete genome of SARS-CoV-2 was determined and reported to GenBank (accession MN908947). Viruses were also isolated from patients to understand the genomic characteristics and mechanism of the viral infection. As revealed by the analysis, the SARS-CoV-2 shared 79% sequence identity to SARS-CoV. In one study, SARS-CoV-2 was found to be closely related to two bat-derived Severe Acute Respiratory Syndrome (SARS)-like coronaviruses, with 87.5% and 87.6% shared identity [1]. In another study, SARS-CoV-2 was 96% identical at the whole-genome level to a bat coronavirus [2].

Despite the high sequence identity between the SARS-CoV-2 and the SARS-CoV in the open reading frame regions, the envelop spike protein (S-protein) [3], which mediates the infection of SARS-CoV via the human host protein ACE-2, has only about 80% shared sequence identity between the SARS-CoV and SARS-CoV-2 [1]. Within the S-protein, the receptor docking domain has a higher divergence, with four out of five critical ACE-2 interacting amino acid residues replaced in the SARS-CoV-2. However, structural modeling indicated that the four residues in the SARS-CoV-2 retain a structural conformation similar to that of SARS-CoV, and the SARS-CoV-2 S-protein should be able to bind ACE-2 with reasonable affinity [4]. Indeed, studies by Zhou et al. using cells expressing human ACE-2 confirmed that the SARS-CoV-2 could infect cells via the same protein on ACE-2 as SARS-CoV did [2]. Thus, one option to treat the infection is to search for an inhibitor that can prevent the interaction of the SARS-CoV-2 S-protein with human ACE-2. The availability of the genome sequence of SARS-CoV-2 allows us to establish structural models for the S-protein [4].

The RNA of coronaviruses encodes polyproteins that can be processed by viral proteases to yield mature proteins. The same mechanism is shared by picornaviruses and retroviruses. Patients treated with protease inhibitors appeared to have much better clinical outcomes than without using the inhibitors (SARS death: 28.8% vs. 2.4%) [5]. Molecular dynamics simulations have revealed that, by molecular docking to the active site of the main protease 3CL of SARS-CoV, both lopinavir and ritonavir could induce conformation changes and potentially interfere with infection by SARS virus [6]. We expect the same will apply for SARS-CoV-2. The crystal structure of the SARS-CoV-2 protease ($3CL^{pro}$) was just recently reported by Liu et al. [7]. Thus, another option to treat the SARS-CoV-2 infection is to search for inhibitors of the SARS-CoV-2 $3CL^{pro}$.

With these models and crystal data, we performed in silico studies of potential inhibitors of the SARS-CoV-2 S-protein and $3CL^{pro}$.

2. Computational Methods

All calculations were operated on Dell PowerEdge C6220 servers. The chemical structures were prepared by AutoDockTools-1.5.6 [8], Chimera 1.14 [9], and Avogadro [10]. The docking studies were performed with Autodock 4.2.6, Autodock4, AutoDockTools4 [11], and Autodock Vina 1.1.2 [12].

2.1. Preparation of Receptor and Ligands

The 3CL protease's three-dimensional crystal structure was retrieved from the Protein Data Bank (PDB ID: 6LU7), and it was applied as the receptor for molecular docking after a cleaning with Chimera. The ligands observed, i.e., FDA-approved drugs (2454 structures in total), were retrieved from the BindingDB (https://www.bindingdb.org), and the structures of the ligands were further optimized with Avogadro. The force field applied for geometry optimization was MMFF94.

The SARS-CoV-2/ACE-2 structure was retrieved using the function of the comparative modeling of the Chimera interface with the modeler (version 9.23) [13]. For the preparation of the SARS-CoV-2/ACE-2 structure, the target template sequence was retrieved from Zhang et al.'s work and the SARS-CoV/ACE-2 (PDB ID: 6ACD) served as a template, as it was also the top candidate from Basic Local Alignment Search Tool (BLAST) results. Because SARS-CoV and SARS-CoV-2 have an 88% similarity, the 3D structure can be predicted with a high accuracy. Next, the sequence alignments were performed using SARS-CoV as a template. Then, the model was built followed by refining the loops, side chain optimization, and model optimization. When the homology model was generated, it was further validated using the WHATCHECK/PROCHECK program [14] for basic parameters like torsion angle, rotational angle, bond length, etc. Finally, this model was used as receptor for docking purposes. The loop refinement and side chain optimization were performed using Chimera 1.14 by selecting the active region; all the parameters were the default of the version.

It is noteworthy that this calculated work was performed before the crystal structure of the COVID-19 S-protein was released (6LZG, 6VW1, etc.). After the crystal structures were released, their structures were compared with ours and the structures overlaid well (Figure 1), with 93.22% of its residues in the allowed region and a minor difference on the top right loop, which was not a site that interacts with the ACE 2, so a re-calculation was not conducted using the new crystal structures.

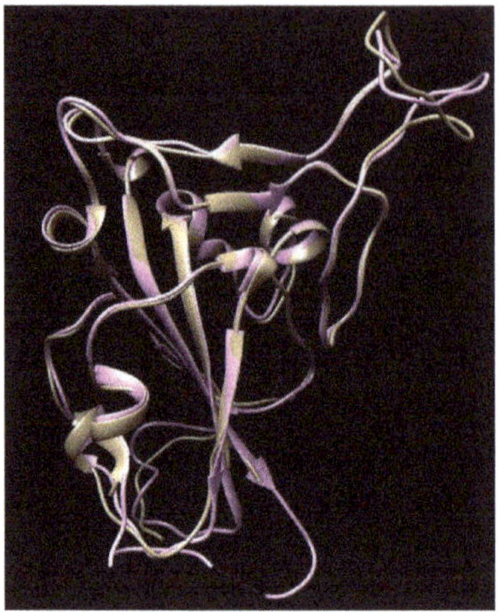

Figure 1. The comparison of the crystal structures of the SARS-2 spike protein (6VW1, pink color) with our homology modeling (light brown color) using the SARS-2 template.

2.2. Molecular Docking with Autodock Vina

For the SARS-CoV-2 3^{CL} inhibition calculation, the input files for Autodock Vina were prepared in the receptor's original file (PDB format) and ligands files (SDF format) using AutoDockTools-1.5.6. After minimizing, the grid box was set at 22.00 Å × 22.00 Å ×22.00 Å along the x, y, and z axis, respectively. The docking site was defined at 1.00 Å when using the Autodock Vina. The grid box was set into the docking site at the H41, C145, and E166 regions according to the docking site of the coronavirus main proteinase (3^{CL}) of Severe Acute Respiratory Syndrome (SARS). Then, the receptor file (PDBQT format, for docking purposes) was prepared by the addition of polar only hydrogen atoms, the removal of all water molecules, and the calculation of the Gasteiger charge. The instructed command prompts were used for the docking process. The docking output file includes the docking energy (in kcal/mol, which is an indication of the binding affinity/efficiency of one specific ligand to the receptor molecule) and the interaction of the ligands with the receptor (hydrogen bond, pi-pi stacking, etc.).

For the SARS-CoV-2 S-protein inhibition calculation, the PDB files of the SARS-CoV-2 S-protein were generated using the homology modeling method in Chimera; the template used for this was the SARS-CoV S-protein. After minimization, the input file was prepared using AutoDockTools-1.5.6. The grid box, which was a rectangular shaped area that covered all the possible docking sites of the SARS-CoV-2 S-protein with its receptor ACE-2, was chosen as 22.00 Å × 42.00 Å ×22.00 Å along the x, y, and z axis, respectively. The docking site was defined at 1.00 Å when using the Autodock Vina. Then, the receptor file (PDBQT format, for docking purposes) was prepared by the addition of polar only hydrogen atoms, the removal of all water molecules, and the calculation of the Gasteiger charge.

2.3. Analyzing the Docking Results with Chimera and BioLuminate

The docking results were ranked in the order from high to low in different modes according to the docking scores (docking energy, kcal/mol). The ligands with the most negative docking scores—i.e., the highest affinities—were selected for the visualization of the docked complexes using Chimera [9].

The docking energies of the SARS-CoV-2 S-protein and human ACE-2 were calculated using BioLuminate [15–17], and then compared to the docking energies of the SARS-CoV S-protein and human ACE-2. To verify whether those ligands can be used for blocking the interaction of the S-protein with human ACE-2, the docking energies of the SARS-CoV-2 S-protein/ligands and human ACE-2 were also calculated. The solvation model used was VSGB [18], and the force field chosen was OPLS_2005 [19] for all the docking energy predictions.

3. Results

3.1. Results of the SARS-CoV-2 3^{CL} Protease

Table 1 shows the binding affinity of several ligands with SARS-CoV-2 3^{CL} protease sorted according to the docking scores (binding affinities) calculated from the Autodock Vina; Figure 2 shows the docking of those with high docking scores—Tadalafil, Dasatinib, and Saquinavir—with the protease in the docking sites of the protease.

Table 1. Different docking scores (binding affinities) of the tested drugs for SARS-CoV-2 proteinase.

Drug Name	Docking Score (kcal/mol)	Usage
Saquinavir	−9.5	Antiretroviral drug to treat or prevent HIV/AIDS [20]
Tadalafil	−9.3	A medication used to treat erectile dysfunction (ED), benign prostatic hyperplasia (BPH), and pulmonary arterial hypertension [21]
Rivaroxaban	−9.2	An anticoagulant medication used to treat and prevent blood clots [22]
Sildenafil	−8.9	A medication used to treat erectile dysfunction and pulmonary arterial hypertension [23]
Dasatinib	−8.8	A targeted therapy used to treat certain cases of chronic myelogenous leukemia (CML) and acute lymphoblastic leukemia (ALL) [24]
Vardenafil	−8.7	A PDE5 inhibitor used to treat erectile dysfunction [25]
Montelukast	−8.5	To treat seasonal and year-round allergies [26]
Indinavir	−8.3	A component antiretroviral therapy to treat HIV/AIDS [27]
Lopinavir	−8.2	Protease inhibitor
Cortisone	−8.2	Can be used for a variety of conditions
celecoxib	−8.1	An anti-inflammation drug
Atazanavir	−8.1	An antiretroviral drug used for HIV treatment
Iressa	−7.9	A drug for cancer treatment
Darunavir	−7.7	An antiretroviral drug used for HIV treatment
Sorafenib	−7.5	A drug for cancer treatment

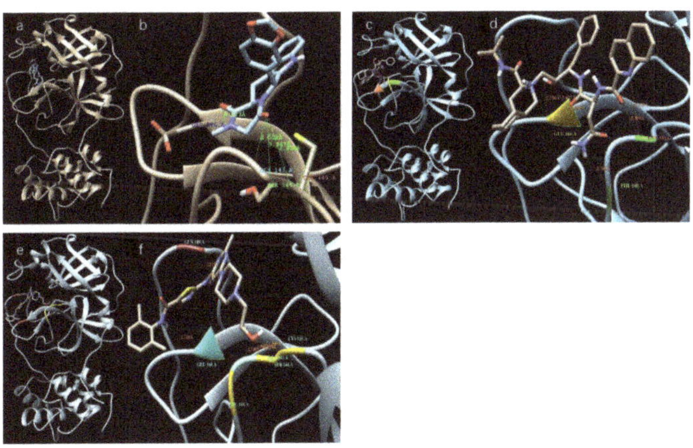

Figure 2. Different ligands in the docking site of SARS-CoV-2 protease. (**a**,**b**) Tadalafil; (**c**,**d**) Saquinavir; (**e**,**f**) Dasatinib.

3.2. Results of SARS-CoV-2 S-Protein

We modeled ligands that may bind at a large docking area on the top of the S-protein that interacts with ACE-2 (red cycle in Figure 3a). Table 2 shows the binding affinities of several ligands with the highest docking scores toward the top docking side of the S-protein. Figure 3b–i shows the dockings of several ligands with the SARS-CoV-2 S-protein.

To understand whether these ligands are reasonably good inhibitors that block the interaction of the SARS-CoV-2 S-protein with ACE-2, the docking energy of the S-protein/ligand complex with ACE-2 was calculated and the results are listed in Table 3. For comparison, the docking energy between the SARS-CoV S-protein and ACE-2 was also calculated and the score was −92.7 kcal/mol, which was close to the −78.6 kcal/mol reported by Xu et al.'s work [4]. The docking energy between the SARS-CoV-2 S-protein and ACE-2 was calculated to be −82.2 kcal/mol, suggesting a slightly weaker interaction than that of the SARS-CoV S-protein with ACE-2. The observation is similar to that reported by Xu et al.'s work [4]. Table 3 shows that more than half of those ligands docking onto the SARS-CoV-2 S-protein do not significantly change the interaction of the SARS-CoV-2 S-protein with ACE-2—i.e., they are not inhibitors to block the interaction of the SARS-CoV-2 S-protein with ACE-2. However, ergotamine, amphotericin B, vancomycin, zafirlukast, and lanicor showed that once they were bound to the S-protein, the interactions of these complex with ACE-2 were no longer energetically favored interactions—i.e., these ligands acted as desired inhibitors that can efficiently block the interaction of the SARS-CoV S-protein with ACE-2. Among these, ergotamine and amphotericin b are most promising, since they demonstrate the highest docking energy to the SARS-CoV-2 S-protein (Table 2). Thus, they are strongly suggested as the core drugs for clinical trials to treat COVID-19 patients. Considering the severe and potentially lethal side effects of amphotericin b [28], ergotamine and vancomycin seem be the top choices.

Table 2. Different docking scores of ligands for the SARS-CoV-2 S-protein.

Drug Name	Docking Score (kcal/mol)	Usage
Ergotamine	−8.8	For treatment of acute migraine attacks [29]
Amphotericin b	−8.3	An antifungal medication used for serious fungal infections and leishmaniasis [30]
Indinavir	−8.1	A component antiretroviral therapy to treat HIV/AIDS
Vancomycin	−7.7	For treatment bacterial infections [31]
Lonpinavir	−7.7	An antiretroviral, often used against HIV infections
Zafirlukast	−7.6	For the chronic treatment of asthma
Lanicor	−7.5	Used to treat heart conditions [32]
PubChem ID: 54098557	−7.5	–
Digitaline Nativelle	−7.5	For treatment of congestive heart failure, also used as angiotensin-converting enzyme (ACE) inhibitor
Rivaroxaban	−7.5	To treat and prevent blood clots [33]
Tadalafil	−7.5	To treat erectile dysfunction
Nelfinavir	−7.3	The treatment of HIV
Montelukast	−7.2	Treatment of asthma
Saquinavir	−7.1	The treatment of HIV
Carfilzomib	−7.1	Anti-cancer drug as proteasome inhibitor
Lapatinib	−7.0	Anti-cancer drug
Atovaquone	−7.0	To treat pneumocystis pneumonia, toxoplasmosis, malaria and babesia
Celecoxib	−7.0	An anti-inflammation drug
Vardenafil	−6.9	For treatment of erectile dysfunction
Dasatinib	−6.8	To treat certain cases of chronic myelogenous leukemia
Cortisone Montelukast	−6.6 −7.2	Can be used for a variety of conditions Treatment of asthma
Saquinavir	−7.1	The treatment of HIV
Carfilzomib	−7.1	Anti-cancer drug as proteasome inhibitor
Lapatinib	−7.0	Anti-cancer drug
Atovaquone	−7.0	To treat pneumocystis pneumonia, toxoplasmosis, malaria and babesia
Celecoxib	−7.0	An anti-inflammation drug
Vardenafil	−6.9	For treatment of erectile dysfunction
Dasatinib	−6.8	To treat certain cases of chronic myelogenous leukemia
Cortisone	−6.6	Can be used for a variety of conditions

Table 3. Docking energy of the SARS-CoV S-protein with and without ligands to human ACE-2.

Interaction of S-Protein and S-Protein/Drug Complex with ACE-2	Docking Energy (kcal/mol)
SARS-CoV S-protein (for comparison)	−92.7
SARS-CoV-2 S-protein	−82.2
SARS-CoV-2 S-protein/Ergotamine	56.4
SARS-CoV-2 S-protein/Amphotericin b	78.6
SARS-CoV-2 S-protein/Indinavir	−61.9
SARS-CoV-2 S-protein/Vancomycin	81.7
SARS-CoV-2 S-protein/Zafirlukast	52.6
SARS-CoV-2 S-protein/Lanicor	4.2
SARS-CoV-2 S-protein/Nelfinavir	−81.5
SARS-CoV-2 S-protein/Montelukast	−71.3
SARS-CoV-2 S-protein/Saquinavir	−48.2
SARS-CoV-2 S-protein/Carfilzomib	−88.1
SARS-CoV-2 S-protein/Lapatinib	−83.1
SARS-CoV-2 S-protein/Atovaquone	−68.2
SARS-CoV-2 S-protein/Celecoxib	−74.2
SARS-CoV-2 S-protein/Dasatinib	−42.3

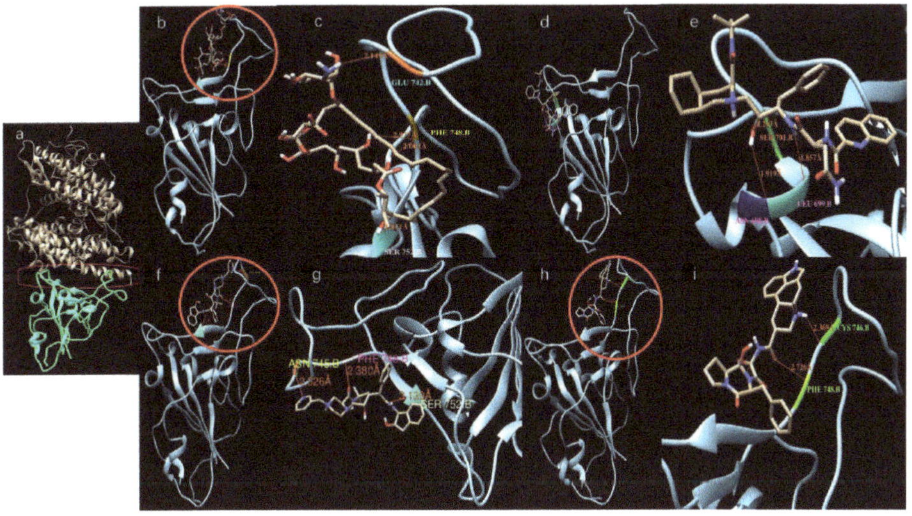

Figure 3. Potential active site selection and ligand-receptor interaction. (**a**) The docking site (inside the red frame) was chosen between the ACE-2 (light brown color) and SARS-CoV-2 S-protein (Cyan color); (**b**,**c**) Amphotericin b docks onto the SARS-CoV-2 S-protein; (**d**,**e**) Saquinavir docks onto the SARS-CoV-2 S-protein; (**f**,**g**) Indinavir docks onto the SARS-CoV-2 S-protein; (**h**,**i**) Ergotamine binds onto the SARS-CoV-2 S-protein.

4. Discussion

Disulfiram, lopinavir, and ritonavir are the three approved and active protease inhibitors against SARS and MERS. Indeed, lopinavir and ritonavir were successfully used to treat a patient in Thailand in January 2020. Our results show that among these ligands, saquinavir, tadalafil, rivaroxaban, sildenafil, dasatinib, vardenafil, montelukast are most promising due to their higher docking scores (<−8.5 kcal/mol, which coresponds to <1 μM IC50) than others. All of these scores appear better than that of the antiviral drug Lopinavir (−8.2 kcal/mol). As a comparison, the docking scores reported for lopinavir with the viral RNA polymerase is −8.3 kcal/mol [18]. It is a remarkable observation that some SARS-CoV-2 inhibitors such as indinavir could not block the interaction of the SARS-CoV-2 S-protein with ACE-2, while other inhibitors, such as ergotamine and amphotericin B, can effectively inhibit such interaction. This is somewhat confusing, since all of these three compounds dock on the same docking site that is marked by the red circles in Figure 3—the grove between an extended insertion that contains the β5/β6 strands and the receptor-binding motif (RBM) loop [28]. To comprehend what caused the significant difference, we overlaid the structures of the three docked compounds and ACE-2 on the SARS-CoV-2 spike protein in Figure 4. The comparison clearly shows that ergotamine (red) and amphotericin B (blue) extend further out toward the ACE-2 and thus effectively block the interaction of the SARS-C oV-2 spike protein with ACE-2 while indinavir (green) clings to the SARS-CoV-2 spike protein, leaving room for ACE-2 to interact with the spike protein.

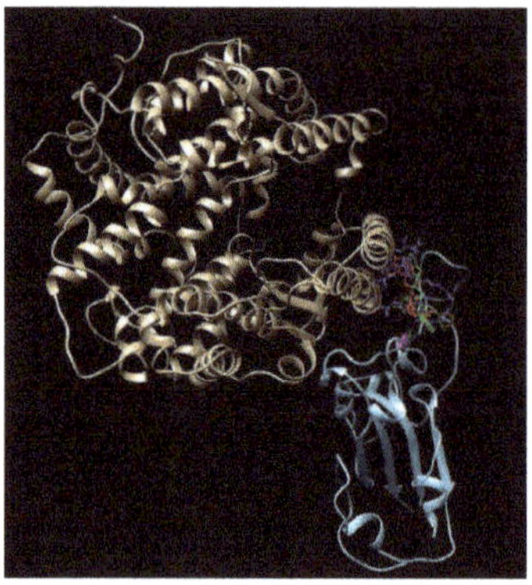

Figure 4. An overlay of the modeled structures of amphotericin, ergotamin, indinavir, and ACE-2 with the SARS-CoV-2 spike protein.

5. Conclusions

For the inhibition of the SARS-CoV-2 3^{CL} protease, saquinavir, tadalafil, rivaroxaban, sildenafil, dasatinib, vardenafil, and montelukast are most promising due to their high docking scores (<-8.5 kcal/mol), which were more negative than those of other ligands.

Among these that showed an excellent inhibiting property to block the interaction of SARS-CoV-2 S-protein with ACE-2 in Table 3, ergotamine, amphotericin b, and vancomycin are the most promising since they are also among the highest to bind to the SARS-CoV-2 S-protein, as shown in Table 2.

For more active results, a combination of 3^{CL} protease inhibitors and ergotamine may be considered.

Author Contributions: Conceptualization, Z.Q., and H.-F.J.; methodology, Z.Q.; software, Z.Q.; validation, Z.Q., H.Z. and H.-F.J.; formal analysis, Z.Q.; writing—original draft preparation, Z.Q. and H.Z.; writing—review and editing, Z.Q., H.Z., Q.C., H.-F.J. All authors have read and agreed to the published version of the manuscript.

Funding: This research was funded by NIH grant number P30 GM122732, H. Zhang was funded by NIH R01HL128895.

Conflicts of Interest: The authors declare no conflict of interest.

References

1. Lu, R.; Zhao, X.; Li, J.; Niu, P.; Yang, B.; Wu, H.; Wang, W.; Song, H.; Huang, B.; Zhu, N.; et al. Genomic characterisation and epidemiology of 2019 novel coronavirus: Implications for virus origins and receptor binding. *Lancet* **2020**, *20*, 30251–30258. [CrossRef]
2. Zhou, P.; Yang, X.; Wang, X.; Hu, B.; Zhang, L.; Zhang, W.; Si, H.; Zhu, Y.; Li, B.; Huang, C.; et al. A pneumonia outbreak associated with a new coronavirus of probable bat origin. *Nature* **2020**, *579*, 270–273. [CrossRef] [PubMed]
3. Li, F. Structure, Function, and Evolution of Coronavirus Spike Proteins. *Annu. Rev. Virol.* **2016**, *3*, 237–261. [CrossRef] [PubMed]

4. Xu, X.; Chen, P.; Wang, J.; Feng, J.; Zhou, H.; Li, X.; Zhong, W.; Hao, P. Evolution of the novel coronavirus from the ongoing Wuhan outbreak and modeling of its spike protein for risk of human transmission. *Sci. China Life Sci.* **2020**, *63*, 457–460. [CrossRef]
5. Chu, C.M.; Cheng, V.C.; Hung, I.F.; Wong, M.M.; Chan, K.H.; Chan, K.S.; Kao, R.Y.; Poon, L.L.; Wong, C.L.; Guan, Y.; et al. Role of lopinavir/ritonavir in the treatment of SARS: Initial virological and clinical findings. *Thorax* **2004**, *59*, 252–256. [CrossRef]
6. Nukoolkarn, V.; Lee, V.S.; Malaisree, M.; Aruksakulwong, O.; Hannongbua, S. Molecular dynamic simulations analysis of ritonavir and lopinavir as SARS-CoV 3CL(pro) inhibitors. *J. Theor. Biol.* **2008**, *254*, 861–867. [CrossRef]
7. Liu, X.; Zhang, B.; Jin, Z.; Yang, H.; Rao, Z. The crytal structure of 2019-nCoV main protease in complex with an inhibitor N3. *Nature* **2020**. [CrossRef]
8. Sanner, M.F. Python: A programming language for software integration and development. *J. Mol. Graph. Model.* **1999**, *17*, 57–61.
9. Pettersen, E.F.; Goddard, T.D.; Huang, C.C.; Couch, G.S.; Greenblatt, D.M.; Meng, E.C.; Ferrin, T.E. UCSF Chimera—A visualization system for exploratory research and analysis. *J. Comput. Chem.* **2004**, *25*, 1605–1612. [CrossRef]
10. Hanwell, M.D.; Curtis, D.E.; Lonie, D.C.; Vandermeersch, T.; Zurek, E.; Hutchison, G.R. Avogadro: An advanced semantic chemical editor, visualization, and analysis platform. *J. Cheminform.* **2012**, *4*. [CrossRef]
11. Morris, G.M.; Huey, R.; Lindstrom, W.; Sanner, M.F.; Belew, R.K.; Goodsell, D.S.; Olson, A.J. Automated docking with selective receptor flexibility. *J. Comput. Chem.* **2009**, *16*, 2785–2791. [CrossRef] [PubMed]
12. Trott, O.; Olson, A.J. AutoDock Vina: Improving the speed and accuracy of docking with a new scoring function, efficient optimization and multithreading. *J. Comput. Chem.* **2010**, *31*, 455–461. [CrossRef]
13. AndrejŠali, A.; Blundell, T.L. Comparative Protein Modelling by Satisfaction of Spatial Restraints. *J. Mol. Biol.* **1993**, *234*, 779–815.
14. Laskowski, R.A.; MacArthur, M.W.; Moss, D.S.; Thornton, J.M. PROCHECK: A program to check the stereochemical quality of protein structures. *J. Appl. Cryst.* **1993**, *26*, 283–291. [CrossRef]
15. Zhu, K.; Day, T.; Warshaviak, D.; Murrett, C.; Friesner, R.; Pearlman, D. Antibody structure determination using a combination of homology modeling, energy-based refinement, and loop prediction. *Proteins* **2014**, *82*, 1646–1655. [CrossRef] [PubMed]
16. Salam, N.K.; Adzhigirey, M.; Sherman, W.; Pearlman, D.A. Structure-based approach to the prediction of disulfide bonds in proteins. *Protein Eng. Des. Sel.* **2014**, *27*, 365–374. [CrossRef]
17. Beard, H.; Cholleti, A.; Pearlman, D.; Sherman, W.; Loving, K.A. Applying Physics-Based Scoring to Calculate Free Energies of docking for Single Amino Acid Mutations in Protein-Protein Complexes. *PLoS ONE* **2013**, *8*, e82849. [CrossRef]
18. Li, J.; Abel, R.; Zhu, K.; Cao, Y.; Zhao, S.; Friesner, R.A. The VSGB 2.0 model: A next generation energy model for high resolution protein structure modeling. *Proteins* **2011**, *79*, 2794–2812. [CrossRef]
19. Banks, J.L.; Beard, H.S.; Cao, Y.; Cho, A.E.; Damm, W.; Farid, R.; Felts, A.K.; Halgren, T.A.; Mainz, D.T.; Maple, J.R.; et al. Integrated Modeling Program, Applied Chemical Theory (IMPACT). *J. Comput. Chem.* **2005**, *26*, 1752–1780. [CrossRef]
20. Kitchen, V.S.; Skinner, C.; Ariyoshi, K.; Lane, E.A.; Duncan, I.B.; Burckhardt, J.; Burger, H.U.; Bragman, K.; Pinching, A.J.; Weber, J.N. Safety and activity of saquinavir in HIV infection. *Lancet* **1995**, *345*, 952–955. [CrossRef]
21. Baumann, M. An Overview of the Key Routes to the Best Selling 5-membered Ring Heterocyclic Pharmaceuticals. *Beilstein J. Org. Chem.* **2011**, *7*, 442–495. [CrossRef] [PubMed]
22. Kakkar, A.K.; Brenner, B.; Dahl, O.E.; Eriksson, B.I.; Mouret, P.; Muntz, J.; Soglian, A.G.; Pap, A.F.; Misselwitz, F.; Haas, S.; et al. Extended duration rivaroxaban versus short-term enoxaparin for the prevention of venous thromboembolism after total hip arthroplasty: A double-blind, randomised controlled trial. *Lancet* **2008**, *372*, 31–39. [CrossRef]
23. Terrett, N.K.; Bell, A.S.; Brown, D.; Ellis, P. Sildenafil (Viagra), a potent and selective inhibitor of type 5 cGMP phosphodiesterase with utility for the treatment of male erectile dysfunction. *Bioorg. Med. Chem. Lett.* **1996**, *6*, 1819–1824. [CrossRef]
24. Talpaz, M.; Shah, N.P.; Kantarjian, H.; Donato, N.; Nicoll, J.; Paquette, R.; Cortes, J.; O'Brien, S.; Nicaise, C.; Bleickardt, E.; et al. Dasatinib in imatinib-resistant Philadelphia chromosome-positive leukemias. *N. Engl. J. Med.* **2006**, *354*, 2531–2541. [CrossRef]

25. Kloner, R.A. Pharmacology and drug interaction effects of the phosphodiesterase 5 inhibitors: Focus on alpha-blocker interactions. *Am. J. Cardiol.* **2005**, *96*, 42M–46M. [CrossRef]
26. De Lepeleire, I.; Reiss, T.F.; Rochette, F.; Botto, A.; Zhang, J.; Kundu, S.; Decramer, M. Montelukast causes prolonged, potent leukotriene D4-receptor antagonism in the airways of patients with asthma. *Clin. Pharmacol. Ther.* **1997**, *61*, 83–92. [CrossRef]
27. Capaldini, L. Protease inhibitors' metabolic side effects: Cholesterol, triglycerides, blood sugar, and "Crix belly. Interview with Lisa Capaldini, M.D. Interview by John S. James". *AIDS Treatment News* **1997**, *277*, 1–4.
28. Park, N.H.; Shin, K.H.; Kang, M.K. Chapter 34. Antifungal and Antiviral Agents. In *Pharmacology and Therapeutics for Dentistry*, 7th ed.; Elsevier: Amsterdam, The Netherlands, 2017; pp. 488–503.
29. Ibraheem, J.J.; Paalzow, L.; Tfelt-Hansen, P. Low bioavailability of ergotamine tartrate after oral and rectal administration in migraine sufferers. *Br. J. Clin. Pharmacol.* **1983**, *16*, 695–699. [CrossRef]
30. World Health Organization. *Control of the Leishmaniasis: Report of a Meeting of the WHO Expert Committee on the Control of Leishmaniases*; World Health Organization: Geneva, Switzerland, 2010.
31. Liu, C.; Bayer, A.; Cosgrove, S.E.; Daum, R.S.; Fridkin, S.K.; Gorwitz, R.J.; Kaplan, S.L.; Karchmer, A.W.; Levine, D.P.; Murray, B.E.; et al. Clinical practice guidelines by the infectious diseases society of America for the treatment of methicillin-resistant Staphylococcus aureus infections in adults and children: Executive summary. *Clin. Infect. Dis.* **2011**, *52*, 285–292. [CrossRef]
32. Wang, W.; Chen, J.S.; Zucker, I.H. Carotid sinus baroreceptor sensitivity in experimental heart failure. *Circulation* **1990**, *81*, 1959–1966. [CrossRef]
33. Turpie, A.G. New oral anticoagulants in atrial fibrillation. *Eur. Heart J.* **2008**, *29*, 155–165. [CrossRef] [PubMed]

computation

MDPI

Article

In Silico Analysis of the Multi-Targeted Mode of Action of Ivermectin and Related Compounds

Maral Aminpour [1], Marco Cannariato [2], Jordane Preto [3], M. Ehsan Safaeeardebili [2], Alexia Moracchiato [2], Domiziano Doria [3], Francesca Donato [2], Eric Adriano Zizzi [2], Marco Agostino Deriu [2], David E. Scheim [4], Alessandro D. Santin [5] and Jack Adam Tuszynski [2,6,*]

1 Department of Biomedical Engineering, University of Alberta, Edmonton, AB T6G 1Z2, Canada; aminpour@ualberta.ca
2 DIMEAS, Politecnico di Torino, Corso Duca degli Abruzzi 24, 10129 Torino, Italy; marco.cannariato@studenti.polito.it (M.C.); s270360@studenti.polito.it (M.E.S.); s281425@studenti.polito.it (A.M.); s198066@studenti.polito.it (F.D.); eric.zizzi@polito.it (E.A.Z.); marco.deriu@polito.it (M.A.D.)
3 Centre de Recherche en Cancérologie de Lyon, Université Claude Bernard Lyon 1, INSERM 1052, CNRS 5286, 69008 Lyon, France; jordane.preto@gmail.com (J.P.); domiziano.doria@studenti.polito.it (D.D.)
4 US Public Health Service, Commissioned Corps, Inactive Reserve, Blacksburg, VA 24060-6367, USA; dscheim@alum.mit.edu
5 Obstetrics, Gynecology & Reproductive Sciences, Yale School of Medicine, P.O. Box 208063, New Haven, CT 06520-8063, USA; alessandro.santin@yale.edu
6 Department of Physics, University of Alberta, Edmonton, AB T6G 1Z2, Canada
* Correspondence: jackt@ualberta.ca

Abstract: Some clinical studies have indicated activity of ivermectin, a macrocyclic lactone, against COVID-19, but a biological mechanism initially proposed for this anti-viral effect is not applicable at physiological concentrations. This in silico investigation explores potential modes of action of ivermectin and 14 related compounds, by which the infectivity and morbidity of the SARS-CoV-2 virus may be limited. Binding affinity computations were performed for these agents on several docking sites each for models of (1) the spike glycoprotein of the virus, (2) the CD147 receptor, which has been identified as a secondary attachment point for the virus, and (3) the alpha-7 nicotinic acetylcholine receptor (α7nAChr), an indicated point of viral penetration of neuronal tissue as well as an activation site for the cholinergic anti-inflammatory pathway controlled by the vagus nerve. Binding affinities were calculated for these multiple docking sites and binding modes of each compound. Our results indicate the high affinity of ivermectin, and even higher affinities for some of the other compounds evaluated, for all three of these molecular targets. These results suggest biological mechanisms by which ivermectin may limit the infectivity and morbidity of the SARS-CoV-2 virus and stimulate an α7nAChr-mediated anti-inflammatory pathway that could limit cytokine production by immune cells.

Keywords: alpha-7 nicotinic receptor; CD147; docking; ivermectin; molecular modeling; SARS-CoV-2

Citation: Aminpour, M.; Cannariato, M.; Preto, J.; Safaeeardebili, M.E.; Moracchiato, A.; Doria, D.; Donato, F.; Zizzi, E.A.; Deriu, M.A.; Scheim, D.E.; et al. In Silico Analysis of the Multi-Targeted Mode of Action of Ivermectin and Related Compounds. *Computation* **2022**, *10*, 51. https://doi.org/10.3390/computation10040051

Academic Editors: Simone Brogi and Vincenzo Calderone

Received: 3 March 2022
Accepted: 16 March 2022
Published: 25 March 2022

1. Introduction

The spread of the SARS-CoV-2 pandemic around the world has spurred on a search for suitable drugs for therapeutic applications against this viral infection. Although vaccination is a proven strategy for containing coronavirus disease 2019 (COVID-19), a new challenge has developed due to the emergence of SARS-CoV-2 variants for which vaccines have offered lesser degrees of protection. Efforts have therefore been focused on the possibility repurposing existing, approved drugs, which do not require de novo design and lengthy testing.

Obstruction of the binding between SARS-CoV-2 spike protein and the ACE2 receptor on target human cells has been one focus of therapeutic intervention [1,2]. Although viral fusion to host cells and replication occur via ACE2, SARS-CoV-2 and several other viral strains, including other betacoronaviruses, initially attach to host cells via more abundantly-distributed glycoconjugate host binding sites [3,4]. Notable among these are sialic acid (SA) and the transmembrane glycoprotein receptor CD147, which are widely distributed in blood, endothelial and several other cells in the human body [4] and provide additional therapeutic targets. Nicotinic acetylcholine receptors, which are densely distributed in neuronal tissue and in cytokine-secreting inflammatory cells (for example macrophages and mast cells), may provide an additional attachment point for SARS-CoV-2 due to a "toxin-like" epitope on the viral spike protein and may thus provide druggable targets as well [5–7].

1.1. Binding of SARS-CoV-2 Spike Protein to Host Cell SA and CD147 Surface Molecules

The SARS-CoV-2 spike protein contains 22 N-linked glycosylation sites on each of its three monomers, with several of these associated glycans capped with terminal SA moieties in various forms [8–11]. Through those glycan bindings, the SARS-CoV-2 spike protein attaches to an SA-coated nanoparticle array, which is the basis for a sensitive viral detection technique [12]. Such a nanoarray is paralleled in densely distributed SA-tipped binding sites of glycophorin A molecules [13–15] and CD147 receptors [16–18] at the surface of red blood cells (RBCs), and SARS-CoV-2 was observed in the hemadsorption assay to clump with human RBCs [19]. Clinical confirmation of viral-RBC attachments was provided by the presence of SARS-CoV-2 spike protein punctae on 41% of RBCs from a series of hospitalized COVID-19 patients [20]. Binding between SARS-CoV-2 spike protein and the host cell receptor CD147 was likewise demonstrated by SPR, Co-IP and ELISA assays, and immuno-electron microscopy [21]. Meplazumab, a humanized anti-CD147 antibody, inhibited SARS-CoV-2 replication in vitro and reduced time to viral clearance from 13 to 3 days for COVID-19 patients in a small clinical study [21,22].

For viruses that bind to SA, including SARS-CoV-2, as noted above, that binding plays a key role in viral infectivity, as SA on host cells typically serves as the initial attachment point for viral spike protein [23–29]. But for such SA-binding viruses, the host limits viral attachment to host cell infectious targets through other entities in the body having SA-rich surfaces, including RBCs, platelets and leukocytes, along with mucins and plasma proteins [30,31]. Some viruses, in turn, dodge that defense through the expression of SA-cleaving enzymes that enable detachment from these blood cells and other snagging substances [28,31–38]. In particular, the SA-cleaving enzyme hemagglutinin esterase (HE) is expressed by the human betacoronaviruses that cause the common cold, OC43 and HKU1, but not by SARS-CoV, SARS-CoV-2 and MERS, the three deadly strains in that viral family [39–41]. It has been proposed that vascular occlusion, central to the morbidities of COVID-19 [42–47], is initially triggered by the clumping and snagging of SARS-CoV-2 with blood and endothelial cells, and that HE expressed by the common cold betacoronaviruses may limit these morbidities [4].

1.2. The Role of CD147 in the Inflammatory Response

In addition to its SARS-CoV-2 binding capability, CD147, a transmembrane glycoprotein receptor encoded in humans by the BSG gene [48], is of interest as a key mediator of inflammatory response, in particular, as related to vascular occlusion. In response to immunogenic stimuli, CD147 is upregulated in T cells [49,50], platelets [51,52] and endothelial cells [53], with upregulation of CD147 in endothelial cells occurring upon exposure to active or UV-deactivated betacoronavirus MHV-4 in vitro. CD147, in turn, has been observed to promote adhesion by RBCs [54–56], leukocytes [52,57–59] and platelets [57,58,60] to other blood cells and endothelial cells. Also, of particular interest are the indicated pro-infectious roles of CD147 and its binding partner cyclophilin A for SARS-CoV-2, SARS-CoV and other viruses [21,61–63]. In a broader clinical framework, the involvement of CD147 in the pathogenesis of a number of diseases, including lung inflammation, atherosclerosis, heart failure, ischemic myocardial injury and stroke [52,56,58,64,65], further suggests that CD147 antagonists or masking agents could mitigate a COVID-19 infection.

1.3. Competitive Binding of Ivermectin to SARS-CoV-2 Spike Protein Binding Sites

Given that the attachment of the SARS-CoV-2 spike protein to host cell targets, including ACE2, SA and CD147, is central to viral infectivity and morbidity, the capability for competitive binding to limit such attachments has been one focus in the search for repurposed COVID-19 therapeutics [3]. Four molecular modeling studies that collectively screened over 800 such molecules were conducted toward that goal [66–69]. The strongest or close to strongest binding affinity in each study was obtained for ivermectin, a macrocyclic lactone with multifaceted antiparasitic and antimicrobial activity which has been distributed in 3.7 billion doses worldwide since 1987 [70–73]. Additional molecular modeling studies of competitive binding to SARS-CoV-2 spike protein sites that focused on ivermectin in particular likewise found strong binding affinities for that agent [74–79].

These findings are of interest given clinical, animal and epidemiological studies, including most of the 20 randomized clinical trials (RCTs) conducted to date, indicating the efficacy of ivermectin against COVID-19 [70,80,81], although interpretations of which of these RCTs are most reliable have been controversial. Ivermectin is suitable for mass use on a global scale, having been the mainstay of two worldwide campaigns to eliminate two devastating scourges affecting millions, onchocerciasis and lymphatic filariasis [82]. It is safe even at much higher doses than the standard dose of 200 μg/kg [83,84], and its limited side effects were noted in the Nobel Committee's 2015 award honoring its discovery and its record of improving the health and wellbeing of millions [85]. However, a biological mechanism initially proposed for ivermectin activity against SARS-CoV-2, entailing blockage of its transport into the host cell nucleus, was proposed in conjunction with in vitro studies conducted at much greater than physiological concentrations and has been questioned [86–88].

1.4. Nicotinic Acetylcholine Receptors: Anti-Inflammatory Modulation and Blockage of Viral Bindings

Another biological mechanism of activity that may underlie the observed clinical benefits of ivermectin treatment of COVID-19 is a potent anti-inflammatory and immune modulatory effect mediated by its action as a positive allosteric modulator of the alpha-7 nicotinic acetylcholine receptor (α7nAChr) [89]. The core receptor of the cholinergic anti-inflammatory pathway is α7nAChr, which is under the control of the vagus nerve [90] and plays a crucial role in balancing of the body's response to inflammation and sepsis [90,91]. This anti-inflammatory pathway connects the involuntary parasympathetic nervous system innervating all major organs to cytokine-producing cells such as TNF, IL1 and IL6-secreting macrophages, lymphocytes and mast cells [90,91], which are reported to play a major role during the inflammatory phase of COVID-19 infection (i.e., the cytokine storm [92]). The ivermectin-induced enhancement of this pathway might rapidly lower pro-inflammatory cytokine levels and decrease expressions of chemokines as well as adhesion molecules at the inflammatory sites [90,91]. Importantly, the marked increase in Ca++ current evoked by acetylcholine (ACh) in the presence of micromolar concentrations of ivermectin (e.g., a 20-fold shift of the affinity of ACh [89]) may also potentially explain the reported clinical activity of ivermectin during the late (i.e., inflammatory), critical phase of severe COVID-19 cases [93].

Recent in silico docking studies have indicated a potential direct interaction between the SARS-CoV-2 spike glycoprotein and α7nAChr, due to a "toxin-like" epitope on the spike glycoprotein, with homology to a sequence of a snake venom toxin [5,6]. Of interest, the α7nAChr receptor, which is densely distributed on neuronal tissue, has previously been shown to serve as the port of entry in the human body for another RNA virus endowed with strong neurotropic action, the rabies virus [7]. The loss of smell (anosmia) and/or taste (ageusia) are considered hallmarks of COVID-19 infection and are likely consequences of the direct SARS-CoV-2 infection of the olfactory and gustatory nerve [94]. Ivermectin high affinity binding to α7nAChr may therefore interfere with the attachment and internalization of SARS-CoV-2 on the olfactory/gustatory nerves, as recently reported in both animal models [94] and human patients [95].

1.5. Subdomains of the SARS-CoV-2 Spike Protein, S1 Region

The SARS-CoV-2 spike protein pockets selected for study as potential ivermectin binding sites were governed by the arrangement of subdomains of interest. The SARS-CoV-2 spike protein is a heavily glycosylated, type I transmembrane protein with 1273 amino acid residues, assembled into trimers and attached on the virion surface, giving the virus its distinctive "corona" or crown-like appearance. Each spike protein trimer contains a central helical stalk consisting of three joined S2 subunits, each capped with an S1 subunit head in a mushroom-like shape [96,97]. The ectodomain of the spike protein S1 attaches to a host cell membrane, after which the S2 stalk engages in fusion, enabling the internalization and replication of the virus [96–99].

That viral-host engagement proceeds, in particular, through fusion of the receptor binding domain (RBD) of the SARS-CoV-2 spike protein S1 to an ACE2 receptor on a host cell [97–99]. As shown in Figure 1, the S1 N-terminal domain (NTD) contains eight of the 22 N-linked glycans on the SARS-CoV-2 spike protein. These eight N-linked glycans on the NTD form initial attachments to host cells' glycoconjugates, including CD147 and others which have SA terminal residues [8,18,100–104]. The RBDs of the virus, one on each spike protein monomer S1 subunit, switch constantly between open ("up") and closed ("down") configurations, with the former enabling both ACE2 binding and immune surveillance and the latter blocking both of those functions [96,105].

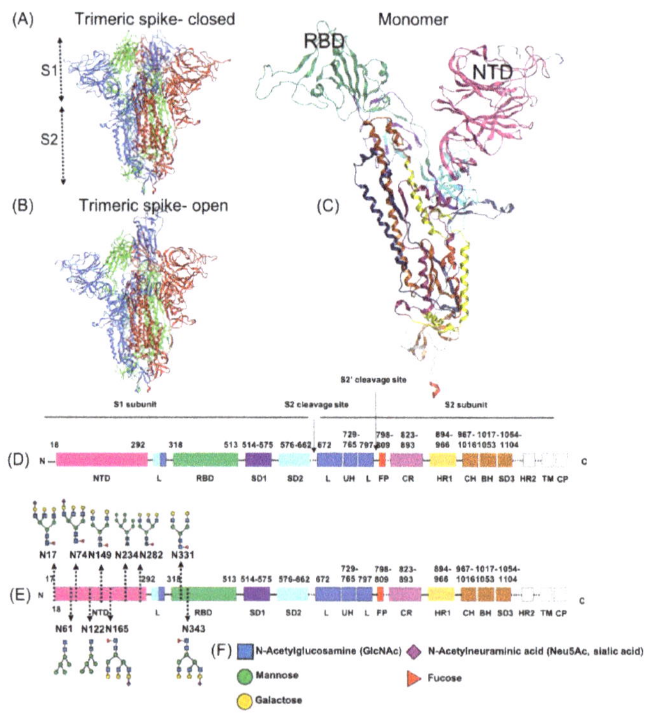

Figure 1. (**A**,**B**) Side view of SARS-CoV-2 spike protein trimer, open (PDB: 6VSB) and closed (PDB: 6VXX) configurations, respectively. Heptad repeats 2 (HR2), HR2 linker, transmembrane domain

(TM), and cytoplasmic domain (CP) is omitted in the S2 subunit represented in A. (**C**) Spike protein monomer color-coded by subdomain. (**D**) Sequence of full-length spike protein with domain assignments, with N-terminal end to the left and C-terminal (stalk) end to the right. (**E**) N-linked glycans are shown, localized in the schematic representation by arrows, for the NTD and RBD domains only. (**F**) A key to the monosaccharides depicted in (**E**). C-E are from Aminpour et al., 2021 [3] (CC-BY 4.0); added glycan representations in E and the glycan key, F, are from Sikora et al., 2021 [106] (CC-BY 4.0).

2. Materials and Methods

2.1. Ligand Database Preparation

A set of 15 test compounds, composed of ivermectin and 14 similar molecules, was collected from the PubChem database and used for the docking studies. Compounds structurally similar to ivermectin were adopted from DrugBank Online under the "similar structures" section of the drug ivermectin [107]. This option provides users the capability to search rapidly for structurally similar small molecules, without having to redraw the molecule and perform additional database searches through the ChemQuery interface. Before evaluating any interaction, the ligand set database was prepared through a "Wash" wizard of the Molecular Operating Environment (MOE) software package. At this stage, the 3D dominant protonation state of each molecule was generated at the physiological pH of 7, followed by a short MOE built-in energy minimization procedure.

2.2. Protein Preparation

The crystal structure of CD147 was obtained from its Protein Data Bank (PDB: 3B5H). Chain A has the strongest electron density and thus was used for analysis. The CHARMM-GUI Archive of COVID-19 Proteins Library [108] was used to collect the structures of the two spike protein conformations, i.e., the closed (PDB: 6VXX) and open (PDB: 6VSB) states [9,109,110]. The NTD (aa 18–292) and RBD (aa 318–513) of one monomer were considered separately in the following analysis. The atomic coordinates of three possible conformations of α7nAChr, namely resting (PDB: 7KOO), desensitized (PDB: 7KOQ) and activated (PDB: 7KOX), were obtained from the PDB [111]. Only the extracellular region of the protein (aa 1–207) was considered in the docking analysis. All proteins were prepared in MOE, adjusting their protonation state according to a physiological pH of 7 and minimizing the potential energy.

2.3. Binding Sites

We employed the Site Finder module in MOE [112] to detect the possible binding sites in the NTD and RBD domains of the spike protein. All the sites we identified using the MOE software had already been reported in the literature, as we summarized them below, so we performed our molecular modeling calculations using the sites specified in Table 1. We manually calculated the center of the binding sites from the residues involved.

Several sialoside-, glycosylation- and ganglioside- binding sites have been reported in the literature. Milanetti et al. [103] proposed a potential sialoside binding site containing three divergent loop regions (site 1). They supported the hypothesis of a structural resemblance between MERS-CoV and SARS-CoV-2 using iso-electron density mapping. Behloul et al. [113] compared the structural features of the SARS-CoV-2 spike protein S1-NTD with BCoV and consequently characterized a binding pocket that has the capability to bind SA species such as Neu5,9Ac2 (site 2). Baker et al. [12] aligned the sequences of the SARS-CoV-2 spike protein, mainly focusing on human coronavirus OC43 as the SA-binding protein. They identified a potential SA binding site, associating its glycan-binding characteristic utilizing glyconanoparticles for the detection (site 3). Gaetano et al. [114] calculated the druggability of all available ligand-binding pockets within the NTD segment of the spike protein S1 using SiteMap of Schrodinger software [115]. As a result, among all of the three hypothesized sialoside-binding pockets in the literature, site 3 by Baker et al. [12] is part of a cavity with a druggable property identified by Gaetano et al. (site 4 in Table 1, or site P1 as Gaetano et al. referenced in their paper [114]). Gaetano et al. also identified an unexpected binding pocket (site 5 in Table 1, or P2 as they referenced in their paper) within S1-NTD. Site 5 (P2) aligns with the recent experimental findings by Bangaru et al. [116].

Table 1. Binding sites of spike NTD and RBD obtained from the literature (as reproduced from Table 3 in Aminpour et al. [3], with reference citation numbers adjusted to this reference list).

Binding Site	Reference	Binding Site Type	Residues	NTD/RBD
Site 1	Milanetti et al. [103]	sialoside	L18-Q23, H66-T78 and G252-S254	NTD
Site 2	Behloul et al. [113]	sialoside	E154, F157, Y160 and the so-called stabilizing loop (N122-N125)	NTD
Site 3	Baker et al. [12]	sialoside	(R21, Q23, L24, H69, F79, P82 and R246)	NTD
Site 4 (P1)	Di Gaetano et al. [114]	sialoside	R21, T22, Q23, L24, P26, R78, P82, V83, L110, F135, C136, N137 and R237	NTD
Site 5 (P2)	Di Gaetano et al. [114]	sialoside	F92, S94, E96, K97, S98, R102, N121, V126, I128, M177, D178, K182, N188, R190, F192, I203, L226, V227 and L229.	NTD
Site 6–14	Watanabe et al. [11]	glycosylation	N122, N149, N165, N17, N61, N74, N234, N282	NTD
Site 15	Fantini et al. [101]	ganglioside	Domain (111–158)- core Q-134 to D-138	NTD
Site 16	Carino et al. [117]	-	F342 N343 A343 T345 R346–W436 N437 S438–L441 D442 S443–G446–N448–Y451 L452	RBD
Site 17	Carino et al. [117]	-	S375–G404 D405–V502 G503–Q506–Y508	RBD
Site 18	Carino et al. [117]	-	E340 V341–F347 A348–N354 R355 K356–S399 F400 V401–V512	RBD
Site 19	Carino et al. [117]	-	F374–N388–Y495 G496 F497	RBD
Site 20	Carino et al. [117]	-	T376 F377 K378 C379 Y380–V407 R408–I410–V433 I444 A445	RBD
Site 21–22	Watanabe et al. [11]	glycosylation	N331–N3443	RBD

Sites 6 to 14 are associated with the glycosylation binding sites proposed by Watanabe et al. [11]. Fantini et al. [101], meanwhile, proposed a new type of ganglioside-binding domain performing molecular dynamics (MD) calculations. The results of his simulations reveal a strong interaction between GM1 ganglioside and S1-NTD (site 15). Finally, Carino et al. [117] utilized the Fpocket server (https://bioserv.rpbs.univ-paris-diderot.fr/services/fpocket/, accessed 22 February 2022) and computationally identified sites 16 to 20 in the RBD fragment of the spike protein. They also studied the binding of several triterpenoids (e.g., glycyrrhetinic and oleanolic acids) and natural bile acids and demonstrated that their semisynthetic derivatives can reduce RBD adhesion to ACE2 in vitro. Sites 21 to 22 belong to the set of glycosylation binding sites proposed by Watanabe et al. [11].

Nine different binding sites in the CD147 dimer identified by MOE (Site Finder) are listed in Table S1 and are illustrated in Figure S1. The three potential *N*-glycosylation sites of CD147 are N44 (site 8), N152 (site 3) and N186 (site 3).

The putative binding pockets of the α7nAChr (desensitized, activated, resting) and CD147 structures were identified using the Site Finder tool in MOE, which computes the possible binding sites of a protein from its 3D structure using a geometrical approach. For each protein, only the sites characterized by a propensity for ligand binding (PLB) greater than or equal to 1 were considered in the docking experiments. The α7nAChr protein is a pentamer with a five-fold symmetry. To have a clear presentation, we summarized the common binding sites between monomers of the α7nAChr protein, excluding the common ones, in Table S2. In total, 37 binding sites with PLB > 1 were identified in all the three conformations of the α7nAChr protein. The binding sites identified by MOE (Site Finder) related to Table S2 are illustrated in Figure S2.

2.4. Molecular Docking Simulations

Docking was performed with a flexible ligand and a rigid receptor approach using the AutoDock Vina program [118] to predict the binding pose of the ligands. In the AutoDock Vina software, receptor–ligand binding affinities were predicted as negative Gibbs free energy (ΔG) values (kcal/mol), which were calculated on the basis of the AutoDock Vina scoring function and classified on the basis of a numerical value referred to as the "Score". The interactions of inhibitors with receptor proteins are predicted on the basis of the Score; the lower the Score (in negative value), the greater the interaction. The Vina scoring function incorporates two features from knowledge-based and empirical potentials. A cubic box with 30.0 Å size, required to delimit the docking area, was used on each binding pocket, centered at their center of geometry. The maximum number of poses to be generated for each docking calculation was set to 20. The minimum root mean square deviation (RMSD) to distinguish between two different poses was 1 Å. Every generated pose was energy-minimized in vacuo using Amber16 by keeping the protein fully rigid [119], with out of box poses then being discarded. Finally, the Vina Score function was used to re-score the poses after the minimization and the pose with the best Score was selected for each compound–receptor pair. The DockBox package was used to facilitate the preparation of docking inputs, the post-processing of the docking results and the rescoring procedure [120]. No constraints were applied in the docking studies. Although we minimized the ligand–protein structures after docking, we double checked the stability of compounds by running 100 ns MD calculations in explicit solvent on the unrestrained ligand–protein complex (see Section 2.5).

To the best of our knowledge, there are no effective therapeutics for COVID-19 which have biological mechanisms similar to those indicated for our 15 test compounds to compare with our docking results, which could be checked for competitive binding to the spike protein or the other host receptors. Therefore, we have the limitation of not being able to usefully check these results against known controls. In order to evaluate docking parameters for a given target prior to undertaking docking calculations on unknown ligands, however, it is always beneficial to perform control docking if the binding of known ligands is available in the crystal structure and if they have a non-covalent nature. Therefore, here, we also performed positive control docking calculation for the ligands that were experimentally available in the crystal structure of the proteins that we used in our study. It was not possible to use the NAG (N-acetyl-D-glucosamine) ligand (PDB: 6VSB) of the spike protein as a positive control since the nature of the binding was covalent. Also, we were not able to perform control docking on the CD147 protein (PDB: 3B5H) since there was no known ligand available in the crystal structure. The ligand Epibatidine (PDB:7KOX) of the alpha-7 nicotinic acetylcholine receptor was used as a control. We were able to successfully generate the same pose (RMSD = 1.2 Å) with a binding affinity of −8.73 kcal/mol (see Figure S3). We used decoys, which are molecules that are physically similar yet chemically dissimilar to the active ligands [121], as a negative control for the docking calculations. We used a state-of-the-art benchmark, the Directory of Useful Decoys (http://dude.docking.org, accessed 22 February 2022), to select decoys for ivermectin [122,123]. The structures of the decoy compounds are presented in the Supplementary Information (Figure S4). The binding affinities of the decoy compounds for the spike protein S1, CD147 and α7nAChr binding sites are in the range of (−3.345 to −5.496 kcal/mol), (−4.217 to −5.137 kcal/mol) and (−4.940 to −6.070 kcal/mol), respectively. The decoy compounds exhibited lower affinities than ivermectin and the most related compounds.

2.5. Molecular Dynamics (MD) Simulations

In order to establish the stability of each docked protein-inhibitor complex, MD simulations were run in explicit solvent using the Amber16 software. The computationally intensive all-atom MD simulations of the Open (6x) and Closed (3x) systems (each tallying ~1.7 million atoms including explicit water, ions and membrane lipids) that was done by another research group on Texas Advanced Computing Center (TACC) achieved benchmarks of ~60 ns/day on 256 GPU nodes [124]. To reduce the computation time, for NTD binding ligands, protein spike S1 was truncated from S698 to D1146, and from P322 to C590. For RBD binding ligands, protein spike S1 is truncated from M1 to E324 and from C590 to D1146. The hydrophobic part of the α7nAChr protein (T207 to L320) was removed in each monomer to prevent the exposure of the hydrophobic area in water. The breaks in all of the structures were capped with MOE's Structure Preparation. All the residues of protein CD147 were kept. MD simulations were carried out on Compute Canada's Graham cluster (V100 GPUs), as well as Cedar (P100 NVIDIA GPUs), depending on their respective availability. Each simulation was carried out on a single GPU. Using the AmberTools 16′ leap program, each complex was solvated in a cubic box with a side length of 12 Å using a three-points (TIP3P) water model. Na^+ and Cl^- ions were added in such a way to adjust the salt concentration to the physiological value of 0.15 M and neutralize the system. The minimization of the complexes was achieved in two steps, using the steepest descent (5000 steps) and conjugate gradient (5000 steps) methods successively. At first, only solvent atoms were minimized, by restraining the protein–ligand complex. Next, the minimization was run with the same parameters without the restraint. After the minimization step, the MD simulations were conducted in three stages: heating, density equilibration and production. At first, each solvated system was heated to 298 K for 500 ps, with weak restraints on all backbone atoms. Next, density equilibration was carried out for 1 ns of constant pressure equilibration at 298 K, with weak restraints. Finally, MD production (one trajectory per complex) were performed without any restraints for all systems for 100 ns. The trajectory of the ligand–protein complex was visually investigated using the VMD package (the University of Illinois at Urbana-Champaign, Urbana, IL, USA). Time-evolutions of the RMSD of top-ranked inhibitors with respect to receptors (spike, CD147 and α7nAChr) were calculated using the CPPTRAJ module of the AMBER16 software. Clustering analysis was carried out on the protein-bound ligand poses where the trajectory reached a plateau using Amber's CPPTRAJ program [67]. Consequently, the representative pose selected from the dominant cluster was considered as a predicted ligand pose.

2.6. Ligand Interaction Fingerprint

The Protein-Ligand Interaction Fingerprint application in the MOE software [112] was used to outline the interactions between ligands and proteins with a fingerprint scheme. Interactions such as hydrogen bonds, ionic interactions and surface contacts are classified in accordance with the residue of the origin and built into a fingerprint scheme which is representative of a given database of protein–ligand complexes.

3. Results

3.1. Molecular Docking Analysis

We docked the 15 test compounds on the SARS-CoV-2 spike protein (open and closed conformations), CD147 and α7nAChr (activated, desensitized and resting states). All docking scores and binding sites are reported in Tables 2 and 3. In the following sections, we discuss the top five spike and α7nAChr and the top six CD147 protein inhibitors, as well as the common inhibitors within the top-ranked inhibitors for all the receptors.

Table 2. Results of the docking analysis for spike protein S1 binding sites on NTD and RBD in open and closed positions. Scores listed are maximum absolute values for the sites listed in Table 1 for NTD or RBD, open or closed, with the maximum for all four combinations shown in column 2. Compounds are sorted in descending order of that maximum |Score| (column 2).

Compound Name	Maximum |Score|		Open				Closed			
			NTD		RDB		NTD		RBD	
	Score (kcal/mol)	At Site	Score (kcal/mol)	Site	Score (kcal/mol)	Site	Score (kcal/mol)	Site	Score (kcal/mol)	Site
Ivermectin	−8.948	NTD-open site 10	−8.948	site 10	−8.256	site 17	−8.205	site 4	−7.735	site 22
Moxidectin	−8.902	NTD-open site 2	−8.902	site 2	−8.218	site 21	−7.659	site 2	−7.989	site 18
Doramectin	−8.885	NTD-open site 2	−8.885	site 2	−8.144	site 21	−8.867	site 9	−8.216	site 19
Oleandrin	−8.787	RBD-closed site 19	−7.787	site 10	−8.051	site 22	−8.083	site 14	−8.787	site 19
Selamectin	−8.774	NTD-closed site 10	−8.476	site 15	−7.432	site 19	−8.774	site 10	−8.142	site 16
Okadaic acid	−8.716	NTD-open site 10	−8.716	site 10	−8.067	site 21	−7.937	site 4	−8.25	site 18
Gitoformate	−8.514	NTD-open site 10	−8.514	site 10	−7.669	site 21	−7.88	site 10	−7.992	site 19
Amphotericin_B	−8.304	NTD-open site 15	−8.304	site 15	−7.516	site 21	−7.931	site 4	−7.332	site 21
P-57AS3	−8.045	NTD-open site 4	−8.045	site 4	−7.663	site 22	−7.704	site 5	−7.627	site 19
Eprinomectin	−7.646	NTD-open site 6	−7.646	site 6	−7.584	site 21	−7.088	site 6	−7.302	site 21
Concanamycin A	−7.564	NTD-open site 10	−7.564	site 10	−7.335	site 19	−7.347	site 3	−7.302	site 21
Natamycin	−7.529	RBD-open site 21	−7.388	site 13	−7.529	site 21	−7.359	site 4	−6.87	site 18
Nystatin	−7.333	RBD-open site 21	−7.226	site 6	−6.845	site 21	−6.867	site 14	−6.773	site 19
beta-Escin	−7.324	NTD-open site 10	−7.324	site 10	−7.333	site 21	−7.264	site 4	−7.296	site 19
Fusicoccin	−6.705	NTD-open site 2	−6.705	site 2	−6.123	site 22	−6.353	site 10	−6.381	site 18

Table 3. Results of the docking analysis on CD147 and α7nAChr. Compounds are sorted in descending order according to |Score| separately for CD147 and α7nAChr.

CD147			α7nAChr		
Compound Name	Score (kcal/mol)	Site	Compound Name	Score (kcal/mol)	Site
Okadaic acid	−8.578	site 5	Ivermectin	−10.636	Activated site 2
Doramectin	−8.253	site 1	Doramectin	−10.243	Activated site 2
Selamectin	−8.082	site 5	Okadaic acid	−10.240	Activated site 2
P-57AS3	−8.010	site 1	Moxidectin	−10.142	Resting site 1
Concanamycin A	−7.847	site 9	Concanamycin A	−9.932	Activated site 2
Ivermectin	−7.527	site 5	P-57AS3	−9.799	Desensitized site 3
Amphotericin_B	−7.481	site 1	Gitoformate	−9.794	Resting site 1
Moxidectin	−7.469	site 1	beta-Escin	−9.711	Resting site 3
Oleandrin	−7.434	site 4	Natamycin	−9.611	Activated site 1
Gitoformate	−7.297	site 8	Oleandrin	−9.465	Activated site 2
Nystatin	−7.038	site 9	Selamectin	−9.397	Activated site 2
Eprinomectin	−6.827	site 9	Nystatin	−9.214	Resting site 3
beta-Escin	−6.755	site 1	Eprinomectin	−8.968	Resting site 3
Natamycin	−6.739	site 7	Fusicoccin	−8.814	Resting site 3
Fusicoccin	−5.872	site 1	Amphotericin_B	−8.811	Resting site 3

All individual docking scores at all sites for ivermectin on the spike, CD147 and α7nAChr are presented in Table S3, Table S4 and Table S5, respectively.

3.2. Selection of the Most Promising Compounds

The top five inhibitors, in descending order of the absolute value of the Score, were found to be as follows: for the spike protein: ivermectin, moxidectin, doramectin, oleandrin and selamectin; for CD147: okadaic acid, doramectin, selamectin, P-57AS3, concanamycin A and ivermectin; and for α7nAChr: ivermectin, doramectin, okadaic acid, moxidectin and concanamycin A. The common inhibitors within the five top-ranked inhibitors were, for the spike and α7nAChr: ivermectin, doramectin, okadaic acid and moxidectin; for the spike and CD147: doramectin; and for CD147 and α7nAChr: okadaic acid, doramectin and concanamycin A. The only common top inhibitor for all the receptors was doramectin. The majority of compounds bound to the spike NTD, while the highest affinity could be observed towards the open conformation. As for α7nAChr, the activated and resting states were preferred with respect to the desensitized state. Moreover, compounds had higher affinities to the activated state of α7nAChr.

All the top inhibitors considered, with the exception of oleandrin, were found to bind to S1-NTD. Both ivermectin and selamectin bound to site 10 of S1-NTD (Figure 2A,C), which is a glycosylation binding site (N61). Moxidectin and doramectin bound to site 2 S1-NTD (Figure 2A), which is a sialoside binding site proposed by Behloul et al. [113]. Oleandrin bound to site 19 of S1-RBD proposed by Carino et al. [117] (Figure 2B). Site 19 includes the N388 glycosylation binding site.

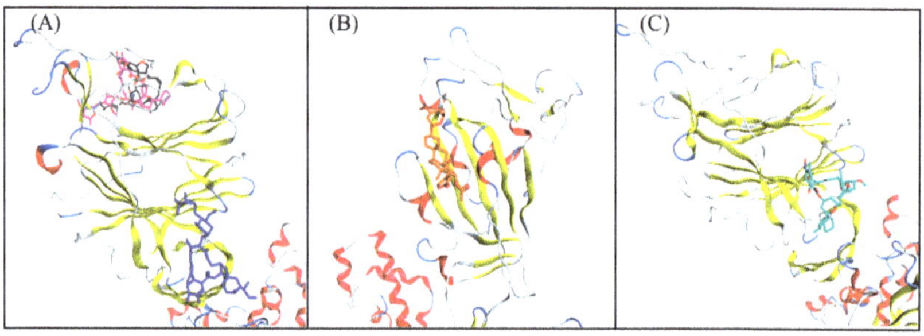

Figure 2. Binding poses of (**A**) ivermectin (dark blue), moxidectin (dark gray), doramectin (purple) on S1-NTD open conformation; (**B**) oleandrin (orange) on S1-RBD closed conformation; (**C**) selamectin (cyan) on S1-NTD closed conformation.

The binding poses of all the compounds with high affinity for CD147 are shown in Figure 3. Okadaic acid, selamectin and ivermectin were found to bind to site 5, which is located in domain A of CD147 protein. Doramectin and P-57AS3 were found to bind to site 1 of CD147, which is in the interface of domain 1 and domain 2 of CD147. Concanamycin A was found to bind to site 9 of CD147.

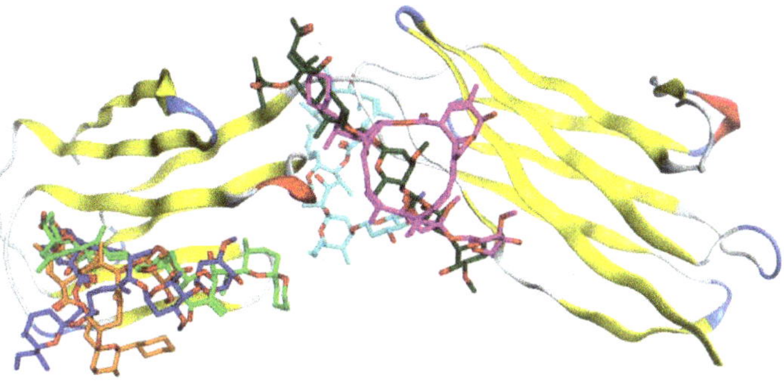

Figure 3. Binding poses of okadaic acid (green), doramectin (purple), selamectin (orange), P-57AS3 (dark green), concanamycin A (cyan) and ivermectin (dark blue) on CD147.

The binding poses of all the compounds with high affinity for α7nAChr are shown in Figure 4. Ivermectin, doramectin, okadaic acid and concanamycin A were found to bind to site 1 of the activated conformation of α7nAChr (Figure 4A,C). Moxidectin was found to bind to site 1 of the resting conformation of α7nAChr (Figure 4B,D). In what follows, the interactions of the top inhibitors for the spike, CD147 and α7nAChr will be discussed.

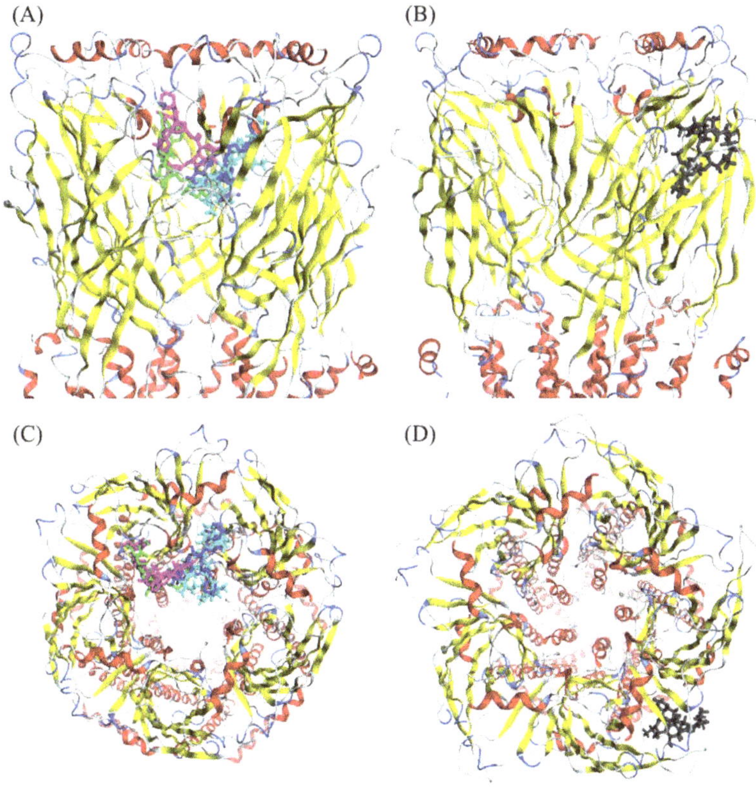

Figure 4. Binding poses of ivermectin (dark blue), doramectin (purple), okadaic acid (green), concanamycin A (cyan) on α7nAChr (**A**) side and (**C**) top view. Binding pose of moxidectin (dark gray) on (**B**) side and (**D**) top view.

3.3. Molecular Dynamics Simulations and RMSD Analysis

A 100ns-long MD simulation was performed to check the stability of each protein-inhibitor complex and to discriminate between stable and unstable docked poses. The top-docked pose (with the lowest docking Score) for each protein–ligand complex was used as an initial structure for the simulations. The binding stability was assessed by following the time evolution of the ligand RMSD in each trajectory, where we used the starting structure as a reference and RMSD alignment was carried out on protein atoms.

From the RMSD analysis and a visual inspection of MD trajectories, we found that, except for salemectin, all of the top five compounds in complex with the spike protein were relatively stable, reaching a RMSD plateau between 2 Å and 4 Å (Figure S5). Conversely, CD147 went through hinge movements during MD (Figure S6), which made it difficult to align the structures and caused fluctuations and higher RMSD values. Visual inspections and ligand–protein interaction analysis (Section 3.4) confirmed that all compounds, except for okadaic acid, maintained their binding to the same binding site during MD simulations (2 Å < RMSD < 6 Å) (Figure S6). Regarding α7nAChr, a common behaviour was observed for almost all of the compounds: before MD, binding to α7nAChr occurred through the interaction between the disaccharide group of each ligand and the activated site 2 of α7nAChr inside the pore (except for moxidectin, which bound to the outer wall of α7nAChr). Benzofuran and spiroketal groups were pointed toward the center of the pore, with no apparent hydrogen bonds with any residue. After conducting MD simulations, the stable structure of compounds tended toward a conformation that maintained its binding with activated site 2, with extra binding through the benzofuran group, by getting close to the pore wall. Ivermectin, okadaic acid and moxidectin manifested a stable RMSD (1.5 < RMSD < 4) (Figure S7). An abrupt shift in the RMSD of doramectin was due to the detachment of the benzofuran group from one monomer and the attachment to another monomer due to the symmetry of the α7nAChr protein. The new conformation still bound, through disaccharide, with the same binding site, and it was as stable as the first conformation. During visual inspection and through ligand–protein interactions, it was confirmed that concanamycin underwent major binding adjustments with regard to its initial docked conformation and ended up leaving the binding site.

3.4. Analysis of the Protein–Ligand Interactions

In stable MD trajectories, the top representative pose of each compound was selected from the populationally dominant cluster using clustering analysis on all the trajectories for further ligand–protein interaction analysis. The Protein-Ligand Interaction Fingerprint module of MOE was used to summarize the interactions between ligands and proteins with a fingerprint scheme. N61, R415, F157 and D40 emerged as main residues of the spike protein due to their interaction with high-affinity compounds (Figure S8). As for CD147, the residues interacting with the selected compounds were L46, K87, R85 and H32 (Figure S9). In case of α7nAChr, four out of the five selected compounds bound to activated site 2 and interacted with P16, N106, W85 and N100, that are exposed on the interior surface of the protein channel. One compound, namely moxidectin, interacted with N110 of resting state α7nAChr, which is exposed on the outer surface of the protein (Figure S10).

The interaction mechanisms of ivermectin with the SARS-CoV-2 spike protein, CD147 and α7nAChr were analysed using MOE software. Binding energies were obtained through the GBVI/WSA forcefield-based scoring function, which uses the AMBER99 forcefield to compute electrostatic, solvation, van der Waals and surface area contributions to the free energy given the ligand pose. Two to four hydrogen bond acceptor interactions were characterized in the best pose of the compounds in all receptors.

Ivermectin remained in the same binding site for all the receptors during MD simulations (2 Å < RMSD < 4 Å). In case of the spike protein (RMSD~2.5 Å), N61 (the main residue of the glycosylation site 10) were involved, with a binding energy of −2.9 kcal/mol, with the benzofuran group of ivermectin and R415 were involved with the lactone group of ivermectin, with a binding energy of −2.9 kcal/mol (Figure 5A).

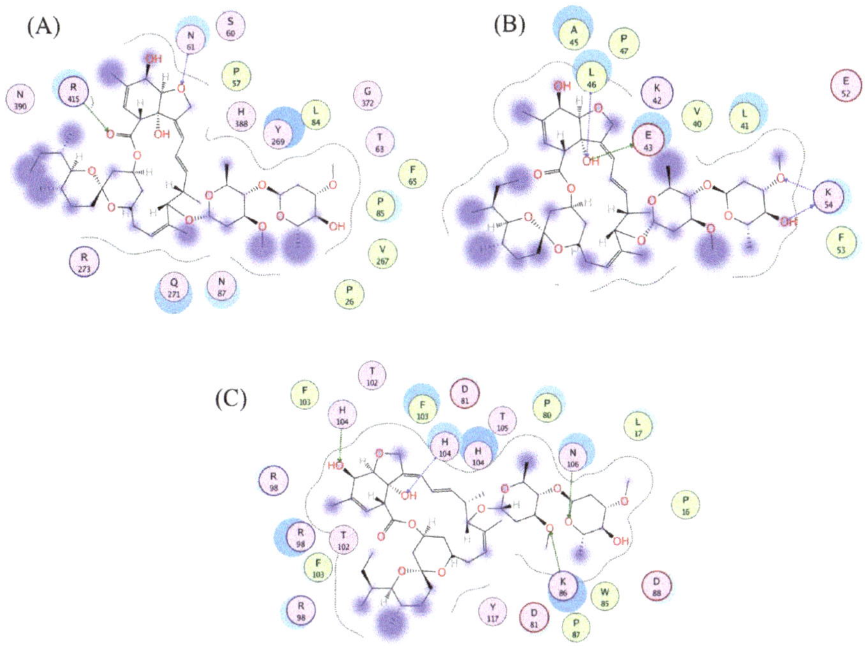

Figure 5. Ligand interaction plots of ivermectin for (**A**) spike, (**B**) CD147 and (**C**) α7nAChr inhibition.

During MD simulations, CD147 went through hinge movements and gave rise to a relatively higher (RMSD < 6 Å) value for ivermectin. Ivermectin stayed stable after 60 ns and strongly bound to CD147 through its disaccharide group, featuring E43 and K54 residues with −2.7 kcal/mol and −4.1 kcal/mol of binding energies, respectively, and a lactone core group featuring L46 residue with −1.1 kcal/mol of binding energy (Figure 5B).

As for α7nAChr, strong hydrogen bond acceptor interactions were found with K86 (−6.8 kcal/mol of binding energy) and N106 (−3 kcal/mol of binding energy). Moreover, it was characterized by an additional hydrophobic interaction with H85 (−2.7 kcal/mol of binding energy) (Figure 5C). In addition to maintaining disaccharide group binding with α7nAChr through K86 and N106, the equilibrated structure formed an extra binding to α7nAChr through its benzofuran group with H85 compared to the initial docking pose. Ivermectin maintained its attachment to α7nAChr at the same binding site with (RMSD < 4 Å). The presence of the same type and number of interactions in the analyzed proteins may support the hypothesis of a multi-targeted action of ivermectin.

A summary of the amino acid mutations of the SARS-CoV-2 Alpha, Beta, Gamma and Delta variants with a focus on the spike protein is presented in Table S6. We do not expect that the given variant will have a significant effect on the binding of the selected compounds, considering that the mutations are not directly involved in the binding sites of ivermectin and related compounds. However, it is noteworthy to mention that allosteric interactions should be taken in consideration for a comprehensive and accurate evaluation.

3.5. Bioactivity of the Test Agents with Greatest Binding Strength

By Lipinski's rule of five, agents with a molecular mass greater than 500 would tend to be suboptimally bioactive as oral agents. However, although among these test agents, ivermectin and doramectin, for example, have molecular masses of 875.1 and 899.1, respectively, both are well-absorbed with similar pharmacokinetics [125]. Ivermectin, in particular, is distributed throughout the human body within eight hours of oral administration [83,126,127], and its success in combatting diseases affecting hundreds of millions of people is well established [70].

3.6. Protein-Protein Interactions

The spike (PDB: 6VSB for open conformation) and α7nAChr (PDB: 7KOX) initial structures were obtained from the RCSB Protein Data Bank. PatchDock software (bioinfo3d.cs.tau.ac.il/PatchDock/ accessed on 2 March 2022) was used for protein–protein docking simulations [128,129]. PatchDock is a geometry-based molecular docking algorithm aimed at finding docking transformations that yield good molecular shape complementarity. Each candidate model is further evaluated by a scoring function that considers both the atomic desolvation energy and the geometric fit. The results obtained from PatchDock were further refined with the associated server FireDock, which delivers a further refinement of both the score function and of the complexes' geometries. We present the highest scoring structure in Figure 6. The best docking pose indicates the interaction between two RBD segments of the spike trimer: from the RBD part of chain B (red) and chain C (green) to the outer surface of two subunits (chain A (cyan) and chain E (gray) of α7nAChr pentamer). We presented the highest scoring spike–α7nAChr complex in Figure S11. The Protein Contacts panel of the MOE software was used to study the interaction between the atoms of proteins. The interaction between the two proteins were evaluated using six types of contacts: Hydrogen bonds (Hbond), metal, ionic, arene, covalent and Van der Waals distance interactions (Distance). We identified the Van der Waals distance interactions between chain E (gray) of α7nAChr and chain C (green) of the spike protein (Figure S11A). There is a main interaction between the receptor-binding motif (aa 437–508) of the spike RBD and aa 186–192 of the extracellular domain of the nAChR 9 subunit. Previously, Farsalinos et al. reported aa 189–192 of the extracellular domain of α7nAChr as part of the a region which forms the core of the "toxin-binding site" of the nAChRs [130]. There is Van der Waals distance interactions between chain A (cyan) of α7nAChr and chain C (green) of the spike protein (Figure S11B). We also identified Hbond interactions between chain E (gray) of α7nAChr and chain B (red) of the spike protein (Figure S11C) and chain A (cyan) α7nAChr and chain B (red) of the spike protein (Figure S11D). Further details relating to the interaction between different chains are presented in Figure S11.

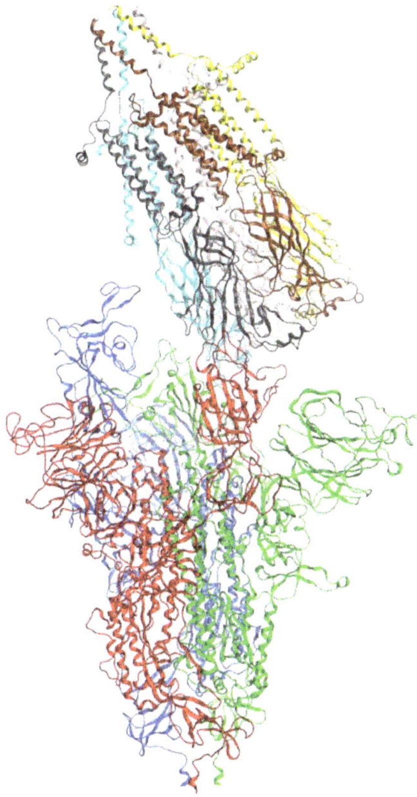

Figure 6. Spike–α7nAChr complex model. Spike protein trimer is colored in dark blue (chain A), red (chain B) and green (chain C). α7nAChr pentamer is colored in cyan (chain A), pink (chain B), yellow (chain C), brown (chain D) and gray (chain E). The yellow and green parts of α7nAChr are interacting with the dark blue and gray monomers from spike protein. Chain B (red) and chain C (green) of α7nAChr are interacting with the chain A(cyan) and chain E(gray) of spike protein.

4. Discussion

Protein–ligand docking is a powerful and popular computational tool to simulate drug–target interactions. Several in silico studies [66–69,74–79] have explored whether competitive binding at subdomains of interest on the SARS-CoV-2 spike protein by ivermectin could explain its efficacy against COVID-19, as indicated in the several RCTs and animal studies related above. In one of these molecular docking studies, Lehrer and Rheinstein (2020) examined potential sites on the SARS-CoV-2 S1 RBD at which ivermectin might bind and competitively block attachment to ACE2, limiting viral replication [74]. They identified one such site at which ivermectin was predicted to dock with high binding energy.

The potential for competitive binding by ivermectin on the spike protein NTD, the subdomain with the highest concentration of glycan binding sites, however, is of interest, especially given the importance of the glycan bindings of SARS-CoV-2 for initial attachments to host cells and the possibilities for hemagglutination, as described above. In particular, the nanometer-scale spacing and the composition of terminal sugar molecules (including SA, galactose, mannose, fucose, N-acetylglucosamine (GlcNAc) and/or N-acetylgalactosamine (GaMAc) for the 22 N-glycosylation sites of the SARS-CoV-2 spike protein [131]) meshes closely with the spacing and terminal sugar composition of glycophorin A [132,133], a ubiquitous molecule on the RBC surface that has no known physiological purpose other than the clearance of viruses and other pathogens [30,31,134].

Here, the AutoDock Vina program was used to perform binding affinity computations for the 15 test compounds (ivermectin and 14 related molecules) for seven binding sites on RBD and 15 on NTD, as identified in the literature as potential SARS-CoV-2 spike protein binding sites of interest for druggability. The inclusion of NTD as well as RBD sites allowed for the consideration of potential competitive binding by ivermectin and related molecules to limit initial viral attachments to host cells and potential hemagglutination-related morbidities. We also examined potential bindings of ivermectin to CD147, an SA-tipped receptor that is densely distributed on RBCs, to provide some indication as to whether ivermectin might limit glycan bindings of SARS-CoV-2 at the host cell end as well. The potential for binding by ivermectin to α7nAChr to inhibit viral attachment to that receptor and activate the cholinergic anti-inflammatory pathway was also explored.

We calculated binding affinities using the Vina Score value, selecting for those bindings most likely to be physically realized. Several of the 15 test compounds, including ivermectin, had bindings of strong or moderate affinity to sites on the spike protein, CD147 and α7nAChr, as detailed here, but since ivermectin, a safe and widely available drug, has been the subject of closest study for COVID-19 treatment among these compounds, the discussion below focuses on the results for that agent and their significance.

As reported in Table S3, docking computations for ivermectin binding to the spike protein found the strongest binding (−8.948 kcal/mol) at site 10 of S1-NTD, which is a glycosylation binding site (N61), in the open position. A study of AutoDock binding energies calculated for a large set of HIV inhibitors and likely non-inhibitors against multiple ligands found that the selection of binding energy <−7.0 kcal/mol identified the inhibitors with 98% sensitivity and 95% specificity [135]. It is thus noteworthy that for the sites at the NTD and RBD in the open position and for NTD in the closed position, most of the binding energies were <−7.0 kcal/mol, and so, per the above, this indicates their capability to be physiologically active. Likewise, binding affinities <−7.0 kcal/mol for ivermectin at five of the 12 sites of CD147 and of 30 of the 37 sites of α7nAChr (for the desensitized, activated and resting states, total) indicate a capability for physiologically-manifested binding to these host receptors as well. Despite the above-cited indication of high sensitivity and specificity for physiological efficacy with binding energies <−7.0 kcal/mol, it is clear that a physiological relevance corresponding to this study's results can only be clearly established through follow-up confirmation with in vitro and/or in vivo findings.

When considering the binding affinities of ivermectin to NTD sites, it is significant that glycan bindings from SARS-CoV-2 and other coronaviruses to host cells are generally weak when univalent but orders of magnitude stronger when multivalent [4,12,29]. Thus, ivermectin, the molecular dimensions of which span approximately 2 × 1 nm [136] (with the length of the spike protein being ~20 nm [137,138]), could block clinically relevant multivalent bindings from the spike protein to host cells by steric interference, even if its actual bindings to some glycan sites on the spike protein were somewhat weaker than predicted. It is relevant, here, that in a rough order of magnitude, considering an average of 0.41 spike protein punctae found attached per RBC in COVID-19 patients [20] and a peak serum concentration of 137.4 nM for ivermectin plus active metabolites after the ingestion of ivermectin with a fatty meal at a dose of 200–350 μg/kg, which is in the range of standard dosing, there would be about 126,000 molecules of ivermectin and active metabolites per spike protein molecule in blood [4].

Two especially significant consequences of these predicted multiple bindings of ivermectin along the spike protein with binding affinities mostly less than −7.0 kcal/mol are, first, that these could provide effective competitive binding for all variants of the virus and, second, since multivalent bindings govern spike protein attachments to host cells, it is noteworthy that competitive inhibitors of such multiple bindings having only moderate binding affinities at individual viral binding sites can have strong inhibitory effects on total attachment strength [139]. Binding affinities as computed <7.0 kcal/mol at five of 12 sites of CD147 further indicate the potential for ivermectin to limit glycan bindings to meshing glycan binding sites on host cells, and also to limit inflammatory pathways mediated by that receptor.

Previous studies using both chick and human cells have demonstrated that a micromolar concentration of ivermectin strongly potentiates the ACh-evoked current of the α7nAChr receptor [89], as expressed on neuronal cells as well as on different airway cells, such as bronchial epithelial cells and type II alveolar epithelial cells, endothelial cells and cytokine secreting immune cells (i.e., macrophages, lymphocytes and mast cells) [140,141]. Importantly, α7nAChr is one of the main receptors under the control of the vagus nerve used by the parasympathetic nervous system to regulate multiple physiological homeostatic mechanisms commonly affected during SARS-CoV-2 infection, including the respiratory rate, heartbeat, blood pressure, vessel tone, hormone secretion, intestinal peristalsis, digestion and inflammation [142]. Our computational studies with ivermectin are consistent with in vitro experimental results obtained with chick and human cells [89] and confirm high affinity bindings to α7nAChr (i.e., Score of -10.636 kcal/mol, the highest of all the 15 test compounds, and Score < -7.0 kcal/mol at 30 of 37 sites for all three states total). Moreover, in agreement with previous reports [5,6], we were also able to demonstrate a potential direct binding of the SARS-CoV-2 spike-1 protein to α7nAChr, suggesting that this ubiquitous cholinergic receptor may represent an additional port of entry for SARS-CoV-2 into human cells. Taken all together, our computational results demonstrating the high-affinity binding of ivermectin to α7nAChr and a potential direct interaction of the cholinergic nicotinic receptor with SARS-CoV-2 spike 1 on neurons, cytokine secreting cells and endothelial cells, which might potentially explain multiple aspects of SARS-CoV-2 infection pathophysiology including but not limited to (a) the typical loss of smell and taste [94,95], (b) the triggering of the life threatening cytokine storm through inactivation of the cholinergic anti-inflammatory α7nAChr pathway on TNF/IL6/IL1 secreting macrophages [90–92] and (c) the impairment of the endothelium dependent acetylcholine-induced vasodilation caused by SARS-CoV-2 spike 1 infection of the lung vasculature [143]. Ivermectin high-affinity binding may therefore potentially shield from infection α7nAChr-expressing host cells while at the same time, through its allosteric agonistic function, potentiate the activation of the cholinergic pathway and attenuate SARS-CoV-2-induced parasympathetic dysregulation by restoring the function of these receptors.

5. Conclusions

In the present study, a computational investigation including molecular docking was conducted to explore the potential bindings of ivermectin and 14 similar compounds to three targets of interest (the spike, CD147 and α7nAChr) that are relevant for drug activity against COVID-19. Strong or moderate affinity bindings were found for ivermectin to multiple sites on the spike protein, CD147 and α7nAChr, which could provide effective competitive bindings to all variants of the virus. According to our calculations, ivermectin binds strongly to a glycosylation binding site (site 10: N61) of the spike protein S1-NTD in the open position and to several other sites on S1 NTD and RBD. We also examined the potential bindings of ivermectin to CD147. Ivermectin was found to bind to site 5, which is located in domain A of the CD147 protein, and to other sites on CD147, indicating that ivermectin might limit glycan bindings of SARS-CoV-2 at the host cell end as well.

Among all the targets, ivermectin has the highest affinity to the α7nAChr receptor. Protein–protein docking results reveal a potential direct binding of the SARS-CoV-2 spike-1 protein to α7nAChr, suggesting that this ubiquitous cholinergic receptor may mediate SARS-CoV-2 entry into cells, shedding light on multiple aspects of SARS-CoV-2 infection pathophysiology (i.e., the loss of smell and taste, the cytokine storm and impairment of the endothelium-dependent acetylcholine-induced vasodilation). In this context, the high affinity of ivermectin and related compounds to α7nAChr may both prevent viral entry and potentiate the activation of the cholinergic pathway and attenuate SARS-CoV-2-induced parasympathetic dysregulation by restoring the function of these receptors. Our preliminary results warrant further in vitro and in vivo testing of the 15 test compounds, in particular ivermectin, an available and safe drug, against SARS-CoV-2, with the hope of containing the virus and limiting its morbidity.

Supplementary Materials: The following supporting information can be downloaded at: https://www.mdpi.com/article/10.3390/computation10040051/s1, Figure S1: Binding sites of CD147 protein obtained from MOE (Site Finder); Figure S2: Binding sites of (A) desensitized, (B) activated and (C) resting conformations of α7nAChr protein obtained from MOE (Site Finder); Figure S3: Positive control docking of epibatidine of (PDB:7KOX) of the α7nAChr receptor; Figure S4: Decoy compounds; Figure S5. Time-evolution of the RMSD of top-ranked inhibitors with respect to spike; Figure S6. Time-evolution of the RMSD of top-ranked inhibitors with respect to the CD147 receptor; Figure S7. Time-evolution of the RMSD of top-ranked inhibitors with respect to the α7nAChr receptor; Figure S8. Ligand interaction plots of compounds selected for spike inhibition; Figure S9. Ligand interaction plots of compounds selected for CD147 inhibition; Figure S10. Ligand interaction plots of compounds selected for α7nAChr inhibition; Figure S11: Protein–protein interaction between (A) chain E (gray) α7nAChr and chain C (green) of spike protein, (B) chain A (cyan) α7nAChr and chain C (green) of spike protein, (C) chain E (gray) α7nAChr and chain B (red) of spike protein and (D) chain A (cyan) α7nAChr and chain B (red) of spike protein; Table S1: Binding sites of CD147 obtained from MOE (Site Finder); Table S2: Binding sites of α7nAChr obtained from MOE (Site Finder); Table S3: Results of the docking analysis of ivermectin for spike protein S1 on all binding sites on NTD and RBD in open and closed positions; Table S4: Results of the docking analysis of ivermectin on all sites of CD147; Table S5: Results of the docking analysis of ivermectin on all sites of α7nAChr; Table S6: Amino acid mutations of SARS-CoV-2 Alpha, Beta, Gamma and Delta variants, with a focus on spike protein.

Author Contributions: Conceptualization, A.D.S., D.E.S. and J.A.T.; methodology, M.A.; validation, M.A., D.E.S., A.D.S. and J.A.T.; formal analysis, M.A., M.C. and J.P.; investigation, M.A., M.C., J.P., M.E.S., A.M., D.D., F.D., D.E.S., A.D.S. and J.A.T.; resources, J.A.T.; data curation, M.A., J.P. and M.C.; writing—original draft preparation, M.A.; writing—review and editing, D.E.S., M.A., A.D.S., and J.A.T.; visualization, M.A. and M.C.; supervision, M.A., E.A.Z., M.A.D., A.D.S., D.E.S. and J.A.T.; project administration, J.A.T. All authors have read and agreed to the published version of the manuscript.

Funding: This work was supported by the Natural Sciences and Engineering Research Council (NSERC) of Canada, grant RES00038219.

Institutional Review Board Statement: Not applicable.

Informed Consent Statement: Not applicable.

Data Availability Statement: The datasets used and analyzed in the current study are available from the corresponding author upon request.

Conflicts of Interest: A.D.S. reports grants from PUMA, grants from IMMUNOMEDICS, grants from GILEAD, grants from SYNTHON, grants and personal fees from MERCK, grants from BOEHINGER-INGELHEIM, grants from GENENTECH, grants and personal fees from TESARO and grants and personal fees from EISAI. The other authors declare no conflict of interest.

Abbreviations

The following abbreviations are used in this manuscript:

α7nAChr	alpha-7 nicotinic acetylcholine receptor
ACE2	angiotensin converting enzyme 2
ACh	acetylcholine
BCov	bovine coronavirus
CD147	cluster of differentiation 147 protein, encoded by the BSG gene
Co-IP	co-immunoprecipitation
COVID-19	coronavirus disease 2019
ELISA	enzyme-linked immunosorbent assay
GPU	graphics processing unit
HE	hemagglutinin esterase
HIV	human immunodeficiency virus
IL-1	interleukin 1
IL-6	interleukin 6
MD	molecular dynamics
MERS	Middle East respiratory syndrome
MHV-4	mouse hepatitis virus 4, JHM strain
MOE	Molecular Operating Environment
NAG	N-acetyl-D-glucosamine
NTD	N-terminal domain
PDB	Protein Data Bank
PLB	propensity for ligand binding
RBC	red blood cell
RBD	receptor binding domain
RCSB	Research Collaboratory for Structural Bioinformatics
RCT	randomized clinical trial
RMSD	root mean square deviation
SA	sialic acid
SARS-CoV-2	severe acute respiratory syndrome coronavirus 2
SPR	surface plasmon resonance
TNF	tumor necrosis factor

References

1. Hoffmann, M.; Kleine-Weber, H.; Schroeder, S.; Krüger, N.; Herrler, T.; Erichsen, S.; Schiergens, T.S.; Herrler, G.; Wu, N.H.; Nitsche, A.; et al. SARS-CoV-2 Cell Entry Depends on ACE2 and TMPRSS2 and Is Blocked by a Clinically Proven Protease Inhibitor. *Cell* **2020**, *181*, 271–280.e8. [CrossRef] [PubMed]
2. Yang, J.; Petitjean, S.J.L.; Koehler, M.; Zhang, Q.; Dumitru, A.C.; Chen, W.; Derclaye, S.; Vincent, S.P.; Soumillion, P.; Alsteens, D. Molecular interaction and inhibition of SARS-CoV-2 binding to the ACE2 receptor. *Nat. Commun.* **2020**, *11*, 4541. [CrossRef] [PubMed]
3. Aminpour, M.; Cannariato, M.; Zucco, A.; Di Gregorio, E.; Israel, S.; Perioli, A.; Tucci, D.; Rossi, F.; Pionato, S.; Marino, S.; et al. Computational Study of Potential Galectin-3 Inhibitors in the Treatment of COVID-19. *Biomedicines* **2021**, *9*, 1208. [CrossRef] [PubMed]
4. Scheim, D.E. A Deadly Embrace: Hemagglutination Mediated by SARS-CoV-2 Spike Protein at its 22 N-Glycosylation Sites, Red Blood Cell Surface Sialoglycoproteins, and Antibody. *Int. J. Mol. Sci.* **2022**, *23*, 2558. [CrossRef] [PubMed]
5. Changeux, J.P.; Amoura, Z.; Rey, F.A.; Miyara, M. A nicotinic hypothesis for COVID-19 with preventive and therapeutic implications. *Comptes Rendus Biol.* **2020**, *343*, 33–39. [CrossRef] [PubMed]
6. Lagoumintzis, G.; Chasapis, C.T.; Alexandris, N.; Kouretas, D.; Tzartos, S.; Eliopoulos, E.; Farsalinos, K.; Poulas, K. Nicotinic cholinergic system and COVID-19: In Silico identification of interactions between alpha7 nicotinic acetylcholine receptor and the cryptic epitopes of SARS-Co-V and SARS-CoV-2 Spike glycoproteins. *Food Chem. Toxicol.* **2021**, *149*, 112009. [CrossRef]
7. Lentz, T.L.; Burrage, T.G.; Smith, A.L.; Crick, J.; Tignor, G.H. Is the acetylcholine receptor a rabies virus receptor? *Science* **1982**, *215*, 182–184. [CrossRef]
8. Chen, W.; Hui, Z.; Ren, X.; Luo, Y.; Shu, J.; Yu, H.; Li, Z. The N-glycosylation sites and Glycan-binding ability of S-protein in SARS-CoV-2 Coronavirus. *bioRxiv* **2020**. [CrossRef]

9. Choi, Y.K.; Cao, Y.; Frank, M.; Woo, H.; Park, S.-J.; Yeom, M.S.; Croll, T.I.; Seok, C.; Im, W. Structure, Dynamics, Receptor Binding, and Antibody Binding of the Fully Glycosylated Full-Length SARS-CoV-2 Spike Protein in a Viral Membrane. *J. Chem. Theory Comput.* **2021**, *17*, 2479–2487. [CrossRef]
10. Shajahan, A.; Supekar, N.T.; Gleinich, A.S.; Azadi, P. Deducing the N- and O-glycosylation profile of the spike protein of novel coronavirus SARS-CoV-2. *Glycobiology* **2020**, *30*, 981–988. [CrossRef]
11. Watanabe, Y.; Allen, J.D.; Wrapp, D.; McLellan, J.S.; Crispin, M. Site-specific glycan analysis of the SARS-CoV-2 spike. *Science* **2020**, *369*, 330–333. [CrossRef] [PubMed]
12. Baker, A.N.; Richards, S.-J.; Guy, C.S.; Congdon, T.R.; Hasan, M.; Zwetsloot, A.J.; Gallo, A.; Lewandowski, J.R.; Stansfeld, P.J.; Straube, A.; et al. The SARS-CoV-2 Spike Protein Binds Sialic Acids and Enables Rapid Detection in a Lateral Flow Point of Care Diagnostic Device. *ACS Cent. Sci.* **2020**, *6*, 2046–2052. [CrossRef] [PubMed]
13. Bharara, R.; Singh, S.; Pattnaik, P.; Chitnis, C.E.; Sharma, A. Structural analogs of sialic acid interfere with the binding of erythrocyte binding antigen-175 to glycophorin A, an interaction crucial for erythrocyte invasion by Plasmodium falciparum. *Mol. Biochem. Parasitol.* **2004**, *138*, 123–129. [CrossRef] [PubMed]
14. Stencel-Baerenwald, J.E.; Reiss, K.; Reiter, D.M.; Stehle, T.; Dermody, T.S. The sweet spot: Defining virus–sialic acid interactions. *Nat. Rev. Microbiol.* **2014**, *12*, 739–749. [CrossRef]
15. Levine, S.; Levine, M.; Sharp, K.A.; Brooks, D.E. Theory of the electrokinetic behavior of human erythrocytes. *Biophys. J.* **1983**, *42*, 127–135. [CrossRef]
16. Odièvre, M.-H.; Bony, V.; Benkerrou, M.; Lapouméroulie, C.; Alberti, C.; Ducrocq, R.; Jacqz-Aigrain, E.; Elion, J.; Cartron, J.-P. Modulation of erythroid adhesion receptor expression by hydroxyurea in children with sickle cell disease. *Haematologica* **2008**, *93*, 502–510. [CrossRef]
17. Bai, Y.; Huang, W.; Ma, L.T.; Jiang, J.L.; Chen, Z.N. Importance of N-glycosylation on CD147 for its biological functions. *Int. J. Mol. Sci.* **2014**, *15*, 6356–6377. [CrossRef]
18. Silva-Filho, J.C.; de Melo, C.G.F.; de Oliveira, J.L. The influence of ABO blood groups on COVID-19 susceptibility and severity: A molecular hypothesis based on carbohydrate-carbohydrate interactions. *Med. Hypotheses* **2020**, *144*, 110155. [CrossRef]
19. Modrof, J.; Kerschbaum, A.; Farcet, M.R.; Niemeyer, D.; Corman, V.M.; Kreil, T.R. SARS-CoV-2 and the safety margins of cell-based biological medicinal products. *Biologicals* **2020**, *68*, 122–124. [CrossRef]
20. Lam, L.M.; Murphy, S.J.; Kuri-Cervantes, L.; Weisman, A.R.; Ittner, C.A.G.; Reilly, J.P.; Pampena, M.B.; Betts, M.R.; Wherry, E.J.; Song, W.-C.; et al. Erythrocytes Reveal Complement Activation in Patients with COVID-19. *MedRxiv* **2020**. [CrossRef]
21. Wang, K.; Chen, W.; Zhang, Z.; Deng, Y.; Lian, J.-Q.; Du, P.; Wei, D.; Zhang, Y.; Sun, X.-X.; Gong, L.; et al. CD147-spike protein is a novel route for SARS-CoV-2 infection to host cells. *Signal Transduct. Target. Ther.* **2020**, *5*, 283. [CrossRef] [PubMed]
22. Bian, H.; Zheng, Z.-H.; Wei, D.; Wen, A.; Zhang, Z.; Lian, J.-Q.; Kang, W.-Z.; Hao, C.-Q.; Wang, J.; Xie, R.-H.; et al. Safety and efficacy of meplazumab in healthy volunteers and COVID-19 patients: A randomized phase 1 and an exploratory phase 2 trial. *Signal Transduct. Target. Ther.* **2021**, *6*, 194. [CrossRef] [PubMed]
23. Hulswit, R.J.G.; Lang, Y.; Bakkers, M.J.G.; Li, W.; Li, Z.; Schouten, A.; Ophorst, B.; van Kuppeveld, F.J.M.; Boons, G.J.; Bosch, B.J.; et al. Human coronaviruses OC43 and HKU1 bind to 9-*O*-acetylated sialic acids via a conserved receptor-binding site in spike protein domain A. *Proc. Natl. Acad. Sci. USA* **2019**, *116*, 2681–2690. [CrossRef] [PubMed]
24. Neu, U.; Bauer, J.; Stehle, T. Viruses and sialic acids: Rules of engagement. *Curr. Opin. Struct. Biol.* **2011**, *21*, 610–618. [CrossRef] [PubMed]
25. Qing, E.; Hantak, M.; Perlman, S.; Gallagher, T. Distinct Roles for Sialoside and Protein Receptors in Coronavirus Infection. *MBio* **2020**, *11*, e02764-19. [CrossRef]
26. Huang, X.; Dong, W.; Milewska, A.; Golda, A.; Qi, Y.; Zhu, Q.K.; Marasco, W.A.; Baric, R.S.; Sims, A.C.; Pyrc, K.; et al. Human Coronavirus HKU1 Spike Protein Uses O-Acetylated Sialic Acid as an Attachment Receptor Determinant and Employs Hemagglutinin-Esterase Protein as a Receptor-Destroying Enzyme. *J. Virol.* **2015**, *89*, 7202–7213. [CrossRef]
27. Li, W.; Hulswit, R.J.G.; Widjaja, I.; Raj, V.S.; McBride, R.; Peng, W.; Widagdo, W.; Tortorici, M.A.; van Dieren, B.; Lang, Y.; et al. Identification of sialic acid-binding function for the Middle East respiratory syndrome coronavirus spike glycoprotein. *Proc. Natl. Acad. Sci. USA* **2017**, *114*, E8508–E8517. [CrossRef]
28. Dai, X.; Zhang, X.; Ostrikov, K.; Abrahamyan, L. Host receptors: The key to establishing cells with broad viral tropism for vaccine production. *Crit. Rev. Microbiol.* **2020**, *46*, 147–168. [CrossRef]
29. Koehler, M.; Delguste, M.; Sieben, C.; Gillet, L.; Alsteens, D. Initial Step of Virus Entry: Virion Binding to Cell-Surface Glycans. *Annu. Rev. Virol.* **2020**, *7*, 143–165. [CrossRef]
30. Baum, J.; Ward, R.H.; Conway, D.J. Natural selection on the erythrocyte surface. *Mol. Biol. Evol.* **2002**, *19*, 223–229. [CrossRef]
31. Varki, A.; Gagneux, P. Multifarious roles of sialic acids in immunity. *Ann. N. Y. Acad. Sci.* **2012**, *1253*, 16–36. [CrossRef] [PubMed]
32. Zeng, Q.; Langereis, M.A.; van Vliet, A.L.W.; Huizinga, E.G.; de Groot, R.J. Structure of coronavirus hemagglutinin-esterase offers insight into corona and influenza virus evolution. *Proc. Natl. Acad. Sci. USA* **2008**, *105*, 9065–9069. [CrossRef] [PubMed]
33. De Groot, R.J. Structure, function and evolution of the hemagglutinin-esterase proteins of corona- and toroviruses. *Glycoconj. J.* **2006**, *23*, 59–72. [CrossRef] [PubMed]
34. Bakkers, M.J.G.; Lang, Y.; Feitsma, L.J.; Hulswit, R.J.G.; de Poot, S.A.H.; van Vliet, A.L.W.; Margine, I.; de Groot-Mijnes, J.D.F.; van Kuppeveld, F.J.M.; Langereis, M.A.; et al. Betacoronavirus Adaptation to Humans Involved Progressive Loss of Hemagglutinin-Esterase Lectin Activity. *Cell Host Microbe* **2017**, *21*, 356–366. [CrossRef]

35. Matrosovich, M.; Herrler, G.; Klenk, H.D. Sialic Acid Receptors of Viruses. In *SialoGlyco Chemistry and Biology II: Tools and Techniques to Identify and Capture Sialoglycans*; Gerardy-Schahn, R., Delannoy, P., von Itzstein, M., Eds.; Springer International Publishing: New York, NY, USA, 2015; pp. 1–28.
36. Miyagi, T.; Yamaguchi, K. 3.17—Sialic Acids. In *Comprehensive Glycoscience*; Kamerling, H., Ed.; Elsevier: Oxford, UK, 2007; pp. 297–323.
37. Wagner, R.; Matrosovich, M.; Klenk, H.D. Functional balance between haemagglutinin and neuraminidase in influenza virus infections. *Rev. Med. Virol.* **2002**, *12*, 159–166. [CrossRef]
38. Lang, Y.; Li, W.; Li, Z.; Koerhuis, D.; van den Burg, A.C.S.; Rozemuller, E.; Bosch, B.-J.; van Kuppeveld, F.J.M.; Boons, G.-J.; Huizinga, E.G.; et al. Coronavirus hemagglutinin-esterase and spike proteins coevolve for functional balance and optimal virion avidity. *Proc. Natl. Acad. Sci. USA* **2020**, *117*, 25759–25770. [CrossRef]
39. Chan, J.F.-W.; Kok, K.-H.; Zhu, Z.; Chu, H.; To, K.K.-W.; Yuan, S.; Yuen, K.-Y. Genomic characterization of the 2019 novel human-pathogenic coronavirus isolated from a patient with atypical pneumonia after visiting Wuhan. *Emerg. Microbes Infect.* **2020**, *9*, 221–236. [CrossRef]
40. Chen, Y.; Liu, Q.; Guo, D. Emerging coronaviruses: Genome structure, replication, and pathogenesis. *J. Med. Virol.* **2020**, *92*, 418–423. [CrossRef]
41. Zaki, A.M.; van Boheemen, S.; Bestebroer, T.M.; Osterhaus, A.D.; Fouchier, R.A. Isolation of a novel coronavirus from a man with pneumonia in Saudi Arabia. *N. Engl. J. Med.* **2012**, *367*, 1814–1820. [CrossRef]
42. Couzin-Frankel, J. The mystery of the pandemic's 'happy hypoxia'. *Science* **2020**, *368*, 455–456. [CrossRef]
43. Rapkiewicz, A.V.; Mai, X.; Carsons, S.E.; Pittaluga, S.; Kleiner, D.E.; Berger, J.S.; Thomas, S.; Adler, N.M.; Charytan, D.M.; Gasmi, B.; et al. Megakaryocytes and platelet-fibrin thrombi characterize multi-organ thrombosis at autopsy in COVID-19: A case series. *EClinicalMedicine* **2020**, *24*, 100434. [CrossRef] [PubMed]
44. Lodigiani, C.; Lapichino, G.; Carenzo, L.; Cecconi, M.; Ferrazzi, P.; Sebastian, T.; Kucher, N.; Studt, J.D.; Sacco, C.; Alexia, B.; et al. Venous and arterial thromboembolic complications in COVID-19 patients admitted to an academic hospital in Milan, Italy. *Thromb. Res.* **2020**, *191*, 9–14. [CrossRef] [PubMed]
45. Price, L.C.; McCabe, C.; Garfield, B.; Wort, S.J. Thrombosis and COVID-19 pneumonia: The clot thickens! *Eur. Respir. J.* **2020**, *56*, 2001608. [CrossRef] [PubMed]
46. Tang, N.; Li, D.; Wang, X.; Sun, Z. Abnormal coagulation parameters are associated with poor prognosis in patients with novel coronavirus pneumonia. *J. Thromb. Haemost.* **2020**, *18*, 844–847. [CrossRef]
47. Marini, J.J.; Gattinoni, L. Management of COVID-19 Respiratory Distress. *JAMA* **2020**, *323*, 2329–2330. [CrossRef]
48. Liao, C.G.; Kong, L.M.; Song, F.; Xing, J.L.; Wang, L.X.; Sun, Z.J.; Tang, H.; Yao, H.; Zhang, Y.; Wang, L.; et al. Characterization of basigin isoforms and the inhibitory function of basigin-3 in human hepatocellular carcinoma proliferation and invasion. *Mol. Cell. Biol.* **2011**, *31*, 2591–2604. [CrossRef]
49. Koch, C.; Staffler, G.; Hüttinger, R.; Hilgert, I.; Prager, E.; Cerný, J.; Steinlein, P.; Majdic, O.; Horejsí, V.; Stockinger, H. T cell activation-associated epitopes of CD147 in regulation of the T cell response, and their definition by antibody affinity and antigen density. *Int. Immunol.* **1999**, *11*, 777–786. [CrossRef]
50. Lv, M.; Miao, J.; Zhao, P.; Luo, X.; Han, Q.; Wu, Z.; Zhang, K.; Zhu, P. CD147-mediated chemotaxis of CD4(+)CD161(+) T cells may contribute to local inflammation in rheumatoid arthritis. *Clin. Rheumatol.* **2018**, *37*, 59–66. [CrossRef]
51. Schmidt, R.; Bültmann, A.; Fischel, S.; Gillitzer, A.; Cullen, P.; Walch, A.; Jost, P.; Ungerer, M.; Tolley, N.D.; Lindemann, S.; et al. Extracellular matrix metalloproteinase inducer (CD147) is a novel receptor on platelets, activates platelets, and augments nuclear factor kappaB-dependent inflammation in monocytes. *Circ. Res.* **2008**, *102*, 302–309. [CrossRef]
52. Loh, D. The potential of melatonin in the prevention and attenuation of oxidative hemolysis and myocardial injury from cd147 SARS-CoV-2 spike protein receptor binding. *Melatonin Res.* **2020**, *3*, 380–416. [CrossRef]
53. Joseph, J.; Knobler, R.L.; Lublin, F.D.; Burns, F.R. Regulation of the expression of intercellular adhesion molecule-1 (ICAM-1) and the putative adhesion molecule Basigin on murine cerebral endothelial cells by MHV-4 (JHM). *Adv. Exp. Med. Biol.* **1993**, *342*, 389–391. [PubMed]
54. De Back, D.Z.; Kostova, E.; Klei, T.; Beuger, B.; van Zwieten, R.; Kuijpers, T.; Juffermans, N.; van den Berg, T.; Korte, D.; van Kraaij, M.; et al. RBC Adhesive Capacity Is Essential for Efficient 'Immune Adherence Clearance' and Provide a Generic Target to Deplete Pathogens from Septic Patients. *Blood* **2016**, *128*, 1031. [CrossRef]
55. Telen, M.J. Red blood cell surface adhesion molecules: Their possible roles in normal human physiology and disease. *Semin. Hematol.* **2000**, *37*, 130–142. [CrossRef]
56. Yurchenko, V.; Constant, S.; Bukrinsky, M. Dealing with the family: CD147 interactions with cyclophilins. *Immunology* **2006**, *117*, 301–309. [CrossRef]
57. Schulz, C.; von Brühl, M.L.; Barocke, V.; Cullen, P.; Mayer, K.; Okrojek, R.; Steinhart, A.; Ahmad, Z.; Kremmer, E.; Nieswandt, B.; et al. EMMPRIN (CD147/basigin) mediates platelet-monocyte interactions in vivo and augments monocyte recruitment to the vascular wall. *J. Thromb. Haemost.* **2011**, *9*, 1007–1019. [CrossRef]
58. Von Ungern-Sternberg, S.N.I.; Zernecke, A.; Seizer, P. Extracellular Matrix Metalloproteinase Inducer EMMPRIN (CD147) in Cardiovascular Disease. *Int. J. Mol. Sci.* **2018**, *19*, 507. [CrossRef]

59. Yee, C.; Main, N.M.; Terry, A.; Stevanovski, I.; Maczurek, A.; Morgan, A.J.; Calabro, S.; Potter, A.J.; Iemma, T.L.; Bowen, D.G.; et al. CD147 mediates intrahepatic leukocyte aggregation and determines the extent of liver injury. *PLoS ONE* **2019**, *14*, e0215557. [CrossRef]
60. Pennings, G.J.; Kritharides, L. CD147 in cardiovascular disease and thrombosis. *Semin. Thromb. Hemost.* **2014**, *40*, 747–755.
61. Carbajo-Lozoya, J.; Ma-Lauer, Y.; Malešević, M.; Theuerkorn, M.; Kahlert, V.; Prell, E.; von Brunn, B.; Muth, D.; Baumert, T.F.; Drosten, C.; et al. Human coronavirus NL63 replication is cyclophilin A-dependent and inhibited by non-immunosuppressive cyclosporine A-derivatives including Alisporivir. *Virus Res.* **2014**, *184*, 44–53. [CrossRef]
62. Chen, Z.; Mi, L.; Xu, J.; Yu, J.; Wang, X.; Jiang, J.; Xing, J.; Shang, P.; Qian, A.; Li, Y.; et al. Function of HAb18G/CD147 in invasion of host cells by severe acute respiratory syndrome coronavirus. *J. Infect. Dis.* **2005**, *191*, 755–760. [CrossRef]
63. Pushkarsky, T.; Zybarth, G.; Dubrovsky, L.; Yurchenko, V.; Tang, H.; Guo, H.; Toole, B.; Sherry, B.; Bukrinsky, M. CD147 facilitates HIV-1 infection by interacting with virus-associated cyclophilin A. *Proc. Natl. Acad. Sci. USA* **2001**, *98*, 6360–6365. [CrossRef] [PubMed]
64. Muramatsu, T. Basigin (CD147), a multifunctional transmembrane glycoprotein with various binding partners. *J. Biochem.* **2016**, *159*, 481–490. [CrossRef] [PubMed]
65. Zhu, X.; Song, Z.; Zhang, S.; Nanda, A.; Li, G. CD147: A Novel Modulator of Inflammatory and Immune Disorders. *Curr. Med. Chem.* **2014**, *21*, 2138–2145. [CrossRef] [PubMed]
66. Dayer, M. Coronavirus (2019-nCoV) Deactivation via Spike Glycoprotein Shielding by Old Drugs, Bioinformatic Study. *Preprints.Org.* **2020**. [CrossRef]
67. Nallusamy, S.; Mannu, J.; Ravikumar, C.; Angamuthu, K.; Nathan, B.; Nachimuthu, K.; Ramasamy, G.; Muthurajan, R.; Subbarayalu, M.; Neelakandan, K. Shortlisting Phytochemicals Exhibiting Inhibitory Activity against Major Proteins of SARS-CoV-2 through Virtual Screening. *Res. Sq.* **2020**. [CrossRef]
68. Suravajhala, R.; Parashar, A.; Malik, B.; Nagaraj, V.A.; Padmanaban, G.; Kavi Kishor, P.B.; Polavarapu, R.; Suravajhala, P. Comparative Docking Studies on Curcumin with COVID-19 Proteins. *Preprints.Org.* **2020**. [CrossRef]
69. Kalhor, H.; Sadeghi, S.; Abolhasani, H.; Kalhor, R.; Rahimi, H. Repurposing of the approved small molecule drugs in order to inhibit SARS-CoV-2 S protein and human ACE2 interaction through virtual screening approaches. *J. Biomol. Struct. Dyn.* **2020**, *40*, 1299–1315. [CrossRef]
70. Yagisawa, M.; Foster, P.J.; Hanaki, H.; Omura, S. Global Trends in Clinical Studies of Ivermectin in COVID-19. *Jpn. J. Antibiot.* **2021**, *74*, 44–95.
71. Campbell, W.C. History of avermectin and ivermectin, with notes on the history of other macrocyclic lactone antiparasitic agents. *Curr. Pharm. Biotechnol.* **2012**, *13*, 853–865. [CrossRef]
72. Juarez, M.; Schcolnik-Cabrera, A.; Dueñas-Gonzalez, A. The multitargeted drug ivermectin: From an antiparasitic agent to a repositioned cancer drug. *Am. J. Cancer Res.* **2018**, *8*, 317–331.
73. Rizzo, E. Ivermectin, antiviral properties and COVID-19: A possible new mechanism of action. *Naunyn Schmiedebergs Arch. Pharm.* **2020**, *393*, 1153–1156.
74. Lehrer, S.; Rheinstein, P.H. Ivermectin Docks to the SARS-CoV-2 Spike Receptor-binding Domain Attached to ACE2. *Vivo* **2020**, *34*, 3023–3026. [CrossRef] [PubMed]
75. Maurya, D. A Combination of Ivermectin and Doxycycline Possibly Blocks the Viral Entry and Modulate the Innate Immune Response in COVID-19 Patients. *ChemRxiv* **2020**. [CrossRef]
76. Dasgupta, J.; Sen, U.; Bakashi, A.; Dasgupta, A. Nsp7 and Spike Glycoprotein of SARS-CoV-2 Are Envisaged as Potential Targets of Vitamin D and Ivermectin. *Preprints* **2020**. [CrossRef]
77. Kaur, H.; Shekhar, N.; Sharma, S.; Sarma, P.; Prakash, A.; Medhi, B. Ivermectin as a potential drug for treatment of COVID-19: An in-sync review with clinical and computational attributes. *Pharmacol. Rep.* **2021**, *73*, 736–749. [CrossRef] [PubMed]
78. Saha, J.K.; Raihan, J. The Binding mechanism of Ivermectin and levosalbutamol with spike protein of SARS-CoV-2. *Struct. Chem.* **2021**, *32*, 1985–1992. [CrossRef] [PubMed]
79. Hussien, M.A.; Abdelaziz, A.E.M. Molecular docking suggests repurposing of brincidofovir as a potential drug targeting SARS-CoV-2 ACE2 receptor and main protease. *Netw. Modeling Anal. Health Inform. Bioinform.* **2020**, *9*, 56. [CrossRef]
80. Santin, A.D.; Scheim, D.E.; McCullough, P.A.; Yagisawa, M.; Borody, T.J. Ivermectin: A multifaceted drug of Nobel prize-honored distinction with indicated efficacy against a new global scourge, COVID-19. *New Microbes New Infect.* **2021**, *43*, 100924. [CrossRef]
81. Kory, P.; Meduri, G.U.; Varon, J.; Iglesias, J.; Marik, P.E. Review of the Emerging Evidence Demonstrating the Efficacy of Ivermectin in the Prophylaxis and Treatment of COVID-19. *Am. J. Ther.* **2021**, *28*, e299–e318. [CrossRef]
82. Crump, A.; Ōmura, S. Ivermectin, 'wonder drug' from Japan: The human use perspective. *Proc. Jpn. Acad. Ser. B Phys. Biol. Sci.* **2011**, *87*, 13–28. [CrossRef]
83. Guzzo, C.A.; Furtek, C.I.; Porras, A.G.; Chen, C.; Tipping, R.; Clineschmidt, C.M.; Sciberras, D.G.; Hsieh, J.Y.; Lasseter, K.C. Safety, tolerability, and pharmacokinetics of escalating high doses of ivermectin in healthy adult subjects. *J. Clin. Pharm.* **2002**, *42*, 1122–1133. [CrossRef] [PubMed]
84. Navarro, M.; Camprubí, D.; Requena-Méndez, A.; Buonfrate, D.; Giorli, G.; Kamgno, J.; Gardon, J.; Boussinesq, M.; Muñoz, J.; Krolewiecki, A. Safety of high-dose ivermectin: A systematic review and meta-analysis. *J. Antimicrob. Chemother.* **2020**, *75*, 827–834. [CrossRef] [PubMed]

85. *The 2015 Nobel Prize in Physiology or Medicine—Press Release*; The Nobel Assembly at Karolinska Institutet: Solna, Sweden, 2015; Available online: https://www.nobelprize.org/prizes/medicine/2015/press-release/ (accessed on 22 February 2022).
86. Caly, L.; Druce, J.D.; Catton, M.G.; Jans, D.A.; Wagstaff, K.M. The FDA-approved drug ivermectin inhibits the replication of SARS-CoV-2 in vitro. *Antivir. Res.* **2020**, *178*, 104787. [CrossRef] [PubMed]
87. Momekov, G.; Momekova, D. Ivermectin as a potential COVID-19 treatment from the pharmacokinetic point of view: Antiviral levels are not likely attainable with known dosing regimens. *Biotechnol. Biotechnol. Equip.* **2020**, *34*, 469–474. [CrossRef]
88. Schmith, V.D.; Zhou, J.J.; Lohmer, L.R.L. The Approved Dose of Ivermectin Alone is not the Ideal Dose for the Treatment of COVID-19. *Clin. Pharm.* **2020**, *108*, 762–765. [CrossRef] [PubMed]
89. Krause, R.M.; Buisson, B.; Bertrand, S.; Corringer, P.J.; Galzi, J.L.; Changeux, J.P.; Bertrand, D. Ivermectin: A positive allosteric effector of the alpha7 neuronal nicotinic acetylcholine receptor. *Mol. Pharm.* **1998**, *53*, 283–294. [CrossRef]
90. Wang, H.; Yu, M.; Ochani, M.; Amella, C.A.; Tanovic, M.; Susarla, S.; Li, J.H.; Wang, H.; Yang, H.; Ulloa, L.; et al. Nicotinic acetylcholine receptor α7 subunit is an essential regulator of inflammation. *Nature* **2003**, *421*, 384–388. [CrossRef]
91. Ren, C.; Tong, Y.L.; Li, J.C.; Lu, Z.Q.; Yao, Y.M. The Protective Effect of Alpha 7 Nicotinic Acetylcholine Receptor Activation on Critical Illness and Its Mechanism. *Int. J. Biol. Sci.* **2017**, *13*, 46–56. [CrossRef]
92. Fajgenbaum, D.C.; June, C.H. Cytokine Storm. *N. Engl. J. Med.* **2020**, *383*, 2255–2273. [CrossRef]
93. Rajter, J.C.; Sherman, M.S.; Fatteh, N.; Vogel, F.; Sacks, J.; Rajter, J.-J. Use of Ivermectin is Associated with Lower Mortality in Hospitalized Patients with COVID-19 (ICON study). *Chest* **2020**, *159*, 85–92. [CrossRef]
94. De Melo, G.D.; Lazarini, F.; Levallois, S.; Hautefort, C.; Michel, V.; Larrous, F.; Verillaud, B.; Aparicio, C.; Wagner, S.; Gheusi, G.; et al. COVID-19-related anosmia is associated with viral persistence and inflammation in human olfactory epithelium and brain infection in hamsters. *Sci. Transl. Med.* **2021**, *13*, eabf8396. [CrossRef] [PubMed]
95. Chaccour, C.; Casellas, A.; Blanco-Di Matteo, A.; Pineda, I.; Fernandez-Montero, A.; Ruiz-Castillo, P.; Richardson, M.-A.; Rodríguez-Mateos, M.; Jordán-Iborra, C.; Brew, J.; et al. The effect of early treatment with ivermectin on viral load, symptoms and humoral response in patients with non-severe COVID-19: A pilot, double-blind, placebo-controlled, randomized clinical trial. *EClinicalMedicine* **2021**, *32*, 100720. [CrossRef] [PubMed]
96. Shang, J.; Wan, Y.; Luo, C.; Ye, G.; Geng, Q.; Auerbach, A.; Li, F. Cell entry mechanisms of SARS-CoV-2. *Proc. Natl. Acad. Sci. USA* **2020**, *117*, 11727–11734. [CrossRef] [PubMed]
97. Li, F. Structure, Function, and Evolution of Coronavirus Spike Proteins. *Annu. Rev. Virol.* **2016**, *3*, 237–261. [CrossRef] [PubMed]
98. Hulswit, R.J.G.; de Haan, C.A.M.; Bosch, B.J. Chapter Two—Coronavirus Spike Protein and Tropism Changes. In *Advances in Virus Research, Ziebuhr, J., Ed.*; Academic Press: New York, NY, USA, 2016; Volume 96, pp. 29–57.
99. Huang, Y.; Yang, C.; Xu, X.-F.; Xu, W.; Liu, S.-W. Structural and functional properties of SARS-CoV-2 spike protein: Potential antivirus drug development for COVID-19. *Acta Pharmacol. Sin.* **2020**, *41*, 1141–1149. [CrossRef] [PubMed]
100. Awasthi, M.; Gulati, S.; Sarkar, D.; Tiwari, S.; Kateriya, S.; Ranjan, P.; Verma, S.K. The Sialoside-Binding Pocket of SARS-CoV-2 Spike Glycoprotein Structurally Resembles MERS-CoV. *Viruses* **2020**, *12*, 909. [CrossRef]
101. Fantini, J.; Di Scala, C.; Chahinian, H.; Yahi, N. Structural and molecular modelling studies reveal a new mechanism of action of chloroquine and hydroxychloroquine against SARS-CoV-2 infection. *Int. J. Antimicrob. Agents* **2020**, *55*, 105960. [CrossRef]
102. Tortorici, M.A.; Walls, A.C.; Lang, Y.; Wang, C.; Li, Z.; Koerhuis, D.; Boons, G.J.; Bosch, B.J.; Rey, F.A.; de Groot, R.J.; et al. Structural basis for human coronavirus attachment to sialic acid receptors. *Nat. Struct. Mol. Biol.* **2019**, *26*, 481–489. [CrossRef]
103. Milanetti, E.; Miotto, M.; Rienzo, L.D.; Monti, M.; Gosti, G.; Ruocco, G. In-Silico evidence for two receptors based strategy of SARS-CoV-2. *bioRxiv* **2020**. [CrossRef]
104. Morniroli, D.; Giannì, M.L.; Consales, A.; Pietrasanta, C.; Mosca, F. Human Sialome and Coronavirus Disease-2019 (COVID-19) Pandemic: An Understated Correlation? *Front. Immunol.* **2020**, *11*, 1480. [CrossRef]
105. Wrapp, D.; Wang, N.; Corbett, K.S.; Goldsmith, J.A.; Hsieh, C.-L.; Abiona, O.; Graham, B.S.; McLellan, J.S. Cryo-EM structure of the 2019-nCoV spike in the prefusion conformation. *Science* **2020**, *367*, 1260–1263. [CrossRef] [PubMed]
106. Sikora, M.; von Bülow, S.; Blanc, F.E.C.; Gecht, M.; Covino, R.; Hummer, G. Computational epitope map of SARS-CoV-2 spike protein. *PLoS Comput. Biol.* **2021**, *17*, e1008790. [CrossRef] [PubMed]
107. DrugBank Online Database, Ivermectin (DB00602). Available online: https://go.drugbank.com/structures/search/small_molecule_drugs/structure?database_id=DB00602&search_type=similarity#results (accessed on 21 February 2022).
108. CHARMM-GUI Archive—COVID-19 Proteins Library. Available online: https://www.charmm-gui.org/?doc=archive&lib=covid19 (accessed on 21 February 2022).
109. Jo, S.; Kim, T.; Iyer, V.G.; Im, W. CHARMM-GUI: A web-based graphical user interface for CHARMM. *J. Comput. Chem.* **2008**, *29*, 1859–1865. [CrossRef] [PubMed]
110. Woo, H.; Park, S.-J.; Choi, Y.K.; Park, T.; Tanveer, M.; Cao, Y.; Kern, N.R.; Lee, J.; Yeom, M.S.; Croll, T.I.; et al. Developing a Fully Glycosylated Full-Length SARS-CoV-2 Spike Protein Model in a Viral Membrane. *J. Phys. Chem. B* **2020**, *124*, 7128–7137. [CrossRef]
111. Noviello, C.M.; Gharpure, A.; Mukhtasimova, N.; Cabuco, R.; Baxter, L.; Borek, D.; Sine, S.M.; Hibbs, R.E. Structure and gating mechanism of the α7 nicotinic acetylcholine receptor. *Cell* **2021**, *184*, 2121–2134.e13. [CrossRef]
112. *Molecular Operating Environment (MOE)*; ULC, Chemical Computing Group: Montreal, QC, Canada, 2019.

113. Behloul, N.; Baha, S.; Shi, R.; Meng, J. Role of the GTNGTKR motif in the N-terminal receptor-binding domain of the SARS-CoV-2 spike protein. *Virus Res.* **2020**, *286*, 198058. [CrossRef]
114. Di Gaetano, S.; Capasso, D.; Delre, P.; Pirone, L.; Saviano, M.; Pedone, E.; Mangiatordi, G.F. More Is Always Better Than One: The N-Terminal Domain of the Spike Protein as Another Emerging Target for Hampering the SARS-CoV-2 Attachment to Host Cells. *Int. J. Mol. Sci.* **2021**, *22*, 6462. [CrossRef]
115. *Schrödinger Release 2019-4: SiteMap*; Schrödinger, LLC.: New York, NY, USA, 2019.
116. Bangaru, S.; Ozorowski, G.; Turner, H.L.; Antanasijevic, A.; Huang, D.; Wang, X.; Torres, J.L.; Diedrich, J.K.; Tian, J.-H.; Portnoff, A.D.; et al. Structural analysis of full-length SARS-CoV-2 spike protein from an advanced vaccine candidate. *bioRxiv* **2020**. [CrossRef]
117. Carino, A.; Moraca, F.; Fiorillo, B.; Marchiano, S.; Sepe, V.; Biagioli, M.; Finamore, C.; Bozza, S.; Francisci, D.; Distrutti, E.; et al. Hijacking SARS-CoV-2/ACE2 Receptor Interaction by Natural and Semi-synthetic Steroidal Agents Acting on Functional Pockets on the Receptor Binding Domain. *Front. Chem.* **2020**, *8*, 572885. [CrossRef]
118. Trott, O.; Olson, A.J. AutoDock Vina: Improving the speed and accuracy of docking with a new scoring function, efficient optimization, and multithreading. *J. Comput. Chem.* **2010**, *31*, 455–461. [CrossRef]
119. Case, D.; Aktulga, H.; Belfon, K.; Ben-Shalom, I.; Brozell, S.; Cerutti, D.; Cheatham, I.T.E.; Cisneros, G. Amber 2021. Available online: https://ambermd.org/index.php (accessed on 21 February 2022).
120. Preto, J.; Gentile, F. Assessing and improving the performance of consensus docking strategies using the DockBox package. *J. Comput. Aided Mol. Des.* **2019**, *33*, 817–829. [CrossRef] [PubMed]
121. Graves, A.P.; Brenk, R.; Shoichet, B.K. Decoys for docking. *J. Med. Chem.* **2005**, *48*, 3714–3728. [CrossRef] [PubMed]
122. Mysinger, M.M.; Carchia, M.; Irwin, J.J.; Shoichet, B.K. Directory of useful decoys, enhanced (DUD-E): Better ligands and decoys for better benchmarking. *J. Med. Chem.* **2012**, *55*, 6582–6594. [CrossRef] [PubMed]
123. Huang, N.; Shoichet, B.K.; Irwin, J.J. Benchmarking sets for molecular docking. *J. Med. Chem.* **2006**, *49*, 6789–6801. [CrossRef]
124. Casalino, L.; Gaieb, Z.; Goldsmith, J.A.; Hjorth, C.K.; Dommer, A.C.; Harbison, A.M.; Fogarty, C.A.; Barros, E.P.; Taylor, B.C.; McLellan, J.S.; et al. Beyond Shielding: The Roles of Glycans in the SARS-CoV-2 Spike Protein. *ACS Cent. Sci.* **2020**, *6*, 1722–1734. [CrossRef]
125. Toutain, P.L.; Upson, D.W.; Terhune, T.N.; McKenzie, M.E. Comparative pharmacokinetics of doramectin and ivermectin in cattle. *Vet. Parasitol.* **1997**, *72*, 3–8. [CrossRef]
126. González Canga, A.; Sahagún Prieto, A.M.; Diez Liébana, M.J.; Fernández Martínez, N.; Sierra Vega, M.; García Vieitez, J.J. The Pharmacokinetics and Interactions of Ivermectin in Humans—A Mini-review. *AAPS J.* **2008**, *10*, 42–46. [CrossRef]
127. Munoz, J.; Ballester, M.R.; Antonijoan, R.M.; Gich, I.; Rodriguez, M.; Colli, E.; Gold, S.; Krolewiecki, A.J. Safety and pharmacokinetic profile of fixed-dose ivermectin with an innovative 18mg tablet in healthy adult volunteers. *PLoS Negl. Trop. Dis.* **2018**, *12*, e0006020. [CrossRef]
128. Duhovny, D.; Nussinov, R.; Wolfson, H.J. *Efficient Unbound Docking of Rigid Molecules*; Springer: Berlin/Heidelberg, Germany, 2002; pp. 185–200.
129. Schneidman-Duhovny, D.; Inbar, Y.; Nussinov, R.; Wolfson, H.J. PatchDock and SymmDock: Servers for rigid and symmetric docking. *Nucleic Acids Res.* **2005**, *33*, W363–W367. [CrossRef]
130. Farsalinos, K.; Eliopoulos, E.; Leonidas, D.D.; Papadopoulos, G.E.; Tzartos, S.; Poulas, K. Nicotinic Cholinergic System and COVID-19: In Silico Identification of an Interaction between SARS-CoV-2 and Nicotinic Receptors with Potential Therapeutic Targeting Implications. *Int. J. Mol. Sci.* **2020**, *21*, 5807. [CrossRef]
131. Gao, C.; Zeng, J.; Jia, N.; Stavenhagen, K.; Matsumoto, Y.; Zhang, H.; Li, J.; Hume, A.J.; Mühlberger, E.; van Die, I.; et al. SARS-CoV-2 Spike Protein Interacts with Multiple Innate Immune Receptors. *bioRxiv* **2020**. [CrossRef]
132. Cohen, M.; Varki, A. Chapter Three–Modulation of Glycan Recognition by Clustered Saccharide Patches. In *International Review of Cell and Molecular Biology*; Jeon, K.W., Ed.; Academic Press: New York, NY, USA, 2014; Volume 308, pp. 75–125.
133. Jaskiewicz, E.; Jodłowska, M.; Kaczmarek, R.; Zerka, A. Erythrocyte glycophorins as receptors for Plasmodium merozoites. *Parasites Vectors* **2019**, *12*, 317. [CrossRef] [PubMed]
134. Anderson, H.L.; Brodsky, I.E.; Mangalmurti, N.S. The Evolving Erythrocyte: Red Blood Cells as Modulators of Innate Immunity. *J. Immunol.* **2018**, *201*, 1343–1351. [CrossRef] [PubMed]
135. Chang, M.W.; Lindstrom, W.; Olson, A.J.; Belew, R.K. Analysis of HIV Wild-Type and Mutant Structures via in Silico Docking against Diverse Ligand Libraries. *J. Chem. Inf. Modeling* **2007**, *47*, 1258–1262. [CrossRef]
136. Mol-Instincts, Structure of IVERMECTIN (C48H74O14), Interactive 3-Dimensional (3D) Visualization. Available online: https://www.molinstincts.com/structure/IVERMECTIN-cstr-CT1079779157.html (accessed on 21 February 2022).
137. Ke, Z.; Oton, J.; Qu, K.; Cortese, M.; Zila, V.; McKeane, L.; Nakane, T.; Zivanov, J.; Neufeldt, C.J.; Cerikan, B.; et al. Structures and distributions of SARS-CoV-2 spike proteins on intact virions. *Nature* **2020**, *588*, 498–502. [CrossRef]
138. Kiss, B.; Kis, Z.; Pályi, B.; Kellermayer, M.S.Z. Topography, Spike Dynamics, and Nanomechanics of Individual Native SARS-CoV-2 Virions. *Nano Lett.* **2021**, *21*, 2675–2680. [CrossRef]
139. Xu, H.; Shaw, D.E. A Simple Model of Multivalent Adhesion and Its Application to Influenza Infection. *Biophys. J.* **2016**, *110*, 218–233. [CrossRef]

140. Cardinale, A.; Nastrucci, C.; Cesario, A.; Russo, P. Nicotine: Specific role in angiogenesis, proliferation and apoptosis. *Crit. Rev. Toxicol.* **2012**, *42*, 68–89. [CrossRef]
141. Macklin, K.D.; Maus, A.D.; Pereira, E.F.; Albuquerque, E.X.; Conti-Fine, B.M. Human vascular endothelial cells express functional nicotinic acetylcholine receptors. *J. Pharm. Exp.* **1998**, *287*, 435–439.
142. Gordan, R.; Gwathmey, J.K.; Xie, L.H. Autonomic and endocrine control of cardiovascular function. *World J. Cardiol.* **2015**, *7*, 204–214. [CrossRef]
143. Lei, Y.; Zhang, J.; Schiavon, C.R.; He, M.; Chen, L.; Shen, H.; Zhang, Y.; Yin, Q.; Cho, Y.; Andrade, L.; et al. SARS-CoV-2 Spike Protein Impairs Endothelial Function via Downregulation of ACE 2. *Circ. Res.* **2021**, *128*, 1323–1326. [CrossRef] [PubMed]

MDPI
St. Alban-Anlage 66
4052 Basel
Switzerland
www.mdpi.com

Computation Editorial Office
E-mail: computation@mdpi.com
www.mdpi.com/journal/computation

www.ingramcontent.com/pod-product-compliance
Lightning Source LLC
Chambersburg PA
CBHW061458220926
51850CB00004B/115
9783725800483